SOCIETÀ ITALIANA DI FISICA

RENDICONTI

DELLA

SCUOLA INTERNAZIONALE DI FISICA
"ENRICO FERMI"

CLXIX Corso

a cura di A. COVELLO, F. IACHELLO e R. A. RICCI

Direttori del Corso

e di

G. MAINO

VARENNA SUL LAGO DI COMO

VILLA MONASTERO

17 – 27 Luglio 2007

Struttura dei nuclei lontano dalla valle di stabilità: nuova fisica e nuova tecnologia

2008

SOCIETÀ ITALIANA DI FISICA
BOLOGNA-ITALY

ITALIAN PHYSICAL SOCIETY

PROCEEDINGS

OF THE

INTERNATIONAL SCHOOL OF PHYSICS
"ENRICO FERMI"

COURSE CLXIX

edited by A. COVELLO, F. IACHELLO and R. A. RICCI

Directors of the Course

and

G. MAINO

VARENNA ON LAKE COMO

VILLA MONASTERO

17 – 27 July 2007

Nuclear Structure far from Stability: New Physics and New Technology

2008

IOS
Press

AMSTERDAM, OXFORD, TOKIO, WASHINGTON DC

ISBN 978-1-58603-885-4 (IOS)
ISBN 978-88-7438-041-1 (SIF)
Library of Congress Control Number: 2008929438

Production Manager
A. OLEANDRI

Copy Editor
M. MISSIROLI

jointly published and distributed by:

IOS PRESS
Nieuwe Hemweg 6B
1013 BG Amsterdam
The Netherlands
fax: +31 20 620 34 19
order@iopress.nl

SOCIETÀ ITALIANA DI FISICA
Via Saragozza 12
40123 Bologna
Italy
fax: +39 051 581340
order@sif.it

Distributor in the UK and Ireland
Gazelle Books Services Ltd.
White Cross Mills
Hightown
Lancaster LA1 4XS
United Kingdom
fax: +44 1524 63232
sales@gazellebooks.co.uk

Distributor in the USA and Canada
IOS Press, Inc.
4502 Rachael Manor Drive
Fairfax, VA 22032
USA
fax: +1 703 323 3668
iosbooks@iospress.com

Proprietà Letteraria Riservata
Printed in Italy

Supported by

Istituto Nazionale di Fisica Nucleare (INFN)

INDICE

H. FELDMEIER and T. NEFF – Fermionic Molecular-Dynamics — clusters, halos, skins and S-factors

A. TUMINO – Tests of clustering in light nuclei and applications to nuclear astrophysics

Preface

The purpose of the CLXIX Course of the "Enrico Fermi" Varenna School was to give an account of recent advances and new perspectives in the study of nuclei far from stability both from the experimental and the theoretical point of view.

Experimental studies of exotic nuclei are currently being performed in several laboratories and new facilities with high-intensity beams are either just completed, or approved and under construction or in their planning stages. At the School an overview of several facilities, RIKEN-RIBF, CERN-ISOLDE, GANIL-SPIRAL2 and GSI-FAIR was given, as well as of the planned FRIB facility in the U.S., with the aim of presenting to the students what will be the experimental scenario in the next 10 years.

The bulk of the School was devoted to nuclear structure models and their derivation from the basic nucleon-nucleon interaction. Three models were extensively discussed: the shell model, the interacting boson model and the cluster model.

In recent years, considerable advance has been made in *ab initio* theories of nuclei, especially of light nuclei. These were also presented at the School thus providing a comprehensive view of the present status of nuclear structure theory.

Another aspect that was discussed was the occurrence of dynamic symmetries and super-symmetries in nuclei, including the newly suggested "critical" symmetries which occur in transitional nuclei when the shape changes from spherical to deformed.

Nuclei far from stability are of particular importance for astrophysics, especially for the r-process. Nuclear Astrophysics was reviewed in detail, including a discussion of energy generation in stars and nucleo-synthesis of elements.

Another aspect of nuclear structure physics is its role in understanding fundamental processes, such as electroweak processes. For example, nuclei offer the opportunity of measuring the neutrino mass through the observation of neutrinoless double-beta decay. A review of the experimental and theoretical situation for double-beta decay both with and without neutrinos was presented at the School.

The main lectures were complemented by seminars on issues of current interest in nuclear structure, some of them being given by students. The students were also given the opportunity to make use of some of the computer programs needed in nuclear structure model calculations.

This CLXIX Course was dedicated to Renato Angelo Ricci on the occasion of his 80th birthday. It was a pleasure to celebrate this joyful event, which gave us the opportunity to recognize the many contributions of Renato to the field of Nuclear Physics, especially those to the structure of fp-shell nuclei.

We gratefully acknowledge the financial support of the Istituto Nazionale di Fisica Nucleare (INFN). We also express our special thanks to Barbara Alzani, Lorenzo Corengia, and Marta Pigazzini for their very efficient and friendly help during the whole School.

A. COVELLO and F. IACHELLO

Acknowledgments

80 years are quite a lot. More than 50 years with nuclear physics and most of them with nuclear structure investigation represent a considerable part of my life and scientific career. This quite a long journey has been gratified not only by interesting activities and results in a pleasant field of research but also by the communion with a lot of nice people, excellent in physics and wonderful as friends.

No doubt that this Course in Varenna, where nuclear spectroscopy, which has been for a long time my preferred field of research, plays an important role, will deserve the status and the perspectives of nuclear structure investigations in a very promising manner.

The fact that the evolution of the understanding of the nuclear structure problems from stable to very exotic nuclei is an essential piece in the history of physics and for the future of the incoming investigations is also very gratifying.

Therefore it has been a pleasure and an honour for me to contribute with Aldo Covello, Franco Iachello and Giuseppe Maino in promoting this Course in the wonderful surrounding of Villa Monastero, a place where I did cover a large part of my engagement in physics and as a President of the Italian Physical Society.

My acknowledgments are therefore also addressed to the President Giuseppe Bassani and to the Council of SIF as well as to the President Roberto Petronzio and the Council of INFN, who have supported and sponsored this School.

Many thanks also to all my friends, pupils and colleagues.

R. A. RICCI

Società Italiana di Fisica
SCUOLA INTERNAZIONALE DI FISICA «E. FERMI »
CLXIX CORSO - VARENNA SUL LAGO DI COMO
VILLA MONASTERO 17 - 27 Luglio 2007

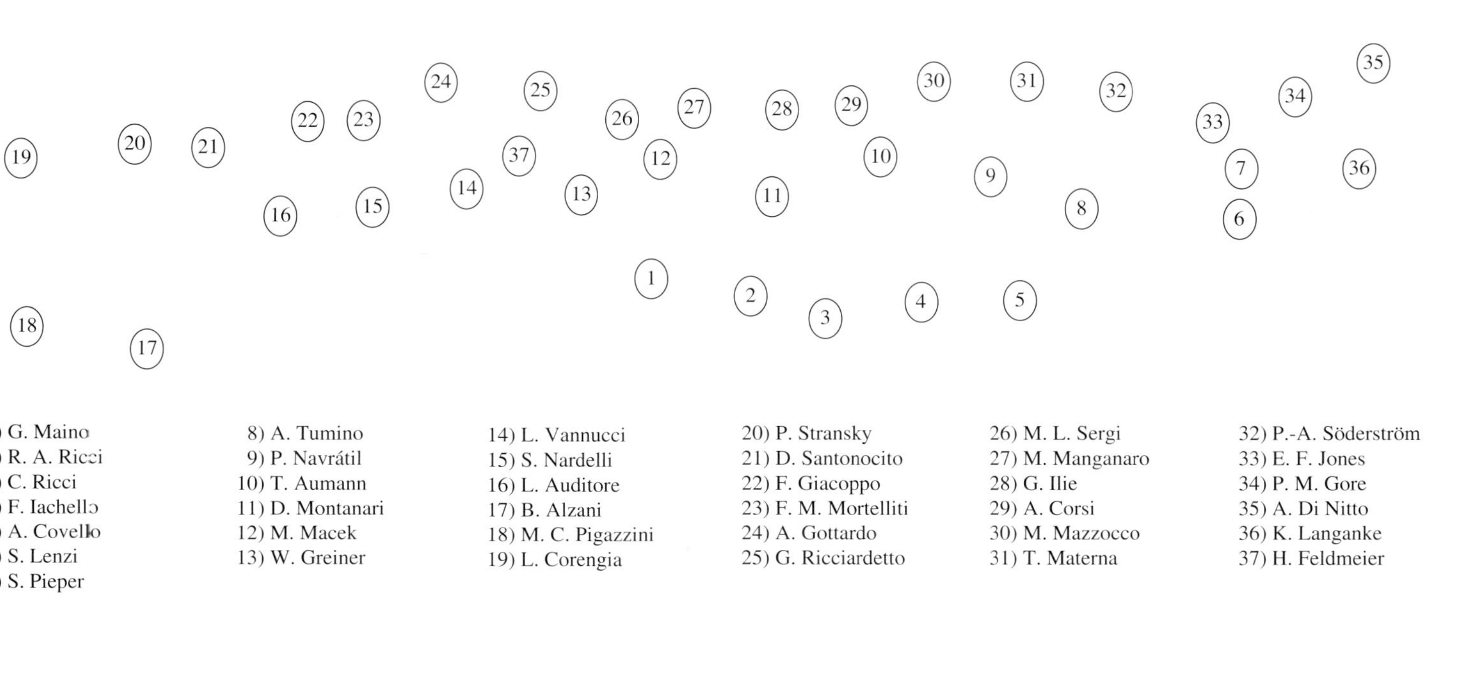

1) G. Maino
2) R. A. Ricci
3) C. Ricci
4) F. Iachello
5) A. Covello
6) S. Lenzi
7) S. Pieper
8) A. Tumino
9) P. Navrátil
10) T. Aumann
11) D. Montanari
12) M. Macek
13) W. Greiner
14) L. Vannucci
15) S. Nardelli
16) L. Auditore
17) B. Alzani
18) M. C. Pigazzini
19) L. Corengia
20) P. Stransky
21) D. Santonocito
22) F. Giacoppo
23) F. M. Mortelliti
24) A. Gottardo
25) G. Ricciardetto
26) M. L. Sergi
27) M. Manganaro
28) G. Ilie
29) A. Corsi
30) M. Mazzocco
31) T. Materna
32) P.-A. Söderström
33) E. F. Jones
34) P. M. Gore
35) A. Di Nitto
36) K. Langanke
37) H. Feldmeier

DOI 10.3254/978-1-58603-885-4-1

Intellectual challenges at radioactive beam facilities

F. IACHELLO

Center for Theoretical Physics, Sloane Physics Laboratory, Yale University
New Haven, CT 06520-8120, USA

Summary. — Three of the fundamental concepts that can be tested at radioactive beam facilities: a) dynamic symmetries and supersymmetries; b) quantum shape phase transitions; c) critical symmetries; are briefly discussed.

1. – Introduction

Nuclear physics is experiencing a *Renaissance* as a result of the possibility of investigating nuclei far from stability, both proton-rich and neutron-rich, by means of radioactive beam facilities. At the same time, the development of computing capabilities has open the way to several theoretical advances, some of which are: i) The possibility to do *ab initio* calculations through the Monte Carlo Green's function method for nuclei with $A \lesssim 12$. ii) The re-visitation of old models, especially the shell model, with the development of a) the no-core shell model for nuclei with $A \lesssim 24$, b) the large-scale shell model for nuclei with $A \lesssim 70$ and c) the Monte Carlo shell model for ground state properties of nuclei. iii) The recasting of old theories, especially mean-field theories in terms of densities functional theories, again for ground-state properties of nuclei. iv) The implementation and improvement of methods used in other fields, such as molecular dynamics, especially for light nuclei. One of the purposes of this school is to give an account of the recent developments both in theory and experiment.

The atomic nucleus can be characterized in various ways: i) a strongly correlated Fermi liquid; ii) a mesoscopic system (2–250 particles) intermediate between a few-body system (2-4 particles) and a condensed-matter system (10^{23} particles); c) a unique many-body system with strong (and short-range) interactions. Because of the nature of the interactions, the nucleus develops a variety of collective modes, three of which are: a) clustering, especially α-clustering; b) surface deformation, especially quadrupole deformation; c) giant oscillations, especially dipole oscillations. In view of these complications, several *models* have been developed in the last 60 years to describe properties of nuclei:

1) the shell model (Jensen, Goeppert-Mayer and Suess, 1947);

2) the collective model (Bohr and Mottelson, 1952);

3) the cluster model (Wildermuth and Kannellopoulos, 1958);

4) the interacting boson model (Arima and Iachello, 1974).

These models, especially 1), 2) and 4), have been used extensively to interpret the enormous amount of data accumulated in these years. This School is intended to give a background on models, methods, theories of nuclear structure and their recent developments.

In addition to providing tests of models, theories and methods, atomic nuclei have given the possibility to test *fundamental concepts* in physics. Radioactive beam facilities will provide a new pool of nuclei where fundamental concepts of physics can be studied. The exploration of fundamental concepts, the eventual introduction of others and their experimental verification are the *intellectual challenges* of radioactive beam facilities. In this paper, three of the fundamental concepts that can be tested at radioactive beam facilities will be briefly discussed:

A) Dynamic symmetries and supersymmetries.

B) Quantum (shape) phase transitions.

C) Critical symmetries.

2. – Dynamic symmetries and supersymmetries

Dynamic symmetries and supersymmetries appear within the context of several areas of physics, ranging from molecular to atomic, from nuclear to particle physics. Here, I will discuss them within the context of the Interacting Boson Model [1-4]. In this model, even-even nuclei are described in terms of correlated pairs of nucleons, treated as bosons, as sketched in fig. 1. This model will be discussed in subsequent lectures by Casten [5] and van Isacker [6]. Here suffices to say that the model, in its simplest version, has dynamic algebra $U(6)$ and three dynamic symmetries:

$$
\begin{array}{llllllllll}
U(6) & \supset & U(5) & \supset & SO(5) & \supset & SO(3) & \supset & SO(2) & \quad\text{(I)} \\
U(6) & \supset & SU(3) & \supset & SO(3) & \supset & SO(2) & & & \quad\text{(II)} \\
U(6) & \supset & SO(6) & \supset & SO(5) & \supset & SO(3) & \supset & SO(2) & \quad\text{(III)}
\end{array}
$$

Dynamic symmetries are situations in which the Hamiltonian operator can be expanded into invariant (Casimir) operators of a chain of algebras $G \supset G' \supset G'' \supset \dots$. When

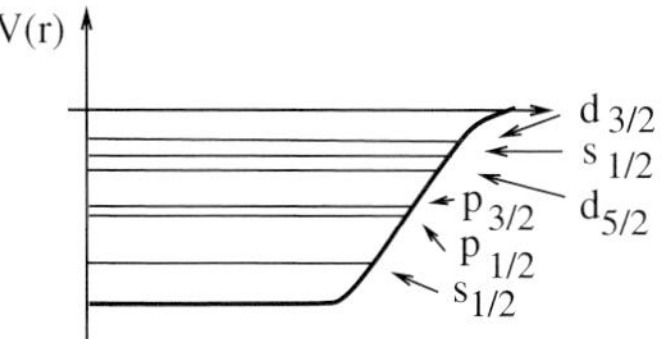

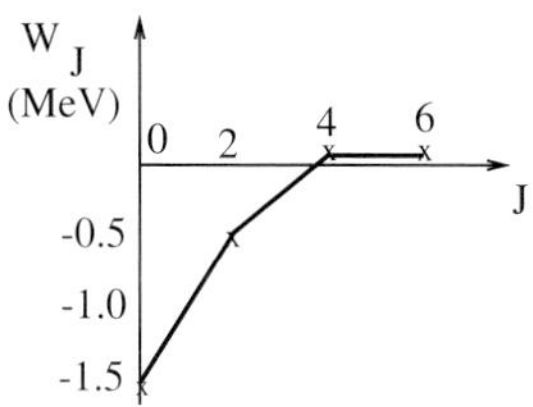

Fig. 1. – A sketch of nuclear structure. Top part: the single-particle potential in nuclei (central field). Central part: the matrix elements of the effective interaction between identical particles in the $f_{7/2}$ shell (correlations). Bottom part: correlated s and d pairs in nuclei (Cooper pairs).

a dynamic symmetry occurs, all properties of the system can be calculated in explicit analytic form. These forms provide a simple way in which experimental data can be analyzed. Dynamic symmetries appear not only in simple systems, such as the hydrogen atom, but also in complex systems, hence the name *simplicity in complexity* given to this program. Many examples of dynamic symmetries in nuclei have been found, three of which are shown in fig. 2 $[U(5)]$, fig. 3 $[SU(3)]$, and fig. 4 $[SO(6)]$. In recent years it has been shown that dynamic symmetries persist in nuclei even to high excitation energies, as shown in fig. 5 [7].

The interacting boson model describes even-even nuclei. For even-odd, odd-even and odd-odd nuclei one needs to consider, in addition to correlated pairs of nucleons, also individual nucleons (fermions), and one is led then to the interacting boson-fermion model [8]. The dynamic algebra of this model is the superalgebra $U(6/\Omega)$, where Ω is the

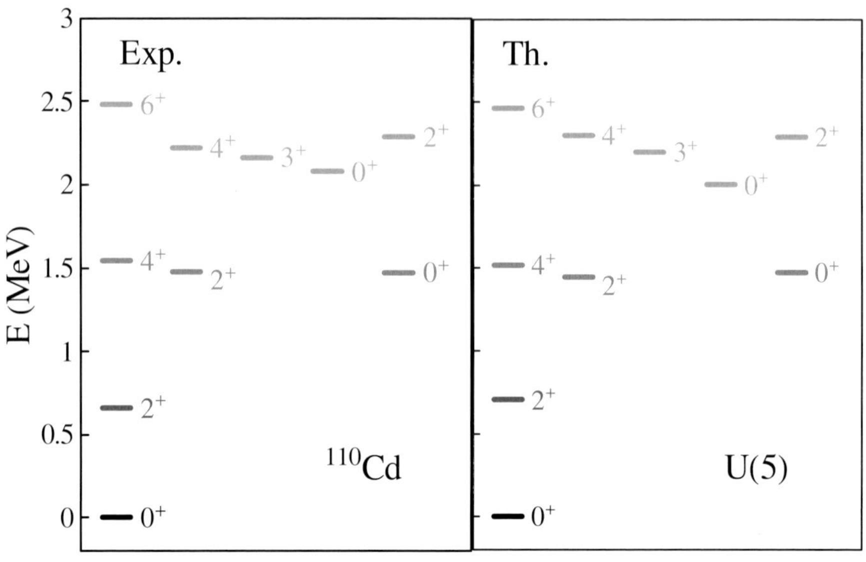

Fig. 2. – An example of $U(5)$ symmetry: $^{110}_{48}\text{Cd}_{62}$. Adapted from [1].

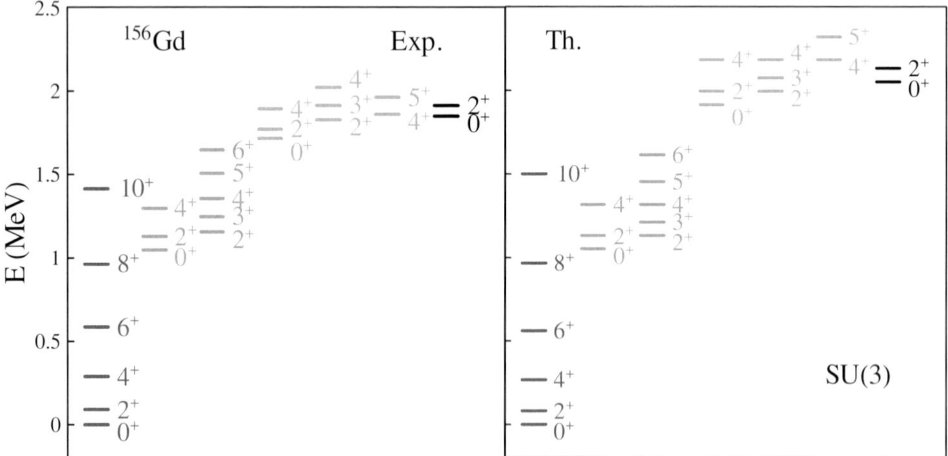

Fig. 3. – An example of $SU(3)$ symmetry: $^{156}_{64}\text{Gd}_{92}$. Adapted from [2].

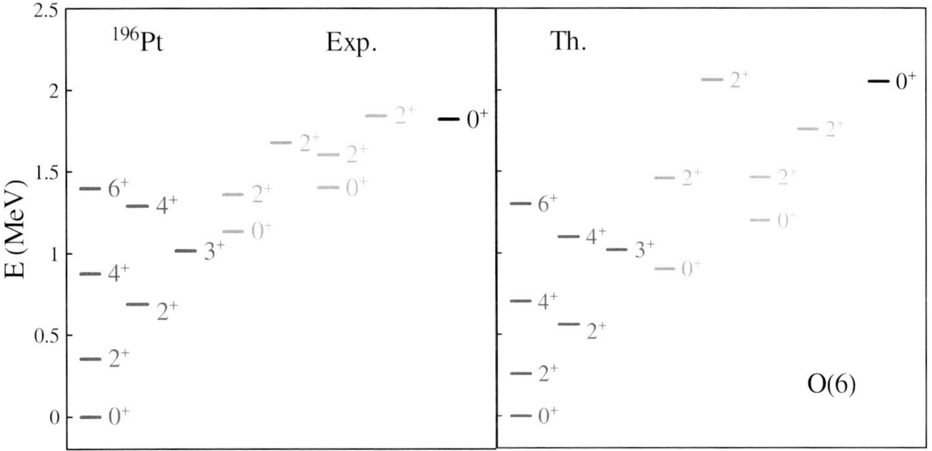

Fig. 4. – An example of $SO(6)$ symmetry: $^{196}_{78}$Pt$_{118}$. Adapted from [3].

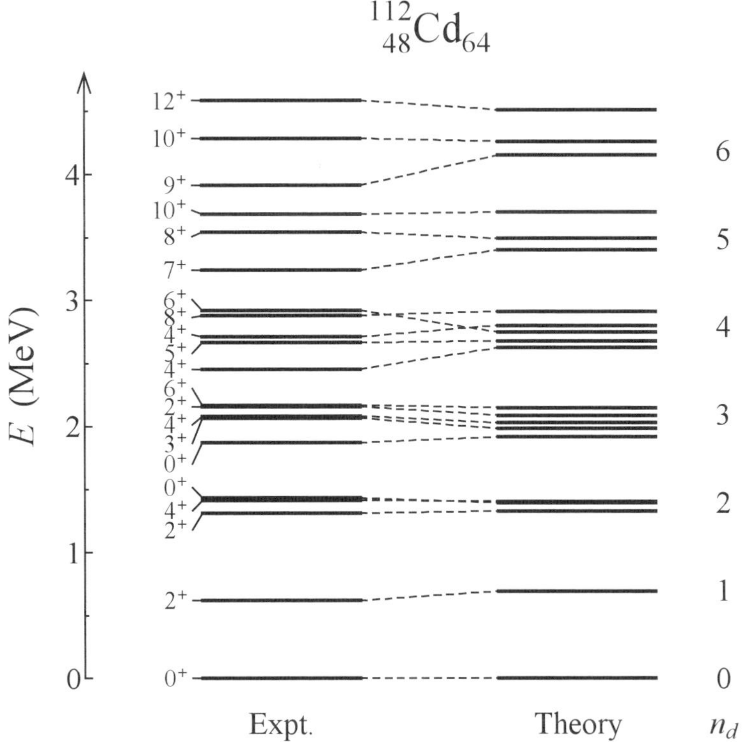

Fig. 5. – A recent example of $U(5)$ symmetry showing the persistance of symmetry at high excitation energy: $^{112}_{48}$Cd$_{64}$ [7].

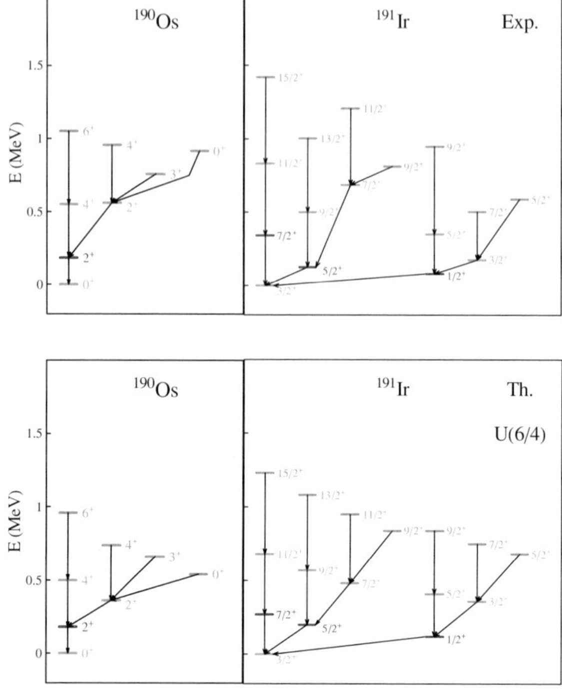

Fig. 6. – An example of $U(6/4)$ supersymmetry: $^{190}_{76}\text{Os}_{114}$-$^{191}_{77}\text{Ir}_{114}$. Adapted from [9].

degeneracy of single-particle orbitals, $\Omega = \sum_j (2j + 1)$. As in the case of the interacting boson model, it is possible also here to construct explicit solutions to the eigenvalue problem for the Hamiltonian

$$(1) \qquad\qquad H = H_{\mathrm{B}} + H_{\mathrm{F}} + V_{\mathrm{BF}} \,,$$

where H_{B} is the Hamiltonian describing the bosons, H_{F} is the Hamiltonian describing the fermions and V_{BF} is their interaction. Within the framework of the interacting boson-fermion model, there are possible several classes of supersymmetry, depending on the symmetry of the bosons and their associated fermions. Some examples of supersymmetry in nuclei have been found, fig. 6 [9], including a very recent example in the spectrum of an odd-odd nucleus, fig. 7 [10]. These are the only examples so far of supersymmetry in physics. The situation is summarized in fig. 8, where a symmetry classification of nuclei is presented [11].

One of the challenges of radioactive beam facilities is to find other examples of *dynamic symmetries and supersymmetries* and to see whether or not other types of symmetry and/or supersymmetry are present in the spectra of nuclei far from stability.

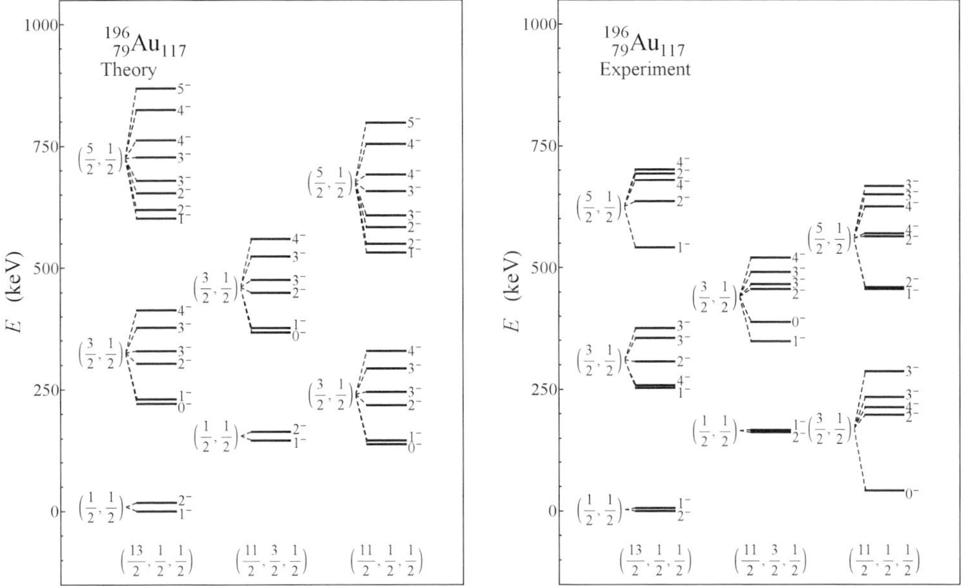

Fig. 7. – An example of supersymmetry in odd-odd nuclei: $^{196}_{79}\text{Au}_{117}$ [10].

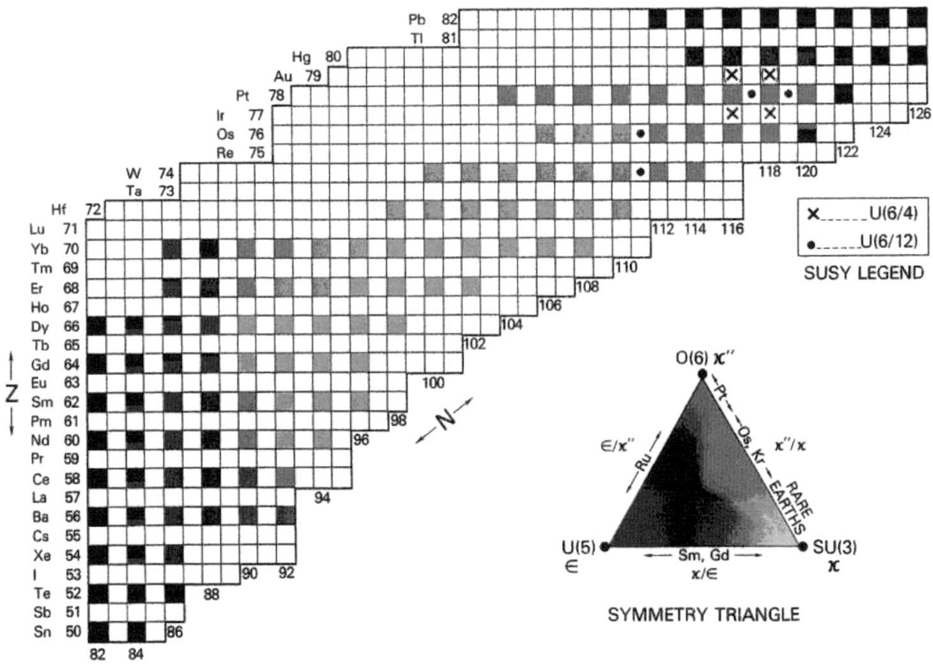

Fig. 8. – A representation of a portion of the nuclear chart in terms of symmetries. Adapted from [11].

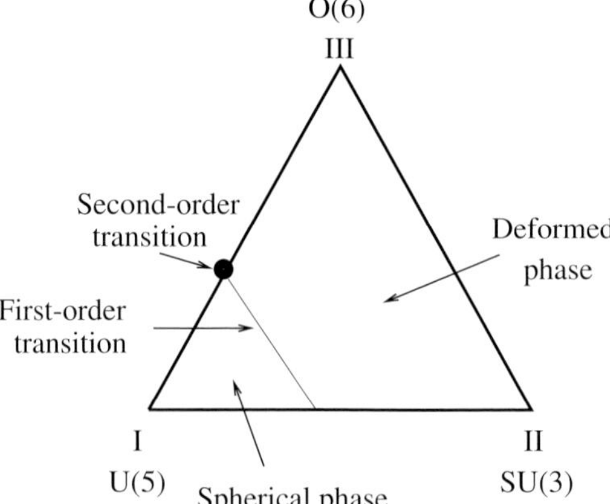

Fig. 9. – Phase diagram of nuclei in the interacting boson model. Adapted from [15].

3. – Shape phase transitions

Atomic nuclei experience several types of phase transitions. One of them is the phase transition associated with the change in shape of the ground state (zero temperature phase transition) [12, 13]. These phase transitions have been termed "quantum phase transitions" since they occur at zero temperature and are driven by quantum fluctuations. The *control* parameter is a coupling constant, g, that appears in the Hamiltonian governing the system

$$(2) \qquad\qquad H = (1 - g)H_1 + gH_2.$$

When $g = 0$, the system is in Phase 1, while when $g = 1$ the system is in Phase 2. The two phases have different symmetry. For some value g_c, the critical value, the system undergoes a phase transition from Phase 1 to Phase 2.

Shape phase transitions can be simply analyzed within the framework of the interacting boson model [14, 15]. The phase diagram of this model is shown in fig. 9. There are two phases, the spherical and the deformed phase, separated by a line of first-order transitions ending in a point of second-order transition. Examples of phase transitions in nuclei have been found. Two signatures of phase transitions have been used: i) Erhenfest signature, that is the study of the ground-state energy and its derivatives as a function of the control parameter and ii) Landau signature, that is the study of the order parameter as a function of the control parameter. It turns out that in nuclei the control parameter is proportional to the number of particles. (This statement is true for any system for which H_1 is a one-body operator and H_2 is a two-body operator.) An example is the Nd-Sm-Gd region where both the Erhenfest and Landau signatures show evidence for a phase transition, figs. 10, 11 [16]. One of the challenges of radioactive beam facilities is understanding whether or not *other phase transitions* are present in nuclei. In particular, in neutron-rich nuclei, a measurement of masses (ground-state energies) and

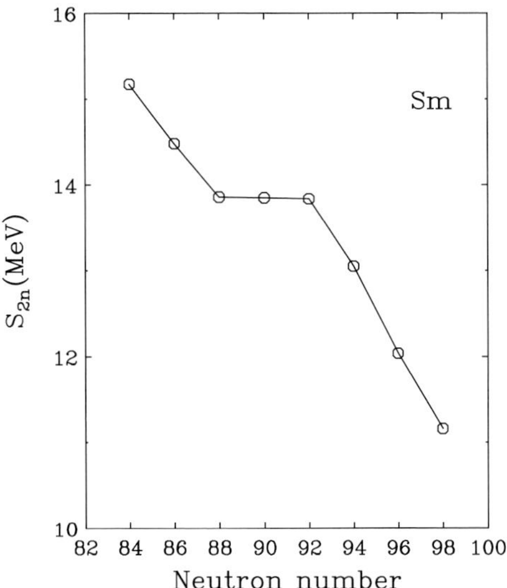

Fig. 10. – Two-neutron separation energies, S_{2n}, as a function of neutron number in the Sm isotopes.

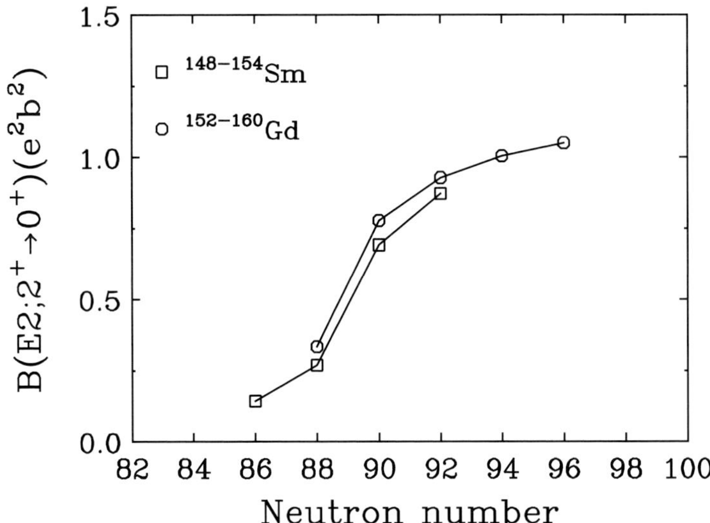

Fig. 11. $B(E2; 2_1^+ \to 0_1^+)$ values as a function of neutron number in Sm and Gd.

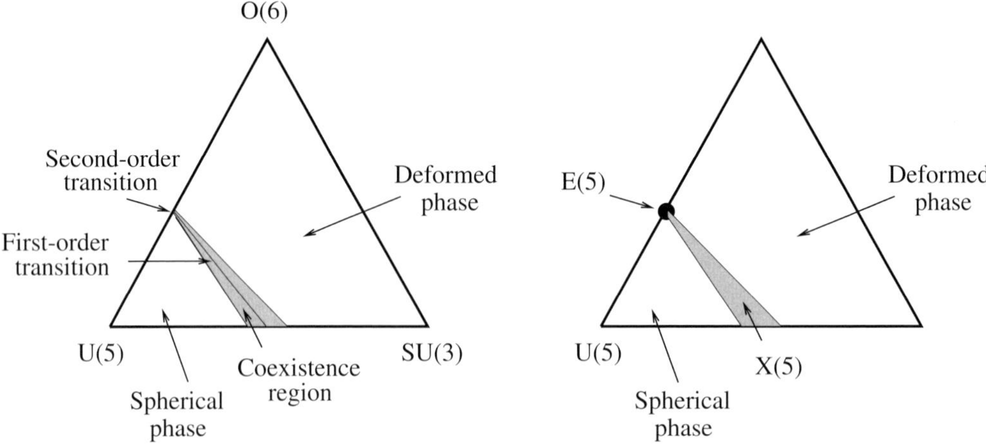

Fig. 12. – Phase diagram of the interacting boson model showing on the right-hand side the location of the analytic descriptions $E(5)$ and $X(5)$.

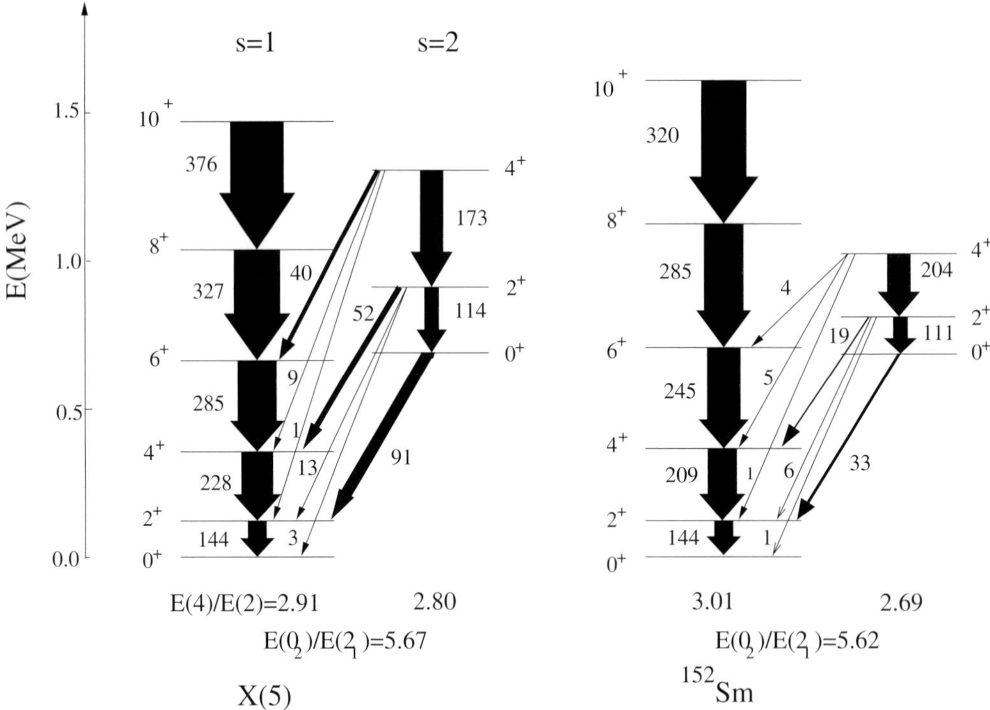

Fig. 13. – Comparison between the experimental spectrum and electromagnetic transition rates in ^{152}Sm with the analytic description $X(5)$. From [23].

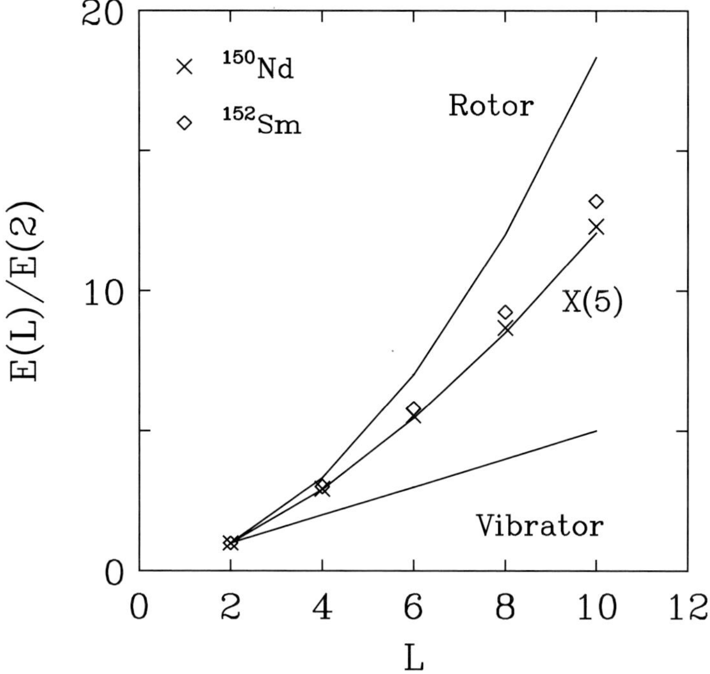

Fig. 14. – Energies in the ground-state band of Sm and Nd as a function of angular momentum L. From [23].

$B(E2; 2_1 \rightarrow 0_1)$ values can determine whether or not a phase transition occurs. Also, because of the large neutron excess, the role of the isospin degree of freedom can be investigated [17, 18].

4. – Critical symmetries

An intriguing (and surprising) result has been found recently: the structure of nuclei at the critical point of a second-order transition and along the critical line of a first-order transition is simple. The energy eigenvalues are well described by the squares of zeros of Bessel functions. These simple analytic descriptions have been termed "critical symmetries" and discussed [19, 20] within the framework of the collective model, as solutions (exact or approximate) of the Bohr Hamiltonian [21]. Since the collective model is formulated in terms of geometric rather than algebraic variables, these symmetries are placed on the right-hand side of fig. 12, which summarizes both the algebraic and geometric symmetries found so far. Examples of both the $E(5)$ and $X(5)$ analytic descriptions have been found [22-24]. One of them is shown in fig. 13. In fig. 14, the behavior of the energies in the ground-state band of the critical nuclei is shown to agree with the simple expression in terms of zeros of Bessel functions. Another challenge of radioactive beam facilities is to find *other examples of critical symmetries.*

5. – Conclusions

Radioactive beam facilities will provide tests of i) microscopic theories of nuclei and of ii) nuclear models. In addition, they will test fundamental concepts in physics, three of which (dynamic symmetries and supersymmetries, quantum phase transitions and critical symmetries) have been discussed here.

$$* * *$$

This work was performed in part under DOE Grant No. DE-FG-02-91ER40608.

REFERENCES

[1] ARIMA A. and IACHELLO F., *Ann. Phys. (N.Y.)*, **99** (1976) 253.
[2] ARIMA A. and IACHELLO F., *Ann. Phys. (N.Y.)*, **111** (1978) 201.
[3] ARIMA A. and IACHELLO F., *Ann. Phys. (N.Y.)*, **123** (1979) 468.
[4] For a review see, IACHELLO F. and ARIMA A., *The Interacting Boson Model* (Cambridge University Press, Cambridge) 1987.
[5] CASTEN R. F., this volume, p.
[6] VAN ISACKER P., this volume p.
[7] DELEZE M., DRISSI S., JOLIE J. *et al.*, *Nucl. Phys. A*, **554** (1993) 1.
[8] For a review see, IACHELLO F. and VAN ISACKER P., *The Interacting Boson-Fermion Model* (Cambridge University Press, Cambridge) 1991.
[9] IACHELLO F., *Phys. Rev. Lett.*, **44** (1980) 772.
[10] METZ A., JOLIE J., GRAW G. *et al.*, *Phys. Rev. Lett.*, **83** (1999) 1542.
[11] CASTEN R. F. and FENG D. H., *Physics Today* **37**, No. 11 (1984) 26.
[12] GILMORE R., *J. Math. Phys.*, **20** (1979) 891.
[13] For a review see, IACHELLO F., *Proceedings of the International School "Enrico Fermi", Course CLIII*, edited by MOLINARI A., RICCATI L., ALBERICO W. M. and MORANDO M. (IOS Press, Amsterdam) (2003) p. 1.
[14] DIEPERINK A. E. L., SCHOLTEN O. and IACHELLO F., *Phys. Rev. Lett.*, **44** (1980) 1747.
[15] FENG D. H., GILMORE R. and DEANS S. R., *Phys. Rev. C*, **23** (1981) 1254.
[16] SCHOLTEN O., IACHELLO F. and ARIMA A., *Ann. Phys. (N.Y.)*, **115** (1978) 325.
[17] CAPRIO M. A. and IACHELLO F., *Phys. Rev. Lett.*, **93** (2004) 242502.
[18] ARIAS J. M., DUKELSKY J. and GARCIA-RAMOS J. E., *Phys. Rev. Lett.*, **93** (2004) 212501.
[19] IACHELLO F., *Phys. Rev. Lett.*, **85** (2000) 3580.
[20] IACHELLO F., *Phys. Rev. Lett.*, **87** (2001) 052502.
[21] BOHR A., *Mat. Fys. Medd. Dan. Vid. Selsk.* **26**, No. 14 (1952).
[22] CASTEN R. F. and ZAMFIR N. V., *Phys. Rev. Lett.*, **85** (2000) 3584.
[23] CASTEN R. F. and ZAMFIR N. V., *Phys. Rev. Lett.*, **87** (2001) 052503.
[24] For a review see, CASTEN R. F. and MCCUTCHAN E. A., *J. Phys. G: Nucl. Part. Phys.*, **34** (2007) R285.

DOI 10.3254/978-1-58603-885-4-13

Superheavy and giant nuclear systems

W. Greiner

Frankfurt Institute for Advanced Studies, J.W. Goethe-Universität
60438 Frankfurt am Main, Germany

V. Zagrebaev

Flerov Laboratory of Nuclear Reaction, JINR - Dubna, 141980, Moscow region, Russia

Summary. — The problem of production and study of superheavy elements is discussed in this paper. Different nuclear reactions leading to formation of superheavy nuclei are analyzed. Collisions of transactinide nuclei are investigated as an alternative way for production of neutron-rich superheavy elements. In many events lifetime of the composite giant nuclear system consisting of two touching nuclei turns out to be rather long ($\geq 10^{-20}$ s); sufficient for observing line structure in spontaneous positron emission from super-strong electric fields, a fundamental QED process.

Dedication

I dedicate this talk to my friend Professor Renato Ricci on occasion of his 80th birthday. I have known him for nearly 40 years. Our first encounter was here at Varenna at the International School on Nuclear Physics, which Renato organized and which was guided by Victor Weisskopf. I had just completed my PhD and was a student at that time. For very many years we met at the island Hvar in Croatia. Nicola Cindro was the third important member of the triplet (Cindro, Ricci and myself). Very important and joyful were our meetings with D. Allan Bromley at Yale. We spend pleasant evenings and dinners together with Mrs. Ricci, Mrs. Bromley and also with my wife Barbara. With the subject of my talk I would like to remind Renato that extremely important physics on time-delay can also be done (for lighter systems) at Legnaro. I urge Renato to stimulate and pave the way for this work there. Dear Renato, stay healthy and be happy for another 20 years. God bless you!

1. – Introduction

Superheavy (SH) nuclei obtained in "cold" fusion reactions with Pb or Bi target [1] are along the proton drip line and very neutron-deficient with a short half-life. In fusion of actinides with ^{48}Ca more neutron-rich SH nuclei are produced [2] with much longer half-life. But they are still far from the center of the predicted "island of stability" formed by the neutron shell around $N = 184$ (see the nuclear map in fig. 1). Unfortunately a small gap between the superheavy nuclei produced in ^{48}Ca-induced fusion reactions and those which were obtained in the "cold" fusion reactions is still remain which should be filled to get a unified nuclear map.

In the "cold" fusion, the cross-sections of SH nuclei formation decrease very fast with increasing charge of the projectile and become less than 1 pb for $Z > 112$ (see fig. 1). Heaviest transactinide, Cf, which can be used as a target in the second method, leads to the SH nucleus with $Z = 118$ being fused with ^{48}Ca. Using the next nearest elements instead of ^{48}Ca (*e.g.*, ^{50}Ti, ^{54}Cr, etc.) in fusion reactions with actinides is expected less encouraging, though experiments of such kind are planned to be performed. In this connection other ways to the production of SH elements in the region of the "island of stability" should be searched for.

In principle, superheavy nuclei may be produced in explosion of supernova [4]. If the half-life of these nuclei is comparable with the age of the Earth they could be searched for

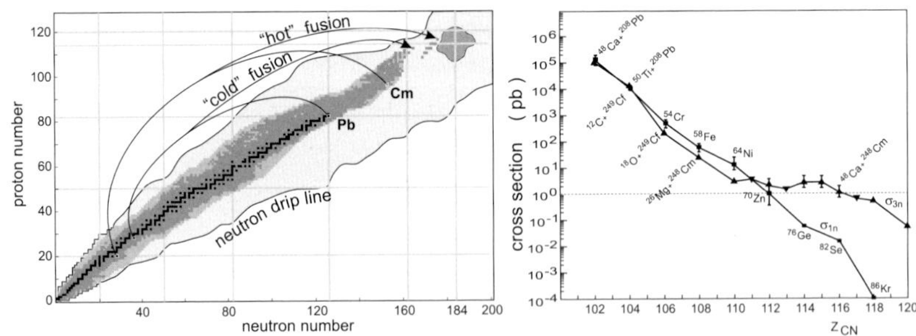

Fig. 1. – Superheavy nuclei produced in "cold" and "hot" fusion reactions. Predicted islands of stability are shown around $Z = 114, 120$ and $N = 184$ (right panel). Experimental and predicted evaporation residue cross-sections for production of superheavy elements produced in "cold" (1n) and "hot" (3n) fusion reactions [3] (left panel).

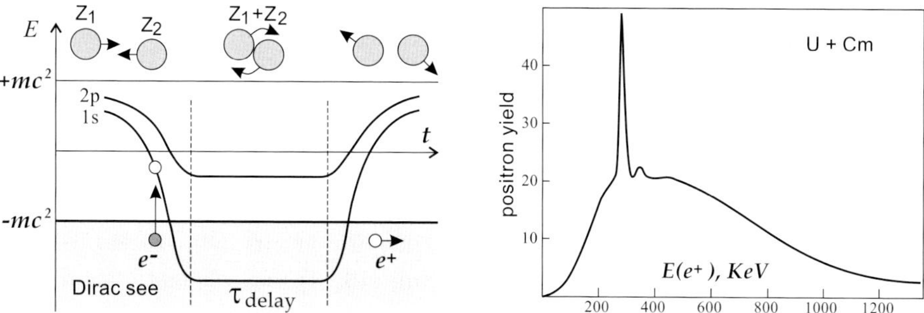

Fig. 2. – Schematic figure of spontaneous decay of the vacuum and spectrum of the positrons formed in supercritical electric field ($Z_1 + Z_2 > 173$).

in nature. However, it is the heightened stability of these nuclei (rare decay) which may hinder from their discovery. To identify these more or less stable superheavy elements supersensitive mass separators should be used. Chemical methods of separation also could be useful here.

About twenty years ago transfer reactions of heavy ions with ^{248}Cm target have been evaluated for their usefulness in producing unknown neutron-rich actinide nuclides [5-7]. The cross-sections were found to decrease very rapidly with increasing atomic number of surviving target-like fragments. However, Fm and Md neutron-rich isotopes have been produced at the level of 0.1 μb. Theoretical estimations for production of primary superheavy fragments in the damped U + U collision have been also performed at this time within the semiphenomenological diffusion model [8]. In spite of obtained high probabilities for the yields of superheavy primary fragments (more than 10^{-2} mb for $Z = 120$), the cross-sections for production of heavy nuclei with low excitation energies were estimated to be rather small: $\sigma_{CN}(Z = 114, E^* = 30\,\text{MeV}) \sim 10^{-6}$ mb for U + Cm collision at 7.5 MeV/nucleon beam energy. The authors concluded, however, that "fluctuations and shell effects not taken into account may conciderably increase the formation probabilities". Such is indeed the case (see below).

Recently a new model has been proposed [9] for *simultaneous* description of all these strongly coupled processes: Deep-Inelastic (DI) scattering, Quasi-Fission (QF), fusion, and regular fission. In this paper we apply this model for analysis of low-energy dynamics of heavy nuclear systems formed in nucleus-nucleus collisions at the energies around the Coulomb barrier. Among others there is the purpose to find an influence of the shell structure of the driving potential (in particular, deep valley caused by the double-shell closure $Z = 82$ and $N = 126$) on formation of compound nucleus (CN) in mass asymmetric collisions and on nucleon rearrangement between primary fragments in more symmetric collisions of actinide nuclei. In the first case, discharge of the system into the lead valley (normal or symmetrizing quasi-fission) is the main reaction channel, which decreases significantly the probability of CN formation. In collisions of heavy transactinide nuclei (U + Cm, etc.), we expect that the existence of this valley may noticeably increase the yield of surviving neutron-rich superheavy nuclei complementary to the Projectile-Like Fragments (PLF) around lead ("inverse" or anti-symmetrizing quasi-fission reaction mechanism).

Direct time analysis of the collision process allows us to estimate also the lifetime of the composite system consisting of two touching heavy nuclei with total charge $Z > 180$. Such "long-living" configurations (if they exist) may lead to spontaneous positron emission from super-strong electric fields of giant quasi-atoms by a static QED process (transition from neutral to charged QED vacuum) [10, 11], see schematic fig. 2.

2. – Nuclear shells

Quantum effects leading to the shell structure of heavy nuclei play a crucial role both in stability of these nuclei and in production of them in fusion reactions. The fission barriers of superheavy nuclei (protecting them from spontaneous fission and, thus, providing their existence) are determined completely by the shell structure. Studies of the shell structure of superheavy nuclei in the framework of the meson field theory and the Skyrme-Hartree-Fock approach show that the magic shells in the superheavy region

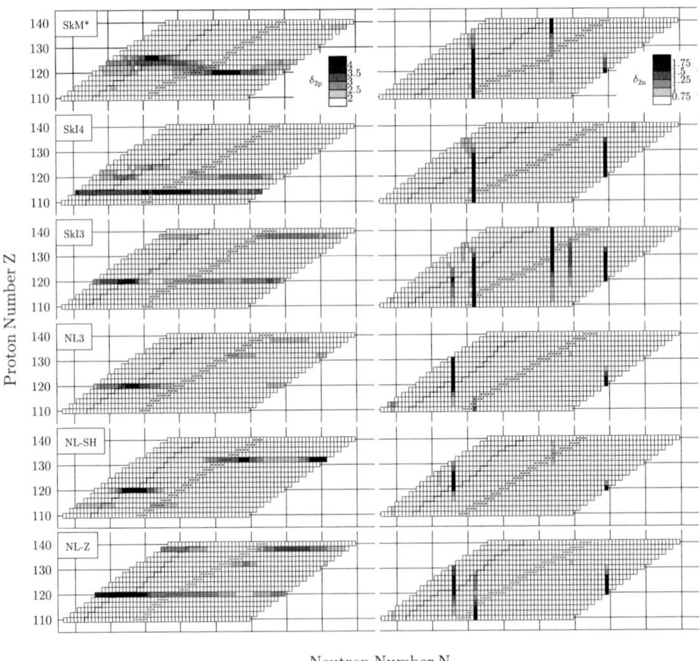

Fig. 3. – Proton (left column) and neutron (right column) gaps in the (N, Z)-plane calculated within the self-consistent Hartree-Fock approach with the forces as indicated [12]. The forces with parameter set SkI4 predict both $Z = 114$ and $Z = 120$ as a magic numbers while the other sets predict only $Z = 120$.

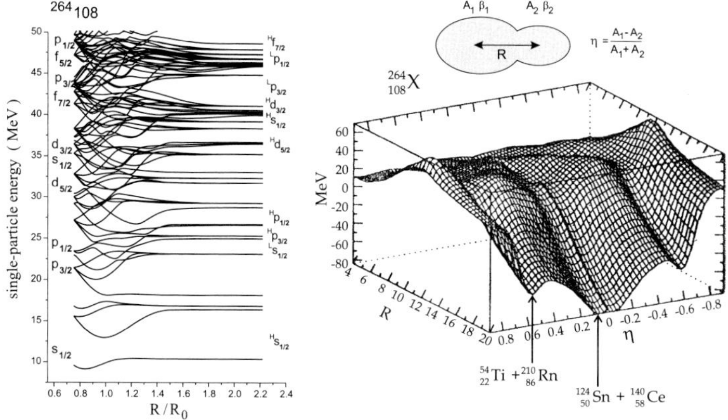

Fig. 4. – Adiabatic energy levels and potential energy surface for the nuclear system $^{264}108$.

are very isotopic dependent [12] (see fig. 3). According to these investigations $Z = 120$ being a magic proton number seems to be as probable as $Z = 114$. Estimated fission barriers for nuclei with $Z = 120$ are rather high though depend strongly on a chosen set of the forces [13].

Interaction dynamics of two heavy nuclei at low (near-barrier) energies is defined mainly by the adiabatic potential energy, which can be calculated, for example, within the two-center shell model [14]. An example of such calculation is shown in fig. 4 for the nuclear system consisting of 108 protons and 156 neutrons. Formation of such heavy nuclear systems in fusion reactions as well as fission and quasi-fission of these systems are regulated by the deep valleys on the potential energy surface (see fig. 4) also caused by the shell effects.

3. – Adiabatic dynamics of heavy nuclear system

At incident energies around the Coulomb barrier in the entrance channel the fusion probability is about 10^{-3} for mass asymmetric reactions induced by ^{48}Ca and much less for more symmetric combinations used in the "cold synthesis". DI scattering and QF are the main reaction channels here, whereas the fusion probability (CN formation) is extremely small. To estimate such a small quantity for CN formation probability, first of all, one needs to be able to describe well the main reaction channels, namely DI and QF. Moreover, the quasi-fission processes are very often indistinguishable from the deep-inelastic scattering and from regular fission, which is the main decay channel of excited heavy compound nucleus.

To describe properly and simultaneously the strongly coupled DI, QF and fusion-fission processes of low-energy heavy-ion collisions we have to choose, first, the unified set of degrees of freedom playing the principal role both at approaching stage and at the stage of separation of reaction fragments. Second, we have to determine the unified potential energy surface (depending on all the degrees of freedom) which regulates all the processes. Finally, the corresponding equations of motion should be formulated to perform numerical analysis of the studied reactions. In contrast with other models, we take into consideration all the degrees of freedom necessary for description of all the reaction stages. Thus, we need not to split artificially the whole reaction into several stages. Moreover, in that case unambiguously defined initial conditions are easily formulated at large distance, where only the Coulomb interaction and zero-vibrations of the nuclei determine the motion. The distance between the nuclear centers R (corresponding to the elongation of a mono-nucleus), dynamic spheroidal-type surface deformations β_1 and β_2, mutual in-plane orientations of deformed nuclei φ_1 and φ_2, and mass asymmetry $\eta = \frac{A_1 - A_2}{A_1 + A_2}$ are probably the relevant degrees of freedom in fusion-fission dynamics.

The two-center shell model [14] seems to be most appropriate for calculation of the adiabatic potential energy surface. A choice of dynamic equations for the considered degrees of freedom is not so evident. The main problem here is a proper description of nucleon transfer and change of the mass asymmetry which is a discrete variable by its nature. The corresponding inertia parameter μ_η, being calculated within the Werner-Wheeler approach, becomes infinite at the contact (scission) point and for separated nuclei. In ref. [9] the inertialess Langevin-type equation for the mass asymmetry has been derived from the corresponding master equation for the distribution function. Finally we use a set of 13 coupled Langevin-type equations for 7 degrees of freedom (relative distance, rotation and dynamic deformations of the nuclei and mass asymmetry) which are solved numerically.

The cross-sections for all the processes are calculated in a simple and natural way. A large number of events (trajectories) are tested for a given impact parameter. Those events, in which the nuclear system overcame the fission barrier from the outside and entered the region of small deformations and elongations, are treated as fusion (CN formation). The other events correspond to quasi-elastic, DI and QF processes. Subsequent decay of the excited CN ($C \rightarrow B + xn + N\gamma$) is described then within the statistical model using an explicit expression for survival probability, which directly takes into account the Maxwell-Boltzmann energy distribution of evaporated neutrons [15]. The double differential cross-sections are calculated as follows

$$(1) \qquad \frac{\mathrm{d}^2\sigma_\eta}{\mathrm{d}\Omega\mathrm{d}E}(E,\theta) = \int_0^\infty b\,\mathrm{d}b \frac{\Delta N_\eta(b,E,\theta)}{N_{\mathrm{tot}}(b)} \frac{1}{\sin(\theta)\Delta\theta\Delta E}.$$

Here $\Delta N_\eta(b, E, \theta)$ is the number of events at a given impact parameter b in which the system enters into the channel η (definite mass asymmetry value) with kinetic energy in the region $(E, E+\Delta E)$ and center-of-mass outgoing angle in the region $(\theta, \theta+\Delta\theta)$, $N_{\mathrm{tot}}(b)$ is the total number of simulated events for a given value of impact parameter. In collisions of deformed nuclei an averaging over initial orientations is performed. Expression (1) describes the mass, energy and angular distributions of the *primary* fragments formed in the binary reaction (both in DI and in QF processes). Subsequent de-excitation cascades of these fragments via emission of light particles and gamma-rays in competition with fission were taken into account explicitly for each event within the statistical model leading to the *final* mass and energy distributions of the reaction fragments. The model allows us to perform also a time analysis of the studied reactions. Each tested event is characterized by the reaction time τ_{int}, which is calculated as a difference between re-separation (scission) and contact times.

4. – Deep inelastic scattering of heavy nuclei

At first we applied the model to describe available experimental data on low-energy damped collision of very heavy nuclei, ^{136}Xe + ^{209}Bi [16], where the DI process should dominate due to expected prevalence of the Coulomb repulsion over nuclear attraction. The adiabatic potential energy surface of this nuclear system is shown on the left panel of fig. 5 in the space of elongation and mass asymmetry at zero dynamic deformations. The colliding nuclei are very compact with almost closed shells and the potential energy has only one deep valley (just in the entrance channel, $\eta \sim 0.21$) giving rather simple mass distribution of the reaction fragments. In that case the reaction mechanism depend mainly on the nucleus-nucleus potential at contact distance, on the friction forces at this region (which determine the energy loss) and on nucleon transfer rate at contact. Note that there is a well-pronounced plateau at contact configuration in the region of zero mass asymmetry (see fig. 5, right panel). It becomes even lower with increasing the deformations and corresponds to formation of the nuclear system consisting of strongly deformed touching fragments ^{172}Er + ^{173}Tm (see fig. 5), which means that a significant mass rearrangement may occur here leading to additional time delay of the reaction.

On the right panel of fig. 5 the landscape of the potential energy is shown at contact configuration depending on mass asymmetry and deformation of the fragments. As can be seen, after contact and before re-separation the nuclei aim to become more deformed. Moreover, beside a regular diffusion (caused by the fluctuations), the final mass distribution is determined also by the two well marked driving paths leading the system to

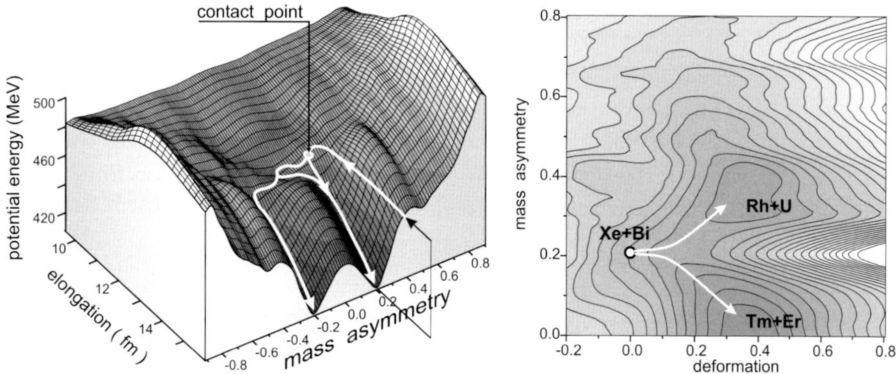

Fig. 5. – Driving potential for the nuclear system formed in ^{136}Xe + ^{209}Bi collision at fixed deformations (left) and at contact configuration (right). The solid lines with arrows show schematically (without fluctuations) most probable trajectories.

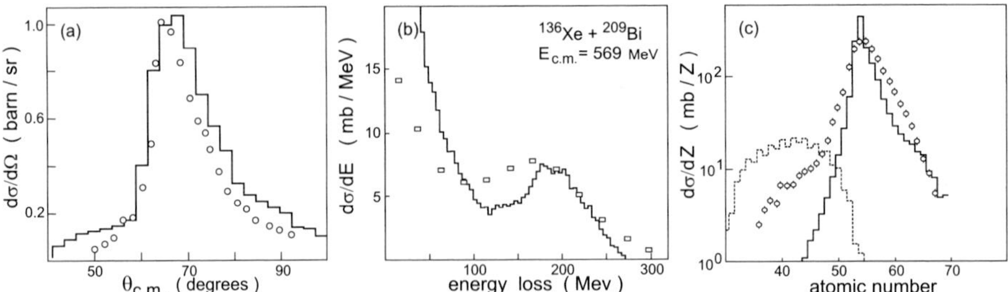

Fig. 6. – Angular (a), energy-loss (b) and charge (c) distributions of the Xe-like fragments obtained in the ^{136}Xe + ^{209}Bi reaction at $E_{c.m.} = 568\,$MeV. Experimental data are taken from ref. [16]. For other notations see the text.

more and to less symmetric configurations. They are not identical and this leads to the asymmetric mass distribution of the primary fragments, see fig. 6(c).

In fig. 6 the angular, energy and charge distributions of the Xe-like fragments are shown comparing with our calculations (histograms). In accordance with experimental conditions only the events with the total kinetic energy in the region of $260 \leq E \leq 546\,$MeV and with the scattering angles in the region of $40° \leq \theta_{c.m.} \leq 100°$ were accumulated. The total cross-section corresponding to all these events is about $2200\,$mb (experimental estimation is $2100\,$mb [16]). Due to the rather high excitation energy sequential fission of the primary heavy fragments may occur in this reaction (mainly those heavier than Bi). In the experiment the yield of the heavy fragments was found to be about 30% less comparing with Xe-like fragments. Our calculation gives $354\,$mb for the cross-section of sequential-fission, which is quite comparable with experimental data. Mass distribution of the fission fragments is shown in fig. 6(c) by the dotted histogram. Note that it is a contamination with sequential fission products of heavy primary fragments leading to the bump around $Z = 40$ in the experimental charge distribution.

At the second step we analyzed the reaction ^{86}Kr + ^{166}Er at $E_{c.m.} = 464\,$MeV [17], in which the nuclear attractive forces may lead, in principle, to the formation of a mononucleus and of CN. The adiabatic potential energy surface, QF and Fusion-Fission (FF) processes should in this case play a more important role. For the analysis of this reaction we used the same value of the nucleon transfer rate and the same friction forces as in the previous case. For the nuclear viscosity we choose the value $\mu_0 = 2 \cdot 10^{-22}\,$MeV s fm^{-3} because of intermediate values of excitation energies available here as compared with the previous reaction.

The interaction time is one of the most important characteristics of nuclear reactions, though it cannot be measured directly. It depends strongly on the reaction channel. The time distribution of all the ^{86}Kr + ^{166}Er collisions at $E_{c.m.} = 464\,$MeV, in which the kinetic energy loss is higher than $35\,$MeV, is shown in fig. 7. The interaction time was calculated starting from $t = 0$ at $R = R_{max} = 40\,$fm up to the moment of scission into two fragments $(R > R_{scission}, p_R > 0)$ or up to CN formation. The approaching time (path from R_{max} to $R_{contact}$) in the entrance channel is very short $(4\text{–}5 \cdot 10^{-22}\,$s depending on

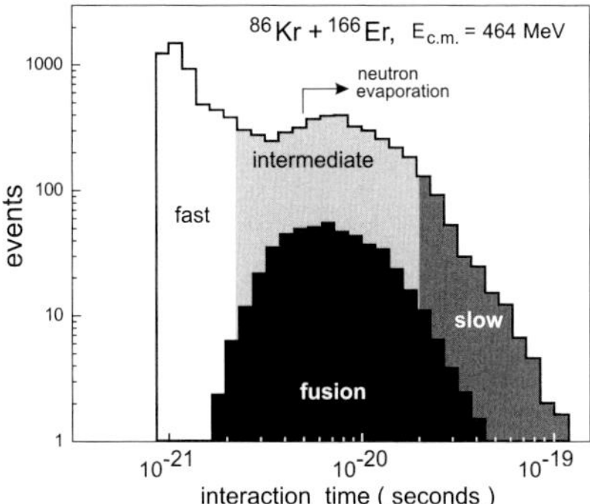

Fig. 7. – Time distribution of all the simulated events for ^{86}Kr + ^{166}Er collisions at $E_{\mathrm{c.m.}}$ = 464 MeV, in which the energy loss was found higher than 35 MeV (totally 10^5 events). Conditionally fast ($< 2 \cdot 10^{-21}$ s), intermediate and slow ($> 2 \cdot 10^{-20}$ s) collisions are marked by the different colors (white, light gray and dark gray, respectively). The black area corresponds to CN formation (estimated cross section is 120 mb), and the arrow shows the interaction time, after which the neutron evaporation may occur.

the impact parameter) and may be ignored here. All the events are divided relatively onto the three groups: fast ($\tau_{\mathrm{int}} < 20 \cdot 10^{-22}$ s), intermediate, and slow ($\tau_{\mathrm{int}} > 200 \cdot 10^{-22}$ s).

A two-dimensional plot of the energy-mass distribution of the primary fragments formed in the ^{86}Kr + ^{166}Er reaction at $E_{\mathrm{c.m.}}$ = 464 MeV is shown in fig. 8. Inclusive angular, charge and energy distributions of these fragments (with energy losses more than 35 MeV) are shown in fig. 9. Rather good agreement with experimental data of all the calculated DI reaction properties can be seen, which was never obtained before in dynamic calculations. Underestimation of the yield of low-Z fragments (fig. 9(c)) could again be due to the contribution of sequential fission of highly excited reaction participants not accounted in the model at the moment.

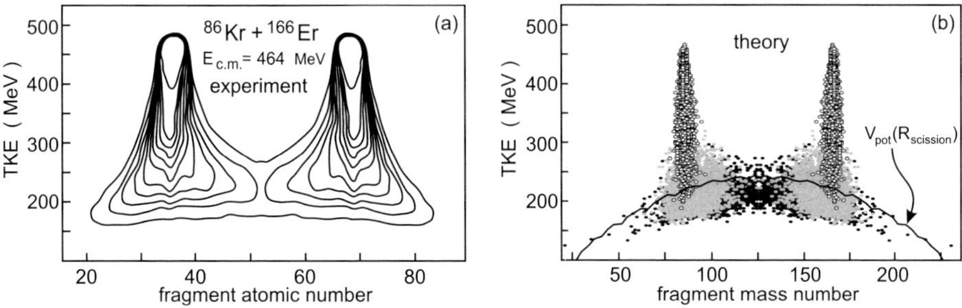

Fig. 8. – (a) TKE-charge distribution of the ^{86}Kr + ^{166}Er reaction products at $E_{\mathrm{c.m.}}$ = 464 MeV [17]. (b) Calculated TKE-mass distribution of the primary fragments. Open, gray and black circles correspond to the fast ($< 2 \cdot 10^{-21}$ s), intermediate and long ($> 2 \cdot 10^{-20}$ s) events (overlapping each other on the plot).

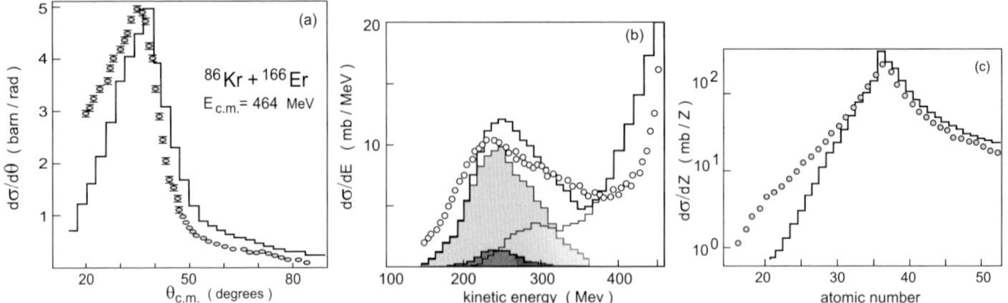

Fig. 9. – Angular (a), energy (b) and charge (c) distributions of the ^{86}Kr + ^{166}Er reaction products at $E_{\text{c.m.}} = 464$ MeV. Experimental data (points) are from [17]. Overlapping white, light and dark gray areas in (b) show the contributions of the fast, intermediate and slow events, respectively (see figs. 7 and 8(b)).

In most of the damped collisions ($E_{\text{loss}} > 35$ MeV) the interaction time is rather short (several units of 10^{-21} s). These fast events correspond to grazing collisions with intermediate impact parameters. They are shown by the white areas in figs. 7 and 9(b) and by the open circles in two-dimensional TKE-mass plot (fig. 8(b)). Note that a large amount of kinetic energy is dissipated here very fast at relatively low-mass transfer (more than 200 MeV during several units of 10^{-21} s).

The other events correspond to much slower collisions with large overlap of nuclear surfaces and significant mass rearrangement. In the TKE-mass plot these events spread over a wide region of mass fragments (including symmetric splitting) with kinetic energies very close to kinetic energy of fission fragments. The solid line in fig. 8(b) corresponds to potential energy at scission point $V(r = R_{\text{scission}}, \beta, \alpha) + Q_{gg}(\alpha)$ minimized over β. Scission point is calculated here as $R_{\text{scission}}(\alpha, \beta) = (1.4/r_0)[R_1(A_1, \beta_1) + R_2(A_2, \beta_2)] + 1$ fm, $Q_{gg}(\alpha) = B(A_1) + B(A_2) - B(^{86}\text{Kr}) - B(^{166}\text{Er})$ and $B(A)$ is the binding energy of a nucleus A. Some gaps between the two groups in the time and energy distributions can also be seen in figs. 7 and 9(b). All these make the second group of slow events quite distinguished from the first one. These events are more similar to fission than to deep-inelastic processes. Formally, they also can be marked as quasi-fission.

5. – Low-energy collisions of transactinide nuclei

Reasonable agreement of our calculations with experimental data on low-energy DI and QF reactions induced by heavy ions stimulated us to study the reaction dynamics of very heavy transactinide nuclei. The purpose was to find an influence of the shell structure of the driving potential (in particular, deep valley caused by the double shell closure $Z = 82$ and $N = 126$) on nucleon rearrangement between primary fragments. In fig. 10 the potential energies are shown depending on mass rearrangement at contact configuration of the nuclear systems formed in ^{48}Ca + ^{248}Cm and ^{232}Th + ^{250}Cf collisions. The lead valley evidently reveals itself in both cases (for ^{48}Ca + ^{248}Cm system there is also

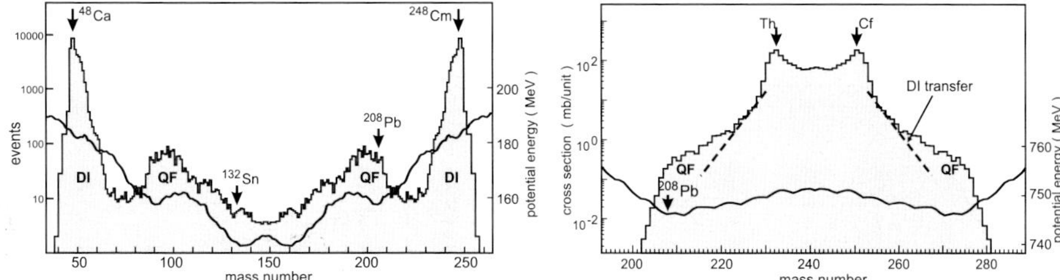

Fig. 10. – Potential energy at contact "nose-to-nose" configuration and mass distribution of primary fragments for the two nuclear systems formed in ^{48}Ca $+$ ^{248}Cm (left) and ^{232}Th $+$ ^{250}Cf (right) collisions.

a tin valley). In the first case (^{48}Ca $+$ ^{248}Cm), discharge of the system into the lead valley (normal or symmetrizing quasi-fission) is the main reaction channel, which decreases significantly the probability of CN formation. In collisions of heavy nuclei (Th $+$ Cf, U $+$ Cm and so on) we expect that the existence of this valley may noticeably increase the yield of surviving neutron-rich superheavy nuclei complementary to the projectile-like fragments around ^{208}Pb ("inverse" or anti-symmetrizing quasi-fission process).

Direct time analysis of the reaction dynamics allows us to estimate also the lifetime of the composite system consisting of two touching heavy nuclei with total charge $Z > 180$. Such "long-living" configurations may lead to spontaneous positron emission from super-strong electric field of giant quasi-atoms by a static QED process (transition from neutral to charged QED vacuum) [10]. About twenty years ago an extended search for this fundamental process was carried out and narrow line structures in the positron spectra were first reported at GSI. Unfortunately these results were not confirmed later, neither at ANL, nor in the last experiments performed at GSI. These negative finding, however, were contradicted by Jack Greenberg (private communication and supervised thesis at Wright Nuclear Structure Laboratory, Yale university). Thus the situation remains unclear, while the experimental efforts in this field have ended. We hope that new experiments and new analysis, performed according to the results of our dynamical model, may shed additional light on this problem and also answer the principal question: are there some reaction features (triggers) testifying a long reaction delays? If they are, new experiments should be planned to detect the spontaneous positrons in the specific reaction channels.

Using the same parameters of nuclear viscosity and nucleon transfer rate as for the system Xe $+$ Bi we calculated the yield of primary and surviving fragments formed in the ^{232}Th $+$ ^{250}Cf collision at 800 MeV center-of mass energy. Low fission barriers of the colliding nuclei and of most of the reaction products jointly with rather high excitation energies of them in the exit channel will lead to very low yield of surviving heavy fragments. Indeed, sequential fission of the projectile-like and target-like fragments dominate in these collisions, see fig. 11. At first sight, there is no chances to get surviving superheavy nuclei in such reactions. However, as mentioned above, the yield of the primary

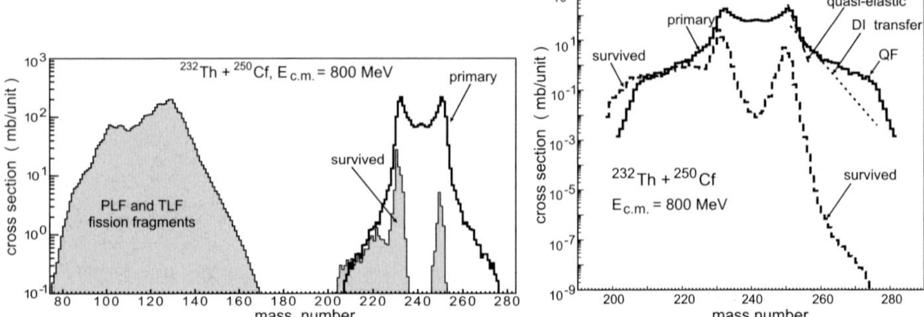

Fig. 11. – Mass distributions of primary (solid histogram), surviving and sequential fission fragments (hatched areas) in the ^{232}Th $+ ^{250}$Cf collision at 800 MeV center-of-mass energy. On the right the result of longer calculation is shown.

fragments will increase due to the QF effect (lead valley) as compared to the gradual monotonic decrease typical for damped mass transfer reactions. Secondly, with increasing neutron number the fission barriers increase on average (also there is the closed sub-shell at $N = 162$). Thus we may expect a non-negligible yield (at the level of 1 pb) of surviving superheavy neutron rich nuclei produced in these reactions [18].

Result of much longer calculations is shown on the right panel of fig. 11. The pronounced shoulder can be seen in the mass distribution of the primary fragments near the mass number $A = 208$ (274). It is explained by the existence of a valley in the potential energy surface (see fig. 10(b)), which corresponds to the formation of doubly magic nucleus ^{208}Pb ($\eta = 0.137$). The emerging of the nuclear system into this valley resembles the well-known quasi-fission process and may be called "inverse (or anti-symmetrizing) quasi-fission" (the final mass asymmetry is larger than the initial one). For $\eta > 0.137$ (one fragment becomes lighter than lead) the potential energy sharply increases and the mass distribution of the primary fragments decreases rapidly at $A < 208$ ($A > 274$).

In fig. 12 the available experimental data on the yield of SH nuclei in collisions of ^{238}U $+ ^{238}$U [5] and ^{238}U $+ ^{248}$Cm [6] are compared with our calculations. The estimated isotopic yields of survived SH nuclei in the ^{232}Th $+ ^{250}$Cf, ^{238}U $+ ^{238}$U and ^{238}U $+ ^{248}$Cm collisions at 800 MeV center-of-mass energy are shown on the right panel of fig. 12. Thus, as we can see, there is a real chance for production of the long-lived neutron-rich SH nuclei in such reactions. As the first step, chemical identification and study of the nuclei up to $^{274}_{107}$Bh produced in the reaction ^{232}Th $+ ^{250}$Cf may be performed.

The time analysis of the reactions studied shows that in spite of absence of an attractive potential pocket the system consisting of two very heavy nuclei may hold in contact rather long in some cases. During this time the giant nuclear system moves over the multidimensional potential energy surface with almost zero kinetic energy (result of large nuclear viscosity). The total reaction time distribution, $\frac{d\sigma}{d\log(\tau)}$ (τ denotes the time after the contact of two nuclei), is shown in fig. 13 for the ^{238}U $+ ^{248}$Cm collision. The dy-

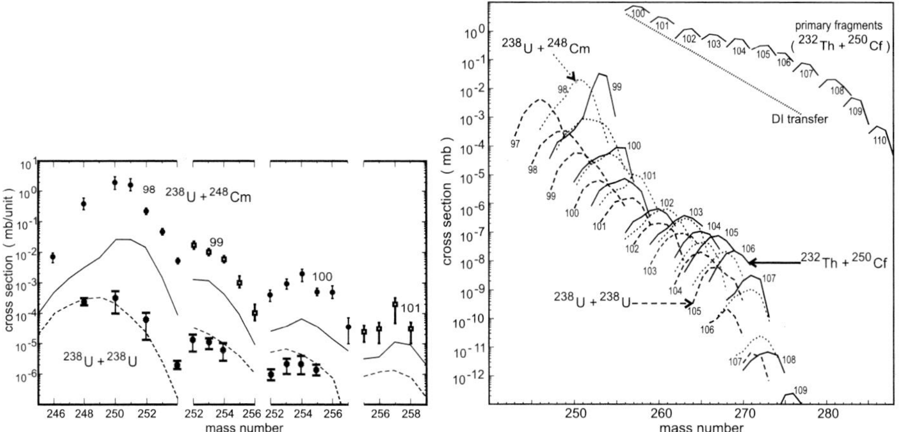

Fig. 12. – Left panel: Experimental and calculated yields of the elements 98–101 in the reactions ^{238}U + ^{238}U (crosses) [5] and ^{238}U + ^{248}Cm (circles and squares) [6]. Right panel: Predicted yields of superheavy nuclei in collisions of ^{238}U + ^{238}U (dashed), ^{238}U + ^{248}Cm (dotted) and ^{232}Th + ^{250}Cf (solid lines) at 800 MeV center-of-mass energy. Solid curves in upper part show isotopic distributions of primary fragments in the Th + Cf reaction.

namic deformations are mainly responsible here for the time delay of the nucleus-nucleus collision. Ignoring the dynamic deformations in the equations of motion significantly decreases the reaction time, see fig. 13(a). With increase of the energy loss and mass transfer the reaction time becomes longer and its distribution becomes more narrow.

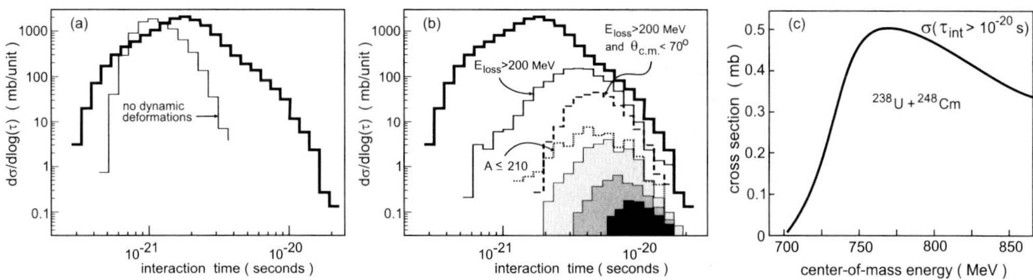

Fig. 13. – Reaction time distributions for the ^{238}U + ^{248}Cm collision at 800 MeV center-of-mass energy. Thick solid histograms correspond to all events with energy loss more than 30 MeV. (a) Thin solid histogram shows the effect of switching-off dynamic deformations. (b) Thin solid, dashed and dotted histograms show reaction time distributions in the channels with formation of primary fragments with $E_{\mathrm{loss}} > 200$ MeV, $E_{\mathrm{loss}} > 200$ MeV and $\theta_{\mathrm{c.m.}} < 70°$ and $A \leq 210$, correspondingly. Hatched areas show time distributions of events with formation of the primary fragments with $A \leq 220$ (light gray), $A \leq 210$ (gray), $A \leq 204$ (dark) having $E_{\mathrm{loss}} > 200$ MeV and $\theta_{\mathrm{c.m.}} < 70°$. (c) Cross-section for events with interaction time longer than 10^{-20} s.

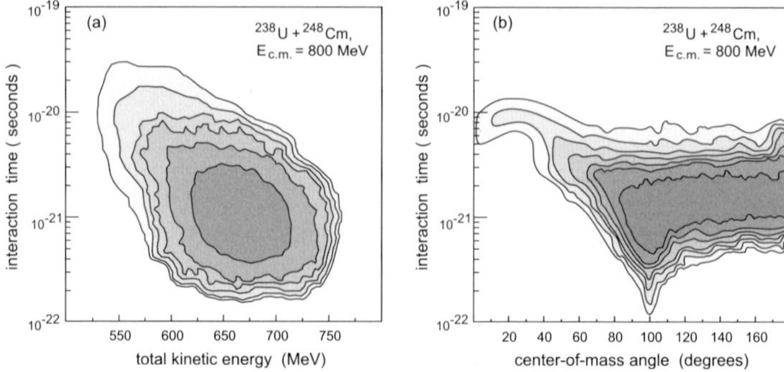

Fig. 14. – Energy-time (a) and angular-time (b) distributions of primary fragments in the ^{238}U + ^{248}Cm collision at 800 MeV ($E_{\text{loss}} > 15$ MeV).

As mentioned earlier, the lifetime of a giant composite system more than 10^{-20} s is quite enough to expect positron line structure emerging on top of the dynamical positron spectrum due to spontaneous e^+e^- production from the supercritical electric fields as a fundamental QED process ("decay of the vacuum") [10]. The absolute cross section for long events is found to be maximal just at the beam energy ensuring the two nuclei to be in contact, see fig. 13(c). The same energy is also optimal for the production of the most neutron-rich SH nuclei. Of course, there are some uncertainties in the used parameters, mostly in the value of nuclear viscosity. However we found only a linear dependence of the reaction time on the strength of nuclear viscosity, which means that the obtained reaction time distribution is rather reliable, see logarithmic scale on both axes in fig. 13(a).

Formation of the background positrons in these reactions forces one to find some additional trigger for the longest events. Such long events correspond to the most damped collisions with formation of mostly excited primary fragments decaying by fission, see fig. 14(a). However there is also a chance for production of the primary fragments in the region of doubly magic nucleus ^{208}Pb, which could survive against fission due to nucleon evaporation. The number of the longest events depends weakly on impact parameter up to some critical value. On the other hand, in the angular distribution of all the excited primary fragments (strongly peaked at the center-of-mass angle slightly larger than $90°$) there is the rapidly decreasing tail at small angles, see fig. 14(b). Time distribution for the most damped events ($E_{\text{loss}} > 150$ MeV), in which a large mass transfer occurs and primary fragments scatter in forward angles ($\theta_{\text{c.m.}} < 70°$), is rather narrow and really shifted to longer time delay, see hatched areas in fig. 13. For the considered case of ^{238}U + ^{248}Cm collision at 800 MeV center-of-mass energy, the detection of the surviving nuclei in the lead region at the laboratory angles of about $25°$ and at the low-energy border of their spectrum (around 1000 MeV for Pb) could be a real trigger for longest reaction time.

6. – Conclusion

For near-barrier collisions of heavy ions it is very important to perform a combined (unified) analysis of all strongly coupled channels: deep-inelastic scattering, quasi-fission, fusion and regular fission. This ambitious goal has now become possible. A unified set of dynamic Langevin type equations is proposed for the *simultaneous* description of DI and fusion-fission processes. For the first time, the whole evolution of the heavy nuclear system can be traced starting from the approaching stage and ending in DI, QF, and/or fusion-fission channels. Good agreement of our calculations with experimental data gives us hope to obtain rather accurate predictions of the probabilities for superheavy element formation and clarify much better than before the mechanisms of quasi-fission and fusion-fission processes. The determination of such fundamental characteristics of nuclear dynamics as the nuclear viscosity and the nucleon transfer rate is now possible. The production of long-lived neutron-rich SH nuclei in the region of the "island of stability" in collisions of transuranium ions seems to be quite possible due to a large mass rearrangement in the inverse (anti-symmetrized) quasi-fission process caused by the $Z = 82$ and $N = 126$ nuclear shells. A search for spontaneous positron emission from a supercritical electric field of long-living giant quasi-atoms formed in these reactions is also quite promising.

REFERENCES

[1] HOFMANN S. and MÜNZENBERG G., *Rev. Mod. Phys.*, **72** (2000) 733.

[2] OGANESSIAN YU. TS., UTYONKOV V. K., LOBANOV YU. V., ABDULLIN F. SH., POLYAKOV A. N., SHIROKOVSKY I. V., TSYGANOV YU. S., GULBEKIAN G. G., BOGOMOLOV S. L., GIKAL B. N., MEZENTSEV A. N., ILIEV S., SUBBOTIN V. G., SUKHOV A. M., VOINOV A. A., BUKLANOV G. V., SUBOTIC K., ZAGREBAEV V. I., ITKIS M. G., PATIN J. B., MOODY K. J., WILD J. F., STOYER M. A., STOYER N. J., SHAUGHNESSY D. A., KENNEALLY J. M., WILK P. A., LOUGHEED R. W., IL'KAEV R. I. and VESNOVSKII S. P., *Phys. Rev. C*, **70** (2004) 064609.

[3] ZAGREBAEV V. I., ITKIS M. G. and OGANESSIAN YU. TS., *Yad. Fiz.*, **66** (2003) 1069.

[4] BOTVINA A. S. and MISHUSTIN I. N., *Phys. Lett. B*, **584** (2004) 233; MISHUSTIN I. N., *Proc. ISHIP Conf., Frankfurt, April 3-6, 2006*, Vol. **16**, No. 4 (World Scientific, Singapore) 2007, pp. 1121-1134.

[5] SCHÄDEL M., KRATZ J. V., AHRENS H., BRÜCHLE W., FRANZ G., GÄGGELER H., WARNECKE I., WIRTH G., HERRMANN G., TRAUTMANN N. and WEIS M., *Phys. Rev. Lett.*, **41** (1978) 469.

[6] SCHÄDEL M., BRÜCHLE W., GÄGGELER H., KRATZ J. V., SÜMMERER K., WIRTH G., HERRMANN G., STAKEMANN R., TITTEL G., TRAUTMANN N., NITSCHKE J. M., HULET E. K., LOUGHEED R. W., HAHN R. L. and FERGUSON R. L., *Phys. Rev. Lett.*, **48** (1982) 852.

[7] MOODY K. J., LEE D., WELCH R. B., GREGORICH K. E., SEABORG G. T., LOUGHEED R. W. and HULET E. K., *Phys. Rev. C*, **33** (1986) 1315.

[8] RIEDEL C. and NÖRENBERG W., *Z. Phys. A*, **290** (1979) 385.

[9] ZAGREBAEV V. and GREINER W., *J. Phys. G*, **31** (2005) 825.

[10] REINHARD J., MÜLLER U. and GREINER W., *Z. Phys. A*, **303** (1981) 173.

[11] Greiner W. (Editor), *Quantum Electrodynamics of Strong Fields* (Plenum Press, New York and London) 1983; Greiner W., Müller B. and Rafelski J., *QED of Strong Fields* (Springer, Berlin and New York) 2nd edition, 1985.

[12] Rutz K., Bender M., Bürvenich T., Schilling T., Reinhard P.-G, Maruhn J. and Greiner W., *Phys. Rev. C*, **56** (1997) 238.

[13] Bürvenich T., Bender M., Maruhn J. and Reinhard P.-G, *Phys. Rev. C*, **69** (2004) 014307.

[14] Maruhn J. and Greiner W., *Z. Phys.*, **251** (1972) 431.

[15] Zagrebaev V. I., Aritomo Y., Itkis M. G., Oganessian Yu. Ts. and Ohta M., *Phys. Rev. C*, **65** (2002) 014607.

[16] Wilcke W. W., Birkelund J. R., Hoover A. D., Huizenga J. R., Schröder W. U., Viola V. E. jr., Wolf K. L. and Mignerey A. C., *Phys. Rev. C*, **22** (1980) 128.

[17] Gobbi A., Lynen U., Olmi A., Rudolf G. and Sann H., in *Proceedings of the International School of Physics "Enrico Fermi", Course LXXVII*, edited by Broglia R. A., Ricci R. A. and Dasso C. H. (North-Holland) 1981, p. 1.

[18] Zagrebaev V. I., Oganessian Yu. Ts., Itkis M. I. and Walter Greiner, *Phys. Rev. C*, **73** (2006) 031602(R).

DOI 10.3254/978-1-58603-885-4-29

Phenomenological nuclear spectroscopy (a personal recollection)

R. A. Ricci

INFN - Laboratori Nazionali di Legnaro
Viale dell'Università 2, 35020 Legnaro (PD), Italy

Summary. — A survey of the nuclear-structure studies through the nuclear-spectroscopy investigations as performed by different tools and techniques is presented. Starting from the simple radioactive decay studies and the associated γ-ray spectroscopy, since the pioneering work with the first scintillation devices, the review will cover the investigation of disintegration schemes and the scintillation spectra analysis performed in the 50's and the early 60's either by radioactive decays or with direct reactions such as stripping and pick-up and inelastic and quasi-free scattering. Examples are the results obtained for determining single particle and collective states in light and medium heavy nuclei. The selection of nuclear states from the *first revolution* in nuclear spectroscopy given by the *scintillation detectors* to the *second revolution* which allowed a more appropriate selection of nuclear states due to the advent of *in-beam γ-ray spectroscopy* and *heavy-ion nuclear reactions* is reviewed. The *third and fourth revolution* in nuclear spectroscopy given by the advent of more sophisticated γ-arrays and by the possibility of accelerating *radioactive beams* are also accounted for. In this context as a typical nuclear spectroscopy investigation the revival of the $1f_{7/2}$ spectroscopy is reported with the aim to show the still attractive future of important aspects of nuclear physics.

Introduction: Nuclear spectroscopy: an old story

In the history of nuclear-structure investigation nuclear spectroscopy plays a major role. The phenomenological description of nuclear states is strongly related to the evolution of experimental investigation of their production and decay via different specific tools starting from the simple α, β radioactive disintegrations and the associated γ-ray decays.

In the '50s, just when I personally started my career as a nuclear physicist, the usual ways of studying the nuclear decay schemes were the determination of the γ-ray spectra arising from excited states of nuclei populated by the disintegration of natural radioactive isotopes and or artificial ones produced by light particle (p, d, α, sometimes n) nuclear reactions.

This of course was accomplished by measuring the β (or α) primary spectrum and (generally in coincidence) the following γ-ray spectra with magnetic and or scintillation spectrometers.

A typical example of such experimental investigations is shown in fig. 1, where a page of my first paper on nuclear physics is reported, concerning the β-decay of ^{214}Bi and the corresponding γ-spectrum of ^{214}Po as measured with a typical β–γ coincidence scintillation technique (a home-made naphthalene crystal for β-rays and a NaI(Tl) counter for γ-rays coupled with RCA photomultipliers, as usual at that time).

On the other hand, the possibility of producing radioactive nuclear species with dedicated proton, deuterons or alpha accelerators or neutron generators has been, during those years, a so powerful tool that an impressive large number of new decay schemes could be investigated.

As typical examples I will report on some of these investigations performed at the Phylips Cyclotron at IKO (Amsterdam) and at the 14 MeV neutron generator in Naples in the years '50s-'60s.

1. – The γ-scintillation era

The scintillation counter, after the discovery in 1947 by H. Kallmann [1] and its coupling with the already invented photomultipliers [2], and the advent of NaI(Tl) detector discovered by R. Hofstadter [3], did develop very rapidly from an instrument for the detection of ionizing radiation to a device with which the energy and intensity of such radiation could be measured with a fair degree of accuracy.

Of course other instruments like magnetic spectrometers were still used, due to their higher precision, for many final analyses of spectra, but the scintillation counter went to be an extraordinary measuring device, especially for γ-rays.

The relative simplicity of the experimental set-up consisting of a scintillator and a photo-cell with electron multiplier made it one of the most manageable instruments in nuclear-physics research. Even the early versions incorporating small crystals and rather crude pulse height discriminators, used for the detection of single spectra only, yielded a surprising large amount of significant data in radioactivity studies.

The comparison with the modern γ-arrays and the computer-assisted analysis of the spectra makes it to appear a museum piece. However, the advent of technical refinements of the instrument components and especially the availability of multichannel pulse height analyzers have made it possible to treat much more complicated spectra, though their

Revision of the β-Ray Spectrum of $^{214}_{83}$Bi(RaC).

R. A. Ricci and G. Trivero

Istituto di Fisica Sperimentale del Politecnico - Torino

(ricevuto il 27 Dicembre 1954)

1. – Several authors have tried to interpret the decay scheme $^{214}_{83}$Bi-(β)-$^{214}_{84}$Po, which presents difficulties due to the complexity of the emitted β and γ ray spectra [1].

In order to test the validity of the different schemes suggested, we have tried to analyze the β spectrum of $^{214}_{83}$Bi with an absorption method, at the same time, using the β-γ coincidences technique [2].

The investigations concern the experimental determination of the absorption curve of $^{226}_{88}$Ra β-spectrum and the absorption curve of the β-γ coincidences.

2. – Fig. 1 shows the experimental apparatus. S is a $^{226}_{88}$Ra sample (about 16 microcurie) in equilibrium with its decay products up to RaD (half-life $T = 22$ years).

F_β is a scintillation counter for β-rays consisting of naphtalene crystal (diameter 1 cm; thickness 0,3 cm) and RCA 1P21 photomultiplier.

F_γ is a scintillation counter for γ-rays (NaI crystal activated with thallium and RCA 913A photomultiplier).

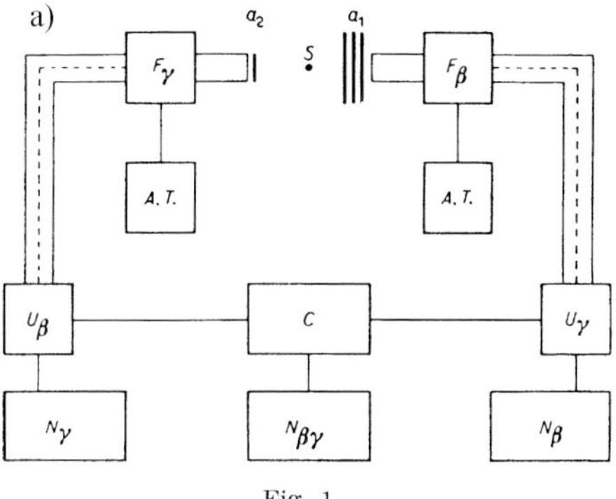

Fig. 1.

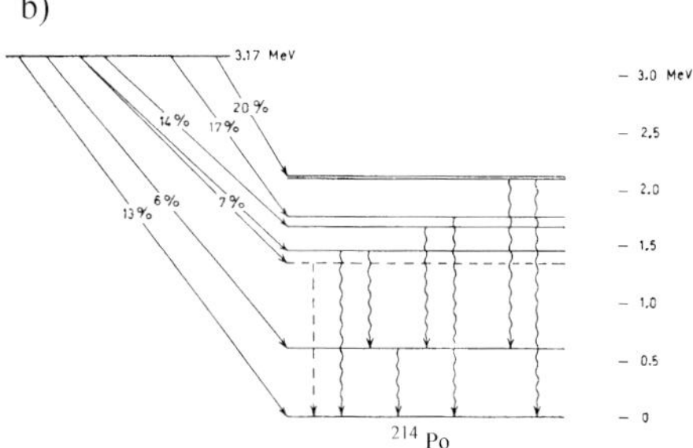

[1] E. Rutherford, W. B. Lewis and B. V. Bowden: *Proc. Roy. Soc.*, A, **142**, 347 (1933); J. Surugue: *Journ. de Phys. et Rad.*, **8**, 7, 145 (1946); G. D. Latyshev: *Rev. Mod. Phys.*, **19**, 132 (1947); A. H. Wapstra: *Int. Conf. β and γ radioactivity* (September 1952), 1247; F. Demichelis and R. Malvano: *Nuovo Cimento*, **10**, 405 (1953).

[2] A. C. G. Mitchell: *Rev. Mod. Phys.*, **20**, 296 (1948).

Fig. 1. – The ^{214}Bi $\rightarrow$ ^{214}Po β-decay.

analysis required development of elaborate techniques to scan the pulse height distribution, in order to extract numerical values for the various experimental data.

Such a survey has been presented in the Chapter devoted to the *"Procedures for the investigation of disintegration schemes"* of the famous book edited by K. Siegbahn in 1965 on *"Alpha, Beta and Gamma-Ray Spectroscopy"* [4].

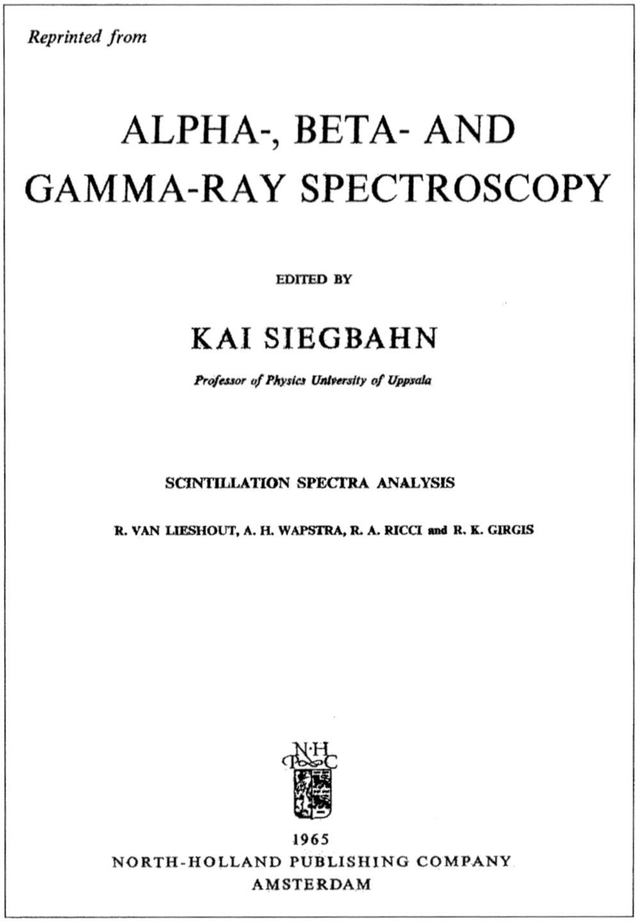

Reprinted from

ALPHA-, BETA- AND GAMMA-RAY SPECTROSCOPY

EDITED BY

KAI SIEGBAHN

Professor of Physics University of Uppsala

SCINTILLATION SPECTRA ANALYSIS

R. VAN LIESHOUT, A. H. WAPSTRA, R. A. RICCI and R. K. GIRGIS

1965

NORTH-HOLLAND PUBLISHING COMPANY
AMSTERDAM

Fig. 2. – Cover of the experimental nuclear spectroscopy "Bible" in the '60s.

The part concerning the *"Scintillation spectra analysis"* constitutes a kind of *"summa"* of the experimental methods used in the investigation of more than 50 radioactive isotopes performed by the IKO (Amsterdam) Group during my stay there in 1957-58 (and for shorter periods in the years '60s).

The cover of such a review is reported in fig. 2.

The evolution of such a methodology is illustrated by some examples of γ-ray spectra of a series of medium-weight nuclei analyzed by the *"peeling"* technique.

A spectrum under study was decomposed into its contributing parts (*i.e.* the different single γ-rays) by comparison with the pulse height distribution of standard calibration sources. The most important aspect of this empirical method lies in the fact that standards could be measured under conditions that were nearly identical to those under which the measurement of the source under study was performed.

Various examples are shown in figs. 3 to 6.

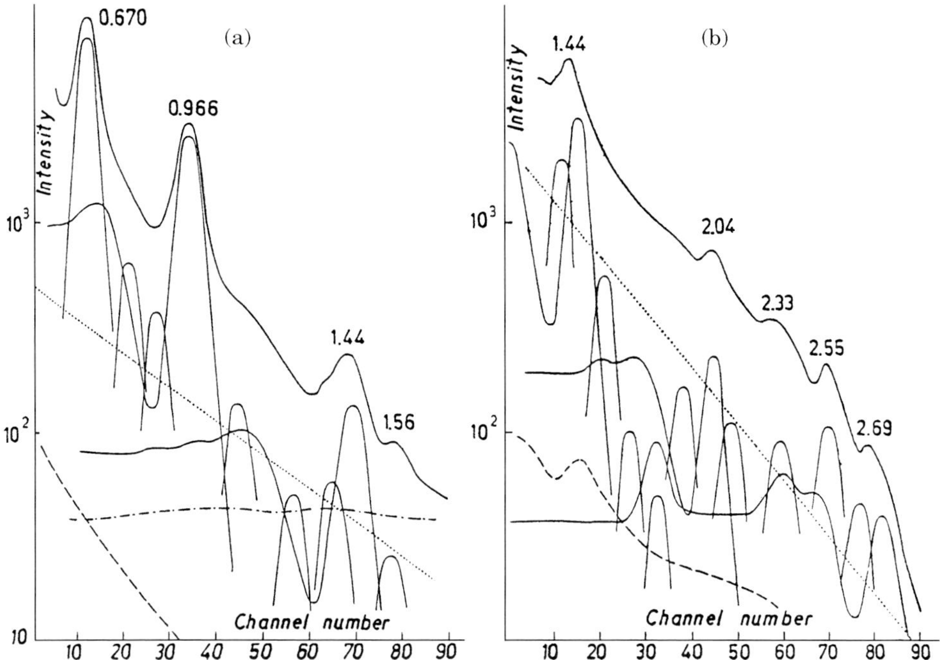

Fig. 3. – a) γ-ray spectrum of ^{63}Zn in the intermediate-energy region. The dotted line represents the contribution due to annihilation in flight. b) γ-ray spectrum of ^{63}Zn in the high-energy region.

Of course, the "crude" pulse-height distribution had to be corrected by some disturbing influences like background effects (with appropriate shielding), bremstrahlung in the presence of strong β-rays or electron capture (thin sources on thin backings or absorbers to saturate the external contribution), positons annihilation in flight (production of 511 kew γ-rays, which were properly accounted for) so as unwanted summing effects.

The final "clean" spectrum was analyzed by peeling off the pulse-height distribution starting from the highest energy peak and subtracting the corresponding shape from the total spectrum.

The shape needed with all the expected features (photopeak, Compton shoulder, pair peaks, ...) is obtained from the library of standard sources, if necessary through interpolation. Normalisation of intensity and small adjustements of pulse heights can be easily accomplished by superimposing a sheet of transparent doubly logarithmic paper with the constructed shape with one on which the spectrum under study is recorded. This procedure is repeated for every peak resolved from the total pulse height distribution, going to lower energies in every step. At every stage the residuals in the higher-energy region and the general appearance of the partially stripped spectrum are inspected. If these are unsatisfactory, it is necessary to go back to an earlier stage in the analysis and make a readjustment in the fitting procedure or, maybe, reinterpret a peak as due to summing effects or re-evaluate background or continuous distributions.

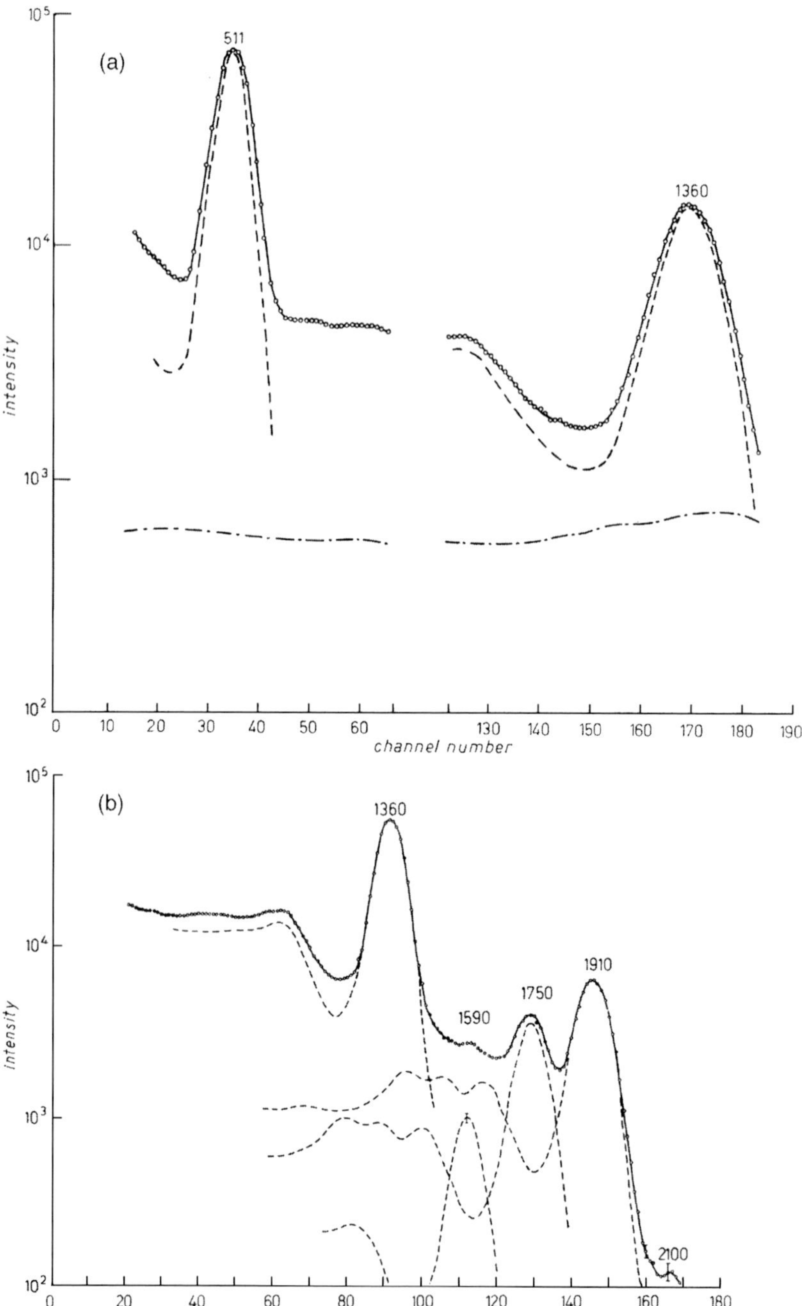

Fig. 4. – a), b) γ-ray spectra from the decay of 37h ^{57}Ni: the dashed curves show the positions of the photopeaks resolved; the dot-dashed curves in a) show the contribution from the high-energy γ-rays.

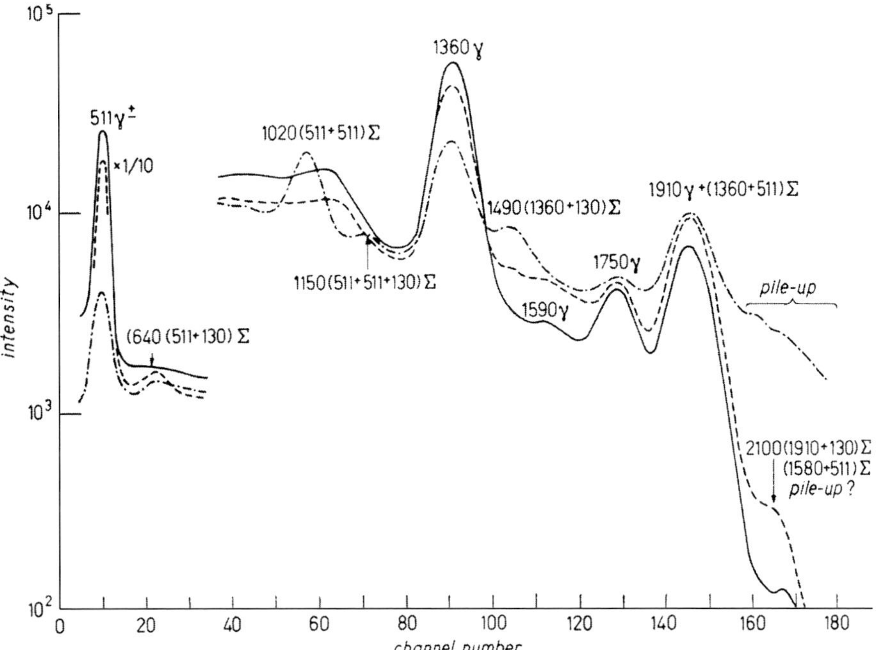

Fig. 5. – Summing spectra of 37h ^{57}Ni with the source in the well of the crystal (dot-dashed curve) and on the top of the crystal (dashed curve), compared to the direct spectrum taken for a source-detector distance of ~ 13 cm (solid curve).

In two cases (^{57}Ni and ^{50}Ti decays) also the "*summing technique*" for detecting γ–γ coincidences is shown: it consists in exploiting in a proper way the so-called "*summing effect*" which is present in the crystal when random and true coincidences occur between the absorbed γ-rays; in such a case the two radiations are detected within the resolving time of the detector arrangement (typically $1\,\mu$s for a NaI(Tl) with a normal pulse-height analyzer). Using an appropriate geometry, close to 4π (well-type crystal with the source inside the well) and comparing the obtained summing spectrum with the spectra corresponding to the source located at various distances from the detector, one obtains a first qualitative (also quantitative with some more elaborated expedient) but good evidence of coincidence events.

An interesting case in point is the lower γ-spectrum of ^{50}Ti following the β-decay of ^{50}Sc as obtained at Naples in 1962 [5] as can be seen in fig. 6 where the comparison of the pulse-height distribution with the source inside and outside the crystal clearly shows the appearance of double and triple γ–γ coincidences corresponding to the level scheme of ^{50}Ti also shown in the figure.

At that time the knowledge of the level sequence given by the interaction of 2 nucleons in the same j shell-model state was far to be clear and the expected sequence $0, 2, 4 \ldots$ $(2j\text{-}1)$ was known only in few cases.

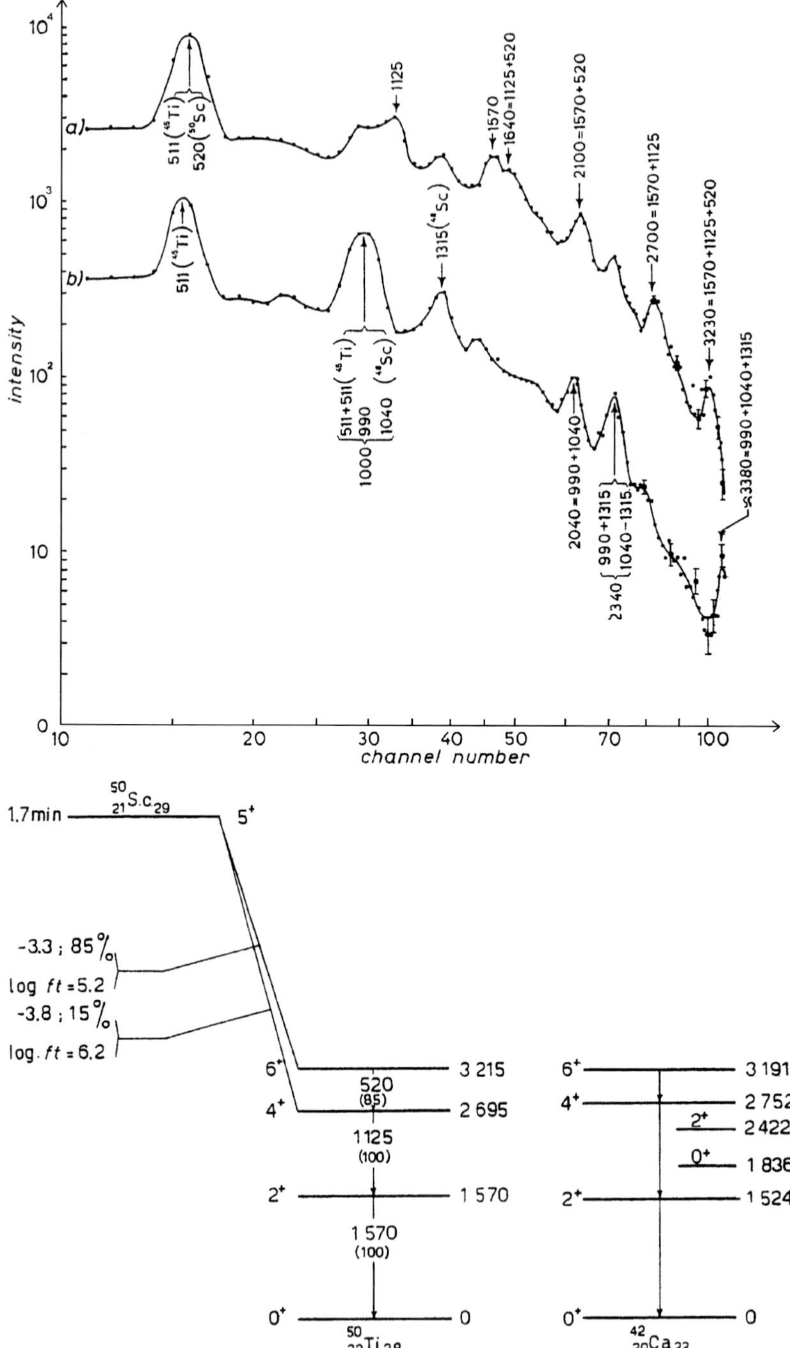

Fig. 6. – The γ-ray and γ-γ summing coincidence spectrum and level scheme of ^{50}Ti compared with ^{42}Ca (see ref. [5], 1963).

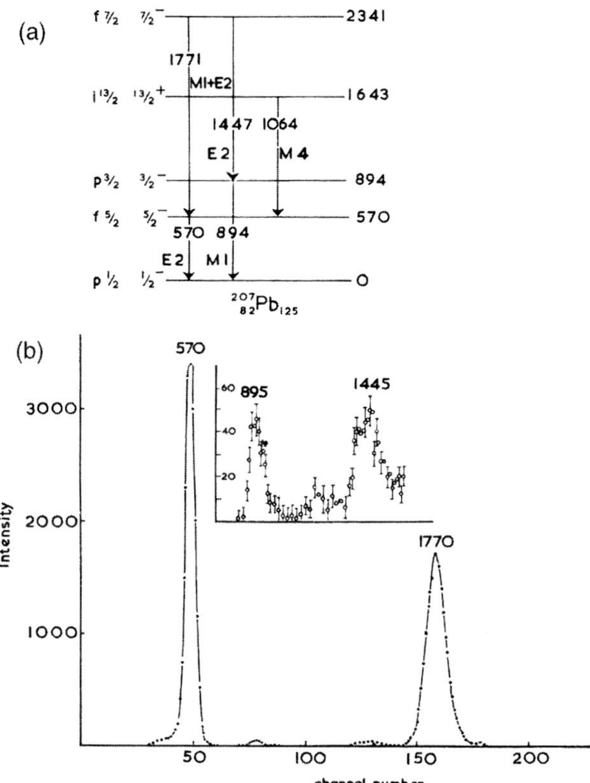

Fig. 7. – a) Decay scheme of ^{207}Pb following the decay of ^{207}Bi. b) Sum coincidences with 2340 keV sum energy. The inset shows the central part enlarged (ref. [6]).

The case of the $1f_{7/2}$ two-body spectra was of particular interest since the 2 nucleons (protons or neutrons) lie in a single j-state in a doubly magic core (^{40}Ca or ^{48}Ca) without important interference of other shells.

Only the 2 neutron spectrum of ^{42}Ca was known.

The 2-proton spectrum of ^{50}Ti was therefore investigated by the nuclear spectroscopy group of Naples in collaboration with H. Morinaga, via the β-decay of the parent isobar ^{50}Sc produced by the ^{50}Ti(n,p) reaction with a 14 MeV neutron generator. The obtained $(1f_{7/2})^2$ proton spectrum was found very similar to the $(1f_{7/2})^2$ neutron spectrum showing without ambiguity the validity of the two-body coupling due to an effective charge-independent nuclear interaction.

Figure 7 also shows another important result concerning the γ transitions of ^{207}Pb as obtained by the decay of ^{207}Bi with the help of the γ-ray scintillation spectrometry [6]. A sum coincidences technique was used in this case (summing discriminator channel connected to a fast-slow concidence system) detecting the cascade γ-rays from a specific level: the $f_{7/2}$ single-particle (hole) state (at 2341 keV) which decays to the $p_{3/2}$ one ($E2$ transition of 1445 keV) and to the $f_{7/2}$ level (pure $M1$ spin-flip transition). The measured

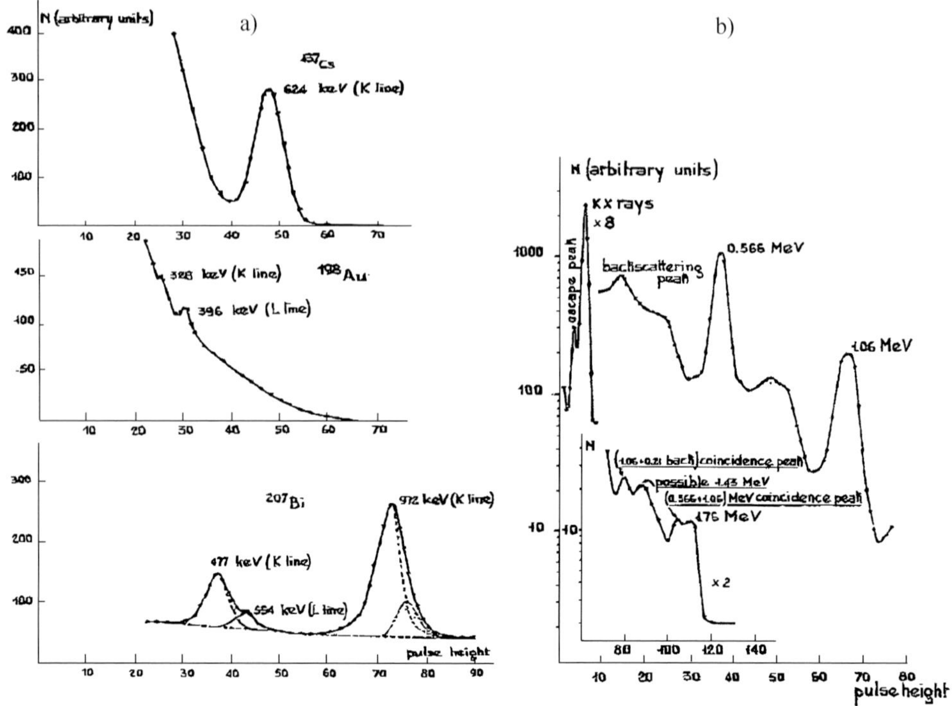

Fig. 8. – a) Electron and conversion spectra of ^{137}Cs, ^{198}Au and ^{207}Bi, as obtained with the scintillation spectrometer. b) Gamma scintillation spectrum of ^{207}Bi.

$M1/E2$ ratio (0.023 ± 0.002) showed that the $M1$ transition is hindered by a factor ~ 4. It is a clear case reported by Bohr and Mottelson [7] concerning the electromagnetic moments with respect to magnetic dipole effects in single-particle configurations.

Coming back to the scintillation spectroscopy methods I should remind that useful information could also be obtained in the case of β-ray spectra especially in the analysis of conversion lines.

Figure 8 shows some cases in point. Electron and conversion spectra of ^{137}Cs (624 keV K line), ^{198}Au (328 keV K line and 396 keV L line) and ^{207}Bi (477 keV and 972 keV K lines, 554 keV L line) were measured with a cylindrical anthracene crystal mounted on a 6292 Du Mont photomultiplier connected with a single-channel analyzer [8]. The γ-ray scintillation spectrum is also shown.

It is interesting to note that such values of the ^{207}Bi K conversion lines have been used to calibrate the liquid scintillation detector adopted in the famous experiment of Reines and Cowan [9] to measure the positron kinetic-energy spectrum from the antineutrino absorption reaction p $(\nu_{\rm e}, \beta+)$n leading to the first neutrino detection.

What is impressive, after all, is the rapid evolution of the γ-ray spectrometry, an essential tool in experimental nuclear physics.

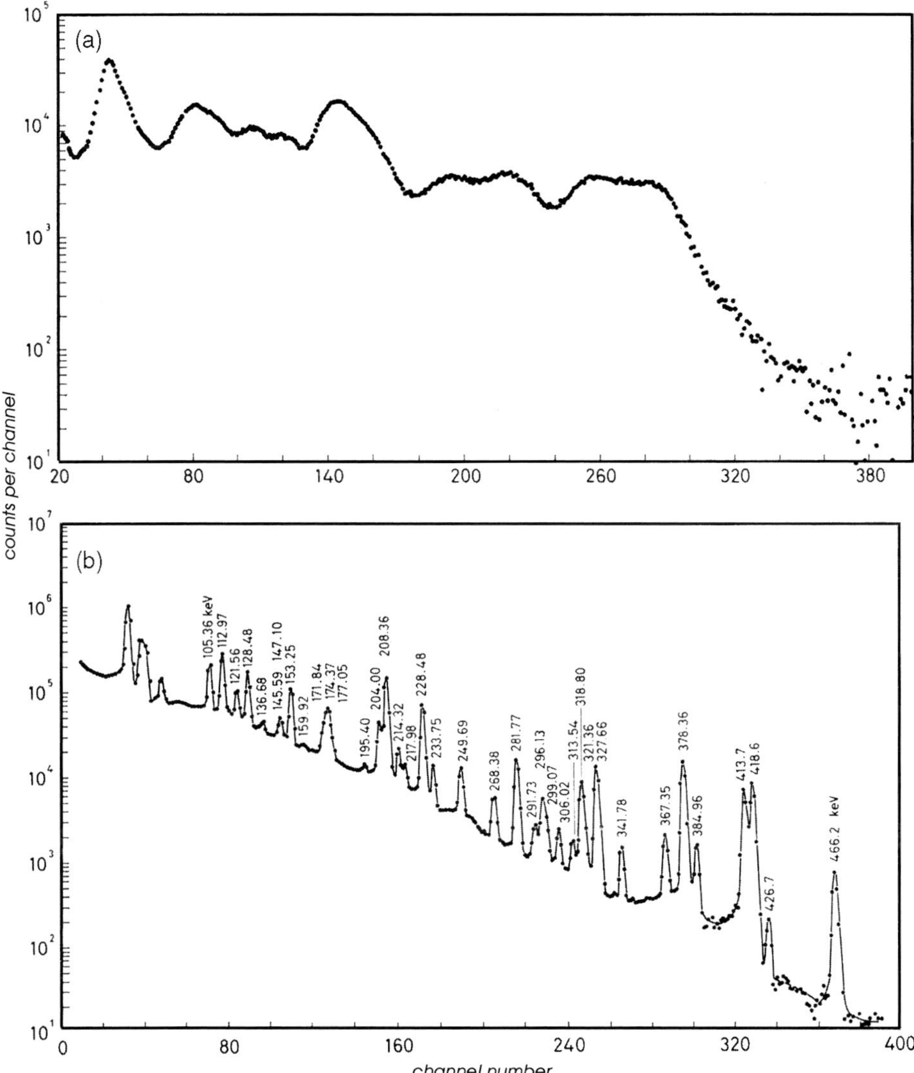

Fig. 9. – a) γ-spectrum of ^{177}Lu measured with a scintillator detector (NaI(Ti), 3 in. $\times$ 3 in.).
b) γ-spectrum of a) measured with a germanium detector ($2\,\mathrm{cm}^2 \times 7\,\mathrm{mm}$).

The advent of more complex and sophisticated γ-arrays is the last step achieved, in the recent years, of such an evolution. Compton suppression, resolution and efficiency have found an impressive improvement.

Figure 9 shows the result of what we could call the *2nd fundamental step* in the γ-ray spectrometry (the first was the advent of scintillation spectroscopy), *i.e.* the advent of germanium detectors.

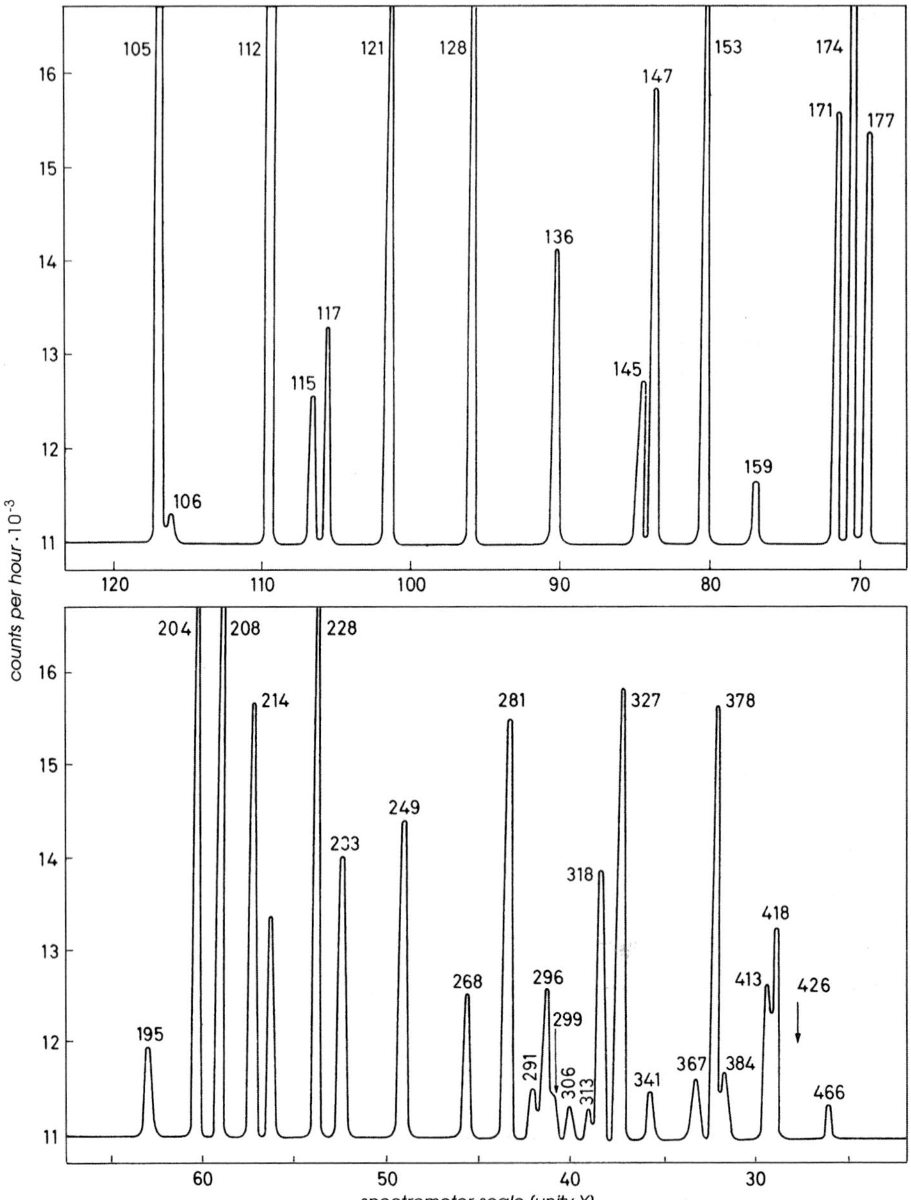

Fig. 10. – Spectrum of fig. 9 as measured with a diffraction spectrometer.

The γ-decay spectrum of ^{177}Lu measured with a NaI/Tl scintillation counter is compared with the one measured with a Ge(Li) $(2\,\text{cm}^2 \times 0.7\,\text{cm})$ detector.

I would like to show also the same spectrum as detected with a diffraction spectrometer (bent crystal) in fig. 10 (cf. ref. [4]).

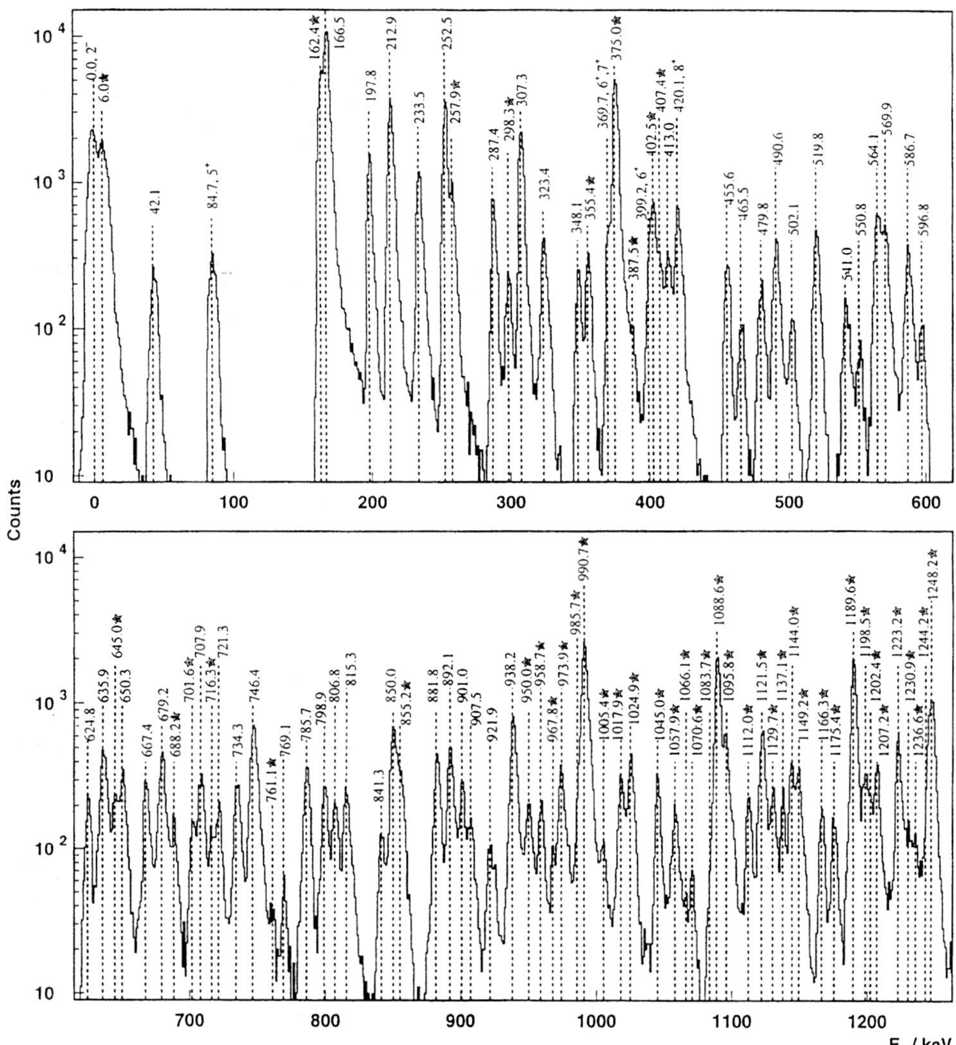

Fig. 11. – The ^{197}Au(p, d) ^{196}Au spectrum measured by Metz *et al.* with 26 MeV protons on a 67 μg/cm^2 target at an angle of 25°. The excitation energy range 0–1.25 MeV is shown. The resolution is 4 keV FWHM. New states, resolved for the first time, are marked by stars. Previous assignments of Nuclear Data Sheets are also included. The spectrum is shown here to emphasize the wealth of information provided by this experiments. With this information it has been possible to test the predictions of supersymmetry unambiguously (from ref. [10]).

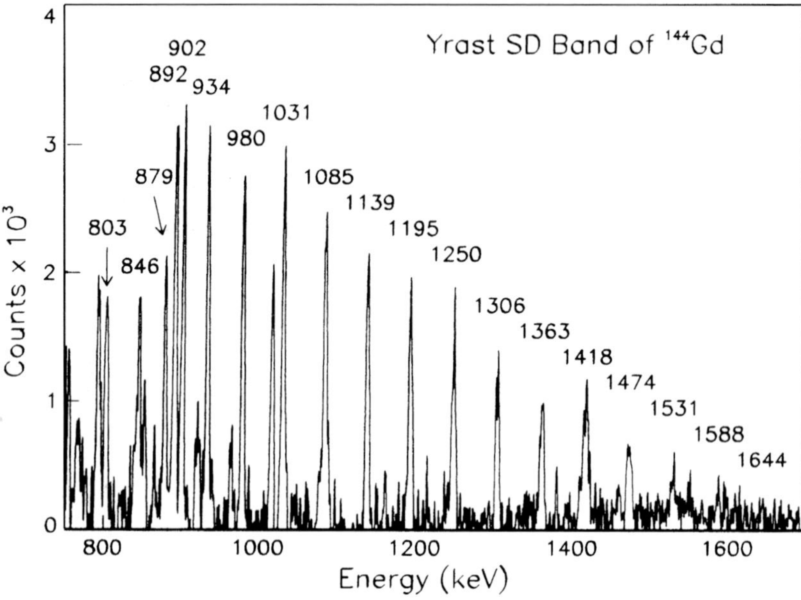

Fig. 12. – The yrast superdeformed band of ^{144}Gd showing a backbending at $E_\gamma = 900\,\mathrm{keV}$. The band numbers are marked by their transition energies in keV (LNL, 1993).

It can be seen that some other techniques can still be very useful as also shown by a more recent case, *i.e.* the magnetic high-resolution spectrometry of ^{196}Au as performed by Metz *et al.* following the ^{197}Au(p, d) ^{196}Au spectrum. The wealth of information provided by this measurement is impressive and could be taken as a test for a clear supersymmetric behaviour in nuclei as predicted by Jachello [10] (see fig. 11).

The third step, and I will show here examples like the γ-spectrum of ^{144}Gd (fig. 12) is given by the more advanced and complex arrays whose employement has opened the new nuclear-spectroscopy era.

2. – Nuclear spectroscopy with direct reactions

An important role in giving essential information about nuclear states has been provided by direct reactions (transfer, inelastic scattering, Coulomb excitation).

Stripping and *pick-up* of one nucleon such as (p,d), (d,He$_3$), (d,p) and (d,He$_3$) reactions were extensively used to determine single-particle (hole) energies so as the corresponding occupation numbers via the *spectroscopic factors* [11].

As an example the sequence of unperturbated single particle and single hole states in the region of the ^{40}Ca and ^{48}Ca cores is reported in fig. 13 as arising from various protons and neutron transfer reaction experiments, including the spin-orbit and the isospin splitting and their evolution in going from ^{40}Ca to ^{48}Ca.

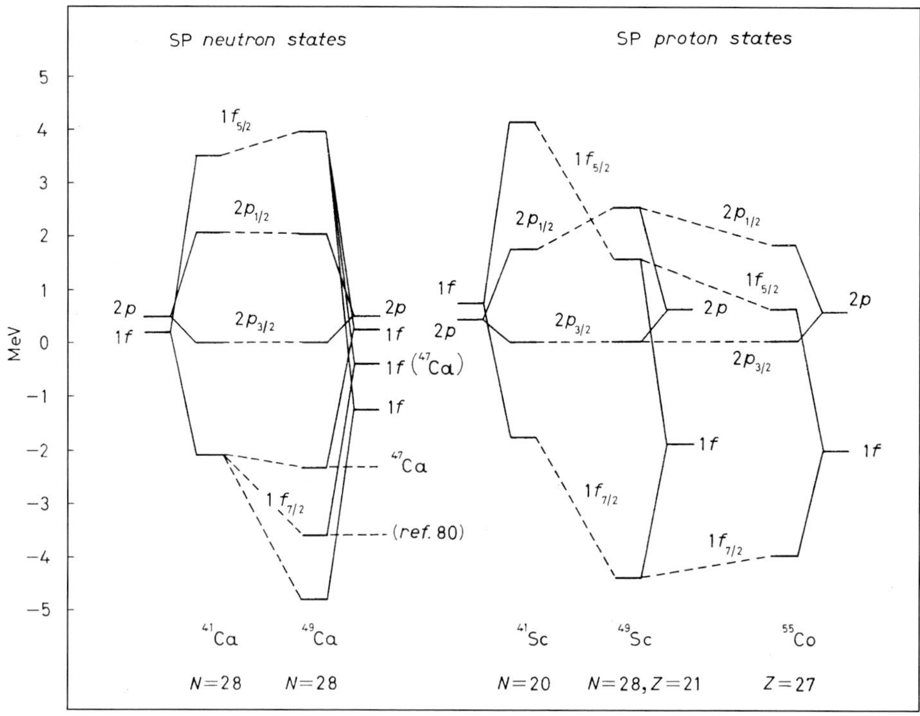

Fig. 13. – Single-particle energies, relative to the $2p_{7/2}$ state of ^{41}Ca, ^{49}Ca (and ^{47}Ca) (neutron states) and of ^{43}Sc, ^{49}Sc and ^{65}Co (proton states). The variations are indicated by dashed lines. The centroids of the $1f$ and $2p$ states are also indicated with the corresponding spin-orbit splitting (ref. [11]).

Another important piece of information in this context was given by the separation energies by (p,2p) *knock-out* and (e,e′p) reactions (*quasi-free scattering*) [12]. An interesting set of data arising from measurements performed at the synchrocyclotron in Orsay at $E_{\mathrm{p}} = 155\,\mathrm{MeV}$ was obtained in the *s-d* and *f-p* nuclear region [13]. As an example in fig. 14 the binding and excitation energies of the $2s$ and $1d$ proton holes in the $1f_{7/2}$ nuclei so obtained, are reported.

One should mention that this kind of investigation has been and still is a very powerful tool for characterizing the nucleon separation energies and therefore in determining the height of the nuclear potential well. Figure 15 reports a summary of such data (proton separation energies as a function of the mass number A) obtained by (e,e′p) and (p,2p) reactions [12, 13].

The other important tool in selecting specific nuclear states has been *inelastic scattering* mainly used to identify *collective states* due to the strong relationship between enhanced $B\,(E2)$ transition probabilities and inelastic scattering cross-sections. This peculiar aspect is emphasized also by *Coulomb excitation* processes as shown in fig. 16,

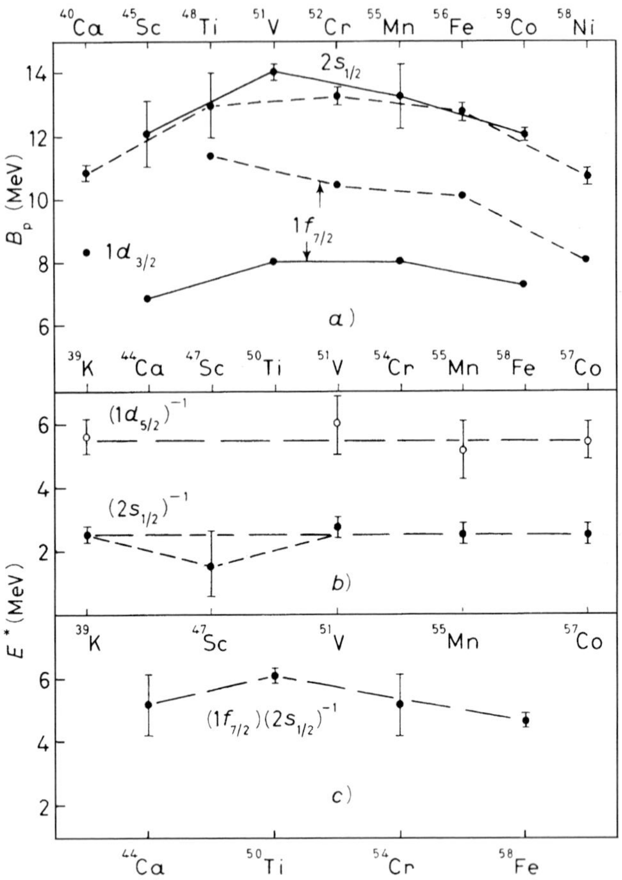

Fig. 14. – a) Binding energies B_p of $2s_{1/2}$ protons in some $1f_{7/2}$ nuclei, from (p, 2p) reactions; the full line connects the data corresponding to even-even residual nuclei. The target nuclei are indicated in the upper part of the figure. The points indicated as $1d_{3/2}$ and $1f_{7/2}$ correspond to the proton binding energies in the ground state of the target nucleus. b) Excitation energies E^* of $2s_{1/2}$ and $1d_{5/2}$ proton-hole states in the odd nuclei reported in a). c) Excitation energies of $2s_{1/2}$ proton-hole states in the even-even nuclei reported in a). The main difference between b) and c) is due to pairing effects.

where the correlation between some $B(E2)$ values determined by Coulomb excitation and the corresponding σ (p,p′) for the same 2^+ excited states is very clearly represented [14].

The numerous inelastic scattering experiments performed in the '60s and '70s such as (e,e′), (p,p′), (α, α') and the measurements of the associate γ-rays leaved a very impressive set of data concerning collective 2^+ and 3^- states in various nuclear regions. Some interesting results are reported in fig. 17, as found in different laboratories and with different probes (electron, protons, alphas) [15].

A detailed set of results concerning nuclear levels excited by 155 MeV proton inelastic scattering at Orsay in 1964 is reported in fig. 18, in the region of $1s$-$2d$ and $1f_{7/2}$ shells.

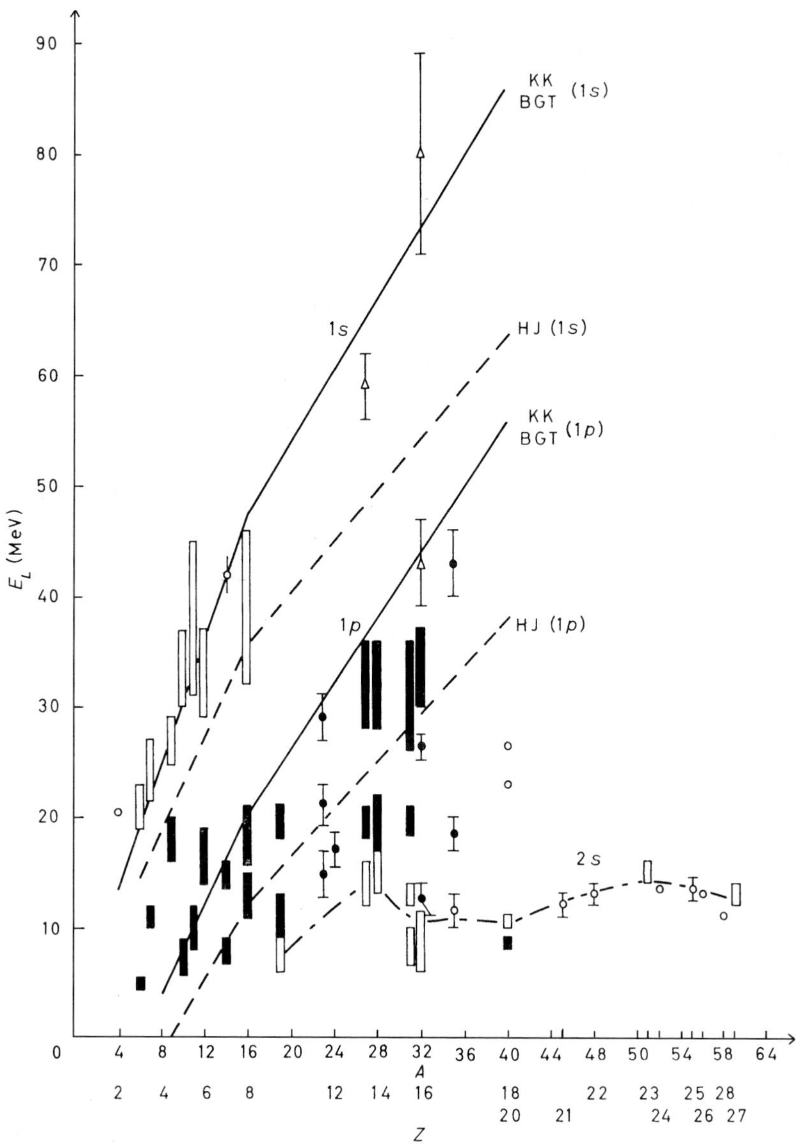

Fig. 15. – Experimental binding energies as a function of A. — KK = Kallio-Kolltveit; BGT = Brueckner-Gammel-Thaler; – – – HJ = Hamada-Johnston; $\square$, $\circ$ $l = 0$ (p, 2p); $\blacksquare$, $\bullet$ $l \neq 0$ (p, 2p); $\triangle$ (e, e$'$, p) (from refs. [12, 13]).

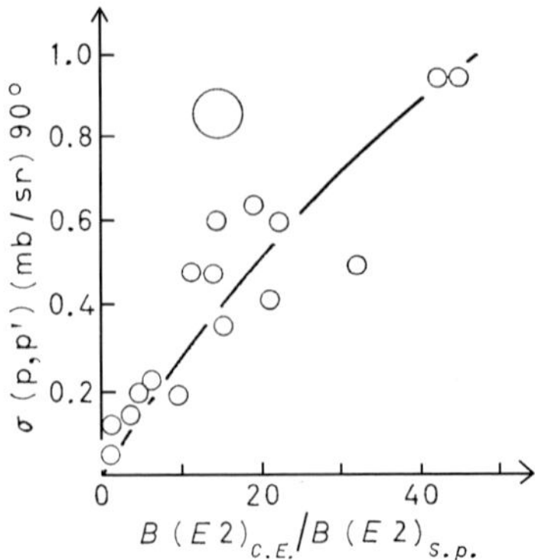

Fig. 16. – Correlation between inelastic scattering σ (p, p') and $B(E2)$ probabilities from Coulomb excitation for the same 2^+ excited states. (Cohen and Rubin, ref. [14]).

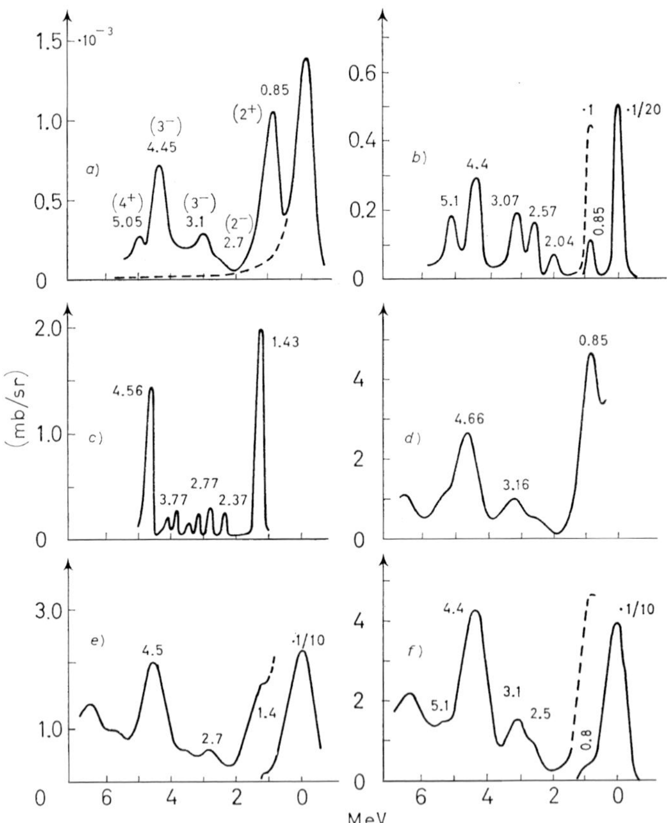

Fig. 17. – Experimental spectra of different scattered particles from various nuclei: a) ^{56}Fe, (ee'), 150 MeV, $\theta_L = 80°$; b) ^{56}Fe, ($\alpha\alpha'$), 44 MeV, $\theta_L = 28°$; c) ^{52}Cr, (pp'), 17 MeV, $\theta_L = 80°$; d) ^{56}Fe, (pp'), 40 MeV, $\theta_L = 25°$; e) ^{52}Cr, (pp'), 163 MeV, $\theta_L = 20°$; f) ^{56}Fe, (pp'), 155 MeV, $\theta_L = 20°$ (see ref. [15]).

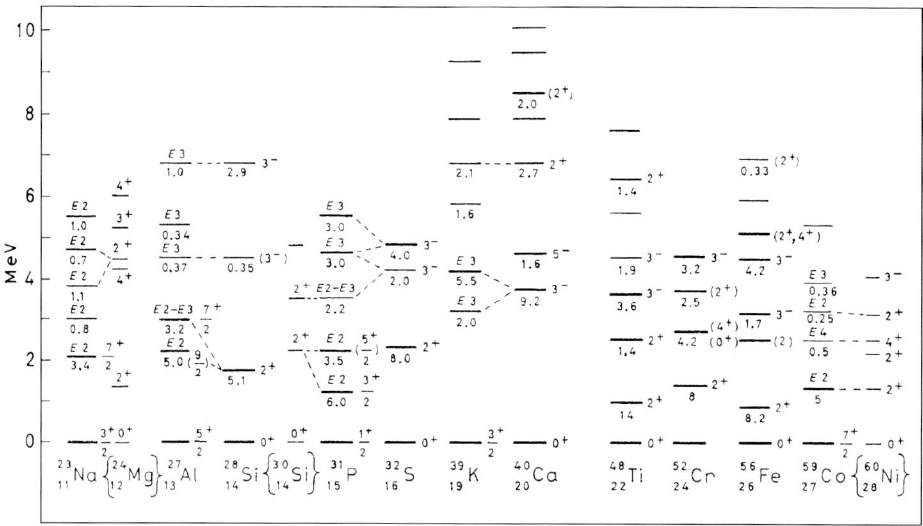

Fig. 18. – Nuclear states excited by inelastic proton scattering at 155 MeV in nuclei of $A = 23$ to $A = 53$ (Orsay, ref. [16]). The strongly excited (collective) levels of even-even nuclei are indicated in bold type. In odd nuclei the levels corresponding to "core excitations" are connected to parent states with dashed lines.

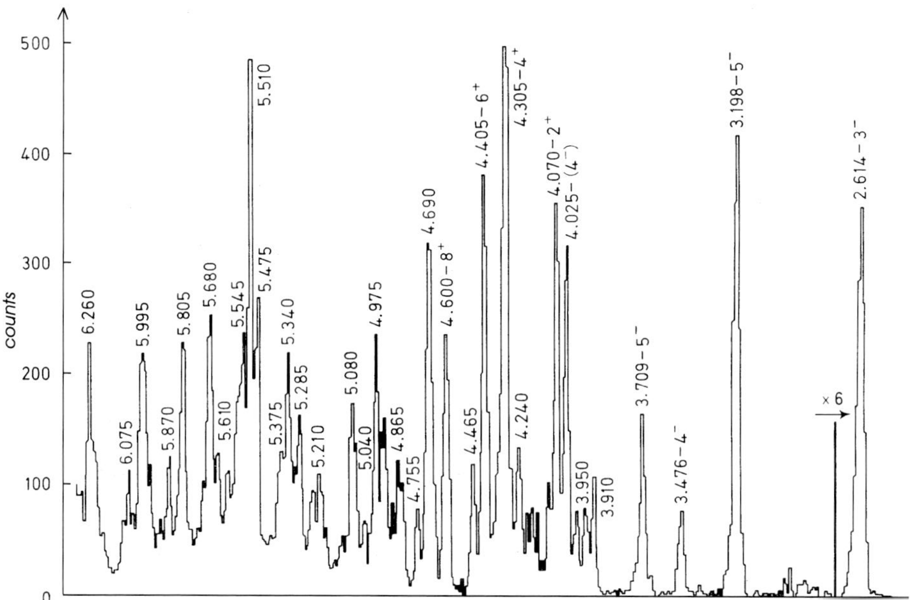

Fig. 19. – Proton inelastic scattering on ^{208}Pb in a very high-resolution experiment (Saclay, ref. [17]).

The collective states (2^+ and 3^-) of even-even nuclei and the *"core-excitations"* in odd-nuclei are clearly shown [16].

On the other hand, fig. 19 displays a very beautiful spectrum of inelastic scattered protons ($E_\mathrm{p} = 24.5\,\mathrm{MeV}$) on ^{208}Pb in a high-resolution experiment (high-efficiency analyzing magnets connected with spark-chambers) performed at Saclay [17].

The sequence of collective 3^-, 5^-, 2^+, 4^+6^+ "collective" states is highlighted by their strong excitation via inelastic scattering.

3. – Selection of nuclear states. In beam γ-ray spectroscopy. The advent of heavy-ion nuclear reactions

Transfer reactions and inelastic scattering so as Coulomb excitations were extensively used especially because of their capability of distinguishing between single particle and collective nuclear states. This specificity of nuclear reactions has been further emphasized by two important steps in nuclear spectroscopy in the '60s-'70s: the in-beam γ-ray spectroscopy and the advent of heavy-ions as powerful tools to select typical nuclear levels.

In-beam γ-ray spectroscopy has been introduced for the first time by Morinaga and Gugelot [18] at IKO (Amsterdam) in 1963.

Figure 20 (a,b) shows the results obtained in their pioneering experiments concerning the level schemes of some Dysprosium even-even isotopes using (α, xn) *reactions* on Gd targets and detecting the on-line cascade γ-rays (rotational and/or yrast bands).

Such a work opened the way to the usual to-day spectrometry with heavy-ions as probing particles.

In fact, the advent of *heavy ions* (together with the *in-beam γ-spectroscopy*) marked the *second revolution* in the nuclear-structure investigation. From the *experimental point* of view the *basic method* is very simple: it consists in observing with a Ge(Li) detector of quite a good resolution the *"prompt γ-rays"* emitted during the bombardment of a target with a heavy-ion beam.

Also the additional measurements usually performed (γ-ray excitation functions, angular distributions, γ–γ coincidences, polarization...) did not require special expedients apart the suitable detection systems [19].

Moreover, as already mentioned, the *very strong selectivity* (very few levels populated) even in light nuclei and at several MeV of excitation gives rise to *impressive specific structure* due to the clean spectrum of cascade γ-rays [19]. This of course is what is observed in the modern nuclear-spectroscopy technology taking also into account the enormous progress achieved in the γ single and coincidence spectra with more and more sophisticated γ-arrays and analyzing detection systems so as new accelerator facilities (radioactive beams).

One of the first examples of this method is the first observation of direct and double Coulomb excitation of the vibrational 2 phonon triplet in ^{76}Se with ^{16}O ions at the Aldermaston Tandem facility in 1962 [20]. The results are shown in fig. 21.

Meanwhile, quite a lot of interesting results were found at the various heavy-ion accelerator facilities worldwide concerning the selection of yrast states, ground-state rotational

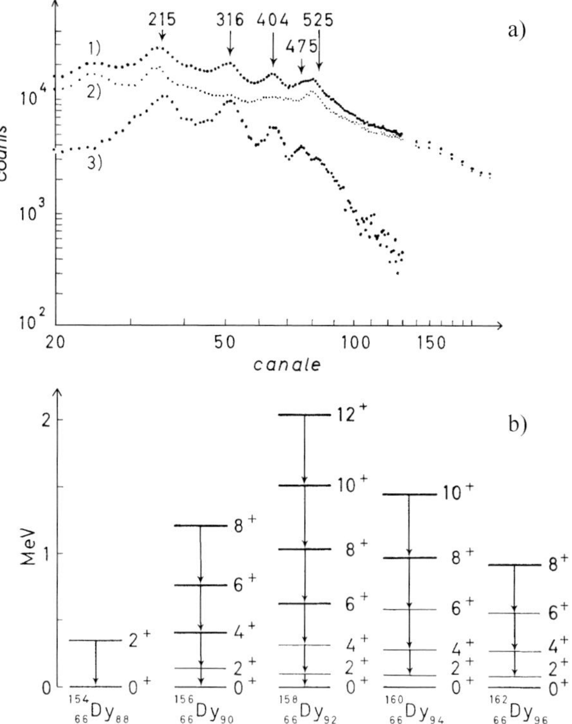

Fig. 20. – a) γ-spectrum observed by (α, xn) reaction on ^{158}Gd$_2$O$_3$ with 52 MeV α: curve 1) 1.5 mm Pb absorber between target and counter; curve 2) 10 mm Pb absorber; curve 3) difference between 1) and 2) (see ref. [18]). b) Level scheme of various even Gd isotopes. The levels found by (α, xn) reaction (ref. [18]) are indicated in bold type.

bands, specific high-excitation shell model states in medium and light nuclei. The discovery of the back-bending and of the super and hyper deformed bands and new nuclear shapes, the production of exotic nuclei far from the stability line are the new frontiers of the nuclear-structure problem opened by this *New Spectroscopy Era*. For instance, with the *TESSA array* in Daresbury it was possible to identify for the first time the famous superdeformed band in ^{152}Dy (until $J = 60\,$h) [21].

Other important achievements were GASP at the Laboratori Nazionali di Legnaro and EUROGAM in Strasbourg. Further developments achieved by the GAMMAS-PHERE at Berkeley (USA) and by EUROBALL in Europe allowed to perform very high-quality γ-spectroscopy.

All these new facilities can be considered as the milestones of the *third revolution of nuclear spectroscopy*.

In Italy we started the heavy-ion physics at the Laboratori Nazionali di Legnaro at the beginning of the '80s after a long period of preparation and assessment, with the advent of the XTU Tandem (High Voltage type with ~ 16 MV at the terminal).

This goal was achieved after 10 years of hard job and confident waiting establishing

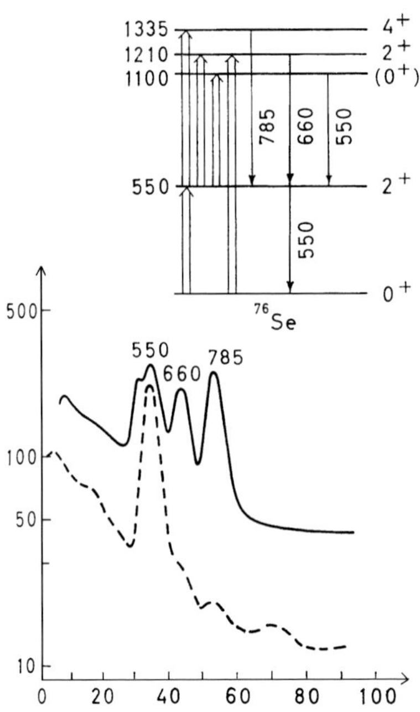

Fig. 21. – Coulomb excitation (direct and double) of the two-phonon triplet in ^{76}Se with ^{16}O ions at 45 MeV at the Aldermaston Tandem. Also the γ scintillation direct spectrum (dashed line) and the one coincident with the 550 γ-ray (continuous line) are reported (from ref. [20]).

a seal milestone for the development of nuclear physics in Italy in the context of the Istituto Nazionale di Fisica Nucleare. The 10 years of hunting this "*White Whale*" were testified by the name "*Moby Dick*" we gave to the accelerator.

The scientific program at LNL Tandem started with fusion-evaporation and deep inelastic experiments in medium-mass nuclei extending the investigation to transfer reaction, dissipative phenomena, in-beam γ-ray spectroscopy, applied nuclear physics together with an impressive development of detection and analysis techniques which are well-known today and competitive at the international level [22].

Moreover, the advent of ALPI (Acceleratore Lineare Per Ioni), the new superconducting LINAC, in 1987 coupled with the Tandem (and today operating also with a ECR source) provided the LNL with more energetic ion beams (^{58}Ni at 347 MeV) [23].

Interesting results obtained in the nuclear-structure studied during the '80s and '90s can be found in ref. [24]. Among them the characterization of shell closure $N = 40$ for ^{68}Ni$_{40}$ (increasing 2^+ state excitation energy) and the investigation of fusion barriers via the excitation functions [24]. Among the others, an illustrative example is given by the already quotedd γ-ray spectrum of ^{144}Gd shown in fig. 12 and measured with GASP at LNL. The Yrast superdeformed band is clearly evidentiated with a backbending at $E_\gamma = 900$ keV.

4. – The evolution of the nuclear-structure problem. The case of $1f_{7/2}$ spectroscopy

The development of nuclear-structure investigations since the already "historical" conception of the nuclear shell-model (the '50s) [25], is under the eyes of everybody. The theoretical (model) description of nuclear states through the establishment of single-particle and collective [26] properties and different coupling schemes and interactions until the more sophisticated large configuration calculations and the symmetry properties of fermions and bosons as specific nuclear characteristics [27], are in continuous impressive progress. From the *3rd revolution* (γ-arrays with high resolution and high efficiency, in beam γ-ray spectroscopy with heavy-ions) we are facing now the *4th revolution in nuclear spectroscopy* with the *advent of radioactive heavy-ion beams* and the production of *exotic nuclei* more and more approaching the *nucleon drip lines*.

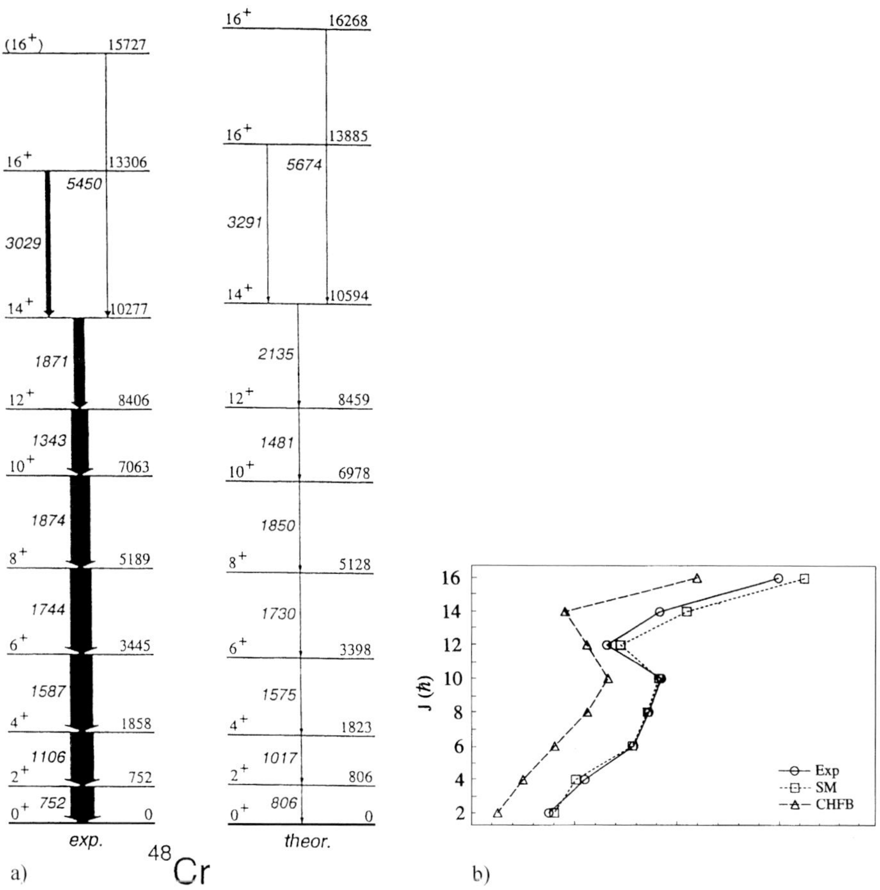

Fig. 22. – a) ^{48}Cr level scheme; experiment *vs.* theory (shell-model with interaction). b) Back-bending plot of the yrast band of ^{48}Cr shell-model (SM), mean field (CHFB) and experiment (Exp) (J *vs.* E_γ) (see ref. [28]).

4`1. *The revival of $1f_{7/2}$ spectroscopy*. – The new possibilities opened by heavy-ion reactions and new detection techniques have a particular application for exploring the structure of medium-mass nuclei. Of particular interest is the case of "$1f_{7/2}$ *nuclei*" (a "*special physics case*" for me, so close to my heart and a large part of my active scientific years). This revival, at Legnaro, has been possible due to the production of new isotopes in the region near the proton drip lines (neutron-deficient nuclei). Exotic $1f_{7/2}$ nuclei are of great interest, of course, and will contribute to new insight into nuclear structure understanding. Among them are $^{48}Cr_{24}$, $^{45}Fe_{19}$, $^{46}Mn_{21}$, $^{48}Ni_{20}$, $^{49}Ni_{21}$, $^{50}Cr_{26}$.

An important issue is the finding of super- and hyper deformed rotational bands displaying back-bending (*i.e.* the inversion of the moment of inertia).

Let me consider some cases in point. I start first with the $^{48}Cr_{24}$ and $^{50}Cr_{26}$ nuclei.

Figure 22 shows the γ-decay scheme of ^{48}Cr, which is a typical ($N = Z$, $T = 0$) nucleus in the middle of $1f_{7/2}$ shell, already far from the stability line. It is found to be a good rotor (Q gives $\beta = 0.3$ where β is the deformation parameter) and its quadrupole properties are well accounted for by the interacting shell-model (full f-p space reducing the valence nucleons to $1f_{7/2}$-$2p_{3/2}$) together with the excellent reproduction of the energy levels [28].

Figure 22 shows also the plot of the yrast band of ^{48}Cr, showing the back-bending at $J > 10\,\text{h}$, as compared with shell-model and mean-field calculations. The good agreement with the first is really interesting.

Even more interesting is the case of ^{50}Cr. Recent investigations [29] did confirm the double backbending as predicted, due to the addition of two neutrons. Also in this case the collective behaviour is accounted for by the shell-model (fig. 23).

A clear progress was made with respect to the previous experiment by the Munich-Padova-Florence group in the '70s, in spite of the fact that the ground-state yrast band was already found with its collective behaviour (see ref. [17]).

Other (interesting) results are shown in fig. 24, where the $T = 1$ isobaric triplet at $A = 50$ is reported. It consists of the $^{50}Cr_{26}$, $^{50}Mn_{25}$ and $^{50}Fe_{24}$. Furthermore the $f_{7/2}$ spin alignment in the mirror ($N \leftrightarrows Z$) nuclei ^{50}Cr and ^{50}Fe is shown indicating the possibility of determining the Coulomb-energy difference as displayed by the corresponding rotational bands (see ref. [28]).

Let me also recall the problem of the ($1f_{7/2}$) two-body spectra which was of paramount importance in the '60s (see ref. [5]) in the frame of the classic ($1f_{7/2}$) nuclear spectroscopy [30].

Figures 25 and 26 show the ^{42}Ca and ^{50}Ti two-nucleon spectra in comparison with a quite complete set of the same spectra for even-even Ti isotopes as more recently found [31]. It is very interesting to note the behaviour of the ($1f_{7/2}$) two-protons spectra as a function of the increasing neutron number. The effective two-body interaction as arising from the ^{42}Ti, ^{50}Ti spectra which are almost identical is well consistent with the $Z = 20$ and $N = 20, 28$ closed shell; moreover the $^{54}Ti_{32}$ spectrum (4 neutron outside the $N = 28$ shell) which is quite similar to those of ^{42}Ti and ^{50}Ti seems to indicate a special "blocking" effect of the $p_{3/2}$ neutron shell.

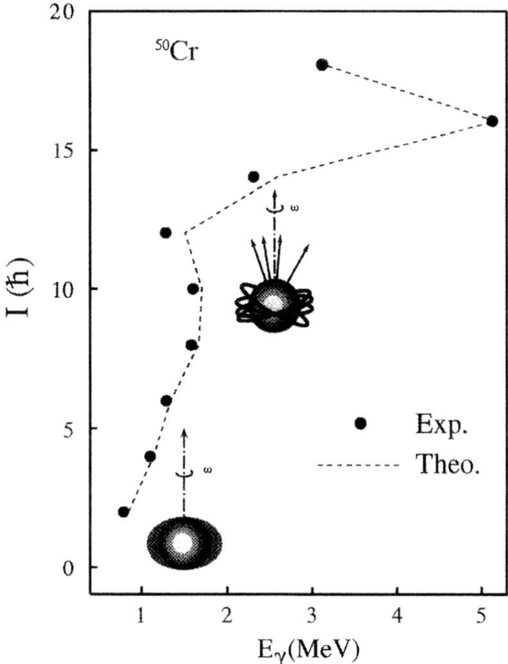

Fig. 23. – Backbending plot (J vs. E_γ) of the yrast band of ^{50}Cr (see text).

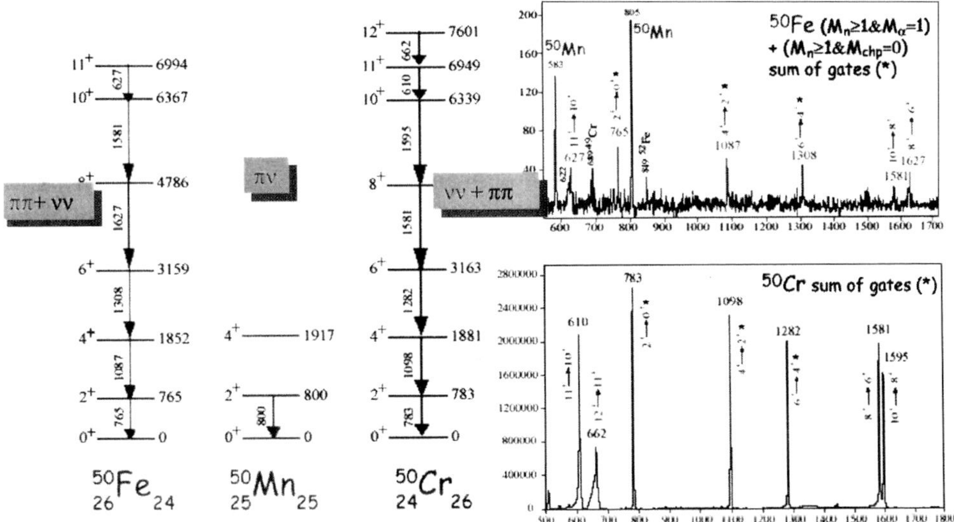

Fig. 24. – $A = 50$, $T = 1$ isobaric triplet.

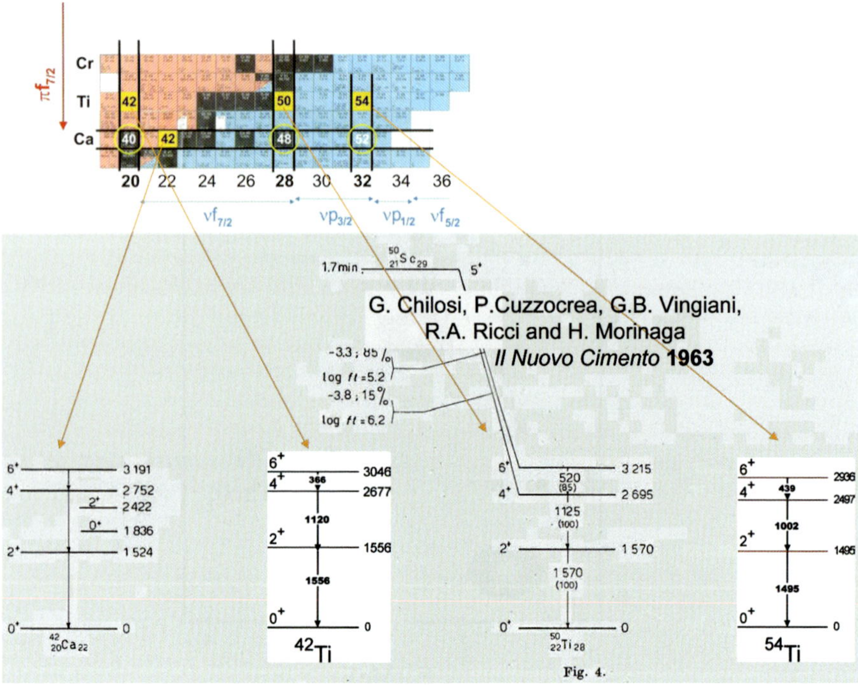

Fig. 25. – Two particle (proton and neutron) spectra in the $1f_{7/2}$ shell (see text).

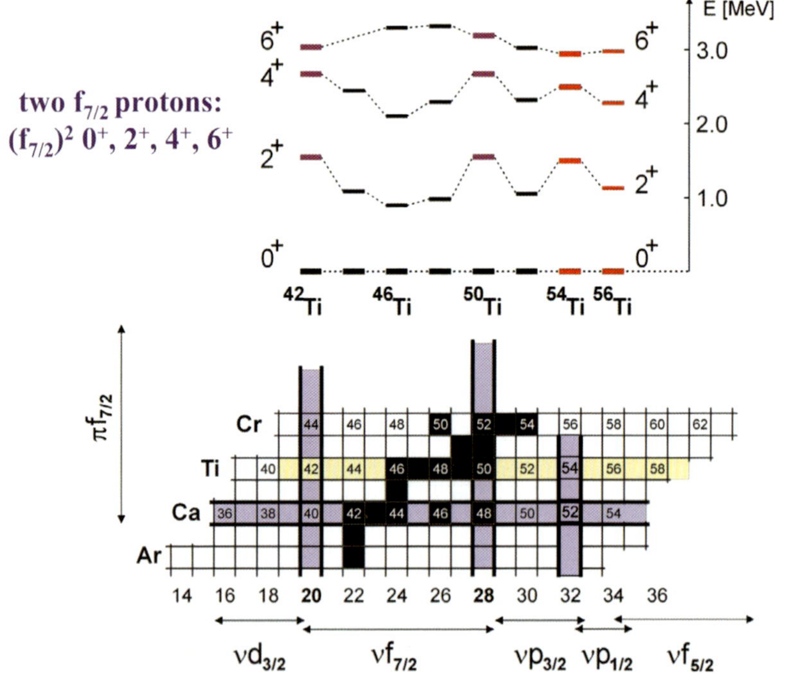

Fig. 26. – Evolution of $1f_{7/2}$ two proton spectra in Ti isotopes (ref. [31]).

5. – Conclusions

The long journey of experimental nuclear spectroscopy as a basic piece of the history of nuclear structure investigations represents a large part of my scientific academic career supported by my "passion" for the field of nuclear physics.

I will conclude with few personal remarks.

First of all it is clear that the *"physics of the nucleus"* will enjoy a real renaissance both in dynamics and structure studies.

The *4th revolution* in nuclear spectroscopy due to the extensive use of *radioactive beams* is opening a new fascinating branch of nuclear investigations.

If one considers that the *"nuclear-structure problem"* could be represented by a diagram with three axis pointing on the direction of the *angular momentum J* (nuclear shapes, super and hyperdeformed bands; magnetic rotations, band terminations); *the nuclear temperature* (collective structures, high excitation modes, chaotic behaviour phase transitions); and the *isospin* (exotic nuclei, halo effects, shell-model structure approaching the drip-lines, new collective states, superheavy elements) so as fundamental *symmetries* and *hypersymmetries* one sees the impressive richness of new experimental and theoretical investigations. This is a good message for young people.

You will allow me, and this is a priviledge I will take, in expressing my deep gratitude for the dedication of this excellent Course to my 80th anniversary, to repeat a message I addressed 5 years ago here in Varenna remembering the 50 years of my *"commitment"* to nuclear physics [32]:

"Physics is a paramount aspect of Science. Science is an intellectual game with crucial impact into society. On the other hand, Science is a school of criticism, freedom and tolerance. Its cultural and social value should be always claimed and defended.
Nuclear Physics is our discipline. Enjoy it"

REFERENCES

[1] KALMANN H., *Nature and Technique*, July 1947: a phosphore of naftalene was used to replace the old zinc sulfide screen of the Rutherford time.
[2] See CURRAN S. C., *Luminesence and the Scintillation Counter* (Academic Press, New York) 1953.
[3] HOFSTADTER R., *Phys. Rev.*, **74** (1948) 100.
[4] MITCHELL A. C. G., VANLIESHOUT R., WAPSTRA A. H., RICCI R. A. and GIRGIS R. K., in *"α–β–γ–ray-Spectroscopy"*, edited by SIEGBAHN K. (North Holland, Amsterdam) 1965; see also RICCI R. A., *From* $1f_{7/2}$ *nuclei to heavy-ion nuclear reactions*, in *Proceedings of the International School of Physics "Enerico Fermi", Course CLIII*, edited by MOLINARI A., RICCATI L., ALBERICO W. M. and MORANDO M. (IOS Press, Amsterdam) 2003, p. 627.
[5] CHILOSI G., CUZZOCREA P., VINGIANI G. B., RICCI R. A. and MORINAGA H., *Nuovo Cimento*, **27** (1963) 407.
[6] CHILOSI G., RICCI R. A., TOUCHARD J. and WAPSTRA A. H., *Nucl. Phys.*, **21** (1964) 235.

[7] Bohr A. and Mottelson B., *Nuclear Structure*, Vol. **1** (New York, Amsterdam) 1969, p. 343, table 3-3.

[8] Ricci R. A., *Performance of a beta scintillation spectrometer*, *Physica*, **XXIII** (1957) 693.

[9] Reines F. and Cowan C. L. jr., *Phys. Rev.*, **113** (1959) 273; see, in particular Nezerick F. A. and Reines F., *Phys. Rev.*, **142** (1966) 142.

[10] See Iachello F., *Nucl. Phys. News*, **10** (2000) 12.

[11] See Ricci R. A., *Experimental Nuclear Spectroscopy in the* $1f_{7/2}$ *Shell*, in *Proceedings of the International School of Physics "Enrico Fermi", Course XL*, edited by Jean M. and Ricci R. A. (Academic Press) 1969; and references quoted herein.

[12] Amaldi U. jr., Campos Venuti G., Cortellessa G., De Sanctis E., Frullani S., Lombard R. and Salvadori P., *Phys. Lett.*, **22** (1966) 593; see also Ricci R. A., *Proceedings of the International School of Physics "Enrico Fermi", Course XXXVI*, edited by Bloch C. L. (Academic Press, London) 1965, p. 566; Riou M., *Rev. Mod. Phys.*, **37** (1965) 385.

[13] Ruhla C., Arditi M., Doubre H., Jacmart J. C., Liu M., Ricci R. A., Riou M. and Roynette J. C., *Nucl. Phys. A*, **95** (1967) 526.

[14] Cohen B. L. and Rubin A. G., *Phys. Rev.*, **114** (1958) 1568; for other references see Ricci R. A., *Suppl. Nuovo Cimento, Ser. I*, **5** (1967) 1385.

[15] See Ricci R. A., *Suppl. Nuovo Cimento, Ser. I*, **3** (1965) 726 and references quoted herein.

[16] See Liu M., Ricci R. A., Riou M. and Ruhla C., *Nucl. Phys.*, **75** (1966) 481; Ricci R. A., Jacmart J. C., Liu M., Riou M. and Ruhla C., *Nucl. Phys. A*, **91** (1967) 609.

[17] Saudinos J., Vallois G., Beer O., Gendrot M. and Lopato P., *International Conference on Progress in Nuclear Physics with Tandems, Heidelberg, July 1966*.

[18] Morinaga H. and Gugelot P. C., *Nucl. Phys.*, **46** (1963) 210.

[19] See Morinaga H., *Proceedings of the International School "Enrico Fermi", Course LXII*, edited by Faraggi H. and Ricci R. A. (North Holland) 1976, p. 351; see also Signorini C., same volume, p. 499.

[20] Bygrave W., Eccleshall D. and Yates M. L., *Nucl. Phys.*, **32** (1962) 190.

[21] See Twin P., *Proceedings of the International School of Physics "Enrico Fermi", Course CIII*, edited by Kienle P., Ricci R. A. and Rubbino A. (North Holland) 1989.

[22] See Ricci R. A., *Nuovo Cimento A*, **81** (1984).

[23] See Stefanini A. M., Lunardi S. and Ricci R. A., *Nucl. Phys. News*, **5** (1995) 9; see also Puglierin G., contribution to this School.

[24] LNL-INFN Rep. 178/2001 and 182/2002.

[25] Mayer M. G., *Phys. Rev.*, **75** (1949) 1969; Axel O., Jensen H. and Suess H., *Phys. Rev.*, **75** (1949) 1766.

[26] Bohr A. and Mottelson B., *Mat. Fys. Medd. Dan. Vid. Selksk*, **27** (1953) 16.

[27] See Iachello F., this volume, p. 1; Covello A., this volume, p. 291.

[28] See Lenzi S. M., this volume, p. 303; see also Lenzi S. M. *et al.*, *Phys. Rev. C*, **60** (1999) 1303.

[29] Brandolini F. *et al.*, LNL-INFN (REP) 2001, p. 14.

[30] See Ricci R. A. and Maurenzig P., *Riv. Nuovo Cimento, Serie I*, vol. I (1999); see also ref. [11].

[31] See Fornal B., to be published.

[32] Ricci R. A., *From* $1f_{7/2}$ *nuclei to heavy-ion nuclear reactions*, in *Proceedings of the International School of Physics "Enrico Fermi", Course CIII*, edited by Molinari A., Riccati L., Alberico W. and Morando M. (IOS Press, Amsterdam) 2003, p. 627.

DOI 10.3254/978-1-58603-885-4-57

Experimental program with rare-isotope beams at FAIR

T. AUMANN

GSI - Gesellschaft für Schwerionenforschung mbH - Darmstadt, Germany

Summary. — The international Facility for Antiproton and Ion Research (FAIR) will provide a combination of accelerators and storage rings for a multidisciplinary scientific program. The FAIR accelerator complex constitutes a substantial extension of the present GSI systems and will use the existing accelerators after several upgrades as an injector. This new facility comprises as an integral part a next-generation rare-isotope beam facility based on the production and separation of secondary beams produced by fragmentation or fission of energetic high-intensity primary ion beams. An advanced experimental program including new and unique concepts promises a new quality in investigations of nuclei far away from stability. A brief overview on the facility is presented with emphasis on the radioactive-beam facility. The experimental program and its related instrumentation is discussed briefly.

1. – Overview

The new Facility for Antiproton and Ion Research (FAIR) which is being built at GSI will provide high-intensity primary and secondary beams for a variety of different experiments enabling a broad range of physics goals to be addressed. The science areas covered include i) the structure of nuclei far from stability and astrophysics, ii) hadron spectroscopy and hadronic matter, iii) compressed nuclear matter, iv) high-energy density in bulk matter, v) quantum electrodynamics, strong fields, and ion-matter interactions.

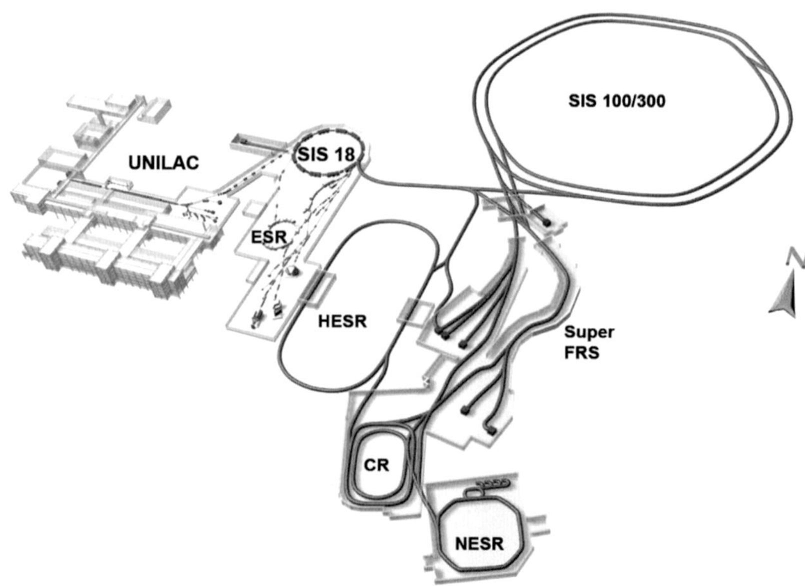

Fig. 1. – Schematic layout of the FAIR facility. The existing GSI accelerators UNILAC and SIS-18 serve as an injector for the two large synchrothon rings SIS-100/300. The beams are distributed to various experimental areas or to the production targets for the anti-proton or radioactive-isotope beam facilities. For beam storage and manipulation several rings are foreseen including the high-energy storage ring HESR for anti-protons, the collector ring CR, and the experimental storage ring NESR. The storage-ring complex provides possibilities of beam cooling as well as target areas for scattering experiments. The superconducting fragment separator Super-FRS for preparation of rare-isotope beams is indicated as well.

The science cases as well as the facility concept and layout have been worked out by a large international community and are described in detail in the Conceptual Design Report (CDR) [1] of the facility. The concept is based on a double-ring synchrotron structure SIS100/300, which utilizes the existing UNILAC/SIS18 accelerators as an injector. Beams of all ions from hydrogen to uranium, as well as antiprotons with a large energy range from rest up to $35 \, \text{GeV}/u$ will be provided by FAIR. A schematic layout of the facility is displayed in fig. 1, showing in addition several storage rings and experimental areas.

Two aspects of the concept are most important to overcome the present intensity limit given by the fact that the space-charge limit is reached in the synchrotron SIS18: i) first, a faster cycling of the synchrotron of $3 \, \text{Hz}$ compared to the present situation of $0.3 \, \text{Hz}$ yields an order of magnitude increase in average intensity, and ii) the acceleration of uranium ions with charge state 28^+ instead of 73^+ will increase the space-charge limit by one order of magnitude. In order to reach the same beam energy (1.5 to $2 \, \text{GeV}/u$) while accelerating a lower charge-state, a synchrotron with higher magnetic rigidity ($100 \, \text{Tm}$) is needed, which is provided by the SIS100 ring. An average intensity of almost 10^{12} ions/s will be reached, $e.g.$, for uranium beams at $1.5 \, \text{GeV}/u$. The second ring, SIS300, serves either as a stretcher ring to provide slow extracted high-duty cycle beams, *e.g.*, for experiments with short-lived nuclei. Alternatively, the two synchrotrons may be used to accelerate ions with higher charge state to higher beam energies (with lower intensity) up to about $30\,\mathrm{GeV}/u$ as required, *e.g.*, for the nucleus-nucleus collisions programme. The rigidity of $100\,\mathrm{Tm}$ does also allow accelerating protons up to $30\,\mathrm{GeV}$, which is optimum for anti-proton production. The experiments utilizing high-energy anti-proton beams are placed around an internal target in the High-Energy Storage Ring HESR, where the antiprotons are circulating and cooled by an electron-cooling system. The system can provide both pulsed beams, *e.g.*, for injection into storage rings or the plasma physics experiments, as well as continuous high-duty cycle beams as required for external target experiments like the Compressed-Barionic Matter (CBM) experiment or several other experiments utilizing secondary beams of short-lived nuclei. A recent collection of articles on the FAIR physics program including a status report can be found in ref. [2].

A high degree of parallel operation of different research programs was an important aspect considered in the design of FAIR. The scheme of multiple synchrotron and storage rings and the resulting possibilities for in-ring experimentation offers itself for a truly parallel and multiple use. For other recent overviews on the FAIR project see refs. [3, 4].

The construction of FAIR is planned in three stages, the first one comprising the superconducting fragment separator Super-FRS linked to SIS-18 for production of rare-isotope beams, the CR and NESR storage rings and experimental areas. Already with completion of this first stage expected for beginning of 2012, the radioactive-beam physics program can be started and will benefit from much higher secondary beam intensities than available presently at GSI, which are provided due to the increased SIS-18 ramping rate, the large phase-space acceptance of the Super-FRS, and the superior functionality of the storage rings. The completed facility is planned being fully operational in 2015.

This paper is focussing on the part of the facility dealing with unstable nuclei. Secondary beams of short-lived nuclei are utilized to investigate the properties of nuclei as a function of isospin in order to illuminate and understand the degrees of freedom governing the properties of nuclei and nuclear matter. Given the possibility of choosing the neutron-proton asymmetry of nuclei enables us to study characteristic effects being enhanced in certain isotopes or being related solely to the neutron-proton asymmetry. The goal of these experimental efforts is to provide stringent tests of theoretical approaches giving guidance towards the fundamental understanding of the many-body quantum system nucleus. In addition, the properties of unstable nuclei as well as reactions involving unstable nuclei are important in many astrophysical scenarios. Such reactions can be studied using radioactive beams, often by measuring the time-reversed process. The astrophysical processes, *e.g.*, the nucleosynthesis via rapid-neutron capture (r-process), involves heavy very neutron-rich nuclei that will be accessible for experiments for the first time at the future next-generation radioactive-beam facility NuSTAR (Nuclear Structure, Astrophysics and Reactions) at FAIR.

A brief overview on the NuSTAR facility is given in the next section. The individual experimental approaches are then discussed in the subsequent sections. The physics program can be grouped essentially into two categories, one concerned with nuclear ground state and decay properties, one devoted to nuclear-structure studies by reactions with various probes. The physics program and the experimental equipment has been proposed and elaborated by a large international collaboration, the NuSTAR collaboration (about 700 scientists), and published as a collection of Letters of Intent, see ref. [5]. Apart from the fact that new regions of the nuclear chart will be in reach due to the higher beam intensities, emphasis has been put to further develop the experimental approaches to substantially increase their potential and quality, as well as to develop new advanced techniques providing access to methods not available so far for radioactive beams. In addition to the physics program, a detailed technical description of the experimental approaches and proposed instrumentation has been presented in individual Technical Proposals of NuSTAR, which have been evaluated by international expert committees. The project proposals of the recommended NuSTAR experiments are published as a separate volume of the Baseline Technical Report (July 2006) [6] of the FAIR facility.

2. – The radioactive beam facility

A schematic view of the layout of the NuSTAR facility is shown in fig. 2. The key requirements for the production of secondary beams of short-lived nuclei by fragmentation and fission are high primary-beam intensities and beam energies of up to $1.5\,\mathrm{GeV}/u$. The latter is important due to several reasons: firstly, an efficient transport of the secondary

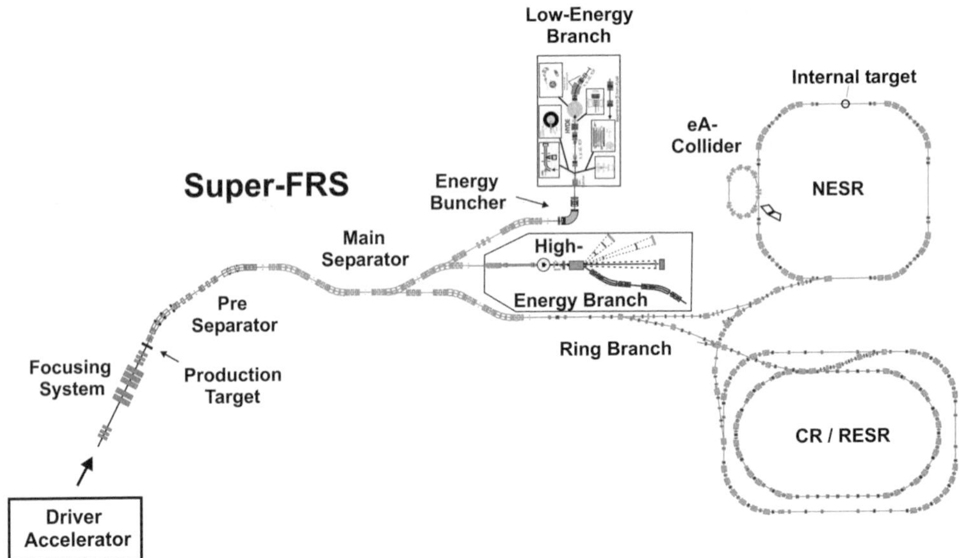

Fig. 2. – Schematic drawing of the two-stage large-acceptance superconducting fragment separator Super-FRS consisting of a pre-separator and a main separator. The Super-FRS serves three experimental areas, the low-energy branch, high-energy branch, and the storage ring complex comprising a large-acceptance Collector Ring (CR), an accumulator ring (RESR), and the experimental storage ring NESR including an internal target and the electron-ion (eA) collider.

beams, in particular those produced by uranium fission, is achievable with reasonable sized magnetic separators only by making use of the kinematic forward focussing due to the high beam energy. Secondly, the $B\rho - \Delta E - B\rho$ separation method allows clean isotopic separation and identification only for one ionic charge state, *i.e.* fully ionized fragments, which can be reached for heavy ions ($Z \approx 80$) only for energies above one GeV/u. Besides the requirements due to the production and separation method, high energies of the secondary beams are also advantageous due to experimental and physics reasons, among those the possibility of using thick targets, high-acceptance measurements, and a clean separation of reaction mechanisms and spectroscopic information to be deduced.

Figure 2 shows a schematic layout of the rare-isotope facility including a superconducting fragment recoil separator Super-FRS [7] and three experimental areas, the low-energy branch, high-energy branch, and ring branch. The ring branch consists of several storage rings. First the collector ring CR, which has a large acceptance to allow efficient injection of secondary beams. Before injection into the new experimental storage ring NESR, the beam is stochastically pre-cooled. Further electron cooling is provided in the NESR, which is equipped with an internal target for reaction studies. A smaller intersecting electron storage ring allows for the first time electron-scattering experiments off radioactive isotopes to be performed in a collider mode (eA-collider). An additional accumulator ring (RESR) is foreseen for fast deceleration with about 1 T/s ramping rate.

The Super-FRS is optimized for efficient transport of fission fragments implying a rather large acceptance of 5% in momentum and ± 40 mrad and ± 20 mrad angular acceptance in the horizontal and vertical planes, respectively. Although the acceptance is largely increased compared to the present FRS, an ion-optical resolving power of 1500 has been retained (for 40 π mm mrad). Besides the resolution, an additional pre-separator ensures low background and clean isotopic separation of the beams of interest. The calculated transmission for beams of neutron-rich medium mass nuclei produced via fission of uranium with primary beam energy of 1.5 GeV/u amounts to about 50%. The gain in transmission compared to the present separator is, as an example, a factor of 30 for the ^{78}Ni region. Together with the gain in primary beam intensity, intensities for separated rare-isotope beams will increase by significant more than three orders of magnitude. Estimated beam intensities are given in fig. 3. Substantially increased beam intensities are achieved in particular for medium- heavy and heavy neutron-rich nuclei far extending into the region of hitherto unknown nuclei (the limits of known nuclei are indicated by the black line). Measurements of masses and decay properties of nuclei important for the understanding of nucleosynthesis of heavy elements in the cosmos by rapid neutron capture processes (r-process) will be possible as indicated in the figure.

In order to achieve a broad and at the same time in-depth insight into the structure of exotic nuclei, it is indispensable to utilize different experimental techniques to measure similar or related observables. Different probes and/or different beam energies are needed in order to optimize the sensitivity to particular nuclear-structure observables. Three experimental areas are foreseen behind the Super-FRS, one housing a variety of experiments with low-energy and stopped beams, a high-energy reaction set-up, and a

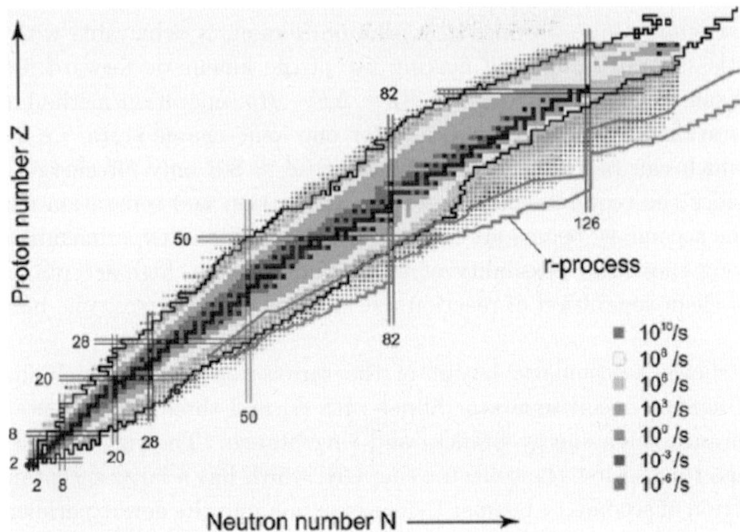

Fig. 3. – Estimated intensities for radioactive beams behind the Super-FRS. The region of the astrophysical r-process path is indicated by the correspondingly labelled line. The black lines indicate the present limits of known nuclei.

storage and cooler ring complex including electron and antiproton colliders. In the following, a few aspects of the different branches will be briefly outlined. This discussion cannot be comprehensive, a more complete and detailed description of the proposed experiments can be found in the Letters of Intent of the NUSTAR collaboration [5].

3. – Experiments with slowed-down and stopped beams

3˙1. *The low-energy branch.* – For experiments requiring low-energy beams or implanted ions, the high-energy radioactive beams have to be slowed down. The key instrument for these experiments is an energy-focusing device [8], which reduces the energy spread of the secondary beams delivered by the Super-FRS. The principle is sketched in fig. 4.

It consists of a high-resolution dispersive separator stage in combination with a set of profiled energy degraders. After the monoenergetic degrader, the momentum spread is greatly reduced allowing, *e.g.*, the ions to be stopped in a 1 m long gas cell, from which they can be extracted for further manipulation and experiments. This method combines the advantages of in-flight separation (no limitations on lifetime and chemistry) with the ISOL-type experimental methods. For gamma spectroscopy studies, beams can be slowed down also to intermediate energies (100 MeV/u) or even energies around the Coulomb barrier. Table I gives an overview on the planned experimental program.

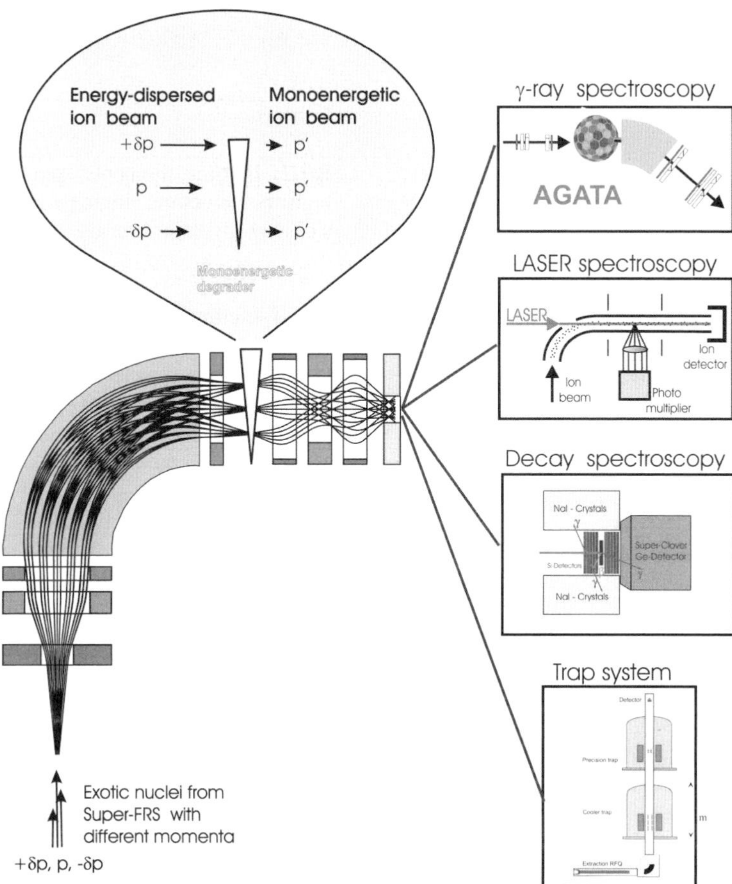

Fig. 4. – Schematic view of the experimental set-up for experiments with low-energy and stopped radioactive beams. Beams from the Super-FRS with large momentum spread are analyzed in a dispersive dipole stage followed by a shaped energy degrader reducing the momentum spread of the slowed-down beam. The beams can be either stopped or directed directly to one of the experimental areas indicated as well.

3˙2. *High-resolution in-flight spectroscopy (HISPEC).* – The core of the HISPEC experimental instrumentation is the AGATA 4π germanium γ-ray spectrometer. Due to its tracking capabilities, the first point of interaction and by that the emission angle of the photon can be determined. This allows to reconstruct the center-of-mass energy of the photons emitted in flight with high resolution even at rather large velocities, *e.g.*, with a resolution of about 0.5% at a velocity of 50% velocity of light. The AGATA array in combination with a magnetic analysis after the target and beam tracking detectors (see right-upper inset in fig. 4) will allow high-resolution gamma spectroscopy experiments using the in-flight technique at medium beam energies (around $100\,\mathrm{MeV}/u$). At higher beam energies, the gamma energy in the laboratory frame becomes rather large due to

TABLE I. – *Summary of experimental opportunities at the low-energy branch. The table is taken from the NuSTAR Letter of Intents* [5].

Field	Experimental method	Subject
Nuclear structure and astrophysics	Multiple Coulex, direct reactions, fusion evaporation	B(E2), lifetime, moments, spins, high spin structure, single particle structure, dynamical properties
	Intermediate energy Coulex, fragmentation	B(E2), lifetime, moments, spins, shape evolution, shell structure
Nuclear structure and astrophysics	α-, β-, γ-, particle-decay spectroscopy	Half-lifes, spins, nuclear moments, GT strength, isomer decay, β-decay branching ratios, β-delayed neutrons
Atomic physics methods applied to nuclear structure	Laser spectroscopy	Nuclear charge radii Magnetic dipole and electric quadrupole moments
Precision experiments	Ion trapping, charge breeding, and spectroscopy	Nuclear binding energies (mass measurements, Q-values)
Tests of the Standard Model	Atom trapping (MOT)	Half-life measurements β-decay branching ratios β-ν correlations
Nuclear structure and antimatter	RIB+pbar trapping	Charge radii, skin- and halo distributions

the Doppler boost in forward direction. In that case, the interaction of the photons in the material are dominated by pair production limiting the tracking capabilities of the detector yielding to larger Doppler broadening effects. For such high energies (at around $500\,\mathrm{MeV}/u$), experiments will be performed more efficiently at the $\mathrm{R}^3\mathrm{B}$ set-up using a high-efficiency gamma calorimeter with high granularity, see subsect. 4'1.

The physics program of HISPEC concentrates on the two energy domains around 50–$100\,\mathrm{MeV}/u$ and also around Coulomb barrier energies. At 50–$100\,\mathrm{MeV}/u$, the first 2^+ state can be excited very efficiently by electromagnetic excitation using heavy targets such as lead. The large cross-sections in combination with the good energy resolution of AGATA allows to determine $B(E2)$ values for short-lived nuclei even with low beam intensities. Shell structure and deformation of exotic nuclei can thus be studied in a wide range. In conjunction with proton inelastic-scattering experiments performed at the EXL facility (see subsect. 5'2), deformations of the neutron and proton densities can be investigated.

Another important part of the experimental program is the γ spectroscopy after fragmentation reactions. Excited states are populated in few-nucleon removal reactions and the γ decays are measured with AGATA in coincidence with the identified fragments after the target. In this reactions, very neutron-rich nuclei can be reached by starting with a neutron-rich beam and observing the excited states in nuclei after few-proton removal reactions. Such reactions promise also access to medium-high spin states of exotic nuclei.

Beam energies reduced to that around the Coulomb barrier are also considered and would allow to access high-spin states by multiple Coulomb excitation or fusion-evaporation reactions.

3‘3. *Decay spectroscopy (DESPEC).* – The set-up proposed by the DESPEC collaboration consists of compact neutron and gamma-ray arrays of high granularity and auxiliary detectors arranged around the active stopper with which time and position signals from the implanted ion and the subsequent beta decay can be correlated. The range straggeling due to the large momentum spread of the beam is largely reduced due to the use of the energy buncher.

From the measurement of the decay of stopped ions, half-lifes, branching ratios, nuclear moments, as well as information on excited states can be determined. The study of isomeric states and exotic decays such as the two-proton decay are further topics of interest. An important characteristic of the decay spectroscopy experiments is its high sensitivity allowing to deduce nuclear-structure information like the lifetime with very low beam intensity, down to about 10^{-3} particles/s. The half lifes of astrophysical relevant nuclei like the very neutron-rich r-process nuclei can thus be determined in a wide range, see fig. 3. This information together with the masses basically determines the complex path of nucleosynthesis. An impressive example was recently reported by the MSU group. The half-life of the r-process waiting-point nucleus ^{78}Ni was determined to be 110^{+100}_{-60} ms [9] measuring a total number of 11 ^{78}Ni isotopes with a beam intensity of about 3×10^{-4}/s.

3‘4. *The advanced trapping system MATS.* – MATS stands for "Precision Measurements of very short-lived nuclei using an Advanced Trapping System for highly-charged ions". Two techniques will be applied: high-accuracy mass measurements and in-trap spectroscopy of conversion-electron and alpha decays. A schematic view of the system is displayed in fig. 5. Radioactive ions from the Super-FRS are energy-bunched and stopped in a gas cell. Figure 5 indicates the manipulation steps from the extraction to the final storage in the precision Penning trap. The MATS collaboration aims for relative mass

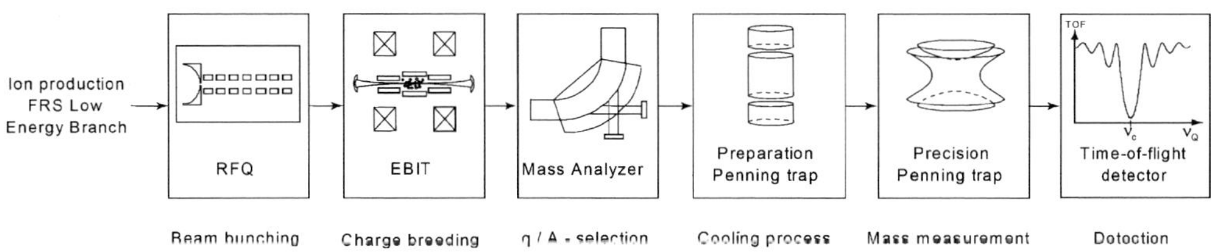

Fig. 5. – Schematic view of the experimental scheme for precision measurements with trapped ions.

precision of better than 10^{-8} even for nuclides with half-lifes of 10 ms only.

The mass-measurement program of MATS is complementary to the ILIMA experiment at the storage rings, see subsect. **5**`1. While the storage ring experiments are able to very effectively measure many masses of radioactive nuclei in a short time, MATS provides a superior resolution. The masses measured by MATS may also serve as calibration points for the survey measurements at the storage rings. Besides its importance for nuclear structure and astrophysics, as discussed above, precise mass values are also needed for tests of the Standard Model, *i.e.* the unitarity test of the CKM matrix in weak decays. Here, the Q values enter and the masses of mother and daughter nuclei have to be known with a precision of about 1×10^{-8}, which will be reached with the MATS Penning-trap system.

3`5. *The Laser-spectroscopy experiment LASPEC.* – A multi-purpose laser-spectroscopy station is proposed by the LASPEC collaboration. The set-up will permit utilizing a variety of optical methods that have been developed and pioneered at ISOL facilities [10], such as fluorescence, resonance ionization and polarization based spectroscopy. These techniques are used to measure isotopic and isomeric nuclear spins, magnetic dipole moments, electric quadrupole moments and changes in mean square charge radii. The latter, *e.g.*, is accomplished by a precise measurement of the isotope shift in an atomic transition. A recent example for the application of this technique is the determination of the charge-radius change for the lithium isotopes ^{9,11}Li [11]. The isotope shift in the $2s \to 3s$ two-photon transition was measured with an uncertainty $< 10^{-5}$ for ^{11}Li (half-life 8.5 ms) with a production rate of $\approx 10^4$ atoms/s only. Together with high-precision calculations for the mass shift, a charge radius of 2.467(37) was deduced. The difference in the two charge radii of ^{9}Li and ^{11}Li is related to polarization of the ^{9}Li core in the two-neutron halo nucleus ^{11}Li, but also involves the correlations among the neutrons in the halo.

The availability of the laser-spectroscopy instrumentation at the NuSTAR low-energy branch offers the unique opportunity of applying such techniques also to isotopes not accessible at ISOL facilities as the refractory elements.

4. – Scattering experiments with high-energy rare-ion beams

4`1. *Reactions with Relativistic Radioactive Beams (R^3B).* – The instrumentation at the high-energy branch is designed for experiments using directly the high-energy secondary beams as delivered from the separator with magnetic rigidities up to 20 Tm, thus taking full advantage of the characteristics of radioactive beams produced by the in-flight separation method including the highest possible transmission to the target. The incoming ions are identified on an event-by-event basis and tracked onto the reaction target enabling a simultaneous use of beams composed out of many isotopes.

The R^{3}B [12] experimental configuration is based on a concept similar to the existing LAND reaction set-up [13] at GSI introducing substantial improvement with respect to resolution and an extended detection scheme, which comprises the additional detection

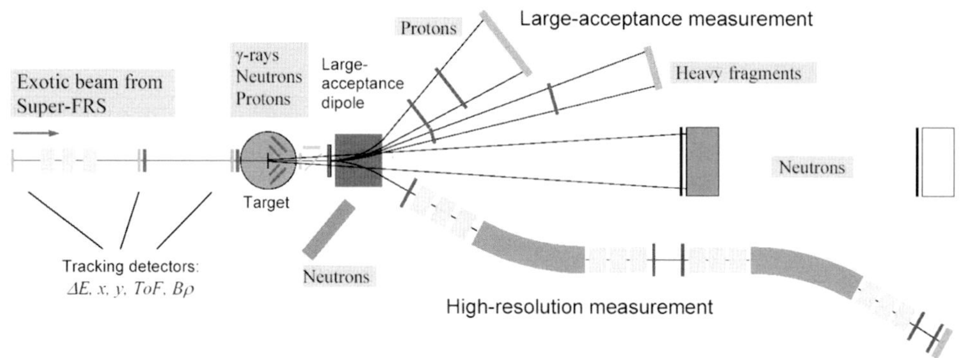

Fig. 6. – Schematic view of the experimental set-up for scattering experiments with relativistic radioactive beams comprising γ-ray and target recoil detection, a large-acceptance dipole magnet, a high-resolution magnetic spectrometer, neutron and light-charged particle detectors, and a variety of heavy-ion detectors.

of light (target-like) recoil particles and a high-resolution fragment spectrometer. The set-up, which is schematically depicted in fig. 6, foresees two operation modes, one for large-acceptance measurements of heavy fragments and light charged particles (left bend in fig. 6), and alternatively for high-resolution momentum measurements ($\Delta p/p \approx 10^{-4}$) using a magnetic spectrometer (right bend in fig. 6). A large-acceptance super-conducting dipole magnet, already under construction and in part funded by the European Commission, serves together with tracking detectors in the momentum analysis of the heavy fragments. The large gap of the dipole allows charged particles and neutrons of projectile rapidity emitted from excited fragments to be detectect behind the magnet by drift chambers and a RPC-based neutron detector at zero degree, respectively, which provide excellent position and time resolution. The target area, with the possibility of inserting a liquid- or frozen-hydrogen target, is surrounded by a silicon vertex tracker and a combined charged particle and gamma calorimeter.

The experimental configuration is suitable for kinematical complete measurements for a wide variety of scattering experiments, *i.e.* such as heavy-ion–induced electromagnetic excitation, knock-out and break-up reactions, or light-ion (in)elastic and quasi-free scattering in inverse kinematics, thus enabling a broad physics programme with rare-isotope beams to be performed. Table II gives an overview on the planned experimental programme. More details can be found in the Letter of Intent of the R^{3}B Collaboration [12, 5].

The use of high beam energies around few hundred MeV/u for studies of nuclear structure and astrophysics is advantageous both from an experimental point of view as well as from theoretical considerations. The high beam energies result in short interaction times and small scattering angles, which allow the use of certain approximations and thus enable a quantitative description of the underlying reaction mechanisms. Experimental merits are the possibility of using relatively thick targets (in the order of g/cm^2) and

TABLE II. – *Reactions with high-energy beams and corresponding achievable information.*

Reaction type	*Physics goals*
Knockout	Shell structure, valence-nucleon wave function, many-particle decay channels unbound states, nuclear resonances beyond the drip lines
Quasi-free scattering	Single-particle spectral functions, shell-occupation probabilities, nucleon-nucleon correlations, cluster structures
Total-absorption measurements	Nuclear matter radii, halo and skin structures
Elastic p scattering	Nuclear matter densities, halo and skin structures
Heavy-ion induced electromagnetic excitation	Low-lying transition strength, single-particle structure, astrophysical S factor, soft coherent modes, low-lying resonances in the continuum, giant dipole (quadrupole) strength
Charge-exchange reactions	Gamow-Teller strength, soft excitation modes, spin-dipole resonance, neutron skin thickness
Fission	Shell structure, dynamical properties
Spallation	Reaction mechanism, astrophysics, applications: nuclear-waste transmutation, neutron spallation sources
Projectile fragmentation and multifragmentation	Equation-of-state, thermal instabilities, structural phenomena in excited nuclei, γ-spectroscopy of exotic nuclei

kinematical forward focusing, which makes full-acceptance measurements feasible with moderately sized detectors. Thus nuclear-structure investigations of very exotic nuclei at the driplines are possible even if such beams are produced with very low rates in the order of one ion per second.

An example is the one-nucleon knock-out reaction. In particular at high beam energies, the reaction mechanism is well understood and spectroscopic factors for single-particle configurations can be obtained from the measured cross-sections. This method has been developed in the past few years as a quantitative spectroscopic tool mainly at the NSCL at Michigan State University, see the review of Hansen and Tostevin [14], and applied to fragmentation beams with low intensities down to a few ions/s. One important aspect of radioactive-beam physics which can be addressed by these reactions is the evolution of the shell structure when going to largely proton-neutron asymmetric and weakly bound nuclei. As already observed experimentally, the shell closures known for stable nuclei might disappear when going away from the valley of β-stability, and new shell closures might appear. The new set-up will allow such studies to be performed with very good resolution, both for the fragment momentum and the coincident γ-decay measurements for energies up to $1\,\mathrm{GeV}/u$ and the heaviest beams available. In addition, related studies using liquid-hydrogen targets are possible. In this case, both nucleons in a (p,2p)- or (p,pn)-type reaction will be tracked and analyzed providing a kinematical complete detection of the quasi-free scattering process.

The heavy-ion–induced electromagnetic-excitation, which was pioneered at GSI using the LAND set-up, is another example. First kinematical complete experiments were performed using stable beams. In such measurements, the two-phonon excitation of the giant dipole resonance [15] was discovered. The method has been applied to high-energy radioactive beams investigating the dipole response of neutron-rich nuclei. A recent result obtained by the LAND collaboration is discussed in the next section. An overview on scattering experiments with fast radioactive beams can be found in the review article [16].

4'2. *Collective multipole response of proton-neutron asymmetric nuclei*. – The multipole response of exotic nuclei might be studied by inelastic scattering of high-energy secondary beams. The question here is, how the excitation spectra change for weakly bound nuclei with asymmetric proton-to-neutron ratios. For stable nuclei we know that the excitation spectra are dominated by the various giant resonances exhausting basically the respective sum-rules. The properties of the giant resonances are closely related to bulk properties of the nucleus. The radius, the deformation, and the asymmetry energy play a role in case of the isovector giant dipole excitation, while for the isoscalar monopole resonance it is the compressibility which is related to the excitation energy.

If we go to nuclei with large isospin, a separation of neutrons and protons energy-wise takes place due to the filling of different single-particle orbits. But also a spatial separation might occur such as the development of halo structures or neutron skins. Such a decoupling of the more weakly bound valence neutrons should manifest itself in the excitation spectra, and the appearance of new collective excitation modes such as a dipole vibration of skin neutrons *vs.* the isospin-saturated core are expected.

Experimentally, dipole excitations are most effectively studied using the electromagnetic-excitation process at high energy, yielding rather large cross-sections. Data for other multipole excitations of radioactive nuclei are very scarce so far. New experimental approaches have to be developed like proton and alpha inelastic scattering in inverse kinematics or electron scattering. These probes, not available at present, will be provided at the future NuSTAR facility at FAIR by utilizing storage rings, as will be discussed in the next section.

A redistribution of dipole strength compared to stable nuclei was observed in several neutron-rich light nuclei. For halo nuclei, a complete decoupling of the dipole response into the part related to the halo and that of the core nucleons is observed. Typically, the neutron halo of a nucleus is visible in the dipole spectrum directly at the threshold at very low excitation energy with large transition probabilities. This part of the dipole strength is of non-resonant character and directly related to the properties of the halo wave function like its spatial extension and its single-particle characteristics. The measurement of the strength function thus can be used as a spectroscopic tool, see ref. [16] for an overview. A recent example is the measurement of the ^{11}Li response at low excitation energy performed at RIKEN [17]. The neutron-neutron spatial correlations in the halo of ^{11}Li enhances the dipole strength at low energies significantly. The correlations in the halo can thus be studied by measuring the dipole response. A similar observation was made for the two-neutron halo nucleus ^{6}He [18].

A first experiment investigating the electromagnetic excitation of heavier neutron-rich nuclei was recently performed at GSI utilizing a mixed secondary beam including ^{132}Sn isotopes produced by fission of a ^{238}U beam with an energy of around 600 MeV/nucleon [19]. The differential cross-section for electromagnetic excitation of

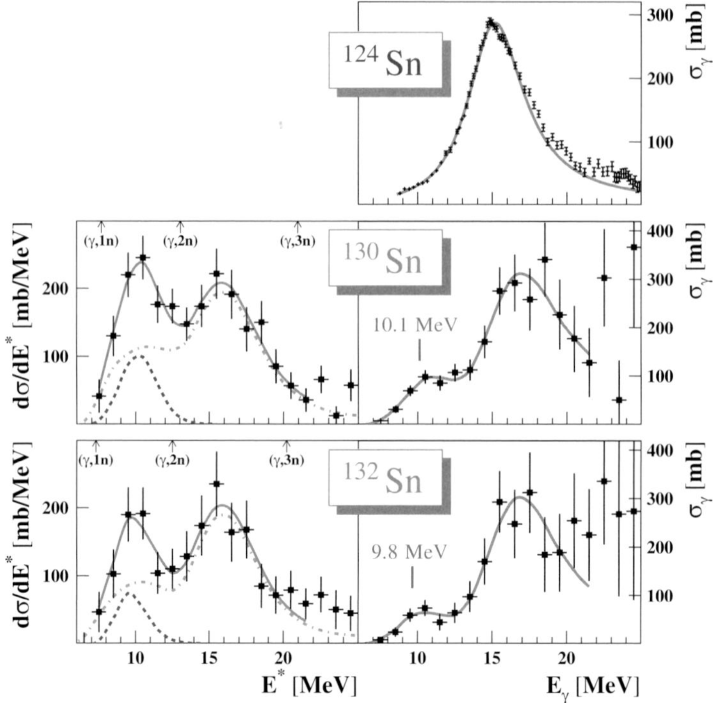

Fig. 7. – Left frames: Differential cross-section for the electromagnetic excitation of ^{130,132}Sn on a lead target at around 500 MeV/nucleon [19]. Corresponding photo-neutron cross-sections are shown in the right panels. For comparison, photo-neutron cross-sections are shown in the upper right frame for the stable isotope ^{124}Sn measured in a real-photon experiment [20]. From Adrich *et al.* [19].

^{130,132}Sn on a lead target is displayed in fig. 7 (left frames). The right frames display the corresponding photo-absorption (γ, n) cross-sections. The upper right panel shows the result from a real-photon experiment [20] for the stable isotope ^{124}Sn for comparison. The spectrum is dominated by the excitation of the giant dipole resonance (GDR). A fit of a Lorentzian plus a Gaussian parametrization for the photo-absorption cross-section (solid curves) yields a position and width of the GDR comparable to those known for stable nuclei in this mass region. The GDR exhausts almost the energy-weighted dipole sum rule. It should be noted that the spectrum shown in fig. 7 was obtained from a measurement with a rather low beam intensity of about 10 ^{132}Sn ions/s only.

An additional peak structure is clearly visible below the GDR energy region. The position of around 10 MeV is close to the predicted energy of the collective soft mode (Pygmy resonance) by the relativistic QRPA calculation of Paar *et al.* [21]. The experimentally observed strength in this peak corresponds to about 4% of the TRK sum rule, also in good agreement with the QRPA prediction [21] as well as with the non-relativistic QRPA calculations of Sarchi *et al.* [22]. The collective character of the low-lying structure is, however, still an open question: while the strength is attributed to a few single-particle excitations in the non-relativistic calculation [22], a coherent superposition of many neutron quasi-particle configurations is found in the relativistic approach [21]. Further theoretical and experimental investigations are called for in order to shed light on the nature of the low-lying dipole strength.

It is interesting to note that the strength of the low-lying mode is related to the neutron-skin thickness and the asymmetry energy. It is observed in mean-field calculations that the slope of the symmetry energy, a quantity related to the pressure of neutron-rich matter, strongly correlates to the neutron radius of heavy nuclei [23]. Similarly, a correlation is observed in the development of a neutron skin and the emergence of low-energy dipole strength [24,25]. Measurements as discussed above thus could provide constraints on the neutron-skin thickness of nuclei and thus on the symmetry energy and the neutron equation of state [26].

5. – Experiments with stored and cooled beams

5`1. *Isomeric Beams, Lifetimes, and Masses (ILIMA)*. – The third branch of the Super-FRS serves a storage- and cooler-ring complex. A schematic layout is shown in the right part of fig. 2. Fragment pulses as short as $50\,$ns are injected into a Collector Ring (CR) with large acceptance, where fast stochastic pre-cooling is applied. A momentum spread of $\Delta p/p \approx 10^{-4}$ is achieved within less than $500\,$ms. The CR may also be used for mass and lifetime measurements of short-lived nuclei applying ToF measurements in the isochronous mode [27] as proposed by the ILIMA Collaboration [5]. In the isochronous mode, the ring optics is tuned such that the differences in velocities are compressed by different trajectories. From the precise measurement of the revolution frequency, masses of short-lived nuclei down to the μs range can be extracted. Nuclei with half-lifes longer than about $1\,$s can be injected after stochastic pre-cooling in the CR into the New Experimental Storage Ring NESR, where electron cooling is applied. Here, the Schottky method [27] is applied to determine the revolution frequency and masses with a resolving power of about 10^6. Both methods for mass- and lifetime measurements have been developed at GSI using the fragment separator FRS and the storage ring ESR at the present GSI facility. The installation at FAIR will benefit not only from the higher beam intensities and larger acceptance of the Super-FRS, but also from the significantly improved acceptance of the CR for hot fragment beams compared to the present storage ring ESR. This will allow mass measurements to be performed in a wide range of the nuclear chart as can be seen in fig. 8, showing the nuclei accessible for mass measurements. The importance of mass measurements for nuclear structure and astrophysics has been discussed in subsect. **3`4**. The new facility will make accessible almost the complete region of nuclei involved in the r-process nucleosynthesis. The expected region of the r-process path is indicated in fig. 8.

5`2. *Reactions at internal targets in the NESR (EXL)*. – The EXL project (EXotic nuclei studied in Light-ion induced reactions at the NESR storage ring) aims at in-

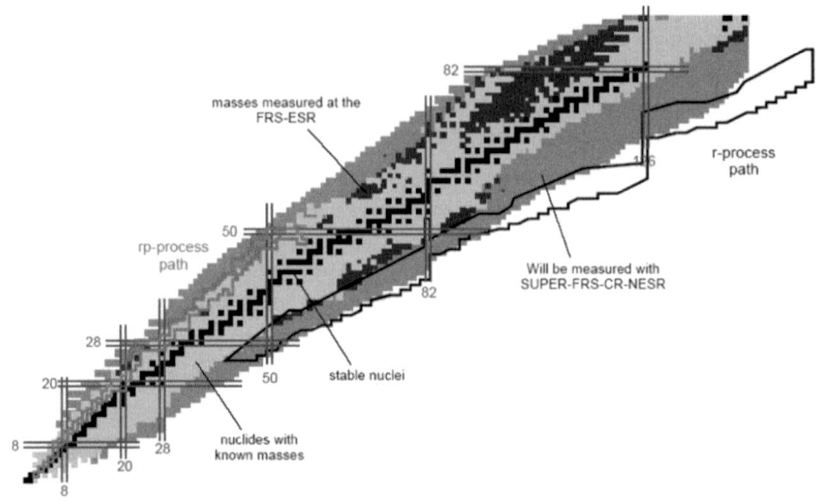

Fig. 8. – Overview on possible mass measurements at the storage rings at NuSTAR@FAIR. The region of the r-process is indicated in the figure showing that mass measurements will be possible for a major part of the r-process nuclei. The figure is taken from the Technical Proposal of the ILIMA collaboration [6].

vestigating light-ion–induced direct reactions in inverse kinematics by using a universal detector system built around the internal target station at the NESR. For studies with the EXL apparatus, the FAIR storage rings are operated in the following scheme: secondary beams of unstable nuclei are produced by fragmentation or fission reactions, are separated in the Super-FRS fragment separator, and then accumulated in the Collector Ring (CR). Bunch rotation and stochastic pre-cooling improves the beam quality to a level already sufficient for most of the envisaged measurements. If required, fast beam energy variation (down to tens of MeV per nucleon) is achieved in the RESR ring. Finally, the beam is transferred into the NESR ring where the measurement is performed. Continuous accumulation of the beam in the NESR can be provided by longitudinal stacking simultaneously to the measurement, and electron cooling compensates an emittance growth from beam-target interactions. In principle, the ions can be decelerated down to energies of a few MeV, the domain of beam energies for transfer and capture reactions, but at the expense of beam losses due to emittance growth.

A schematic layout of the set-up is displayed in fig. 9. The cooled radioactive-ion beam passes a gas-jet target while circulating in the storage ring. Recoiling target ions, e.g., protons or α-particles, are detected by the recoil detector surrounding the target. Particles are tracked by position sensitive Si-strip detectors and their energy is measured in a scintillator (see right part of fig. 9). In forward direction, charged ejectiles and neutrons are measured by detectors placed at small angles around the beam pipe. Heavy fragments can be analyzed by using the first arc of the storage ring as a magnetic spectrometer. The experimental approach allows thus a kinematical complete measurement of the reaction products.

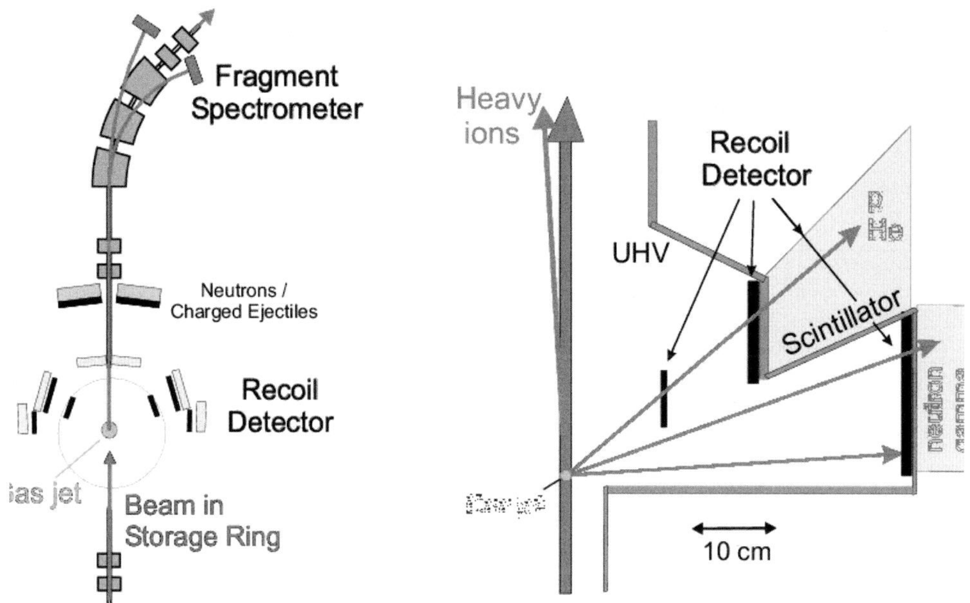

Fig. 9. – Detection scheme for reaction studies at the internal target in the NESR.

Table III gives an overview on the physics programme of the EXl experiment. This program has some overlap with the physics aims of the R^{3}B experiment discussed in subsect. 4.1. The experimental approach of EXL, however, is complementary to the R^{3}B one since it covers the low-momentum transfer part of the scattering processes, which cannot be measured efficiently at R^{3}B. The kinematics of the scattering process in inverse kinematics is illustrated in fig. 10.

TABLE III. – *Nuclear structure information obtainable from scattering off light nuclei* [5].

Method (reactions)	Physical observables	Related effects in exotic nuclei
elastic scattering (p,p); (^{4}He,^{4}He);	nuclear matter radii and distributions	halo; neutron skin; central density; optical potential
inelastic scattering (p,p');($\bar{p}$,p'); (^{4}He,^{4}He')	surface collective states; electric giant resonances; isovector magnetic excitation for (p,p'); analyzing powers	bulk properties in N-Z asymmetric matter; proton/neutron deformation; nuclear compressibility; threshold strength; soft modes
charge exchange (p,n); (d,^{2}He); (^{3}He,t)	spin-isospin excitations; Gamow-Teller; Spin-dipole resonance	(stellar) weak interaction rates; spin excitations; neutron skin
transfer reaction (p,d); (d,^{3}He); (p,t); (d,p)	Spectroscopic factors; Single particle (hole) states; Pair transfer	single-particle structure; spin-orbit; pairing interaction
quasi-free scattering (p,2p);(p,np); (p,p ^{4}He)	single-particle spectral function; cluster knockout	single-particle structure; nucleon-nucleon and cluster correlations; in-medium interactions

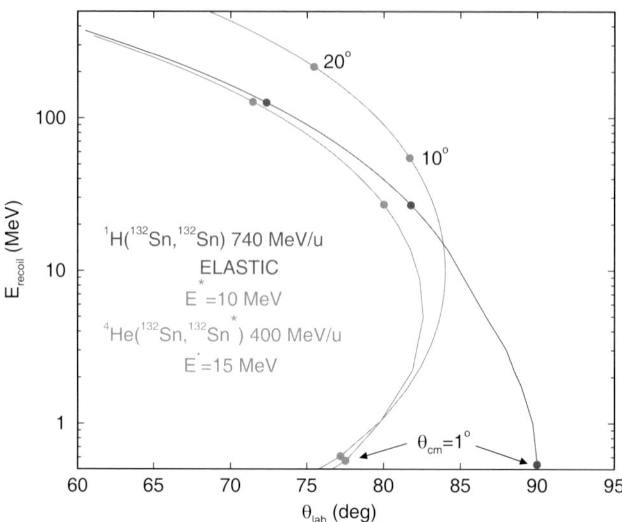

Fig. 10. – Kinetic energy *versus* scattering angle in the laboratory frame of recoiling ^{1}H and ^{4}He target nuclei. Center-of-mass angles θ_{cm} are indicated at the curves.

The kinetic energy of the recoiling (target) nuclei is shown versus the scattering angle in the laboratory frame. An example is shown for inelastic scattering of α particles off ^{132}Sn nuclei at $400\,\mathrm{MeV}/u$, a typical reaction used to study isoscalar excitations such as the giant monopole resonance. The angular distribution, which is sensitive to the multipolarity, has to be measured down to low center-of-mass angels, for monopole transitions preferentially close to 0 degree. This means that recoiling particles will have very little kinetic energy as can be seen from fig. 10, implying the need of using thin targets. External target experiments would thus yield very low luminosities. The advantage of performing such experiments in a storage ring arises mainly from the fact that thin gas targets can be used while gaining luminosity by the revolution frequency of $\approx 10^6\,\mathrm{s}^{-1}$ and by continued accumulation of the secondary beams. EXL will thus allow to extend the applicability of high-energy reactions towards the low-momentum transfer region not accessible so far, or only with rather low luminosity.

5$^.$**3.** *Electron scattering with short-lived nuclei (ELISe).* – Electron scattering has been a very powerful tool for decades to extract charge densities and radii of nuclei as well as to investigate collective states by means of inelastic scattering. In addition, electron-induced knock-out reactions give access to the single-particle structure as well as nucleon-nucleon correlations in nuclei. The advantages of using electrons in nuclear-structure investigations essentially arise from the fact that electrons are point-like particles (no formfactor) and that the electron-nucleus interaction is relatively weak and theoretically well described. Charge distributions and transition densities can thus be reliably deduced from measured cross-sections. Independent variation of energy and momentum transfer allows selecting certain scattering processes.

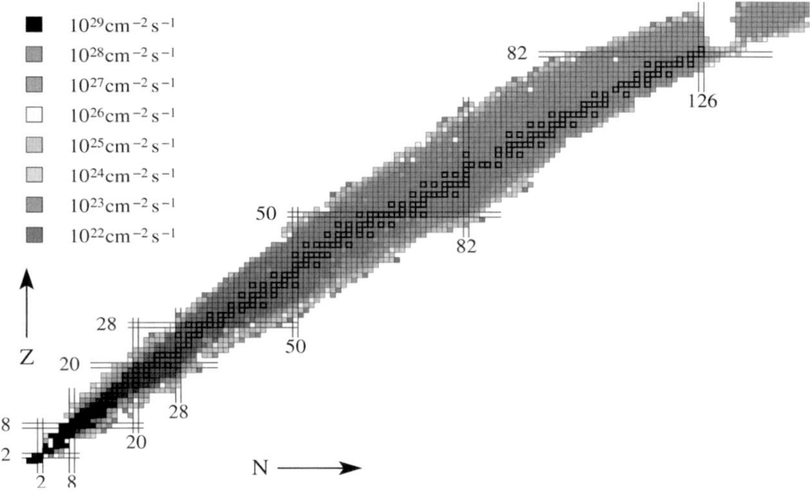

Fig. 11. – Estimated luminosities for electron-scattering experiments off radioactive nuclei. The figure is taken from the Technical Proposal of the ELISe collaboration [6].

So far, the application of the electron-scattering process to short-lived nuclei was impossible. The ELISe collaboration proposes to utilize two intersecting storage rings, the NESR and a smaller storage ring for electrons, to scatter electrons in a collider mode off radioactive nuclei circulating in the NESR, see fig. 2 (E-A collider). The stored electrons with variable energies from 125 to 500 MeV are contra-propagating to heavy-ion bunches with energies typically between 200 and 740 MeV/u. Scattered electrons will be momentum analyzed by a large-acceptance spectrometer. The crucial quantity first of all is the reachable luminosity. The number of stored ions is apart from production limitations and short lifetimes limited by space-charge effects. A full simulation of expected luminosities including the production as well as transmission losses, nuclear and beam (storage) life-time is shown in fig. 11. This has to be compared to the luminosity requirements of the different experiments. Giant resonance studies, for instance, can be performed with luminosities around 10^{28} cm^{-2}/s. Clearly, a large number of rare isotopes potentially become accessible for such studies, and even more for measurements of elastic scattering.

5`4. *The Antiproton-Ion-Collider AIC*. – The AIC Collaboration proposes to use the electron storage ring, or a modified version of it, for storing antiprotons. This would allow to study antiproton-ion scattering in a collider mode similar to the eA collider. The total antiproton absorption cross-section is the interesting quantity from which the neutron-skin thickness of nuclei can be determined by measuring the individual cross-sections for annihilations on proton and neutrons. The antiproton will annihilate on an individual nucleon. The resulting A-1 fragments are circulating in the storage ring where they can be individually detected via the Schottky method. In head-on collisions with the unstable nuclei stored in the NESR, luminosities of 10^{23} cm^{-2} s^{-1} are estimated in case of 10^6 stored ions. Since antiproton absorption cross-sections are of the order of a barn, systematic studies over extended isotopic or isotonic chains are enabled.

6. – Conclusion

The NuSTAR facility will allow the extraction of the basic properties of nuclei at the extremes of nuclear existence while at the same time allowing in-depth investigations of exotic nuclei closer to stability. Exploring nuclear structure and nuclear stability under extreme conditions is essential for a comprehensive understanding of the nuclear many-body system. It is also the basis for understanding the various aspects of nuclear astrophysics and for many applications of nuclear physics. High beam intensities provided by the FAIR accelerators, effective radioactive-beam production and transportation by the Super-FRS, in combination with advanced experimental instrumentation of high sensitivity and efficiency are the most important ingredients for the next-generation rare-isotope beam facility described in the present paper. The NuSTAR facility will span the full range of experimental techniques needed to pursue the intellectual challenges related to the many-body system nucleus, including novel experimental techniques not available at present for radioactive nuclei like, *e.g.*, light-ion and electron scattering off exotic nuclei.

REFERENCES

[1] *An International Accelerator Facility for Beams of Ions and Antiprotons*, Conceptual Design Report, GSI (2001), `http://www.gsi.de/GSI-Future/cdr/`.
[2] Körner G.-E. (Editor), *Nuclear Physics News*, Vol. **16**, No. 1 (2006).
[3] Henning W. F., *Nucl. Phys. A*, **734** (2004) 654.
[4] Gutbrod H. H., *Nucl. Phys. A*, **752** (2005) 457c.
[5] *Letters of Intent of the NUSTAR Collaboration at FAIR (April 2004)*, `http://www.gsi.de/forschung/kp/kp2/nustar_e.html/`.
[6] *FAIR, An International Accelerator Facility for Beams of Ions and Antiprotons*, Baseline Technical Report, Vol. **4**, July 2006, `http://www.gsi.de/fair/reports/btr.html`.
[7] Geissel H., Weick H., Winkler M., Münzenberg G., Chichkine V., Yavor M., Aumann T., Behr K. H., Böhmer M., Brünle A., Burkard K., Benlliure J., Cortina-Gil D., Chulkov L., Dael A., Ducret J.-E., Emling H., Franczak B., Friese J., Gastineau B., Gerl J., Gernhäuser R., Hellström M., Jonson B., Kojouharova J., Kulessa R., Kindler B., Kurz N., Lommel B., Mittig W., Moritz G., Mühle C., Nolen J. A., Nyman G., Roussel-Chomaz P., Scheidenberger C., Schmidt K.-H., Schrieder G., Sherrill B. M., Simon H., Sümmerer K., Tahir N. A., Vysotsky V., Wollnik H. and Zeller A. F., *Nucl. Instrum. Methods Phys. Res. B*, **204** (2003) 71.
[8] Scheidenberger C., Geissel H., Maier M., Münzenberg G., Portillo M., Savard G., Van Duppen P., Weick H., Winkler M., Yavor M., Attallah F., Behr K.-H., Chichkine V., Eliseev S., Hausmann M., Hellström M., Kaza E., Kindler B., Litvinov Y., Lommel B., Marx G., Matos M., Nankov N., Ohtsubo T., Sümmerer K., Sun Z.-Y. and Zhou Z., *Nucl. Instrum. Methods B*, **204** (2003) 119.

[9] HOSMER P. T., SCHATZ H., APRAHAMIAN A., ARNDT O., CLEMENT R. R. C., ESTRADE A., KRATZ K.-L., LIDDICK S. N., MANTICA P. F., MUELLER W. F., MONTES F., MORTON A. C., OUELLETTE M., PELLEGRINI E., PFEIFFER B., REEDER P., SANTI P., STEINER M., STOLZ A., TOMLIN B. E., WALTERS W. B. and WÖHR A., *Phys. Rev. Lett.*, **94** (2005) 112501.

[10] OTTEN E. W., *Nuclear radii and moments of unstable isotopes*, in *Treatise on Heavy-Ion Science*, edited by BROMLEY D. A. (Plenum, New York) 1990, pp. 517-638.

[11] SÁNCHEZ R., NÖRTERSHÄUSER W., EWALD G., ALBERS D., BEHR J., BRICAULT P., BUSHAW B. A., DAX A., DILLING J., DOMBSKY M., DRAKE G. W. F., GÖTTE S., KIRCHNER R., KLUGE H.-J., KÜHL TH., LASSEN J., LEVY C. D. P., PEARSON M. R., PRIME E. J., RYJKOV V., WOJTASZEK A., YAN Z.-C. and ZIMMERMANN C., *Phys. Rev. Lett.*, **96** (2006) 033002.

[12] Reactions with Relativistic Radioactive Beams (R3B), http://www-land.gsi.de/r3b/.

[13] PALIT R., ADRICH P., AUMANN T., BORETZKY K., CARLSON B. V., CORTINA D., ELZE TH. W., EMLING H., GEISSEL H., HELLSTRÖM M., JONES K. L., KRATZ J. V., KULESSA R., LEIFELS Y., LEISTENSCHNEIDER A., MÜNZENBERG G., NOCIFORO C., REITER P., SIMON H., SÜMMERER K. and WALUS W., *Phys. Rev. C*, **68** (2003) 034318.

[14] HANSEN P. G. and TOSTEVIN J. A., *Annu. Rev. Nucl. Part. Sci.*, **53** (2003) 219.

[15] AUMANN T., BORTIGNON P. F. and EMLING H., *Annu. Rev. Nucl. Part. Sci.*, **48** (1998) 351.

[16] AUMANN T., *Eur. Phys. J. A*, **26** (2005) 441.

[17] NAKAMURA T., VINODKUMAR A. M., SUGIMOTO T., AOI N., BABA H., BAZIN D., FUKUDA N., GOMI T., HASEGAWA H., IMAI N., ISHIHARA M., KOBAYASHI T., KONDO Y., KUBO T., MIURA M., MOTOBAYASHI T., OTSU H., SAITO A., SAKURAI H., SHIMOURA S., WATANABE K., WATANABE Y. X., YAKUSHIJI T., YANAGISAWA Y. and YONEDA K., *Phys. Rev. Lett.*, **96** (2006) 252502.

[18] AUMANN T., ALEKSANDROV D., AXELSSON L., BAUMANN T., BORGE M. J. G., CHULKOV L. V., CUB J., DOSTAL W., EBERLEIN B., ELZE TH. W., EMLING H., GEISSEL H., GOLDBERG V. Z., GOLOVKOV M., GRÜNSCHLOSS A., HELLSTRÖM M., HENCKEN K., HOLECZEK J., HOLZMANN R., JONSON B., KORSHENNINIKOV A. A., KRATZ J. V., KRAUS G., KULESSA R., LEIFELS Y., LEISTENSCHNEIDER A., LETH T., MUKHA I., MÜNZENBERG G., NICKEL F., NILSSON T., NYMAN G., PETERSEN B., PFÜTZNER M., RICHTER A., RIISAGER K., SCHEIDENBERGER C., SCHRIEDER G., SCHWAB W., SIMON H., SMEDBERG M. H., STEINER M., STROTH J., SUROWIEC A., SUZUKI T., TENGBLAD O. and ZHUKOV M. V., *Phys. Rev. C*, **59** (1999) 1252.

[19] ADRICH P., KLIMKIEWICZ A., FALLOT M., BORETZKY K., AUMANN T., CORTINA-GIL D., DATTA PRAMANIK U., ELZE TH. W., EMLING H., GEISSEL H., HELLSTRÖM M., JONES K. L., KRATZ J. V., KULESSA R., LEIFELS Y., NOCIFORO C., PALIT R., SIMON H., SURÓWKA G., SÜMMERER K. and WALUS W., *Phys. Rev. Lett.*, **95** (2005) 132501.

[20] FULTZ S. C., BERMAN B. L., CALDWELL J. T., BRAMBLETT R. L. and KELLEY M. A., *Phys. Rev.*, **186** (1969) 1255.

[21] PAAR N., RING P., NIKŠIĆ T. and VRETENAR D., *Phys. Rev. C*, **67** (2003) 034312.

[22] SARCHI D., BORTIGNON P. F. and COLÒ G., *Phys. Lett. B*, **601** (2004) 27.

[23] TYPEL S. and BROWN B. A., *Phys. Rev. C*, **64** (2001) 027302.

[24] PIEKAREWICZ J., *Phys. Rev. C*, **73** (2006) 044325.

[25] PAAR N., personal communication.

[26] LAND COLLABORATION (KLIMKIEWICZ A. *et al.*), *Phys. Rev. C*, **76** (2007) 051603(R).

[27] GEISSEL H. and WOLLNIK H., *Nucl. Phys. A*, **693** (2001) 19.

DOI 10.3254/978-1-58603-885-4-79

SPIRAL2 at GANIL: A world leading ISOL facility for the next decade

S. GALES

GANIL (DSM-CEA/IN2P3-CNRS) - Blvd. Henri Becquerel, F-14076 Caen cedex, France

Summary. — During the last two decades, RIB has allowed the investigation of a new territory of nuclei with extreme N/Z called "terra incognita". Due to technical limitations, existing facilities were able to cover some part of the light mass region of this "terra incognita". The main goal of SPIRAL2 is clearly to extend our knowledge of the limit of existence and the structure of nuclei deeply in the medium and heavy mass region ($A = 60$ to 140) which is to day an almost unexplored continent. SPIRAL2 is based on a high power, CW, superconducting driver LINAC, delivering $5\,\mathrm{mA}$ of deuteron beams at $40\,\mathrm{MeV}$ ($200\,\mathrm{KW}$) directed on a C converter+ Uranium target and producing therefore more 10^{13} fissions/s. The expected radioactive beams intensities for exotic species in the mass range from $A = 60$ to $A = 140$, of the order of 10^6 to 10^{10} pps will surpass by two orders of magnitude any existing facilities in the world. These unstable atoms will be available at energies between few keV/n to $15\,\mathrm{MeV/n}$. The same driver will accelerate high intensity ($100\,\mu\mathrm{A}$ to $1\,\mathrm{mA}$), heavier ions up to Ar at $14\,\mathrm{MeV/n}$ producing also proton-rich exotic nuclei. In applied areas SPIRAL2 is considered as a powerful variable energy neutron source, a must to study the impact of nuclear fission and fusion on materials. The intensities of these unstable species are excellent opportunities for new tracers and diagnostics either for solid state, material or for radiobiological science and medicine.

The "Go" decision has been taken in May 2005. The investments and personnel costs amount to 190 M€, for the construction period 2006-2012. The project group has been completed in 2006, the construction of the accelerator started in the beginning of 2007 whereas detail design of the RIB production processes are underway. Construction of the SPIRAL2 facility is shared by ten French laboratories and a network of international partners. Under the 7FP program of European Union called "Preparatory phase for the construction of new facilities", the SPIRAL2 project has been granted a budget of about 4 M€ to build up an international consortium around this new venture. Regarding the future physics program a call for Letter of intents has been launched in Oct 2006. A very positive and enthusiastic response is the basis of new large international collaborations. Proposals for innovative new instrumentation and methods for the for SPIRAL2 facility are being examined by an International Scientific Advisory Committee. The status of the construction of SPIRAL2 accelerator and technical R&D programs for physics instrumentation (detectors, spectrometers) in collaboration with EU and International partners will be presented.

1. – Introduction

The search for a better understanding of nuclei and interactions between them as well as the way that matter is synthesised in the Universe, all depend crucially on our knowledge of the physics of the nucleus. In particular, the modern basic nuclear physics research is aiming to study exotic nuclei and the underlying physics, beyond the realm of the stable nuclei, even reaching the "drip-lines" at the very edge of the nuclear chart. The physicists are seeking today to explore the entire chart.

New exotic shapes and excitation modes, halo-like and molecular structures, new modes of nuclear decay have been recently observed, while tests of fundamental symmetries, testing and refinement of the Standard Model of fundamental interactions, and exploration of the "magic" numbers of protons and nuclei in very exotic nuclei are all enticing avenues of discovery. The study of radioactive nuclei, involved as they are in nucleosynthesis in the stars, leading to the creation of nuclei, atoms, molecules and all other complex structures our world is made of, has until now been strongly restricted by their short lifetimes and the limited production yields. By using new technologies, we are now able to produce accelerators and ingenious systems for producing beams of very exotic ions in quantities, which will permit their properties to be measured and understood. The quest for Rare Isotope Beams (RIB), which are orders of magnitude more intense than those currently available, is the main motivation behind the SPIRAL2 project [1-6].

The importance of the RIBs has been underlined by NuPECC (Nuclear Physics European Collaboration Committee—an expert committee of the European Science Foundation) which recommend the construction of two new-generation complementary RIB facilities in Europe. One based on the In-Flight Fragmentation (IFF) as proposed for the FAIR facility at GSI (Darmstadt, Germany) [7] and the other on Isotope-Separation

On-Line (ISOL) method. In ISOL method, a thick target is bombarded with a primary light or heavy-ion beam or with a secondary neutron beam in case of fission targets. The produced radioactive nuclei are post-accelerated to energies from keV to tens of MeV. The ISOL method provides high-intensity beams of high optical quality comparable to stable beams.

The French Ministry of Research took a decision on the construction of SPIRAL2 in the end of May 2005. Its construction cost, estimated to $190\,\text{M}€$ (including personnel and contingency), will be shared by French founding agencies CNRS/IN2P3 and CEA/DSM, the Region of Basse Normandie and the European partners. Funding from EU 7th framework and from others partnership countries are expected to contribute for about 20% to this budget. The construction will last about seven years (2006–2012) and operation of the facility will cost $8.5\,\text{M}€$ per year.

In October 2006, ESFRI (European Strategy Forum on Research Infrastructures) published a list of selected 34 Research Infrastructure projects corresponding to major needs of the European scientific community (all fields of science included) in the coming years. The FAIR and SPIRAL2 are among the selected projects and were recommended as Research Infrastructures necessary to maintain Europe's position at the cutting-edge of world research.

As a result of the ESFRI selection process, under the 7FP program of European Union called "Preparatory phase for the construction of new facilities", the SPIRAL2 project has been granted a budget of about $4\,\text{M}€$ to build up an international consortium around this new venture.

The first beams are expected by the end of 2011, beginning of 2012. As an added bonus, the facility will also produce over 10^{15} neutrons per second, making it the world's most powerful source of fast neutrons for several years. SPIRAL2 will reinforce European leadership in the field of exotic nuclei and will serve a community of about 1000 scientists.

2. – Description of the project

The quest for Rare Isotope Beams (RIB), which are orders of magnitude more intense than those currently available, is the main motivation behind the SPIRAL2 project.

The new SPIRAL2 facility is a major improvement of the SPIRAL1 facility. More than 20 laboratories from 10 countries from 2001 to 2005 have developed the conceptual and the technical design of SPIRAL2. More than 200 scientists and engineers have contributed to the scientific case and the design of the facility [1,2].

The SPIRAL2 facility (fig. 1) is based on a high power, superconducting driver LINAC, which will deliver a high-intensity, $40\,\text{MeV}$ deuteron beam as well as a variety of heavy-ion beams with mass over charge ratio equals to 3 and energy up to $14.5\,\text{MeV/nucl}$. Using a carbon converter, the $5\,\text{mA}$ deuteron beam and a uranium carbide target, a fast-neutron–induced fission is expected to reach a rate of up to 10^{14} fissions/s. The RIB intensities in the mass range from $A = 60$ to $A = 140$ will surpass by one or two orders of magnitude any existing facilities in the world. A direct irradiation of the UC_x target with beams of deuterons, $^{3,4}\text{He}$, $^{6,7}\text{Li}$, or ^{12}C can be used if higher excitation energy

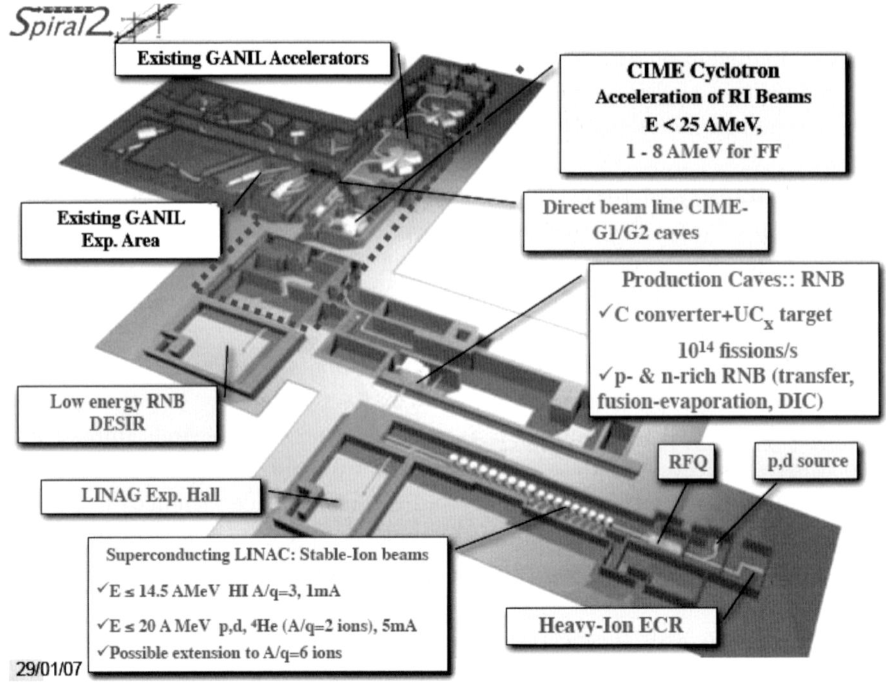

Fig. 1. – Schematic layout of the SPIRAL2 facility at GANIL.

leads to higher production rate for a nucleus of interest.

The neutron-rich fission RIB could be complemented by beams of nuclei near the proton drip line provided by fusion-evaporation or transfer reactions. For example, a production of up to 8×10^4 atoms of ^{80}Zr per second using a 200 μA ^{24}Mg^{8+} beam on a ^{58}Ni target should be possible. Similarly, the heavy- and light-ion beams from LINAG on different production targets can be used to produce high-intensity light RIB with ISOL technique.

The extracted RIB will be subsequently accelerated to energies of up to 20 MeV/nucl. (Typically 6-8 MeV/nucl. for fission fragments) by the existing CIME cyclotron. Thus using different production mechanisms and techniques SPIRAL2 would allow performing experiments in a wide range of neutron- and proton-rich nuclei far from the line of stability (fig. 2).

Recent papers describing the technical and instrumentation developments related to the SPIRAL2 facility can be found in refs. [8-11].

SPIRAL2 has also a remarkable potential for neutron-based research both for fundamental physics and various applications. In particular, in the neutron energy range from a few MeV to about 35 MeV this research would have a leading position for the next 10–15 years if compared to other neutron facilities in operation or under construction worldwide.

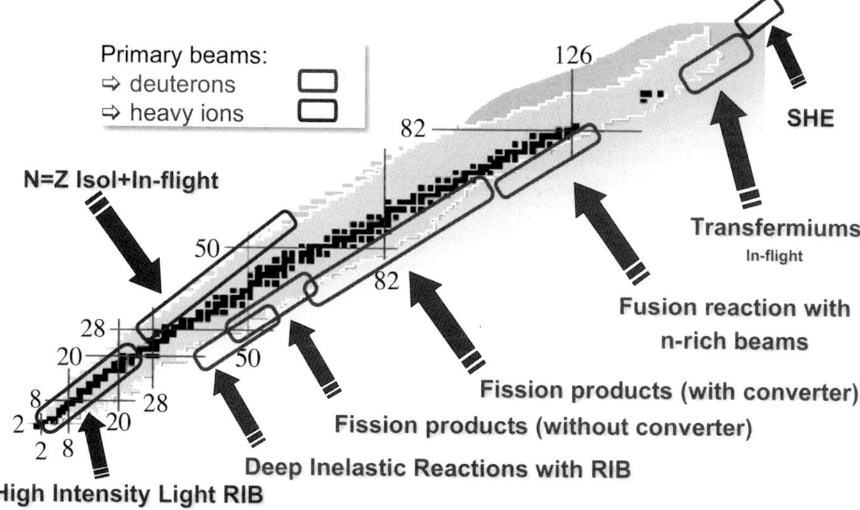

Fig. 2. – Regions of the chart of nuclei accessible for research of nuclei far from stability at SPIRAL2.

3. – Performances of the SPIRAL2 facility

3`1. *Intense stable beams from LINAG*. – Very intense stable heavy-ions beams will be available at the exit of the LINAG accelerator. These heavy ions (from He to Ca in a first step, then up to U), will reach intensities from $1\,\mathrm{p}\mu\mathrm{A}$ up to $1\,\mathrm{pmA}$ using new PHOENIX ion sources in combination with $A/Q = 1/3$ RFQ injector and later $A/Q = 1/6$ for heavier ions (Kr to U) as illustrated in fig. 3.

The construction of the PHOENIX $A/q = 3$ ion source has started at Grenoble (LPSC Grenoble-GANIL collaboration) [12-14]. In addition a recent MoU has been signed between Argonne National Laboratory, Accelerator Division and GANIL-SPIRAL2 project. Under the agreed terms ANL has started the design of a new RFQ injector with $A/Q = 6$, devoted to the production of intense stable beams with mass higher than $A = 40$.

Reaching such beams intensities will open a wide range of physics experiments not accessible to day. Super Heavy Elements synthesis and spectroscopy, spectroscopy at or beyond the proton drip line, multinuclear transfer and deep inelastic scattering, study of mechanisms and reaction product distributions, production and study of isomers, ground state properties of rare nuclei.

A new device, S3 (for Super Separator Spectrometer) has been proposed for experiments with the very high intensity stable beams of SPIRAL2 coming from the LINAG accelerator [15-18].

Most of the physics cases listed above can be seen as either the study of rare processes or as the study of a secondary reaction with exotic ions produced in a first step reaction.

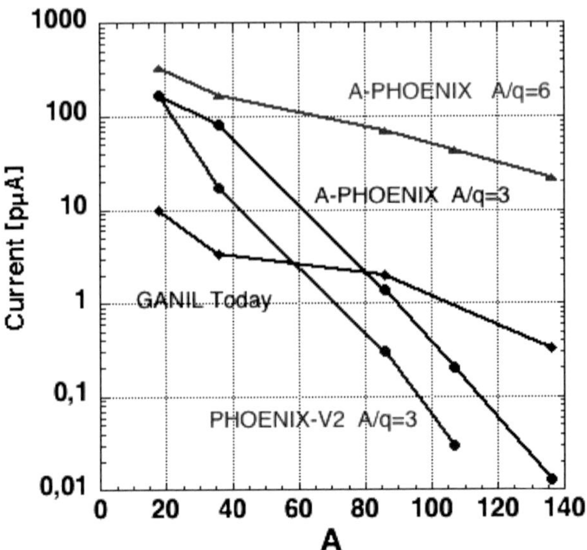

Fig. 3. – Expected intensities of stable heavy-ion beams at the exit of the LINAG accelerator for various versions of PHOENIX ion sources.

S3 shall be designed to answer most experimental needs of these physics cases. The crucial point of the device is to cope with the very high intensity beams provided by the LINAG accelerator. That means:

- the targets should be able to cope with very high power dissipation,

- the reaction products should be separated from the beam ions before they can be detected,

- the full optics of the spectrometer should reduce the background coming from scattered ions and activation products.

Meanwhile, the vast array of different physics that could be possible with S3 implies that it should be compatible with various detection settings. The basic design considered could be the following (see fig. 4).

The primary beam will impinges on a primary target which should be made resistant to the high fluxes (rotating cooled targets, gas or liquid targets...).

The separation of the beam can proceed in two steps, the first being a high rejection device and an appropriate beam stopper.

After this first step, ions of interest could, if necessary, react on a secondary target. As they deal with moderate fluxes, this secondary target point could have gamma or particle prompt detection.

Next to this point, a secondary spectrometer helps with separation if very high purity is needed. It has some mass/momentum resolution and could rotate for reactions products at large angles.

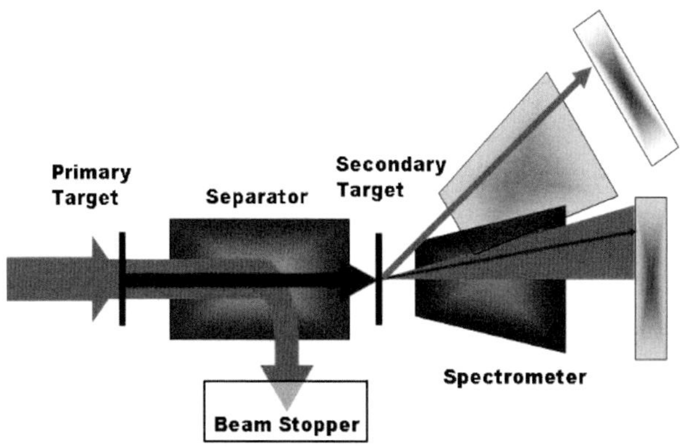

Fig. 4. – Basic scheme for S3 design.

The final focal plane can be equipped with the relevant detection system.

3˙2. *The radioactive ion production system.* – The SPIRAL2 facility at GANIL will produce radioactive isotopes ranging from the lightest to very heavy elements beyond uranium. Different production mechanisms will be utilised.

A layout of the radioactive production vaults is shown in fig. 5.

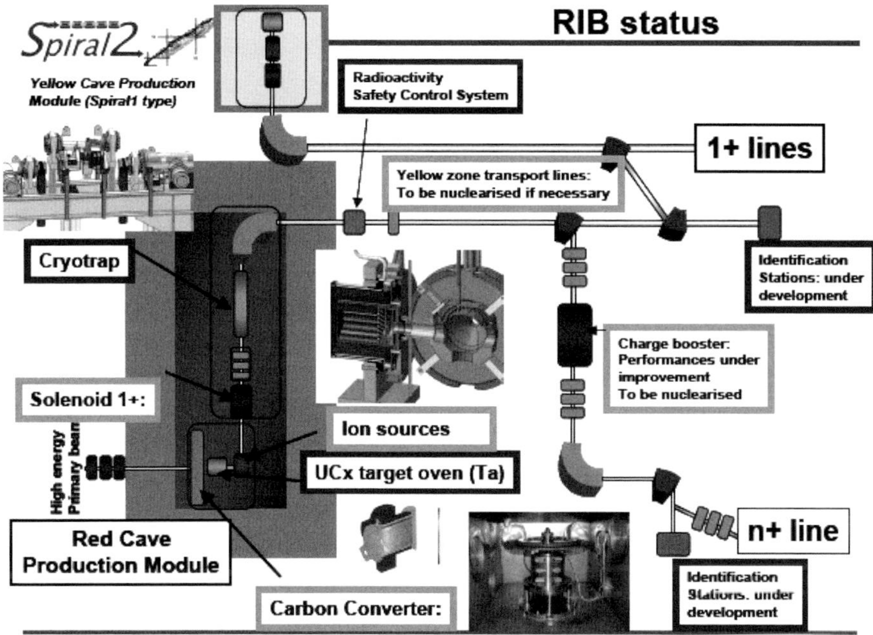

Fig. 5. – Layout of the radioactive production halls for the SPIRAL2 facility.

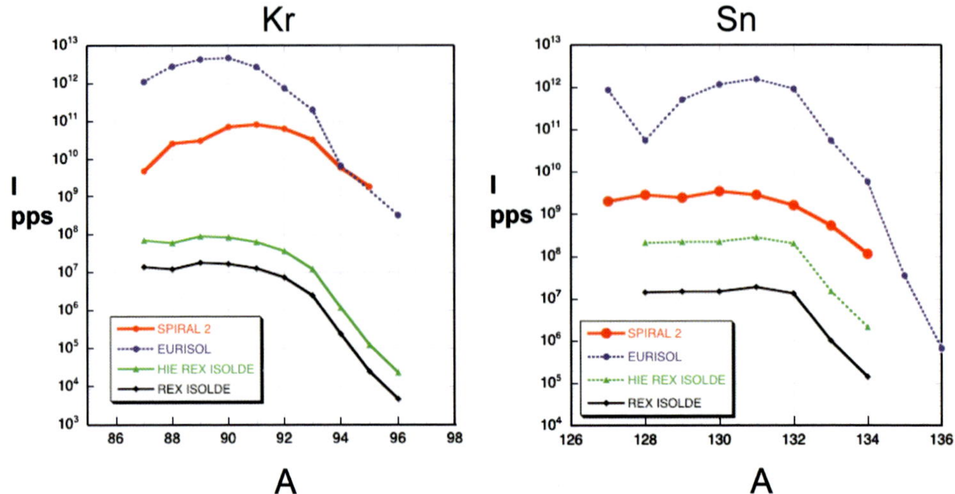

Fig. 6. – Expected intensities of Kr and Sn isotopes at SPIRAL2 after acceleration compared to REX-ISOLDE present and future (HE-REX ISOLDE) and to EURISOL predictions.

The radioactive neutron rich beams will be mainly produced via the fission process [19], with the aim of 5×10^{13} fissions/s. The retained solution is the fast neutrons induced fission in a depleted UCx target. The neutrons are generated by the beak-up of 40 MeV deuterons in a thick C target (converter). For the aimed high fission rate the converter should subtend a beam power of 200 kW (see fig. 1). With a length of 80 mm and at about 40 mm from the entrance of the converter, 5×10^{13} fissions/s can be reached in the target volume with an UC depleted target of 11 g/cm^3. The resulting fission products accumulate in the target, diffuse from the surface from which they evaporate towards an ion source.

Few target-ion sources systems are being developed, often in the framework of international collaborations. Let us mention an ECR 1+ "MONOBOB" for gaseous elements in test at GANIL [20], a FEBIAD, "IRENA" ion source well suited for short lifetimes at IPN Orsay [12], a Laser Ion source, particularly efficient for its selectivity and for metallic beams. An international collaboration around the Laser Ion Source will be built under the 7th European FP.

For neutron-rich fission products the whole area will be confined in a red zone (see fig. 5) with only remote controls and maintenance. At the exit of the target ion source, Cryotraps are being developed in order to catch volatile gaseous atoms which are not ionized. Then, the 1+ ions are separated and charge bred in a ECR ion source in order to be accelerated by the existing CIME cyclotron up to 5–9 MeV/n for masses $A = 70$ to $A = 140$.

The expected intensities (to be validated by measurements) of typical fission fragment beams of Kr and Sn isotopes from SPIRAL2 are compared to existing or planned RIB facilities in fig. 6.

TABLE I. – *Calculated production yields for light and $N = Z$ exotic nuclei using transfer or fusion evaporation reactions.*

RI Beam	Reaction	Production method	Yield (min.–max.) pps
^{6}He	^{9}Be$(n, \alpha)^6$He	ISOL	5×10^7–10^{12}
^{11}C	^{14}N$(p, \alpha)^{11}$C	ISOL	10^7–3×10^{11}
^{15}O	^{15}N$(d, 2n)^{15}$O	ISOL	3×10^7–10^{10}
^{18}Ne	^{19}F$(p, 2n)^{18}$Ne	ISOL	6×10^6–7×10^9
^{34}Ar	^{35}Cl$(p, 2n)^{34}$Ar	ISOL	2×10^6–2×10^8
^{56}Ni	^{58}Ni$(p, p2n)^{56}$Ni	Batch mode	2×10^4–10^8
^{58}Cu	^{58}Ni$(p, n)^{58}$Cu	Batch mode	10^4–10^8
^{80}Zn	^{24}Mg $+ {}^{58}$Ni	In-flight	$< 3 \times 10^4$

For SPIRAL2 the RIB intensities in the mass range from $A = 60$ to $A = 140$ will be of the order of 10^6 to 10^{11} part/s. For example, the intensities should reach $10^{9\text{-}10}$ part/s for ^{132}Sn and 10^{11} pps for ^{92}Kr surpassing by two orders of magnitude any existing facilities in the world (see fig. 6).

Production of rare isotopes are also possible using a second target station (yellow cave production see fig. 5) similar to SPIRAL1 cave where intense heavy-ion beams from LINAG can induce fusion evaporation reactions leading to rare isotopes for medium-mass, neutron-deficient nuclei. In addition, using transfer and deep-inelastic reactions, SPIRAL2 facility will produce intense secondary beams of light-to-medium mass nuclei closer to the line of stability. Examples of reactions leading to the production of light neutron-rich and $N = Z$ exotic nuclei are listed in table I below.

4. – Selected examples of the scientific opportunities at SPIRAL2

As discussed in previous section, the multi-beam driver LINAC will deliver intense heavy-ion stable beams and may be dedicated to the synthesis of very heavy and super-heavy nuclei.

A new experimental Hall (AEL) will host the new S3 "Super-Separator-Spectrometer" and a wide range of physics experiments not accessible today will be at reach.

In the same experimental hall a neutron-TOF facility is planned, thanks to the high energy and high intensity neutron flux available at SPIRAL2, the facility will offer a unique opportunity for material irradiations and cross-section measurements both for fission- (notably Accelerator-Driven Systems (ADS) and Gen-IV fast reactors) and fusion-related research, tests of various detection systems and of resistance of electronics components to irradiations, etc.

The DESIR collaboration [21], formed after the SPIRAL2 workshop on low-energy physics at GANIL in July 2005, proposes the construction of an experimental facility to exploit the low-energy "exotic" beams from SPIRAL2. Experiments performed at other ISOL facilities (*e.g.*, ISOLDE or ISAC) show that a high degree of purity is necessary for the ions beams. The most efficient and universal way of achieving isotopic ally pure beams is a high-resolution mass separator.

At SPIRAL2 and its low-energy beam facility DESIR, there is an opportunity of building three kinds of set-up to try to cover the whole nuclear chart. A ß-NMR set-up will be suitable to study precisely the moments of key cases and, in combination with collinear hyperfine spectroscopy, get unambiguous spin assignments. With a collinear spectroscopy set-up, we will have the possibility to study series of isotopes in the intermediate and heavy mass regions. For the heavy elements, a double laser + RF spectroscopy in a Paul trap will be used to study the hyperfine anomaly and higher-order moments up to very high precision.

Shell structure of neutron-rich isotopes near the $N = 20$ and 28 shell are today the subject of intensive research all around the world. Shell quenching, weakening of the spin-orbit interaction, tensor component in the n-n force are investigated using the rich and powerful tools of nuclear reactions from nucleon transfer [22], in beam gamma spectroscopy [23], Coulomb excitation, fusion-evaporation and deep inelastic processes.

SPIRAL2 will be well suited to address these fundamental questions, thanks to the expected large coverage of RIB species, very high intensities and energy range covered by the facility. Two main regions of the mass chart in the vicinity of the $N = 50$, and 82 shell closures are of special interest.

a) The $N = 50$ regions

One important issue is to determine how thick can be the neutron skin for nuclei far from stability? From the measurement of the charge radius, one has access to the neutron radius. A key element would be germanium which is produced in large quantities in neutron stars. So the spectroscopy of this element and its neighbours (Zn, As, Ga, Se) over the magic shell $N = 50$ using collinear laser spectroscopy and reactions will answer this question.

Another typical example of an element that has to be studied is of course nickel as close as possible to ^{78}Ni. Is this doubly-magic nucleus still spherical?

b) The $N = 82$ regions

A lot of nuclear models predict strong evolution of shapes and/or change of shell structure around neutron number $N = 60$ and mass number $A = 110$. Neutron skin is predicted to occur for the very neutron-rich Sn isotopes. Studying tin isotopes very far from stability will be a very stringent test for these nuclear models.

Nuclei were created during major events that made the evolution of the universe and are the witness of the origin of matter: while the lightest elements were created in the first minutes following the big bang, heavier ones are synthesized in stars. Understanding our world, its origin and its evolution is one goal of nuclear physics. However, these various nucleosynthesis processes involve radioactive nuclei. SPIRAL2 will abundantly produce some of them opening new investigation field.

The future SPIRAL2 facility will contribute prominently to several areas of active research in nuclear astrophysics, such as explosive hydrogen burning, s-process and r-process nucleosynthesis, which are linked to astrophysical observations (novae, X-ray bursts, type-2 supernovae, etc.,...). For example, radiative capture cross-sections can be determined from spectroscopic factors (or Asymptotic Normalization Coefficients) obtained in transfer reaction or elastic resonant scattering.

The (d, p) reaction can be used to simulate the (n, γ) capture on medium-mass nuclei at SPIRAL2 near the r-process possible paths.

The way the energy of theses s or p states evolve from $Z = 50$ down to $Z = 40$ determines how neutron capture cross-sections will evolve along the $N = 82$ shell closure and how the r process synthesis of nuclei will be blocked at this shell closure. The neutron rich isotopes of Sn and Cd nuclei are excellent test cases. The neutron-capture through the soft Pigmy resonance could also play an important role if the energy of the soft pigmy vibration mode is peaked around the S_n value. The way this energy is shifted when moving from Sn to Cd isotopic is certainly an important issue to consider.

The scientific opportunities discussed above are just a few examples among many exciting avenues which can be investigated with SPIRAL2. New exotic shapes and excitation modes, halo-like and molecular structures, single particle properties and new magic numbers, new modes of nuclear decay have been recently observed, while studies of fundamental symmetries, testing and refinement of the Standard Model and of fundamental interactions in very exotic nuclei have all a huge potential of discovery.

5. – Construction of the facility and International Collaborations

Since September 2005, a project construction group has been formed and the whole project has been organized around three main sections:

- the driver accelerator,

- the secondary radioactive beams (production and transport),

- the infrastructure (buildings, vacuum and cryogenics, power supplies,...).

The project has been divided in each section into work packages.

For the driver accelerator, the work packages are under the responsibility of IPN Orsay and DAPNIA/DSM/Saclay laboratories [24-27]. Figure 7 shows the driver accelerator divided into main components and the associated target dates for completion. The R&D on prototypes' for the RFQ, SC cavities and RF couplers are completed and order of all driver components will be completed at the end of 2007.

A SPIRAL2-France consortium had been set to coordinate the efforts of 10 French nuclear physics laboratories from CNRS/IN2P3 and CEA. Moreover SPIRAL2 facility will be built by a large European and International collaboration. Several MoU have been already signed or are in final negotiation stages with JINR Dubna, Israel, Romania,

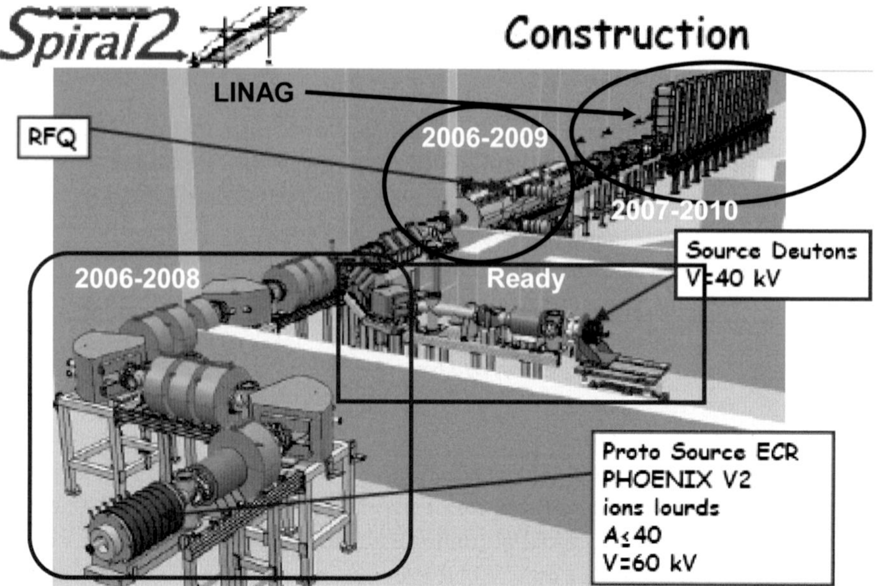

Fig. 7. – SPIRAL2 driver accelerator divided into main components.

Isolde (CERN), Italy, UK, Germany, Poland, Canada, Japan and US, both for accelerator components (control system, beam loss monitors, beam lines) and/or construction of components of the RIB production units (neutron converter, beam lines, target-ions sources, laser ion sources, high-resolution magnetic separator, ...).

At the EU research organisation level, SPIRAL2 as well as FAIR have been selected by the ESFRI Committee and a proposal for the construction phase in the framework of the 7th FPRD has been recently granted $4\,\text{M}\text{€}$ to coordinate the contributions of more than 20 Laboratories from EU, USA, Japan and India.

The process of formation of international collaborations around the SPIRAL2 research programme has started in 2004 and pursued the next years with the organization of 5 specialised workshops held in 2004-2006 (Neutrons for science at SPIRAL2, 13-14/12/2004, Physics with separated low energy beams at SPIRAL2, 4-5/07/2005, Future prospects for high resolution gamma spectroscopy at GANIL, 4-6/10/2005, Nuclear Astrophysics with SPIRAL2, 17-18/10/2005, SPIRAL2 Reactions, 19-21/10/2005).

In the beginning of 2006, under the umbrella of SPIRAL2 Scientific Advisory Committee (SAC) a Call for letters of intent has been launched in March 2006. A very positive and enthusiastic response from the International community has been obtained. 19 Letters of Intents (see refs. [1, 18]), coming from 600 physicists and 34 countries have been received and evaluated by the SAC in October 2006.

After this evaluation, the formation of about 9 large international collaborations around major instruments is under progress. Next important step is the call for Technical proposals around the spring of 2008.

In terms of schedule for the new facility, the ambition is to start operation of the driver accelerator with intense stable beams before the end of 2011 and to complete the whole project including RIB production and post acceleration of fission fragments before the beginning of 2013.

6. – Conclusion

Nuclear physics has been revolutionized by the recent development of the ability to produce accelerated beams of radioactive nuclei. For the first time it will be possible to study reactions to produce the 6000 to 7000 nuclei we believe exist rather than the 300 stable ones that nature provides. At the larger scale one wants to understand the limits of nuclear stability by producing exotic nuclei with vastly different numbers of neutrons and protons.

SPIRAL2 is major expansion of the SPIRAL facility at GANIL, which will help maintain European leadership in ISOL (Isotope separation on line) development, aiming at two orders of magnitude increase of the secondary beams available for nuclear physics studies.

The technical challenges of the acceleration, targetry and experimental equipments will provide essential knowledge on the road toward EURISOL [28], the most advanced nuclear physics research facility presently imaginable and based on the ISOL principle. It is expected, that the realisation of SPIRAL2 will substantially increase the know-how of technical solutions to be applied not only for EURISOL but also in a number of other European/ world projects. The scientific programme, proposes the investigation of the most challenging nuclear and astrophysics questions aiming at the deeper understanding of the nature of matter.

SPIRAL2 will contribute to the physics of nuclear fission and fusion based on the collection of unprecedented detailed basic nuclear data, to the production of rare radioisotopes for medicine, to radiobiology and to material science.

REFERENCES

[1] www.ganil.fr/research/developments/spiral2/.
[2] "The scientific objectives of the SPIRAL2 project", rapport WB_SP2_Final-1.pdf June 2006. http://www.ganil.fr/research/developments/spiral2/index.html and Letter of Intention therein.
[3] MITTIG W. *et al.*, "LINAG Phase1", internal report GANIL (2002).
[4] SAINT-LAURENT M.-G. *et al.*, "SPIRAL phase II, European RTT", R0103 internal report GANIL (2001).
[5] MOSCATELLO M.-H., *SPIRAL2 at GANIL*, in *Seventeenth International Conference on the cyclotrons and their applications, Tokyo, 2004 October 18-22* (Particle Accelerator Society of Japan) 2005, p. 518.
[6] JUNQUERA T. *et al.*, *High Intensity Linac Driver for the SPIRAL-2 Project: Design of Superconducting 88 MHz Quarter Wave Resonators (beta 0.12), Power Couplers and Cryomodules*, in *9th European Particle Accelerator Conference (EPAC 2004) Luzern, July 2004*, TUPLT058 (European Physical Society Accelerator Group (EPS-AG)) 2004, p. 1285.

[7] http://www.gsi.de/fair/reports/btr.html, FAIR Baseline Technical report (Sept 2006).

[8] LEWITOWICZ M., *The SPIRAL2 project*, in *Tours Symposium on Nuclear Physics VI, Tours (FR) 5th-8th September 2006*, AIP Conf. Proc., **891** (2007) 91.

[9] JACQUEMET M. and PETIT E. *et al.*, *Status report on the SPIRAL2 project*, to be published in *XV International Conference on Electromagnetic Isotopes Separators and Techniques related to their applications EMIS 2007, Deauville, 24th–29th June 2007*.

[10] GALES S., *SPIRAL2 at GANIL: a world leading ISOL facility at the dawn of the next decade*, in *International Workshop on Nuclear Physics 28th Course radioactive Beams, Nuclear Dynamics and Astrophysics, Erice, September 16th-24th, 2006*, Progr. Part. Nucl. Phys., **59** (2007) 22.

[11] SAINT-LAURENT M.-G., *Future opportunities with SPIRAL2 at GANIL*, to be published.

[12] CHEIK MHAMED M. *et al.*, Rev. Sci. Instrum., **77** (2006) 03A702.

[13] LAMY T. *et al.*, Rev. Sci. Instrum., **77** (2006) 03B101.

[14] THULLIER T. *et al.*, Rev. Sci. Instrum., **77** (2006) 03A323.

[15] VILLARI A. C. C. *et al.*, *S3: the Super Separator Spectrometer for LINAG*, in *International Symposium on Exotic Nuclei, Khanty-Mansiysk, (RU) July 17-22 j 2006*, AIP Conf. Proc., **912** (2007) 436.

[16] DROUART A. *et al.*, *Design Study of a Pre-Separator for the LINAG Super Separator Spectrometer*, in *XV International Conference on Electromagnetic Isotope Separators and Techniques related to their Applications (EMIS2007), France (2007, June 24th-27th)*, to be published.

[17] SAVARD G. *et al.*, *Radioactive beams from gas catchers: RIBS with and without driver*, to be published in *XV International Conference on Electromagnetic Isotopes Separators and Techniques related to their applications, Deauville, 24th-29th June 2007*.

[18] S3 Collaboration LOI-SPI2-03, http://www.ganil.fr/research/developments/spiral2/loi_texts.html.

[19] FADIL M. *et al.*, *About the production rates and the activation of the uranium carbide target in SPIRAL2*, in *XV International Conference on Electromagnetic Isotopes Separators and Techniques related to their applications EMIS 2007, Deauville, 24th-29th June 2007*, to be published.

[20] HUET-EQUIBEC C. *et al.*, Nucl. Instrum. Methods Phys. Res. Sect. B, **240** (2005) 752.

[21] DESIR Collaboration LOI-SPI2, see ref. [18].

[22] GAUDEFROY L. *et al.*, Phys. Rev. Lett., **97** (2006) 092501.

[23] BASTIN B. *et al.*, Phys. Rev. Lett., **99** (2007) 022503.

[24] FERDINAND R. *et al.*, *SPIRAL2 RFQ prototype - First results, 10th European Particle Accelerator Conference (EPAC'06), Edinburgh, 26-30 June 2006*, MOPCH103 (European Physical Society Accelerator Group (EPS-AG)) 2006.

[25] BERNAUDIN P. E. *et al.*, *Design of the Low-Beta, quarter-wave resonator and its cryomodule for the SPIRAL2 project*, in *9th European Particle Accelerator Conference (EPAC 2004), Luzern, July 5-9 2004*, TUPLT054 (European Physical Society Accelerator Group (EPS-AG)) 2004, p. 1276.

[26] OLRY G. *et al.*, *Status of the beta 0.12 superconducting cryomodule development for SPIRAL2 project*, in *Proceedings of EPAC 2006, Edinburgh, 26-30 June 2006*, MOPCH146 (European Physical Society Accelerator Group (EPS-AG)) 2006.

[27] GOMEZ MARTINEZ Y. *et al.*, *Theoretical study and experimental result of the RF coupler prototypes of Spiral2*, in *Proceedings of EPAC 2006, Edinburgh*, THPCH160 (European Physical Society Accelerator Group (EPS-AG)) 2006.

[28] http://www.eurisol.org/site01/index.php.

DOI 10.3254/978-1-58603-885-4-93

The ISOLDE facility and HIE-ISOLDE

K. Riisager

PH Department, CERN - CH-1211 Geneve 23, Switzerland

Summary. — Selected examples illustrate the present physics possibilities of the ISOLDE facility at CERN. The future upgrades are discussed.

1. – Introduction

Nuclear structure physics has been revitalized by the new experimental information obtained with radioactive beams. The high level of activity is reflected in many overview papers, a brief introduction to radioactive-beam research can be found in [1]. The ISOLDE radioactive beam facility at CERN [2] has been operating for close to 40 years and has played a key role in many of the developments in the ISOL (Isotope Separation On-Line) field. This paper describes the present ISOLDE facility and gives selected examples that illustrate the present physics possibilities. Many different projects exist for construction of new facilities to bring the field of radioactive beams even further. The upgrade plans for ISOLDE, the HIE-ISOLDE project, will be discussed at the end.

2. – Production of radioactive beams

There are two basic methods used today for the production of radioactive beams, the in-flight separation and the ISOL method [3]. As revealed by its name, ISOLDE belongs to the latter category. A general overview of the ISOL method can be found in [4,5],

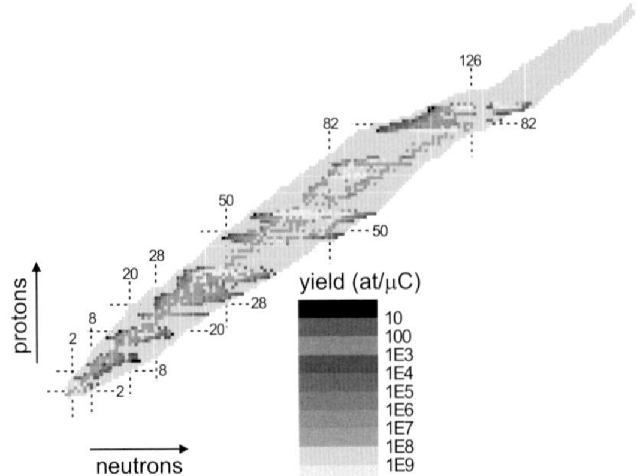

Fig. 1. – Yields of radioactive isotopes available at ISOLDE [13].

this section will concentrate on some aspects specific to ISOLDE. A major advantage for ISOLDE made possible by its being at CERN is the use of high-energy (1–1.4 GeV) protons for the production of radioactive nuclei, this is recognized as being the optimum choice for a single driver accelerator facility [6] and is also the choice made for the ambitious next generation European ISOL facility, EURISOL [7]. ISOLDE is situated after the PS Booster accelerator at the beginning of the CERN accelerator complex, the proton pulses can contain up to $3 \cdot 10^{13}$ protons with a spacing of 1.2 s or more, giving a typical average proton current of $2\,\mu$A.

The PSB-ISOLDE facility is described in [8]. The radioactive nuclei can here be produced in two different target stations (GPS and HRS), they are accelerated to typically 60 keV and after mass separation led into a common experimental hall. The knowledge accumulated over decades on how to construct targets and ion sources tailored to release pure beams of specific elements is one of ISOLDE's strong points, more than 1000 different isotopes of around 70 elements are available today. An overview of the different types of ion sources used is given in [9], the present "workhorse" for the facility, the selective RILIS laser ion source, is described in detail in [10,11]. The release from the target can be rather complex [12] and is modelled in terms of diffusion and effusion processes within the target and ion source, but can in the best cases take place on timescales around 100 ms. Very short-lived products will therefore suffer decay losses. There is an ongoing target development programme aiming at optimizing the efficiency of the production and at finding improved designs that can withstand the increased primary beam intensity of the next generation facilities. The present status of the available yields as collected by the target group is given in fig. 1.

2˙1. *REX-ISOLDE*. – The present worldwide increase of activities at ISOL-facilities is to a large extent caused by the possibility of post-accelerating the low-energy ISOL beams, a development first completed at Louvain-la-Neuve and in the present decade followed up at many laboratories, *e.g.*, at GANIL (SPIRAL), Oak Ridge and Triumf. At ISOLDE the REX-ISOLDE post-accelerator [14, 15] was constructed combining resources from several countries and started operation in 2001. A thorough technical description of the REX accelerator has been published recently [16], put briefly the singly charged ions from ISOLDE are captured and bunched in a large Penning trap (REXTRAP) and charge bred in the REXEBIS ion source to a A/q ratio between 3 and 4.5. This allows them to be subsequently efficiently accelerated in a compact linear accelerator, in the first years the maximum final energy was $2.2\,\mathrm{MeV}/u$ but additional acceleration cavities has brought this to $3.1\,\mathrm{MeV}/u$ (see sect. **5** for the plans of increasing this further). The overall efficiency, *i.e.* the intensity after acceleration relative to the intensity out of the ISOLDE target, reaches 5% for many elements. The breeding time is typically $20\,\mathrm{ms}$ for the lighter ions for which REX was designed, but it has recently been shown [17] that one can charge breed and accelerate ions all the way up to $^{238}\mathrm{U}$, although breeding times then become longer.

3. – Physics with low-energy beams

Many science questions within nuclear physics and neighbouring areas—atomic physics, fundamental physics, nuclear astrophysics, condensed-matter physics and biophysics—can be addressed with low-energy radioactive beams. At the moment about 35 different experiments take place every year covering all the mentioned subjects and more than 400 experimentalists perform experiments at ISOLDE. It is not possible in a single presentation to cover the complete range of results coming out of these experiments, so I will restrict myself to giving examples from some of the major research lines, but will first give a few brief comments and references to some others.

Traps and other devices for ion and atom manipulation have become very important tools not only for the general preparation of the radioactive ions for experiments, but also as key ingredients in specific experimental set-ups. The use of Penning traps in mass measurements is an obvious example, but traps are now also contributing importantly to investigations of the structure of fundamental interactions [18], *e.g.*, in the WITCH experiment. The older on-line low-temperature set-up NICOLE [19] that can reach temperatures down to $10\,\mathrm{mK}$ still contributes to the field of fundamental interactions, but its capability of polarizing nuclear spins gives it a wider range of applicability. The non-nuclear physics programmes within solid-state physics [20] and life science [21, 22] are important, but will not be treated further here.

3'1. *Nuclear masses.* – The mass of a nucleus is along with its halflife and main decay modes typically the first of its properties to be measured. The total binding energy of a nucleus contains much physics information [23] and precision mass measurements therefore often provide important tests of nuclear models. The ISOLTRAP mass measurement [24, 25] has during the last two decades successfully pioneered precision measurements of nuclei with halflives extending down to a fraction of a second and has thereby prompted the construction of many similar Penning trap set-ups at other radioactive beam facilities, see [26, 27] for general reviews of Penning traps and their use in mass measurements.

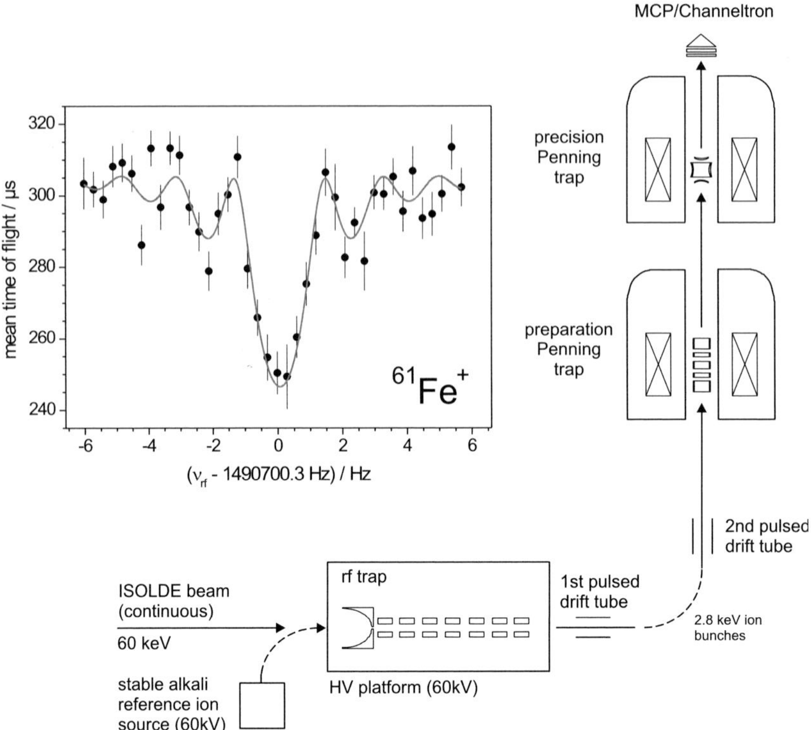

Fig. 2. – Sketch of the experimental setup of ISOLTRAP. The inset shows a time-of-flight cyclotron resonance for ^{61}Fe$^+$ (rf excitation duration of 600 ms). The solid line is a fit of the theoretical line shape to the data.

The set-up at ISOLDE is shown in fig. 2, the continuous beam from ISOLDE is collected and bunched before being sent to the preparation trap where purification can take place if needed. A detailed analysis of the ion motion in the precision trap shows that the cyclotron resonance of the ion can be found through measurement of the time of flight of the ions when ejected from the trap, this is illustrated in the inset of the figure. New developments still take place, among the latest is the use of in-trap decay: the nucleus of interest, in the figure the example of Fe is shown, may not be accessible directly as an ISOL beam but is instead produced via in-trap decay in the preparation Penning trap [28].

3˙2. Nuclear moments. – The experimental determination of nuclear radii and nuclear moments have for many years been a major activity at ISOLDE [29, 30]. The nuclear properties are deduced from accurate laser spectroscopy measurements of the atomic level spectrum. The application of lasers in nuclear physics is reviewed in [31]; an overview of the various approaches used in measurements of nuclear magnetic and electric moments can be found in [32].

As in the case of mass measurements, the systematic measurement of nuclear moments along an isotopic, isotonic or isobaric series is a powerful tool in extracting nuclear structure changes. One example that also demonstrates the versatility of ISOLDE is the measurement of radii of the light Pb isotopes [33] in which the isotope shifts were measured by tuning of the lasers in the RILIS ion source. This method can—and has—been extended to provide relatively pure beams of isomeric states in nuclei.

3'3. *Beta decay*. – Beta decay not only allows the nuclear physicist to contribute to tests of the weak interaction, it is also a well-tested probe of nuclear structure and benefits from the selection rules governing the decays. When moving away from the line of beta-stability, beta decay actually increases in importance due to the higher Q-values which allows more states to be fed in the decays. In many cases one finds that beta-decay experiment countrates remain competitive with those of nuclear reactions for the same final nucleus in spite of the decay starting in a more exotic nucleus. A general overview of beta decay of exotic nuclei can be found through [34] and references therein; more specific reviews of ISOLDE work on beta-delayed two-proton emission [35] and on halo nuclei [36] are also available.

Two recent examples that deserve mentioning are the studies of ^{30,31,32}Mg [37] (including fast timing measurements) and $^{74-78}$Cu [38], both cases relevant for the question of how stable nuclear shells remain as we move to exotic nuclei. The $N = 20$ shell, the first case, is known to disapper, whereas the answer is still unknown for the second case in the region of $Z = 28$.

3'4. *Nuclear astrophysics*. – Other lectures at this school introduce this exciting interdisciplinary field; more information can be obtained through a recent compilation [39]. Several experiments at ISOLDE have contributed to the field, one of the major sustained efforts being beta-decay experiments of neutron-rich nuclei relevant for the r-process path [40], *e.g.*, reaching from ^{132}Sn down to ^{129}Ag. For the r-process mass measurements are of course also needed. Low-energy reactions have been studied in great detail at Louvain-la-Neuve and Triumf and may become important at REX-ISOLDE in the near future.

Another recent result is the investigation [41] of the excited level structure in ^{12}C through beta-decays of the isobaric nuclei. The astrophysical triple-alpha process depends crucially on the nuclear structure here, and the improved nuclear physics knowledge reduced the systematic uncertainty of it in several temperature ranges of interest.

4. – Physics with accelerated beams

The physics programme at ISOLDE with accelerated beams only started a few years ago and the possibilities are far from being exploited fully. I will give examples from the two major types of experiments performed so far.

4'1. *Miniball experiments*. – In the energy range up to 3.1 MeV/u available at present at ISOLDE many reactions will take place at or below the Coulomb barrier, so the

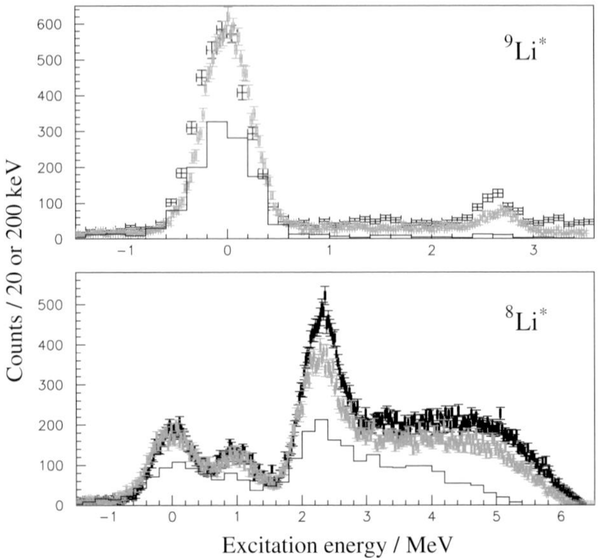

Fig. 3. – Excited states in ^{9}Li and ^{8}Li populated in the ^{9}Li + ^{2}H reaction. The histogram (200 keV bins) show data at 2.2 MeV/u published in [50, 52], the data points (20 keV bins) are from a recent experiment at 2.8 MeV/u.

main reaction mechanism is Coulomb excitation. The electromagnetic interaction is well understood and is in general used as a probe of nuclear structure from the subbarrier energies in Coulomb excitation experiments to high-energy dissociation experiments [42].

The key ingredient in these experiments is of course the gamma detection, in ISOLDE's case this is done in the Miniball gamma-detector array [43]. The first result was the "safe" Coulomb excitation of ^{30}Mg [44], again probing the $N = 20$ shell disappearance, but many experiments have been performed by now. Three recent examples that illustrate the broadness of the experimental programme are the determination of the quadrupole moment for the first 2^+ state in ^{70}Se [45], the Coulomb (de)excitation of isomerically purified beams on neutron-rich Cu isotopes [46] and the B(E2) measurement of light Sn isotopes [47] probing the structural evolution towards ^{100}Sn.

4˙2. *Transfer experiments*. – Elastic and inelastic scattering, transfer reactions, etc. will also yield important information on the structure of exotic nuclei [48] and have been initiated at several laboratories. At REX-ISOLDE the present beam energies restrict studies to the light nuclei and experiments have been performed near the driplines where many interesting results have been uncovered during the last decade [49]. In the ^{9}Li + ^{2}H reaction neutron transfer will populate either ^{8}Li or the unbound nucleus ^{10}Li (relevant for the understanding of the famous halo nucleus ^{11}Li) [50, 51]. For the ^{8}Li case the results can be compared to *ab initio* calculations [52]. The data shown in fig. 3 are from two runs performed just after REX-ISOLDE started operating and less than a year ago, respectively, and illustrate the rapid improvement in the experimental capabilities.

5. – The HIE-ISOLDE project

Radioactive nuclear beams is given a high priority in the present planning of future nuclear physics facilities on all continents. I shall concentrate here on the developments leading towards EURISOL, other contributions to this school cover other aspects, in particular the physics motivation for pursuing this field. Development work at ISOLDE and elsewhere is already focussed on the technical challenges [53] that need to be resolved before EURISOL can be constructed, most likely about a decade from now.

To bridge the time until EURISOL is ready and to prepare the user community we need upgrades of the present facilities, the two main projects presently appear to be SPIRAL2 at GANIL and HIE-ISOLDE at CERN. The HIE-ISOLDE project intends to improve the experimental capabilities at ISOLDE over a wide front, the technical description was published last year [54] and a detailed physics case is presently being finalized. The main features will be an increase of the energy of REX-ISOLDE to $5.5\,\mathrm{MeV}/u$ (with the option of a later boost to $10\,\mathrm{MeV}/u$), an increase of the primary-beam intensity and improvements in several respects of the secondary beam properties (purity, ionization efficiency and beam optical quality). The superconducting acceleration technology will be introduced in REX and is planned to gradually replace the present machine. Some of the secondary beam improvements have already started, *e.g.*, the construction of a RFQ ion cooler and a major upgrade of the RILIS set-up including an off-line laboratory. The HIE-ISOLDE project is naturally staged and already ongoing, this promises well for the physics outcomes of the facility also in the coming years.

$$* \quad * \quad *$$

I would like to thank L. FRAILE, A. HERLERT, H. JEPPESEN and M. LINDROOS for help with material presented here. This work was in part supported by the European Union Sixth Framework through RII3-EURONS (contract no. 506065).

REFERENCES

[1] HUYSE M., in *The Euroschool Lectures on Physics with Exotic Beams*, Vol. **I**, edited by AL-KHALILI J. and ROECKL E., *Lect. Notes Phys.*, **651** (2004) 1.
[2] http://www.cern.ch/isolde
[3] GEISSEL H., MÜNZENBERG G. and RIISAGER K., *Annu. Rev. Nucl. Part. Sci.*, **45** (1995) 163.
[4] RAVN H. L. and ALLARDYCE B. A., in *Treatise on Heavy-Ion Science*, Vol. **8**, edited by BROMLEY D. A. (Plenum) 1989, p. 363.
[5] VAN DUPPEN P., in *The Euroschool Lectures on Physics with Exotic Beams*, Vol. **II**, edited by AL-KHALILI J. and ROECKL E., *Lect. Notes Phys.*, **700** (2006) 37.
[6] RAVN H. L. *et al.*, *Nucl. Instrum. Methods B*, **88** (1994) 441.
[7] http://www.eurisol.org
[8] KUGLER E., *Hyperfine Interact.*, **129** (2000) 23.
[9] VAN DUPPEN P. *et al.*, *Rev. Sci. Instrum.*, **63** (1992) 2381.
[10] FEDOSSEEV V. *et al.*, *Nucl. Instrum. Methods A*, **204** (2003) 353.
[11] CATHERALL R. *et al.*, *Rev. Sci. Instrum.*, **75** (2004) 1614.

[12] LETTRY J. *et al.*, *Nucl. Instrum. Methods B*, **126** (1997) 130.

[13] TURRION M., private communication.

[14] HABS D. *et al.*, *Hyperfine Interact.*, **129** (2000) 43.

[15] KESTER O. *et al.*, *Nucl. Instrum. Methods A*, **204** (2003) 20.

[16] CEDERKALL J. *et al.* (Editors), *The REX-ISOLDE Facility*, CERN Report, CERN-2005-009.

[17] WENANDER F., *ISOLDE Workshop, February 2007*.

[18] SEVERIJNS N., BECK M. and NAVILIAT-CUNCIC O., *Rev. Mod. Phys.*, **78** (2006) 991.

[19] SCHLÖSSER K. *et al.*, *Hyperfine Interact.*, **43** (1988) 141.

[20] FORKEL-WIRTH D., *Rep. Prog. Phys.*, **62** (1999) 527.

[21] TRÖGER W. and BUTZ T., *Hyperfine Interact.*, **129** (2000) 511.

[22] BEYER G. J., *Hyperfine Interact.*, **129** (2000) 529.

[23] LUNNEY D., PEARSON J. M. and THIBAULT C., *Rev. Mod. Phys.*, **75** (2003) 1021.

[24] BOLLEN G. *et al.*, *Nucl. Instrum. Methods A*, **368** (1996) 675.

[25] HERFURTH F. *et al.*, *J. Phys. B*, **36** (2003) 931.

[26] BOLLEN G., in *The Euroschool Lectures on Physics with Exotic Beams*, Vol. **I**, edited by AL-KHALILI J. and ROECKL E., *Lect. Notes Phys.*, **651** (2004) 169.

[27] BLAUM K., *Phys. Rep.*, **425** (2006) 1.

[28] HERLERT A. *et al.*, *New J. Phys.*, **7** (2005) 44.

[29] OTTEN E. W., in *Treatise on heavy-ion science*, Vol. **8**, edited by BROMLEY D. A. (Plenum) 1989, p. 517.

[30] OTTEN E. W., *Phys. Rep.*, **225** (1993) 157.

[31] KLUGE H.-J. and NÖRTERSHÄUSER W., *Spectrochim. Acta, Part B*, **58** (2003) 1031.

[32] NEYENS G., *Rep. Prog. Phys.*, **66** (2003) 633.

[33] DE WITTE H. *et al.*, *Phys. Rev. Lett.*, **98** (2007) 112502.

[34] JONSON B. and RIISAGER K., *Nucl. Phys. A*, **693** (2001) 77.

[35] BORGE M. J. G. and FYNBO H. O. U., *Hyperfine Interact.*, **129** (2000) 97.

[36] NILSSON T., NYMAN G. and RIISAGER K., *Hyperfine Interact.*, **129** (2000) 67.

[37] MACH H. *et al.*, *Eur. Phys. J. A*, **25**, Supp. 1 (2005) 105.

[38] VAN ROOSBROECK J. *et al.*, *Phys. Rev. C*, **71** (2005) 054307.

[39] *Special issue on Nuclear Astrophysics*, edited by LANGANKE K., THIELEMANN F.-K. and WIESCHER M., *Nucl. Phys. A*, **777** (2006) 1.

[40] KRATZ K.-L. *et al.*, *Hyperfine Interact.*, **129** (2000) 185.

[41] FYNBO H. O. U. *et al.*, *Nature*, **433** (2005) 136.

[42] BAUR G., HENCKEN K. and TRAUTMANN D., *Prog. Part. Nucl. Phys.*, **51** (2003) 487.

[43] EBERTH J. *et al.*, *Prog. Part. Nucl. Phys.*, **46** (2001) 389.

[44] NIEDERMAIER O. *et al.*, *Phys. Rev. Lett.*, **94** (2005) 172501.

[45] HURST A. M. *et al.*, *Phys. Rev. Lett.*, **98** (2007) 072501.

[46] STEFANESCU I. *et al.*, *Phys. Rev. Lett.*, **98** (2007) 122701.

[47] CEDERKALL J. *et al.*, *Phys. Rev. Lett.*, **98** (2007) 172501.

[48] ALAMANOS N. and GILLIBERT A., in *The Euroschool Lectures on Physics with Exotic Beams*, Vol. **I**, edited by AL-KHALILI J. and ROECKL E., *Lect. Notes Phys.*, **651** (2004) 295.

[49] JONSON B., *Phys. Rep.*, **389** (2004) 1.

[50] JEPPESEN H. B. *et al.*, *Nucl. Phys. A*, **748** (2005) 374.

[51] JEPPESEN H. B. *et al.*, *Phys. Lett. B*, **642** (2006) 449.

[52] JEPPESEN H. B. *et al.*, *Phys. Lett. B*, **635** (2006) 17.

[53] VAN DUPPEN P., *Nucl. Instrum. Methods A*, **204** (2003) 9.

[54] LINDROOS M. and NILSSON T. (Editors), *HIE-ISOLDE: the technical options*, CERN Report, CERN-2006-013.

DOI 10.3254/978-1-58603-885-4-101

RIKEN RI Beam Factory and its research opportunities

T. Motobayashi

RIKEN Nishina Center - 2-1 Hirosawa, Wako, Saitama 351-0198, Japan

Summary. — RIKEN RI Beam Factory (RIBF) is one of the new-generation facilities for Radioactive Isotope (RI) beam, which are dedicated to provide various nuclei very far from the stability valley served as beam with high-power driver accelerators. The RIKEN RIBF is their first realization. The accelerator complex has already started operation, and the first result of new isotope production has been performed. Research opportunities with RI beams, especially for the ones with intermediate energies, are discussed.

1. – Nuclei far from the stability

Figure 1 shows a current nuclear chart. The solid squares represent stable isotopes, which are counted to about 300, and the nuclei experimentally known so far are about 3000 as shown by the open squares. Theoretical estimates give approximately 10000 isotopes that have lifetimes longer than micro seconds. They are stable against particle decay including proton and neutron emission and fission with a very short lifetime. Nuclear physicists have made continuous effort to extend the territory of known nuclei. About 500 nuclei were already known in 1940, and the number of known nuclei has increased steadily until now. As mentioned later, the RIKEN RI Beam Factory (RIBF) and other new-generation RI beam facilities as FAIR in Germany, FRIB in US and EURISOL

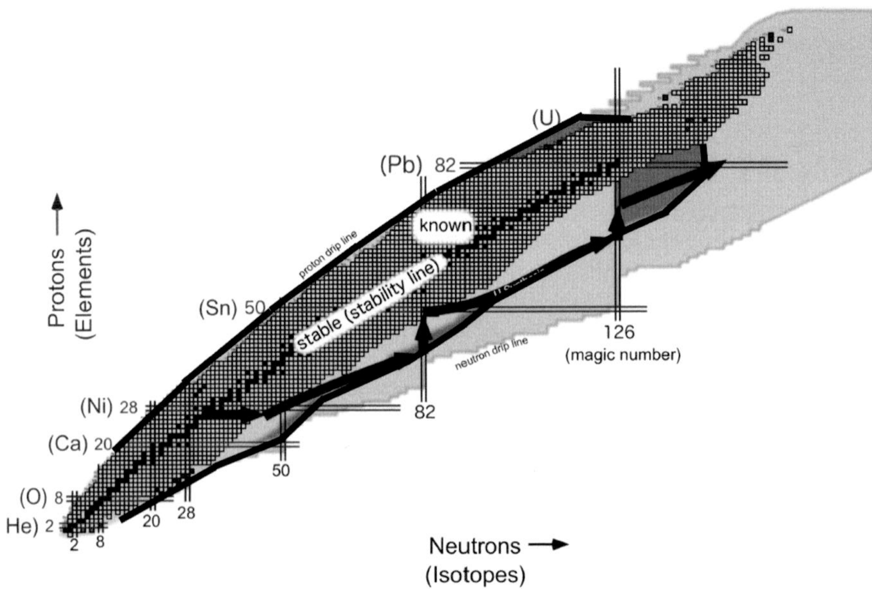

Fig. 1. – Nuclear chart. The expected r-process path is indicated by the lines with arrows. The thick solid curves indicate the limit of RI productions of 1 particle per day with $1\,\mathrm{p}\mu\mathrm{A}$ primary beam intensity at RIBF.

in Europe are designed to have capability of increasing the number of known nuclei to 4000 or more.

Nuclear physics research is not only to extend the territory in the nuclear chart. To understand the interaction and structure of atomic nuclei, a very important issue is to study properties of excited states of nuclei, which are less known for nuclei far from the stability line. It is also known that nuclear reactions play important roles in evolution of the Universe by creating energies in astrophysical objects and generating chemical elements. Reactions with unstable nuclei are important for explosive nuclear burning in stars and early Universe.

2. – Fast RI beam and new experimental methods

The word "RI beams" is abbreviation of "Radioactive Isotopes beams", and is equivalent to "radioactive ion beams" or beams with unstable nuclei. RI beams at RIKEN have been produced by the projectile fragmentation scheme since 1990. Heavy-ion beams with intermediate energy (typically $100\,\mathrm{MeV/nucleon}$) bombard a production target, and products of projectile fragmentation reactions are separated by a magnetic analyzer system and served as beams. Due to the high performance of the RIken Projectile fragment Separator (RIPS) [1], the world-highest-intensity RI beams are available for many light neutron-rich nuclei. For example, the very neutron-rich fluorine isotope ^{31}F could be

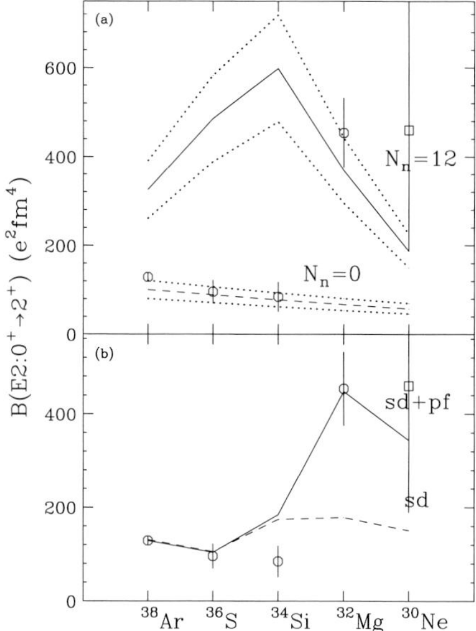

Fig. 2. – Reduced transition probability $B(E2)$ of $N = 20$ even-even nuclei.

found to be particle-stable [2], owing to this high-intensity capability. Besides these intermediate-energy RI beams, beams of light unstable nuclei at lower energies, typically 5 MeV/nucleon, are also available in the CRIB facility constructed by CNS [3]. The CRIB uses in-flight direct reactions, such as (p, n) and (^{3}He, n), to produce RI beams, which are mainly used for studying low-energy reactions of astrophysical interest.

The intermediate-energy RI beams have been used for various experiments. Studies of interaction cross-section have been performed, and neutron halo and neutron skin structures have been established in some light neutron-rich nuclei [4]. The neutron halo or the neutron skin indicates that some neutrons can be decoupled from protons despite of the strong p-n interaction.

Disappearance of magic numbers is another phenomena of interest. The first Coulomb excitation experiment with fast RI beams has been performed at RIKEN for the neutron-rich ^{32}Mg nucleus with the $N = 20$ magic number [5]. The extracted large $B(E2)$ value in contrast to heavier $N = 20$ nuclei (see fig. 2) indicates that the $N = 20$ shell closure disappears in ^{32}Mg.

A new method for studying properties of unstable nuclei, measurement of γ-rays from fast-moving excited nuclei in coincidence with reaction products with particle identification, was applied to this experiment, and many experiments using this technique have been performed so far for studying nuclear structures of nuclei around the shell closure at $N = 8$ and $N = 20$.

Recently, the decoupling of protons and neutrons has been revealed also in excitation of the ^{16}C nucleus to its 2^+ state by measuring the 2^+ lifetime [6, 7], inelastic scattering

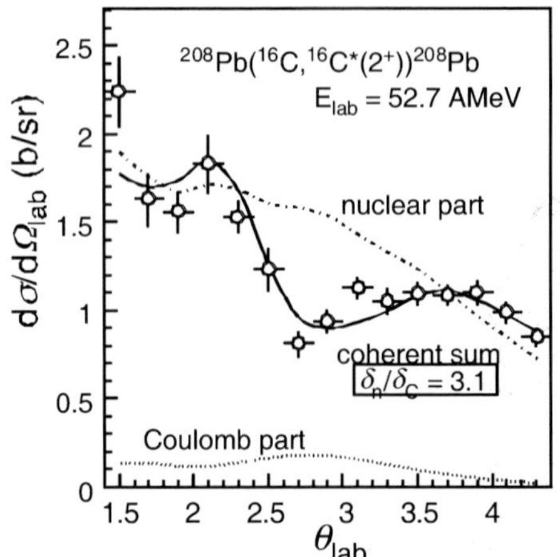

Fig. 3. – Angular distribution of the ^{16}C + Pb inelastic scattering feeding the first 2^+ state in ^{16}C.

with ^{1}H [8], and the one with ^{208}Pb [9]. Analysis of the interference between the nuclear and Coulomb excitation amplitudes in the angular distribution for the ^{16}C + ^{208}Pb inelastic scattering (see fig. 3) could separate the proton and neutron contributions to the 2^+ state excitation. The separation is also possible by comparing the lifetime and proton inelastic data. Both the results indicates that neutrons strongly contribute to the 2^+ state excitation, while protons have little contribution.

This decoupling picture might be related to the low electric quadrupole moments measured for ^{15}B and ^{17}B using the β-NMR technique [10]. Studies of particle-unbound states in unstable nuclei are another highlight. The Coulomb dissociation to simulate astrophysical (p, γ) reactions [11] and the neutron halo structure [12] have been extensively studied.

The nuclear reactions used in these studies involve no mass transfer and match the fast RI beam, since their good kinematical matching persists even in the relevant energy range.

3. – RI Beam Factory

3`1. *Overview*. – Encouraged by the success of research using the fast RI beams, the RI Beam Factory (RIBF) project was planned. The RIBF new facility consists of three cyclotrons, the fixed-frequency Ring Cyclotron (fRC) with $K = 570$ MeV, Intermediate stage Ring Cyclotron (IRC) with $K = 980$ MeV, and Superconducting Ring Cyclotron (SRC) with $K = 2600$ MeV, which successively boosts the beam energy up to 345 MeV/nucleon [13]. The 16 MV variable-frequency linear accelerator (RILAC) and four-sector ring cyclotron (RIKEN Ring Cyclotron, RRC) with $K = 540$ MeV, which have been in operation since 1987, are used as their injectors. A schematic view of the facility is shown in fig. 4.

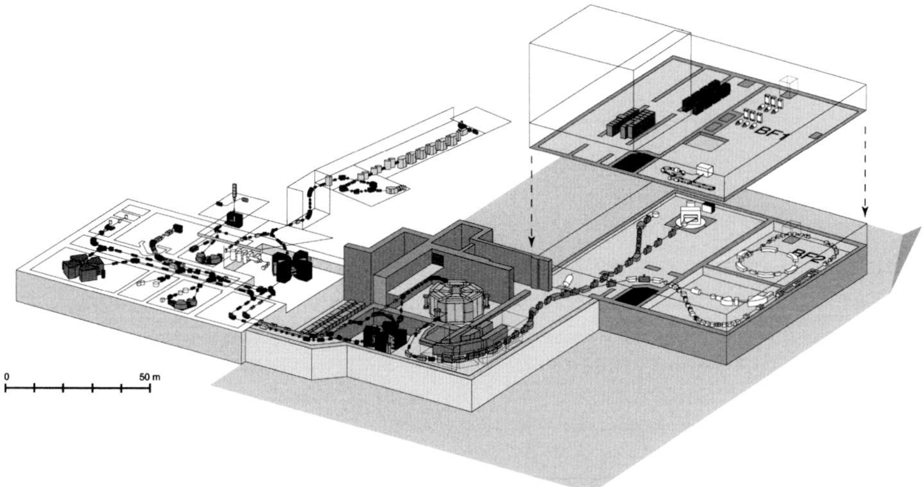

Fig. 4. – Schematic view of the RIKEN accelerator complex.

To provide intense beams the new RIBF accelerators, several improvements have been made around the injector RILAC. Thus, intense ions at several MeV/nucleon energies became available. In 2004, the first event indicating the production of the isotope $^{278}113$ by the $^{209}\mathrm{Bi}(^{70}\mathrm{Zn,n})$ fusion reaction has been observed [14]. The beams have been used also for various applications to nuclear chemistry, bio and medical science, and materials science.

RI beams are produced via the projectile fragmentation of heavy ions or in-flight fission of uranium ions by a superconducting fragment separator BigRIPS [15] shown in fig. 5. It consists of fourteen magnets of superconducting quadrupole triplets and six room-temperature dipoles magnets. Separation of reaction products is made by a two-

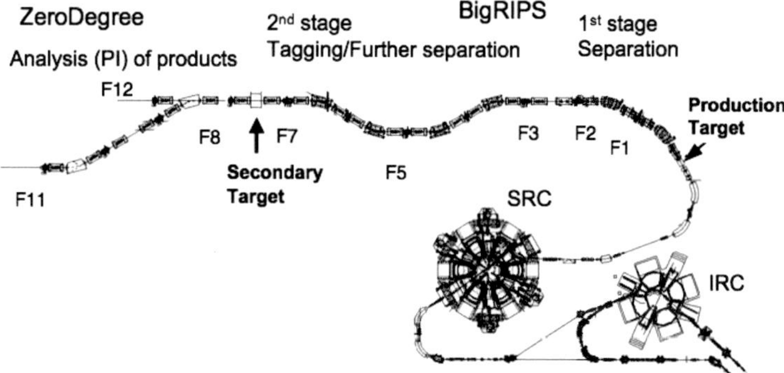

Fig. 5. – Plan view of the BigRIPS and zero degree spectrometer together with the two cyclotrons IRC and SRC of the RIBF accelerate complex.

stage separation scheme. The first stage serves to produce and separate RI beams with a wedge-shaped degrader inserted at the momentum-dispersive focus F1. The second-stage identifies RI beam species event-by-event and tags the secondary beam that still contains various different ions. The horizontal and vertical angular acceptances are designed, respectively, to 80 mrad 100 mrad vertically, while the momentum acceptance is 6%. These angular and momentum acceptances enable one to collect about half of the fission fragments produced by a 350 MeV/nucleon uranium beam. With the goal intensity of the primary beams, 1 pμA for ions up to uranium, the RI beam yield estimated by the code EPAX2 [16] is shown in fig. 1. For example, the intensities of the doubly magic nuclei ^{78}Ni, ^{132}Sn and ^{100}Sn are expected to be 10, 100 and 1 particles/s, respectively, encouraging detailed studies of nuclei far from the stability valley. Most of the expected path of the r-process nucleosynthesis can be reached with the intensity higher than one particle per day.

The Zero Degree Spectrometer (ZDS), also shown in fig. 5, analyzes secondary reaction products emitted in the beam direction. Its typical use is for γ-ray measurements in coincidence with fast-moving excited nuclei. The ZDS is used to identify the reaction product. For other applications, such as β-decay measurements with stopped RIs, the ZDS is also useful.

3`2. *Status of RIBF*. – The parts completed so far are the cyclotron complex, the Big RIPS, and the zero degree spectrometer. After the first extraction of the primary beam of 345 MeV/nucleon ^{27}Al^{10+}, on December 28 in 2006, the accelerators have been tuned, and ^{238}U ions were successfully accelerated in March 2007.

Attempts of RI beam production started in March 2007 and the first experiment of new-isotope production [17] was performed with a 345 MeV/nucleon ^{238}U^{86+} beam delivered to a beryllium production target of 7 mm thick. The parameters of the BigRIPS were matched to neutron-rich fission products with the atomic number around 50. The atomic number Z and the mass-to-charge ratio A/Q were obtained event-by-event by measuring the velocity (or time-of-flight), energy loss, total energy, and magnetic rigidity using beam-line counters. The yield for $Z = 46$ (palladium) is plotted as a function of A/Q in fig. 6. Observed double-peak structures are due to the mixture of ions not fully stripped. A peak corresponding to the new isotope ^{125}Pd, indicated by the arrow, is clearly seen with a good statistics, and some indication for ^{126}Pd is also seen. The data were taken with the primary beam current of 4×10^7 particle/s on an average, which is about 10^{-5} of the goal intensity, 1 particle μA, and the data acquisition time is about one-day. This indicates high potential of the BigRIPS separator coupled with the RIBF accelerator complex, and future possibility in accessing a large amount of unknown unstable nuclei is foreseen. Since the RI beam intensities achieved so far is not enough for secondary reaction studies, various efforts are on-going including the improvement of the beam transmission in every stage of the accelerator complex. The use of the ^{48}Ca and ^{86}Kr beams is currently assumed for the first series of experiments.

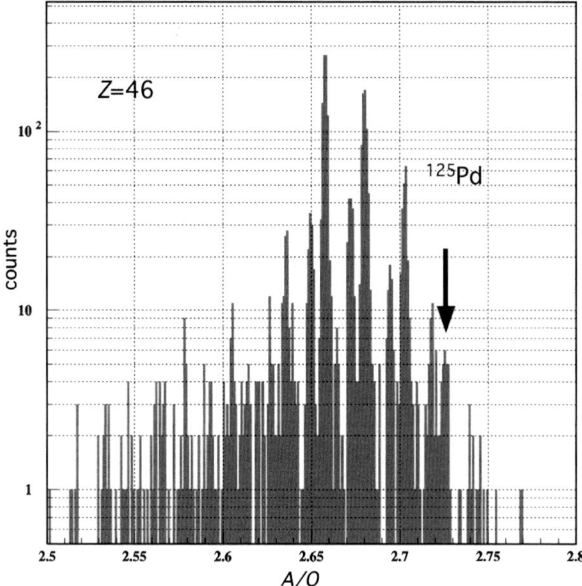

Fig. 6. – Yield distribution for the atomic number $Z = 46$ obtained by the in-flight fission on uranium beams with the energy of 345 MeV/nucleon. The arrow indicates the peak position for the new isotope ^{125}Pd.

3˙3. *New experimental installations.* – Besides experimental studies in the first periods using the Big RIPS and the zero degree Spectrometer, the nuclear-physics and nuclear-astrophysics programs in the RIBF includes construction of various kinds of new apparatuses.

The SHARAQ spectrometer [18] is of a QQ-D-Q-D configuration. By employing dispersion-matching optics with a specially designed beam line from the BigRIPS, the momentum resolution of 15000 will be achieved. The SHARAQ spectrometer is under construction and will be commissioned in 2008.

The SLOWRI [19] aims at conducting various experiments using slow RI beams. Fast secondary beams from the BigRIPS will be efficiently stopped and extracted by a gas-catcher system with the RF ion-guide technique.

The SAMURAI is a spectrometer with a large solid angle and a large momentum acceptance dedicated to particle-correlation studies. It is of a QQQ-D configuration. Its dipole magnet (D) is superconducting one with 6.7 Tm rigidity. The large gap of 80 cm is useful for measurements of projectile-rapidity neutrons. The SAMURAI spectrometer will be completed in the year 2011.

The system for electron-RI scattering experiments employs a Self-Confining Radioactive Ion Target (SCRIT [20]). RI ions are transversely confined by the attractive force caused by the electron beam itself. A mirror potential is applied externally to achieve longitudinal confinement. Test experiments to examine the confinement mechanism are successfully made at an existing electron ring.

The rare RI mass ring [21] is to measure the mass of rarely produced (1 particle per day, for example) exotic nuclides in 10^{-6} accuracy. Each ion is injected individually to the ring by a trigger signal provided from a counter in the BigRIPS. The ring is tuned to achieve the isochronous condition, and a time-of-flight of the ion in the ring is measured.

To allow for running an RI-beam–based experiment and a superheavy element search simultaneously, construction of a new injector [22] is being considered. Another possibility to extend the research opportunity is to build a beam line that brings back the IRC beam to the RIPS separator. Various nuclear moment studies and condensed-matter research are planned.

4. – Conclusions

The RIKEN RI Beam Factory (RIBF), one of the new-generation facilities for beams of unstable nuclei, has started its operation. RI-beams are produced by the RIBF accelerator complex with three newly-built cyclotrons coupled with the superconducting RI beam separator BigRIPS. Their potential was demonstrated by production of the new isotope ^{125}Pd. When the RIBF reaches to its full performance, almost all nuclides along the r-process nuclear synthesis can be created and experimentally studied. Continuous attempts to improve the primary beam intensity together with construction of various experimental equipments will open a new domain of nuclear physics, and will create a new view on atomic nucleus as well as on the element genesis in the universe, together with the world efforts for realizing RI-beam facilities being constructed or planned.

REFERENCES

[1] Kubo T. *et al.*, *Nucl. Instrum. Methods B*, **70** (1992) 309.
[2] Sakurai H. *et al.*, *Phys. Lett. B*, **448** (1999) 180.
[3] Yanagisawa Y. *et al.*, *Nucl. Instrum. Methods A*, **539** (2005) 74.
[4] Suzuki T. *et al.*, *Phys. Rev. Lett.*, **75** (1995) 3241.
[5] Motobayashi T. *et al.*, *Phys. Lett. B*, **346** (1995) 9.
[6] Imai N. *et al.*, *Phys. Rev. Lett.*, **92** (2004) 062501.
[7] Ong H. J. *et al.*, submitted to *Phys. Rev. C*.
[8] Ong H. J. *et al.*, *Phys. Rev. C*, **73** (2006) 024610.
[9] Elekes Z. *et al.*, *Phys. Lett. B*, **686** (2004) 34.
[10] Ueno H. *et al.*, *Nucl. Phys. A*, **738** (2004) 211.
[11] For example, Motobayashi T., *Nucl. Phys. A*, **693** (2001) 258.
[12] For example, Nakamura T. *et al.*, *Phys. Rev. Lett.*, **96** (2006) 252502.
[13] Yano Y., *Nucl. Instrum. Methods B*, **261** (2007) 1009.
[14] Morita K. *et al.*, *J. Phys. Soc. Jpn.*, **73** (2004) 2593.
[15] Kubo T. *et al.*, *IEEE Trans. Appl. Superconductiv.*, **17** (2007) 1069.
[16] Suemmerer K. and Blank B., *Phys. Rev. C*, **61** (2000) 034607.
[17] Yano Y., presented at the *International Symposium on Nuclear Physics (INCP07), Tokyo, June 2007*.
[18] Uesaka T. *et al.*, CNS Annual Report 2004, (2005) 42.

[19] WADA M. *et al.*, *Nucl. Instrum. Methods A*, **532** (2004) 40.
[20] WAKASUGI M. *et al.*, *Nucl. Instrum. Methods A*, **532** (2004) 216.
[21] YAMAGUCHI Y. *et al.*, CNS Annual Report 2004, (2005) 83.
[22] KAMIGAITO O. *et al.*, *RIKEN Accel. Prog. Rep.*, **39** (2005) 261.

DOI 10.3254/978-1-58603-885-4-111

Quantum Monte Carlo calculations of light nuclei

STEVEN C. PIEPER

Physics Division, Argonne National Laboratory - Argonne, IL 60439, USA

Summary. — During the last 15 years, there has been much progress in defining the nuclear Hamiltonian and applying quantum Monte Carlo methods to the calculation of light nuclei. I describe both aspects of this work and some recent results.

1. – Introduction

The goal of *ab initio* light-nuclei calculations is to understand nuclei as collections of nucleons interacting with realistic (bare) potentials through reliable solutions of the many-nucleon Schrödinger equation. Such calculations can study binding energies, excitation spectra, relative stability, densities, transition amplitudes, cluster-cluster overlaps, low-energy astrophysical reactions, and other aspects of nuclei. Such calculations are also essential to claims of sub-nucleonic effects, such as medium modifications of the nuclear force or nucleon form factors; if a reliable pure nucleonic degrees-of-freedom calculation can reproduce experiment, then there is no basis for claims of seeing sub-nucleonic degrees of freedom in that experiment (beyond the obvious fact that the free-space nucleon interactions are a result of sub-nucleonic degrees of freedom).

There are two problems in microscopic few- and many-nucleon calculations: 1) determining the Hamiltonian, and 2) given H, accurately solving the Schrödinger equation for A nucleons; I will discuss both of these in this contribution. The two-nucleon (NN) force is determined by fitting the large body of NN scattering data. Several modern NN

potentials are in common use. The Argonne v_{18} is a local potential written in operator format; this potential is used in the calculations described here, and is presented in some detail below. Other modern potentials are generally non-local; some of them are discussed in other contributions to this school.

It has long been known that calculations with just realistic NN potentials fail to reproduce the binding energies of nuclei; three-nucleon (NNN) potentials are also required. These arise naturally from an underlying meson-exchange picture of the nuclear forces or from chiral effective field theories. Unfortunately, much NNN scattering data is well reproduced by calculations using just NN forces, so the NNN force must determined from properties of light nuclei. In this contribution the recent Illinois models with 2π and 3π rings are used.

Our understanding of nuclear forces has evolved over the last 70 years:

– Meson-exchange theory of Yukawa (1935).

– Fujita-Miyazawa three-nucleon potential (1955).

– First phase-shift analysis of NN scattering data (1957).

– Gammel-Thaler, Hamda-Johnston and Reid phenomenological potentials (1957–1968).

– Bonn, Nijmegen and Paris field-theoretic models (1970s).

– Tuscon-Melbourne and Urbana NNN potential models (late 70's–early 80's).

– Nijmegen partial wave analysis (PWA93) with $\chi^2/\text{dof} \sim 1$ (1993).

– Nijm I, Nijm II, Reid93, Argonne v_{18} and CD-Bonn (1990s).

– Effective field theory at N^{3}LO (2004).

References for a number of these developments are given in the following sections.

Accurate solutions of the many-nucleon Schrödinger equation have also evolved over many decades:

– ^{2}H by numerical integration (1952) – a pair of coupled second-order differential equations in 1 variable. At the time this took "between 5 and 20 minutes for the calculation and the printout another 5 minutes" [1]!

– ^{3}H by Faddeev (1975–1985).

– ^{4}He by Green's function Monte Carlo (GFMC) (1988).

– $A = 6$ by GFMC and No-core shell model (NCSM) (1994-95).

– $A = 7$ by GFMC and NCSM (1997-98).

– $A = 8$ by GFMC and NCSM (2000).

- ^{4}He benchmark by 7 methods to 0.1% (2001).

- $A = 9, 10$ by GFMC and NCSM (2002).

- ^{12}C by GFMC and NCSM (2004–).

- ^{16}O by Coupled Cluster (CC) (2005–).

References for the $A = 3, 4$ calculations may be found in ref. [2]; the GFMC calculations are the subject of this paper; the NCSM are discussed in Petr Navrátil's contribution to this Course; and CC results may be found in ref. [3].

This contribution is limited to Variational Monte Carlo (VMC) and GFMC calculations of light nuclei. Section **2** describes the Hamiltonians used and sects. **3** through **5** describe the computation methods. Section **6** gives a number of results for energies of nuclear states; sect. **7** describes GFMC calculations of scattering states; and sect. **8** gives some results for densities. Finally some conclusions and prospects for the future are presented in sect. **9**.

2. – Hamiltonians

The nuclear Hamiltonian used here has the form

$$(1) \qquad H = \sum_i K_i + \sum_{i<j} v_{ij} + \sum_{i<j<k} V_{ijk}.$$

Here K_i is the non-relativistic kinetic energy, including $m_n - m_p$ effects, v_{ij} is the NN potential and V_{ijk} is the NNN potential.

2$^{.}$1. *Argonne v_{ij}.* – The NN potential (v_{ij}) is Argonne v_{18} [4] (AV18) which has the form

$$(2) \qquad v_{ij} = v_{ij}^{\gamma} + v_{ij}^{\pi} + v_{ij}^{R} + v_{ij}^{\mathrm{CIB}}.$$

The v_{ij}^{γ} is a very complete representation of the pp, pn and nn electromagnetic terms, including first- and second-order Coulomb, magnetic, vacuum polarization, etc., components with form factors. (Reference [5] provides a heuristic introduction to AV18.)

The v_{ij}^{π} is the isoscalar one-pion exchange potential represented as a local operator:

$$(3) \qquad v_{ij}^{\pi} = \frac{1}{3} \frac{f_{\pi NN}^2}{4\pi} m_\pi \, X_{ij} \, \boldsymbol{\tau}_i \cdot \boldsymbol{\tau}_j,$$

$$(4) \qquad X_{ij} = T(m_\pi r_{ij}) \, S_{ij} + Y(m_\pi r_{ij}) \, \boldsymbol{\sigma}_i \cdot \boldsymbol{\sigma}_j,$$

$$(5) \qquad Y(x) = \frac{e^{-x}}{x} \, \xi(r),$$

$$(6) \qquad T(x) = \left(\frac{3}{x^2} + \frac{3}{x} + 1 \right) Y(x) \, \xi(r),$$

$$(7) \qquad \xi(r) = (1 - e^{-c_\pi r^2}),$$

where $\boldsymbol{\tau}_i$, $\boldsymbol{\sigma}_i$ and S_{ij} are isospin, spin and tensor operators, respectively. In light nuclei, $\langle v_{ij}^\pi \rangle$ contributes $\sim 85\%$ of $\langle v_{ij} \rangle$.

The remaining isospin-conserving terms are

$$(8) \qquad v_{ij}^R = \sum_{p=1,14} v_p(r_{ij}) O_{ij}^p,$$

$$(9) \qquad O_{ij}^{p=1,14} = \left[1, \boldsymbol{\sigma}_i \cdot \boldsymbol{\sigma}_j, S_{ij}, \mathbf{L} \cdot \mathbf{S}, \mathbf{L}^2, \mathbf{L}^2 \boldsymbol{\sigma}_i \cdot \boldsymbol{\sigma}_j, (\mathbf{L} \cdot \mathbf{S})^2\right] \otimes [1, \boldsymbol{\tau}_i \cdot \boldsymbol{\tau}_j],$$

where $v_p(r)$ has short-, intermediate-, and long-range components. The long-range components are just the $Y(r)$ and $T(r)$ of the one-pion potential and are present only for those operators that have contributions from one-pion exchange. The intermediate-range components are proportional to $T^2(r)$ and the short-range component is of the Woods-Saxon form.

Finally, v_{ij}^{CIB} is the strong charge independence breaking part of the potential and consists of four operators:

$$(10) \qquad O_{ij}^{p=15,18} = [1, (\boldsymbol{\sigma}_i \cdot \boldsymbol{\sigma}_j), S_{ij}] \otimes T_{ij}, (\tau_{zi} + \tau_{zj}).$$

The long-range part of $O^{p=15,17}$ comes from one-pion exchange by inserting $m_{\pi+-}$ or m_{π^0} in eqs. (3) and (4) and using $f_{\pi NN}^2 \propto m_\pi$.

The parameters in the short- and intermediate-range components were determined by making a direct fit to the 1993 Nijmegen data base [6, 7] containing 1787 pp and 2514 np data in the range 0–350 MeV, the nn scattering length, and deuteron binding energy. The fit of approximately 40 parameters results in a $\chi^2/\text{d.o.f.}$ of 1.09, which is typical of 1990's NN potentials.

2`2. *Illinois V_{ijk}*. – The three-nucleon potential used for most of the examples presented here is the Illinois-2 [8]. It consists of two- and three-pion terms and a simple phenomenological repulsive term:

$$(11) \qquad V_{ijk} = V_{ijk}^{2\pi} + V_{ijk}^{3\pi} + V_{ijk}^R.$$

The two-pion term, illustrated in fig. 1, contains P- and S-wave πN-scattering terms:

$$(12) \qquad V_{ijk}^{2\pi} = V_{ijk}^{2\pi,P} + V_{ijk}^{2\pi,S}.$$

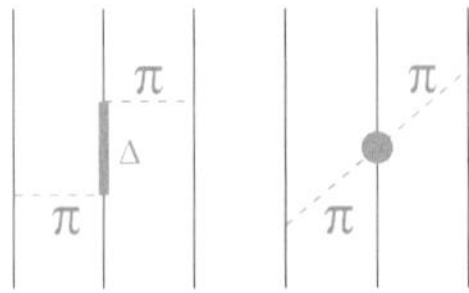

Fig. 1. – Two-pion exchange terms in the Illinois NNN potentials.

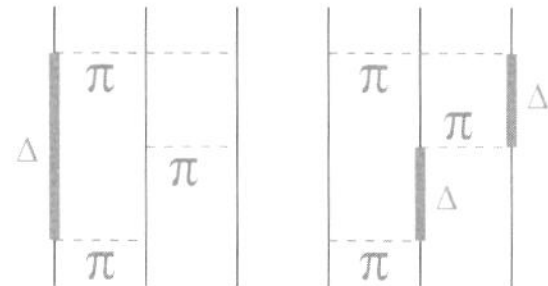

Fig. 2. – Three-pion ring terms in the Illinois NNN potentials.

The P-wave term (left panel of fig. 1) is the well-known Fujita-Miyazawa [9, 10] term which is present in all realistic NNN potentials. It has the form

$$(13) \qquad V_{ijk}^{2\pi,P} = A_{2\pi,P} \sum_{\text{cyclic}} \{X_{ij}, X_{jk}\}\{\boldsymbol{\tau}_i \cdot \boldsymbol{\tau}_j, \boldsymbol{\tau}_j \cdot \boldsymbol{\tau}_k\} + \frac{1}{4}[X_{ij}, X_{jk}][\boldsymbol{\tau}_i \cdot \boldsymbol{\tau}_j, \boldsymbol{\tau}_j \cdot \boldsymbol{\tau}_k],$$

where X_{ij} is defined in eq. (4). This is the longest-ranged nuclear NNN potential and is attractive in all nuclei and nuclear matter. However it is very small or even slightly repulsive in pure neutron systems.

The second panel of fig. 1 represents the S-wave part of $V_{ijk}^{2\pi}$. This term was introduced in the Tuscon-Melbourne NNN potential [11] and is required by chiral perturbation theory. However, in practice it is only 3–4% of $V_{ijk}^{2\pi,P}$ in light nuclei.

The three-pion term (fig. 2) was introduced in the Illinois potentials. It consists of the subset of three-pion rings that contain only one Δ mass in the energy denominators. Even so it has a quite complicated form which is given in ref. [8]. An important aspect of this structure is that there is a significant attractive term which acts only in $T = 3/2$ triples. In most light nuclei $\langle V_{ijk}^{3\pi}\rangle \lesssim 0.1\langle V_{ijk}^{2\pi}\rangle$.

The final term in the NNN potential, V_{ijk}^{R}, represents all other diagrams including relativistic effects. It is strictly phenomenological and purely central and repulsive:

$$(14) \qquad V_{ijk}^{R} = A_R \sum_{\text{cyclic}} T^2(m_\pi r_{ij}) T^2(m_\pi r_{jk}).$$

This repulsive term is principally needed to make nuclear matter saturate at the proper density instead of a too-high density and to obtain a hard enough equation of state for neutron matter.

The coupling constants $A_{2\pi,P}$, $A_{3\pi}$, and A_R were adjusted to fit 17 nuclear levels for $A \leq 8$. The $V_{ijk}^{2\pi,S}$ is too weak to be determined by fitting and its coupling was left at the value predicted by chiral perturbation theory.

In light nuclei we find

$$(15) \qquad \langle V_{ijk}\rangle \sim (0.02 \text{ to } 0.09)\langle v_{ij}\rangle \sim (0.15 \text{ to } 0.6)\langle H\rangle,$$

where the large fraction of $\langle H\rangle$ is due to a large cancellation of K and v_{ij}. From this

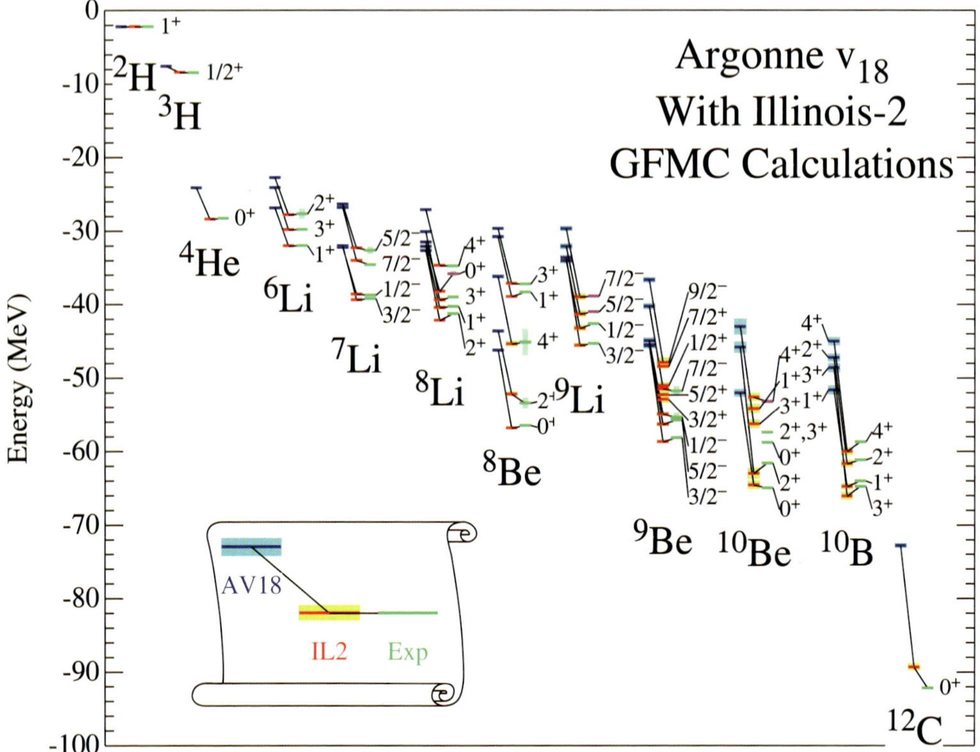

Fig. 3. – GFMC computations of energies for the AV18 and AV18+IL2 Hamiltonians compared with experiment.

we expect

$$\langle V_{4N} \rangle \sim 0.06 \langle V_{ijk} \rangle \ \sim (0.02 \text{ to } 0.04) \langle H \rangle \sim (0.5 \text{ to } 2.) \text{ MeV}. \tag{16}$$

This is comparable to the accuracy of our calculations. Even if more accurate calculations could be made, it would probably not be possible to disentangle four-nucleon potential effects from uncertainties in the fitted parameters of V_{ijk}.

2′3. *What makes nuclear structure?* – We have defined a very complicated nuclear Hamiltonian and it is reasonable to ask if it is all necessary to reproduce the structure of light nuclei. A study [12] was made of this in which features of the nuclear Hamiltonian were systematically removed and the effects on nuclear level energies investigated. For each simplification of the two-nucleon part of H, the remaining terms were readjusted to continue reproducing as many low partial-wave phase shifts, and the deuteron, as possible.

Figure 3 shows the energies of various nuclear states. For each isotope there are three sets of energies: the right-most are the experimental values, the left-most are the results

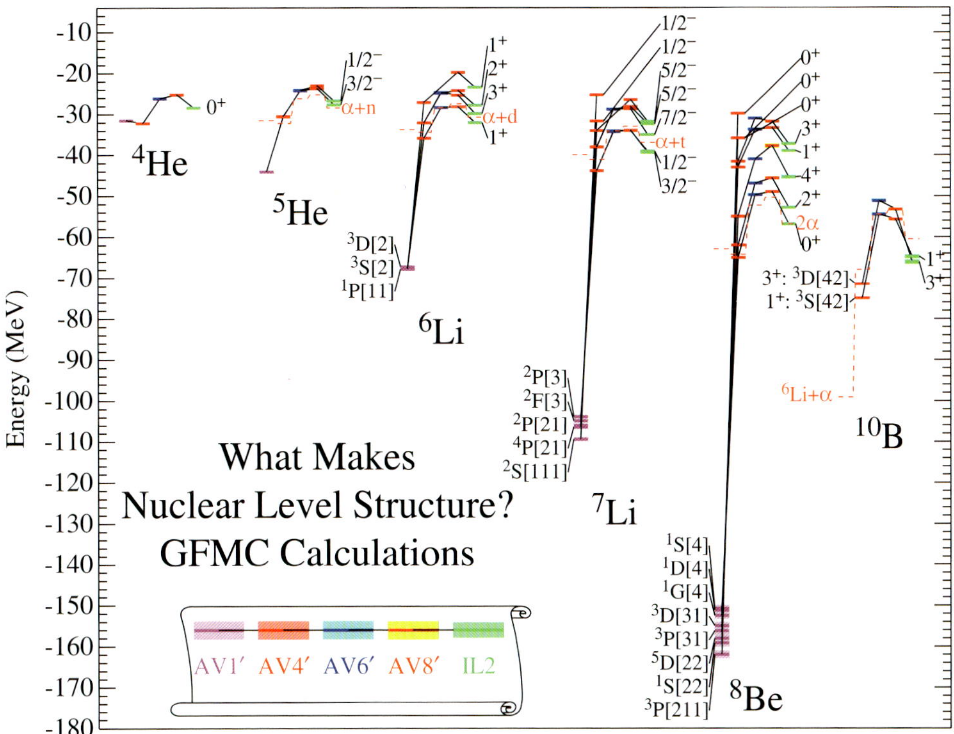

Fig. 4. – Nuclear energy levels for various simplifications of the Hamiltonian.

of the GFMC calculations to be described using just the NN potential AV18, and the middle ones are GFMC calculations using the AV18+IL2 Hamiltonian. The AV18+IL2 results are generally in good agreement with the data; the r.m.s. deviation is $\sim 0.75\,\mathrm{MeV}$. However without the IL2 NNN potential the comparison to data gets steadily worse as the number of nucleons increases. This is a general result that has also been obtained by others using different many-body methods and different NN potentials.

Figure 4 shows the effects of making further simplifications to H beyond removing the NNN potential. Here the right-most results are again for the full AV18+IL2 Hamiltonian and thus are close to the data. The next set of results to left are for the AV8′ NN potential [13] with no NNN force. With eight operators $([1, \boldsymbol{\sigma}_i \cdot \boldsymbol{\sigma}_j, S_{ij}, \mathbf{L} \cdot \mathbf{S}] \otimes [1, \boldsymbol{\tau}_i \cdot \boldsymbol{\tau}_j])$, AV8′ can reproduce AV18 results for eight partial waves; these are chosen to be $^1S_0$, $^3S_1$, $^3D_1$, ϵ_1, $^1P_1$, and $^3P_{0,1,2}$ (ϵ_1 is the $^3S_1$-$^3D_1$ mixing angle). (Strictly speaking, the $^3P_2$ potential of AV18 is reproduced but, because the $^3F_2$ and ϵ_2 potentials are different from those of AV18, the $^3P_2$ phase shifts are not reproduced.) This potential is more attractive in nuclei than AV18 and more than makes up for the lost binding due to the removal of the NNN potential. In general it gives a good qualitative picture of nuclear energies.

To the left of the AV8′ results are results for AV6′ which does not have $L \cdot S$ terms.

Not surprisingly, these have only negligible spin-orbit splittings. In addition ^{6,7}Li are essentially unbound to break up into $\alpha + $d or $\alpha + $t (the dashed lines show the indicated thresholds for each Hamiltonian). The next simplification is AV4$'$ which contains no tensor force. With this force, the deuteron no longer has a D state but still has the correct binding energy. The $^{1}S_0$, $^{3}S_1$, $^{1}P_1$, and an average $^{3}P_J$ partial waves of AV18 are reproduced. This simplified force results in spurious degeneracies of nuclear levels and somewhat overbinds all the shown nuclei. In particular ^{8}Be is bound against breakup into two alpha particles – an important failure because bound ^{8}Be would result in very different stellar nucleosynthesis.

Finally the left-most results are for AV1$'$, a pure central force that is an average of the $^{1}S_0$ and $^{3}S_1$ potentials of AV18. This produces an "upside-down" spectrum in which states with the least spatial symmetry are most bound. More importantly, there is no nuclear saturation; each increase in A results in much more binding and there are no $A = 5, 8$ mass gaps which are essential to big-bang nucleosynthesis and stellar evolution.

Reference [12] shows results for several other Hamiltonians including an AV2$'$ that contains only central and space-exchange terms and thus is very similar to the popular Volkov potentials [14]. Besides erroneously binding the dineutron, this potential has the strange feature of binding ^{6}He but not ^{6}Li so that $A = 6$ beta decay would be in the wrong direction. The conclusion of this study is that one needs almost the full, complicated, Hamiltonian to do realistic nuclear physics.

3. – Quantum Monte Carlo methods

The many-body problem with the full Hamiltonian described above is very difficult as is indicated by the slow progress over the last half-century that is outlined in the introduction. We need to solve

$$(17) \qquad \mathcal{H}\,\Psi\left(\vec{r}_1, \vec{r}_2, \cdots, \vec{r}_A; s_1, s_2, \cdots, s_A; t_1, t_2, \cdots, t_A\right)$$
$$= E\,\Psi\left(\vec{r}_1, \vec{r}_2, \cdots, \vec{r}_A; s_1, s_2, \cdots, s_A; t_1, t_2, \cdots, t_A\right),$$

where $s_i = \pm\frac{1}{2}$ are nucleon spins, and $t_i = \pm\frac{1}{2}$ are nucleon isospins (proton or neutron). Thus we need to solve the equivalent of $2^A \times \binom{A}{Z}$ complex coupled second-order equations in $3A - 3$ variables (the number of isospin states can be reduced; see subsect. 4'2). For ^{12}C this corresponds to 270336 coupled equations in 33 variables.

Furthermore, the coupling is strong; the expectation value of the tensor component of v^{π} (eq. (3)) is approximately 60% of the total $\langle v_{ij} \rangle$ but it is identically zero if there are no tensor correlations. Thus we cannot perturbatively introduce the couplings.

We use two successive quantum Monte Carlo (QMC) methods to solve this problem. The first is variational Monte Carlo (VMC) in which a trial wave function, containing variational parameters, is posited and the expectation value of the Hamiltonian computed using Monte Carlo integration. In practice we have not been able to formulate accurate enough trial wave functions and so the second step, Green's function Monte Carlo (GFMC), is needed to iteratively project the exact eigenfunction out of the trial wave function. These two methods are described in the following sections. School and review articles on these methods are refs. [15-17]. Detailed descriptions may be found in refs. [13, 18-21].

4. – Variational Monte Carlo

In VMC we start with a trial wave function, Ψ_T, which contains a number of variational parameters. We vary these parameters to minimize the expectation value of H,

$$(18) \qquad E_T = \frac{\langle \Psi_T | H | \Psi_T \rangle}{\langle \Psi_T | \Psi_T \rangle} \geq E_0.$$

As indicated, the resulting E_T is, by the Raleigh-Ritz variational principle, greater than the true ground-state energy for the quantum numbers (J^π, J_z, T, and T_z) of Ψ_T. A simplified form of our trial wave functions is

$$(19) \qquad |\Psi_T\rangle = \left[\mathcal{S} \prod_{i<j} (1 + U_{ij} + \Sigma_k U_{ijk}) \right] \prod_{i<j} f_c(r_{ij}) |\Phi\rangle.$$

Here $f_c(r)$ is a central (mostly short-ranged repulsion) correlation, U_{ij} are non-commuting two-body correlations induced by v_{ij}, and U_{ijk} is a simplified three-body correlation from V_{ijk}.

More specifically,

$$(20) \qquad U_{ij} = \sum_{p=2,6} u_p(r_{ij}) O^p_{ij},$$

contains $\boldsymbol{\tau}_i \cdot \boldsymbol{\tau}_j$, $\boldsymbol{\sigma}_i \cdot \boldsymbol{\sigma}_j$, $\boldsymbol{\sigma}_i \cdot \boldsymbol{\sigma}_j \, \boldsymbol{\tau}_i \cdot \boldsymbol{\tau}_j$, S_{ij}, and $S_{ij} \, \boldsymbol{\tau}_i \cdot \boldsymbol{\tau}$ operators, of which the $S_{ij} \, \boldsymbol{\tau}_i \cdot \boldsymbol{\tau}$ is most important due to the already noted strong tensor contribution from v^π. The $f_c(r)$ and $u_p(r)$ are solutions of coupled differential equations with v_{ij} as input [18].

The Φ (see below) is fully antisymmetric; hence the rest of eq. (19) must be symmetric. But the U_{ij} do not commute; for example

$$(21) \qquad [\boldsymbol{\sigma}_1 \cdot \boldsymbol{\sigma}_2 , \boldsymbol{\sigma}_1 \cdot \boldsymbol{\sigma}_3] = 2i \, \boldsymbol{\sigma}_1 \cdot (\boldsymbol{\sigma}_2 \times \boldsymbol{\sigma}_3).$$

The symmetrizer $\mathcal{S}$ fixes this by summing over all $[\frac{A(A-1)}{2}]!$ permutations of the ordering in $\prod_{i<j}$. In practice this is done by using just one Monte Carlo chosen ordering per wave function evaluation.

4·1. *The one-body part of* Ψ_T, Φ. – The one-body part of Ψ_T, Φ, is a $1\hbar\omega$ shell-model wave function. It determines the quantum numbers of the state being computed and is fully antisymmetric. For ^{3}H and ^{3,4}He, Φ can be antisymmetrized in just spin-isospin space, for example

$$(22) \quad \left| \Phi \left({}^3H, M_J = \frac{1}{2} \right) \right\rangle = \frac{1}{\sqrt{6}} \left(|p \uparrow n \uparrow n \downarrow\rangle - |p \uparrow n \downarrow n \uparrow\rangle + |n \downarrow p \uparrow n \uparrow\rangle \right.$$
$$\left. - |n \uparrow p \uparrow n \downarrow\rangle + |n \uparrow n \downarrow p \uparrow\rangle - |n \downarrow n \uparrow p \uparrow\rangle \right).$$

For $A > 4$ we need P-wave radial wave functions in order to antisymmetrize Φ; the antisymmetrization is achieved by summing over all partitions of the A nucleons into four S-shell nucleons (the α core) and A-4 P-shell nucleons which are antisymmetrically coupled to J^π and T. To make Φ translationally invariant, we express all functions of single-particle positions as functions of position relative to the center of mass of the A nucleons or of some sub-cluster of them. The one-body wave functions are solutions of Woods Saxon potentials containing several variational parameters. If desired the separation energy of these one-body wave functions can be fixed at the experimental value to guarantee that the Ψ_T has the correct asymptotic form. In general Φ has several spatial-symmetry components depending on how many ways a state of the desired quantum numbers can be constructed in the P-shell basis. For example the ${}^6\mathrm{Li}$ Φ have the form

$$(23) \quad |\Phi\rangle = \mathcal{A} \sum_{LS} \beta_{LS} \left| \Phi_6(LSJMTT_3)_{1234:56} \right\rangle,$$

$$(24) \quad \Phi_6(LSJMTT_3)_{1234:56} = \Phi_4(0000)_{1234} \phi_p^{LS}(R_{\alpha 5}) \phi_p^{LS}(R_{\alpha 6})$$
$$\times \left\{ \left[Y_{1m_l}(\Omega_{\alpha 5}) Y_{1m_l'}(\Omega_{\alpha 6}) \right]_{LM_L} \times \left[\chi_5 \left(\frac{1}{2} m_s \right) \chi_6 \left(\frac{1}{2} m_s' \right) \right]_{SM_S} \right\}_{JM}$$
$$\times \left[\nu_5 \left(\frac{1}{2} t_3 \right) \nu_6 \left(\frac{1}{2} t_3' \right) \right]_{TT_3}.$$

A $1\hbar\omega$ LS-basis diagonalization determines the β_{LS}.

$4\,{}^\cdot2.$ *Representing* Ψ_T *in the computer.* – The wave function, $\Psi_T(\vec{r}_1, \vec{r}_2, \cdots, \vec{r}_A)$, is a complex vector in spin-isospin space with dimension $[N_S$ components for spin$] \times [N_T$ components for isospin$]$. The number of spin states is 2^A. However for even A, if we choose to use $M_J = 0$, we can calculate and retain only half the spin vector, say that part with positive spin for the last nucleon, and obtain the other half of the vector by time-reversal symmetry. The number of isospin states, N_T, depends on the isospin basis being used:

$$(25) \quad N_T = \binom{A}{Z} \qquad \text{for a proton-neutron basis,}$$

$$(26) \quad = \frac{2T+1}{A/2+T+1} \binom{A}{A/2+T} \qquad \text{for a good isospin basis.}$$

Potentials (v_{ij}, V_{ijk}) and correlations (u_{ij}, U_{ijk}) involve repeated operations on Ψ. For example $\boldsymbol{\sigma}_i \cdot \boldsymbol{\sigma}_j$ may be written as

$$\boldsymbol{\sigma}_i \cdot \boldsymbol{\sigma}_j = 2(\sigma_i^+ \sigma_j^- + \sigma_i^- \sigma_j^+) + \sigma_i^z \sigma_j^z \tag{27}$$

$$= 2P_{ij}^\sigma - 1 \tag{28}$$

$$= \begin{pmatrix} 1 & 0 & 0 & 0 \\ 0 & -1 & 2 & 0 \\ 0 & 2 & -1 & 0 \\ 0 & 0 & 0 & 1 \end{pmatrix} \text{ acting on } \begin{pmatrix} \uparrow\uparrow \\ \uparrow\downarrow \\ \downarrow\uparrow \\ \downarrow\downarrow \end{pmatrix}. \tag{29}$$

Here P_{ij}^σ exchanges the spin of i and j. Consider the spin part of an $A = 3$ wave function; $\boldsymbol{\sigma}_i \cdot \boldsymbol{\sigma}_j$ will not mix different isospin components and, for different i and j, will separately act on different, non-contiguous, 4-element blocks of Ψ:

$$\Psi = \begin{pmatrix} a_{\uparrow\uparrow\uparrow} \\ a_{\uparrow\uparrow\downarrow} \\ a_{\uparrow\downarrow\uparrow} \\ a_{\uparrow\downarrow\downarrow} \\ a_{\downarrow\uparrow\uparrow} \\ a_{\downarrow\uparrow\downarrow} \\ a_{\downarrow\downarrow\uparrow} \\ a_{\downarrow\downarrow\downarrow} \end{pmatrix} ; \qquad \boldsymbol{\sigma}_1 \cdot \boldsymbol{\sigma}_2 \Psi = \begin{pmatrix} a_{\uparrow\uparrow\uparrow} \\ a_{\uparrow\uparrow\downarrow} \\ 2a_{\downarrow\uparrow\uparrow} - a_{\uparrow\downarrow\uparrow} \\ 2a_{\downarrow\uparrow\downarrow} - a_{\uparrow\downarrow\downarrow} \\ 2a_{\uparrow\downarrow\uparrow} - a_{\downarrow\uparrow\uparrow} \\ 2a_{\uparrow\downarrow\downarrow} - a_{\downarrow\uparrow\downarrow} \\ a_{\downarrow\downarrow\uparrow} \\ a_{\downarrow\downarrow\downarrow} \end{pmatrix} ; \tag{30}$$

$$\boldsymbol{\sigma}_2 \cdot \boldsymbol{\sigma}_3 \Psi = \begin{pmatrix} a_{\uparrow\uparrow\uparrow} \\ 2a_{\uparrow\downarrow\uparrow} - a_{\uparrow\uparrow\downarrow} \\ 2a_{\uparrow\uparrow\downarrow} - a_{\uparrow\downarrow\uparrow} \\ a_{\uparrow\downarrow\downarrow} \\ a_{\downarrow\uparrow\uparrow} \\ 2a_{\downarrow\downarrow\uparrow} - a_{\downarrow\uparrow\downarrow} \\ 2a_{\downarrow\uparrow\downarrow} - a_{\downarrow\downarrow\uparrow} \\ a_{\downarrow\downarrow\downarrow} \end{pmatrix} ; \qquad \boldsymbol{\sigma}_3 \cdot \boldsymbol{\sigma}_1 \Psi = \begin{pmatrix} a_{\uparrow\uparrow\uparrow} \\ 2a_{\downarrow\uparrow\uparrow} - a_{\uparrow\uparrow\downarrow} \\ a_{\uparrow\downarrow\uparrow} \\ 2a_{\downarrow\downarrow\uparrow} - a_{\uparrow\downarrow\downarrow} \\ 2a_{\uparrow\uparrow\downarrow} - a_{\downarrow\uparrow\uparrow} \\ a_{\downarrow\uparrow\downarrow} \\ 2a_{\uparrow\downarrow\downarrow} - a_{\downarrow\downarrow\uparrow} \\ a_{\downarrow\downarrow\downarrow} \end{pmatrix} . \tag{31}$$

Similarly, the tensor operator is

$$S_{ij} = 3 \, \boldsymbol{\sigma}_i \cdot \hat{r}_{ij} \, \boldsymbol{\sigma}_j \cdot \hat{r}_{ij} - \boldsymbol{\sigma}_i \cdot \boldsymbol{\sigma}_j \tag{32}$$

$$= 3 \begin{pmatrix} z^2 - 1/3 & z(x - iy) & z(x - iy) & (x - iy)^2 \\ z(x + iy) & -z^2 - 1/3 & x^2 + y^2 - 2/3 & -z(x - iy) \\ z(x_i y) & x^2 + y^2 - 2/3 & -z^2 + 1/3 & -z(x - iy) \\ (x + iy)^2 & -z(x + iy) & -z(x + iy) & z^2 - 1/3 \end{pmatrix} , \tag{33}$$

where $x = x_i - x_j$, etc. As shown in eqs. (30) and (31), these 4×4 matrices form a sparse matrix of (non-contiguous) 4×4 blocks in the A-body problem. Specially coded subroutines are used to efficiently perform these operations.

TABLE I. – *Scaling of wave function computation time.*

	A	Pairs	Spin $\times$ Isospin	$\prod(/^8\mathrm{Be})$
^{4}He	4	6	8×2	0.001
^{5}He	5	10	32×5	0.020
^{6}Li	6	15	32×5	0.036
^{7}Li	7	21	128×14	0.66
^{8}Be	8	28	128×14	1.
^{8}Li	8	28	128×28	2.
^{9}Be	9	36	512×42	18.
^{10}B	10	45	512×42	24.
^{10}Be	10	45	512×90	51.
^{11}B	11	55	2048×132	400.
^{12}C	12	66	2048×132	530.
^{16}O	16	120	32768×1430	224,000.
^{40}Ca	40	780	$3.6 \times 10^{21} \times 6.6 \times 10^9$	2.8×10^{20}
^{8}n	8	28	128×1	0.071
^{14}n	14	91	8192×1	26.

Most of the time in VMC or GFMC calculations is spent evaluating wave functions (or in GFMC making a propagation step which is equivalent). The pair operators dominate this time. The evaluation of a kinetic energy involves numerical second derivatives which require $6A$ wave function computations. Hence the product of A, the number of pairs, and of the length of the spin-isospin vector is a good indication of how the total computational time scales with A. Table I shows this scaling for various nuclei, assuming $M = 0$ for even J nuclei and that good-isospin bases are being used. The final column shows the product of the first three columns relative to ^{8}Be. We can do calculations up to $A = 10$ routinely and a few ^{12}C calculations have been done. It is clear that this approach is not reasonable for ^{16}O. The last two lines are for "neutron drops" for which isospin does not have to be considered. This allows somewhat bigger A to be reached.

4‘3. *A variational Monte Carlo calculation.* – The basic steps in a variational calculation are

– Generate a random position: $\mathbf{R} = \vec{r}_1, \vec{r}_2, \cdots, \vec{r}_A$.

– Make many (1000's) random steps based on the probability $P = |\Psi_T(\mathbf{R})|^2$.

– Start integration loop:

    – Make order 10 steps based on P.

    – Compute and sum $H_{\mathrm{local}}(\mathbf{R}) = [\Psi_T(\mathbf{R})^\dagger H \Psi_T(\mathbf{R})]/|\Psi_T(\mathbf{R})|^2$. Gradients and Laplacians are computed by differences: $6A$ evaluations of $\Psi_T(\mathbf{R} + \delta_j \vec{r}_i)$.

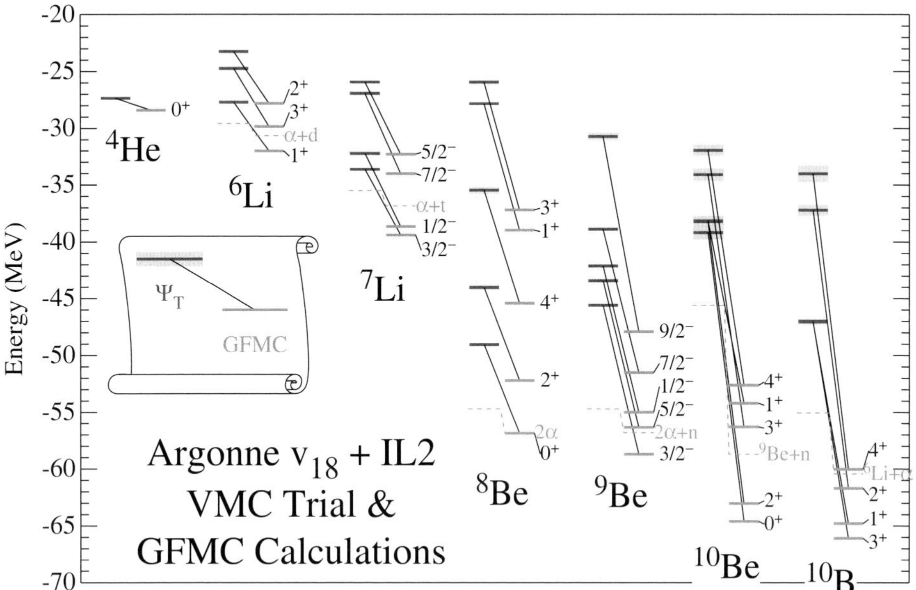

Fig. 5. – Comparison of VMC and GFMC energies for the AV18+IL2 Hamiltonian. The light shading shows Monte Carlo statistical errors.

- $\langle \Psi_T | H | \Psi_T \rangle / \langle \Psi_T | \Psi_T \rangle = \text{average}(H_{\text{local}})$.

A random step from a given position, $\mathbf{R}$, to a new position $\mathbf{R}'$ is made using the Metropolis method:

- Use $3A$ uniform random numbers on $(0,1)$, $\{w_j\}$, to make $\Delta\mathbf{R}$; $\Delta x_i = 2\delta r(w_j - 1)$.

- Set $\mathbf{R}' = \mathbf{R} + \Delta\mathbf{R}$, and compute $P(\Delta\mathbf{R}) = |\Psi_T(\mathbf{R}')|^2 / |\Psi_T(\mathbf{R})|^2$.

- Make another random number on $(0,1)$: p

- If $P > p$, the step is accepted; replace $\mathbf{R}$ with $\mathbf{R}'$.
  if $P < p$, the step is rejected; discard $\mathbf{R}'$ and stay at $\mathbf{R}$.

$4\dot{}4$. *Accuracy of VMC energies*. – Figure 5 compares VMC energies of various nuclear states with the corresponding GFMC values for the AV18+IL2 Hamiltonian. As is described in the next section, the GFMC results are believed to be accurate to 1–2%. For ^{4}He the VMC result is quite close to the GFMC. However as we move into the P shell, the VMC results get steadily worse. In fact, although the GFMC calculations show that this Hamiltonian binds the nuclei shown with the exception of ^{8}Be, the VMC energies are all above the VMC energies for the subclusters that the nuclei can break up into. Furthermore the ^{8}Be VMC energy is actually lower than those of 9- and 10-body nuclei. Calculations with simpler Hamiltonians show that these failures of the VMC energies are related to the tensor force; VMC calculations for the simple AV4$'$ potential discussed in subsect. **2**$\cdot$**3** are quite accurate, while those for AV6$'$ have significant errors.

5. – Green's function Monte Carlo—General description

As shown above, our VMC trial wave functions are not good enough for P-shell nuclei. This means that they contain admixtures of excited-state components in addition to the desired exact ground-state component, Ψ_0;

$$\Psi_T = \Psi_0 + \sum \alpha_i \Psi_i. \tag{34}$$

We use Green's function Monte Carlo to project Ψ_0 out of Ψ_T by propagating in imaginary time, τ:

$$\Psi(\tau) = \exp\left[-(H - \tilde{E}_0)\tau\right]\Psi_T, \tag{35}$$

$$= e^{-(E_0 - \tilde{E}_0)\tau} \times \left[\Psi_0 + \sum \alpha_i e^{-(E_i - E_0)\tau}\Psi_i\right], \tag{36}$$

$$\lim_{\tau \to \infty} \Psi(\tau) \propto \Psi_0, \tag{37}$$

where $\tilde{E}_0$ is a guess for the exact E_0.

The eigenvalue E_0 is calculated exactly while other expectation values are generally calculated neglecting terms of order $|\Psi_0 - \Psi_T|^2$ and higher. In contrast, the error in the variational energy, E_T, is of order $|\Psi_0 - \Psi_T|^2$, and other expectation values calculated with Ψ_T have errors of order $|\Psi_0 - \Psi_T|$.

The evaluation of $\Psi(\tau)$ is made by introducing a small time step, $\Delta\tau$, $\tau = n\Delta\tau$,

$$\Psi(\tau) = \left[e^{-(H - E_0)\Delta\tau}\right]^n \Psi_T = G^n \Psi_T, \tag{38}$$

where G is the short-time Green's function. The $\Psi(\tau)$ is represented by a vector function of $\mathbf{R}$, and the Green's function, $G_{\alpha\beta}(\mathbf{R}', \mathbf{R})$ is a matrix function of $\mathbf{R}'$ and $\mathbf{R}$ in spin-isospin space, defined as

$$G_{\alpha\beta}(\mathbf{R}', \mathbf{R}) = \left\langle \mathbf{R}', \alpha \left| e^{-(H - E_0)\Delta\tau} \right| \mathbf{R}, \beta \right\rangle. \tag{39}$$

It is calculated with leading errors of order $(\Delta\tau)^3$ as discussed below. Omitting spin-isospin indices for brevity, $\Psi(\mathbf{R}_n, \tau)$ is given by

$$\Psi(\mathbf{R}_n, \tau) = \int G(\mathbf{R}_n, \mathbf{R}_{n-1}) \cdots G(\mathbf{R}_1, \mathbf{R}_0)\Psi_T(\mathbf{R}_0)\mathrm{d}\mathbf{P}, \tag{40}$$

and

$$E(\tau) = \frac{\int \Psi_T^\dagger(\mathbf{R}_n)\, G^\dagger(\mathbf{R}_n, \mathbf{R}_{n-1}) \cdots G^\dagger(\mathbf{R}_1, \mathbf{R}_0)\, H\, \Psi_T(\mathbf{R}_0)\mathrm{d}\mathbf{P}}{\int \Psi_T^\dagger(\mathbf{R}_n)\, G^\dagger(\mathbf{R}_n, \mathbf{R}_{n-1}) \cdots G^\dagger(\mathbf{R}_1, \mathbf{R}_0)\, \Psi_T(\mathbf{R}_0)\mathrm{d}\mathbf{P}}, \tag{41}$$

where $d\mathbf{P} = d\mathbf{R}_0 d\mathbf{R}_1 \cdots d\mathbf{R}_n$. Here we have placed the $\Psi(\tau)$ to the left side of H because the derivatives in H may be evaluated only on Ψ_T; we cannot compute gradients or Laplacians of $\Psi(\tau)$. This $3An$-dimensional integral is computed by Monte Carlo.

5`1. *The short-time propagator.* – The success of a GFMC calculation depends on an accurate and fast evaluation of the short-time propagator, $G_{\alpha\beta}(\mathbf{R}',\mathbf{R})$. One wants to be able to do this for the largest possible value of $\Delta\tau$ to reduce the number of steps, n, needed to reach some asymptotic value of τ. The most important features of $\Psi(\tau)$ are induced by the *NN* potential, so consider first $G_{\alpha\beta}(\mathbf{R}',\mathbf{R})$ for a Hamiltonian with no *NNN* potential. This can be written as

$$(42) \qquad G_{\alpha\beta}(\mathbf{R}',\mathbf{R}) = e^{\tilde{E}_0\Delta\tau} G_0(\mathbf{R}',\mathbf{R}) \langle\alpha| \left[S \prod_{i<j} \frac{g_{ij}(\mathbf{r}'_{ij},\mathbf{r}_{ij})}{g_{0,ij}(\mathbf{r}'_{ij},\mathbf{r}_{ij})} \right] |\beta\rangle,$$

where

$$(43) \qquad G_0(\mathbf{R}',\mathbf{R}) = \langle\mathbf{R}'|e^{-K\Delta\tau}|\mathbf{R}\rangle = \left[\sqrt{\frac{m}{2\pi\hbar^2\Delta\tau}} \right]^{3A} \exp\left[\frac{-(\mathbf{R}'-\mathbf{R})^2}{2\hbar^2\Delta\tau/m} \right],$$

is the many-nucleon free propagator and $g_{0,ij}$ is the corresponding two-nucleon free propagator,

$$(44) \qquad g_{0,ij}(\mathbf{r}'_{ij},\mathbf{r}_{ij}) = \left[\sqrt{\frac{\mu}{2\pi\hbar^2\Delta\tau}} \right]^3 \exp\left[-\frac{(\mathbf{r}'_{ij}-\mathbf{r}_{ij})^2}{2\hbar^2\Delta\tau/\mu} \right],$$

and $\mu = m/2$ is the reduced mass.

The $G_0(\mathbf{R}',\mathbf{R})$ is included in the Monte Carlo integration (eq. (41)) by using it to make the step from $\mathbf{R}$ to $\mathbf{R}'$. The magnitudes of the $3A$ steps (x, y, and z for each nucleon) are determined by sampling a Gaussian of the width given in eq. (43) and the directions of the steps are picked by importance sampling; see ref. [13] for details.

Equation (42) introduces the exact two-body propagator,

$$(45) \qquad g_{ij}(\mathbf{r}'_{ij},\mathbf{r}_{ij}) = \langle\mathbf{r}'_{ij}|e^{-H_{ij}\Delta\tau}|\mathbf{r}_{ij}\rangle,$$

$$(46) \qquad H_{ij} = -\frac{\hbar^2}{m}\nabla^2_{ij} + v_{ij}.$$

All terms containing any number of the same v_{ij} and K are treated exactly in this propagator, as we have included the imaginary-time equivalent of the full two-body scattering amplitude. Equation (42) still has errors of order $(\Delta\tau)^3$, however they are from commutators of terms like $v_{ij}Kv_{ik}(\Delta\tau)^3$ which become large only when both pairs ij and ik are close.

To calculate g_{ij}, we use the techniques developed by Schmidt and Lee [22] for scalar interactions. These allow g_{ij} to be calculated with high (~ 10 digit) accuracy. However, this calculation is quite time consuming. Therefore, prior to the GFMC calculation, we compute and store the the propagator on a grid. For a spin-independent interaction, the propagator g_{ij} would depend only upon the two magnitudes r' and r and the angle $\cos(\theta) = \hat{\mathbf{r}}' \cdot \hat{\mathbf{r}}$ between them. Here, though, there is also a dependence upon the spin quantization axis. Rotational symmetry allows one to calculate the spin-isospin components of $g_{ij}(\mathbf{r}', \mathbf{r})$ for any $\mathbf{r}'$ and $\mathbf{r}$ by simple $SU3$ spin rotations and values of g_{ij} on a grid of initial points $\mathbf{r} = (0, 0, z)$ and final points $\mathbf{r}' = (x', 0, z')$. In addition, the fact that the propagator is Hermitian allows us to store only the values for $z > z'$.

Returning to the full Hamiltonian including NNN forces, the complete propagator is given by

$$(47) \qquad G_{\alpha\beta}(\mathbf{R}', \mathbf{R}) = e^{E_0 \Delta\tau} G_0(\mathbf{R}', \mathbf{R}) \exp\left[-\sum \left(V_{ijk}^R(\mathbf{R}') + V_{ijk}^R(\mathbf{R}) \right) \frac{\Delta\tau}{2} \right]$$

$$\times \langle \alpha | I_3(\mathbf{R}') | \gamma \rangle \langle \gamma | \left[\mathcal{S} \prod_{i<j} \frac{g_{ij}(\mathbf{r}'_{ij}, \mathbf{r}_{ij})}{g_{0,ij}(\mathbf{r}'_{ij}, \mathbf{r}_{ij})} \right] | \delta \rangle \langle \delta | I_3(\mathbf{R}) | \beta \rangle,$$

with

$$(48) \qquad I_3(\mathbf{R}) = \left[1 - \frac{\Delta\tau}{2} \sum V_{ijk}^{2\pi}(\mathbf{R}) \right].$$

The exponential of $V_{ijk}^{2\pi}$ is expanded to first order in $\Delta\tau$ thus, there are additional error terms of the form $V_{ijk}^{2\pi} V_{i'j'k'}^{2\pi} (\Delta\tau)^2$. However, they have negligible effect since $V_{ijk}^{2\pi}$ has a magnitude of only a few MeV. It was verified that the results for ^{4}He do not show any change, outside of statistical errors, when $\Delta\tau$ is decreased from $0.5\,\mathrm{GeV}^{-1}$.

5'**2.** *Problems with nuclear GFMC.* – While GFMC is in principal exact for the $\langle H \rangle$, there are several practical difficulties that make it only approximate. We have made many tests of the accuracy of the GFMC energies, both by comparison to other methods and by comparing calculations with different $\Delta\tau$, starting Ψ_T, and other computational parameters. These tests show that our results for energy are good to $\sim 1\%$ up to $\sim 2\%$ for larger A or $N - Z$ (^{8}He is particularly difficult). Some of the problems are listed below.

5'**2.1.** Limitation on H. The exact propagator of eq. (46) can be computed for the full v_{18} potential, however the $\mathbf{L}^2$ and $(\mathbf{L} \cdot \mathbf{S})^2$ terms in the potential correspond to state-dependent changes of the mass appearing in the free Green's function. Since we do not know how to sample such a free Green's function, we cannot use the exact g_{ij} for the full potential, but rather must use one constructed for an approximately equivalent potential that does not contain quadratic $\mathbf{L}$ terms, namely the AV8$'$ introduced in subsect. **2**'**3**. The difference between the desired and approximate potentials is computed perturbatively. Comparisons with more accurate calculations for ^{3}H and ^{4}He suggest that this introduces errors of less than 1%.

5˙2.2. Fermion sign problem. The $G(\mathbf{R}_i, \mathbf{R}_{i-1})$ is a local operator and can mix in the boson solution. This has a (much) lower energy than the fermion solution and thus is exponentially amplified in subsequent propagations. In the final integration with the antisymmetric Ψ_T, the desired fermionic part is projected out in eq. (41), but in the presence of large statistical errors that grow exponentially with τ. Because the number of pairs that can be exchanged grows with A, the sign problem also grows exponentially with increasing A. For $A \geq 8$, the errors grow so fast that convergence in τ cannot be achieved.

For simple scalar wave functions, the fermion sign problem can be controlled by not allowing the propagation to move across a node of the wave function. Such "fixed-node" GFMC provides an approximate solution which is the best possible variational wave function with the same nodal structure as Ψ_T. However, a more complicated solution is necessary for the spin- and isospin-dependent wave functions of nuclei. This is provided by "constrained-path" propagation in which those configurations that, in future generations, will contribute only noise to expectation values are discarded. If the exact ground state $|\Psi_0\rangle$ were known, any configuration at time step n for which

$$(49) \qquad \Psi(\mathbf{R}_n)^\dagger \Psi_0(\mathbf{R}_n) = 0,$$

where a sum over spin-isospin states is implied, could be discarded. The sum of these discarded configurations can be written as a state $|\Psi_d\rangle$, which obviously has zero overlap with the ground state. The Ψ_d contains only excited states and should decay away as $\tau \to \infty$, thus discarding it is justified. Of course the exact Ψ_0 is not known, and so configurations are discarded with a probability such that the average overlap with the trial wave function,

$$(50) \qquad \langle \Psi_d | \Psi_T \rangle = 0.$$

Many tests of this procedure have been made [19] and it usually gives results that are consistent with unconstrained propagation, within statistical errors. However a few cases in which the constrained propagation converges to the wrong energy (either above or below the correct energy) have been found. Therefore a small number, $n_u = 10$ to 20, of unconstrained steps are made before evaluating expectation values. These few unconstrained steps, out of typically 400 total steps, appear to be enough to damp out errors introduced by the constraint, but do not greatly increase the statistical error. Unfortunately, the constrained-path $E(\tau)$ are not upper bounds to the true E_0; examples have been found in which the constrained energies evaluated with inadequate n_u are below E_0.

5˙2.3. Mixed estimates extrapolation. As shown in eq. (41), GFMC computes "mixed" expectation values between Ψ_T and $\Psi(\tau)$ of operators,

$$(51) \qquad \langle O \rangle_{\text{Mixed}} = \frac{\langle \Psi(\tau) | O | \Psi_T \rangle}{\langle \Psi(\tau) | \Psi_T \rangle} \, .$$

TABLE II. – *Contributions to $\langle H \rangle$ for ^{6}Li (MeV).*

	$\langle O \rangle_T$	$\langle O \rangle_{\text{Mixed}}$	$\langle O \rangle_{\text{Mixed}} - \langle O \rangle_T$	$\langle O \rangle$
K	146.60	153.49	6.89	160.39
v_{nuc}	−171.64	−180.49	−8.85	−189.34
v_C	1.53	1.54	0.01	1.55
V_{ijk}	−4.10	−6.43	−2.34	−8.77
Sum	−27.61	−31.90	−4.28	−36.17
H	−27.61	−31.90	—	−31.90

The desired expectation values, of course, have $\Psi(\tau)$ on both sides. By writing $\Psi(\tau) = \Psi_T + \delta\Psi(\tau)$ and neglecting terms of order $[\delta\Psi(\tau)]^2$, we obtain the approximate expression

$$(52) \qquad \langle O(\tau) \rangle = \frac{\langle \Psi(\tau) | O | \Psi(\tau) \rangle}{\langle \Psi(\tau) | \Psi(\tau) \rangle} \approx \langle O(\tau) \rangle_{\text{Mixed}} + [\langle O(\tau) \rangle_{\text{Mixed}} - \langle O \rangle_T],$$

where $\langle O \rangle_T$ is the variational expectation value. More accurate evaluations of $\langle O(\tau) \rangle$ are possible, essentially by measuring the observable at the mid-point of the path. However, such estimates require a propagation twice as long as the mixed estimate and require separate propagations for every $\langle O \rangle$ to be evaluated.

The expectation value of the Hamiltonian is a special case. The $\langle H(\tau) \rangle_{\text{Mixed}}$ can be re-expressed as [23]

$$(53) \qquad \langle H(\tau) \rangle_{\text{Mixed}} = \frac{\langle \Psi_T | e^{-(H-E_0)\tau/2} H e^{-(H-E_0)\tau/2} | \Psi_T \rangle}{\langle \Psi_T | e^{-(H-E_0)\tau/2} e^{-(H-E_0)\tau/2} | \Psi_T \rangle} \geq E_0,$$

since the propagator $\exp[-(H-E_0)\tau]$ commutes with the Hamiltonian. Thus $\langle H(\tau) \rangle_{\text{Mixed}}$ is already the correct expectation value and must not be extrapolated. This results in the unfortunate circumstance that the sum of the pieces of $\langle H \rangle$ is not equal to the full GFMC value of $\langle H \rangle$. An example is shown in table II which shows VMC, mixed, and extrapolated energies for ^{6}Li computed with the AV18+IL2 Hamiltonian. The sum of the extrapolated kinetic and potential energy values is 4.3 MeV different from the total energy. This means that the extrapolated values of the pieces have errors whose absolute sum is at least this big.

Instead of the linear extrapolation of eq. (52), one can also use a ratio extrapolation:

$$(54) \qquad \langle O(\tau) \rangle \approx \frac{\langle O(\tau) \rangle_{\text{Mixed}}^2}{\langle O \rangle_T},$$

which is the same as eq. (52) to lowest order in $[\langle O(\tau) \rangle_{\text{Mixed}} - \langle O \rangle_T]$, but obviously has different quadratic errors. This method has the feature that if both $\langle O(\tau) \rangle_{\text{Mixed}}$ and $\langle O \rangle_T$ have the same sign, then the extrapolated $\langle O(\tau) \rangle$ will also have that sign. This is an advantage for quantities such as densities (see sect. **8**) which must be positive; if the GFMC is reducing the density at large r where it is exponentially falling, linear extrapolation can result in negative values. Of course such large extrapolations by either method are uncertain.

5`3. *A simplified GFMC calculation.* – The basic steps in a GFMC calculation are

– start with collection of $\Psi(\tau{=}0, \mathbf{R}_j) = \Psi_T(\mathbf{R}_j)$ from a VMC calculation,

– Loop over time steps $\tau_n = n\Delta\tau$:

  – loop over configurations j
  a) Make a random step to $\mathbf{R}'_j = \mathbf{R}_j + \Delta\mathbf{R}_j$ by sampling $G_0(\mathbf{R}', \mathbf{R})$.
  b) Sample several directions based on simplified Ψ_T and potential.
  c) Compute $\Psi(\tau_{n+1}, \mathbf{R}'_j) = G(\mathbf{R}'_j, \mathbf{R}_j)\Psi(\tau_n, \mathbf{R}_j)$.
  d) Possibly mark as killed due to the constraint $\Psi^\dagger(\tau_{n+1}, \mathbf{R}'_j) \cdot \Psi_T(\mathbf{R}'_j)$.
  e) Use importance sampling to kill or replicate the configuration $\Psi(\tau_{n+1}, \mathbf{R}'_j)$.

  – Every 20–40 time steps
  a) Compute the local energy $\Psi^\dagger(\tau_n, \mathbf{R}_j)H\Psi_T(\mathbf{R}_j)/\Psi^\dagger(\tau_n, \mathbf{R}_j)\Psi_T(\mathbf{R}_j)$.
  b) Check that total number of configurations is staying reasonably constant.

GFMC calculations are quite computer intensive. For example, a typical ^{8}Li calculation requires 300 processor hours running at a delivered (not theoretical-peak) speed of one GigaFLOPS. As shown in table I, a ^{10}B calculation will need about 10 times this and ^{12}C 250 times it. Clearly such calculations are practical only on highly parallel computers. Our GFMC program uses a master-slave structure in which each slave gets a number of configurations to propagate as outlined in the preceding paragraph. The computed energy results are sent back to master for averaging as they are generated.

Because configurations are multiplied or killed during propagation, the work load fluctuates. It is important to periodically rebalance the work load—otherwise slaves will wind up with nothing to do while the last slave with the biggest work load finishes its calculations. To do this, the master periodically collects load statistics and then tells slaves to redistribute some of their configurations. The slaves have work (energy calculations left from previous time steps) set aside to do during this synchronization. This method results in parallelization efficiencies of typically 95% on up to 2000 processors. However, the next generation of large computers will have 10000 to 100000 processors, which is more than the number of configurations to be propagated for a large nucleus like ^{12}C. Thus the program has to be made parallel at a finer level; this is being worked on.

5`4. *Examples of GFMC propagation.* – Figure 6 shows the $E(\tau)$ as a function of the imaginary time, τ, for the beginning of GFMC propagation for ^{4}He. The propagation starts from the VMC value of $-26.92\,\mathrm{MeV}$, and initially decreases rapidly with increasing

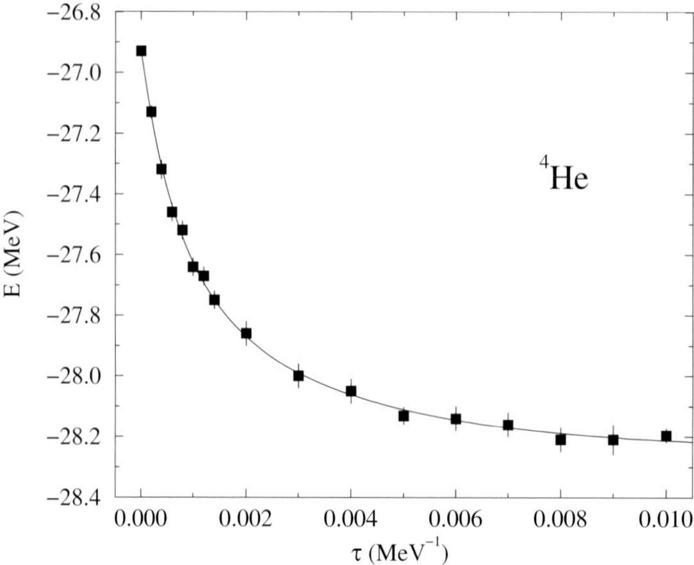

Fig. 6. – GFMC propagation for ^{4}He. The $E(\tau)$ is shown as a function of imaginary time, τ.

τ; essentially converged values are achieved by $\tau = 0.01\,\mathrm{MeV}^{-1}$. The propagation is continued another factor of 10 to $\tau = 0.1\,\mathrm{MeV}^{-1}$ and the results averaged over the last half of the propagation to get the converged result of $-28.300(15)\,\mathrm{MeV}$. The curve is a fit of the form

$$(55) \qquad E(\tau) = E_0 + \frac{\sum_i \alpha_i^2 E_i^\star \exp[-E_i^\star \tau]}{1 + \sum_i \alpha_i^2 \exp[-E_i^\star \tau]},$$

to the computed $E(\tau)$ using three terms. The $E_1^\star$ was fixed at the first 0^+ excitation energy of $20.2\,\mathrm{MeV}$ of ^{4}He, and the other two $E_i^\star$ and the three α_i were varied in the fit. The fitted $E_i^\star$ turn out to be very large, 340 and $1480\,\mathrm{MeV}$, with small α_i, 0.0018 and 0.00046, respectively. Thus the errors in the VMC Ψ_T correspond to small amounts of extremely high excitation energy; GFMC is particularly efficient at filtering out such errors.

Figure 7 shows GFMC propagation, using the AV18+IL2 Hamiltonian, of the ground, first 3^+, and 2^+ states of ^{6}Li. The propagation for the ground state (which is particle stable with this H) and the 3^+ (which is only slightly above the d + α threshold and experimentally has a narrow width) is stable after $\tau = 0.2\,\mathrm{MeV}^{-1}$. However the 2^+ (a broad resonance) never becomes stable; the $E(\tau)$ are decaying to the threshold energy of separated α and d clusters. Because the 3^+ state $E(\tau)$ stops decreasing around $\tau = 0.2$, the $E(\tau = 0.2)$, shown by the star, is best GFMC estimate we can currently make of the resonance energy. However it is now possible to make GFMC calculations using scattering-wave boundary conditions (see sect. **7**) and this method will be applied to states such as ^{6}Li(2^+).

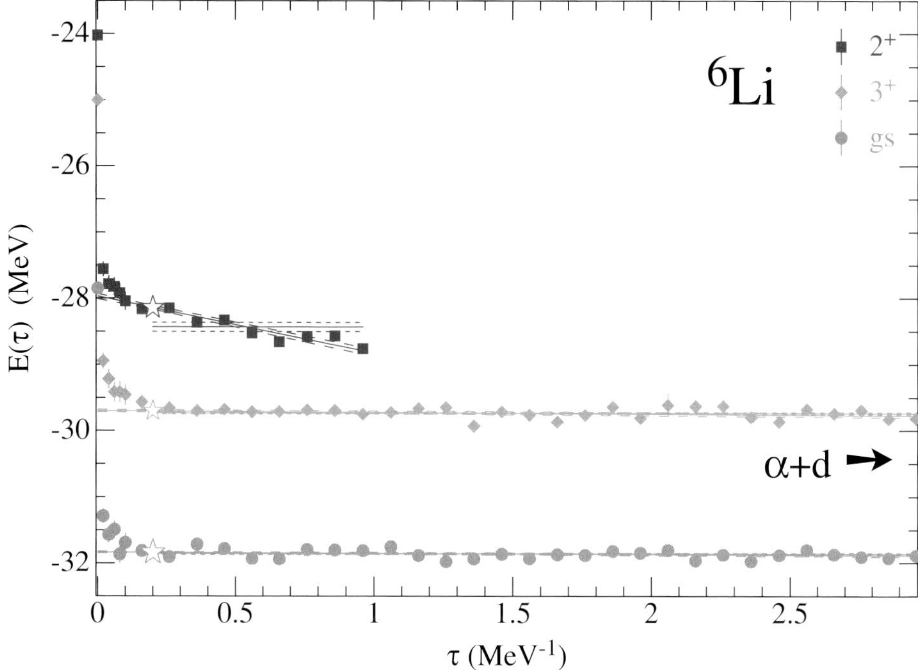

Fig. 7. – GFMC propagation for three states of ^{6}Li.

6. – Results for energies of nuclear states

Figure 3 compares the GFMC energies of various nuclear states with experiment and fig. 8 does the same for excitation energies. In both cases the left set of bars for each isotope shows results using just the AV18 NN potential while the middle set of bars is for the full AV18+IL2 Hamiltonian. As has already been observed, AV18 alone significantly underbinds all nuclei except the deuteron; including IL2 results in fairly good agreement with the experimental values. The excitation spectra in fig. 8 show that IL2 also fixes other problems that arise when just a NN potential is used. For example, spin-orbit splittings are usually too small without the NNN potential (note the $\frac{1}{2}^- - \frac{3}{2}^-$ and $\frac{5}{2}^- - \frac{7}{2}^-$ splittings in ^{7}Li and the $\frac{1}{2}^- - \frac{3}{2}^-$ splitting in ^{9}Li). As is discussed in the next subsection, even the ordering of states can be changed by the NNN potential.

The discussion of sect. **5** implies that GFMC can be used only for the lowest state of each set of quantum numbers but fig. 3 shows several states with the same J^π. The ability of GFMC to provide such results was demonstrated in ref. [21].

6`1. *Ordering of states in* ^{10}Be *and* ^{10}B. – Figure 9 shows the beginning of the computed and experimental excitation spectra of ^{10}Be and ^{10}B. We see that NN potentials with no NNN predict a 1^+ ground state for ^{10}B while the Illinois-2 NNN potential fixes this and gives the correct 3^+ ground state. No-core shell model calculations show that

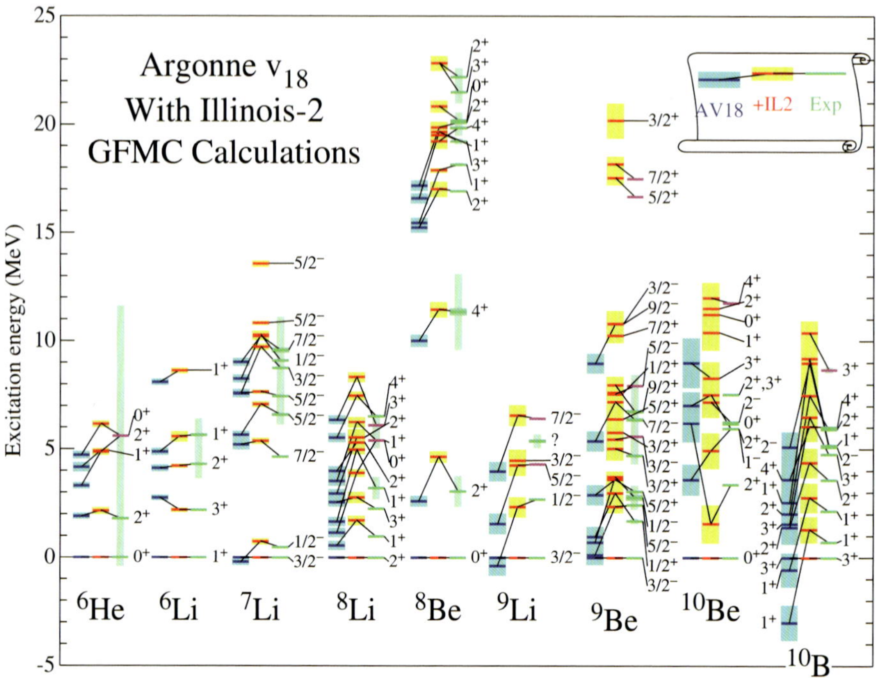

Fig. 8. – GFMC computations of excitation energies for the AV18 and AV18+IL2 Hamiltonians compared with experiment. The shading on the experimental energies shows the widths of resonances.

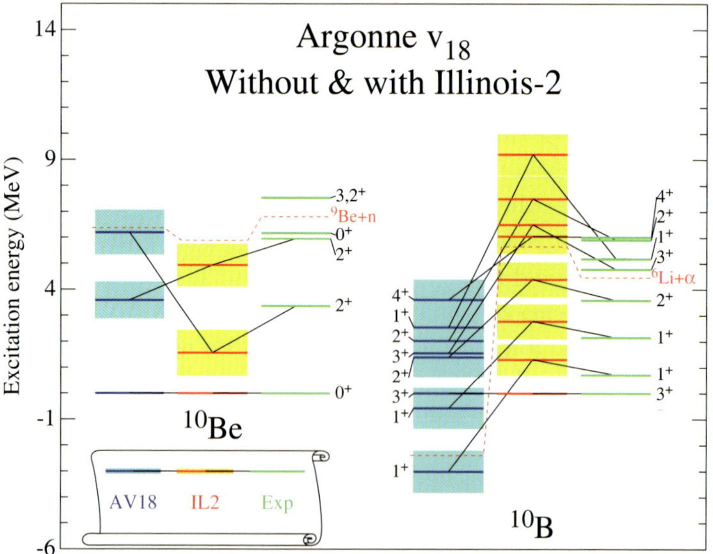

Fig. 9. – Excitation energies of ^{10}Be and ^{10}B.

other NN potentials without NNN potentials also give a 1^+ ground state for ^{10}B [24], so this is not a failure of just AV18. This incorrect ground-state prediction is another manifestation of too-small spin-orbit splitting using just NN potentials; in 1956 D. Kurath showed that the relative positions of the 3^+ and 1^+ levels depends on the amount of spin-orbit strength in a shell-model calculation [25].

The first two excited states in ^{10}Be are both 2^+ and fig. 9 shows that including the IL2 NNN potential reverses their order. VMC and GFMC calculations predict large positive and negative quadrupole moments (Q) for these states; with no NNN potential, the GFMC energy of the $Q > 0$ state is the lower, while adding the IL2 NNN potential reverses this. VMC also predicts a large $B(E2)$ to the g.s. for only the state with $Q > 0$. An ATLAS experiment for the $B(E2)$ and quadrupole moments of these states will be made to determine if the reversal of order given by IL2 is correct.

$6`2. Charge dependence and isospin mixing. –$ The differences of the energies of states in the same isomultiplet are a probe of isospin-breaking components of the Hamiltonian. The largest such component is the Coulomb potential between protons, but it has been known since 1969 that this does not fully account for the measured differences [26]; the discrepancy is referred to as the Nolen-Schiffer anomaly. It is convenient to parametrize the energies of the isomultiplets by the coefficients $a_{A,T}^{(n)}$,

$$(56) \qquad E_{A,T}(T_z) = \sum_{n \leq 2T} a_{A,T}^{(n)} Q_n(T, T_z),$$

where $Q_0 = 1$ is the isoscalar component, $Q_1 = T_z$ the isovector, and $Q_2 = \frac{1}{2}(3T_z^2 - T^2)$ the isotensor. Term 18 of AV18 contributes to the isovector component and terms 15 to 17 to the isotensor component; both terms receive contributions from the various electro-magnetic terms in AV18. Table III shows computed and experimental values of the coefficients for several isomultiplets; the $v^{\text{CSB+CD}}$ column gives contribution of the nuclear

TABLE III. – *Computed* (AV18 + IL2) *and experimental Nolen-Schiffer energies for several isomultiplets.*

	T	n	v^{Coul}	v^{otherEM}	$v^{\text{CSB+CD}}$	K^{CSB}	Total	Expt.
^{3}H–^{3}He	$\frac{1}{2}$	1	649	29	64	14	757	764
^{7}Li–^{7}Be	$\frac{1}{2}$	1	1458	40	83	23	1605	1644
^{7}He, ^{7}Li*, ^{7}Be*, ^{7}B	$\frac{3}{2}$	1	1286	14	49	17	1366	1373
	$\frac{3}{2}$	2	132	7	34		174	175
^{8}Li, ^{8}Be*, ^{8}B	1	1	1692	24	78	24	1818	1770
	1	2	140	5	−5		140	145
^{8}He, ^{8}Li*, ^{8}Be*, ^{8}B*, ^{8}C	2	1	1719	13	83	26	1840	1659
	2	2	153	7	42		203	153

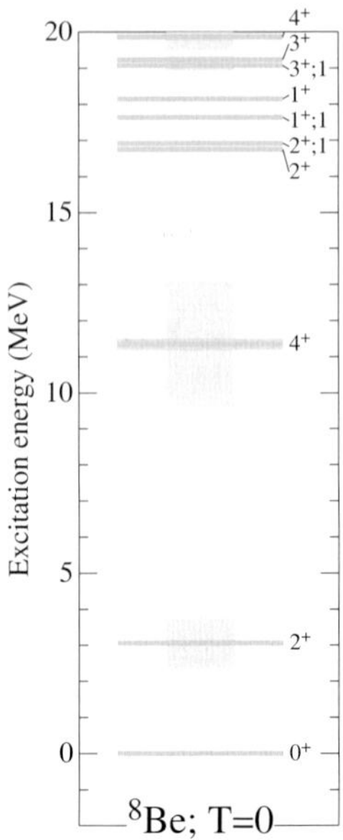

Fig. 10. – Experimental spectrum of ^{8}Be.

(strong-interaction) terms in AV18, while the K^{CSB} column shows that resulting from the difference of the proton and neutron masses. In general the non-Coulomb electromagnetic and strong charge symmetry breaking and charge dependent terms result in good agreement with the experimental values and thus resolve the Nolen-Schiffer anomaly.

Figure 10 shows the beginning of the experimental excitation spectrum of ^{8}Be. From 16 to 19 MeV excitation there are pairs of 2^+, 1^+, and 3^+ levels with isospin 0 and 1. The two 2^+ levels are very close and hence strongly isospin mixed; the 3^+ levels also have significant mixing. The mixing has been experimentally known from the decay properties of the states since 1966 [27]. However, as with the Nolen Schiffer anomaly, calculations using just Coulomb mixing underestimate the amount of mixing. Table IV shows GFMC calculations using AV18+IL2 of the mixing matrix elements for the three

TABLE IV. – *Isospin mixing matrix elements for* ^{8}Be *in* keV.

J^P	GFMC				Expt
	Coulomb	Strong CSB	Other	Total	
2nd 2^+	78	21	16	115	144
1^+	80	18	4	102	120
3^+	61	15	14	90	63
1st 2^+	4	0.4	1	6	–

pairs [28]. The contribution of the nuclear CSB term is relatively more important here than for the Nolen Schiffer anomaly. The agreement of the predictions with the data is not as satisfactory as for the Nolen Schiffer anomaly. The final line of the table shows the mixing matrix element between the first 2^+ (at $3\,\mathrm{MeV}$) state and the isospin-1 17 MeV state. The small value of this matrix element, and the large energy denominator, shows that the first 2^+ state has very little $T = 1$ contamination; this is important to the possibility of using $^8\mathrm{Li}(\beta^-)^8\mathrm{Be}(1^{\mathrm{st}}\ 2^+)$ decay as a test of $V - A$.

$6\!\cdot\!3$. *Can modern nuclear Hamiltonians tolerate a bound tetraneutron?* – In 2002 a claimed observation of a bound tetraneutron was published [29, 30]. The experiment did not produce a definite binding energy, just the statement that four neutrons are weakly bound. It is well known that the dineutron is not bound, but rather has low-energy pole on the second sheet (pseudo bound state); AV18 along with other realistic potentials has this feature. A set of GFMC calculations were made to see if the AV18+IL2 Hamiltonian, or acceptable modifications of it, could reproduce the tetraneutron claim [31]. It was clearly established that the unmodified AV18+IL2 does not bind $^4\mathrm{n}$; at most there is some weak resonance at $E \sim +2.\,\mathrm{MeV}$.

Minimal modifications to AV18+IL2 to give $E(^4\mathrm{n}) \sim -0.5\,\mathrm{MeV}$ were then made and the effects of such modifications on the energies of other, well established, nuclei were computed. Figure 11 shows some of these. In the first case (the left-hand bars),

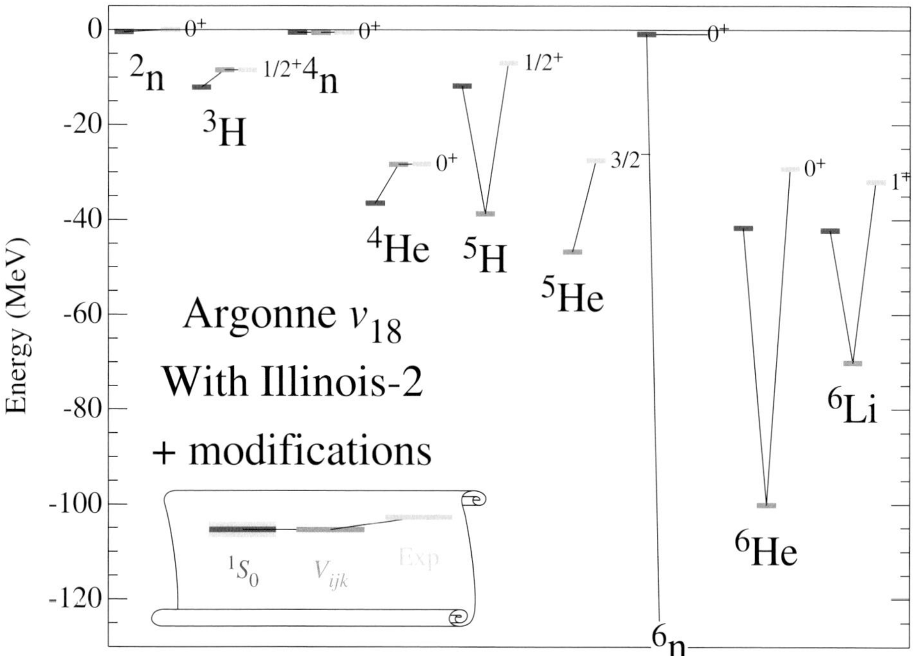

Fig. 11. – Nuclear energies with Hamiltonians modified to bind a tetraneutron.

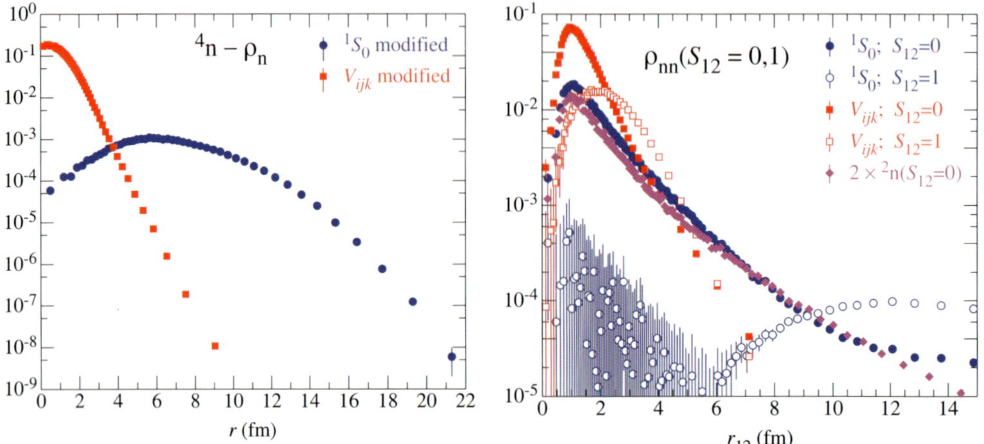

Fig. 12. – One-body (left) and two-body (right) densities of tetraneutrons. Circles are for the $^1S_0$ NN and squares for the $T = \frac{3}{2}$ NNN modifications; the diamonds are for ^{2}n.

the intermediate-range part of the $^1S_0$ partial-wave potential in of AV18 was increased enough to bind four neutrons. This increase results in the dineutron also being bound, in fact the ^{4}n is still unbound against breakup into two dineutrons in this model. As the figure shows, other existing nuclei (^{3}H, ^{4}He, etc.) all become significantly over bound with this increased $^1S_0$ potential. Also ^{6}n and ^{8}n (not shown) become bound. An attempt to bind the ^{4}n by changing the ^3P_J part of AV18 was also made, but this requires a huge change which very strongly overbinds other nuclei.

A second attempt was to add an attractive $V_{ijk}(T = 3/2)$ to H. This has two advantages: 1) it has no effect on NN scattering and does not make a bound ^{2}n; 2) because the modification is made only in isospin-$\frac{3}{2}$ triples, it has no effect on ^{3}H, ^{3}He, or ^{4}He; these have only isospin-$\frac{1}{2}$ triples. However, as can be seen in the figure, as soon as this potential can act (*i.e.* in nuclei with $T = \frac{3}{2}$ triples), it produces huge overbinding. The most dramatic effect is in pure neutron systems: ^{6}n is bound by 220 MeV and ^{8}n by 650 MeV. These are the most stable 6- and 8-nucleon systems with this Hamiltonian, so all other 6- and 8-body nuclei would beta decay to them!

Finally an attractive four-nucleon, $T = 2$, potential was added (not shown). This does not effect ^{6}Li but does very strongly overbind ^{6}He, larger nuclei, and pure neutron systems with more than four neutrons. The conclusion of the study is that a bound ^{4}n is incompatible with our understanding of nuclear forces. In the meantime the experiment has not been successfully reproduced.

The $^1S_0$ NN and $T = \frac{3}{2}$ NNN modifications have very different effects on $A > 4$ binding energies, even though both have been adjusted to bind ^{4}n by only 0.5 MeV; the NNN modification results in much more severe overbinding. Figure 12 helps to explain this. The left panel shows the one-body densities of the two bound ^{4}n systems; the $^1S_0$ NN modification results in a very diffuse ^{4}n with a r.m.s. radius of 8.9 fm. The right panel shows the pair density which is proportional to finding two neutrons a given distance apart. For the $^1S_0$ NN modification, the density for a $S = 0$ pair is peaked around 1 fm and is very similar to the pair distribution of the isolated ^{2}n which is bound by this potential. The $S = 1$ pair distribution is peaked at 12 fm; it arises from neutrons in different dineutron clusters. Thus this ^{4}n looks like two widely separated dineutrons. The binding comes from the small tails where neutrons from each dineutron get close enough to interact. The change of $v_{NN}(^1S_0)$ needed to achieve this is not that big and hence bigger nuclei overbound only somewhat.

On the other hand, a V_{NNN} requires all three neutrons in a triple to be close together to be effective. Thus the $T = \frac{3}{2}$ NNN modification must bring the two dineutrons close together. This results in the much more compact one-body density, with a r.m.s. radius of only 1.9 fm, shown in the left panel and a two-body density that is much sharper than the isolated ^{2}n. The $V_{NNN}(T = \frac{3}{2})$ must be large to achieve this high density and it thus has a large effect in all bigger nuclei.

7. – GFMC for scattering states

The GFMC calculations presented so far have treated the nucleus as a particle-stable system; that is the starting wave function, Ψ_T, is exponentially decaying as any nucleon is removed to a large distance from the center of mass. However many of the states of interest are particle unstable; they are above the threshold for emission of a single nucleon or, as is often the case for light nuclei, the threshold for break-up into subclusters. This approximation appears to be adequate for narrow resonances which have only a small scattering-wave component. However broad resonances should really be computed with proper scattering-wave boundary conditions; as shown in fig. 7 we do not achieve a converged energy for a broad state with the bound-state GFMC. In addition to being the correct approach, scattering solutions allow one to compute the phase shift as a function of energy and thus obtain the width of the resonance. Finally, one might also be interested in the phase shifts for partial waves that have no resonance.

Preliminary GFMC calculations of neutron-alpha scattering were made in 1991 [32], but detailed, high-statistics results have been obtained only recently [33]. Instead of using exponentially decaying wave functions, we construct Ψ_T to have a specified logarithmic derivative, γ, at some large boundary radius ($R \geq 7\,\mathrm{fm}$). Here R is the maximum distance that any nucleon is allowed to get from the other $A - 1$ nucleons. The GFMC propagation uses a method of images to preserve γ at R, and thus finds $E(R,\gamma)$; the eigenenergy that corresponds to the boundary condition. The phase shift, $\delta(E)$, can then be computed from R, γ, and $E(R,\gamma)$. This procedure is repeated for a number of γ and $\delta(E)$ is mapped out parametrically.

Figure 13 shows an example of this for the broad $\frac{1}{2}^-$ resonance in ^{5}He. The bound-state boundary condition does not give a stable energy, but is decaying to the $n + \,^4$He threshold energy. The scattering boundary condition produces a stable energy; the value of γ used in this example results in an energy slightly above the resonance energy.

Figure 14 compares the calculations of $n+\alpha$ scattering with a R-matrix analysis of the data [34] (solid curves). The left panel shows the partial-wave phase shifts computed for

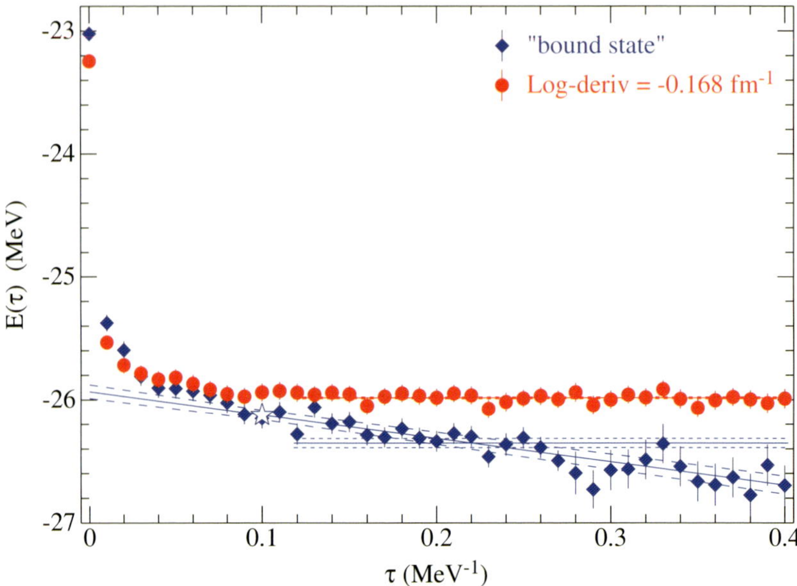

Fig. 13. – GFMC propagation for ^{5}He($\frac{1}{2}^-$) using bound- and scattering-state boundary conditions (diamonds and circles, respectively).

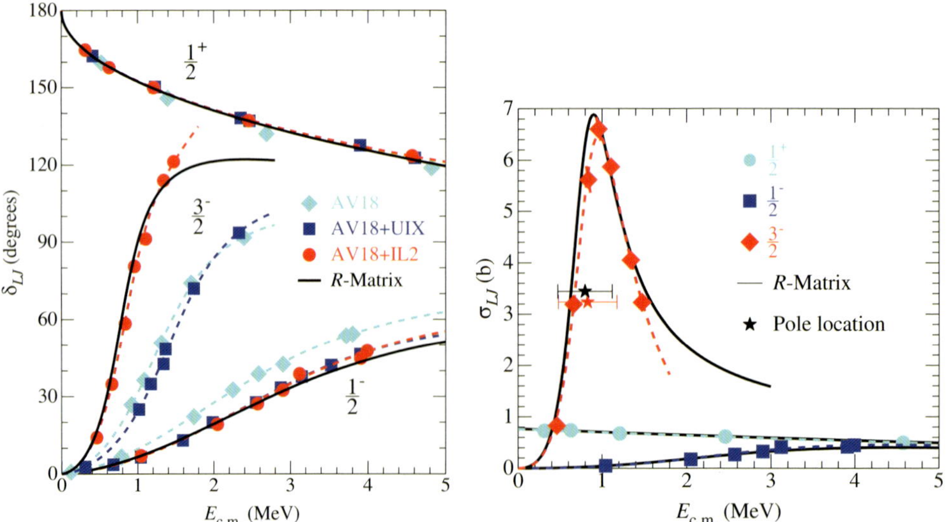

Fig. 14. – GFMC calculations of $n + \alpha$ scattering in the three principal partial waves. The left panel shows partial-wave phase shifts for three Hamiltonians. The right panel shows corresponding partial-wave cross-sections for the AV18+IL2 Hamiltonian. The experimental data is represented by the solid curves.

three different Hamiltonians: AV18 with no V_{ijk}, AV18+IL2, and AV18+UIX (UIX is an older NNN potential [35] that, with AV18, correctly binds ^{3}H and ^{4}He, but underbinds P-shell nuclei). All three Hamiltonians give very similar results for the $\frac{1}{2}^+$ partial wave which has no resonance. However only the AV18+IL2 correctly reproduces the two P-wave partial waves; AV18 alone misses both of them and AV18+UIX fits the $\frac{1}{2}^-$ partial wave but has too-small spin-orbit splitting and misses the $\frac{3}{2}^-$ one. The right panel shows the partial-wave cross-sections for the AV18+IL2 Hamiltonian. The strong resonance in the $\frac{3}{2}^-$ channel is very well reproduced showing that both the position and width of the resonance agree with the data; the much broader $\frac{1}{2}^-$ resonance is also well reproduced. In addition the good agreement with the low-energy $\frac{1}{2}^+$ cross-section data shows that the scattering length is reproduced.

This first study is very promising; the GFMC method, with its ability to have correct asymptotic forms, should be applied to other scattering calculations including a number of broad resonances [^{7,9}He, ^{6}Li(2^+), ^{8}Be($2^+, 4^+$), etc.] and the initial states of astrophysically interesting capture reactions [^{4}He(d, γ)^{6}Li, ^{7}Be(p, γ)^{8}B, etc.].

8. – Coordinate- and momentum-space densities

Up to now we have been concentrating on GFMC calculations of energies of nuclear states. However matrix elements of any operator may be evaluated in the GFMC propagation by using the extrapolation formulas, eqs. (52) or (54). I discuss some recent work on the charge radii, the corresponding densities, and momentum-space densities in this section.

8·1. *RMS radii and one-body densities of helium isotopes.* – A few years ago, a group at Argonne measured the RMS charge radius of the radioactive nucleus ^{6}He (β-decay half-life 0.8 s) with the remarkable accuracy of 0.7% [36] and the corresponding measurement for ^{8}He (β-decay half-life 0.1 s) will be published soon [37]. This has led us to attempt equally precise GFMC calculations of the corresponding point proton RMS radii. Such calculations are very difficult because of the small separation energies of the two valence neutrons in these isotopes ($E_{\text{sep}} = 0.97$ MeV for ^{6}He and 2.14 MeV for ^{8}He). Changes in the starting Ψ_T and other aspects of the GFMC calculations can result in changes of 200 keV (400 keV for ^{8}He) in the computed energy (and hence E_{sep}). The r.m.s. radius depends strongly on E_{sep}; as E_{sep} goes to zero, the radius goes to infinity. Thus we cannot give a precise value for the computed RMS radius for a specific Hamiltonian. Instead we find that the computed values for the same Hamiltonian with different GFMC calculations, or even for different Hamiltonians, all lie in a band of radius *vs.* separation energy. This is shown in fig. 15 which shows results for two Hamiltonians, AV18+IL2 and AV18+IL6, each with several GFMC calculations (IL6 is a newer, unpublished, version of IL2). The stars in each panel show the experimental point radii at the experimental separation energies; they are clearly consistent with our calculations which give $1.92(4)$ fm for ^{6}He and $1.82(2)$ fm, for ^{8}He. These numbers are both significantly bigger than the RMS point radius of ^{4}He which is 1.46 fm.

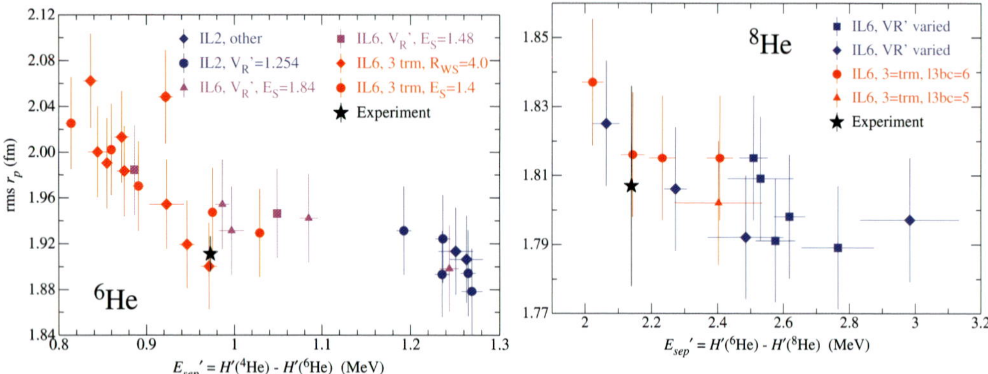

Fig. 15. – GFMC calculations of point proton RMS radii of ^{6,8}He plotted as a function of the two-neutron separation energy obtained in the calculation.

Figure 16 shows the point proton and neutron densities of ^{4,6,8}He. The alpha-particle is extremely compact; its central density is twice that of nuclear matter. In these calculations it has identical proton and neutron densities which is a very good approximation. As is shown below, the valence neutrons in ^{6,8}He do not seriously distort the ^{4}He core, rather they just drag the ^{4}He center of mass around. This results in the proton density being spread out which is why the charge radii of ^{6,8}He are so much greater than that of ^{4}He even though all three nuclei have just two protons. The right panel of the figures clearly shows that ^{6,8}He have large neutron halos due to the weak binding of the extra neutrons. The neutron halo of ^{6}He is more diffuse than that of ^{8}He as is expected from the smaller E_{sep} of ^{6}He.

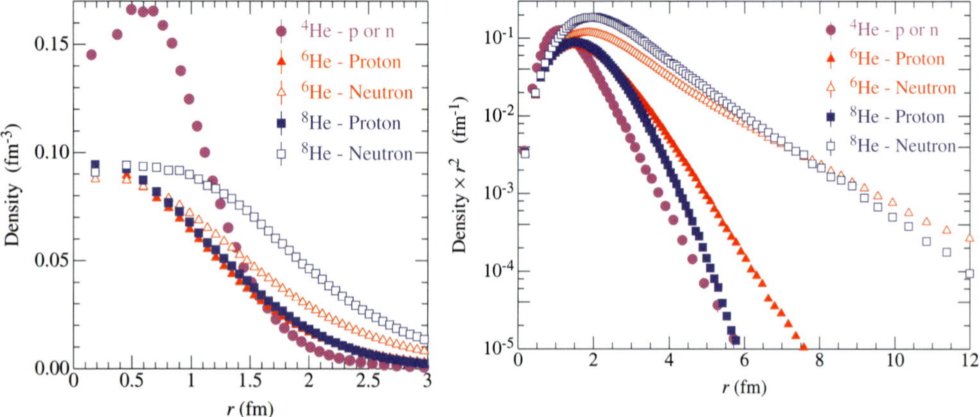

Fig. 16. – GFMC calculations, using AV18+IL2, of proton and neutron point densities for helium isotopes. The left panel shows density on a linear scale; the right panel $r^2\rho$ on a logarithmic scale.

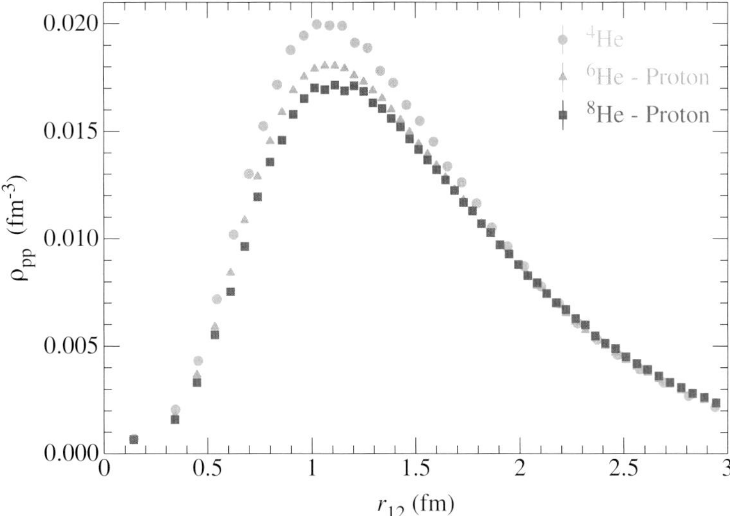

Fig. 17. – GFMC calculations, using AV18+IL2, of two-proton densities of helium isotopes.

8˙2. *Is an alpha-particle in a sea of neutrons still an alpha-particle?* – The previous subsection showed that the proton density in ^{6,8}He is much more spread out than the density of ^{4}He, even though ^{6,8}He have only extra neutrons added to a ^{4}He core. This might first be thought to indicate that the core of ^{6,8}He has been considerably enlarged by the neutrons. This can be studied by computing ρ_{pp}, the pair density which is proportional to the probability for finding two protons a given density apart. These distribution functions are shown in fig. 17, again calculated with GFMC for the AV18+IL2 model. These nuclei each have just one pp pair which presumably is in the "alpha core" of ^{6,8}He. Unlike the one-body densities, these distributions are not sensitive to center of mass effects, and thus if the alpha core of ^{6,8}He is not distorted by the surrounding neutrons, all three ρ_{pp} distributions in the figure should be the same.

We see that the pp distribution spreads out slightly with neutron number in the helium isotopes, with an increase of the pair r.m.s. radius of approximately 4% in going from ^{4}He to ^{6}He, and 8% to ^{8}He. While this could be interpreted as a swelling of the alpha core, it might also be due to the charge-exchange $(\tau_i \cdot \tau_j)$ correlations which can transfer charge from the core to the valence nucleons. Since these correlations are rather long-ranged, they can have a significant effect on the pp distribution. VMC calculations of ^{4}He with wave functions modified to give ρ_{pp} distributions close to those of ^{6,8}He suggest that the alpha cores of ^{6,8}He are excited by ~ 80 and ~ 350 keV, respectively, which corresponds to only a 0.4–2% admixture of the first 0^+ excited state of ^{4}He at 20 MeV. Thus almost all of the increased RMS radius of the proton density is due to the α core of ^{6,8}He being pushed around by the neutrons and not distortions of the core.

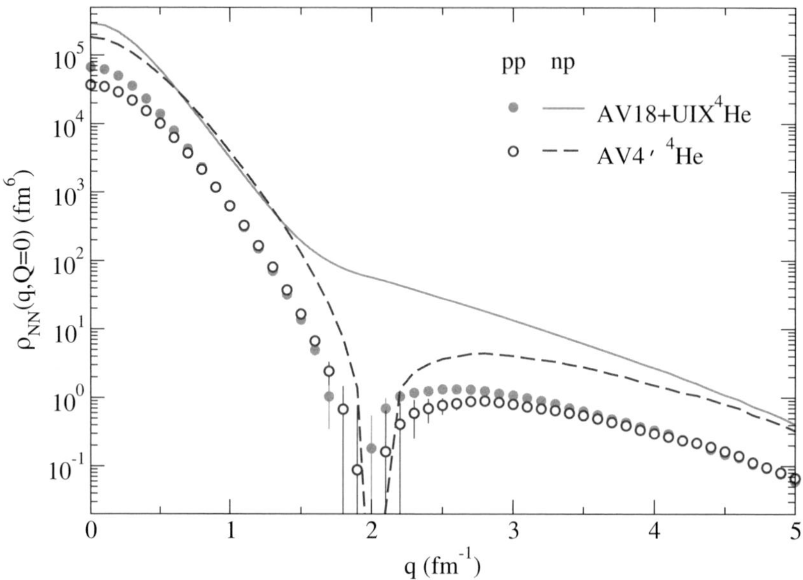

Fig. 18. – Two-nucleon momentum distributions in ^{4}He, computed by VMC for the AV18+UIX Hamiltonian. The symbols show pp distributions; the curves are for np.

8˙3. *Two-nucleon knockout—$(e, e'pN)$. –* A recent JLAB experiment for ^{12}C$(e, e'pN)$ measured back to back pp and np pairs; that is pairs with total c.m. momentum $Q = 0$, as a function of their relative momentum, q [38]. They found that the cross-section for np pairs with q in the range 2–3 fm^{-1} is 10–20 times larger than that for pp pairs in the same range. To study this we made VMC calculations of the corresponding pair momentum distributions (ρ_{NN}) in several nuclei from ^{3}H to ^{8}Be [39]. The calculations for ^{4}He are shown in fig. 18. Results for the AV18+UIX Hamiltonian (which for $A = 3, 4$ is a good approximation to AV18+IL2) are shown as the solid line (np pairs) and solid circles (pp pairs). Around $q = 2$ fm^{-1} there is a deep minimum in the pp density which results in large values for ρ_{np}/ρ_{pp}. The dashed curve and open circles show the corresponding quantities computed for the AV4$'$ NN potential (subsect. **2**˙3) with no NNN potential; in this case both densities have a deep dip and there is no enhancement of the ratio.

The AV4$'$ potential has no tensor force and thus np pairs are just S-wave while, for the full Hamiltonian, isospin-0 np pairs, like the deuteron, have a D-wave admixture. The S-wave deuteron momentum distribution has a zero at 2 fm^{-1} which is filled in by the D-wave contribution, the same as is seen here for np pairs in ^{4}He. The tensor force is much smaller in the pure $T = 1$ pp pairs, so the deep S-wave minimum is not filled in for those pairs, even with the AV18+UIX Hamiltonian. Calculations for ^{3}He, ^{6}Li, and ^{8}Be all show this effect although the deep minimum is somewhat filled in for ^{8}Be. Thus the JLAB experiment shows the importance of tensor correlations up to > 3 fm^{-1}.

9. – Conclusions

Quantum Monte Carlo methods are powerful tools for studying light nuclei with realistic nuclear interactions. Calculations of $A = 6$–12 nuclear energies with accuracies of 1–2% are possible and the AV18+IL2 reproduces binding energies with an average error of order $0.7\,\text{MeV}$ for $A = 3$–12. The NNN potential is required for overall P-shell energies and for spin-orbit splittings and several level orderings.

The QMC methods allow matrix elements of many operators of interest to be computed. This contribution presents rms radii, one- and two-body densities and two-body momentum distributions. These are generally in good agreement with experiment. Recently GFMC values of $A = 6, 7$ electromagnetic and weak transitions have also been computed; these improve on older VMC calculations and also generally agree with experiment [40]. Another topic not covered here is overlap functions and the related spectroscopic factors; these are used as input to calculations (such as distorted-wave Born approximation) of nuclear reactions; recent results are presented in refs. [41,42]. A just-finished interesting study used VMC calculations to investigate the effects on nuclear binding energies of changes in the Hamiltonian induced by changes of the fundamental constants [43].

GFMC calculations of are very computer intensive and at present ^{12}C can just barely be done. However a new generation of extremely parallel computers is becoming available and we are working with computer scientists to enable the GFMC program to make use of these machines. This should lead to the possibility of detailed studies of ^{12}C including second 0^+ (Hoyle) state which is the doorway for triple-alpha burning. This state has resisted precise calculation by shell-model based methods; we hope that our more flexible variational wave functions, combined with GFMC propagation, will overcome these difficulties.

But perhaps the most important advance in nuclear GFMC is the computation of scattering states. In these calculations the correct scattering-wave boundary condition is achieved. The resulting wave functions will be used to compute reactions of astrophysical interest such as $^{3}\text{He} + \alpha \to {}^{7}\text{Be}$, $\text{p} + {}^{7}\text{Be} \to {}^{8}\text{B}$, and $\text{n} + (\alpha + \alpha) \to {}^{9}\text{Be}$. Indeed all big-bang nucleosynthesis, solar neutrino, and some r-process seeding reactions are accessible.

$$* \ * \ *$$

As can be seen from the author lists of the citations, the work reported in this contribution is the result of long-term collaborations with J. CARLSON (who invented nuclear GFMC), K. M. NOLLETT (who is doing the GFMC scattering), V. R. PANDHARIPANDE (who for many years guided and inspired our group), R. SCHIAVILLA (who has provided the expertise on the electroweak currents) and R. B. WIRINGA (who has developed the VMC wave functions). I thank R. B. WIRINGA for also making a critical reading of this MS. The work would not have been possible without extensive computer resources provided over the years by Argonne's Mathematics and Computer Science Division (most recently on the IBM Blue Gene), Argonne's Laboratory Computing Resource Center, and the US Department of Energy's National Energy Research Scientific Computing Center.

This work is supported by the US Department of Energy, Office of Nuclear Physics, under contract DE-AC02-06CH11357.

REFERENCES

[1] BLATT J. M. and KALOS M. H., *Phys. Rev.*, **91** (1953) 444.

[2] CARLSON J. and SCHIAVILLA R., *Rev. Mod. Phys.*, **70** (1998) 743.

[3] HAGEN G., DEAN D. J., HJORTH-JENSEN M., PAPENBROCK T. and SCHWENK A., *Phys. Rev. C*, **76** (2007) 044305.

[4] WIRINGA R. B., STOKS V. G. J. and SCHIAVILLA R., *Phys. Rev. C*, **51** (1995) 38.

[5] WIRINGA R. B., *Nucleon-nucleon interactions*, presented at *Contemporary Nuclear Shell Models*, edited by PAN X.-W., FENG D. H. and VALLIÈRES M. (Springer-Verlag, Berlin) 1997.

[6] BERGERVOET J. R., VAN CAMPEN P. C., KLOMP R. A. M., DE KOK J.-L., RIJKEN T. A., STOKS V. G. J. and DE SWART J. J., *Phys. Rev. C*, **41** (1990) 1435.

[7] STOKS V. G. J., KLOMP R. A. M., RENTMEESTER M. C. M. and DE SWART J. J., *Phys. Rev. C*, **48** (1993) 792.

[8] PIEPER S. C., PANDHARIPANDE V. R., WIRINGA R. B. and CARLSON J., *Phys. Rev. C*, **64** (2001) 014001.

[9] FUJITA J. and MIYAZAWA H., *Prog. Theor. Phys.*, **17** (1957) 360.

[10] FUJITA J. and MIYAZAWA H., *Prog. Theor. Phys.*, **17** (1957) 366.

[11] COON S. A. *et al.*, *Nucl. Phys. A*, **317** (1979) 242.

[12] WIRINGA R. B. and PIEPER S., *Phys. Rev. Lett.*, **89** (2002) 182501.

[13] PUDLINER B. S., PANDHARIPANDE V. R., CARLSON J., PIEPER S. C. and WIRINGA R. B., *Phys. Rev. C*, **56** (1997) 1720.

[14] VOLKOV A. B., *Nucl. Phys. A*, **74** (1965) 33.

[15] CARLSON J. A. and WIRINGA R. B., *Variational Monte-Carlo techniques in nuclear physics*, in *Computational Nuclear Physics*, Vol. **1**, edited by LANGANKE K., MARUHN J. A. and KOONIN S. E. (Springer-Verlag, Berlin) 1990, Ch. 9, the source and input files are available at http://www.phys.washington.edu/users/bulgac/Koonin/Monte_Carlo/.

[16] PIEPER S. C., *Monte Carlo calculations of nuclei*, in *Lecture Notes in Physics – Microscopic Quantum Many-Body Theories and Their Applications*, Vol. **510**, edited by NAVARRO J. and POLLS A. (Springer-Verlag, Berlin) 1998.

[17] PIEPER S. C. and WIRINGA R. B., *Annu. Rev. Nucl. Part. Sci.*, **51** (2001) 53.

[18] WIRINGA R. B., *Phys. Rev. C*, **43** (1991) 1585.

[19] WIRINGA R. B., PIEPER S. C., CARLSON J. and PANDHARIPANDE V. R., *Phys. Rev. C*, **62** (2000) 014001.

[20] PIEPER S. C., VARGA K. and WIRINGA R. B., *Phys. Rev. C*, **66** (2002) 044310.

[21] PIEPER S. C., WIRINGA R. B. and CARLSON J., *Phys. Rev. C*, **70** (2004) 054325.

[22] SCHMIDT K. E. and LEE M. A., *Phys. Rev. E*, **51** (1995) 5495.

[23] CEPERLEY D. M. and KALOS M. H., in *Monte Carlo Methods in Statistical Physics*, Vol. **510**, edited by BINDER K. (Springer, Heidelberg) 1979.

[24] NAVRÁTIL P. and ORMAND W. E., *Phys. Rev. C*, **68** (2003) 034305.

[25] KURATH D., *Phys. Rev.*, **101** (1956) 216.

[26] NOLEN J. A. and SCHIFFER J., *Annu. Rev. Nucl. Sci.*, **19** (1969) 471.

[27] BARKER F. C., *Nucl. Phys.*, **83** (1966) 418.

[28] WIRINGA R. B., PIEPER S. C. and PERVIN M., *Bull. Am. Phys. Soc.*, **52** (2007) 9, and in preparation.

[29] Marqués F. M. *et al.*, *Phys. Rev. C*, **65** (2002) 044006.

[30] Marqués F. M. and Orr N. A., nucl-ex/03030005 (2003).

[31] Pieper S. C., *Phys. Rev. Lett.*, **90** (2003) 252501.

[32] Carlson J., *Nucl. Phys. A*, **522** (1991) 185c.

[33] Nollett K., Pieper S., Wiringa R., Carlson J. and Hale G. M., *Phys. Rev. Lett.*, **99** (2007) 022502.

[34] Hale G. M., Dodder D. C. and Witte K., unpublished (1995).

[35] Pudliner B. S., Pandharipande V. R., Carlson J. and Wiringa R. B., *Phys. Rev. Lett.*, **74** (1995) 4396.

[36] Wang L.-B. *et al.*, *Phys. Rev. Lett.*, **93** (2003) 142501.

[37] Mueller P. *et al.*, in preparation (2007).

[38] Piasetzky E. and Higinbotham D., private communication (2007).

[39] Schiavilla R., Wiringa R. B., Pieper S. C. and Carlson J., *Phys. Rev. Lett.*, **98** (2007) 132501.

[40] Pervin M. and Pieper S. C. and Wiringa R. B., arXiv:0710.1265 (2007).

[41] Wuosmaa A. H. *et al.*, *Phys. Rev. Lett.*, **94** (2005) 082502.

[42] Wuosmaa A. H. *et al.*, *Phys. Rev. C*, **72** (2005) 061301(R).

[43] Flambaum V. V. and Wiringa R. B., *Phys. Rev. C.*, **76** (2007) 054002.

DOI 10.3254/978-1-58603-885-4-147

Ab initio no-core shell model calculations for light nuclei

P. Navrátil

Lawrence Livermore National Laboratory, L-414
P.O. Box 808, Livermore, CA 94551, USA

Summary. — An overview of the *ab initio* no-core shell model is presented. Recent results for light nuclei obtained with the chiral two-nucleon and three-nucleon interactions are highlighted. Cross-section calculations of capture reactions important for astrophysics are discussed. The extension of the *ab initio* no-core shell model to the description of nuclear reactions by the resonating group method technique is outlined.

1. – Introduction

The major outstanding problem in nuclear physics is to calculate properties of finite nuclei starting from the basic interactions among nucleons. This problem has two parts. First, the basic interactions among nucleons are complicated, they are not uniquely defined and there is evidence that more than just two-nucleon forces are important. Second, the nuclear many-body problem is very difficult to solve. This is a direct consequence of the complex nature of the inter-nucleon interactions. Both short-range and medium-range correlations among nucleons are important and for some observables long-range correlations also play a significant role. Various methods have been used to solve the few-nucleon problem in the past. The Faddeev method [1] has been successfully applied to solve the three-nucleon bound-state problem for different nucleon-nucleon potentials [2-4]. For the solution of the four-nucleon problem one can employ Yakubovsky's generalization of the Faddeev formalism [5] as done, *e.g.*, in refs. [6] or [7]. Alternatively, other methods have also been succesfully used, such as, the correlated hyperspherical harmonics expansion method [8,9] or the Green's function Monte Carlo method [10]. Recently, a benchmark calculation by seven different methods was performed for a four-nucleon bound-state problem [11]. However, there are few approaches that can be successfully applied to systems of more than four nucleons when realistic inter-nucleon interactions are used. Apart from the coupled cluster method [12-15] applicable typically to closed-shell nuclei, the Green's function Monte Carlo method is a prominent approach capable to solve the nuclear many-body problem with realistic interactions for systems of up to $A = 12$. Another method developed recently applicable to light nuclei up to $A = 16$ and beyond is the *ab initio* no-core shell model (NCSM) [16]. In this paper, an overview of this approach is given and results obtained very recently are presented. Also, future developments, in particular applications to nuclear reactions, are outlined.

2. – *Ab initio* no-core shell model

In the *ab initio* no-core shell model, we consider a system of A point-like non-relativistic nucleons that interact by realistic two- or two- plus three-nucleon (NNN) interactions. Under the term realistic two-nucleon (NN) interactions we mean NN potentials that fit nucleon-nucleon phase shifts with high precision up to certain energy, typically up to 350 MeV. A realistic NNN interaction includes terms related to two-pion exchanges with an intermediate delta excitation. In the NCSM, all the nucleons are considered active, there is no inert core like in standard shell model calculations. Therefore the "no-core" in the name of the approach. There are two other major features in addition to the employment of realistic NN or NN+NNN interactions. The first one is the use of the harmonic-oscillator (HO) basis truncated by a chosen maximal total HO energy of the A-nucleon system. The reason behind the choice of the HO basis is the fact that this is the only basis that allows to use single-nucleon coordinates and consequently the second-quantization representation without violating the translational invariance of the system. The powerful techniques based on the second quantization and developed for standard shell model calculations can then be utilized. Therefore the "shell model" in the name of the approach. As a downside, one has to face the consequences of the incorrect asymptotic behavior of the HO basis. The second feature comes as a consequence of the basis truncation. In order to speed up convergence with the basis enlargement, we construct an effective interaction from the original realistic NN or NN+NNN potentials by means of a unitary transformation. The effective interaction depends on the basis truncation and by construction becomes the original realistic NN or NN+NNN interaction as the size of the basis approaches infinity. In principle, one can also perform calculations with the unmodified, "bare", original interactions. Such calculations are then variational with the HO frequency and the basis truncation parameter as variational parameters.

2˙1. *Hamiltonian*. – The starting Hamiltonian of the *ab initio* NCSM is

$$(1) \qquad H_A = \frac{1}{A}\sum_{i<j}^{A}\frac{(\vec{p}_i-\vec{p}_j)^2}{2m} + \sum_{i<j}^{A} V_{\mathrm{NN},ij} + \sum_{i<j<k}^{A} V_{\mathrm{NNN},ijk}\,,$$

where m is the nucleon mass, $V_{\mathrm{NN},ij}$ the NN interaction, $V_{\mathrm{NNN},ijk}$ the three-nucleon interaction. In the NCSM, we employ a large but finite HO basis. Due to properties of the realistic nuclear interaction in eq. (1), we must derive an effective interaction appropriate for the basis truncation. To facilitate the derivation of the effective interaction, we modify the Hamiltonian (1) by adding to it the center-of-mass (CM) HO Hamiltonian $H_{\mathrm{CM}} = T_{\mathrm{CM}} + U_{\mathrm{CM}}$, where $U_{\mathrm{CM}} = (\frac{1}{2})Am\Omega^2\vec{R}^2$, $\vec{R} = (\frac{1}{A})\sum_{i=1}^{A}\vec{r}_i$. The effect of the HO CM Hamiltonian will later be subtracted out in the final many-body calculation. Due to the translational invariance of the Hamiltonian (1) the HO CM Hamiltonian has in fact no effect on the intrinsic properties of the system. The modified Hamiltonian can be cast into the form

$$(2) \quad H_A^{\Omega} = H_A + H_{\mathrm{CM}} = \sum_{i=1}^{A} h_i + \sum_{i<j}^{A} V_{ij}^{\Omega,A} + \sum_{i<j<k}^{A} V_{\mathrm{NNN},ijk} = \sum_{i=1}^{A}\left[\frac{\vec{p}_i^{\,2}}{2m} + \frac{1}{2}m\Omega^2\vec{r}_i^{\,2}\right]$$

$$+ \sum_{i<j}^{A}\left[V_{\mathrm{NN},ij} - \frac{m\Omega^2}{2A}(\vec{r}_i-\vec{r}_j)^2\right] + \sum_{i<j<k}^{A} V_{\mathrm{NNN},ijk}\,.$$

2˙2. *Basis*. – In the *ab initio* NCSM, we use a HO basis. A single-nucleon HO wave function can be written as

$$(3) \qquad \varphi_{nlm}(\vec{r};b) = R_{nl}(r;b)Y_{lm}(\hat{r}),$$

with $R_{nl}(r,b)$ the radial HO wave function and b the HO length parameter related to the HO frequency Ω as $b = \sqrt{\hbar/m\Omega}$, with m the nucleon mass. The HO length parameter b is often dropped in R_{nl} and φ_{nlm} in the following text to simplify notation.

The HO wave functions have important transformation properties that we utilize frequently. Let us consider two particles with different masses of ratio $d = m_2/m_1$ moving in a HO well. Their relative and center-of-mass coordinates can be defined by an orthogonal transformation

$$(4) \qquad \vec{r} = \sqrt{\frac{d}{1+d}}\vec{r}_1 - \sqrt{\frac{1}{1+d}}\vec{r}_2\,,$$

$$(5) \qquad \vec{R} = \sqrt{\frac{1}{1+d}}\vec{r}_1 + \sqrt{\frac{d}{1+d}}\vec{r}_2\,,$$

where all the vectors in (4) and (5) are defined as products of square root of the respective mass and the position vector: $\vec{r} = \sqrt{m}\vec{x}$. The product of the single-particle HO wave functions can then be expressed as a linear combination of the relative-coordinate HO wave function and the CM coordinate HO wave function:

$$(6) \qquad \left[\varphi_{n_1 l_1}(\vec{r}_1)\varphi_{n_2 l_2}(\vec{r}_2)\right]_k^{(K)} = \sum_{nlNL} \langle nlNLK | n_1 l_1 n_2 l_2 K \rangle_d \left[\varphi_{nl}(\vec{r})\varphi_{NL}(\vec{R})\right]_k^{(K)},$$

with $\langle nlNLK | n_1 l_1 n_2 l_2 K \rangle_d$ a generalized HO bracket that can be evaluated according to the algorithm in, *e.g.*, ref. [17].

As the NN and NNN interactions depend on relative coordinates and/or momenta, the natural coordinates in the nuclear problem are the relative, or Jacobi, coordinates.

We work in the isospin formalism and consider nucleons with the mass m. A generalization to the proton-neutron formalism with unequal masses for the proton and the neutron is straightforward. We will use Jacobi coordinates that are introduced as an orthogonal transformation of the single-nucleon coordinates. In general, Jacobi coordinates are proportional to differences of centers of mass of nucleon sub-clusters.

For the present purposes we consider just a single set of Jacobi coordinates. More general discussion can be found in ref. [18]. The following set

$$(7) \qquad \vec{\xi}_0 = \sqrt{\frac{1}{A}}\left[\vec{r}_1 + \vec{r}_2 + \ldots + \vec{r}_A\right],$$

$$(8) \qquad \vec{\xi}_1 = \sqrt{\frac{1}{2}}\left[\vec{r}_1 - \vec{r}_2\right],$$

$$(9) \qquad \vec{\xi}_2 = \sqrt{\frac{2}{3}}\left[\frac{1}{2}\left(\vec{r}_1 + \vec{r}_2\right) - \vec{r}_3\right],$$

$$\ldots$$

$$(10) \qquad \vec{\xi}_{A-2} = \sqrt{\frac{A-2}{A-1}}\left[\frac{1}{A-2}\left(\vec{r}_1 + \vec{r}_2 + \ldots + \vec{r}_{A-2}\right) - \vec{r}_{A-1}\right],$$

$$(11) \qquad \vec{\xi}_{A-1} = \sqrt{\frac{A-1}{A}}\left[\frac{1}{A-1}\left(\vec{r}_1 + \vec{r}_2 + \ldots + \vec{r}_{A-1}\right) - \vec{r}_A\right],$$

is useful for the construction of the antisymmetrized HO basis. Here, $\vec{\xi}_0$ is proportional to the center of mass of the A-nucleon system. On the other hand, $\vec{\xi}_\rho$ is proportional to the relative position of the $\rho + 1$-st nucleon and the center of mass of the ρ nucleons.

As nucleons are fermions, we need to construct an antisymmetrized basis. Here we illustrate how to do this for the simplest case of three nucleons. One starts by introducing a HO basis that depends on Jacobi coordinates $\vec{\xi}_1$ and $\vec{\xi}_2$, defined in eqs. (8) and (9), *e.g.*,

$$(12) \qquad |(nlsjt;\mathcal{NLJ})JT\rangle.$$

Here n, l and $\mathcal{N}, \mathcal{L}$ are the HO quantum numbers corresponding to the harmonic oscillators associated with the coordinates (and the corresponding momenta) $\vec{\xi}_1$ and $\vec{\xi}_2$, respectively. The quantum numbers s, t, j describe the spin, isospin and angular momentum of the relative-coordinate two-nucleon channel of nucleons 1 and 2, while $\mathcal{J}$ is the angular momentum of the third nucleon relative to the center of mass of nucleons 1 and 2. The J and T are the total angular momentum and the total isospin, respectively. Note that the basis (12) is antisymmetrized with respect to the exchanges of nucleons 1 and 2, as the two-nucleon channel quantum numbers are restricted by the condition $(-1)^{l+s+t} = -1$. It is not, however, antisymmetrized with respect to the exchanges of nucleons $1 \leftrightarrow 3$ and $2 \leftrightarrow 3$. In order to construct a completely antisymmetrized basis, one needs to obtain eigenvectors of the antisymmetrizer

$$(13) \qquad \mathcal{X} = \frac{1}{3}\left(1 + \mathcal{T}^{(-)} + \mathcal{T}^{(+)}\right),$$

where $\mathcal{T}^{(+)}$ and $\mathcal{T}^{(-)}$ are the cyclic and the anti-cyclic permutation operators, respectively. The antisymmetrizer $\mathcal{X}$ is a projector satisfying $\mathcal{X}\mathcal{X} = \mathcal{X}$. When diagonalized in the basis (12), its eigenvectors span two eigenspaces. One, corresponding to the eigenvalue 1, is formed by physical, completely antisymmetrized states and the other, corresponding to the eigenvalue 0, is formed by spurious states. There are about twice as many spurious states as the physical ones [19].

Due to the antisymmetry with respect to the exchanges $1 \leftrightarrow 2$, the matrix elements in the basis (12) of the antisymmetrizer $\mathcal{X}$ can be evaluated simply as $\langle \mathcal{X} \rangle = (\frac{1}{3})(1 - 2\langle P_{2,3} \rangle)$, where $P_{2,3}$ is the transposition operator corresponding to the exchange of nucleons 2 and 3. Its matrix element can be evaluated in a straightforward way, *e.g.*,

$$(14) \qquad \langle (n_1 l_1 s_1 j_1 t_1; \mathcal{N}_1 \mathcal{L}_1 \mathcal{J}_1) JT | P_{2,3} | (n_2 l_2 s_2 j_2 t_2; \mathcal{N}_2 \mathcal{L}_2 \mathcal{J}_2) JT \rangle =$$

$$= \delta_{N_1, N_2} \hat{t}_1 \hat{t}_2 \begin{Bmatrix} \frac{1}{2} & \frac{1}{2} & t_1 \\ \frac{1}{2} & T & t_2 \end{Bmatrix}$$

$$\times \sum_{LS} \hat{L}^2 \hat{S}^2 \hat{j}_1 \hat{j}_2 \hat{\mathcal{J}}_1 \hat{\mathcal{J}}_2 \hat{s}_1 \hat{s}_2 (-1)^L \begin{Bmatrix} l_1 & s_1 & j_1 \\ \mathcal{L}_1 & \frac{1}{2} & \mathcal{J}_1 \\ L & S & J \end{Bmatrix} \begin{Bmatrix} l_2 & s_2 & j_2 \\ \mathcal{L}_2 & \frac{1}{2} & \mathcal{J}_2 \\ L & S & J \end{Bmatrix}$$

$$\times \begin{Bmatrix} \frac{1}{2} & \frac{1}{2} & s_1 \\ \frac{1}{2} & S & s_2 \end{Bmatrix} \langle n_1 l_1 \mathcal{N}_1 \mathcal{L}_1 L | \mathcal{N}_2 \mathcal{L}_2 n_2 l_2 L \rangle_3 \,,$$

where $N_i = 2n_i + l_i + 2\mathcal{N}_i + \mathcal{L}_i, i = 1, 2$; $\hat{j} = \sqrt{2j+1}$; and $\langle n_1 l_1 \mathcal{N}_1 \mathcal{L}_1 L | \mathcal{N}_2 \mathcal{L}_2 n_2 l_2 L \rangle_3$ is the general HO bracket for two particles with mass ratio 3 as defined, *e.g.*, in ref. [17]. The expression (14) can be derived by examining the action of $P_{2,3}$ on the basis states (12). That operator changes the state $|nl(\vec{\xi}_1), \mathcal{N}\mathcal{L}(\vec{\xi}_2), L\rangle$ to $|nl(\vec{\xi'}_1), \mathcal{N}\mathcal{L}(\vec{\xi'}_2), L\rangle$, where $\vec{\xi'}_i, i = 1, 2$ are defined as $\vec{\xi}_i, i = 1, 2$ but with the single-nucleon indexes 2 and 3 exchanged. The primed Jacobi coordinates can be expressed as an orthogonal transformation of the unprimed ones, see eq. (6). Consequently, the HO wave functions depending on the primed Jacobi coordinates can be expressed as an orthogonal transformation of the original HO wave functions. Elements of the transformation are the generalized HO brackets for two particles with the mass ratio d, with d determined from the orthogonal transformation of the coordinates, see eq. (4).

The resulting antisymmetrized states can be classified and expanded in terms of the original basis (12) as follows:

$$(15) \qquad |NiJT\rangle = \sum \langle nlsjt; \mathcal{NLJ} || NiJT \rangle | (nlsjt; \mathcal{NLJ})JT \rangle,$$

where $N = 2n+l+2\mathcal{N}+\mathcal{L}$ and where we introduced an additional quantum number i that distinguishes states with the same set of quantum numbers N, J, T, $e.g.$, $i = 1, 2, \ldots r$ with r the total number of antisymmetrized states for a given N, J, T. The symbol $\langle nlsjt; \mathcal{NLJ} || NiJT \rangle$ is a coefficient of fractional parentage.

A generalization to systems of more than three nucleons can be done as shown, $e.g.$, in ref. [18]. It is obvious, however, that as we increase the number of nucleons, the antisymmetrization becomes more and more involved. Consequently, in the standard shell model calculations one utilizes antisymmetrized wave functions constructed in a straightforward way as Slater determinants of single-nucleon wave functions depending on single-nucleon coordinates $\varphi_i(\vec{r}_i)$. It follows from the transformations (6) that a use of a Slater determinant basis constructed from single nucleon HO wave functions such as

$$(16) \qquad \varphi_{nljmm_t}(\vec{r}, \sigma, \tau; b) = R_{nl}(r; b)(Y_l(\hat{r})\chi(\sigma))_m^{(j)}\chi(\tau)_{m_t},$$

results in eigenstates of a translationally invariant Hamiltonian that factorize as products of a wave function depending on relative coordinates and a wave function depending on the CM coordinates. This is true as long as the basis truncation is done by a chosen maximum of the sum of all HO excitations, $i.e.$: $\sum_{i=1}^{A}(2n_i + l_i) \leq N_{\text{totmax}}$. In eq. (16), σ and τ are spin and isospin coordinates of the nucleon. The physical eigenstates of an translationally invariant Hamiltonian can then be selected as eigenstates with the CM in the $0\hbar\Omega$ state:

$$(17) \qquad \langle \vec{r}_1 \ldots \vec{r}_A \sigma_1 \ldots \sigma_A \tau_1 \ldots \tau_A | A\lambda JMTM_T \rangle_{\text{SD}} =$$
$$= \langle \vec{\xi}_1 \ldots \vec{\xi}_{A-1}\sigma_1 \ldots \sigma_A \tau_1 \ldots \tau_A | A\lambda JMTM_T \rangle \varphi_{000}(\vec{\xi}_0; b).$$

For a general single-nucleon wave function this factorization is not possible. A use of any other single nucleon wave function than the HO wave function will result in mixing of CM and internal motion.

In the *ab initio* NCSM calculations, we use both the Jacobi-coordinate HO basis and the single-nucleon Slater determinant HO basis. One can choose, whichever is more convenient for the problem to be solved. One can also mix the two types of bases. In general for systems of $A \leq 4$, the Jacobi coordinate basis is more efficient as one can perform the antisymmetrization easily. The CM degrees of freedom can be explicitly removed and a coupled $J^\pi T$ basis can be utilized with matrix dimensions of the order of thousands. For systems with $A > 4$, it is in general more efficient to use the Slater determinant HO basis. In fact, we use so-called m-scheme basis with conserved quantum numbers $M = \sum_{i=1}^{A} m_i$, parity π and $M_T = \sum_{i=1}^{A} m_{ti}$. The antisymmetrization is trivial, but the dimensions can be huge as the CM degrees of freedom are present and no JT coupling is considered. The advantage is the possibility to utilize the powerful second quantization technique, shell model codes, transition density codes and so on.

As mentioned above, the model space truncation is always done using the condition $\sum_{i=1}^{A}(2n_i + l_i) \leq N_{\text{totmax}}$. Often, instead of N_{totmax} we introduce the parameter N_{max} that measures the maximal allowed HO excitation energy above the unperturbed ground state. For $A = 3, 4$ systems $N_{\text{max}} = N_{\text{totmax}}$. For the p-shell nuclei they differ, *e.g.* for ^{6}Li, $N_{\text{max}} = N_{\text{totmax}} - 2$, for ^{12}C, $N_{\text{max}} = N_{\text{totmax}} - 8$, etc.

2˙3. *Effective interaction.* – In the *ab initio* NCSM calculations we use a truncated HO basis as discussed in previous sections. The inter-nucleon interactions act, however, in the full space. In order to obtain meaningful results in the truncated space, or model space, the inter-nucleon interactions need to be renormalized. We need to construct an effective Hamiltonian with the inter-nucleon interactions replaced by effective interactions. By meaningful results we understand results as close as possible to the full space exact results for a subset of eigenstates. Mathematically we can construct an effective Hamiltonian that exactly reproduces the full space results for a subset of eigenstates. In practice, we cannot in general construct this exact effective Hamiltonian for the A-nucleon problem we want to solve. However, we can construct an effective Hamiltonian that is exact for a two-nucleon system or for a three-nucleon system or even for a four-nucleon system. The corresponding effective interactions can then be used in the A-nucleon calculations. Their use in general improves the convergence of the problem to the exact full space result with the increase of the basis size. By construction, these effective interactions converge to the full-space inter-nucleon interactions therefore guaranteeing convergence to exact solution when the basis size approaches the infinite full space.

In our approach we employ the so-called Lee-Suzuki similarity transformation method [20, 21], which yields a starting-energy–independent Hermitian effective interaction. We first recapitulate general formulation and basic results of this method. Applications of this method for computation of two- or three-body effective interactions are described afterwards.

Let us consider an *arbitrary* Hamiltonian H with the eigensystem $E_k, |k\rangle$, *i.e.*,

$$(18) \qquad\qquad H|k\rangle = E_k|k\rangle.$$

Let us further divide the full space into the model space defined by a projector P and the complementary space defined by a projector Q, $P+Q = 1$. A similarity transformation of the Hamiltonian $e^{-\omega}He^{\omega}$ can be introduced with a transformation operator ω satisfying the condition $\omega = Q\omega P$. The transformation operator is then determined from the requirement of decoupling of the Q-space and the model space as follows:

$$(19) \qquad\qquad Qe^{-\omega}He^{\omega}P = 0.$$

If we denote the model space basis states as $|\alpha_P\rangle$, and those which belong to the Q-space, as $|\alpha_Q\rangle$, then the relation $Qe^{-\omega}He^{\omega}P|k\rangle = 0$, following from eq. (19), will be satisfied for a particular eigenvector $|k\rangle$ of the Hamiltonian (18), if its Q-space components can be expressed as a combination of its P-space components with the help of the transformation operator ω, *i.e.*

$$(20) \qquad \langle\alpha_Q|k\rangle = \sum_{\alpha_P}\langle\alpha_Q|\omega|\alpha_P\rangle\langle\alpha_P|k\rangle.$$

If the dimension of the model space is d_P, we may choose a set $\mathcal{K}$ of d_P eigenevectors, for which the relation (20) will be satisfied. Under the condition that the $d_P \times d_P$ matrix defined by matrix elements $\langle\alpha_P|k\rangle$ for $|k\rangle \in \mathcal{K}$ is invertible, the operator ω can be determined from (20) as

$$(21) \qquad \langle\alpha_Q|\omega|\alpha_P\rangle = \sum_{k\in\mathcal{K}}\langle\alpha_Q|k\rangle\langle\tilde{k}|\alpha_P\rangle,$$

where we denote by tilde the inverted matrix of $\langle\alpha_P|k\rangle$, *e.g.*, $\sum_{\alpha_P}\langle\tilde{k}|\alpha_P\rangle\langle\alpha_P|k'\rangle = \delta_{k,k'}$, for $k,k' \in \mathcal{K}$.

The Hermitian effective Hamiltonian defined on the model space P is then given by [21]

$$(22) \qquad \bar{H}_{\text{eff}} = \left[P(1+\omega^{\dagger}\omega)P\right]^{1/2} PH(P+Q\omega P)\left[P(1+\omega^{\dagger}\omega)P\right]^{-1/2}.$$

By making use of the properties of the operator ω, the effective Hamiltonian $\bar{H}_{\text{eff}}$ can be rewritten in an explicitly Hermitian form as

$$(23) \qquad \bar{H}_{\text{eff}} = \left[P(1+\omega^{\dagger}\omega)P\right]^{-1/2}(P+P\omega^{\dagger}Q)H(Q\omega P+P)\left[P(1+\omega^{\dagger}\omega)P\right]^{-1/2}.$$

With the help of the solution for ω (21) we obtain a simple expression for the matrix elements of the effective Hamiltonian

$$(24) \qquad \langle\alpha_P|\bar{H}_{\text{eff}}|\alpha_{P'}\rangle = \sum_{k\in\mathcal{K}}\sum_{\alpha_{P''}}\sum_{\alpha_{P'''}}\langle\alpha_P|(1+\omega^{\dagger}\omega)^{-1/2}|\alpha_{P''}\rangle\langle\alpha_{P''}|\tilde{k}\rangle E_k\langle\tilde{k}|\alpha_{P'''}\rangle$$
$$\times\langle\alpha_{P'''}|(1+\omega^{\dagger}\omega)^{-1/2}|\alpha_{P'}\rangle.$$

For computation of the matrix elements of $(1+\omega^{\dagger}\omega)^{-1/2}$, we can use the relation

$$(25) \qquad \langle\alpha_P|(1+\omega^{\dagger}\omega)|\alpha_{P''}\rangle = \sum_{k\in\mathcal{K}}\langle\alpha_P|\tilde{k}\rangle\langle\tilde{k}|\alpha_{P''}\rangle,$$

to remove the summation over the Q-space basis states. The effective Hamiltonian (24) reproduces the eigenenergies $E_k, k \in \mathcal{K}$ in the model space.

It has been shown [22] that the Hermitian effective Hamiltonian (23) can be obtained directly by a unitary transformation of the original Hamiltonian:

$$(26) \qquad \bar{H}_{\text{eff}} = P e^{-S} H e^{S} P,$$

with an anti-Hermitian operator $S = \text{arctanh}(\omega - \omega^{\dagger})$. The transformed Hamiltonian then satisfies decoupling conditions $Q e^{-S} H e^{S} P = P e^{-S} H e^{S} Q = 0$.

We can see from eqs. (24) and (25) that in order to construct the effective Hamiltonian we need to know a subset of exact eigenvalues and model space projections of a subset of exact eigenvectors. This may suggest that the method is rather impractical. Also, it follows from eq. (24) that the effective Hamiltonian contains many-body terms, in fact for an A-nucleon system all terms up to A-body will in general appear in the effective Hamiltonian even if the original Hamiltonian consisted of just two-body or two- plus three-body terms.

In the *ab initio* NCSM we use the above effective interaction theory as follows. Since the two-body part dominates the A-nucleon Hamiltonian (2), it is reasonable to expect that a two-body effective interaction that takes into account full space two-nucleon correlations would be the most important part of the exact effective interaction. If the NNN interaction is taken into account, a three-body effective interaction that takes into account full space three-nucleon correlations would be a good approximation to the exact A-body effective interaction. We construct the two-body or three-body effective interaction by application of the above-described Lee-Suzuki procedure to a two-nucleon or three-nucleon system. The resulting effective interaction is then exact for the two- or three-nucleon system. It is an approximation of the exact A-nucleon effective interaction.

Using the notation of eq. (2), the two-nucleon effective interaction is obtained as

$$(27) \qquad V_{2\text{eff},12} = P_2 \left[e^{-S_{12}} (h_1 + h_2 + V_{12}^{\Omega,A}) e^{S_{12}} - (h_1 + h_2) \right] P_2,$$

with $S_{12} = \text{arctanh}(\omega_{12} - \omega_{12}^{\dagger})$ and P_2 is a two-nucleon model space projector. The two-nucleon model space is defined by a truncation $N_{12\text{max}}$ corresponding to the A-nucleon N_{max}. For example, for $A = 3, 4$, $N_{12\text{max}} = N_{\text{max}}$, for p-shell nuclei with $A > 5$ $N_{12\text{max}} = N_{\text{max}} + 2$. The operator ω_{12} is obtained with the help of eq. (21) from exact solutions of the Hamiltonian $h_1 + h_2 + V_{12}^{\Omega,A}$ which are straightforward to find. In practice, we actually do not need to calculate ω_{12}, rather we apply eqs. (24) and (25) with the two-nucleon solutions to directly calculate $P_2 e^{-S_{12}} (h_1 + h_2 + V_{12}^{\Omega,A}) e^{S_{12}} P_2$. To be explicit, the two-nucleon calculation is done with

$$(28) \qquad H_2^{\Omega} = H_{02} + V_{12}^{\Omega,A} = \frac{\vec{p}^2}{2m} + \frac{1}{2} m\Omega^2 \vec{r}^2 + V_{NN}(\sqrt{2}\vec{r}) - \frac{m\Omega^2}{A} \vec{r}^2,$$

where $\vec{r} = \sqrt{(1/2)}(\vec{r}_1 - \vec{r}_2)$ and $\vec{p} = \sqrt{(1/2)}(\vec{p}_1 - \vec{p}_2)$ and where H_{02} differs from $h_1 + h_2$ by the omission of the center-of-mass HO term of nucleons 1 and 2. Since $V_{12}^{\Omega,A}$ acts on relative coordinate, the S_{12} is independent of the two-nucleon center of mass and the two-nucleon center-of-mass Hamiltonian cancels out in eq. (27). We can see that for $A > 2$ the solutions of (28) are bound. The relative-coordinate two-nucleon HO states used in the calculation are characterized by quantum numbers $|nlsjt\rangle$ with the radial and orbital HO quantum numbers corresponding to coordinate $\vec{r}$ and momentum $\vec{p}$. Typically, we solve the two-nucleon Hamiltonian (28) for all two-nucleon channels up to $j = 8$. For the channels with higher j only the kinetic-energy term is used in the many-nucleon calculation. The model space P_2 is defined by the maximal number of allowed HO excitations $N_{12\text{max}}$ from the condition $2n + l \leq N_{12\text{max}}$. In order to construct the operator ω (21) we need to select the set of eigenvectors $\mathcal{K}$. We select the lowest states obtained in each channel. It turns out that these states also have the largest overlap with the P_2 model space. Their number is given by the number of basis states satisfying $2n + l \leq N_{12\text{max}}$.

An improvement over the two-body effective interaction approximation is the use of three-body effective interaction that takes into account the full-space three-nucleon correlations. If the NNN interaction is included, the three-body effective interaction approximation is rather essential for $A > 3$ systems. First, let us consider the case with no NNN interaction. The three-body effective interaction can be calculated as

$$(29) \quad V_{3\text{eff},123}^{\text{NN}} =$$
$$= P_3 \left[e^{-S_{123}^{\text{NN}}}(h_1 + h_2 + h_3 + V_{12}^{\Omega,A} + V_{13}^{\Omega,A} + V_{23}^{\Omega,A})e^{S_{123}^{\text{NN}}} - (h_1 + h_2 + h_3) \right] P_3.$$

Here, $S_{123}^{\text{NN}} = \text{arctanh}(\omega_{123} - \omega_{123}^{\dagger})$ and P_3 is a three-nucleon model space projector. The P_3 space contains all three-nucleon states up to the highest possible three-nucleon excitation, which can be found in the P space of the A-nucleon system. For example, for $A = 6$ and $N_{\text{max}} = 6$ ($6\hbar\Omega$) space we have P_3 defined by $N_{123\text{max}} = 8$. Similarly, for the p-shell nuclei with $A \geq 7$ and $N_{\text{max}} = 6$ ($6\hbar\Omega$) space we have $N_{123\text{max}} = 9$. The operator ω_{123} is obtained with the help of eq. (21) from exact solutions of the Hamiltonian $h_1 + h_2 + h_3 + V_{12}^{\Omega,A} + V_{13}^{\Omega,A} + V_{23}^{\Omega,A}$, which are found using the antisymmetrized three-nucleon Jacobi coordinate HO basis. In practice, we again do not need to calculate ω_{123}, rather we apply eqs. (24) and (25) with the three-nucleon solutions. The three-body effective interaction is then used in A-nucleon calculations using the effective Hamiltonian

$$(30) \qquad H_{A,\text{eff}}^{\Omega} = \sum_{i=1}^{A} h_i + \frac{1}{A-2} \sum_{i<j<k}^{A} V_{3\text{eff},ijk}^{\text{NN}} \,,$$

where the $\frac{1}{A-2}$ factor takes care of over-counting the contribution from the two-nucleon interaction.

If the NNN interaction is included, we need to calculate in addition to (29) the following effective interaction:

$$(31) \quad V_{3\text{eff},123}^{\text{NN}+\text{NNN}} =$$

$$= P_3 \left[e^{-S_{123}^{\text{NN}+\text{NNN}}} (h_1 + h_2 + h_3 + V_{12}^{\Omega,A} + V_{13}^{\Omega,A} + V_{23}^{\Omega,A} + V_{\text{NNN},123}) e^{S_{123}^{\text{NN}+\text{NNN}}} \right.$$

$$\left. - (h_1 + h_2 + h_3) \right] P_3.$$

This three-body effective interaction is obtained using full-space solutions of the Hamiltonian $h_1 + h_2 + h_3 + V_{12}^{\Omega,A} + V_{13}^{\Omega,A} + V_{23}^{\Omega,A} + V_{\text{NNN},123}$. The three-body effective interaction contribution from the NNN interaction we then define as

$$(32) \qquad V_{3\text{eff},123}^{\text{NNN}} \equiv V_{3\text{eff},123}^{\text{NN}+\text{NNN}} - V_{3\text{eff},123}^{\text{NN}}.$$

The effective Hamiltonian used in the A-nucleon calculation is then

$$(33) \qquad H_{A,\text{eff}}^{\Omega} = \sum_{i=1}^{A} h_i + \frac{1}{A-2} \sum_{i<j<k}^{A} V_{3\text{eff},ijk}^{\text{NN}} + \sum_{i<j<k}^{A} V_{3\text{eff},ijk}^{\text{NNN}}.$$

At this point we also subtract the H_{CM} and, if the Slater determinant basis is to be used, we add the Lawson projection term $\beta(H_{\text{CM}} - (\frac{3}{2})\hbar\Omega)$ to shift the spurious CM excitations.

It should be noted that all the effective interaction calculations are performed in the Jacobi coordinate HO basis. As discussed above, the two-body effective interaction is performed in the $|nlsjt\rangle$ basis and the three-body effective interaction in the $|NiJT\rangle$ basis (15). In order to perform the A-nucleon calculation in the Slater determinant HO basis as is typically done for $A > 4$, the effective interaction needs to be transformed to single-nucleon HO basis. This is done with the help of the HO wave function transformations (6). The details for the three-body case in particular are given in refs. [23] and [24].

As a final remark, we note that the unitary transformation performed on the Hamiltonian should be also applied to other operators that are used to calculate observables. If this is done, the model-space-size convergence of observables improves. More details on calculation of effective operators are given in ref. [25].

2`4. *Convergence tests.* – In this subsection, we give examples of convergence of *ab initio* NCSM calculations. In fig. 1, we show the convergence of the ^{3}H ground-state energy with the size of the basis. Thin lines correspond to results obtained with the NN interaction only. Thick lines correspond to calculations that also include the NNN interaction. The full lines correspond to calculations with two-body effective interaction derived from the chiral effective field theory (EFT) NN interaction of ref. [26] discussed in more details in the next section. The dashed lines correspond to calculations with the bare, that is the original unrenormalized, chiral EFT NN interaction. The bare NNN interaction is added to either the bare NN or to the effective NN interaction in calculations depicted

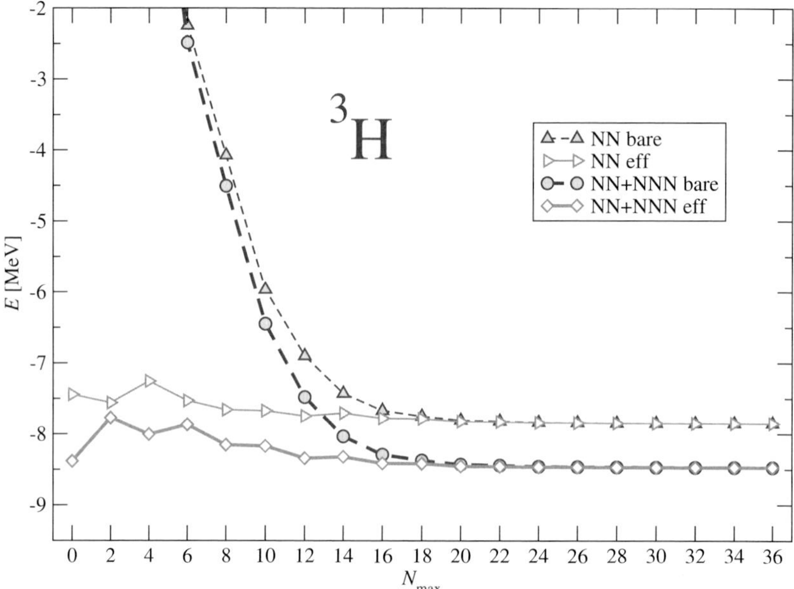

Fig. 1. – ^{3}H ground-state energy dependence on the size of the basis. The HO frequency of $\hbar\Omega = 28\,\text{MeV}$ was employed. Results with (thick lines) and without (thin lines) the NNN interaction are shown. The full lines correspond to calculations with two-body effective interaction derived from the chiral NN interaction, the dashed lines to calculations with the bare chiral NN interaction.

by thick lines. Here, we use the chiral EFT NNN interaction that will be discussed in details in the next section. We observe that the convergence is faster when the two-body effective interaction is used. However, starting at about $N_{\max} = 24$ the convergence is reached also in calculations with the bare NN interaction. The rate of convergence also depends on the choice of the HO frequency. In general, it is always advantageous to use the effective interaction in order to improve the convergence rate. It should be noted that in calculations with the effective interaction, the effective Hamiltonian is different at each point as the effective interaction depends on the size of the model space given by $N_{\max}$. The calculation with the bare interaction is a variational calculation converging from above with $N_{\max}$ and HO frequency Ω as variational parameters. The calculation with the effective interaction is not variational. The convergence can be from above, from below or oscillatory. This is because a part of the exact effective Hamiltonian is omitted. The calculation without NNN interaction converges to the ^{3}H ground-state energy $-7.852(5)\,\text{MeV}$, well above the experimental $-8.482\,\text{MeV}$. Once the NNN interaction is added, we obtain $-8.473(5)\,\text{MeV}$, close to experiment. As discussed in the next section, the NNN parameters were tuned to reproduce ^{3}H and ^{3}He binding energies.

In fig. 2, we show convergence of the ^{4}He ground-state energy. The NCSM calculations are performed in basis spaces up to $N_{\max} = 20$. Thin lines correspond to results

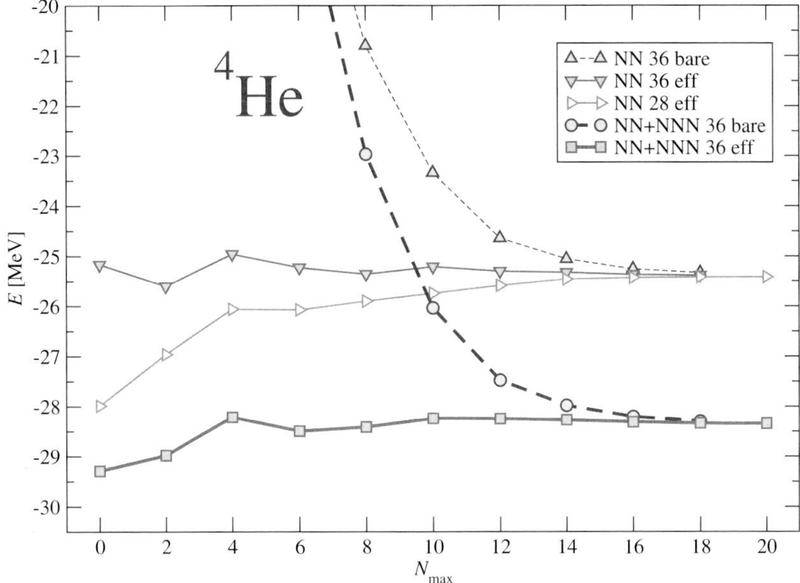

Fig. 2. – ^{4}He ground-state energy dependence on the size of the basis. The HO frequencies of $\hbar\Omega = 28$ and 36 MeV was employed. Results with (thick lines) and without (thin lines) the NNN interaction are shown. The full lines correspond to calculations with three-body effective interaction, the dashed lines to calculations with the bare interaction. For further details see the text.

obtained with the NN interaction only, while thick lines correspond to calculations that also include the NNN interaction. The dashed lines correspond to results obtained with bare interactions. The full lines correspond to results obtained using three-body effective interaction (the NCSM three-body cluster approximation). It is apparent that the use of the three-body effective interaction improves the convergence rate dramatically. We can see that at about $N_{\mathrm{max}} = 18$ the bare interaction calculation reaches convergence as well. It should be noted, however, that p-shell calculations with the NNN interactions are presently feasible in model spaces up to $N_{\mathrm{max}} = 6$ or $N_{\mathrm{max}} = 8$. The use of the three-body effective interaction is then essential in the p-shell calculations. The calculation without NNN interaction was done for two different HO frequencies. It is apparent that convergence to the same result is in both cases. We note that in the case of no NNN interaction, we may use just the two-body effective interaction (two-body cluster approximation), which is much simpler. The convergence is slower, however, see discussion in ref. [27]. We also note that ^{4}He properties with the chiral EFT NN interaction that we employ here were calculated using two-body cluster approximation in ref. [28] and present results are in agreement with results found there. Our ^{4}He results ground state energy results are $-25.39(1)$ MeV in the NN case and $-28.34(2)$ MeV in the NN+NNN case. The experimental value is -28.296 MeV. We note that the present *ab initio* NCSM ^{3}H

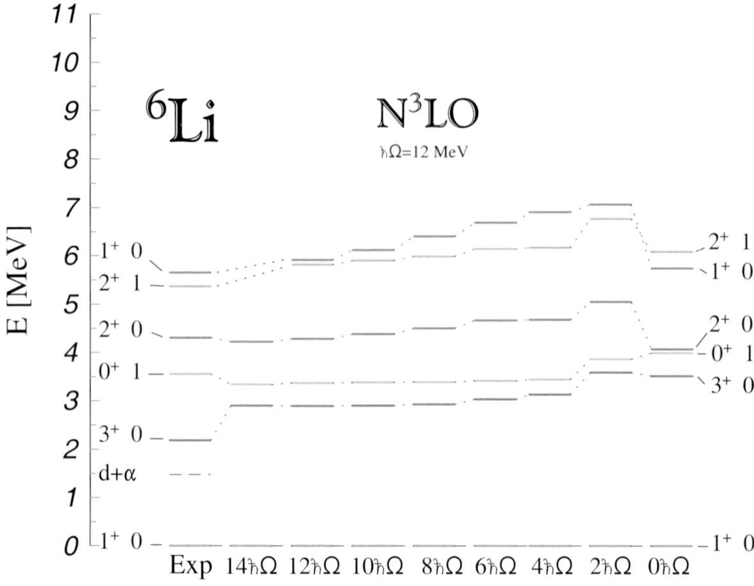

Fig. 3. – Calculated positive-parity excitation spectra of ^{6}Li obtained in $0\hbar\Omega$-$14\hbar\Omega$ basis spaces using two-body effective interactions derived from the chiral EFT NN potential are compared to experiment. The HO frequency of $\hbar\Omega = 12\,\mathrm{MeV}$ was used.

and ^{4}He results obtained with the chiral EFT NN interaction are in a perfect agreement with results obtained using the variational calculations in the hyperspherical harmonics basis as well as with the Faddeev-Yakubovsky calculations published in ref. [29]. A satisfying feature of the present NCSM calculation is the fact that the rate of convergence is not affected in any significant way by inclusion of the NNN interaction.

As yet another example of convergence of *ab initio* NCSM calculations, we present the excitation energy calculations of the five lowest excited states of ^{6}Li using the chiral EFT NN potential. The NCSM excitation energy dependence on the basis size is presented in fig. 3 for the HO frequency of $\hbar\Omega = 12\,\mathrm{MeV}$. Due to the complexity of the calculations, the lowest four states were obtained in basis spaces up to $N_{\mathrm{max}} = 14$ while we stopped at $N_{\mathrm{max}} = 12$ for the 2^+1 and the 1_2^+0 state. The calculations were performed using the two-body effective interaction in the Slater determinant HO basis with the shell model code Antoine [30]. Results for other HO frequencies were published in ref. [28]. We observe that the convergence rate with N_{max} is different for different states. In particular, the 3^+0 state and the 0^+1 state converge faster in the higher-frequency calculations ($\hbar\Omega = 12, 13\,\mathrm{MeV}$), while the higher-lying states converge faster in the lower frequency calculations ($\hbar\Omega = 8, 10\,\text{MeV}$). The results in fig. 3 demonstrate a good convergence of the excitation energies in particular for the 3^+0 and 0^+1 states. An interesting result is the overestimation of the 3^+0 excitation energy compared to experiment. It turns out that this problem is resolved once the NNN interaction is included in the Hamiltonian.

3. – Light nuclei from chiral EFT interactions

Interactions among nucleons are governed by quantum chromodynamics (QCD). In the low-energy regime relevant to nuclear structure, QCD is non-perturbative, and, therefore, hard to solve. Thus, theory has been forced to resort to models for the interaction, which have limited physical basis. New theoretical developments, however, allow us connect QCD with low-energy nuclear physics. The chiral effective field theory (χEFT) [31] provides a promising bridge. Beginning with the pionic or the nucleon-pion system [32] one works consistently with systems of increasing nucleon number [33-35]. One makes use of spontaneous breaking of chiral symmetry to systematically expand the strong interaction in terms of a generic small momentum and takes the explicit breaking of chiral symmetry into account by expanding in the pion mass. Thereby, the NN interaction, the NNN interaction and also πN scattering are related to each other. The χEFT predicts, along with the NN interaction at the leading order, an NNN interaction at the 3rd order (next-to-next-to-leading order or N^{2}LO) [31, 36, 37], and even an NNNN interaction at the 4th order (N^{3}LO) [38]. The details of QCD dynamics are contained in parameters, low-energy constants (LECs), not fixed by the symmetry. These parameters can be constrained by experiment. At present, high-quality NN potentials have been determined at order N^{3}LO [26]. A crucial feature of χEFT is the consistency between the NN, NNN and NNNN parts. As a consequence, at N^{2}LO and N^{3}LO, except for two LECs, assigned to two NNN diagrams, the potential is fully constrained by the parameters defining the NN interaction.

We adopt the potentials of the χEFT at the orders presently available, the NN at N^{3}LO of ref. [26] and the NNN interaction at N^{2}LO [36, 37]. Since the NN interaction is non-local, the *ab initio* NCSM is the only approach currently available to solve the resulting many-body Schrödinger equation for mid-p-shell nuclei. We are in a position to use the *ab initio* NCSM calculations in two ways. One of them is the determination of the LECs assigned to two NNN diagrams that must be determined in $A \geq 3$ systems. The other is testing predictions of the chiral NN and NNN interactions for light nuclei.

The NNN interaction at N^{2}LO of the χEFT comprises of three parts: i) the two-pion exchange, ii) the one-pion exchange plus contact and the three-nucleon contact, see fig. 4. The LECs associated with the two-pion exchange also appear in the NN interaction and are therefore determined in the $A = 2$ system. The one-pion exchange plus contact term (D-term) is associated with the LEC c_D and the three-nucleon contact term (E-term) is associated with the LEC c_E. The c_D and c_E LECs, expected to be of order one, can be constrained by the $A = 3$ binding energy. We then still need additional observables to determine the two parameters. Their determination from three-nucleon scattering data is difficult due to a correlation of the ^{3}H binding energy and, *e.g.*, the *nd* doublet scattering

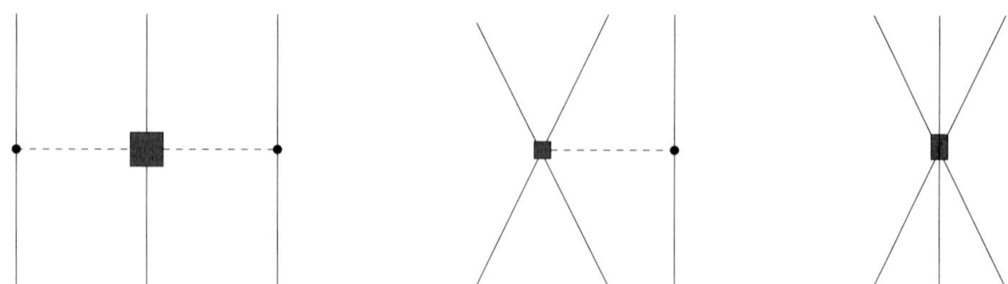

Fig. 4. – Terms of the N^2LO χEFT NNN interaction.

length [37] and, in general, due to the lack of an in-depth three-nucleon scattering phase shift analysis. We therefore investigate sensitivity of the $A > 3$ nuclei properties to the variation of the constrained LECs.

Before presenting results of the *ab initio* NCSM calculations with the χEFT NN+NNN interactions, most of which were published in ref. [39], let us discuss briefly a few technical details of the calculations with the NNN interaction. The NNN interaction is symmetric under permutation of the three nucleon indexes. It can be written as a sum of three pieces related by particle permutations:

$$(34) \qquad W = W_1 + W_2 + W_3.$$

To obtain its matrix element in an antisymmetrized three-nucleon basis we need to consider just a single term, *e.g.* W_1. Using the basis introduced in eq. (15), a general matrix element can be written as

$$(35) \qquad \langle NiJT|W|N'i'JT\rangle = 3\langle NiJT|W_1|N'i'JT\rangle$$
$$= 3\sum \langle nlsjt, \mathcal{NLJ}||NiJT\rangle\langle n'l's'j't', \mathcal{N'L'J'}||N'i'JT\rangle$$
$$\times \langle (nlsjt, \mathcal{NLJ})JT|W_1|(n'l's'j't', \mathcal{N'L'J'})JT\rangle.$$

We consider just the most trivial part of the χEFT N^2LO NNN interaction, the three-nucleon contact term and evaluate its matrix element to demonstrate that non-trivial effort is needed to include the NNN interactions in many-body calculations. The three-nucleon contact term can be written as

$$(36) \qquad W_1^{\text{cont}} = E\vec{\tau}_2 \cdot \vec{\tau}_3 \delta(\vec{r}_1 - \vec{r}_2)\delta(\vec{r}_3 - \vec{r}_1)$$
$$= E\vec{\tau}_2 \cdot \vec{\tau}_3 \frac{1}{(2\pi)^6} \frac{1}{(\sqrt{3})^3} \int d\vec{\pi}_1 d\vec{\pi}_2 d\vec{\pi}_1' d\vec{\pi}_2' |\vec{\pi}_1 \vec{\pi}_2\rangle\langle \vec{\pi}_1' \vec{\pi}_2'|,$$

with $E = \frac{c_E}{F_\pi^4 \Lambda_\chi}$, where Λ_χ is the chiral symmetry-breaking scale of the order of the ρ meson mass and $F_\pi = 92.4\,\text{MeV}$ is the weak pion decay constant. The $\vec{\pi}_1$ and $\vec{\pi}_2$ are

Jacobi momenta associated with the the Jacobi coordinates $\vec{\xi}_1$ and $\vec{\xi}_2$ defined in eqs. (8) and (9).

The contact term must be regulated before it can be used in many-body calculations. We consider a regulator depending on momentum transfer:

$$(37) \quad W_1^{\text{cont,Q}} = E\vec{\tau}_2 \cdot \vec{\tau}_3 \frac{1}{(2\pi)^6} \frac{1}{(\sqrt{3})^3} \int d\vec{\pi}_1 d\vec{\pi}_2 d\vec{\pi}_1' d\vec{\pi}_2' |\vec{\pi}_1 \vec{\pi}_2\rangle F(\vec{Q}^2;\Lambda) F(\vec{Q}'^2;\Lambda) \langle \vec{\pi}_1' \vec{\pi}_2'|$$

$$= E\vec{\tau}_2 \cdot \vec{\tau}_3 \int d\vec{\xi}_1 d\vec{\xi}_2 |\vec{\xi}_1 \vec{\xi}_2\rangle Z_0(\sqrt{2}\xi_1;\Lambda) Z_0\left(\left|\frac{1}{\sqrt{2}}\vec{\xi}_1 + \sqrt{\frac{3}{2}}\vec{\xi}_2\right|;\Lambda\right) \langle \vec{\xi}_1 \vec{\xi}_2|,$$

with the regulator function $F(q^2;\Lambda) = \exp(-q^4/\Lambda^4)$. We defined momenta transferred by nucleon 2 and nucleon 3: $\vec{Q} = \vec{p}_2' - \vec{p}_2 = -\frac{1}{\sqrt{2}}(\vec{\pi}_1' - \vec{\pi}_1) + \frac{1}{\sqrt{6}}(\vec{\pi}_2' - \vec{\pi}_2)$ and $\vec{Q}' = \vec{p}_3' - \vec{p}_3 = -\sqrt{\frac{2}{3}}(\vec{\pi}_2' - \vec{\pi}_2)$. Also, we introduced the function

$$(38) \qquad Z_0(r;\Lambda) = \frac{1}{2\pi^2} \int dq\, q^2 j_0(qr) F(q^2;\Lambda).$$

This results in an interaction local in coordinate space because of the dependence of the regulator function on differences of initial and final Jacobi momenta. We can see that after the regulation, the form of the contact interaction becomes much more complicated. The three-nucleon matrix element of the regulated term is then obtained in the form

$$(39) \qquad \langle (nlsjt, \mathcal{NLJ})JT | W_1^{\text{cont,Q}} | (n'l's'j't', \mathcal{N'L'J'})JT \rangle =$$

$$= E6\delta_{ss'} \hat{t}\hat{t}'(-1)^{t+t'+T+\frac{1}{2}} \left\{ \begin{matrix} t & t' & 1 \\ \frac{1}{2} & \frac{1}{2} & \frac{1}{2} \end{matrix} \right\} \left\{ \begin{matrix} t & t' & 1 \\ \frac{1}{2} & \frac{1}{2} & T \end{matrix} \right\}$$

$$\times \hat{j}\hat{j}'\hat{\mathcal{J}}\hat{\mathcal{J}}'\hat{l}'\hat{\mathcal{L}}'(-1)^{\mathcal{J}-\frac{1}{2}+\mathcal{J}'-\mathcal{J}+l+\mathcal{L}+s}$$

$$\times \sum_X (-1)^X \hat{X}^2 \left\{ \begin{matrix} l' & l & X \\ j & j' & s \end{matrix} \right\} \left\{ \begin{matrix} j & j' & X \\ \mathcal{J}' & \mathcal{J} & J \end{matrix} \right\} \left\{ \begin{matrix} \mathcal{J}' & \mathcal{J} & X \\ \mathcal{L} & \mathcal{L}' & \frac{1}{2} \end{matrix} \right\}$$

$$\times (l'0X0|l0)(\mathcal{L}'0X0|\mathcal{L}0)$$

$$\times \int d\xi_1 d\xi_2 \xi_1^2 \xi_2^2 R_{nl}(\xi_1, b) R_{\mathcal{NL}}(\xi_2, b) R_{n'l'}(\xi_1, b) R_{\mathcal{N'L'}}(\xi_2, b)$$

$$\times Z_0\left(\sqrt{2}\xi_1;\Lambda\right) Z_{0,X}\left(\sqrt{\frac{1}{2}}\xi_1, \sqrt{\frac{3}{2}}\xi_2;\Lambda\right),$$

with a new function

$$(40) \qquad Z_{0,X}(r_1, r_2;\Lambda) = \frac{1}{2\pi^2} \int dq\, q^2 j_X(qr_1) j_X(qr_2) F(q^2;\Lambda).$$

and customary abbreviation $\hat{l} = \sqrt{2l+1}$. Evaluation of the other N^{2}LO NNN terms is still more complicated.

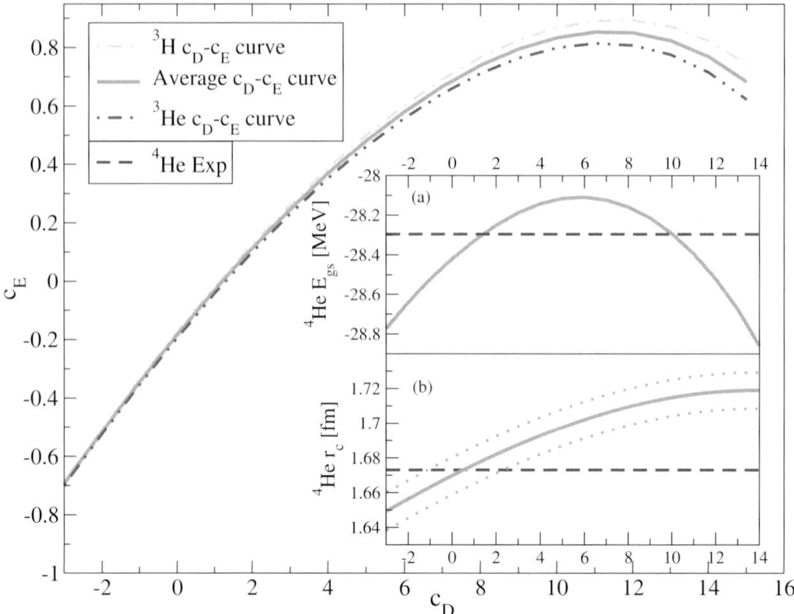

Fig. 5. – Relations between c_D and c_E for which the binding energy of ^{3}H (8.482 MeV) and ^{3}He (7.718 MeV) are reproduced. (a) ^{4}He ground-state energy along the averaged curve. (b) ^{4}He charge radius r_c along the averaged curve. Dotted lines represent the r_c uncertainty due to the uncertainties in the proton charge radius.

It is important to note that our NCSM results through $A = 4$ are fully converged in that they are independent of the N_{max} cut-off and the $\hbar\Omega$ HO energy. This was demonstrated in the subsection on the *ab initio* NCSM convergence tests in particular for the chiral EFT interactions we are investigating here. For heavier systems, we characterize the approach to convergence by the dependence of results on N_{max} and $\hbar\Omega$.

Figure 5 shows the trajectories of the two LECs $c_D - c_E$ that are determined from fitting the binding energies of the $A = 3$ systems. Separate curves are shown for ^{3}H and ^{3}He fits, as well as their average. There are two points where the binding of ^{4}He is reproduced exactly. We observe, however, that in the whole investigated range of $c_D - c_E$, the calculated ^{4}He binding energy is within a few hundred keV of experiment. Consequently, the determination of the LECs in this way is likely not very stringent. We therefore investigate the sensitivity of the p-shell nuclear properties to the choice of the $c_D - c_E$ LECs. First, we maintain the $A = 3$ binding energy constraint. Second, we limit ourselves to the c_D values in the vicinity of the point $c_D \sim 1$ since the values close to the point $c_D \sim 10$ overestimate the ^{4}He radius. Also this large value might be considered "unnatural" from the χEFT point of view.

While most of the p-shell nuclear properties, *e.g.*, excitation spectra, are not very sensitive to variations of c_D in the vicinity of the $c_D \sim 1$ point, we were able to identify

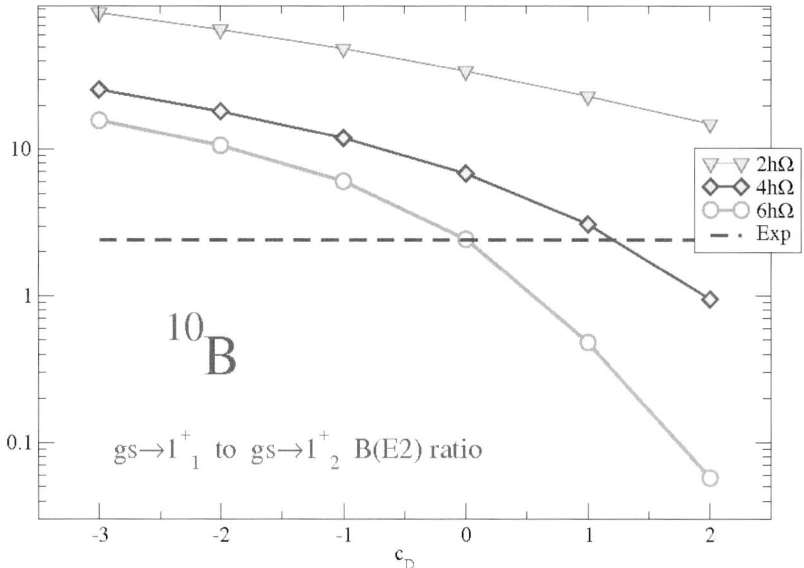

Fig. 6. – Dependence on the c_D with the c_E constrained by the $A = 3$ binding energy fit for different basis sizes for ^{10}B $B(E2; 3_1^+0 \to 1_1^+0)/B(E2; 3_1^+0 \to 1_2^+0)$ ratio. The HO frequency of $\hbar\Omega = 14\,$MeV was employed.

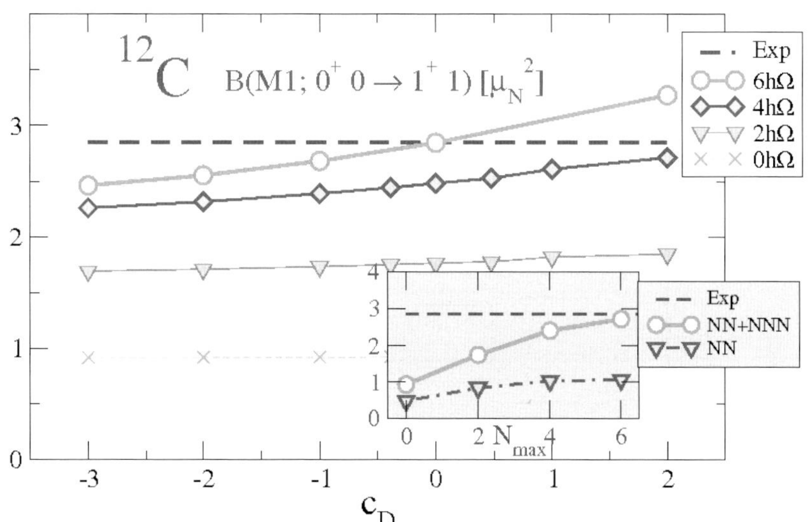

Fig. 7. – Dependence on the c_D with the c_E constrained by the $A = 3$ binding energy fit for different basis sizes for the ^{12}C $B(M1; 0^+0 \to 1^+1)$. The HO frequency of $\hbar\Omega = 15\,$MeV was employed. In the inset, the convergence of the $B(M1; 0^+0 \to 1^+1)$ is presented for calculations with (using $c_D = -1$) and without the NNN interaction.

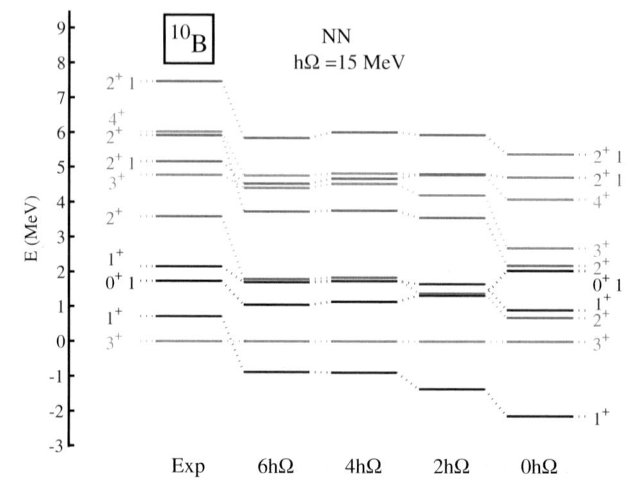

Fig. 8. – ^{10}B excitation spectra as function of the basis space size N_{max} at $\hbar\Omega = 15\,\mathrm{MeV}$ using the chiral NN interaction and comparison with experiment. The isospin of the states not explicitly depicted is $T = 0$.

several observables that do demonstrate strong dependence on c_D. For example, the ^{6}Li quadrupole moment that changes sign depending on the choice of c_D. In fig. 6, we display the ratio of the $B(E2)$ transitions from the ^{10}B ground state to the first and the second 1^+0 state. This ratio changes by several orders of magnitude depending on the c_D variation. This is due to the fact that the structure of the two 1^+0 states is exchanged depending on c_D.

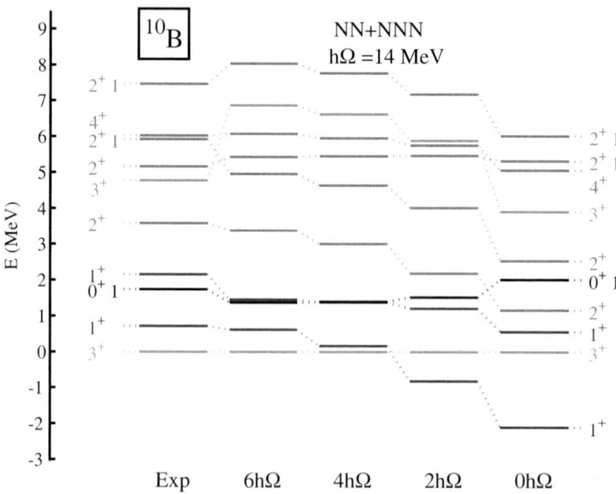

Fig. 9. – ^{10}B excitation spectra as function of the basis space size N_{max} at $\hbar\Omega = 14\,\mathrm{MeV}$ using the chiral NN+NNN interaction and comparison with experiment. The isospin of the states not explicitly depicted is $T = 0$.

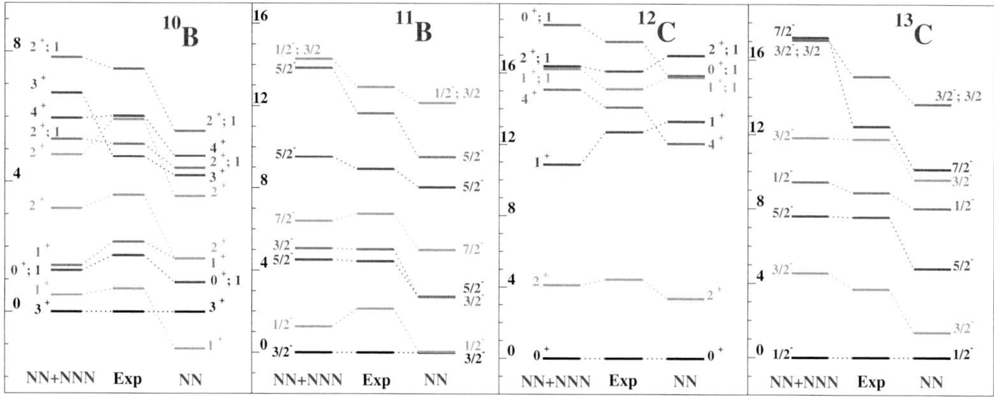

Fig. 10. – States dominated by p-shell configurations for ^{10}B, ^{11}B, ^{12}C, and ^{13}C calculated at $N_{\max} = 6$ using $\hbar\Omega = 15\,\mathrm{MeV}$ (14 MeV for ^{10}B). Most of the eigenstates are isospin $T = 0$ or $1/2$, the isospin label is explicitly shown only for states with $T = 1$ or $3/2$. The excitation energy scales are in MeV.

In fig. 7, we present the ^{12}C B(M1) transition from the ground state to the 1^+1 state. The B(M1) transition inset illustrates the importance of the NNN interaction in reproducing the experimental value [40]. Overall our results show that for $c_D < -2$ the ^{4}He radius and the ^{6}Li quadrupole moment underestimate experiment while for $c_D > 0$ the lowest two 1^+ states of ^{10}B are reversed and the ^{12}C $B(M1; 0^+0 \to 1^+1)$ is overestimated. We therefore select $c_D = -1$ as globally the best choice and use it for our further investigation.

We present in figs. 8 and 9 the excitation spectra of ^{10}B as a function of $N_{\max}$ for both the chiral NN+NNN, as well as with the chiral NN interaction alone. In both cases, the convergence with increasing $N_{\max}$ is quite quite reasonable for the low-lying states. Similar convergence rates are obtained for our other p-shell nuclei.

A remarkable feature of the ^{10}B results is the observation that the chiral NN interaction alone predicts incorrect ground-state spin of ^{10}B. The experimental value is 3^+0 while the calculated one is 1^+0. On the other hand, once we also include the chiral NNN interaction in the Hamiltonian, which is actually required by the χEFT, the correct ground-state spin is predicted. Further, once we select the c_D value as discussed above, *i.e.* $c_D = -1$, we also obtain the two lowest 1^+0 states in the experimental order.

We display in fig. 10 the natural parity excitation spectra of four nuclei in the middle of the p-shell with both the NN and the NN+NNN effective interactions from χEFT. The results shown are obtained in the largest basis spaces achieved to date for these nuclei with the NNN interactions, $N_{\max} = 6$ $(6\hbar\Omega)$. Overall, the NNN interaction contributes significantly to improve theory in comparison with experiment. This is especially well demonstrated in the odd mass nuclei for the lowest few excited states. The case of the ground-state spin of ^{10}B and its sensitivity to the presence of the NNN interaction discussed also in figs. 8 and 9 is clearly evident. We note that the ^{10}B results with the

NN interaction only in fig. 8 were obtained with the HO frequency of $\hbar\Omega = 15\,\text{MeV}$, while those in fig. 10 with $\hbar\Omega = 14\,\text{MeV}$. A weak HO frequency dependence of the $N_{\max} = 6$ results is evident. The ^{10}B results with NN+NNN interaction presented in figs. 9 and 10 were obtained using the same HO frequency. Still, one may notice small differences of the $N_{\max} = 6$ results. The reason behind those differences is the use of two alternative D-term regularizations (both depending on the momentum transfer, details will be discussed elsewhere). As the dependence on the regulator is a higher-order effect than the χEFT expansion order used to derive the NNN interaction, these differences should have only minor effect. It is satisfying that the present ^{10}B results appear to support this expectation.

Concerning the ^{12}C results, there is an initial indication that the chiral NNN interaction is somewhat over-correcting the inadequacies of the NN interaction since, *e.g.*, 1^+0 and the 4^+0 states in ^{12}C are not only interchanged but they are also spread apart more than the experimentally observed separation. In the ^{13}C results, we can also identify an indication of an overly strong correction arising from the chiral NNN interaction as seen in the upward shift of the $\frac{7}{2}^-$ state. However, the experimental $\frac{7}{2}^-$ may have significant intruder components and is not well matched with our state. In addition, convergence for some higher-lying states is affected by incomplete treatment of clustering in the NCSM. This point will be elaborated upon later. These results required substantial computer resources. A typical $N_{\max} = 6$ spectrum shown in fig. 10 and a set of additional experimental observables, takes 4 hours on 3500 processors of the LLNL's Thunder machine. The A-nucleon calculations were performed in the Slater determinant HO basis using the shell model code MFD [41]. More details on some of the results discussed here are published ref. [39].

The calculations presented in this section demonstrate that the chiral NNN interaction makes substantial contributions to improving the spectra and other observables. However, there is room for further improvement in comparison with experiment. In these calculations we used a strength of the 2π-exchange piece of the NNN interaction, which is consistent with the NN interaction that we employed (*i.e.* from ref. [26]). This strength is somewhat uncertain (see, *e.g.*, ref. [24]). Therefore, it will be important to study the sensitivity of our results with respect to this strength. Further on, it will be interesting to incorporate sub-leading NNN interactions and also four-nucleon interactions, which are also order N^{3}LO [38]. Finally, it will be useful to extend the basis spaces to $N_{\max} = 8$ ($8\hbar\Omega$) for $A > 6$ to further improve convergence.

4. – Cluster overlap functions and S-factors of capture reactions

In the *ab initio* NCSM calculations, we are able to obtain wave functions of low-lying states of light nuclei in large model spaces. An interesting and important question is, what is the cluster structure of these wave functions. That is we want to understand, how much, *e.g.*, a ^6Li eigenstate looks like ^4He plus deuteron, a ^7Be eigenstate looks like ^4He plus ^3He, a ^8B eigenstate looks like ^7Be plus proton and so on. This information is important for the description of low-energy nuclear reactions. To gain insight, one introduces channel cluster form factors (or overlap integrals, overlap functions). The formalism for calculating the channel cluster form factors from the NCSM wave functions was developed in ref. [42]. Here we just briefly repeat a part of the formalism relevant to the simplest case when the lighter of the two clusters is a single-nucleon.

We consider a composite system of A nucleons, *e.g.* ^{8}B, a nucleon projectile, here a proton, and an $(A-1)$-nucleon target, *e.g.* ^{7}Be. Both nuclei are assumed to be described by eigenstates of the NCSM effective Hamiltonians expanded in the HO basis with identical HO frequency and the same (for the eigenstates of the same parity) or differing by one unit of the HO excitation (for the eigenstates of opposite parity) definitions of the model space. The target and the composite system is assumed to be described by wave functions expanded in Slater determinant single-particle HO basis (that is obtained from a calculation using a shell model code like Antoine).

Let us introduce a projectile-target wave function

$$(41) \quad \left\langle \vec{\xi}_1 \ldots \vec{\xi}_{A-2} r' \hat{r} \middle| \Phi^{(A-1,1)JM}_{(l\frac{1}{2})j;\alpha I_1}; \delta_r \right\rangle = \sum (jmI_1 M_1 | JM)(lm_l \tfrac{1}{2}m_s|jm)\frac{\delta(r-r')}{rr'}$$
$$\times Y_{lm_l}(\hat{r})\chi_{m_s}\left\langle \vec{\xi}_1 \ldots \vec{\xi}_{A-2} \middle| A-1\alpha I_1 M_1 \right\rangle,$$

where $\left\langle \vec{\xi}_1 \ldots \vec{\xi}_{A-2}|A-1\alpha I_1 M_1\right\rangle$ and χ_{m_s} are the target and the nucleon wave function, respectively. Here, l is the channel relative orbital angular momentum, $\vec{\xi}$ are the target Jacobi coordinates defined in eqs. (8)–(10) and $\vec{r} = [\frac{1}{A-1}(\vec{r}_1 + \vec{r}_2 + \ldots + \vec{r}_{A-1}) - \vec{r}_A]$ describes the relative distance between the nucleon and the center of mass of the target. The spin and isospin coordinates were omitted for simplicity.

The channel cluster form factor is then defined by

$$(42) \quad g^{A\lambda J}_{(l\frac{1}{2})j;A-1\alpha I_1}(r) = \left\langle A\lambda J \middle| \mathcal{A}\Phi^{(A-1,1)J}_{(l\frac{1}{2})j;\alpha I_1}; \delta_r \right\rangle,$$

with $\mathcal{A}$ the antisymmetrizer and $|A\lambda J\rangle$ an eigenstate of the A-nucleon composite system (here ^{8}B). It can be calculated from the NCSM eigenstates obtained in the Slater-determinant basis from a reduced matrix element of the creation operator. The derivation is as follows. First, we use the relation (17) for both the composite A-nucleon and the target $(A-1)$-nucleon eigenstate. With the help of relations analogous to (6):

$$(43) \quad \sum_{Mm}(LMlm|Qq)\varphi_{NLM}(\vec{R}^{A-1}_{\mathrm{CM}})\varphi_{nlm}(\vec{r}_A) =$$
$$\sum_{n'l'm'N'L'M'} \langle n'l'N'L'Q|NLnlQ\rangle_{\frac{1}{A-1}}(l'm'L'M'|Qq)\varphi_{n'l'm'}(\vec{\xi}_{A-1})\varphi_{N'L'M'}(\vec{\xi}_0),$$

we obtain

$$(44) \quad {}_{\mathrm{SD}}\!\left\langle A\lambda J \middle| \mathcal{A}\Phi^{(A-1,1)J}_{(l\frac{1}{2})j;\alpha I_1}; nl \right\rangle_{\mathrm{SD}} = \langle nl00l|00nll\rangle_{\frac{1}{A-1}} \left\langle A\lambda J \middle| \mathcal{A}\Phi^{(A-1,1)J}_{(l\frac{1}{2})j;\alpha I_1}; nl \right\rangle,$$

with a general HO bracket due to the CM motion. The nl in (44) refers to a replacement of δ_r by the HO $R_{nl}(r)$ radial wave function. Second, we relate the SD overlap to a linear combination of matrix elements of a creation operator between the target and the composite eigenstates $_{\mathrm{SD}}\langle A\lambda J|a^{\dagger}_{nlj}|A-1\alpha I_1\rangle_{\mathrm{SD}}$. The subscript SD refers to the fact that these states were obtained in the Slater determinant basis. Such matrix elements are easily calculated by shell model codes. The result is

$$(45) \qquad \left\langle A\lambda J \middle| \mathcal{A}\Phi^{(A-1,1)J}_{(l\frac{1}{2},j);\alpha I_1}; \delta_r \right\rangle = \sum_n R_{nl}(r)\frac{1}{\langle nl00l|00nll\rangle_{\frac{1}{A-1}}}\frac{1}{\hat{j}}(-1)^{I_1-J-j}$$
$$\times_{\mathrm{SD}}\langle A\lambda J||a^{\dagger}_{nlj}||A-1\alpha I_1\rangle_{\mathrm{SD}}.$$

The eigenstates expanded in the Slater determinant basis contain CM components. A general HO bracket, which value is simply given by

$$(46) \qquad \langle nl00l|00nll\rangle_{\frac{1}{A-1}} = (-1)^l \left(\frac{A-1}{A}\right)^{\frac{2n+l}{2}},$$

then appears in eq. (45) in order to remove these components. The $R_{nl}(r)$ in eq. (45) is the radial HO wave function with the oscillator length parameter $b = \sqrt{\frac{\hbar}{\frac{A-1}{A}m\Omega}}$, where m is the nucleon mass.

A conventional spectroscopic factor is obtained by integrating the square of the cluster form factor:

$$(47) \qquad S^{A\lambda J}_{(l\frac{1}{2})j;A-1\alpha I_1} = \int \mathrm{d}r r^2 \left|g^{A\lambda J}_{(l\frac{1}{2})j;A-1\alpha I_1}(r)\right|^2.$$

A generalization for projectiles (= the lighter of the two clusters) with 2, 3 or 4 nucleons is straightforward, although the expressions become more involved. In all cases, the projectile is described by wave function expanded in Jacobi coordinate HO basis, while the composite and the target eigenstates are expanded in the Slater determinant HO basis. Full details are given in ref. [42].

As an example, in fig. 11 we present the cluster overlap function of ^{10}B ground state with ^{6}Li $+$ ^{4}He. Results are given for ^{6}Li in the 1^+0 ground state and in the 3^+0 excited state. We can see that the ^{6}Li 1^+0 ground-state component is rather small. The ^{10}B ground state is dominated by a superposition of S-, D- and G-waves of relative motion of ^{4}He and ^{6}Li in the 3^+0 state.

The overlap functions introduced in this subsection are relevant for description of low-energy γ-capture reactions important for nuclear astrophysics. Next, we investigate three reactions of this type.

4.1. ^{7}Be(p, γ)^{8}B. – The ^{7}Be(p, γ)^{8}B capture reaction serves as an important input for understanding the solar neutrino flux [43]. Recent experiments have determined the neutrino flux emitted from ^{8}B with a precision of 9% [44]. On the other hand, theoretical predictions have uncertainties of the order of 20% [45, 46]. The theoretical neutrino flux

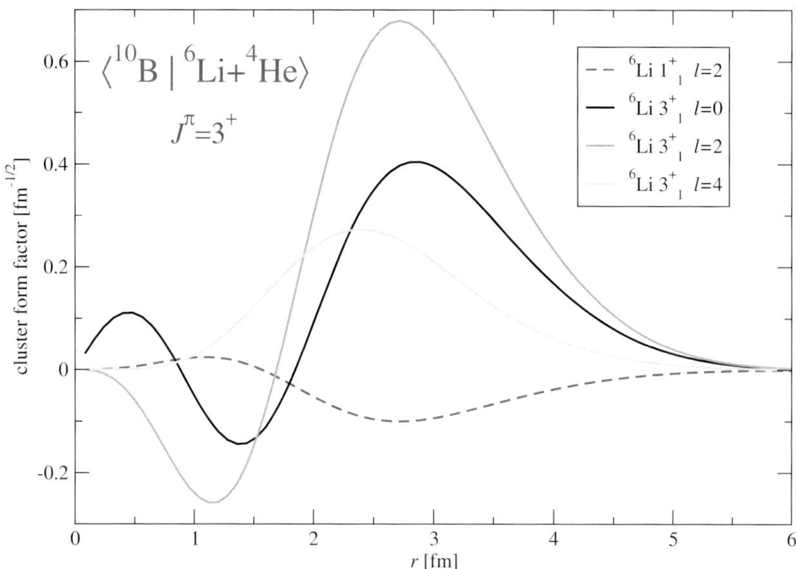

Fig. 11. – Overlap integral of ^{10}B ground state with ^{6}Li $+$ ^{4}He as a function of separation between the ^{4}He and the ^{6}Li. Results for the ^{6}Li $1^{+}0$ ground state and the first excited $3^{+}0$ state are compared. The χEFT NN+NNN interaction and the $N_{\mathrm{max}} = 6$ model space for ^{10}B and ^{6}Li were used.

depends on the ^{7}Be(p, γ)^{8}B S-factor. Many experimental and theoretical investigations studied this reaction.

In this subsection, we discuss a calculation of the ^{7}Be(p, γ)^{8}B S-factor starting from *ab initio* wave functions of ^{8}B and ^{7}Be. It should be noted that the aim of *ab initio* approaches is to predict correctly absolute cross-sections (S-factors), not only relative cross-sections. The full details of our ^{7}Be(p, γ)^{8}B investigation were published in refs. [47, 48].

Our calculations for both ^{7}Be and ^{8}B nuclei were performed using the high-precision CD-Bonn 2000 NN potential [49] in model spaces up to $10\hbar\Omega$ ($N_{\mathrm{max}} = 10$) for a wide range of HO frequencies. From the obtained ^{8}B and ^{7}Be wave functions, we calculate the channel cluster form factors (overlap functions, overlap integrals) $g^{A\lambda J}_{(l\frac{1}{2})j;A-1\alpha I_{1}}(r)$ as discussed in the previous subsection. Here, $A = 8$, l is the channel relative orbital angular momentum and $\vec{r} = [\frac{1}{A-1}(\vec{r}_{1}+\vec{r}_{2}+\ldots+\vec{r}_{A-1})-\vec{r}_{A}]$ describes the relative distance between the proton and the center of mass of ^{7}Be. The two most important channels are the p-waves, $l = 1$, with the proton in the $j = 3/2$ and $j = 1/2$ states, $\vec{j} = \vec{l}+\vec{s}, s = 1/2$. In these channels, we obtain the spectroscopic factors of 0.96 and 0.10, respectively. The dominant $j = 3/2$ overlap integral is presented in fig. 12 by the full line. The $10\hbar\Omega$

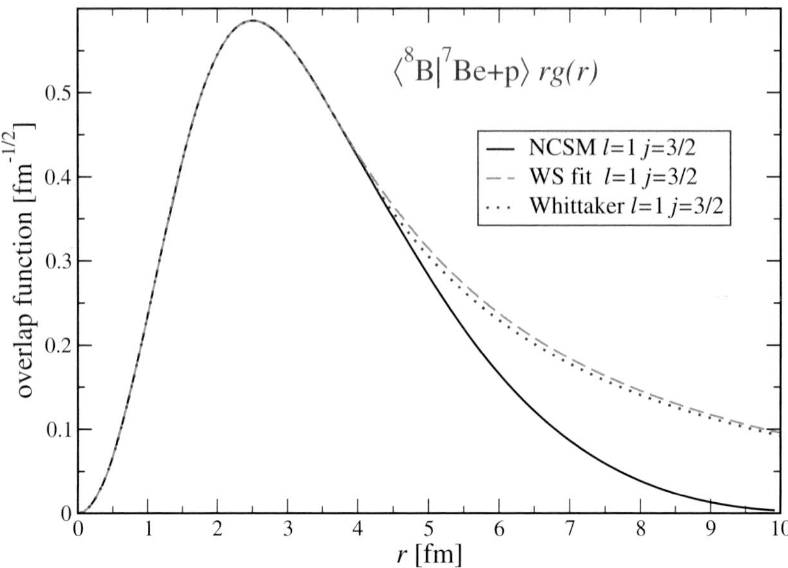

Fig. 12. – Overlap function, $rg(r)$, for the ground state of ^{8}B with the ground state of ^{7}Be plus proton as a dependence on separation between the ^{7}Be and the proton. The p-wave channel with $j = 3/2$ is shown. The full line represents the NCSM result obtained using the CD-Bonn 2000 NN potential, the $10\hbar\Omega$ model space and the HO frequency of $\hbar\Omega = 12$ MeV. The dashed lines represent corrected overlaps obtained from a Woods-Saxon potential whose parameters were fit to the NCSM overlaps up to 4.0 fm under the constraint to reproduce the experimental separation energy. The dotted lines represent overlap corrections by the direct Whittaker function matching.

model space and the HO frequency of $\hbar\Omega = 12$ MeV were used. Despite the fact, that a very large basis was employed in the present calculation, it is apparent that the overlap function is nearly zero at about 10 fm. This is a consequence of the HO basis asymptotic behavior. As already discussed, in the *ab initio* NCSM, the short-range correlations are taken into account by means of the effective interaction. The medium-range correlations are then included by using a large, multi-$\hbar\Omega$ HO basis. The long-range behavior is not treated correctly, however. The proton capture on ^{7}Be to the weakly bound ground state of ^{8}B associated dominantly by the $E1$ radiation is a peripheral process. In order to calculate the S-factor of this process we need to go beyond the *ab initio* NCSM as done up to this point. We expect, however, that the interior part of the overlap function is realistic. It is then straightforward to find a quick fix and correct the asymptotic behavior of the overlap functions, which should be proportional to the Whittaker function.

One possibility we explored utilizes solutions of a Woods-Saxon (WS) potential. In particular, we performed a least-square fit of a WS potential solution to the interior of the NCSM overlap in the range of 0–4 fm. The WS potential parameters were varied in the fit under the constraint that the experimental separation energy of ^{7}Be + p, $E_0 = 0.137\,\text{MeV}$, was reproduced. In this way we obtain a perfect fit to the interior of the overlap integral and a correct asymptotic behavior at the same time. The result is shown in fig. 12 by the dashed line.

Another possibility is a direct matching of logarithmic derivatives of the NCSM overlap integral and the Whittaker function: $\frac{\mathrm{d}}{\mathrm{d}r}\ln(rg_{lj}(r)) = \frac{\mathrm{d}}{\mathrm{d}r}\ln(C_{lj}W_{-\eta,l+1/2}(2k_0r))$, where η is the Sommerfeld parameter, $k_0 = \sqrt{2\mu E_0}/\hbar$ with μ the reduced mass and E_0 the separation energy. Since asymptotic normalization constant (ANC) C_{lj} cancels out, there is a unique solution at $r = R_m$. For the discussed overlap presented in fig. 12, we found $R_m = 4.05\,\text{fm}$. The corrected overlap using the Whittaker function matching is shown in fig. 12 by a dotted line. In general, we observe that the approach using the WS fit leads to deviations from the original NCSM overlap starting at a smaller radius. In addition, the WS solution fit introduces an intermediate range from about $4\,\text{fm}$ to about $6\,\text{fm}$, where the corrected overlap deviates from both the original NCSM overlap and the Whittaker function. Perhaps, this is a more realistic approach compared to the direct Whittaker function matching. In any case, by considering the two alternative procedures we are in a better position to estimate uncertainties in our S-factor results.

In the end, we re-scale the corrected overlap functions to preserve the original NCSM spectroscopic factors (table 2 of ref. [47]). In general, we observe a faster convergence of the spectroscopic factors than that of the overlap functions. The corrected overlap function should represent the infinite space result. By re-scaling a corrected overlap function obtained at a finite $N_{\max}$, we approach faster the infinite space result. At the same time, by re-scaling we preserve the spectroscopic factor sum rules.

The S-factor for the reaction ^{7}Be(p$,\gamma)^8$B also depends on the continuum wave function, $R_{lj}^{(c)}$. As we have not yet developed an extension of the NCSM to describe continuum wave functions (see, however, the discussion in sect. **5**), we obtain $R_{lj}^{(c)}$ for s and d waves from a WS potential model. Since the largest part of the integrand stays outside the nuclear interior, one expects that the continuum wave functions are well described in this way. In order to have the same scattering wave function in all the calculations, we chose a WS potential from ref. [50] that was fitted to reproduce the p-wave 1^+ resonance in ^{8}B. It was argued [51] that such a potential is also suitable for the description of s- and d-waves. We note that the S-factor is very weakly dependent on the choice of the scattering-state potential (using our fitted potential for the scattering state instead changes the S-factor by less than $1.5\,\text{eV}$ b at $1.6\,\text{MeV}$ with no change at $0\,\text{MeV}$).

Our obtained S-factor is presented in fig. 13 where contributions from the two partial waves are shown together with the total result. It is interesting to note a good agreement of our calculated S-factor with the recent Seattle direct measurement [52].

In order to judge the convergence of our S-factor calculation, we performed a detailed investigation of the model-space-size and the HO frequency dependences. We used the HO frequencies in the range from $\hbar\Omega = 11\,\text{MeV}$ to $\hbar\Omega = 15\,\text{MeV}$ and the model spaces from $6\hbar\Omega$ to $10\hbar\Omega$. By analysing these results, we arrived at the S-factor value of $S_{17}(10\,\text{keV}) = 22.1 \pm 1.0\,\text{eV}$ b.

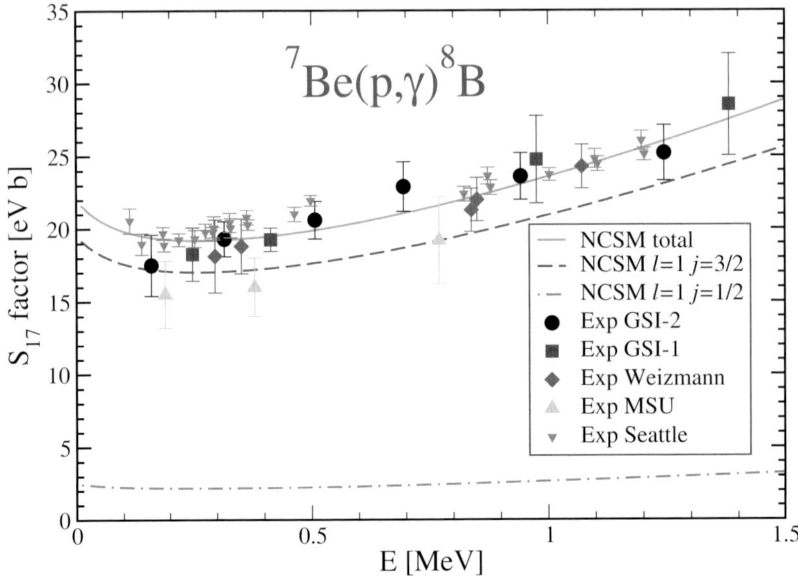

Fig. 13. – The $^7\mathrm{Be}(p,\gamma)^8\mathrm{B}$ S-factor obtained using the NCSM overlap functions with corrected asymptotics as described in the text. The dashed and dash-dotted lines show the contribution due to the $l = 1$, $j = 3/2$ and $j = 1/2$ partial waves, respectively. Experimental values are from refs. [52,53].

4˙2. $^3\mathrm{He}(\alpha,\gamma)^7\mathrm{Be}$. – The $^3\mathrm{He}(\alpha,\gamma)^7\mathrm{Be}$ capture reaction cross-section was identified the most important uncertainty in the solar model predictions of the neutrino fluxes in the p-p chain [46]. We investigated the bound states of $^7\mathrm{Be}$, $^3\mathrm{He}$ and $^4\mathrm{He}$ within the *ab initio* NCSM and calculated the overlap functions of $^7\mathrm{Be}$ bound states with the ground states of $^3\mathrm{He}$ plus $^4\mathrm{He}$ as a function of separation between the $^3\mathrm{He}$ and the α-particle. The obtained p-wave overlap functions of the $^7\mathrm{Be}$ $3/2^-$ ground state excited state are presented in fig. 14 by the full line. The dashed lines show the corrected overlap function obtained by the least-square fits of the WS parameters done in the same way as in the $^8\mathrm{B} \leftrightarrow {}^7\mathrm{Be} + \mathrm{p}$ case. The corresponding NCSM spectroscopic factors obtained using the CD-Bonn 2000 in the $10\hbar\Omega$ model space for $^7\mathrm{Be}$ ($12\hbar\Omega$ for $^{3,4}\mathrm{He}$) and HO frequency of $\hbar\Omega = 13\,\mathrm{MeV}$ are 0.93 and 0.91 for the ground state and the first excited state of $^7\mathrm{Be}$, respectively. We note that contrary to the $^8\mathrm{B} \leftrightarrow {}^7\mathrm{Be} + \mathrm{p}$ case, the $^7\mathrm{Be} \leftrightarrow {}^3\mathrm{He} + \alpha$ p-wave overlap functions have a node.

Using the corrected overlap functions and a $^3\mathrm{He} + \alpha$ scattering state obtained using the potential model of ref. [54] we calculated the $^3\mathrm{He}(\alpha,\gamma)^7\mathrm{Be}$ S-factor. Our $10\hbar\Omega$ result is presented in the left panel of fig. 15. We show the total S-factor as well as the contributions from the capture to the ground state and the first excited state of $^7\mathrm{Be}$.

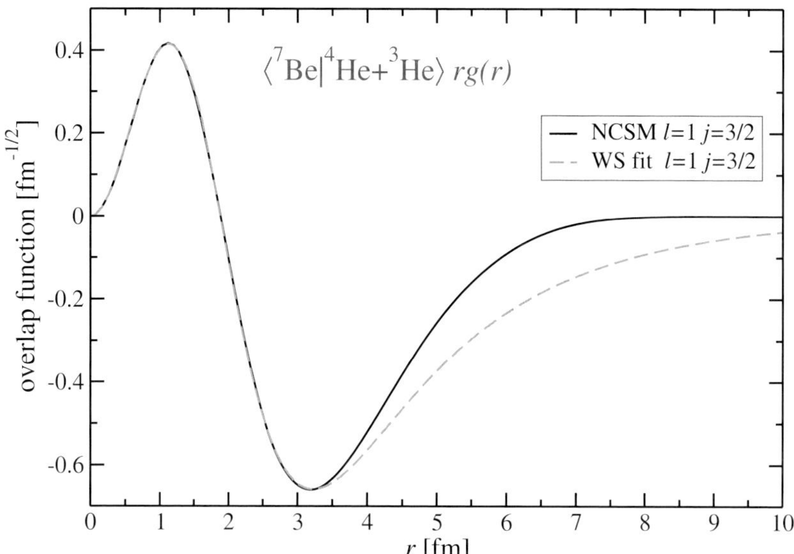

Fig. 14. – The overlap function, $rg(r)$, for the first excited state of ${}^{7}\mathrm{Be}$ with the ground state of ${}^{3}\mathrm{He}$ plus α as a dependence on separation between the ${}^{3}\mathrm{He}$ and the α-particle. The p-wave channel overlap function with $j = 3/2$ is shown. The full line represents the NCSM result obtained using the CD-Bonn 2000 NN potential and the $10\hbar\Omega$ model space for ${}^{7}\mathrm{Be}$ ($12\hbar\Omega$ for ${}^{3,4}\mathrm{He}$) with the HO frequency of $\hbar\Omega = 13\,\mathrm{MeV}$. The dashed line represents a corrected overlap obtained with a Woods-Saxon potential whose parameters were fit to the NCSM overlap up to 3.4 fm under the constraint to reproduce the experimental separation energy.

By investigating the model space dependence for $8\hbar\Omega$ and $10\hbar\Omega$ spaces we estimate the ${}^{3}\mathrm{He}(\alpha,\gamma){}^{7}\mathrm{Be}$ S-factor at zero energy to be higher than $0.44\,\mathrm{keV}\,\mathrm{b}$, the value that we obtained in the discussed case shown in fig. 15. Our results are similar to those obtained by Nollett [55] using the variational Monte Carlo wave functions for the bound states and potential model wave functions for the scattering state.

4˙3. ${}^{3}\mathrm{H}(\alpha,\gamma){}^{7}\mathrm{Li}$. – An important check on the consistency of the ${}^{3}\mathrm{He}(\alpha,\gamma){}^{7}\mathrm{Be}$ S-factor calculation is the investigation of the mirror reaction ${}^{3}\mathrm{H}(\alpha,\gamma){}^{7}\mathrm{Li}$, for which more accurate data exist [56]. Our results obtained using the CD-Bonn 2000 NN potential are shown in fig. 16. It is apparent that our ${}^{3}\mathrm{H}(\alpha,\gamma){}^{7}\mathrm{Li}$ results are consistent with our ${}^{3}\mathrm{He}(\alpha,\gamma){}^{7}\mathrm{Be}$ calculation. We are on the lower side of the data and we find an increase of the S-factor as we increase the size of our basis.

More details on the *ab initio* NCSM investigation of the ${}^{3}\mathrm{He}(\alpha,\gamma){}^{7}\mathrm{Be}$ and ${}^{3}\mathrm{H}(\alpha,\gamma){}^{7}\mathrm{Li}$ S-factors are given in ref. [57].

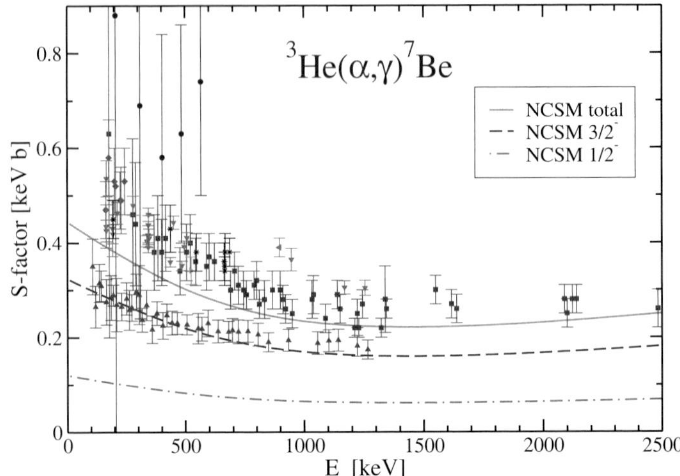

Fig. 15. – The full line shows the ^{3}He$(\alpha,\gamma)^7$Be S-factor obtained using the NCSM overlap functions with corrected asymptotics. The dashed lines show the ^{7}Be ground- and the first-excited-state contributions. The calculation was done using the CD-Bonn 2000 NN potential and the $10\hbar\Omega$ model space for ^{7}Be ($12\hbar\Omega$ for ^{3,4}He) with the HO frequency of $\hbar\Omega = 13$ MeV.

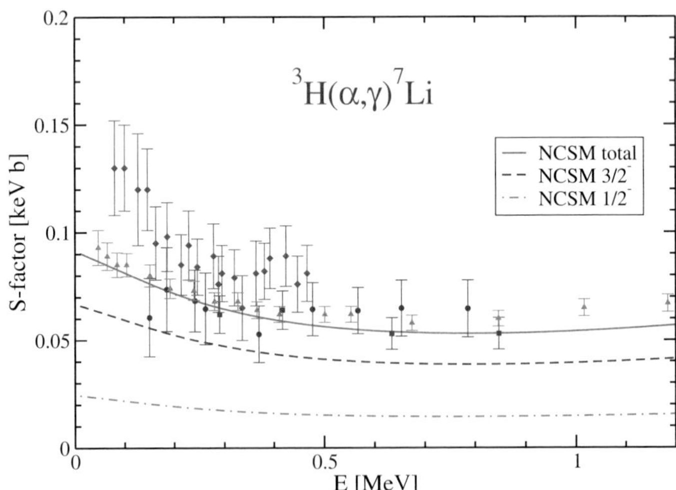

Fig. 16. – The full line shows the ^{3}H$(\alpha,\gamma)^7$Li S-factor obtained using the NCSM overlap functions with corrected asymptotics. The dashed lines show the ^{7}Li ground- and the first-excited-state contributions. The calculation was done using the CD-Bonn 2000 NN potential and the $10\hbar\Omega$ model space for ^{7}Li ($12\hbar\Omega$ for ^{3}H and ^{4}He) with the HO frequency of $\hbar\Omega = 13$ MeV.

5. – Towards the *ab initio* NCSM with continuum

In the previous section, we highlighted shortcomings of the *ab initio* NCSM, its incorrect description of long-range correlations and its lack of coupling to continuum. If we want to build upon the *ab initio* NCSM to microscopically describe loosely bound systems as well as nuclear reactions, the approach must be augmented by explicitly including cluster states such as, *e.g.*, those given in eq. (41), and solve for their relative motion while imposing the proper boundary conditions. This can be done by extending the *ab initio* NCSM HO basis through the addition of the cluster states. This would result in an over-complete basis with the cluster relative motion wave functions as amplitudes that need to be determined. The first step in this direction is to consider the cluster basis alone. This approach is very much in the spirit of the resonating group method (RGM) [58], a technique that considers clusters with fixed internal degrees of freedom, treats the Pauli principle exactly and solves the many-body problem by determining the relative motion between the various clusters. In our approach, we use the *ab initio* NCSM wave functions for the clusters involved and the *ab initio* NCSM effective interactions derived from realistic NN (and eventually also from NNN) potentials.

The general outline of the formalism is as follows. The many-body wave function is approximated by a superposition of binary cluster channel wave functions

$$(48) \qquad \Psi^{(A)} = \sum_\nu \hat{\mathcal{A}} \left[\psi^{(A-a)}_{1\nu} \psi^{(a)}_{2\nu} \varphi_\nu(\vec{r}_{A-a,a}) \right] = \sum_\nu \int d\vec{r}\, \varphi_\nu(\vec{r})\, \hat{\mathcal{A}}\, \Phi^{(A-a,a)}_{\nu\vec{r}},$$

with

$$(49) \qquad \Phi^{(A-a,a)}_{\nu\vec{r}} = \psi^{(A-a)}_{1\nu} \psi^{(a)}_{2\nu} \delta(\vec{r} - \vec{r}_{A-a,a}).$$

Here, $\hat{\mathcal{A}}$ is the antisymmetrizer accounting for the exchanges of nucleons between the two clusters (which are already antisymmetric with respect to exchanges of internal nucleons). The relative-motion wave functions φ_ν depend on the relative-distance between the center of masses of the two clusters in channel ν. They can be determined by solving the many-body Schrödinger equation in the Hilbert space spanned by the basis functions (49):

$$(50) \qquad H\Psi^{(A)} = E\Psi^{(A)} \longrightarrow \sum_\nu \int d\vec{r} \left[\mathcal{H}^{(A-a,a)}_{\mu\nu}(\vec{r}\,',\vec{r}) - E\mathcal{N}^{(A-a,a)}_{\mu\nu}(\vec{r}\,',\vec{r}) \right] \varphi_\nu(\vec{r}),$$

where the Hamiltonian and norm kernels are defined as

$$(51) \qquad \mathcal{H}^{(A-a,a)}_{\mu\nu}(\vec{r}\,',\vec{r}) = \left\langle \Phi^{(A-a,a)}_{\mu\vec{r}\,'} \left| \hat{\mathcal{A}} H \hat{\mathcal{A}} \right| \Phi^{(A-a,a)}_{\nu\vec{r}} \right\rangle,$$

$$(52) \qquad \mathcal{N}^{(A-a,a)}_{\mu\nu}(\vec{r}\,',\vec{r}) = \left\langle \Phi^{(A-a,a)}_{\mu\vec{r}\,'} \left| \hat{\mathcal{A}}^2 \right| \Phi^{(A-a,a)}_{\nu\vec{r}} \right\rangle.$$

The most challenging task is to evaluate the Hamiltonian kernel and the norm kernel. We now briefly outline, how this is done when *ab initio* NCSM wave functions are used for the binary cluster states. From now on, let us consider the cluster states with a single-nucleon projectile ($a = 1$ in eq. (48)). A generalization is straightforward. Using an alternative coupling scheme compared to eq. (41), we introduce

$$(53) \qquad \left\langle \vec{\xi}_1 \ldots \vec{\xi}_{A-2} \xi'_{A-1} \hat{\xi}_{A-1} \Big| \Phi^{(A-1,1)JMTM_T}_{(\alpha I_1 T_1, \frac{1}{2}\frac{1}{2});sl}; \delta_{\xi_{A-1}} \right\rangle =$$

$$= \sum (I_1 M_1 \tfrac{1}{2} m_s | s m)(s m l m_l | J M)(T_1 M_{T_1} \tfrac{1}{2} m_t | T M_T) \frac{\delta(\xi_{A-1} - \xi'_{A-1})}{\xi_{A-1} \xi'_{A-1}}$$

$$\times Y_{l m_l}(\hat{\xi}_{A-1}) \chi_{m_s} \chi_{m_t} \left\langle \vec{\xi}_1 \ldots \vec{\xi}_{A-2} \Big| A - 1 \alpha I_1 M_1 T_1 M_{T_1} \right\rangle,$$

with the spin and isospin coordinates omitted to simplify the notation. The Jacobi coordinates were defined in eq. (8)–(11). Using the latter cluster basis and the following definition of the antisymmetrizer $\hat{\mathcal{A}} = 1/\sqrt{A}(1 - \sum_{j=1}^{A-1} P_{j,A})$ with $P_{j,A}$ the transposition operator of nucleons j and A, the norm kernel can be expressed as

$$(54) \qquad \mathcal{N}^{(A-1,1)}_{\mu\nu}(r',r) = \delta_{\mu\nu} \frac{\delta(r'-r)}{r'r} - (A-1) \sum_{n'n} R_{n'l'}(r$$

$$\times \left\langle \Phi^{(A-1,1)JT}_{(\alpha' I'_1 T'_1, \frac{1}{2}\frac{1}{2})s'l'}; n'l' \Big| P_{A,A-1} \Big| \Phi^{(A-1,1)JT}_{(\alpha I_1 T_1, \frac{1}{2}\frac{1}{2})sl}; nl \right\rangle R_{nl}(r),$$

with $\mu \equiv (\alpha' I'_1 T'_1, \frac{1}{2}\frac{1}{2})s'$, $\nu \equiv (\alpha I_1 T_1, \frac{1}{2}\frac{1}{2})s$ and $P_{A,A-1}$ the transposition operator of nucleons A and $A - 1$. The coordinates r are related to ξ_{A-1} by $r = \sqrt{\frac{A}{A-1}}\xi_{A-1}$ and the HO length parameter of the radial HO wave functions is $b = \sqrt{\frac{\hbar}{\frac{A-1}{A}m\Omega}}$. The matrix element of the transposition operator $P_{A,A-1}$ can be directly evaluated using the *ab initio* NCSM wave functions expanded in Jacobi coordinate HO basis following a procedure analogous to the derivation of eq. (14). However, a crucial feature of the *ab initio* NCSM approach is that the matrix elements that enter the norm kernel and the Hamiltonian kernel can be equivalently evaluated using the *ab initio* NCSM wave functions expanded in the Slater determinant HO basis. This is achieved in two stages. First, we calculate the SD matrix element as

$$(55) \qquad {}_{\mathrm{SD}}\left\langle \Phi^{(A-1,1)JT}_{(\alpha' I'_1 T'_1, \frac{1}{2}\frac{1}{2})s'l'}; n'l' \Big| P_{A,A-1} \Big| \Phi^{(A-1,1)JT}_{(\alpha I_1 T_1, \frac{1}{2}\frac{1}{2})sl}; nl \right\rangle_{\mathrm{SD}} =$$

$$= \frac{1}{A-1} \sum_{jj'K\tau} \begin{Bmatrix} I_1 & \frac{1}{2} & s \\ l & J & j \end{Bmatrix} \begin{Bmatrix} I'_1 & \frac{1}{2} & s' \\ l' & J & j' \end{Bmatrix} \begin{Bmatrix} I_1 & K & I'_1 \\ j' & J & j \end{Bmatrix} \begin{Bmatrix} T_1 & \tau & T'_1 \\ \frac{1}{2} & T & \frac{1}{2} \end{Bmatrix}$$

$$\times \hat{s}\hat{s}'\hat{j}\hat{j}'\hat{K}\hat{\tau}(-1)^{I'_1+j'+J}(-1)^{T_1+\frac{1}{2}+T}$$

$$\times {}_{\mathrm{SD}}\left\langle A-1\alpha' I'_1 T'_1 \Big|\Big|\Big| \left(a^\dagger_{nlj\frac{1}{2}} \tilde{a}_{n'l'j'\frac{1}{2}}\right)^{(K\tau)} \Big|\Big|\Big| A-1\alpha I_1 T_1 \right\rangle_{\mathrm{SD}}.$$

Second, it is possible to show that the matrix element in the SD basis is related to the

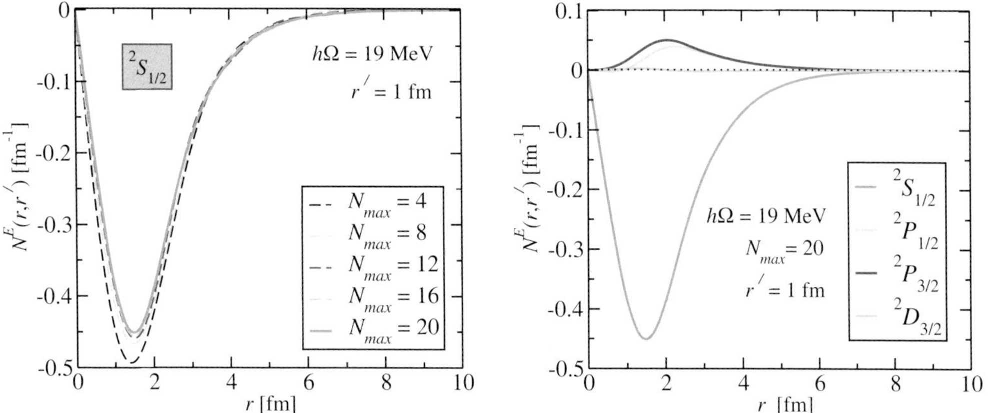

Fig. 17. – The exchange part of the norm kernel of the n + ^{4}He system. Left, the convergence with the size of the basis of the ^{4}He wave function for the $^2S_{1/2}$ channel. Right, results for channels are compared. The chiral EFT NN potential was used.

one in the Jacobi coordinate basis:

$$
(56) \quad _{\rm SD}\left\langle \Phi^{(A-1,1)JT}_{(\alpha' I_1' T_1', \frac{1}{2}\frac{1}{2})s'l'}; n'l' \middle| P_{A,A-1} \middle| \Phi^{(A-1,1)JT}_{(\alpha I_1 T_1, \frac{1}{2}\frac{1}{2})sl}; nl \right\rangle_{\rm SD} =
$$

$$
= \sum_{n_r l_r n_r' l_r' J_r} \left\langle \Phi^{(A-1,1)J_r T}_{(\alpha' I_1' T_1', \frac{1}{2}\frac{1}{2})s'l_r'}; n_r' l_r' \middle| P_{A,A-1} \middle| \Phi^{(A-1,1)J_r T}_{(\alpha I_1 T_1, \frac{1}{2}\frac{1}{2})sl_r}; n_r l_r \right\rangle
$$

$$
\times \sum_{NL} \hat{l}\hat{l}'\hat{J}_r^2 (-1)^{s+l_r-s-l_r'}
\left\{ \begin{array}{ccc} s & l_r & J_r \\ L & J & l \end{array} \right\}
\left\{ \begin{array}{ccc} s' & l_r' & J_r \\ L & J & l' \end{array} \right\}
$$

$$
\times \left\langle n_r l_r N L l | 00 n l l \right\rangle_{\frac{1}{A-1}} \left\langle n_r' l_r' N L l' | 00 n' l' l' \right\rangle_{\frac{1}{A-1}} .
$$

This relation then defines a matrix that one inverts to get the Jacobi-coordinate matrix element. This is analogous to what was done to obtain the translationally invariant density in ref. [59]. The Hamiltonian kernel can be evaluated in a similar yet more involved way. It consists of a kinetic term, a NN potential direct term associated with the operator $V_{A,A-1}(1 - P_{A,A-1})$ and a NN potential exchange term associated with the operator $V_{A,A-2}P_{A,A-1}$ (plus terms arising from the NNN interaction). The ability to employ wave functions expanded in the SD basis opens the possibility to apply this formalism for nuclei with $A > 5$.

In fig. 17, we show the exchange part of the norm kernel for the n + ^{4}He system, in particular the second term of eq. (54) multiplied by rr'. It is apparent that we are able to reach convergence for the kernel. Furthemore, the $^2S_{1/2}$ channel shows, as it should the effect of the Pauli principle. Indeed, the ^{4}He wave function is dominated by the four nucleon s-shell configuration. The Pauli principle prevents adding the fifth nucleon to the same shell.

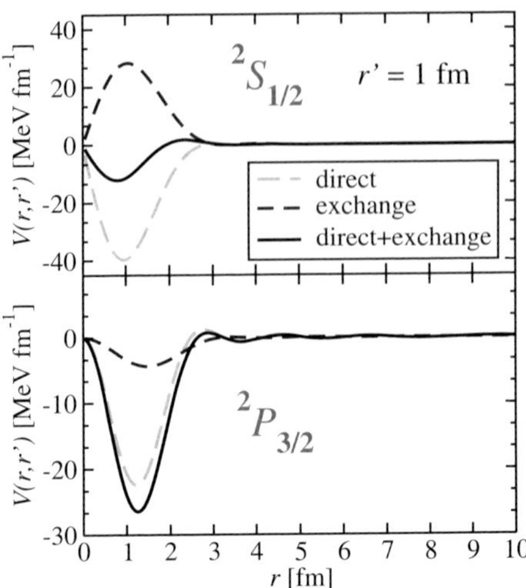

Fig. 18. – The NN potential direct and exchange terms of the Hamiltonian kernel of the n + ^{4}He system. The $^2S_{1/2}$ and $^2P_{3/2}$ channels are compared. The low-momentum NN potential of ref. [60] was used in the $14\hbar\Omega$ model space with the HO frequency of $\hbar\Omega = 18\,\mathrm{MeV}$.

In fig. 18, we show the direct and the exchange contributions of the NN potential to the Hamiltonian kernel as well as their sum for the n + ^{4}He system. Again, the Pauli principle is manifest in the $^2S_{1/2}$ channel. We were able to obtain the presented results using wave functions expanded both in the Jacobi-coordinate and the SD basis. The two

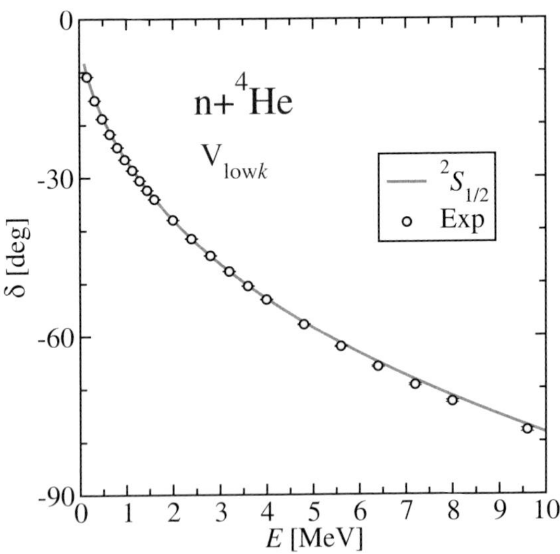

Fig. 19. – The calculated $^2S_{1/2}$ phase shift for the n + ^{4}He system compared with experimental data. The low-momentum NN potential of ref. [60] was used in the $16\hbar\Omega$ model space with the HO frequency of $\hbar\Omega = 18\,\mathrm{MeV}$.

independent calculations gave identical results as expected. A converged calculation of the $^2S_{1/2}$ phase shift together with experimental data is presented in fig. 19. Full details regarding this approach are given in ref. [61].

6. – Conclusions

The *ab initio* NCSM evolved into a powerful many-body technique. Presently, it is the only method capable to use interactions derived within the chiral EFT for systems of more than four nucleons, in particular for mid-p-shell nuclei. Among its successes is the demonstration of importance of the NNN interaction for nuclear structure. Applications to nuclear reactions with a proper treatment of long-range properties are under development. Extension to heavier nuclei is achieved through the importance-truncated NCSM [62]. Within this approach, *ab initio* calculations for nuclei as heavy as ^{40}Ca become possible.

$$* \ * \ *$$

I would like to thank all the collaborators that contributed to the cited papers and, in particular, SONIA QUAGLIONI for useful discussions and input for sect. **5**. This work performed under the auspices of the U.S. Department of Energy by Lawrence Livermore National Laboratory under Contract DE-AC52-07NA27344. Support from the LDRD contract No. 04-ERD-058 and from U.S. DOE/SC/NP (Work Proposal Number SCW0498) is acknowledged. This work was also supported in part by the Department of Energy under Grant DE-FC02-07ER41457.

REFERENCES

[1] FADDEEV L. D., *Zh. Eksp. Teor. Fiz.*, **39** (1960) 1459 (*Sov. Phys. JETP*, **12** (1961) 1014).
[2] CHEN C. R., PAYNE G. L., FRIAR J. L. and GIBSON B. F., *Phys. Rev. C*, **31** (1985) 2266.
[3] FRIAR J. L., PAYNE G. L., STOKS V. G. J. and DE SWART J. J., *Phys. Lett. B*, **311** (1993) 4.
[4] NOGGA A., HÜBER D., KAMADA H. and GLÖCKLE W., *Phys. Lett. B*, **409** (1997) 19.
[5] YAKUBOVSKY O. A., *Sov. J. Nucl. Phys.*, **5** (1967) 937.
[6] GLÖCKLE W. and KAMADA H., *Phys. Rev. Lett.*, **71** (1993) 971.
[7] CIESIELSKI F. and CARBONELL J., *Phys. Rev. C*, **58** (1998) 58; CIESIELSKI F., CARBONELL J. and GIGNOUX C., *Nucl. Phys. A*, **631** (1998) 653c.
[8] VIVIANI M., KIEVSKY A. and ROSATI S., *Few-Body Systems*, **18** (1995) 25.
[9] BARNEA N., LEIDEMANN W. and ORLANDINI G., *Nucl. Phys. A*, **650** (1999) 427.
[10] PUDLINER B. S., PANDHARIPANDE V. R., CARLSON J., PIEPER S. C. and WIRINGA R. B., *Phys. Rev. C*, **56** (1997) 1720; WIRINGA R. B., PIEPER S. C., CARLSON J. and PANDHARIPANDE V. R., *Phys. Rev. C*, **62** (2000) 014001; PIEPER S. C. and WIRINGA R. B., *Annu. Rev. Nucl. Part. Sci.*, **51** (2001) 53; PIEPER S. C., VARGA K. and WIRINGA R. B., *Phys. Rev. C*, **66** (2002) 044310.
[11] KAMADA H. *et al.*, *Phys. Rev. C*, **64** (2001) 044001.
[12] BISHOP R. F., FLYNN M. F., BOSCÁ M. C., BUENDÍA E. and GUARDIOLA R., *Phys. Rev. C*, **42** (1990) 1341.

[13] HEISENBERG J. H. and MIHAILA B., *Phys. Rev. C*, **59** (1999) 1440.

[14] KOWALSKI K. *et al.*, *Phys. Rev. Lett.*, **92** (2004) 132501.

[15] WŁOCH M. *et al.*, *Phys. Rev. Lett.*, **94** (2005) 212501.

[16] NAVRÁTIL P., VARY J. P. and BARRETT B. R., *Phys. Rev. Lett.*, **84** (2000) 5728.

[17] TRLIFAJ L., *Phys. Rev. C*, **5** (1972) 1534.

[18] NAVRÁTIL P., KAMUNTAVIČIUS G. P. and BARRETT B. R., *Phys. Rev. C*, **61** (2000) 044001.

[19] NAVRÁTIL P., BARRETT B. R. and GLÖCKLE W., *Phys. Rev. C*, **59** (1999) 611.

[20] SUZUKI K. and LEE S. Y., *Prog. Theor. Phys.*, **64** (1980) 2091.

[21] SUZUKI K., *Prog. Theor. Phys.*, **68** (1982) 246; SUZUKI K. and OKAMOTO R., *Prog. Theor. Phys.*, **70** (1983) 439.

[22] SUZUKI K., *Prog. Theor. Phys.*, **68** (1982) 1999; SUZUKI K. and OKAMOTO R., *Prog. Theor. Phys.*, **92** (1994) 1045.

[23] NAVRÁTIL P. and ORMAND W. E., *Phys. Rev. C*, **68** (2003) 034305.

[24] NOGGA A., NAVRÁTIL P., BARRETT B. R. and VARY J. P., *Phys. Rev. C*, **73** (2006) 064002.

[25] STETCU I., BARRETT B. R., NAVRATIL P. and VARY J. P., *Phys. Rev. C*, **71** (2005) 044325.

[26] ENTEM D. R. and MACHLEIDT R., *Phys. Rev. C*, **68** (2003) 041001(R).

[27] NAVRÁTIL P. and ORMAND W. E., *Phys. Rev. Lett.*, **88** (2002) 152502.

[28] NAVRÁTIL P. and CAURIER E., *Phys. Rev. C*, **69** (2004) 014311.

[29] VIVIANI M., MARCUCCI L. E., ROSATI S., KIEVSKY A. and GIRLANDA L., *Few-Body Systems*, **39** (2006) 159.

[30] CAURIER E., MARTINEZ-PINEDO G., NOWACKI F., POVES A., RETAMOSA J. and ZUKER A. P., *Phys. Rev. C*, **59** (1999) 2033; CAURIER E. and NOWACKI F., *Acta Phys. Pol. B*, **30** (1999) 705.

[31] WEINBERG S., *Physica A*, **96** (1979) 327; *Phys. Lett. B*, **251** (1990) 228; *Nucl. Phys. B*, **363** (1991) 3; GASSER J. *et al.*, *Ann. Phys. (N.Y.)*, **158** (1984) 142; *Nucl. Phys. B*, **250** (1985) 465.

[32] BERNARD V., KAISER N. and MEISSNER ULF-G., *Int. J. Mod. Phys. E*, **4** (1995) 193.

[33] ORDONEZ C., RAY L. and VAN KOLCK U., *Phys. Rev. Lett.*, **72** (1994) 1982; *Phys. Rev. C*, **53** (1996) 2086.

[34] VAN KOLCK U., *Prog. Part. Nucl. Phys.*, **43** (1999) 337.

[35] BEDAQUE P. F. and VAN KOLCK U., *Annu. Rev. Nucl. Part. Sci.*, **52** (2002) 339; EPELBAUM E., *Prog. Part. Nucl. Phys.*, **57** (2006) 654.

[36] VAN KOLCK U., *Phys. Rev. C*, **49** (1994) 2932.

[37] EPELBAUM E., NOGGA A., GLÖCKLE W., KAMADA H., MEISSNER ULF-G. and WITALA H., *Phys. Rev. C*, **66** (064001) 2002.

[38] EPELBAUM E., *Phys. Lett. B*, **639** (2006) 456.

[39] NAVRÁTIL P., GUEORGUIEV V. G., VARY J. P., ORMAND W. E. and NOGGA A., *Phys. Rev. Lett.*, **99** (2007) 042501.

[40] HAYES A. C., NAVRÁTIL P. and VARY J. P., *Phys. Rev. Lett.*, **91** (2003) 012502.

[41] VARY J. P., "The Many-Fermion-Dynamics Shell-Model Code", Iowa State University, 1992, unpublished.

[42] NAVRÁTIL P., *Phys. Rev. C*, **70** (2004) 054324.

[43] ADELBERGER E. *et al.*, *Rev. Mod. Phys.*, **70** (1998) 1265.

[44] SNO COLLABORATION (AHMED S. N. *et al.*), *Phys. Rev. Lett.*, **92** (2004) 181301.

[45] COUVIDAT S., TURCK-CHIÈZE S. and KOSOVICHEV A. G., *Astrophys. J.*, **599** (2003) 1434.

[46] BAHCALL J. N. and PINSONNEAULT M. H., *Phys. Rev. Lett.*, **92** (2004) 121301.

[47] NAVRÁTIL P., BERTULANI C. A. and CAURIER E., *Phys. Lett. B*, **634** (2006) 191.

[48] NAVRÁTIL P., BERTULANI C. A. and CAURIER E., *Phys. Rev. C*, **73** (2006) 065801.

[49] MACHLEIDT R., *Phys. Rev. C*, **63** (2001) 024001.

[50] ESBENSEN H. and BERTSCH G. F., *Nucl. Phys. A*, **600** (1996) 37.

[51] ROBERTSON R. G. H., *Phys. Rev. C*, **7** (1973) 543.

[52] JUNGHANS A. R., MOHRMANN E. C., SNOVER K. A., STEIGER T. D., ADELBERGER E. G., CASANDJIAN J. M., SWANSON H. E., BUCHMANN L., PARK S. H., ZYUZIN A. and LAIRD A. M., *Phys. Rev. C*, **68** (2003) 065803.

[53] IWASA N. *et al.*, *Phys. Rev. Lett.*, **83** (1999) 2910; DAVIDS B. *et al.*, *Phys. Rev. Lett.*, **86** (2001) 2750; SCHUMANN F. *et al.*, *Phys. Rev. Lett.*, **90** (2003) 232501.

[54] KIM B. T., IZUMUTO T. and NAGATANI K., *Phys. Rev. C*, **23** (1981) 33.

[55] NOLLETT K. M., *Phys. Rev. C*, **63** (2001) 054002.

[56] BRUNE C. R., KAVANAGH R. W. and ROLFS C., *Phys. Rev. C*, **50** (1994) 2205.

[57] NAVRÁTIL P., BERTULANI C. A. and CAURIER E., *Nucl. Phys. A*, **787** (2007) 539c.

[58] TANG Y. C., LeMERE M. and THOMPSON D. R., *Phys. Rep.*, **47** (1978) 167; LANGANKE K. and FRIEDRICH H., *Advances in Nuclear Physics* (Plenum, New York) 1987, chapt. 4; LOVAS R. G., LIOTTA R. J., INSOLIA A., VARGA K. and DELION D. S., *Phys. Rep.*, **294** (1998) 265.

[59] NAVRÁTIL P., *Phys. Rev. C*, **70** (2004) 014317.

[60] BOGNER S. K., KUO T. T. S. and SCHWENK A., *Phys. Rep.*, **386** (2003) 1.

[61] QUAGLIONI S. and NAVRÁTIL P., to be published.

[62] ROTH R. and NAVRÁTIL P., *Phys. Rev. Lett*, **99** (2007) 092501.

DOI 10.3254/978-1-58603-885-4-185

Fermionic Molecular-Dynamics — clusters, halos, skins and S-factors

H. FELDMEIER and T. NEFF

Gesellschaft für Schwerionenforschung mbH - Planckstr. 1, D-64291 Darmstadt, Germany

Summary. — The nuclear many-body problem is discussed. Starting from realistic nuclear forces the problem of induced short-range correlations is treated with the unitary correlation operator method. Fermionic molecular dynamics many-body states are used to describe phenomena such as clustering of nucleons inside light nuclei, weakly bound two-neutron halos and nucleus-nucleus reactions at low energies.

1. – Introduction

On the way to understanding the structure of atomic nuclei two major obstacles have to be overcome. The first is the fact that the interaction between the constituents, the protons and neutrons, is rather complex and cannot be deduced in a unique fashion from data or from the underlying quantum chromodynamics. The second challenge is the numerical solution of the many-body problem. After a period of stagnation new developments in the past ten years have advanced the field considerably, both in the derivation of realistic nucleon-nucleon forces and in the many-body methods.

In this lecture we address the problem of short- and long-range correlations induced by the nucleon-nucleon forces. For the short-range correlations we discuss the Unitary Correlation Operator Method in sect. **3**. Long-range correlations are supposed to be represented by explicit superposition of many-body states. Here we concentrate on cluster structures and halos for which the Fermionic Molecular Dynamics (FMD) model is especially suited. This model utilizes localized single-particle states to set up antisymmetrized many-body states that are well adapted to the considered physical phenomena. FMD is explained in detail in sect. **4**.

Many illustrative examples are presented: for cluster structures in sect. **5**, for halos in sect. **6** and for neutron skins and astrophysical S-factors in sect. **7**.

2. – The nuclear many-body problem

2'1. *Degrees of freedom*. – At low energies the relevant degrees of freedom are the center-of-mass positions, the spins and the isospins of the nucleons. Therefore one represents the many-body state $|\Psi\rangle$ of A particles

$$(1) \qquad \Psi(\vec{r}_1\vec{\sigma}_1\vec{\tau}_1,\cdots,\vec{r}_A\vec{\sigma}_A\vec{\tau}_A) = \langle\vec{r}_1\vec{\sigma}_1\vec{\tau}_1,\cdots,\vec{r}_A\vec{\sigma}_A\vec{\tau}_A|\Psi\rangle$$

in terms of these degrees of freedom. This is by no means trivial as the nucleons themselves are complex many-body objects composed of quarks and gluons, the constituents of quantum chromodynamics. The chosen degrees of freedom will be appropriate if the internal structure of a nucleon is rigid enough so that it can be regarded as an entity. However, at short distances between the nucleons polarization or virtual excitation of intrinsic excited states is an important issue.

The general task of nuclear structure theory is to solve the many-body Schrödinger equation

$$(2) \qquad \underset{\sim}{H}\,|\widehat{\Psi};\alpha\rangle = E_\alpha\,|\widehat{\Psi};\alpha\rangle$$

which provides the many-body eigenstates $|\widehat{\Psi};\alpha\rangle$ that contain all possible information about the structure of the nuclear system and the corresponding energy eigenvalues E_α. (In these lectures operators acting in the Hilbert space are marked by an underlining $\sim$.)

It will turn out that the nuclear many-body states $|\widehat{\Psi};\alpha\rangle$ contain many correlations of various kinds that cannot be represented in a simple way by many-body basis states like Slater determinants.

First one has to consider the Hamilton operator $\underset{\sim}{H}$. After having decided about the degrees of freedom, one has to construct $\underset{\sim}{H}$ in terms of the corresponding operators.

$$(3) \qquad \underset{\sim}{H} = \underset{\sim}{T}_{\text{int}} + \underset{\sim}{V}_{NN} + \underset{\sim}{V}_{NNN}$$

$$= \underset{\sim}{T}_{\text{int}} + \sum_{i<j=1}^{A} V_{NN}\left(\underset{\sim}{\vec{r}}_{ij},\underset{\sim}{\vec{p}}_{ij},\underset{\sim}{\vec{\sigma}}_i,\underset{\sim}{\vec{\sigma}}_j,\underset{\sim}{\vec{\tau}}_i,\underset{\sim}{\vec{\tau}}_j\right) + \sum_{i<j<k=1}^{A} \underset{\sim}{V}_{NNN}(ijk).$$

The center-of-mass kinetic energy is subtracted from the total kinetic energy so that the boost-invariant intrinsic kinetic energy $\underset{\sim}{T}_{\text{int}}$ is left.

$$(4) \qquad \underset{\sim}{T}_{\text{int}} = \sum_{i=1}^{A} \frac{\underset{\sim}{\vec{p}}_i^{\,2}}{2m_i} - \frac{\underset{\sim}{\vec{P}}_{\text{CM}}^2}{2M} = \frac{1}{M} \sum_{i<j=1}^{A} \frac{(m_j \underset{\sim}{\vec{p}}_i - m_i \underset{\sim}{\vec{p}}_j)^2}{2m_i m_j}$$

$$(5) \qquad \underset{\sim}{\vec{P}}_{\text{CM}} = \sum_{i=1}^{A} \underset{\sim}{\vec{p}}_i \,; \quad M = \sum_{i=1}^{A} m_i$$

$\underset{\sim}{\vec{P}}_{\text{CM}}$ is the total momentum and M is the total mass of the system.

It has been realised that realistic nucleon-nucleon interactions $\underset{\sim}{V}_{NN}$ depend not only on the relative distance $\vec{r}_{ij} = \underset{\sim}{\vec{r}}_i - \underset{\sim}{\vec{r}}_j$ of particles i and j, but also on their relative momentum $\underset{\sim}{\vec{p}}_{ij} = (m_j \underset{\sim}{\vec{p}}_i - m_i \underset{\sim}{\vec{p}}_j)/(m_i + m_j)$. Furthermore, the exchange of charged pseudoscalar pions leads to a strong tensor component of the type $(\underset{\sim}{\vec{r}}_{ij}\underset{\sim}{\vec{\sigma}}_i)(\underset{\sim}{\vec{r}}_{ij}\underset{\sim}{\vec{\sigma}}_j)(\underset{\sim}{\vec{\tau}}_i\underset{\sim}{\vec{\tau}}_j)$. Details of the Hamiltonian will be discussed later.

In recent years it has become clear that one also needs three-body potentials $\underset{\sim}{V}_{NNN}$ in order to achieve high precision in the reproduction of many-body data. In this lecture three-body forces are not used, for more information on this subject see the lectures of S. Pieper and P. Navrátil.

The second object one has to consider is the many-body Hilbert space which the many-body states $|\widehat{\Psi}; \alpha\rangle$ inhabit and in which the Hamiltonian acts. For that one has to find an appropriate basis to represent the correlations among the particles which are induced by the nuclear interaction.

A convenient many-body basis is composed of Slater determinants

$$(6) \qquad |n_1, n_2, \ldots, n_A\rangle = \sqrt{A!}\, \underset{\sim}{\mathcal{A}} |n_1\rangle \otimes |n_2\rangle \otimes \cdots \otimes |n_A\rangle$$

$$= \frac{1}{\sqrt{A!}} \sum_{\text{all } \rho} \text{sgn}(\rho) |n_{\rho(1)}\rangle \otimes |n_{\rho(2)}\rangle \cdots \otimes |n_{\rho(A)}\rangle \,,$$

where $|n_i\rangle$, $i = 1, \ldots, A$, denote the occupied single-particle states and $\text{sgn}(\rho)$ is the sign of the permutation ρ. The antisymmetrizer $\underset{\sim}{\mathcal{A}}$ projects on the subspace of the A-body Hilbert space which is antisymmetric under permutations of the particles.

An often used single-particle representation is the harmonic-oscillator shell-model basis. The many-body state $|\widehat{\Psi}; \alpha\rangle$ is then represented by a very large sum of Slater determinants

$$(7) \qquad |\widehat{\Psi}; \alpha\rangle = \sum_{n_1 < n_2 < \cdots < n_A} \widehat{\Psi}^{\alpha}_{n_1, n_2, \ldots, n_A} |n_1, n_2, \ldots, n_A\rangle .$$

If the summations over n_i are restricted such that all many-body states up to a maximum number of oscillator quanta are included the model is called "No-Core Shell Model" (see lecture of P. Navrátil), if only a selected set of single-particle orbits around the Fermi edge are included one speaks of "Configuration Mixing Shell Model" or simply "Shell

Model". If the resulting many-body Hilbert space reaches dimensions that challenge modern computers the phrase "Large Scale Shell Model" is often used.

Another representation is the coordinate representation given in eq. (1) which is used in Green's function Monte Carlo calculations discussed in the lecture of S. Pieper.

In this lecture we will concentrate on the many-body space spanned by Fermionic Molecular Dynamics (FMD) states that are especially well suited to describing cluster structures, halos, and nucleus-nucleus scattering. The details are explained in sect. **4**.

2'2. *The nucleon-nucleon potential.* – The nuclear two-body system has only one bound state, the deuteron. Thus the main information on the nucleon-nucleon potential comes from measurements of the elastic-scattering cross-section. As there is precise data available, one can deduce the phase shifts and mixing angles as a function of energy for all partial waves that contribute.

Potentials that describe phase shifts, mixing angle and the deuteron are called realistic potentials. At large inter-nucleon distances, beyond about 1.4 fm, all realistic potentials are based on the exchange of the lightest meson, the pion. For intermediate and shorter distances the exchange of heavier mesons like ω or ρ is the underlying concept in the Bonn-type potentials [1]. However, at short distances form factors have to be employed to take into account the finite size of the nucleons.

A new class of so-called chiral potentials [2-5] has been developed in recent years. They are based on the approximate chiral symmetry of the underlying Quantum Chromodynamics (QCD). Their advantage is that one has expansion parameters that allow different contributions to be sorted according to their expected size. Furthermore, a consistent treatment of three-body forces is possible. However, even this appealing scheme needs *ad hoc* form factors and the number of parameters that have to be adjusted to data increases with the addition of higher-order terms.

Another successful potential is the Argonne V18 potential (AV18) that treats the medium- and short-range part of the interaction in a more phenomenological way. It is expressed in coordinate space representation and is almost local, in the sense that the momentum dependence is reduced to an unavoidable minimum.

In fig. 1 we display the AV18-potential as a function of the distance $r_{12} = |\vec{r}_1 - \vec{r}_2|$ between two protons, or two neutrons, with total spin $S = 0$ and relative momentum $\vec{p}_{12} = 0$. The potential extends to about 2.5 fm and becomes strongly repulsive below $r_{12} \approx 0.5$ fm. Although it is 100 MeV deep the attraction is not sufficient to form a bound di-proton or di-neutron, in accordance with observations. Along with the potential, we show in the insets how much two protons overlap at the three indicated distances. At 0.5 fm the nucleons overlap to a large degree which explains the repulsive core of the potential. At maximum attraction around 1 fm there is still an appreciable overlap and at 1.8 fm, which is the mean distance in the interior of large nuclei (or nuclear matter at saturation density), the nucleons are almost separated. This picture shows: first, that nucleons are relatively large objects, their diameter is not really small compared to their mean distance, and second, the ansatz of a potential is certainly a strong simplification. When the overlap of the nucleons, which are complex many-body systems composed of

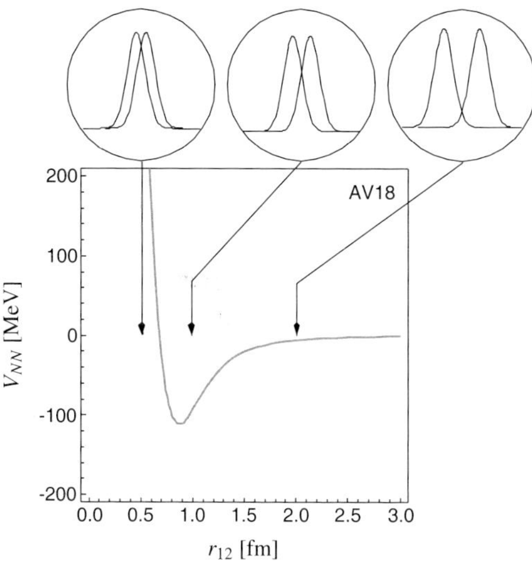

Fig. 1. – Argonne V18 potential in the $S = 0$, $T = 1$ channel for relative momentum $p_{12} = 0$ as a function of r_{12}. The insets display (on the same length scale as that used for r_{12}) the charge distribution of two protons at the three indicated relative distances.

quarks and gluons, is so big, one cannot expect that their intrinsic structure remains unaffected. One would also not expect the potential to be an adiabatic one. Inside nuclei the Fermi velocity is about a third of the speed of light, thus a momentum dependence is plausible.

Even if all the dynamic polarizations between two nucleons could be cast into a momentum-dependent two-body potential, the interaction of three nucleons that are in close proximity need not be just the sum of three pairwise interactions. The presence of a third nucleon, when two nucleons are interacting and polarizing each other, can induce additional polarizations which alter the original interaction between the pair. This effect causes a three-body interaction V_{NNN} not present in the two-body system. The simplest case is that nucleon 1 sends a pion to nucleon 2 which is virtually excited (polarized) to a Δ-resonance and subsequently emits another pion which is absorbed by nucleon 3.

The probability to find a pair of nucleons at short distances is highly reduced by the strong repulsion. This short-range correlation in the relative distance cannot be represented by Slater determinants as they are antisymmetrized products of single-particle states. Thus, Hartree-Fock calculations, where the trial state is a single Slater determinant, are not feasible with realistic interactions. But also the superposition of a huge number of them (cf. eq. (7)) as done in the No-Core Shell Model (NCSM) is not sufficient. Therefore, the NCSM employs the so-called Lee-Suzuki transformation to resolve this problem (see lecture of P. Navrátil).

If one looks at nucleon pairs with total spin $S = 1$, the situation is more complex as the tensor force (not present between $S = 0$ states) induces correlations between the spin direction and the direction of the relative distance vector that again cannot be represented by a finite number of Slater determinants. These correlations will be discussed in more detail in the following section, where we introduce a unitary correlation operator to address the short-range correlations.

3. – The Unitary Correlation Operator Method (UCOM)

Our aim is to perform calculations of nuclei with mass numbers larger than those typically accessible to exact few-body models but still starting from realistic interactions like the Bonn or Argonne potentials. However, the repulsive core and the tensor force of the nuclear interaction induce strong short-range central and tensor correlations in the nuclear many-body system. These correlations are in the relative coordinates $\vec{r}_{ij} = \vec{r}_i - \vec{r}_j$ and thus cannot be represented by products of single-particle states like Slater determinants, eq. (6), that are usually used as many-body states in Hartree-Fock or as a basis in shell model calculations.

We want to treat the short-range correlations explicitly by a unitary correlation operator $\underset{\sim}{C}$ that imprints them into uncorrelated states $|\Psi\rangle$

$$|\widehat{\Psi}\rangle = \underset{\sim}{C}\,|\Psi\rangle \tag{8}$$

such that the many-body state $|\widehat{\Psi}\rangle$ contains the short-ranged correlations [6-8]. For the correlator we make the following ansatz:

$$\underset{\sim}{C} = \underset{\sim}{C}_\Omega\,\underset{\sim}{C}_r = \exp\left[-i\sum_{i<j}\underset{\sim}{g}_{\Omega ij}\right]\exp\left[-i\sum_{i<j}\underset{\sim}{g}_{r ij}\right]. \tag{9}$$

It is the product of a radial correlator $\underset{\sim}{C}_r$ and a tensor correlator $\underset{\sim}{C}_\Omega$, both having a Hermitian two-body generator in the exponent.

The important advantage of treating the correlations in a unitary fashion compared to a Jastrow ansatz [9] is that one can either work with correlated states or correlated operators just by a similarity transformation:

$$\langle\widehat{\Psi}|\underset{\sim}{A}|\widehat{\Phi}\rangle = \langle\Psi|\underset{\sim}{C}^\dagger\underset{\sim}{A}\,\underset{\sim}{C}|\Phi\rangle = \langle\Psi|\underset{\sim}{\widehat{A}}|\Phi\rangle, \tag{10}$$

where the correlated operator is defined by

$$\underset{\sim}{\widehat{A}} = \underset{\sim}{C}^\dagger\underset{\sim}{A}\,\underset{\sim}{C} = \underset{\sim}{C}^{-1}\underset{\sim}{A}\,\underset{\sim}{C}. \tag{11}$$

Unitarity conserves the overlaps $\langle\widehat{\Psi}|\widehat{\Phi}\rangle = \langle\Psi|\Phi\rangle$ and the correlated product of operators equals the product of correlated operators:

$$\underset{\sim}{\widehat{AB}} = \underset{\sim}{C}^{-1}\underset{\sim}{AB}\,\underset{\sim}{C} = \underset{\sim}{C}^{-1}\underset{\sim}{A}\,\underset{\sim}{C}\,\underset{\sim}{C}^{-1}\underset{\sim}{B}\,\underset{\sim}{C} = \underset{\sim}{\widehat{A}}\,\underset{\sim}{\widehat{B}}, \tag{12}$$

which will be very helpful later on.

3˙1. *Cluster expansion.* – As the ansatz for the correlator contains a two-body operator in the exponent, any correlated operator will contain many-body parts. For example a Hamiltonian consisting of one- and two-body parts will turn into

$$(13) \qquad \widehat{H} = \underset{\sim}{C}^{\dagger}\underset{\sim}{H}\underset{\sim}{C} = \underset{\sim}{C}^{\dagger}\left(\sum_i \underset{\sim}{T}_i + \sum_{i<j} \underset{\sim}{V}_{ij}\right)\underset{\sim}{C}$$

$$= \sum_i \underset{\sim}{T}_i + \sum_{i<j} \widehat{\underset{\sim}{T}}^{[2]}_{ij} + \sum_{i<j<k} \widehat{\underset{\sim}{T}}^{[3]}_{ijk} + \cdots + \sum_{i<j} \widehat{\underset{\sim}{V}}^{[2]}_{ij} + \sum_{i<j<k} \widehat{\underset{\sim}{V}}^{[3]}_{ijk} + \cdots ,$$

where $^{[n]}$ indicates irreducible n-body operators. Here we introduce an approximation by keeping terms only up to two-body operators. This approximation should be good for systems where the range R_c of the correlators is short compared to the mean particle distance. $\underset{\sim}{C}_r$ and $\underset{\sim}{C}_\Omega$ act only at small distances because $g_{r\,ij} = 0$ and $g_{\Omega\,ij} = 0$ for $r_{ij} > R_c$ and hence $\underset{\sim}{C}_r = \underset{\sim}{1}$ and $\underset{\sim}{C}_\Omega = \underset{\sim}{1}$ for $r_{ij} > R_c$. The strength of the induced three-body terms is governed by the probability to find three particles simultaneously within the correlation range R_c. The aim is to keep these contributions small.

3˙2. *Radial correlator.* – The radial correlator $\underset{\sim}{C}_r$ (described in detail in [6]) shifts a pair of particles away from each other in the radial direction so that they avoid the repulsive core. To perform these shifts the generator of the radial correlator uses the relative radial momentum operator $\underset{\sim}{p}_r$ together with a shift function $s(\underset{\sim}{r})$ that depends on the distance between of the two nucleons:

$$(14) \qquad\qquad \underset{\sim}{g}_{rij} = \frac{1}{2}\left(\underset{\sim}{p}_{rij}\, s(\underset{\sim}{r}_{ij}) + s(\underset{\sim}{r}_{ij})\, \underset{\sim}{p}_{rij}\right).$$

The shift function $s(\underset{\sim}{r})$ is optimized for the potential under consideration. It is large for short distances and vanishes at large distances.

The effect of the transformation $|\Psi\rangle \rightarrow \underset{\sim}{C}_r|\Psi\rangle$ is shown in the upper part of fig. 2 where the two-body density $\rho^{(2)}_{S,T}$ is displayed as a function of the distance vector $(\vec{r}_1 - \vec{r}_2)$ between two nucleons in ^{4}He. On the l.h.s. $\rho^{(2)}_{S,T}$ is calculated with the shell model state $|(0s_{1/2})^4\rangle$ that is just a product of four Gaussians. It has a maximum at zero distance which is in contradiction with the short-ranged repulsion of the interaction. This inconsistency is removed by the action of the radial correlator $\underset{\sim}{C}_r$ that moves nuclear density out of the repulsive region of the potential. The corresponding kinetic, potential and total energies are displayed in the lower part of the figure for three doubly magic nuclei. The radially correlated kinetic energy $\langle \underset{\sim}{C}_r^{\dagger}\underset{\sim}{T}\underset{\sim}{C}_r\rangle$ increases somewhat compared to $\langle \underset{\sim}{T}\rangle$ but this is more than compensated for by the gain of binding energy of about 25 MeV per particle due to the correlated potential. Nevertheless the nuclei are still unbound.

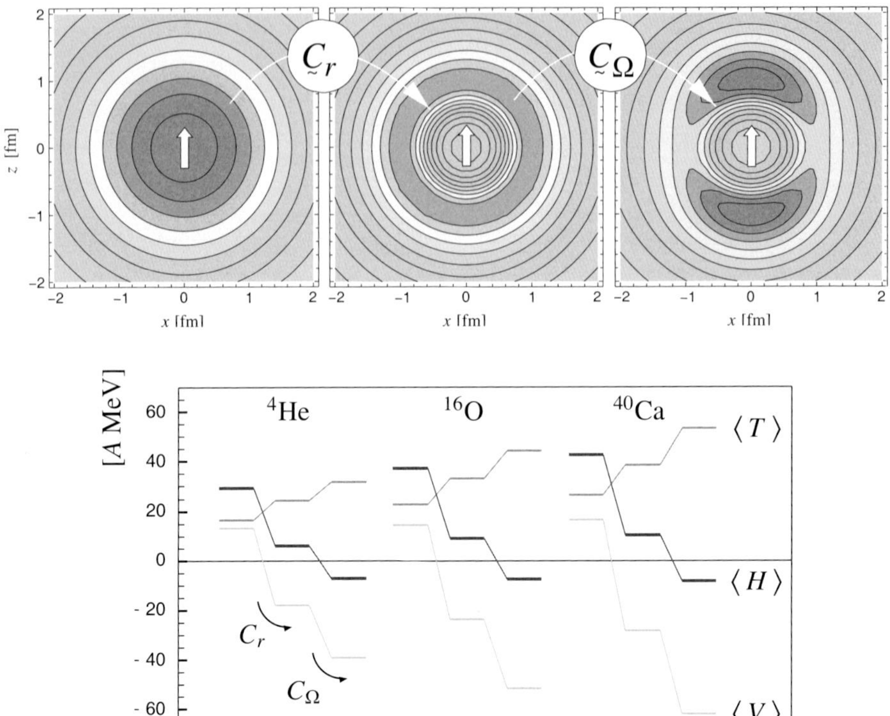

Fig. 2. – Upper part: two-body density $\rho^{(2)}_{S,T}(\vec{r}_1 - \vec{r}_2)$ of ^{4}He for a pair of nucleons with isospin $T = 0$ and parallel spins, $S = M_S = 1$. The arrow indicates the spin direction and $(x, y, z) = (\vec{r}_1 - \vec{r}_2)$ are the components of the relative coordinate vector. Lower part: corresponding kinetic, potential and total energies per particle of ^{4}He, ^{16}O and ^{40}Ca, without correlations $\langle T \rangle$, $\langle V \rangle$, $\langle H \rangle$; with radial correlations $\langle C_r^\dagger T C_r \rangle$, $\langle C_r^\dagger T C_r \rangle$, $\langle C_r^\dagger T C_r \rangle$; and with radial and tensor correlations $\langle C_r^\dagger C_\Omega^\dagger T C_\Omega C_r \rangle$, $\langle C_r^\dagger C_\Omega^\dagger V C_\Omega C_r \rangle$, $\langle C_r^\dagger C_\Omega^\dagger H C_\Omega C_r \rangle$ (AV18 potential).

3'3. *Tensor correlator*. – The tensor force in the $S = 1$ channels of the nuclear interaction depends on the spins and the spatial orientation $\hat{r}_{12} = (\vec{r}_1 - \vec{r}_2)/(|\vec{r}_1 - \vec{r}_2|)$ of the nucleons according to the tensor operator

$$(15) \qquad S_{12} = 3(\vec{\sigma}_1 \hat{r}_{12})(\vec{\sigma}_2 \hat{r}_{12}) - (\vec{\sigma}_1 \vec{\sigma}_2) = 6(\vec{S} \hat{r}_{12})^2 - 2\vec{S}^2 .$$

An alignment of $\hat{r}$ with the direction of total spin $\vec{S} = \frac{1}{2}(\vec{\sigma}_1 + \vec{\sigma}_2)$ is favoured energetically. This can be seen clearly in fig. 3 where the Argonne V18 potential is plotted for two orientations of the total spin $\vec{S}$. If $\vec{S}$ is parallel (or antiparallel) to the orientation of $\hat{r}_{12}$ the strongest attraction is encountered, while in situations where $\hat{r}_{12}$ is perpendicular to $\vec{S}$ the proton-neutron pair does not feel any attraction anymore, only the short-range repulsion. The tensor force alone is repulsive but the other attractive central parts counterbalance the repulsion. It is obvious that this strong dependence on the spin

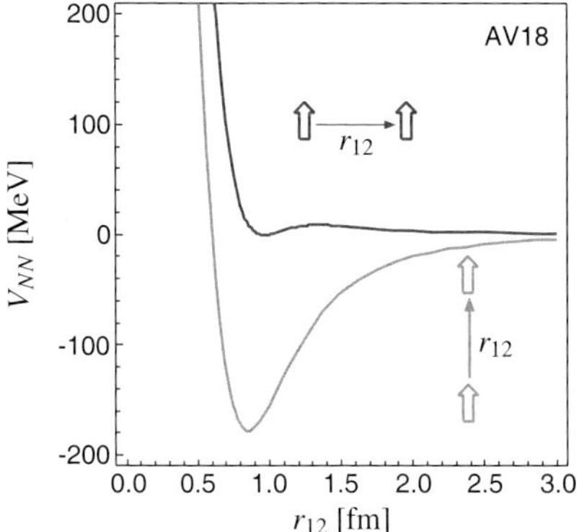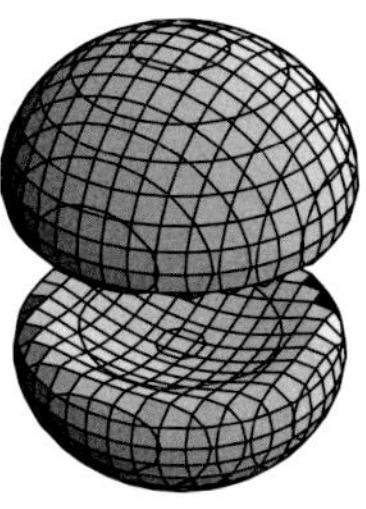

Fig. 3. – L.h.s.: Argonne V18 potential for two nucleons at rest with isospin $T = 0$ and parallel spins, $S = 1$, as a function of relative distance. The behaviour resembles the interaction between two bar magnets: largest attraction when the magnets are aligned with the line connecting them. R.h.s.: the equidensity surface of the deuteron for $J = M = 1$ shows the resulting tensor correlations, the proton-neutron pair density is aligned with the spin direction. At small distances around $\vec{r}_{12} = 0$ the density is suppressed due to the strong repulsive core in the interaction.

orientation will induce corresponding correlations not only in the deuteron but also in many-body states. Spin $S = 1$, $T = 0$ pairs of nucleons will arrange themselves such that they can be found with a higher probability in areas where their spins are aligned with the orientation of their relative coordinate $\vec{r}_{12}$ than elsewhere. This leads to the typical dumbbell shape of the two-body correlation function known already from the deuteron, displayed on the r.h.s. of fig. 3. These tensor correlations cannot be represented by product states like Slater determinants.

In order to take them into account a tensor correlator $\underset{\sim}{C}_\Omega = \exp[-i \sum_{i<j} \underset{\sim}{g}_{\Omega\,ij}]$ defined by the generator

$$(16) \qquad \underset{\sim}{g}_{\Omega ij} = \vartheta(\underset{\sim}{r}_{ij})\frac{3}{2}\left((\vec{\sigma}_i\,\vec{\underset{\sim}{p}}_{\Omega ij})(\vec{\sigma}_j\,\vec{\underset{\sim}{r}}_{ij}) + (\vec{\sigma}_i\,\vec{\underset{\sim}{r}}_{ij})(\vec{\sigma}_j\,\vec{\underset{\sim}{p}}_{\Omega ij}) \right),$$

is introduced, which imprints this alignment on the uncorrelated states by shifts perpendicular to the relative orientation $\hat{r}_{ij}$. To achieve this the generator $\underset{\sim}{g}_{\Omega ij}$ is a tensor operator constructed with the orbital part of the relative momentum operator $\vec{\underset{\sim}{p}}_{\Omega ij} = \vec{\underset{\sim}{p}}_{ij} - \vec{\underset{\sim}{p}}_{rij}$. The r-dependent strength and the range of the tensor correlations is controlled by $\vartheta(r)$, for details see ref. [7].

The application of the tensor correlator $\underset{\sim}{C}_\Omega$ leads to the two-body density of ^{4}He depicted in the right-hand contour plot of fig. 2. One may visualize the action of $\underset{\sim}{C}_\Omega$ as a displacement of probability density from the "equator" to both "poles", where the spin of the $S = 1$ component of the nucleon pair defines the "south-north" direction. Again this costs kinetic energy due to the increased curvature in the wave functions, but now the many-body state is in accord with the tensor interaction and one gains the binding needed to end up with about $-8\,\mathrm{MeV}$ per particle as seen in the lower part of fig. 2.

3'4. *The effective interaction V_{UCOM}.* – The correlated Hamiltonian

$$(17) \qquad \widehat{\underset{\sim}{H}} = \underset{\sim}{C}^{-1}\,\underset{\sim}{H}\,\underset{\sim}{C} = \underset{\sim}{T}_{\mathrm{int}} + \big(\widehat{\underset{\sim}{T}}^{[2]} + \widehat{\underset{\sim}{V}}^{[2]}\big) + \big(\widehat{\underset{\sim}{T}}^{[3]} + \widehat{\underset{\sim}{V}}^{[3]}\big) + \dots$$

is by construction phase-shift equivalent to the realistic Hamiltonian $\underset{\sim}{H}$ one starts from. Therefore the distinction between genuine and correlation induced 3-body forces is not so obvious. In order to keep the induced $n = 3$- and higher-body parts $\widehat{\underset{\sim}{T}}^{[n]}$ and $\widehat{\underset{\sim}{V}}^{[n]}$ small the range of the correlation fucntions $s(r)$ and $\vartheta(r)$ should be smaller than the mean particle distance. While the central correlation functions could be determined from the two-body system, the tensor correlation has a long range and thus $\vartheta(r)$, which is determined from the exact deuteron wave function, also has a long range. Therefore the range of the tensor correlator has to be confined in order to avoid induced many-body forces.

To test the two-body approximation no-core shell model calculations were performed for ^{4}He and ^{3}He and compared with exact results [8]. It turned out, that by neglecting the induced $n = 3$- and 4-body potentials of the correlated Hamiltonian, both ground-state energies could be reproduced, when a specific range of the tensor correlator was chosen. This choice defines the phase shift equivalent nucleon-nucleon potential V_{UCOM}

$$(18) \qquad V_{\mathrm{UCOM}} = \widehat{\underset{\sim}{T}}^{[2]} + \widehat{\underset{\sim}{V}}^{[2]} \, .$$

3'5. *Effective operators.* – Not only the Hamiltonian but also all other observables have to be correlated. Due to the unitarity of the correlator $\underset{\sim}{C}$ matrix elements of transitions or expectation values of some observable $\underset{\sim}{A}$ can be calculated either with correlated states or correlated operators.

$$(19) \qquad \langle\widehat{\Psi}|\underset{\sim}{A}|\widehat{\Phi}\rangle = \langle\Psi|\underset{\sim}{C}^{\dagger}\underset{\sim}{A}\,\underset{\sim}{C}|\Phi\rangle = \langle\Psi|\widehat{\underset{\sim}{A}}|\Phi\rangle$$

In actual applications effective operators

$$(20) \qquad \widehat{\underset{\sim}{A}} = \underset{\sim}{C}^{-1}\underset{\sim}{A}\,\underset{\sim}{C} \approx \underset{\sim}{A}^{[1]} + \underset{\sim}{A}^{[2]}$$

are introduced by again neglecting induced three- and higher-body parts.

4. – Fermionic Molecular Dynamics (FMD)

4'1. *FMD many-body states.* – In Fermionic Molecular Dynamics (FMD) [10-13] the A-body Hilbert space is spanned by a set of non-orthogonal Slater determinants

$$(21) \qquad |Q\rangle = \underset{\sim}{\mathcal{A}}\,|q_1\rangle \otimes |q_2\rangle \cdots \otimes |q_A\rangle = \frac{1}{A!}\sum_{\mathrm{all}\,\rho}\mathrm{sgn}(\rho)\,|q_{\rho(1)}\rangle \otimes |q_{\rho(2)}\rangle \cdots \otimes |q_{\rho(A)}\rangle$$

which are antisymmetrized Kronecker products of single-particle states

$$(22) \qquad |q_k\rangle = \sum_j c_{kj} \, |a_{kj}, \vec{b}_{kj}\rangle \otimes |\chi_{kj}\rangle \otimes |\xi_k\rangle$$

consisting of a spatial, spin and isospin part. The spatial parts $|a_k, \vec{b}_k\rangle$ are Gaussians represented in coordinate space as

$$(23) \qquad \langle \vec{x} | a_k, \vec{b}_k \rangle = \exp\left[-\frac{(\vec{x} - \vec{b}_k)^2}{2a_k} \right].$$

The complex parameters $\vec{b}_k$ localize the nucleon in coordinate and momentum space, the complex width parameters a_k control its spread in position and momentum. The spin part $|\chi_k\rangle$

$$(24) \qquad |\chi_k\rangle = \chi_k^{\uparrow} \, |\!\uparrow\rangle + \chi_k^{\downarrow} \, |\!\downarrow\rangle$$

denotes the most general two-spinor for a non-relativistic spin-(1/2) fermion and the isospin part $|\xi_k\rangle$ distinguishes between proton and neutron. The sum is usually only over one or two Gaussians. By superimposing two Gaussians with different widths one can for example improve the description of the surface of the nucleus, especially for halo nuclei.

The eigenstates of the nuclear Hamiltonian are eigenstates of parity, total spin, and magnetic quantum number. The total center-of-mass motion separates out. The FMD representation (21) does not possess these symmetries in the general case. Therefore the symmetries of the Hamiltonian are implemented by appropriate projections.

$\mathbf{4}`2$. *Center-of-mass projection.* – In the special case that all width parameters $a_k = a$ are identical, each Slater determinant $|Q\rangle$ can be factorised into intrinsic and centre-of-mass motion. Identical widths occur for example in Antisymmetric Molecular Dynamics (AMD) [14-16] or in Brink's cluster model [17]. However, a more precise description of nuclear structure demands different width parameters. Therefore, in the general case we perform a CM projection on the momentum zero eigenstate of total momentum $\underset{\sim}{\vec{P}}_{\text{CM}} = \sum_{i=1}^{A} \underset{\sim}{\vec{p}}_i$ by means of the operator

$$(25) \qquad \underset{\sim}{P}_{\text{CM}} = \frac{1}{(2\pi)^3} \int \mathrm{d}^3 X \, \exp\left[-i \, \underset{\sim}{\vec{P}}_{\text{CM}} \vec{X} \right].$$

Applying $\underset{\sim}{P}_{\text{CM}}$ to an arbitrary many-body state causes it to become translationally invariant. Furthermore $\underset{\sim}{P}_{\text{CM}}$ commutes with any rotation and the total angular momentum of a CM projected state is the intrinsic one.

4'3. *Angular-momentum projection.* – The Hamiltonian is invariant under rotations and the parity operation. Therefore, we also project all FMD states $|Q^{(i)}\rangle$ on total angular momentum J and parity $\pi = \pm 1$

$$(26) \qquad |Q^{(i)}; J^\pi MK\rangle := \underset{\sim}{P}{}^J_{MK} \frac{1}{2}(1 + \pi\underset{\sim}{\Pi}) \underset{\sim}{P}_{\mathrm{CM}} |Q^{(i)}\rangle = \underset{\sim}{P}{}^{J^\pi}_{MK} \underset{\sim}{P}_{\mathrm{CM}} |Q^{(i)}\rangle .$$

$\underset{\sim}{\Pi}$ denotes the parity operation and $\underset{\sim}{P}{}^J_{MK}$ projects on angular momentum:

$$(27) \qquad \underset{\sim}{P}{}^J_{MK} = \frac{2J+1}{8\pi^2} \int \mathrm{d}\alpha \, \sin\beta \, \mathrm{d}\beta \, \mathrm{d}\gamma \, D^{J\,*}_{MK}(\alpha,\beta,\gamma) \, \underset{\sim}{R}(\alpha,\beta,\gamma),$$

with the rotation operator

$$(28) \qquad \underset{\sim}{R}(\alpha,\beta,\gamma) = \exp\left[-i\alpha \underset{\sim}{J}_z\right] \exp\left[-i\beta \underset{\sim}{J}_y\right] \exp\left[-i\gamma \underset{\sim}{J}_z\right]$$

and the Wigner functions

$$(29) \qquad D^J_{MK}(\alpha,\beta,\gamma) = \langle JM|\underset{\sim}{R}(\alpha,\beta,\gamma)|JK\rangle = e^{-iM\alpha} \, d^J_{MK}(\beta) \, e^{-iK\gamma} .$$

Details can be found in [18], here we quote only the property

$$(30) \qquad \left(\underset{\sim}{P}{}^J_{MK}\right)^\dagger \underset{\sim}{P}{}^{J'}_{M'K'} = \delta_{JJ'} \, \delta_{MM'} \, \underset{\sim}{P}{}^J_{KK'} ,$$

which shows that $\underset{\sim}{P}{}^J_{MK}$ is not a projection operator in the strict sense. This property is used to reduce the numerical effort required to calculate the matrix elements of tensor operators. For example, the Hamiltonian is a tensor of rank zero and commutes with rotations, thus

$$(31) \qquad \langle Q^{(i)}; J^\pi MK|\underset{\sim}{H}|Q^{(j)}; J^\pi M'K'\rangle = \langle Q^{(i)}|\underset{\sim}{H}\underset{\sim}{P}{}^{J^\pi}_{KK'}\underset{\sim}{P}_{\mathrm{CM}}|Q^{(j)}\rangle \, \delta_{MM'}$$

4'4. *K-mixing.* – The dimension $n^{(i)}$ of the subspace spanned by the $2J+1$ projected states

$$(32) \qquad \left\{ |Q^{(i)}; J^\pi MK\rangle; \ K = -J, -J+1, \ldots, J \right\}$$

depends on the symmetries of the intrinsic state $\underset{\sim}{P}_{\mathrm{CM}}|Q^{(i)}\rangle$. For example, a spherically symmetric intrinsic state has $n^{(i)} = 1$ for $J = M = K = 0$ and $n^{(i)} = 0$ for all other J, M, K combinations. For an axially symmetric intrinsic state the dimension is not $2J+1$ but $n^{(i)} = 1$ as rotation about the symmetry axis does not yield new states.

In the general case we first perform so-called K-mixing by solving the eigenvalue problem for the intrinsic Hamiltonian $\underset{\sim}{H}$ in the subspace (32) employing a singular value decomposition. We find the lowest $n^{(i)}$ eigenvalues whilst excluding states with insignificant norms:

$$(33) \qquad |Q^{(i)}; J^\pi M; \kappa_i\rangle = \sum_{K=-J}^{J} |Q^{(i)}; J^\pi M K\rangle \, C^{(i)}_{K\kappa_i} \, ; \quad \kappa_i = 1, \ldots, n^{(i)} \, .$$

4'5. *Many-body Hilbert space.* – For given total angular momentum and parity J^π the FMD many-body Hilbert space is spanned by the non-orthogonal states $|Q^{(i)}; J^\pi M; \kappa_i\rangle$ given in (33). In order to keep the notation concise, the following short notation is introduced for matrix elements:

$$(34) \qquad \mathsf{N}_{ij} = \langle Q^{(i)}; J^\pi M; \kappa_i | Q^{(j)}; J^\pi M; \kappa_j\rangle,$$

$$(35) \qquad \mathsf{H}_{ij} = \langle Q^{(i)}; J^\pi M; \kappa_i | \underset{\sim}{H} | Q^{(j)}; J^\pi M; \kappa_j\rangle,$$

$$(36) \qquad \mathsf{B}_{ij} = \langle Q^{(i)}; J^\pi M; \kappa_i | \underset{\sim}{B} | Q^{(j)}; J^\pi M; \kappa_j\rangle, \text{ etc.}$$

The index i stands for both the number (i) of the FMD parameter set $Q^{(i)}$ and the quantum number κ_i from the K-mixing procedure. Thus, a sum over i means a sum over (i) and κ_i, $i = 1, \ldots, n^{(i)}$. The matrix element of a translationally and rotationally invariant operator $\underset{\sim}{B}$ is represented in long notation by

$$(37) \qquad \mathsf{B}_{ij} = \sum_{K,K'=-J}^{J} C^{(i)\,*}_{K\kappa_i} \langle Q^{(i)} | \underset{\sim}{B} \, \underset{\sim}{P}_{\mathrm{CM}} \underset{\sim}{P}^{J^\pi}_{KK'} | Q^{(j)} \rangle C^{(j)}_{K'\kappa_j} \, .$$

In short notation the energy eigenvalue problem reads

$$(38) \qquad \sum_j \mathsf{H}_{ij}\, \Psi_j^\alpha = E_\alpha \sum_j \mathsf{N}_{ij}\, \Psi_j^\alpha \, ,$$

or, even more concisely,

$$(39) \qquad \mathsf{H} \cdot \vec{\Psi}^\alpha = E_\alpha\, \mathsf{N} \cdot \vec{\Psi}^\alpha \, .$$

It is understood that all matrices N, H etc. and all eigenvalues E_α and coefficients $\vec{\Psi}_\alpha$ should also carry the label J^π, as they are different in the different J^π many-body Hilbert spaces. The basis states $|Q^{(i)}; J^\pi M; \kappa_i\rangle$ are not orthonormalized, therefore, in order to avoid numerical inaccuracies, a singular value decomposition is employed to eliminate eigenstates of the overlap matrix N with eigenvalues below a threshold.

The energy eigenvalue problem formulated in the many-body Hilbert space

$$(40) \qquad \underset{\sim}{H} |J^\pi M; \alpha\rangle = E_\alpha |J^\pi M; \alpha\rangle$$

leads to eigenstates of the Hamiltonian given in long notation by

$$(41) \qquad |J^\pi M; \alpha\rangle = \sum_{j=1}^{n} \sum_{\kappa_j=1}^{n^{(j)}} \Psi^\alpha_{(j)\kappa_j} \sum_{K=-J}^{J} |Q^{(j)}; J^\pi MK\rangle \, C^{(j)}_{K\kappa_j} \, .$$

4'6. *Ritz variational principle.* – Let us denote the exact ground state of a Hamiltonian $\underset{\sim}{H}$ by $|\Psi; 0\rangle$ and the corresponding eigenvalue by E_0. Let a many-body state $|\Phi\rangle$ be specified by a manifold of parameters $\Phi = \{\phi_1, \phi_2, \ldots, \phi_n\}$. Minimizing the energy

$$(42) \qquad E[\Phi] = \frac{\langle \Phi | \underset{\sim}{H} | \Phi \rangle}{\langle \Phi | \Phi \rangle}$$

of the trial state $|\Phi\rangle$ with respect to all parameters ϕ_k leads to an approximation of the true ground state $|\Psi; 0\rangle$. If the manifold of parameters $\Phi 1 = \{\phi_1, \phi_2, \ldots, \phi_n, \phi_{n+1}\}$ is enlarged, the minimum in the energy of the trial state $|\Phi 1\rangle$ is lowered or unaltered

$$(43) \qquad E[\Phi] \geq E[\Phi 1] \geq E_0$$

and approaches the true ground-state energy from above. At the same time, the overlap with the ground state increases indicating a better approximation to the exact ground state.

$$(44) \qquad \frac{|\langle \Phi | \Psi; 0 \rangle|^2}{\langle \Phi | \Phi \rangle \langle \Psi; 0 | \Psi; 0 \rangle} \leq \frac{|\langle \Phi 1 | \Psi; 0 \rangle|^2}{\langle \Phi 1 | \Phi 1 \rangle \langle \Psi; 0 | \Psi; 0 \rangle} \leq 1$$

These easy to prove relations constitute the variational principle of Ritz [19].

A variety of approximations arise from the Ritz variational principle.

– The Hartree-Fock (HF) approximation is obtained when $|\Phi\rangle = \mathcal{A} |\phi_1\rangle \otimes \cdots \otimes |\phi_A\rangle$ is a Slater determinant and the variational manifold consists of the most general single-particle states $|\phi_k\rangle$. If the form of the single-particle states is restricted, for example to the FMD states $|q_k\rangle$, one obtains an approximation to Hartree-Fock.

– Variation after parity projection denotes the variation of an FMD many-body basis state that is projected on parity. The minimization of the energy

$$(45) \qquad E^\pi[Q] = \frac{\langle Q | \underset{\sim}{H} \, (1 + \pi \underset{\sim}{\Pi}) | Q \rangle}{\langle Q | 1 + \pi \underset{\sim}{\Pi} | Q \rangle}$$

leads to an FMD basis state $|Q^\pi\rangle$ that is to be regarded as an intrinsic state. The resulting parameter set Q^π is different for $\pi = +1$ and $\pi = -1$.

This variation after projection on parity eigenstates is a simple example of restoration of a symmetry, here the parity symmetry. If the representation of many-body states $|Q\rangle$ allows for a symmetry to be broken one can enforce this symmetry by projecting on its eigenstates. The variation after projection yields energies which are lower or equal to the variation without projection. Even in cases where a HF minimization yields a state $|Q\rangle$ which is already a parity eigenstate, the variation after projection on parity usually gives an intrinsic state $|Q^\pi\rangle$ that is different from $|Q\rangle$ and $(1 + \pi\underset{\sim}{\Pi})|Q^\pi\rangle$ has a lower energy than $|Q\rangle$.

It is important to note that the energies finally quoted in this lecture are calculated with the angular-momentum projected state $|Q; J^\pi M; \kappa\rangle$ defined in eq. (33), therefore the method is called **P**rojection **A**fter **V**ariation of a **P**arity eigenstate (PAV$^\pi$).

– VAP denotes **V**ariation **A**fter angular momentum and parity **P**rojection. The trial state is projected on angular momentum and the energy to be minimized is

$$(46) \qquad E^{J^\pi}[Q] = \frac{\langle Q|\underset{\sim}{H}\,\underset{\sim}{P}{}^{J^\pi}_{KK}|Q\rangle}{\langle Q|\underset{\sim}{P}{}^{J^\pi}_{KK}|Q\rangle} \,.$$

The value of K is chosen according to the largest contribution in K-mixing. This variation is by far the most numerically demanding one.

– Configuration mixing can also be derived from the Ritz variational principle. In this case the trial state is a linear combination of selected states with a given J^π, see eq. (33).

$$(47) \qquad |\Phi\rangle = \sum_i \Psi_i\,|Q^{(i)}; J^\pi M; \kappa_i\rangle \,.$$

The variation manifold consists of the coefficients Ψ_i. Determining the extrema of the energy function

$$(48) \qquad E^{J^\pi}\left[\{\Psi_j^*, \Psi_j;\ j = 1,\ldots n\}\right] = \frac{\sum_{ij} \Psi_i^*\,\mathsf{H}_{ij}\,\Psi_j}{\sum_{ij} \Psi_i^*\,\mathsf{N}_{ij}\,\Psi_j} \,,$$

i.e. $\partial E^{J^\pi}/\partial\Psi_i^* = 0$ for $i = 1,\ldots,n$, leads to the eigenvalue problem (38) which has as solutions the $\vec{\Psi}^\alpha$ and their corresponding eigenvalues E_α.

4'7. *Generator Coordinate Method (GCM).* – The minimum in the energy surface $E[Q]$ is often located in a narrow valley with small second derivatives in certain directions along the multidimensional valley. This indicates that the nuclear system has small restoring forces in this submanifold. Therefore one expects quantum zero point oscillations around the minimum. Or, in other words, a superposition (41) of different angular-momentum projected states $|Q^{(i)}; J^\pi M; \kappa_i\rangle$ along the valley should improve the quality of the many-body state because these ground state correlations are then included.

The simplest examples for flat valleys are translations by a vector $\vec{X}$ and rotations about the three Euler angles α, β, γ. Both are contained in the FMD manifold Q and the derivative of $E[Q]$ with respect to these parameters is zero. A superposition of all translations and all rotations with the appropriate coefficients is just the CM and angular-momentum projection discussed in subsects. 4'2 and 4'3.

In principle one could find soft directions by looking at the major axes of the second derivatives of $E[Q]$ at the minimum. As this is rather time consuming, in practice one varies $E[Q]$ under some constraints $C[Q] = c_i$. Examples are usually collective observables like the square radius or the dipole moment $\vec{\mathcal{D}}$

$$(49) \qquad \vec{\mathcal{D}} = \frac{\langle Q | \sum_{i=1}^{A} (\vec{r}_i - \vec{R}_{\mathrm{CM}}) \tau_{3i} | Q \rangle}{\langle Q | Q \rangle} \quad \text{with } C[Q] = (\vec{\mathcal{D}})^2 .$$

The constraint should be invariant under translations and rotations, thus $(\vec{\mathcal{D}})^2$ is used. In the case of the quadrupole moment, the trace of the quadrupole tensor squared or the determinant are invariant quantities.

By minimizing the energy under these constraints one finds new configurations $Q^{(i)}$ as a function of the value c_i of the constraint which is called generator coordinate. The projected many-body states $P^{J^\pi}_{MK} P_{\mathrm{CM}} | Q^{(i)} \rangle$ may then serve to span the Hilbert space for a configuration mixing calculation.

4'8. *FMD Hamiltonian.* – Comparison of FMD with no-core shell model calculations using V_{UCOM} as the two-body potential show that even after treating the short-range correlations a numerically feasible FMD Hilbert space is not sufficient to treat the remaining long-range correlations. To benefit from the special properties of FMD states without increasing the numerical effort beyond practicability we decided to add a phenomenological correction term consisting of a momentum-dependent central and spin-orbit correction H_{corr} with parameters adjusted to ^{4}He, ^{16}O, ^{40}Ca, ^{24}O, ^{34}Si, and ^{48}Ca. The resulting effective interaction $H = T_{\mathrm{int}} + V_{\mathrm{UCOM}} + H_{\mathrm{corr}}$ is henceforth used for all nuclei. The expectation value of H_{corr} is typically 15% of the correlated interaction energy $\langle V_{\mathrm{UCOM}} \rangle$.

5. – Cluster structure

5'1. *Be isotopes.* – In fig. 4 the densities of the intrinsic states $|Q^\pi\rangle$ obtained by minimizing the energy $E^\pi[Q]$, eq. (45), after parity projection are shown for the Beryllium isotopes. For ^{7}Be and ^{8}Be pronounced cluster structures can be seen and we find a strong overlap with Brink cluster wave functions.

The ^{8}Be density is a nice example of the restoration of a symmetry. Minimization of $\langle Q | H | Q \rangle / \langle Q | Q \rangle$ leads to an intrinsic state $|Q\rangle$ which is an eigenstate of parity. However, minimizing the energy $E^+[Q]$ of the parity projected state $(1 + \Pi)|Q^+\rangle$ leads to an FMD state $|Q^+\rangle$ that has broken parity. Its density is not mirror symmetric although ^{8}Be consists of two identical α-clusters. The density of the parity projected state is of course symmetric. The energy $E^+[Q^+]$ in which the off-diagonal matrix element $\langle Q^+ | H \Pi | Q^+ \rangle$ contributes, see eq. (45), is lower than $\langle Q | H | Q \rangle / \langle Q | Q \rangle$ which in turn is lower than $\langle Q^+ | H | Q^+ \rangle / \langle Q^+ | Q^+ \rangle$. The physical reason is that compared to $|Q\rangle$ the state $(1 + \Pi)|Q^+\rangle$ with restored parity has more freedom to describe the surface and the mutual polarization of the α-clusters. If the manifold of the parameters Q already has good parity, the Gaussians of one α-cluster are the same as the ones of the other except

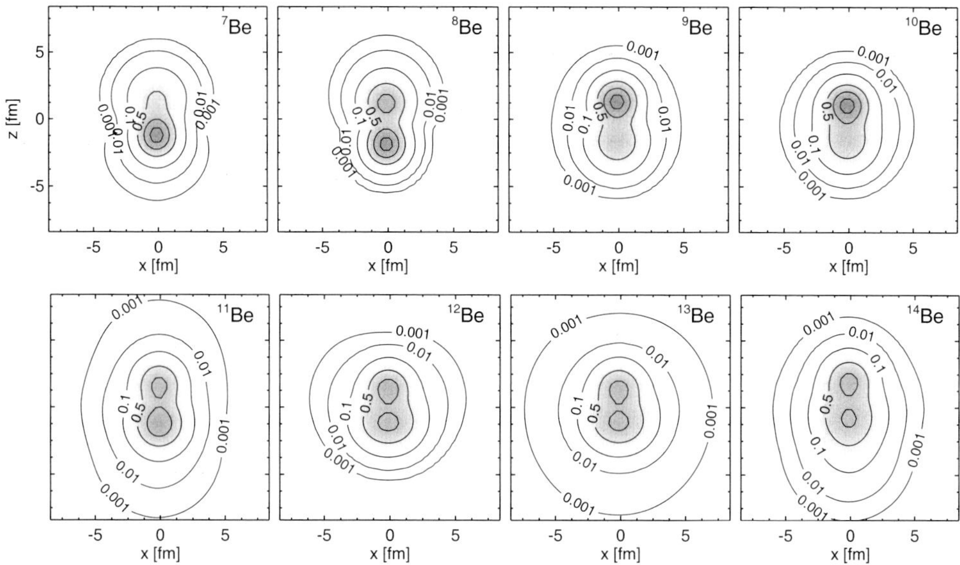

Fig. 4. – Density of intrinsic states $|Q^-\rangle$ for ^{7,9,13}Be and $|Q^+\rangle$ for ^{8,10,12,14}Be. In the case of ^{11}Be the density of the unnatural positive parity state $|Q^+\rangle$ is shown.

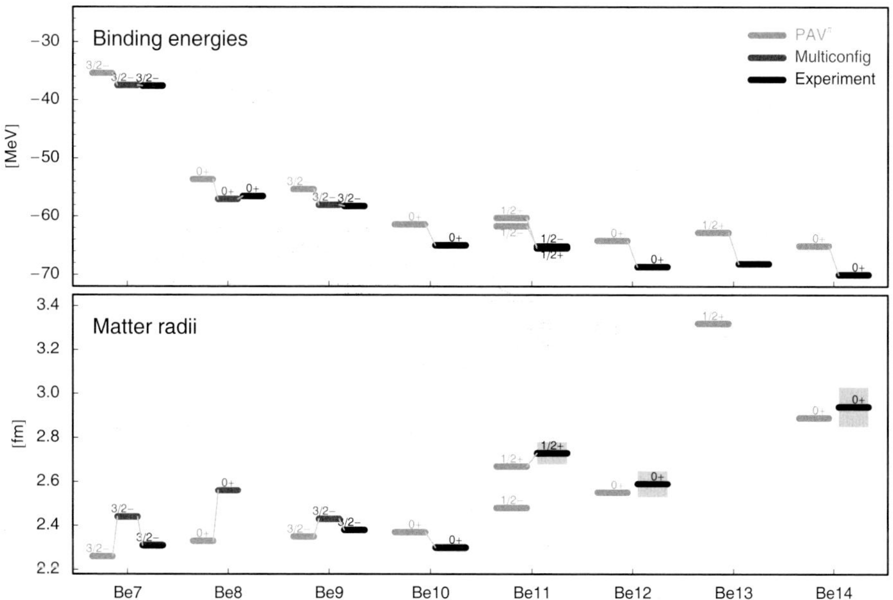

Fig. 5. – Energies and matter radii of beryllium isotopes. So far multiconfiguration calculations have only been done for ^{7}Be, ^{8}Be and ^{9}Be.

for mirroring them at the origin. Thus only half of the parameters can actually be chosen independently.

In the heavier isotopes the α-cluster structure survives but is modified by the additional neutrons. In ^{9}Be and ^{10}Be the additional neutrons occupy the $0p_{3/2}$ single-particle states. In experiment a parity inversion in ^{11}Be is observed. The $0d_{5/2}$ and $1s_{1/2}$ single-particle states come down in energy and the $1/2^+$ ground state is almost degenerate with the $1/2^-$ excited state. In the FMD PAV$^\pi$ calculations we find the $1/2^+$ state to be 1.5 MeV higher in energy than the $1/2^-$ state (see fig. 5). The large matter radius of the $1/2^+$ state is in good agreement with the experimental value. The unnatural parity states in ^{11}Be are discussed in more detail in the following subsection. The general behaviour of ^{12}Be being bound, ^{13}Be being unbound and ^{14}Be being bound again as well as the matter radii are well reproduced. For the heavier ^{12}Be, ^{13}Be and ^{14}Be nuclei the $0d_{5/2}$ and also the $1s_{1/2}$ single-particle states are important.

$5^\cdot 2.$ *Unnatural parity ground state in* ^{11}Be. – Let us have a closer look at the $J^\pi = 1/2^+$ intruder state in ^{11}Be. A PAV$^\pi$ calculation results in the energy spectrum displayed in fig. 6. The first observation is that ^{10}Be is underbound by 3.6 MeV which indicates that a single PAV$^\pi$ state is not sufficient, configuration mixing and/or variation after projection (VAP) needs to be done to improve the many-body states. The energies relative to the ^{10}Be ground state are however close to the data.

In fig. 7 the intrinsic densities are displayed for the three cases. The upper row shows the proton and neutron densities of the intrinsic states $|Q^+\rangle$ for ^{10}Be, and $|Q^-\rangle$ for the expected parity as well as $|Q^+\rangle$ for the unnatural parity in the case of ^{11}Be. The row below shows the corresponding densities after projection on parity. The intruder state with unnatural parity develops a much broader neutron distribution. This is even more visible in the densities of the angular-momentum projected states displayed in the lower right corner. The $1/2^-$ states looks like a normal p-shell nucleus while the $1/2^+$ state has a neutron halo. This is in accord with the large measured matter radius (see

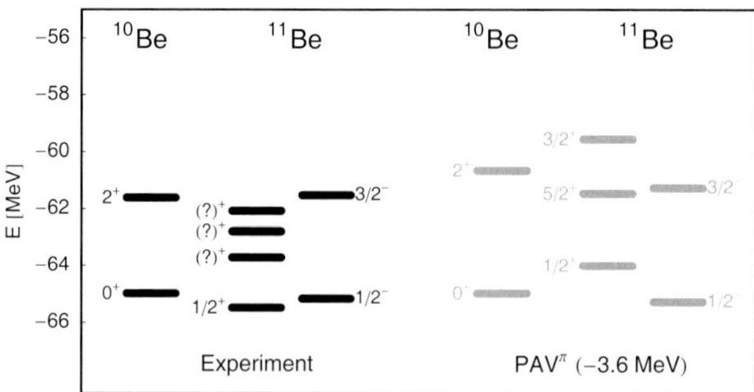

Fig. 6. – Experimental and calculated energy spectra of ^{10}Be and ^{11}Be. The theoretical spectra obtained from a PAV$^\pi$ calculation are shifted down as a whole by 3.6 MeV to align with the experimental ^{10}Be ground state.

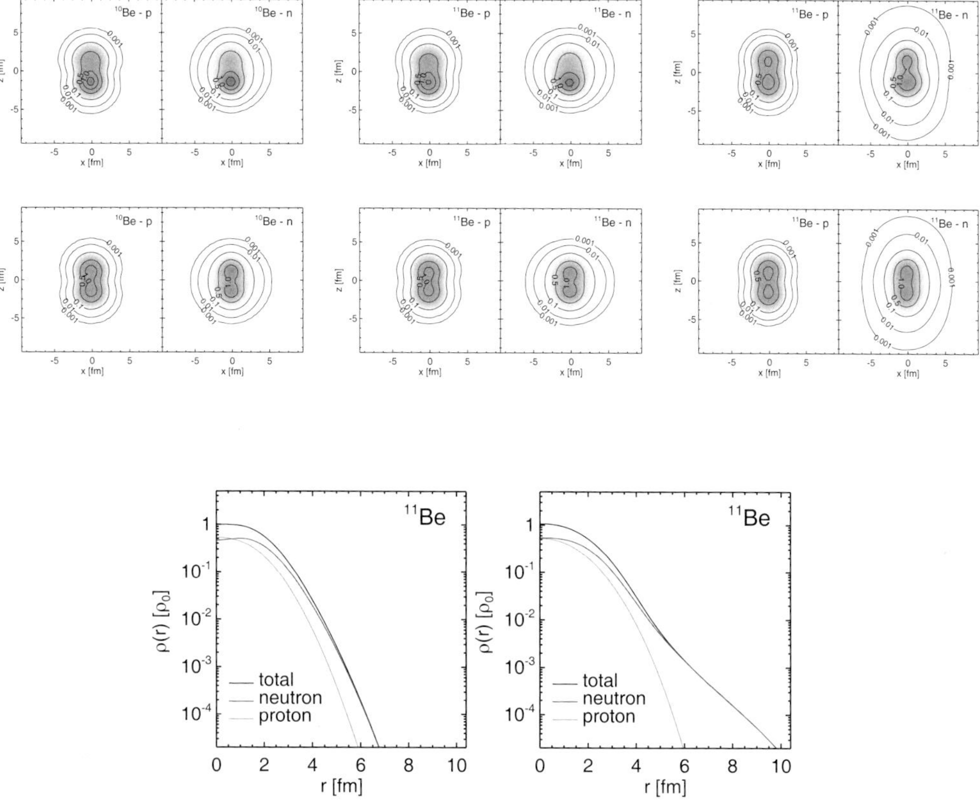

Fig. 7. – Densities of ^{10}Be, ^{11}Be negative and positive parity. First row: proton and neutron densities of intrinsic states $|Q^+\rangle$ for ^{10}Be, $|Q^-\rangle$ and $|Q^+\rangle$ for ^{11}Be. Second row: densities after parity projection. R.h.s: total, proton, and neutron densities after angular momentum projection for $J^\pi = 1/2^-$ and the unnatural parity state $J^\pi = 1/2^+$. All densities are for pointlike nucleons in units of $\rho_0 = 0.17\,\mathrm{fm}^{-3}$.

fig. 5). It remains to be seen but it is quite conceivable that like in ^{7}Be, ^{8}Be, and ^{9}Be a configuration mixing calculation would lower all energy levels of interest so that the missing 3.6 MeV would be gained. It also has to be checked whether the level ordering of the $1/2^-$ and $1/2^+$ states will change in the multiconfiguration calculation.

5'3. ^{12}C. – In no-core shell model calculations some of the excited states at energies around the three α breakup threshold cannot be described [20, 21]. The reason is that these states consist essentially of three α-particles loosely bound in resonances with small widths. A representation in terms of harmonic oscillator states requires huge Hilbert spaces. The first excited 0^+ state just above the 3α threshold, see fig. 8, has gained fame because it was predicted by the astronomer Fred Hoyle who concluded that the abundance of ^{12}C in the universe could only be understood if there was a resonance in

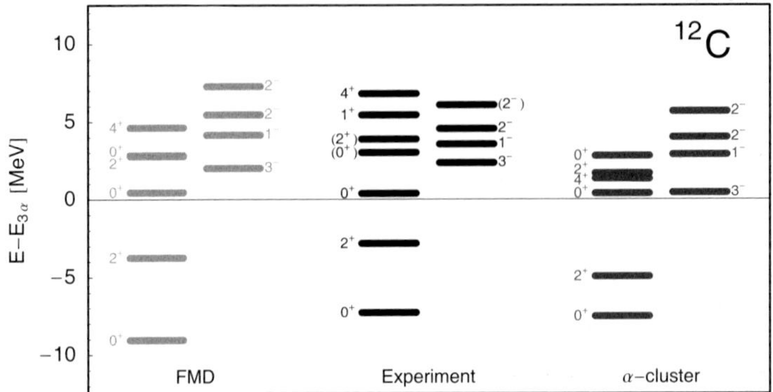

Fig. 8. – Energy spectra of ^{12}C with respect to 3α threshold. α-cluster model calculated with modified Volkov V2 interaction.

^{12}C close to the ^{8}Be + ^{4}He or ^{4}He + ^{4}He + ^{4}He threshold [22], otherwise there would be much less ^{12}C. This state at 7.65 MeV excitation was later found and is often referred to as the Hoyle state.

The FMD many-body states are very flexible and can describe cluster states as well as shell model like configurations [23]. Therefore it is interesting to see to what extent FMD can reproduce the spectrum of ^{12}C. In the present multiconfiguration calculation the many-body basis consists of 16 intrinsic states obtained in a variation after angular-momentum projection procedure (projecting on 0^+ and 2^+ states) with constraints on the radii and additional 57 states that have been iteratively selected to minimize the energies of the first three 0^+ states. These states are chosen out of a set of 42 FMD states obtained in variation after parity projection calculations with constraints on radii and quadrupole deformation and 165 explicit α-cluster triangle configurations. An α-cluster is defined here as a product of four Gaussian single-particle states with total spin and isospin equal to zero. The Antisymmetrized Molecular Dynamics (AMD) model (see [24] for a recent discussion of ^{12}C) uses similar wave functions but imposes a fixed width parameter on the Gaussian wave packets. As cluster states and shell model like states prefer different widths in ^{12}C this is an important downside when compared to the FMD approach. In fig. 8 the resulting energy spectrum is compared with the experimental one. The FMD calculation can reproduce the energies of the Hoyle state and the neighbouring resonances quite well.

In a second model (labeled α-cluster) we restrict ourselves to the α-cluster triangle configurations. Convergence for the first three 0^+ states is achieved with a subset of 55 states. In this case we essentially implement a microscopic α-cluster model using Brink-type [17] wave functions. However, with the α-cluster states alone, a significant underbinding is observed when the FMD Hamiltonian is used. Therefore, we employ the modified Volkov V2 interaction proposed in [25, 26] which is fine-tuned to reproduce the ground and Hoyle state energies in ^{12}C within an α-cluster model, see fig. 8. One has to keep in mind that this interaction is especially tailored and cannot be used in other nuclei, for example for ^{16}O it already leads to an overbinding of about 25 MeV. The

Table I. – *Energies, radii and transition strengths. Units of energies are* MeV, *of radii,* fm; $M(E0)$, e fm^2 *and* $B(E2)$, e^2 fm^4. *Data are from* [29], *"BEC" results from* [27].

	Exp	FMD	α-cluster	"BEC"
$E(0_1^+)$	-92.16	-92.64	-89.56	-89.52
$E^*(0_2^+)$	7.65	9.50	7.89	7.73
$E(0_2^+) - E(3\alpha)$	0.38	0.44	0.38	0.26
$E^*(0_3^+)$	(10.3)	11.90	10.33	
$E^*(2_1^+)$	4.44	5.31	2.56	2.81
$E^*(2_2^+)$	(11.16)	11.83	9.21	
$E(3\alpha)$	-84.89	-83.59	-82.05	-82.05
$r_{\text{charge}}(0_1^+)$	2.47 ± 0.02	2.53	2.54	
$r(0_1^+)$		2.39	2.40	2.40
$r(0_2^+)$		3.38	3.71	3.83
$r(0_3^+)$		4.62	4.75	
$r(2_1^+)$		2.50	2.37	2.38
$r(2_2^+)$		4.43	4.02	
$M(E0, 0_1^+ \to 0_2^+)$	5.4 ± 0.2	6.53	6.52	6.45
$B(E2, 2_1^+ \to 0_1^+)$	7.6 ± 0.4	8.69	9.16	
$B(E2, 2_1^+ \to 0_2^+)$	2.6 ± 0.4	3.83	0.84	

addition of a spin-orbit force would destroy the reproduction of the ^{12}C ground-state properties. Therefore the predictive power is limited.

The same Volkov interaction is used in the third model (labelled "BEC") by Funaki *et al.* [27]. Here the number of degrees of freedom is reduced even further by using basis states where the center-of-mass coordinates of all the α-clusters are given by the same (deformed) wave function like in a Bose-Einstein condensate. Of course the state must in the end be antisymmetrized. The bosonic nature of the wave function therefore only survives when the density of the α-clusters is low enough that antisymmetrization is not important. This is certainly not the case for the ground state and only to a certain extent for the Hoyle state. A detailed analysis [28] within an α-cluster model, using a slightly different interaction, shows that the probability to find all α-clusters in the same s-wave orbit is about 30% for the ground state and about 70% for the Hoyle state. Thus the attribute "Bose-Einstein condensate" should not be taken too literally.

A comparison of the three models for energies, radii and transition strengths in ^{12}C is shown in table I. The α-cluster results agree very well with the "BEC" approach and also with resonating group method (RGM) calculations [25]. All models give very large radii for the Hoyle state as well as for the 0_3^+ and the 2_2^+ state. In the cluster models the absence of spin-orbit forces leads to the well-known underestimation of the 2_1^+ energy indicating again their schematic nature.

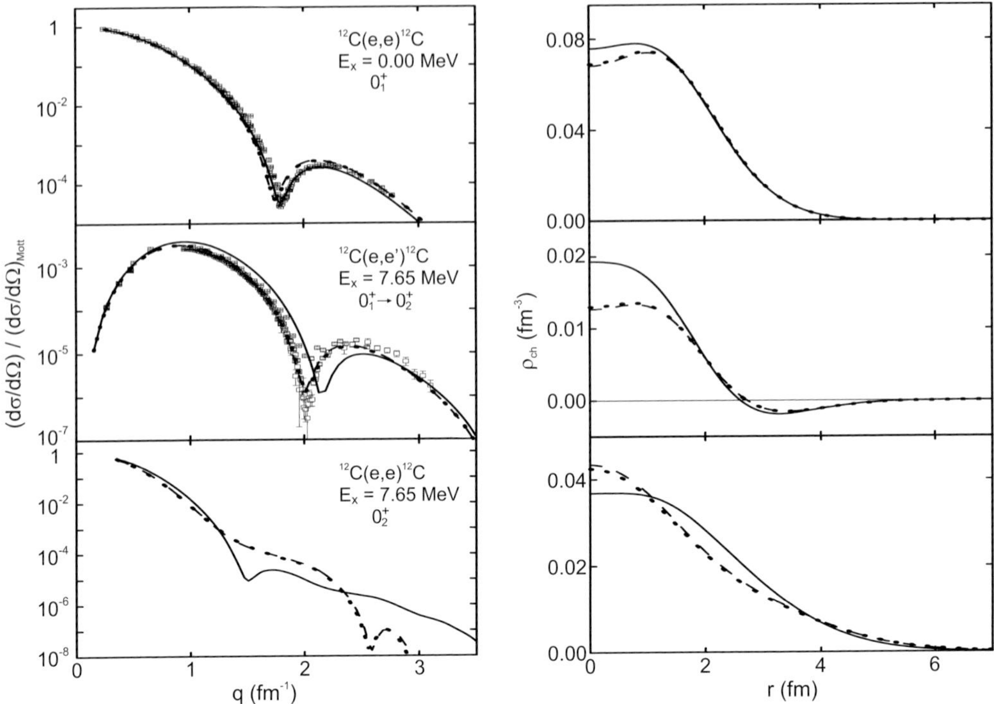

Fig. 9. – L.h.s.: FMD (solid lines), α-cluster (dashed lines) and "BEC" (dotted lines) predictions of the charge form factors in ^{12}C in comparison to experimental data (open squares). Elastic scattering on g.s. (top), transition to the Hoyle state (middle) and elastic scattering on the Hoyle state (bottom). R.h.s.: corresponding charge density distributions. "BEC" results are from [30].

To quantify the degree of α-clustering within the FMD wave functions, which are obtained by a multiconfiguration mixing calculation containing shell model like and cluster states, we calculate the overlap of the eigenstates with the α-cluster model space. For that we construct a projection operator $\underset{\sim}{P}_\alpha$ using the 165 α-cluster triangle configurations (see eq. (26)):

$$(50) \qquad \underset{\sim}{P}_\alpha = \sum_{J^\pi, M} \sum_{i,j=1}^{165} |Q^{(i)}; J^\pi M \kappa_i\rangle \, \mathsf{O}_{ij} \, \langle Q^{(j)}; J^\pi M \kappa_j|, \quad \text{where } \mathsf{O}_{ij} = (\mathsf{N}^{-1})_{ij}.$$

We obtain $\langle 0_1^+|\underset{\sim}{P}_\alpha|0_1^+\rangle = 0.52$, $\langle 0_2^+|\underset{\sim}{P}_\alpha|0_2^+\rangle = 0.85$, $\langle 0_3^+|\underset{\sim}{P}_\alpha|0_3^+\rangle = 0.92$, $\langle 2_1^+|\underset{\sim}{P}_\alpha|2_1^+\rangle = 0.67$ and $\langle 2_2^+|\underset{\sim}{P}_\alpha|2_2^+\rangle = 0.99$. A restriction to α-cluster configurations is obviously not sufficient for a description of the ground state $|0_1^+\rangle$. The spin-orbit force breaks the α-clusters and a large shell model component is found in the FMD ground state. The Hoyle state $|0_2^+\rangle$ on the other hand is dominated by α-cluster contributions but still has a sizeable component of shell model nature.

In fig. 9 we compare calculated electron scattering form factors with measured data and show the corresponding charge densities for the ground state, the Hoyle state and the transition between them. The data are given as the ratio of the measured cross-section to the Mott cross-section. The comparison between experimental and theoretical cross sections is performed in distorted wave Born approximation (DWBA) [31, 32].

In the FMD and the α-cluster model we calculate first the matter densities of point-like protons and neutrons which are then folded with the charge densities of proton and neutron, respectively. Thus we obtain the charge densities of ^{12}C shown in fig. 9. The same procedure is used to calculate the densities from the matter densities obtained within the "BEC" model [30].

A good reproduction of the ground-state form factor is a prerequisite to draw sound conclusions about the charge distribution of the Hoyle state from the transition form factor because both states enter the transition matrix element on an equal footing. As can be seen from fig. 9 the ground-state form factor is well described by the FMD model. The results for the α-cluster and "BEC" models are almost identical and show a slightly worse agreement with the data.

The α-cluster model and the "BEC" reproduce the shape of the transition form factor well. The FMD model on the other hand somewhat overestimates the data in the region of the first maximum and has its node at $q = 2.2\,\mathrm{fm}^{-1}$ while the experimental minimum is at $q = 2.0\,\mathrm{fm}^{-1}$. The differences in the transition form factors are mainly due to differences in the Hoyle state. Compared to the α-cluster models the FMD charge density of the Hoyle state has a smaller surface thickness and a lower central density, leading to a stronger oscillation in the transition density. These differences also show up in the form factors of the Hoyle state where the models show noticeable differences at medium and high momentum transfers. We suspect that minor modifications to the FMD interaction, taking α-α scattering data into account, could result in an improved description—investigations are under way.

Charge densities and form factors are essentially one-body observables and do not reflect many-body correlations existing in the many-body state. Therefore the form factors

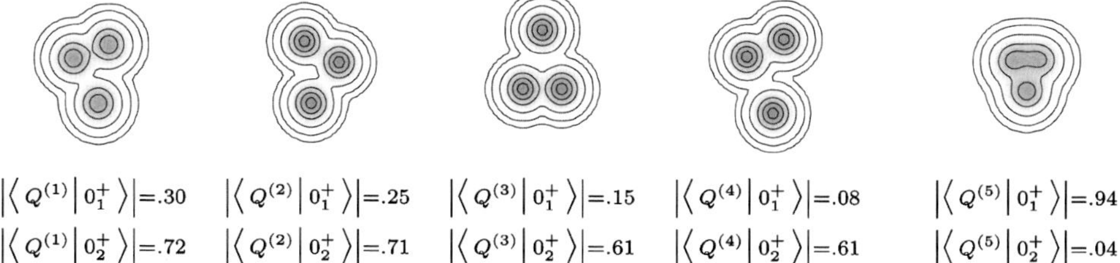

$$\left|\left\langle Q^{(1)}\middle|0_1^+\right\rangle\right|=.30 \qquad \left|\left\langle Q^{(2)}\middle|0_1^+\right\rangle\right|=.25 \qquad \left|\left\langle Q^{(3)}\middle|0_1^+\right\rangle\right|=.15 \qquad \left|\left\langle Q^{(4)}\middle|0_1^+\right\rangle\right|=.08 \qquad \left|\left\langle Q^{(5)}\middle|0_1^+\right\rangle\right|=.94$$

$$\left|\left\langle Q^{(1)}\middle|0_2^+\right\rangle\right|=.72 \qquad \left|\left\langle Q^{(2)}\middle|0_2^+\right\rangle\right|=.71 \qquad \left|\left\langle Q^{(3)}\middle|0_2^+\right\rangle\right|=.61 \qquad \left|\left\langle Q^{(4)}\middle|0_2^+\right\rangle\right|=.61 \qquad \left|\left\langle Q^{(5)}\middle|0_2^+\right\rangle\right|=.04$$

Fig. 10. – Intrinsic one-body densities of the four FMD states which contribute most to the Hoyle state and their respective amplitudes for the ground state $|0_1^+\rangle$ and the Hoyle state $|0_2^+\rangle$. The fifth state, obtained by variation after projection on angular momentum, is the leading component in the ground state. Note that the FMD states are not orthogonal.

provide no direct information on the α-cluster structure, neither in the ground state nor in the Hoyle state. However, a Hoyle state exhibiting cluster nature is also supported by the FMD calculations, where the Hamiltonian can choose between shell model like and cluster configurations. An analysis of the FMD Hoyle state shows that its leading components displayed in fig. 10 are cluster-like and resemble $^8\mathrm{Be}+\alpha$ configurations. Two of the α-particles are typically close to each other and the third one is further away. The ground state is dominated by more compact configurations which have a large overlap with shell model states (see right most configuration in fig. 10). In the 0_3^+ and 2_2^+ states we also find the leading components to be of $^8\mathrm{Be}+\alpha$ nature but featuring more prolate open triangle configurations.

6. – Halos

6`1. *He isotopes.* – Figure 11 in the upper row shows the one-body densities of the intrinsic states $|Q^+\rangle$ for even and $|Q^-\rangle$ for odd mass numbers, respectively. The FMD states were obtained by minimizing the energy $\langle Q^\pm|\underset{\sim}{H}(\underset{\sim}{1}\pm\underset{\sim}{\Pi})|Q^\pm\rangle/\langle Q^\pm|(\underset{\sim}{1}\pm\underset{\sim}{\Pi})|Q^\pm\rangle$ of the parity projected states with respect to all parameters in $Q^\pm$. In all helium isotopes a dipole deformation caused by a displacement of the neutrons against the α-core is found. In $^6\mathrm{He}$ the configuration with two neutrons on the same side of the core is preferred to configurations with the two neutrons located at opposite sides of the core. In $^8\mathrm{He}$ one approaches the $0p_{3/2}$ neutron shell closure with an almost spherical neutron distribution

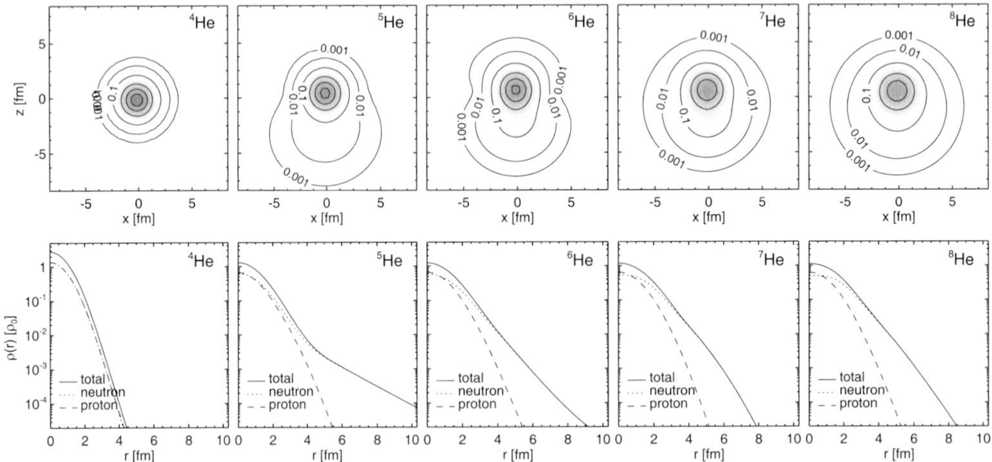

Fig. 11. – Upper part: intrinsic shapes of helium isotopes corresponding to the minima in a variation after parity projection calculation. Shown are cuts through the nucleon density calculated with the intrinsic states $|Q^+\rangle$ or $|Q^-\rangle$ before parity projection. Lower part: radial density distributions for the helium isotopes using angular-momentum projected multiconfiguration states. The center-of-mass motion has been removed. Densities are given in units of nuclear matter density $\rho_0 = 0.17\,\mathrm{fm}^{-3}$.

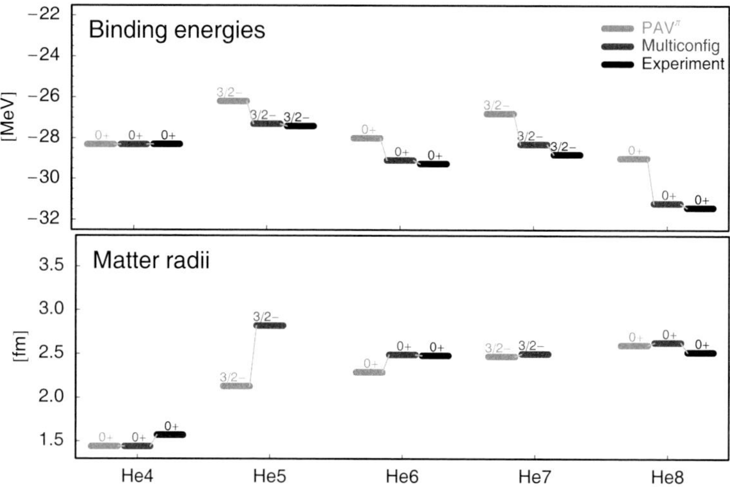

Fig. 12. – Binding energies and matter radii for the helium isotopes. Results are given for the PAV^π and the multiconfiguration calculations. Experimental matter radii are taken from [33].

but the displacement is still visible. In fig. 12 the binding energies and matter radii obtained after angular-momentum projection (PAV^π) are compared to the experimental binding energies and radii.

To improve the many-body states we create additional configurations using the dipole moment as a generator coordinate, an example is shown in fig. 13 for ^{6}He. The multiconfiguration calculations, which consists of diagonalizing the Hamiltonian in the many-body space spanned by these configurations, reproduce the experimental binding energies and radii very well (see fig. 12). This illustrates the importance of the soft-dipole mode for the understanding of the borromean nature of ^{6}He and ^{8}He. In the ground state it is realized in the form of a zero-point oscillation (or ground-state correlations) that correspond to a pygmy resonance. However the resonance itself cannot be excited any more because it decays at once by emitting the two loosely bound neutrons. The proton and neutron densities of the ground states calculated by multiconfiguration mixing are shown in the lower part of fig. 11. Besides the neutron halo in ^{6}He we can see the broadened proton distribution that is caused by the motion of the α-core against the center of mass of the nucleus. For ^{6}He we calculate a charge radius of 2.02 fm that should be compared with the recently measured value of 2.054 ± 0.014 fm [34].

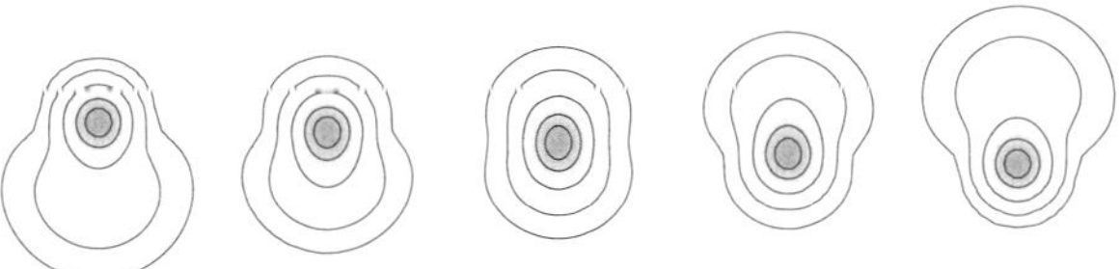

Fig. 13. – Typical configurations used for multiconfiguration calculations of ^{6}He.

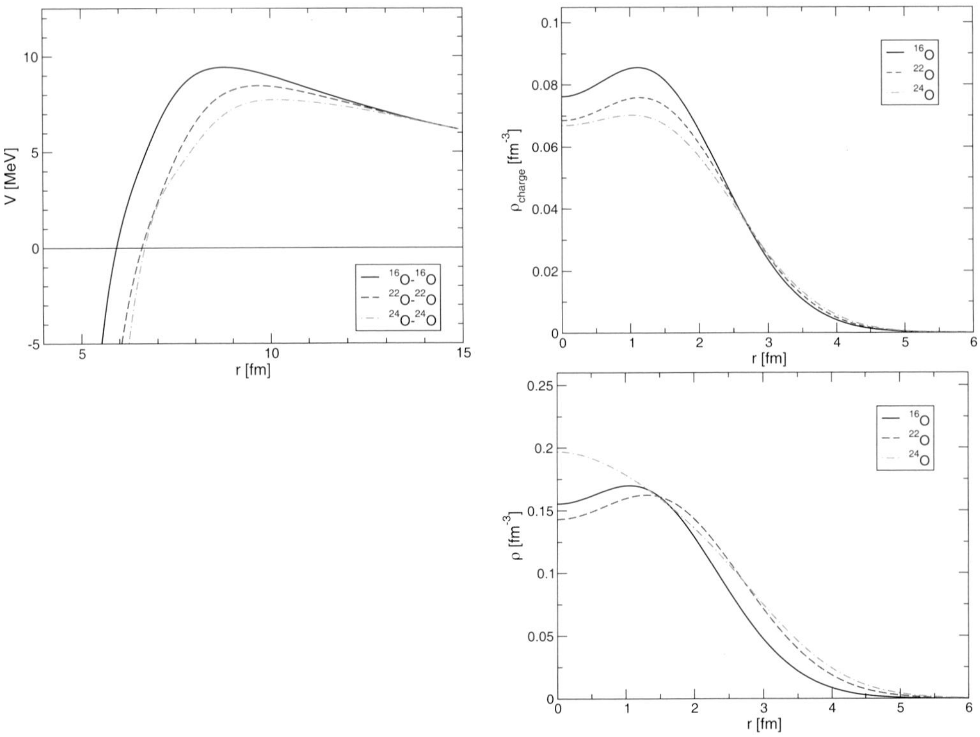

Fig. 14. – Nucleus-nucleus potentials calculated in the frozen state approximation for $^{16}\text{O} + {}^{16}\text{O}$, $^{22}\text{O} + {}^{22}\text{O}$ and $^{24}\text{O} + {}^{24}\text{O}$. Top right: charge densities, bottom right: total densities of oxygen isotopes. Finite size of protons and neutrons is taken into account.

7. – S-factor and neutron skins

The fusion of neutron-rich, unstable isotopes is expected to take place in pycno-nuclear reactions in the outer layers of accreting neutron stars. At densities larger than about 10^{12}g/cm^3 nuclei are densely packed in lattice configurations. The height of the Coulomb barrier is lowered by neutron skins and there is enhancement of tunneling due to the high electron density between neighbouring nuclei [35, 36].

By positioning the FMD ground states of two nuclei on a grid at different distances and antisymmetrizing the product state one can construct a basis. By projecting on various J^π one can span a many-body Hilbert space that describes the relative motion of the nuclei in the J^π channel. From the Hamiltonian matrix represented in this space one can determine the nucleus-nucleus potential. From that the probability for fusion can be calculated at energies above and below the Coulomb barrier where the nuclei have to tunnel. Energies which occur in astrophysical scenarios are usually very low so that the tunneling probability is very small. The resulting tiny cross-sections make measurements very difficult or impossible. Therefore one needs reliable models to extrapolate to these very low energies.

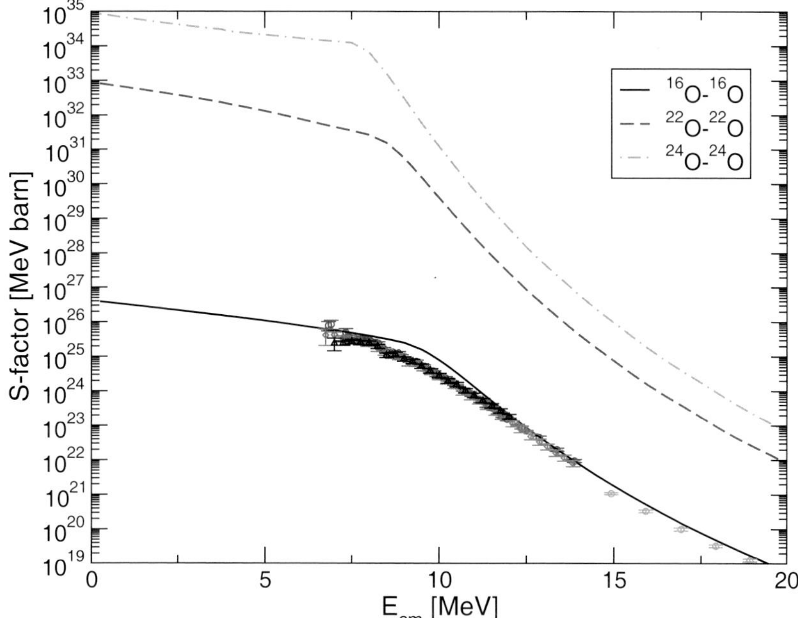

Fig. 15. – S-factor as function of relative energy for $^{16}O + ^{16}O$, $^{22}O + ^{22}O$ and $^{24}O + ^{24}O$. In the case of $^{16}O + ^{16}O$ available data are shown.

In a microscopic model like FMD there are no adjustable parameters in calculating the nucleus-nucleus potential, in particular, the isospin dependence comes from the isospin dependence of the nucleon-nucleon interaction and the neutron to proton ratios in the surface of the nuclei. As an example, in fig. 14 we display the ground-state densities of ^{16}O, ^{22}O, and ^{24}O together with the calculated nucleus-nucleus potential for $^{16}O + ^{16}O$, $^{22}O + ^{22}O$ and $^{24}O + ^{24}O$ [37]. As ^{22}O and ^{24}O are radioactive isotopes it is impossible to measure their fusion S-factor.

Compared to ^{16}O the charge densities of ^{22}O and ^{24}O (see fig. 14 top right) are somewhat broadened due to the additional neutrons which arrange themselves mostly at the surface. The total densities (proton plus neutron) displayed below the charge densities clearly show the formation of a neutron skin. One can also see that the additional two neutrons in ^{24}O occupy the $1s_{1/2}$ shell as they contribute to the density in the interior of the nucleus.

These properties are reflected in the nucleus-nucleus potentials displayed in fig. 14. Comparing the potential between two ^{16}O nuclei with that for $^{22}O + ^{22}O$ one sees a lowered barrier and a shortened tunneling distance at low energies. For ^{24}O the barrier is lowered further but the tunneling path at energies close to zero gets longer again. This leads to fusion cross-sections displayed in fig. 15 in terms of S-factors. The S-factor is defined as

$$(51) \qquad S(E) = \sigma(E)\, E\, \exp\left[\frac{2\pi Z_1 Z_2 e^2}{\hbar^2\, v_\infty}\right],$$

where $Z_{1,2}$ are the proton numbers and v_∞ is the relative velocity of the nuclei at an infinitely large distance. One multiplies the cross-section by an energy-dependent factor which removes the effect of the exponentially dropping tunneling probability due to the Coulomb barrier. The remaining S-factor is supposed to vary smoothly with energy so that an extrapolation to zero energy is more reliable. (See lecture of K. Langanke.)

At energies far below the Coulomb barrier one sees a steeper increase in the fusion probability for the neutron rich isotope ^{22}O, reflecting the lower barrier and the shorter distance for tunneling. The microscopically calculated S-factor for ^{16}O $+$ ^{16}O compares nicely with the one measured around the Coulomb barrier (hinted at by the kink in the S-factor), especially considering that no adjustable parameters are at our disposal.

REFERENCES

[1] MACHLEIDT R., *Phys. Rev. C*, **63** (2001) 024001.
[2] ENTEM D. R. and MACHLEIDT R., *Phys. Lett. B*, **524** (2001) 93.
[3] ENTEM D. R. and MACHLEIDT R., *Phys. Rev. C*, **68** (2003) 041001.
[4] EPELBAUM E., GLÖCKLE W. and MEISSNER U.-G., *Nucl. Phys. A*, **747** (2005) 362.
[5] EPELBAUM E., *Prog. Part. Nucl. Phys.*, **57** (2006) 654.
[6] FELDMEIER H., NEFF T., ROTH R. and SCHNACK J., *Nucl. Phys. A*, **632** (1998) 61.
[7] NEFF T. and FELDMEIER H., *Nucl. Phys. A*, **713** (2003) 311.
[8] ROTH R., HERGERT H., PAPAKONSTANTINOU P., NEFF T. and FELDMEIER H., *Phys. Rev. C*, **72** (2005) 034002.
[9] JASTROW R., *Phys. Rev.*, **98** (1955) 1479.
[10] FELDMEIER H. and SCHNACK J., *Rev. Mod. Phys.*, **72** (2000) 655, and references therein.
[11] ROTH R., NEFF T., HERGERT H. and FELDMEIER H., *Nucl. Phys. A*, **745** (2004) 3.
[12] NEFF T. and FELDMEIER H., *Nucl. Phys. A*, **738** (2004) 357.
[13] NEFF T., FELDMEIER H. and ROTH R., *Nucl. Phys. A*, **752** (2005) 321c.
[14] KANADA-EN'YO Y. and HORIUCHI H., *Prog. Theor. Phys. Suppl.*, **142** (2001) 205.
[15] KANADA-EN'YO Y., KIMURA M. and HORIUCHI H., *C. R. Phys.*, **4** (2003) 497.
[16] KANADA-EN'YO Y. and HORIUCHI H., *Phys. Rev. C*, **68** (2003) 014319.
[17] BRINK D. M., *Proceedings of the International School of Physics "Enrico Fermi", Course XXXVI* (Academic Press, New York/London) 1996, p. 247.
[18] RING P. and SCHUCK P., *The Nuclear Many-Body Problem* (Springer) 2000.
[19] COHEN-TANNOUDJI C., DIU B. and LALOË F., *Quantum Mechanics* (Wiley & sons, Herman) 1977.
[20] NAVRÁTIL P. and ORMAND W. E., *Phys. Rev. C*, **68** (2003) 034305.
[21] NAVRÁTIL P., GUEORGUIEV V. G., VARY J., ORMAND W. E. and NOGGA A., *Phys. Rev. Lett.*, **99** (2007) 042501.
[22] HOYLE F., *Astrophys. J. Suppl.*, **1** (1954) 121.
[23] CHERNYK M., FELDMEIER H., NEFF T., VON NEUMANN-COSEL P. and RICHTER A., *Phys. Rev. Lett.*, **98** (2007) 032501.
[24] KANADA-EN'YO Y., *Prog. Theor. Phys.*, **117** (2007) 655.

[25] Fukushima Y. and Kamimura M., *J. Phys. Soc. Jpn. Suppl.*, **44** (1978) 225.

[26] Kamimura M., *Nucl. Phys. A*, **351** (1981) 456.

[27] Funaki Y., Tohsaki A., Horiuchi H., Schuck P. and Röpke G., *Phys. Rev. C*, **67** (2003) 051306.

[28] Matsumara H. and Suzuki Y., *Nucl. Phys. A*, **739** (2004) 238.

[29] Ajzenberg-Selove F., *Nucl. Phys. A*, **506** (1990) 1.

[30] Funaki Y., Tohsaki A., Horiuchi H., Schuck P. and Röpke G., *Eur. Phys. J. A*, **28** (2006) 259.

[31] Heisenberg J. and Blok H. P., *Annu. Rev. Nucl. Part. Sci.*, **33** (1983) 569.

[32] Bähr C., code PHASHI, unpublished TU Darmstadt.

[33] Ozawa A., Suzuki T. and Tanihata I., *Nucl. Phys. A*, **693** (2001) 32.

[34] Wang L.-B., Mueller P., Bailey K., Drake G. W. F., Greene J. P., Henderson D., Holt R. J., Janssens R. V. F., Jiang C. L., Lu Z.-T., O'Connor T. P., Pardo R. C., Rehm K. E., Schiffer J. P. and Tang X. D., *Phys. Rev. Lett.*, **93** (2004) 142501.

[35] Haensel P. and Zudnik J., *Astron. Astrophys.*, **404** (2003) 33.

[36] Haensel P. and Zudnik J., *Astron. Astrophys.*, **227** (1990) 431.

[37] Feldmeier H., Neff T. and Langanke K., nucl-th/0703030v1 (2007).

DOI 10.3254/978-1-58603-885-4-215

Tests of clustering in light nuclei and applications to nuclear astrophysics

A. Tumino

Laboratori Nazionali del Sud, INFN and Dipartimento di Metodologie Fisiche e Chimiche,
Università di Catania - 95123 Catania, Italy

Summary. — Starting from the earliest ideas of clustering in light nuclei, the signatures traditionally supporting the existence of clusters will be discussed. In particular the procedure to perform an angular correlation analysis in order to infer a dominant angular momentum to a resonance will be shown. Then, moving from the α-chains in ^{8}Be and ^{12}C, the nuclear dimers and polymers will be introduced. Moreover, applications of cluster physics to nuclear astrophysics by means of quasi-free scattering and reactions will be considered.

1. – Introduction

Clustering is a particular structural mode identified in many nuclei and a manifestation of symmetries appearing in the nuclear interaction.

The concept of clustering in nuclei has a long and distinguished history dating back to the birth of nuclear physics itself. Speculations about the existence of clusters such as alpha-particles have been around even before the discovery of the neutron. They were stimulated initially by the observation of alpha-particle decay. However, for several decades after the discovery of the neutron this topic went out of favor mainly due to the lack of experimental results with which to test those original ideas. Around the 1950's,

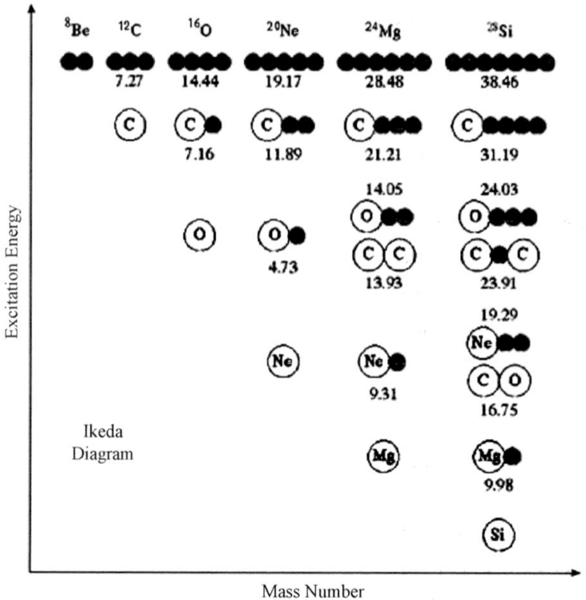

Fig. 1. – Ikeda Diagram showing possible cluster configuration in 4N self-conjugated nuclei.

the incapability of the neutron proton model in describing peculiar nuclear properties resulting from the earliest spectroscopy experiments, triggered a renewed interest on the idea of clustering. The Hoyle speculation on the ^{12}C nucleus described in terms of 3α-particles dates back to that period. In order to account for the observed ^{12}C-abundance in stars, he suggested the existence of the 7.66 MeV level in ^{12}C as responsible for greatly enhancing the small probability of α-capture by the short-lived ^{8}Be during the ^{4}He burning phase. During the 1960's a number of theoretical groups began to work on the problem of describing the nucleus in terms of clusters of alpha-particles. Two results from the period are quite illustrative. The figure below shows what has become known as the "Ikeda Diagram" [1], and it shows possible cluster structures in a number of light nuclei (fig. 1). Ikeda believed that each of the different cluster structures would be mostly noticeable in the energy region just above the relevant separation energy for the fragments. In the light of our recent discoveries, the predictions of this model seem remarkably far sighted. Figure 2 gives some predictions using the Bloch-Brink cluster model [2]. It is in essence a multi-centre shell model in which a maximum of four nucleons is associated with any given centre and constrained to move inside the same well. Each member of such a quartet has the same spatial function, which is taken to be that of a $0s$ harmonic-oscillator state with origin at a fixed centre. Centre positions and widths of the wells are allowed to vary until the energy of the system is minimized and this is taken as the nuclear ground-state configuration. It was formally shown that in the limit of zero separation between the wells the wave function reduces to that of the shell model.

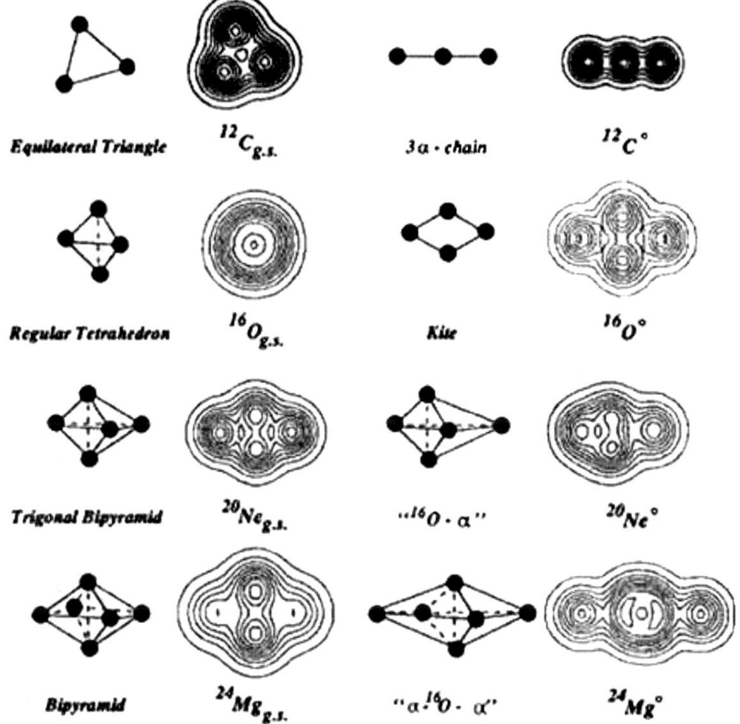

Fig. 2. – Clusters in nuclei as crystalline structures and density probability from the Bloch-Brink Cluster Model.

The Bloch-Brink Cluster Model (BB) has recently been used to study the role of clustering in 4N (*sd*)-shell nuclei and to correlate large bodies of experimental data; it gives intuitively appealing and physically transparent prescriptions of clustering in light and medium-heavy nuclei. The light alpha conjugate nuclei are seen to be almost crystalline structures with specific arrangements of the alpha clusters (see fig. 2) as originally proposed by Brink at about the same period as Ikeda.

Soon after that Ikeda and Brink were developing their early models of alpha clustering, the first experiments which suggested the existence of larger clusters were just beginning. The evidence came from some of the earliest experiments involving the collision of two heavy ions [3]. When the yield of reaction products from the collision of two ^{12}C nuclei was measured, striking, narrow resonances appeared in the excitation function. These were correlated in a number of reaction channels (see fig. 3), leading to the speculation that the two ^{12}C nuclei were forming an intermediate "nuclear molecule" in the reaction. This, then, would be an even more extreme form of cluster structure in the nucleus.

Of course the observation of a complex fragment (or cluster) emerging from a nucleus need not necessarily imply that the fragment existed as a pre-formed entity in the

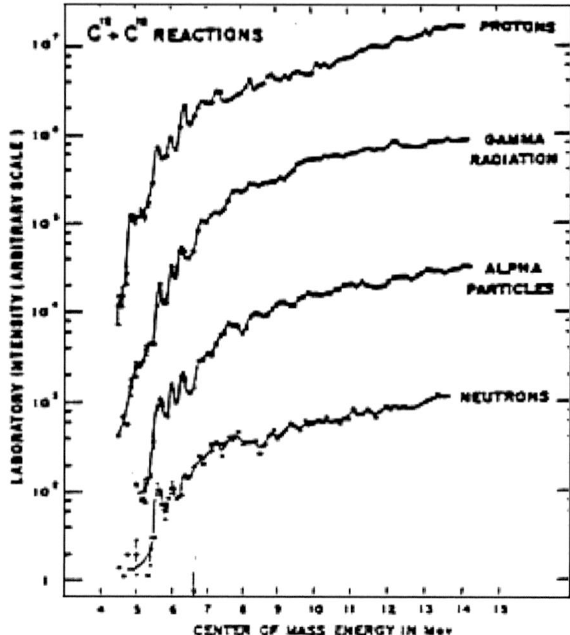

Fig. 3. – Excitation functions showing ^{12}C + ^{12}C scattering resonances around the Coulomb barrier for several exit channels.

nucleus. However, extensive investigations over the next couple of decades by a large number of groups resulted in the development of new experimental techniques and in the discovery of several signatures to probe cluster structure. These results have rekindled interest in the topic.

2. – Evidence of clustering

After a wealth of experimental as well as theoretical work, it is possible to list the experimental signatures which traditionally support the existence of cluster states:

– Selective excitation in many nucleon-transfer reactions.

– Enhanced electromagnetic moments and transition strengths; the most important physical quantities pointing to a collective behaviour are the quadrupole moment $Q(J)$ and the reduced transition probability $B(E2; J \to J')$. $Q(J)$ is responsible for measuring excursions away from spherical symmetry embodying various types of "deformation". The $B(E2)$ magnitude is also closely related to the deformation, in particular the larger the bigger the deformation, both $Q(J)$ and $B(E2)$ depending on the intrinsic quadrupole moment. $B(E2)$ is important when $B(E1)$ is forbidden. Indeed, in a collective representation of the nucleus, the center of electric charges has the same coordinates of the center of mass, whose motion under the action of internal forces is prohibited by the momentum conservation law. The almost disappearance of the dipole radiation is a common situation in many transitions at low energy.

– Appreciable cluster emission widths for resonant states.

– Rotationally spaced energy levels; the feature is the appearance of a rotational band-like structure with energies within a band that vary as $J(J+1)$. This comes indeed from the expression of the quantum-symmetrical top rotor

$$(1) \qquad E(J) = \frac{\hbar^2}{2I} J(J+1) + E_0 \,,$$

where I is the moment of inertia of the cluster configuration which is expected to describe those levels, and $E_0 = E_b + E_c$ with the first term representing the binding energy of the nuclear system in the exit channel, while the second one the Coulomb barrier. More complex formula exist to account for perturbations to this ideal picture of pure rotation in an axially symmetric nucleus. For high J values the centrifugal force stretches the rotor and deviations from eq. (1) are quite well accounted for by using

$$(2) \qquad E(J) = \frac{\hbar^2}{2I} [J(J+1) + \beta J^2 (J+1)^2].$$

– Quasi-free scattering and reactions, extensively studied in the past in order to investigate the cluster structure of light nuclei as will be shown later.

2$`1. The $^{12}C + {}^{12}C$ scattering resonances._ – These are behaviours already clearly seen in the mentioned excitation curves of the $^{12}C + {}^{12}C$ experiments. Studies of the on-resonance angular distributions revealed that most of the observed structures are characterised by good angular-momentum quantum numbers, as required by a true isolated resonance. Furthermore, as is apparent from fig. 4, states of the same J spin were clustered in the same excitation region with the spin value increasing with excitation energy.

The correlation between resonance energies and dominant J values is reproduced by an expression like eq. (1) [4-6]. This evidence up-helps the above suggestions that states resulted from two ^{12}C nuclei orbiting one another in a molecular configuration. The multiplicity of states for a given spin was attributed to the fragmentation of each rotational state into a series of vibrational states. The vibration-rotation model applied to the data gave a value of 76 keV for the rotational parameter $(\hbar^2/2I)$, close to the 70 keV predicted for a highly deformed ^{12}C-^{12}C molecule [7]. Similar features have been observed in a number of other systems. They also speak for extreme deformations in the intermediate nuclear system, hence the notion of molecular states led to much discussion on the nature of the interaction between the two nuclei.

^{24}Mg state is assumed to exist for an interaction time of about 10^{-21} s (of the same order of the collision time associated with a direct reaction) as two ^{12}C nuclei bonded

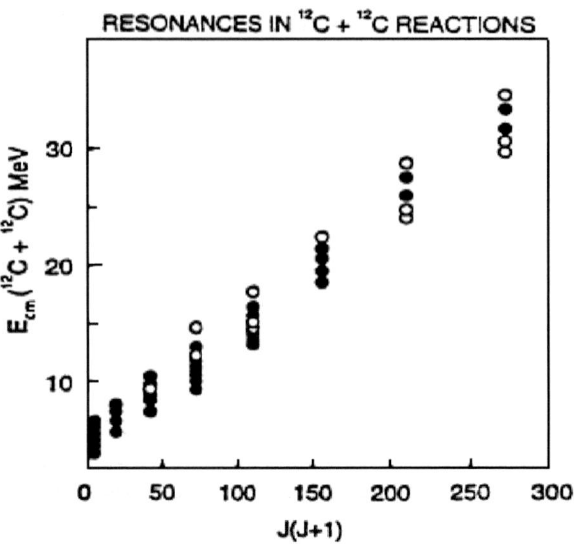

Fig. 4. – Energy-spin relation for the ^{12}C $+$ ^{12}C scattering resonances. Open circles indicate uncertain spin.

together in a fashion reminiscent of the saddle point in fission. This state de-excites either by re-emission of the ^{12}C nuclei or by collapse into a normal compound nucleus from which α-particles and nucleons are emitted. Many more resonances have since been discovered in ^{12}C-^{12}C reaction, not only near the Coulomb barrier with typical widths of about 100 keV, but also at higher energies where intermediate and gross-structure resonances with much greater widths have been found. The scattering potential was drawn to display a pocket at large radial separation, formed by the sum of the nuclear, Coulomb and centrifugal components. The choice of nuclear potential was at first purely phenomenological as in the optical potential, where its radial form is simply parameterized, for example, as a Wood-Saxon for

$$(3) \qquad V_{\text{real}}(r) = -V_0\left[1 + e^{(r-R_r)/a_r}\right], \qquad V_{\text{imag}}(r) = -W_0\left[1 + e^{(r-R_i)/a_i}\right]$$

with radius and diffuseness parameters R and a.

2'1.1. Couplings. One of the first hypotheses made to justify the survival of molecular configurations, concerned the strength of the coupling potential. The weakness of this potential was considered to be crucial for the observation of the intermediate structure in the excitation functions associated with these configurations. This so-called weak-coupling approach offers an understanding of the reaction process in terms of the relative motion of the nuclei in an appropriate nucleus-nucleus potential. By contrast, a different idea has been developed, which relates the molecular configurations to the proper structure of the compound system. The strong-coupling method stems to this idea. The weak coupling is characterized by a short reaction time because of the weak interaction between the two nuclei, while a longer interaction period is involved when the strength of coupling increases. The width of resonances reflects the interaction time. Several models have been proposed to describe the intrinsic nature of nuclear molecular configurations, based on the two different approaches. The weak-coupling idea is the basis of the double-excitation mechanism of Nogami and Imanishi [8], developed in the Band Crossing Model (BCM) by Abe and co-workers. The model describes the reaction features through the interaction and excitation of some collective states of the colliding nuclei. Details can be found in [9, 10]. As regards the strong coupling the α-cluster models and others based on deformed shell model calculations assume the existence of correlation between resonance structures and intrinsic configurations in high-lying states of the intermediate system as said before. See [11, 12].

3. – Resonant particle spectroscopy and angular correlation analysis

With the introduction of multistrip and position-sensitive charged particle detectors, new experimental techniques, such as the resonant particle spectroscopy, became available for studying cluster structures in light nuclei. This method involves the coincidence detection of the decay products of highly excited nuclei making it possible a measurement of the partial widths for different decay channels as well as a model-independent determination of the spins of the decaying states. The angular correlation analysis of the final-state particles allows the angular dependence of the primary excitation and subsequent decay processes to be unravelled. In such measurements the resonant particle of interest may be excited via inelastic scattering or transfer reactions. In particular, the advantage of this last method consists in the possibility of exploring an extended excitation energy range with a single beam energy.

The decay of the resonant particle into the cluster components may then be used as possible indicator of cluster structure, *i.e.* states possessing large reduced widths for cluster decay should appear most strongly in the decay spectrum. The reconstruction of the emission angles of break-up fragments, using position sensitive detectors permits the reconstruction of the excitation energy of the decaying state and a measurement of the angular distributions with which the decay process proceeds. Before showing some results of such analysis, some details regarding the general angular correlation method are given.

3`1. *Angular correlation: general method*. – The first assumption of this method is concerned with the decay process, occurring through a particular intermediate state of one of the two first emitted nuclei. In general a sequential break-up reaction can be described using the notation $a(A, B^* \rightarrow c + C)b$ as illustrated in fig. 5. The correlation between the three final-state particles c, C and b can be parameterized in terms of four spherical polar angles [13, 14]. The angle θ^* represents the polar centre-of-mass scattering angle of the resonant particle B^*, and Ψ is defined as the polar emission angle of the decay products in the B^* centre-of-mass frame. Both angles are measured with respect

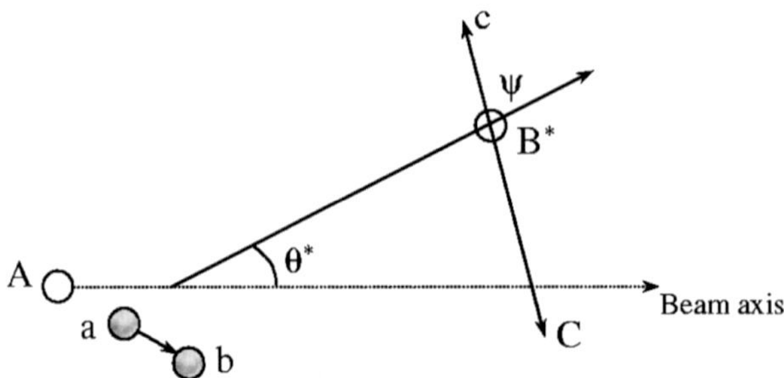

Fig. 5. – A sequential break-up reaction $a(A, B^* \to c + C)b$ is schematically described. The diagram shows moreover the correlation angles θ^*, Ψ.

to the beam axis which is taken as the quantization axis. The other two angles ϕ^* and χ are the azimuthal angular components associated with θ^* and Ψ, respectively. The horizontal plane is defined by the conditions of the azimuthal angles to be zero.

$$(4) \qquad \frac{d\sigma}{d\Omega d\Omega_\Psi} = \sum_{m_a, m_A, m_c, m_C, m_b} \left| \sum_{m_{B^*}} T^{m_a, m_A}_{m_b, m_{B^*}}(\Omega^*) T^{m_{B^*}}_{m_c, m_{C^*}}(\Omega_\Psi) \right|^2 .$$

The first term in the product represents the formation transition amplitude between states with magnetic quantum numbers m_a, m_A and m_b, m_{B^*}, and the second term is the decay amplitude from the state formed in the substate m_{B^*} to states in the two fragments with quantum numbers m_c, m_C. If the two final nuclei coming from the resonant particle break-up have spin zero, the decay transition amplitudes are simply expressed in terms of spherical harmonics $Y^{m_{B^*}}_{J_{B^*}}(\Omega_\Psi)$. Moreover, if the angular correlation is restricted to the region around $\theta^* = 0$, only the $m = 0$ substates can contribute. Under this condition the double differential cross-section is expressed in the following form:

$$(5) \qquad \frac{d\sigma}{d\Omega d\Omega_\Psi} = \frac{2J_{B^*} + 1}{4\pi} |P_{J_{B^*}}(\cos \Psi)|^2 .$$

A similar simplified formula is obtained when θ^* is unrestricted and Ψ is constrained such that $\Psi = 0°$ or $180°$. The presence of all the magnetic substates far from these particular conditions may make the angular correlation analysis complex. However if the analysis is restricted to the in-plane reaction, that means $\phi^* = 0$ and $\chi = 0$, the eventual correlation structure appearing at $\theta^* = 0$ is not removed, but it simply appears shifted such that

$$(6) \qquad \frac{d\sigma}{d\Omega d\Omega_\Psi} = \frac{2J_{B^*} + 1}{4\pi} |P_{J_{B^*}}(\cos(\Psi + \Delta\Psi))|^2 .$$

Under the assumption that a single entrance channel partial wave l_i dominates in the reaction, that is the reaction proceeds through a resonance in the intermediate system, the phase shift can be related to this l_i through the relation

$$(7) \qquad \frac{\Delta\theta^*}{\Delta\Psi} = \frac{J_{B^*}}{l_i - J_{B^*}} .$$

As a consequence, the θ^*-Ψ correlation appears as a series of diagonal ridges whose periodicity and gradient can be used to infer the angular momenta contributing to the reaction process. Equation (7) allows to obtain an estimation of the dominant l_i once the resonant angular momentum is known. Moreover, if the stretched configuration corresponding to the lowest centrifugal barrier is considered, the relationship $l_i = J_B^* + l_f$ gives the value of the final-state grazing angular momentum l_f. The in-plane constraint demands that only a limited part of available data can be used to perform the correlation analysis, and the low statistics and the limited detection efficiency very often prevents any meaningful spin assignment. However, in special cases when the direction of the emitting nucleus is a symmetry axis for the correlation function, an alternative way to obtain the information on the spin J can be used. The correlation function is thus expressed in a simplified form, which no longer depends on both θ^* and Ψ angles, but it is proportional to a Legendre polynomial with argument $\Psi' = \Psi - \theta_L$, θ_L being the angle of the emitting B^* nucleus in the laboratory system. The angular distribution of this new angle simply carries out the required information on the J in a way which avoids the problem of the $\theta^* - \Psi$ locus distortion allowing for using all the available data to carry out the information on the spin.

This method was successfully applied in the $^{16}\mathrm{O} + ^{12}\mathrm{C}$ reaction [15] to study the high-lying states of $^{24}\mathrm{Mg}$ decaying into α-particles and/or their subsystems. In particular quite original results have been obtained by selecting the $^{16}\mathrm{O} + ^{8}\mathrm{Be}$ decay channel of the excited $^{24}\mathrm{Mg}$, which is seen to be excited up to about 35 MeV. The segmented detection set-up, consisting of a number of position sensitive detectors in coincidence with a telescope to identify the residue, allowed to perform the "Resonant particle spectroscopy" with the coincidence detection of the decay products of the highly excited $^{24}\mathrm{Mg}$. Results are reported in figs. 6 ($^{16}\mathrm{O}_{\mathrm{g.s.}}$-$^{8}\mathrm{Be}_{\mathrm{g.s.}}$ relative energy spectrum) and 7 (θ^* vs. Ψ and θ^* vs. Ψ' correlation diagrams for the 34.5 MeV structure observed in fig. 6) Another original way to infer the dominant J value to a resonance consists in the integration of the excitation function over small angular ranges around the zeros of Legendre polynomials of given orders, expected in the excitation energy region investigated. If a peak is dominated by one of these values, it is expected to disappear (or to be strongly depressed) in the excitation function integrated over its zeros. Such kind of analysis brings to assignment of a dominant spin value to several structures populated in the $^{12}\mathrm{C}(^{12}\mathrm{C}, ^{8}\mathrm{Be})^{16}\mathrm{O}$ reaction in the relative energy range from 20 to 30 MeV, as reported in [16]. These results together with those obtained on the same reaction at lower energies [17-19] and the more recent reported in fig. 6, show a fair consistency with the linear rule typical of the rotational bands. The deduced resonance energy vs. $J(J+1)$ slope coefficient of about 90 keV gives a moment of inertia of the system which corresponds to the one calculated for the D1 (α-$^{16}\mathrm{O}$-α) configuration of $^{24}\mathrm{Mg}$.

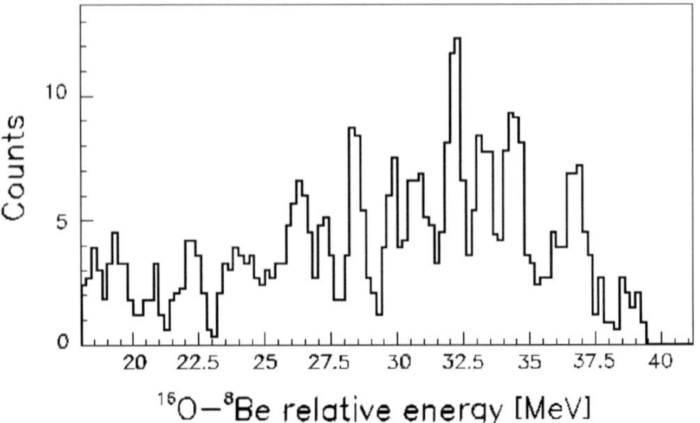

Fig. 6. – $^{16}\mathrm{O}_{\text{g.s.}}$-$^{8}\mathrm{Be}_{\text{g.s.}}$ relative energy spectrum from the $^{16}\mathrm{O} + {}^{12}\mathrm{C}$ interaction with the in-plane data selection.

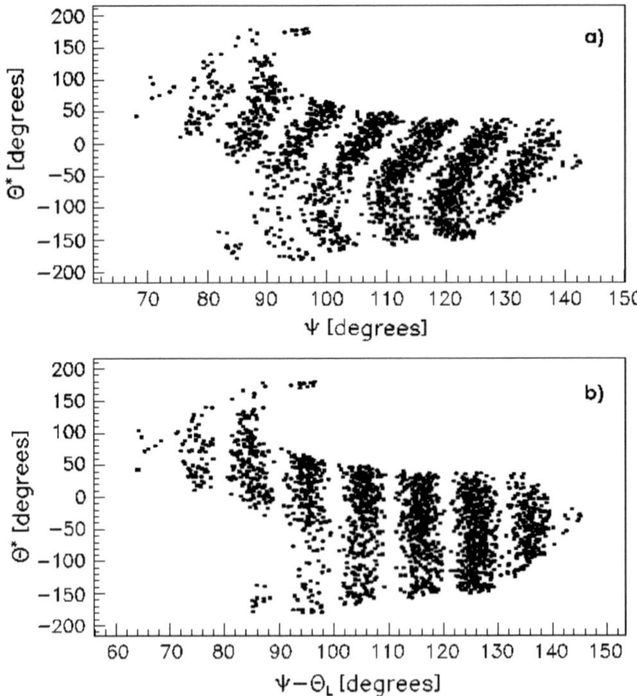

Fig. 7. – θ^* *vs.* Ψ (a) and θ^* *vs.* Ψ' (b) correlation diagrams for the 34.5 MeV structure observed in fig. 6.

4. – α-chains in light nuclei

For light nuclei an α-particle is considered as a possible building block because of its stable and inert behaviour, due to the strong binding of two protons and two neutrons.

4$^{\cdot}$1. ^{8}Be. – It provides the most concrete example for the existence of an α-chain structure in the ground and low-lying excited states, albeit only two units in length. The ^{8}Be ground state is believed to possess a configuration with a very large axis ratio of approximately $2 : 1$. Experimentally, the rotational band built on the ground state is known up to spin $J = 4^+$, and the moment of inertia of this band is consistent with a superdeformed configuration, which can be described in terms of two touching α-particles. Nilsson-Strutinsky and Alpha Cluster Model calculations [20, 21] reproduce this rotational behaviour extremely well. Further evidence for its cluster structure is the significantly larger α-decay width of the ^{8}Be ground state, than simple shell model expectations. Horiuchi [22] managed to reproduce the experimental α-decay width of this state, within a microscopic treatment which takes into account the influence of the Pauli principle on the structure of the ^{8}Be nucleus. A phase shift analysis of resonances observed in ^{4}He $+ {}^4$He elastic scattering has been performed by Buck *et al.* using the resonating group formalism [23], further demonstrating the importance of the cluster structure and the associated antisymmetrization effects in the $\alpha - \alpha$ wave function.

4$^{\cdot}$2. ^{12}C. – Several models have been applied to ^{12}C [24, 25]; from these studies, the ground rotational band, consisting of the ground 0_1^+, the 2_1^+ at 4.44 MeV and the 4_1^+ at 14.08 MeV states, was regarded as having a strong shell-model–like character, although α-clustering effect is important. But none of the models based on the assumption of stable average nuclear field are able to describe satisfactory the low-lying positive-parity excited states, 0_2^+ at 7.66 MeV and 2_2^+ (or 0_3^+) at 10.3 MeV. Morinaga [26] assigned a 3α linear chain structure to these levels to account for their anomalously low excitation energies. Following Morinaga, the structure of the ground band was re-examined by the use of the BB Model [2]. An equilateral triangular configuration was assumed for this band. Microscopic calculations revealed a dominant shell-like structure for the ground state; at the same time the structure of the excited 0_2^+ state was found to be well-developed cluster one. Uegaki and collaborators found within a GCM formalism [27, 28], that in the positive-parity excited states, α-clusters are largely separated with each other and are not confined to a specific geometrical arrangement. Hence the structure of loosely coupled 3α-particles system was fully microscopically confirmed. The BB Model ruled out also the hypothesis of the linear chain; this structure cannot explain the large α-decay width as near the Wigner limit value for the 0_2^+ state. Furthermore the energy difference between this state and the 2_2^+ one does not allow them to belong to a rotational band built on the 3α chain. A 2^+ member of such a rotational band, is predicted by the BB Model to lie at around 8.4 MeV of excitation energy. At present the most credible structure is that one coming from the microscopic calculations, which is pictured in terms of a system of 3 loosely bound α-particles. If a spin 0^+ is assigned to the 10.3 MeV state, it may be better candidate for the aligned 3α chain configuration; it possesses a significantly enhanced reduced α-decay width. There is a possible 2^+ state at $11.2\,\text{MeV}$ and a 4^+ at $14.08\,\text{MeV}$ although this latter state is predicted to be the 4^+ member of the ground rotational band. Some years ago we have performed a study on ^{12}C by using a treatment of the n-body problem in terms of the $U(3n\text{-}2)$ spectrum generating algebra [29, 30]. Starting from an equilateral triangle 3α-configuration, with this formulation we could add to the rotational band built on the ground state the first 3^- state at $9.63\,\text{MeV}$ of excitation energy. The symmetry of triangular configuration allows the negative parity states to be present, unlike the one of the α-chain. Moreover the 0_2^+ and $1^-(10.84\,\text{MeV})$ states were assumed to represent the stretching and bending vibrations respectively, and with this interpretation energy and form factors for those levels could be reproduced. However, present experimental data do not allow to exclude that this state could be the band head of another configuration as the one consisting of three *alpha*-particles in a line. A dedicated kinematically complete experiment in order to look for the second 2^+ state predicted at $8.4\,\text{MeV}$ would be likely.

Detailed measurements have shown that α-chain configurations are somehow unlikely to exist in heavier nuclei, in contrast to earlier experimental evidence. Rather, these measurements demonstrated that the resonances that had previously been associated with the chain-configuration were actually connected with a planar cluster structure as revealed by a precise angular correlation analysis.

4$^\cdot$3. *Nuclear dimers and polymers.* – In the last decade we have started to extend the well-developed cluster structure of α-conjugate nuclei to neutron-rich nuclei and to study the role of neutrons in stabilizing the bond. In particular, due to the nature of the α-α potential, which looks more like a molecular one with an attractive part at distances larger than 3–$4\,\text{fm}$ and a repulsive core, the additional neutrons have been described in terms of covalent particles. Molecular structure has been suggested to appear in neutron-rich Be, B and C isotopes, based on the concept of covalent molecular orbitals of neutrons [31-33]. α-cluster in Be isotopes was reproduced by model independent Antisymmeterised Molecular Dynamics (AMD) calculations [34, 35]. The Molecular Orbit (MO) model was successfully used for the calculation of the structure of Be and C isotopes [36]. The population of orbitals in Be isotopes was considered at the minimum (3–$4\,\text{fm}$) in the α-α potential. Figure 8 shows the relevant information on binding energies for Be, including the heavier isotopes. The excitation energies are shifted so as to have a common line for the relevant threshold ($2\alpha + xn$) for all isotopes. The first possible isomeric two-center states are indicated in boxes. From the experimental point of view, the evidence for rotational bands is based mostly on inelastic α-scattering $(\alpha, {}^3\text{He})$ reactions as well as (d, p) reactions on ^{9}Be. Rotational bands are summarized in fig. 9 [31]. The evidence for states in ^{11}Be corresponding to dimers can be obtained by means of a two-neutron transfer reaction on ^{9}Be, instead of a one-neutron stripping on ^{10}Be, the ground state having not the appropriate structure. For rotational states belonging to the $K = 1/2$ (projection of the total angular momentum J on the symmetry axis) the back-bending effect due to the Coriolis decoupling is evident.

Starting from the 0_2^+ state of ^{12}C the energy diagrams of molecular chain states in

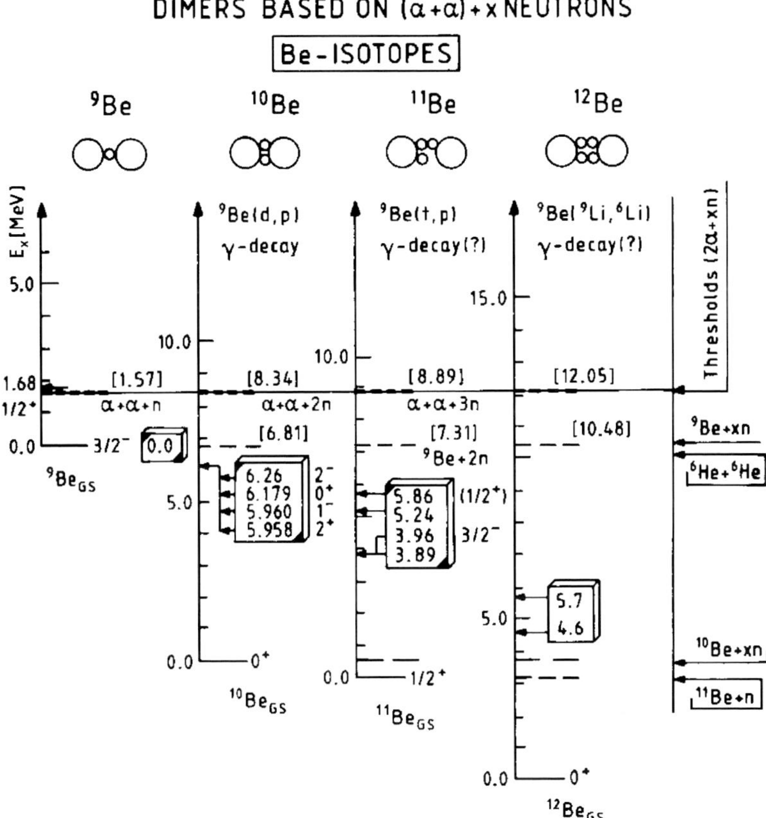

Fig. 8. – Energy diagrams for dimers of the beryllium isotopes.

carbon isotopes was built (see fig. 10). The energy scales are aligned to the same level for the decay threshold into 3α-particles and x-neutrons. The binding energy for the isomeric chain states can be read from this common reference line. Isomeric states are indicated in boxes with black corners, if properties (excitation energies, populating reactions, etc.) are known. Predictions for rotational bandheads are given in boxes without black corners. In C isotopes the stability of the linear 3α-chain state has been examined in terms of the α-α distance and bending angle and the calculations revealed that the addition of valence neutron helps to stabilize the chain structure, suggesting that in neutron-rich carbon isotopes molecular chain may exist. For example ^{16}C with the combination of the π^- and σ^- orbits occupied by neutrons has the simultaneous stabilities against the breathing-like break-up path and bending-like one. Recent calculations show that this is the case for ^{14}C also. The threshold for the ^{12}C$^* + 2n$ channel is quite high (20.7 MeV) (see fig. 10). Similarly the threshold for the "unbound chain" (^{10}Be$^* + 1\alpha$) is high at 18.0 MeV. We may thus expect resonances in the ^{10}Be$^* + 1\alpha$ system just above 18.0 MeV; these would correspond to a chain state where the two valence neutrons are kept in one

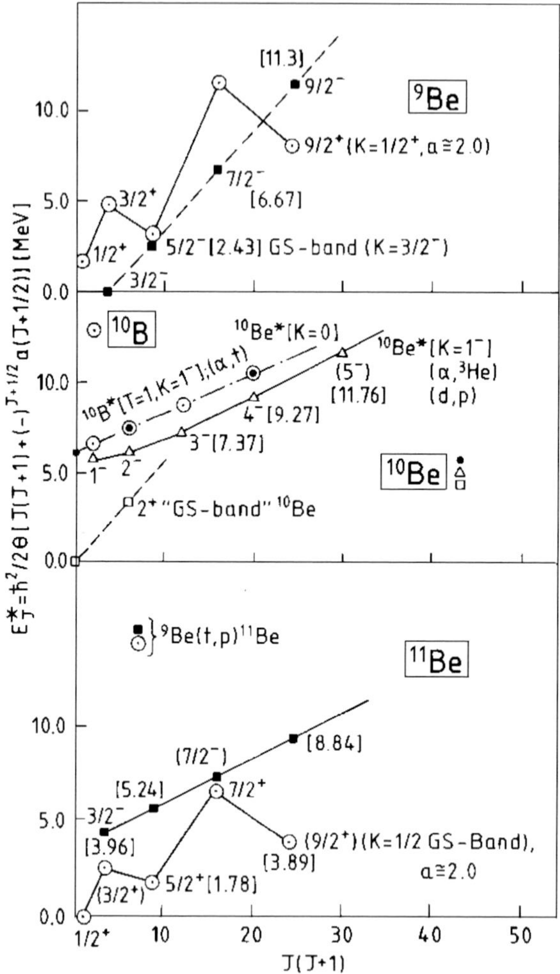

Fig. 9. – Resonance energy *vs.* $J(J+1)$ for Be isotopes.

covalent bond (configuration "Y" in fig. 10). We expect a series of states connected to the group of states at 6.0 MeV in ^{10}Be*$(0^1, 2^1, 1^2, 2^2)$ ranging up to 19.0 MeV. There is a clear possibility to observe γ-decay branches from and between those states just like for the 0_2^+ in ^{12}C, because the $(^9\text{Be} + 1\alpha + 1n)$ channel is closed. The decay into other channels like $(^{12}\text{C} + 2n)$ or $(^{13}\text{C} + 1n)$ would imply a very strong structural change because of the very different shapes. The configuration where the valence bond is equally distributed (configuration X) is expected to come below the configurations (Y), because of the stronger binding effect for each valence neutron $(2 \times 1.66\,\text{MeV})$. This state was predicted at about 17.4 MeV. Another important aspect enters in the case ^{14}C in the Y configuration as well as in ^{13,15}C isotopes, *i.e.* the occurrence of intrinsically reflection-asymmetric chain states which gives rise to parity doublets. For details see [37, 38].

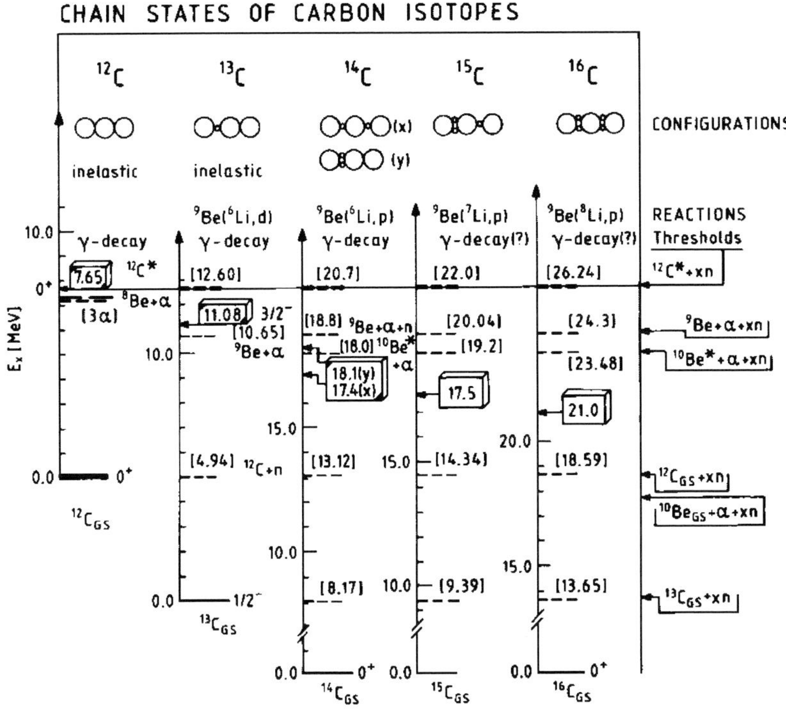

Fig. 10. – Energy diagrams for polymers of carbon isotopes.

5. – Quasi-free processes to probe clustering in light nuclei

Quasi-free scattering and reactions have been extensively studied in the past in order to investigate the cluster structure of light nuclei [39-42]. A number of theoretical approaches based on the Impulse Approximation (IA) were developed [5, 6], which describe a quasi-free $A + a \to c + C + s$ reaction (a being the nucleus described in terms of $x \oplus s$ clusters) by a pseudo-Feynman diagram where only the first term of the Feynman series is retained (see fig. 11). A pole of the diagram refers to the break-up of the target nucleus a into the clusters x and s, and the other one contains the information on the virtual $A + x \to c + C$ two-body process, which leaves the cluster s as a spectator. The IA factorizes the three-body cross-section as

$$(8) \qquad \frac{\mathrm{d}^3\sigma}{\mathrm{d}E_1\mathrm{d}\Omega_1\mathrm{d}\Omega_2} \propto KF|\Phi(\vec{p}_s)|^2 \left(\frac{\mathrm{d}\sigma}{\mathrm{d}\Omega_{\mathrm{cm}}}\right)^{\mathrm{off}},$$

where $\mathrm{d}\sigma/\mathrm{d}\Omega_{\mathrm{cm}}^{\mathrm{off}}$ is the off-shell differential cross-section for the two-body process, KF is a kinematical factor depending on masses, momenta and angles of the outgoing particles and containing the final phase-space factor [40], and $\Phi(\vec{p}_s)$ is essentially the Fourier

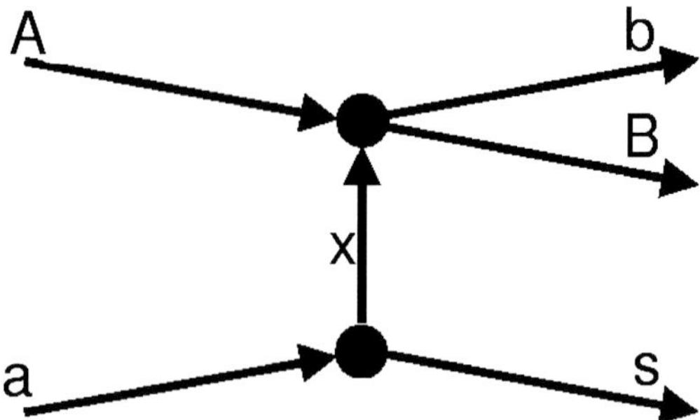

Fig. 11. – Pole diagram describing the quasi-free mechanism.

transform of the radial wave function for the $x - s$ intercluster motion inside a [40-42]. Once we know the three-body as well as the two-body cross-sections from independent experiments and we calculate KF, we can get the shape of the momentum distribution $\Phi(\vec{p}_s)$, which can be compared with the expected behaviour obtained by means of the Fourier transform. A kind of clustering probability can be obtained by integrating the full momentum distribution over p_s. In the 80's this technique was applied to the study of ^{6}Li [43], considered the simplest nucleus that could be described in the framework of cluster models. Its low dissociation energy into a deuteron and an α-particle (1.47 MeV) suggested its description in terms of these two clusters: ^{6}Li $= \alpha \oplus d$. This description was then confirmed with the agreement between theoretical and experimental momentum distributions, shown in fig. 12.

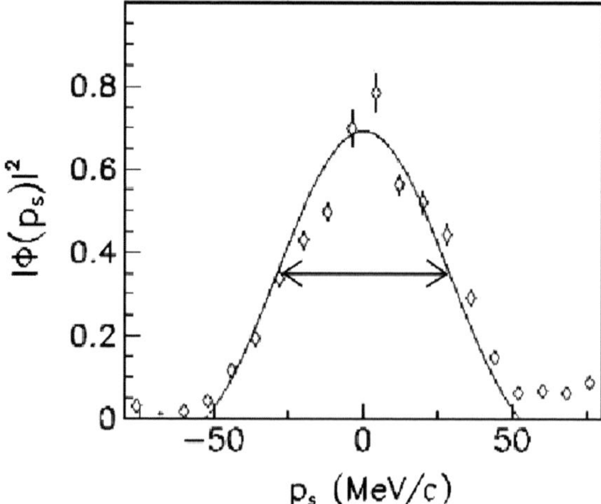

Fig. 12. – Theoretical (full line) momentum distribution of α and deuteron inside ^{6}Li and experimental one obtained from the ^{6}Li(^{6}Li, $\alpha\alpha$)^{4}He quasi-free reaction.

6. – Nuclear clusters as virtual projectiles/targets for nuclear astrophysics: the Trojan Horse Method

Recently the quasi-free mechanism has been applied in the framework of the Trojan Horse Method (THM), which is a successful indirect technique for nuclear astrophysics. The idea is to use nuclear clusters as virtual projectiles/targets in order to measure cross-sections at ultra-low energies overcoming the main problems of direct measurements. Indeed, when studying charged particle reactions at sub-Coulomb energies, the Coulomb barrier causes a strong suppression of the cross-section, which drops exponentially with decreasing energy. This complicates a lot the experimental study. In addition, the electron screening effect due to the electrons surrounding the interacting ions prevents one to measure the bare nucleus cross-section. Up to date, the only way to reach the ultra-low energy is by extrapolating the behaviour of the higher energy data. This is done by means of the definition of the astrophysical $S(E)$ factor which represents essentially the cross-section free of Coulomb suppression:

$$(9) \qquad S_b(E) = E\sigma_b(E)\exp[2\pi\eta],$$

where $\exp[2\pi\eta]$ is the inverse of the Gamow factor (η is the Sommerfeld parameter), which removes the dominant energy dependence of $\sigma_b(E)$, due to the barrier penetrability.

Recently the Trojan Horse Method (THM) [44-47] has provided a successful alternative path. The method selects the quasi-free contribution of an appropriate three-body $A + a \rightarrow c + C + s$ reaction performed at energies well above the Coulomb barrier to extract the relevant two-body $A + x \rightarrow c + C$ reaction cross-section down to sub-Coulomb energies. The wave function of a, the so-called Trojan Horse nucleus, is required to have a large amplitude for a $x \oplus s$ cluster configuration. Thanks to the high energy in the $A + a$ entrance channel, the two-body interaction can be considered as taking place inside the nuclear field, without experiencing either Coulomb suppression or electron screening effects. In order to compensate for the $A + a$ relative motion, kinematical conditions are chosen in a way that the $x - s$ binding energy can play this role and particle s remains as spectator to the $A + x$ process. The THM was successfully applied to a number of reactions of astrophysical interest [44-47], and it represents nowadays the only way to get the bare nucleus astrophysical $S(E)$-factor without extrapolation. In a typical THM experiment the decay products (c and C) of the virtual two-body reaction of interest are detected and identified by means of telescopes (silicon detector or ionization chamber as ΔE step and position sensitive detector as E step) placed at the so-called quasi-free angles. After the selection of the reaction channel, the most critic point is to disentangle the quasi-free mechanism from other reaction mechanisms feeding the same particles in the final state, *e.g.*, sequential decay and direct break-up. An observable which turns out to be very sensitive to the reaction mechanism is the shape of the experimental momentum distribution of the spectator. In order to reconstruct the experimental p_s distribution, the energy sharing method [40] was applied for each pair of coincidence QF angles, selecting c-C relative energy windows of $100\,\mathrm{keV}$.

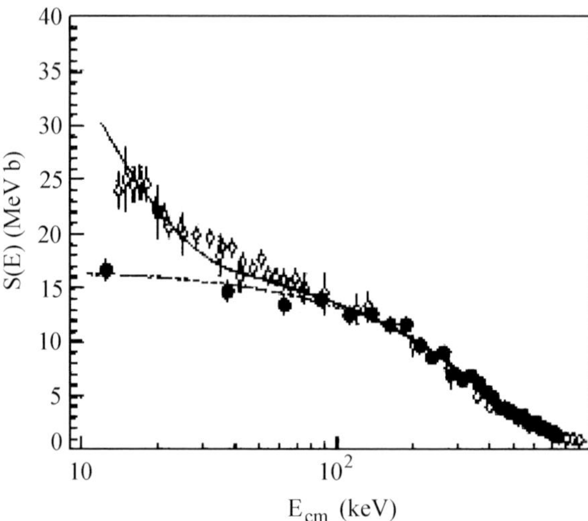

Fig. 13. – THM S(E) factor for the ^{6}Li(d, α)α reaction (full dots) and direct behaviour (open diamonds).

The extracted experimental momentum distribution is then compared with the theoretical one and further data analysis is limited to the data lying in the region where the agreement between the two distributions exists (usually within few tens of MeV/c). In fig. 13 the THM astrophysical $S(E)$ factor for the ^{6}Li(d, α)α reaction is reported as full dots [44] and compared with the screened behaviour from direct experiment [48], shown as open diamonds. According to the inhomogeneous big-bang model this is a reaction responsible for the destruction of ^{6}Li in neutron-rich sites.

Another nice example of THM $S(E)$ factor is reported in fig. 14. It refers to the ^{15}N(p, α)^{12}C reaction studied by performing the ^{2}H(^{15}N, α^{12}C)n reaction. The reaction rate of this two-body process is considered among the primary sources of uncertainty in predicting the fluorine abundance in AGB stars, whose chemical evolution is strongly influenced by the reactions belonging to the production/destruction path of ^{19}F.

The astrophysical $S(E)$ factor ^{15}N(p, α)^{12}C reaction has been directly measured down to 73 keV (open symbols) [49] and an extrapolation has been performed to derive its trend down to zero energy [49]. This extrapolation has also been employed to calculate the reaction rate [50] to be introduced in the stellar evolutionary codes. Of course it does not account for either the contribution of subthreshold resonances, or the enhancement of the $S(E)$ factor due to the electron screening. Indeed the electron screening effect introduces at least a 10% increase of the S-factor at 80 keV for the ^{15}N + p system, which can modify the low-energy behaviour of the $S(E)$ factor. The THM data are reported as red dots and they go down to 20 keV. In order to strengthen this result an independent half-off-energy-shell R-matrix approach has been applied [51] which accounts also for possible interference between intermediate ^{16}O levels with $J^\pi = 1^-$ feeding the ^{12}C + α channel. The theoretical curve is reported as full line and it appears clearly in agreement with the data.

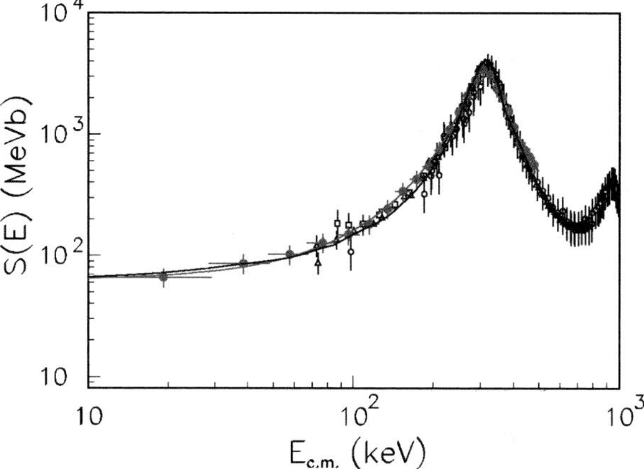

Fig. 14. – THM $S(E)$ factor for the $^{15}\mathrm{N}(\mathrm{p},\alpha)^{12}\mathrm{C}$ reaction (solid dots) compared with the direct behaviour (open symbols). The full line is the result of a half-off-shell R-Matrix calculation.

6ʹ1. *$p-p$ scattering via the THM.* – Recently we have extended the application of the THM to scattering processes. In particular, we have address the study of the $p+p$ elastic scattering through the $^{2}\mathrm{H}(p,pp)n$ reaction [52]. The aim of the experiment was to investigate the suppression of the Coulomb amplitude also for scattering. Indeed the $p+p$ system is the simplest frame for this important test: nuclear and Coulomb scatterings coexist and are coherent, so that interference terms between the two effects are expected to contribute to the cross-section. In particular, there is a region at low $p-p$ relative energy (E_{pp}) in which the nuclear scattering amplitude in the ^{1}S-state and the even part of the Coulomb scattering amplitude give destructive interference [53], leading to the deep minimum in the $p-p$ cross-section (about $1\,\mathrm{mbarn/sr}$) which occurs at $E_{pp} = 191.2\,\mathrm{keV}$ and $\theta_{\mathrm{c.m.}} = 90°$ [54]. The test was done by checking whether there is remaining evidence of the nuclear plus Coulomb interference pattern in the half-off-energy-shell (HOES) $p+p$ cross-section extracted via the THM. The extracted $p-p$ HOES cross-section is presented in fig. 15 as a function of E_{pp} (black dots) and compared to the free p-p cross-section (solid line) where the $l = 0$ phase shift is calculated by using the formalisms reported in [53] with effective length $a_p = -7.806\,\mathrm{fm}$ and scattering radius $r_0 = 2.794\,\mathrm{fm}$. The dashed-dotted line represents the calculated HOES $p-p$ cross-section [52]. Both calculated curves are integrated over $80° \leq \theta_{\mathrm{c.m.}} \leq 100°$ and averaged over an energy bin of $20\,\mathrm{keV}$. Experimental and calculated HOES cross-sections are normalized to the calculated OES one at E_{pp} close to the Coulomb barrier ($500\,\mathrm{keV}$). We observe a striking disagreement between the THM (HOES) and the free $p-p$ (OES) cross-sections throughout the region of the interference minimum, which is missing in the THM data. Instead, the calculated HOES $p-p$ nicely fits the THM data.

In order to strengthen this result, the THM $p-p$ cross-section was compared with the on-shell $n-n$, $p-n$ and pure nuclear $p-p$ ones [53]. The expression of the nuclear two-body cross-section at low energy was considered for different values of the a and r_0

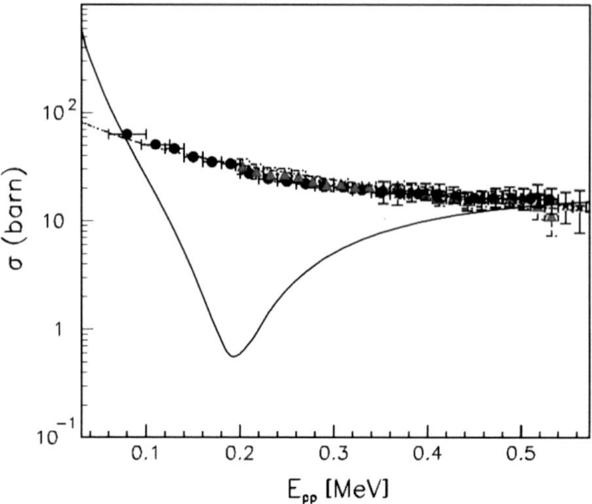

Fig. 15. – THM two-body cross-section (symbols) *vs.* pp relative energy. Solid line represents the theoretical OES $p-p$ cross-section calculated as explained in the text. The dash-dotted line is the calculated HOES cross-section.

parameters: $a = -18.5\,\mathrm{fm}$, $r_0 = 2.75\,\mathrm{fm}$ for the $n-n$, $a = -23.748\,\mathrm{fm}$, $r_0 = 2.75\,\mathrm{fm}$ for the $n-p$, $a = -17.3\,\mathrm{fm}$, $r_0 = 2.85\,\mathrm{fm}$ for the nuclear $p-p$ scattering [55]. Results are reported in fig. 16 (solid line for the $n-n$, dashed line for the $p-n$ and dotted line for the pure nuclear $p-p$ cross-section). The calculated HOES cross-section is again shown

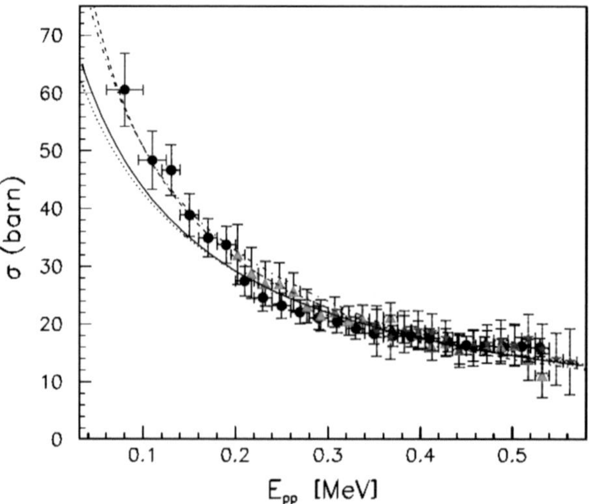

Fig. 16. – THM two-body cross-section (symbols) *vs.* pp relative energy compared with the on-shell $n-n$ (solid line), $p-n$ (dashed line) and pure nuclear $p-p$ (dotted line) ones. The HOES calculated cross-section is also reported as dash-dotted line.

for completeness (dash-dotted line). There is clearly good agreement between the OES nucleon-nucleon nuclear cross-sections and the THM data. This agreement represents compelling evidence of the validity of the THM method for elastic scattering: the HOES $p + p$ scattering cross-section has the same energy dependence as pure nuclear OES nucleon-nucleon cross-sections. Through a mechanism different from that of nuclear rearrangement reactions, these results strongly confirm the THM hypothesis, *i.e.* the suppression of Coulomb effects in the two-body cross-section at sub-Coulomb energies. It turns out to be an universal effect whether we consider binary elastic or rearrangement processes. This is the hypothesis which makes the THM so powerful in the framework of nuclear astrophysics as well as in all physics contexts where it can be important to investigate nuclear effects at low energies.

7. – Conclusions

Many years ago the idea that the atomic nucleus need not remain a tightly bound, compact, spherical object, but might show a cluster substructure, has strongly supported the development of the physics of clusters. Recent experimental advances, backed by new quantum-mechanical models, are now revealing a wide variety of clustering ranging from nuclear molecules to possible chains of alpha-particles. We are also beginning to understand how valence nucleons can form molecular-type orbits between these clusters to provide additional stability. We have many observables which can provide us strong signatures in favor of clustering. From the discussion above, we can draw the conclusion that this physics is still very attractive. In particular, we need to understand a lot at high excitation energy, where the selective feeding of states at such high level density might pick-out new insights into the most intriguing cluster mode, *i.e.* the α conglomerates. This study has been regarded as a way of testing the Bose-Einstein condensation of α-clusters in the atomic nucleus.

We have considered also a possible application of cluster physics to nuclear astrophysics and we have seen how the idea of virtual projectiles/targets may help nuclear physicists to overcome the main problems encountered when performing nuclear reactions at astrophysical energies.

REFERENCES

[1] IKEDA K. *et al.*, *Prog. Theor. Phys. Suppl.*, Extra Number (1968) 464.
[2] BRINK D., *Proceedings of the International School of Physics "Enrico Fermi", Course XXXVI*, edited by BLOCH C. L. (Academic Press, New York and London) 1966, p. 247.
[3] ALM *et al.*, *Phys. Rev. Lett.*, **4** (1960) 515.
[4] COSMAN T. M. *et al.*, *Phys. Rev. Lett.*, **35** (1975) 265.
[5] CORMER T. M. *et al.*, *Phys. Rev. Lett.*, **38** (1977) 17.
[6] CORMER T. M. *et al.*, *Phys. Rev. Lett.*, **40** (1978) 14.
[7] CINDRO N., *Ann. Phys. (Paris)*, **24** (1988) 289.
[8] IMANISHI B., *Phys. Lett. B*, **27** (1968) 267.
[9] MATSUSE T. *et al.*, *Progr. Theor. Phys.*, **59** (1978) 1904.

[10] Kato K. *et al.*, *Progr. Theor. Phys.*, **68** (1982) 1794.

[11] Hiura J. *et al.*, *Progr. Theor. Phys.*, **42** (1969) 555.

[12] Hiura J. *et al.*, *Progr. Theor. Phys.*, **52** (1972) 173.

[13] da Silveira E. F., Contribution to *the XIV Winter Meeting on Nuclear Physics, Bormio*, edited by Iori I. (Università di Milano) 1976, p. 293.

[14] Freer M., *Nucl. Instrum. Methods Phys. Res A*, **383** (1996) 463.

[15] Tumino A. *et al.*, *Eur. Phys. J. A*, **12** (2001) 327.

[16] Lattuada M. *et al.*, *Nuovo Cimento A*, **110** (1997) 1007.

[17] Fletcher N. R. *et al.*, *Phys. Rev. C*, **13** (1976) 1173.

[18] James D. R. and Fletcher N. R., *Phys. Rev. C*, **17** (1978) 2248.

[19] Weidinger A. *et al.*, *Nucl. Phys. A*, **257** (1976) 144.

[20] Ragnarsson I. *et al.*, *Physica Scripta*, **24** (1981) 215.

[21] Merchant A. C. *et al.*, *Nucl. Phys. A*, **549** (1992) 431.

[22] Horiuchi H., *Prog. Theor. Phys.*, **43** (1970) 375.

[23] Buck B. *et al.*, *Phys. Rev. C*, **11** (1975) 1803.

[24] Cohen S. and Kurath D., *Nucl. Phys.*, **73** (1965) 1.

[25] Jaeger H. U. and Kirchbach M., *Nucl. Phys. A*, **291** (1977) 32.

[26] Morinaga H., *Phys. Lett.*, **21** (1966) 78.

[27] Uegaki E. *et al.*, *Prog. Theor. Phys.*, **57** (1977) 1262.

[28] Uegaki E. *et al.*, *Prog. Theor. Phys.*, **62** (1979) 1621.

[29] Bijker R. and Iachello F., *Phys. Rev. C*, **61** (2000) 067305.

[30] Tumino A., *Proceedings of the 7th International Conference on Clustering Aspects of Nuclear Structure and Dynamics*, edited by Korolija M., Basrak Z. and Caplar R. (World Scientific Singapore) 2000.

[31] von Oertzen W., *Z. Phys. A*, **354** (1996) 37.

[32] von Oertzen W., *Z. Phys. A*, **357** (1997) 355.

[33] von Oertzen W., *Nuovo Cimento A*, **110** (1997) 895.

[34] Kanada-En'yo Y. and Horiuchi H., *Phys. Rev. C*, **52** (1995) 647.

[35] Kanada-En'yo Y. and Horiuchi H., *Phys. Rev. C*, **60** (1999) 64304.

[36] Itagaki N. and Okabe S., *Phys. Rev. C*, **61** (2000) 044306.

[37] Milin M. and von Oertzen W., *Eur. Phys. J. A*, **14** (2002) 295.

[38] von Oertzen W. *et al.*, *Eur. Phys. J. A*, **21** (2005) 193.

[39] Miljanic Dj. *et al.*, *Nucl. Phys. A*, **215** (1973) 221.

[40] Kasagi J. *et al.*, *Nucl. Phys. A*, **239** (1975) 233.

[41] Lattuada M. *et al.*, *Nucl. Phys. A*, **458** (1986) 493.

[42] Zadro M. *et al.*, *Nucl. Phys. A*, **474** (1987) 373.

[43] Lattuada M. *et al.*, *Nuovo Cimento A*, **83** (1984) 151 and references therein.

[44] Spitaleri C. *et al.*, *Phys. Rev. C*, **63** (2001) 005801 and references therein.

[45] Tumino A. *et al.*, *Phys. Rev. C*, **67** (2003) 065803 and references therein.

[46] Spitaleri C. *et al.*, *Phys. Rev. C*, **69** (2004) 055806.

[47] La Cognata M. *et al.*, *Phys. Rev. C*, **72** (2005) 065802 and references therein.

[48] Engstler S. *et al.*, *Z. Phys. A*, **342** (1992) 471.

[49] Redder A. *et al.*, *Z. Phys. A*, **305** (1982) 325.

[50] Angulo C. *et al.*, *Nucl. Phys. A*, **656** (1999) 3.

[51] La Cognata M. *et al.*, *Phys. Rev. C*, unpublished (2007).

[52] Tumino A. *et al.*, *Phys. Rev. Lett.*, **98** (2007) 252502.

[53] Jackson J. D. and Blatt J. M., *Rev. Mod. Phys.*, **22** (1950) 77.

[54] Dombrovski H. *et al.*, *Nucl. Phys. A*, **619** (1997) 97.

[55] Miller G. A. *et al.*, *Phys. Rep.*, **194** (1990) 1.

DOI 10.3254/978-1-58603-885-4-237

Borromean halo nuclei: Continuum structures and reactions

J. S. VAAGEN and Ø. JENSEN

Department of Physics and Technology, University of Bergen - N-5007 Bergen, Norway

B. V. DANILIN

Russian Research Center "The Kurchatov Institute" - 123182 Moscow, Russia

S. N. ERSHOV

Joint Institute for Nuclear Research - 141980 Dubna, Russia

G. HAGEN

Physics Division, Oak Ridge National Laboratory - P.O. Box 2008, Oak Ridge, TN 37831, USA

Department of Physics and Astronomy, University of Tennessee - Knoxville, TN 37996, USA

Summary. — Halo nuclei exhibit a new type of structure found in extremely neutron-rich light nuclei, at the limits of nuclear existence. Of particular interest are Borromean nuclei, where none of the binary substructures can bind, demonstrating features of universality. Nuclear physics has in recent years taken further steps to explore the nature of the halo continuum, this being in fact the major part of the spectrum since halo nuclei support only one or a few bound states. Since $3 \rightarrow 3$ scattering is prohibitively difficult to perform, the halo continuum has so far been excited in binary collisions, proceeding via the exotic ground state which to various degrees puts its imprint on the result. We discuss via examples how to disentangle continuum structures, comparing with recent correlation data, and the challenges of linking reaction theory and modern structure calculations.

1. – Borromean physics—Dreams and realization

Dripline physics, currently being—or to be explored with ever larger and more sophisticated experimental tools, can be described as "fundamental matter under extreme conditions". The new physics intertwines structure and reactions more than ever before. This feature is clearly exhibited by the exotic halo structures discovered and identified over the last two decades at the neutron dripline. These are characterized by self-organized clustering into a central object of normal density surrounded by a loosely bound veil of neutron matter with extremely low density and correspondingly abnormal extension. Of particular conceptual importance are the light three-body–like Borromean nuclei, the name being inspired by the Borromean rings, an analogue model [1]. The exotics of such systems are not only contained in the bound state, but also in the correlated continuum structures near the three-body threshold. Since 3-3 scattering is prohibitively difficult to carry out, a major challenge is to find ways to explore the nature of the continuum without being swamped by the exotic structure of the ground state.

Visions about Borromean structures or related Efimov states, the latter predicted in 1970 [2], have also been described in other fields, in particular by Zhen and Macek [3]. Recently evidence for Efimov quantum states was found in ultracold (10 nK) gas of caesium atoms [4]. A couple of years ago molecular Borromean ring structures (literally!) were addressed in Science [5] as an upcoming nanotech field. Few-body systems, famous for being "intractable" have thus become beacons for exciting new physics more widely, providing fundamental tools for a variety of important fundamental questions such as showing that the hydrogen-antihydrogen molecule is unstable. As tool for hypernuclei, hyperon "glue" may be used for i) tugging beyond-dripline-nuclei back to the dripline, or vice versa, ii) structurally well-understood light dripline nuclei may be used as laboratory for exploring fundamental YN and YY interactions, see [6] and references therein.

Other speakers at this Fermi School have, in great detail, discussed various *ab initio* approaches to modern nuclear many-body structure calculations, based on realistic nuclear interactions, and with the ambition to cover nuclei more generally all the way to the driplines. Being based on the nucleon degrees of freedom, this carries hope of understanding of the origin of halo structures, *i.e.* how this exotic self-organization arises. Significant progress toward this end has been made, giving clear evidence for the cluster sub-structures that underlie the emergent few-body degrees of freedom historically used to understand Borromean halo nuclei.

The characteristic features of the three-body Borromean halo structure is that it is bound in spite of lack of bound states between any of the three binary subsystems; for example, between the "clusters" $n - n$, $\alpha - n$ in ^{6}He, described as $\alpha + n + n$. These binary sub-systems are all characterized by large scattering lengths which have to be reproduced when tuning the intra-cluster interactions in a three-body formulation. Granted the role of the large scattering lengths, we may ask how much the details of the interaction matter. Thus the physics of the Borromean halo belongs to what is referred to as "Universality in few body systems with large scattering length" (see for example [7] for a comprehensive discussion).

The so-called Efimov states for three identical bosons provide the cardinal example. The nuclear cases that we discuss, say $^6\text{He} = \alpha + n + n$, contain fermions. Thus the Pauli principle must be obeyed and the halo neutrons kept outside the α cluster. The universality property explains the successes of the few-body approach in explaining a vast amount of data that depend primarily on the cluster degrees of freedom. *Ab initio* calculations for such quantities are not expected to be sensitive to short-range features of the nucleon-nucleon force and a low-momentum approximation may be applied.

An additional remark may, however, be made concerning *ab initio* approaches. *Ab initio* in this context means starting from nucleons and their interactions (possibly also three-body) resulting in self-organized halo structures, *i.e.* not putting clustering in by hand as is done in cluster models (which, however, use realistic interactions between the clusters). Substantial progress toward this challenging goal is currently being made. A remaining challenge is however to obtain the large width of the continuum structures and a description of the very nature of the reaction scenarios which contain colliding clusters (nuclei) producing new outgoing clusters (residual nuclei). Reactions obviously favor cluster formulations and the most advanced current reaction theory is within this picture, although it consistently invokes interactions at the nucleonic level between the reaction partners. The true *ab initio* approach has quite some way to go to also encompass reactions.

2. – Emergent degrees of freedom—Few-body modelling

In the early days of halo physics two decades ago, a system like ^6He was thought of as a pointlike dineutron moving with respect to an alpha-particle core. A decade later, after a more proper treatment of the emergent three-body degrees of freedom had proven to capture the leading halo physics [1], the spatial correlation plot also contained an enhancement corresponding to a halo neutron on each side of the alpha core, in addition to the dineutron peak (see fig. 1). This gross structure simply reflects the Pauli principle, which prevents the halo neutrons from entering the s-motion already occupied in the alpha cluster. Thus they are pushed out to form an angular momentum zero pair, in which each halo neutron is predominantly in p-motion with respect to the core. The gross features of the spatial correlation plot emerge by transforming this orbital motion to Jacobi **T** coordinates, the dineutron separation and the coordinate of the dineutron center of mass (CM) relative to the core. The finer details of the landscape given by the plot, essential for the correct halo physics, are caused by the interactions and dynamics of the three-body system [1]. The spatial correlation plot attracted great attention, keep in mind the chaotic nature of similar classical systems, and attempts were made to probe the peak structures. Two neutron transfer was used as a dissection device indicated a dominant role of the dineutron peak in one step transfer [8]. In the context of the present paper, where the halo continuum is probed in processes that proceed via the halo ground state, this state together with the transition operators act as a filter, that could lead to deceptive conclusions about the genuine structure of the continuum states. Thus we have to look for observables that amplify the very nature of the continuum states.

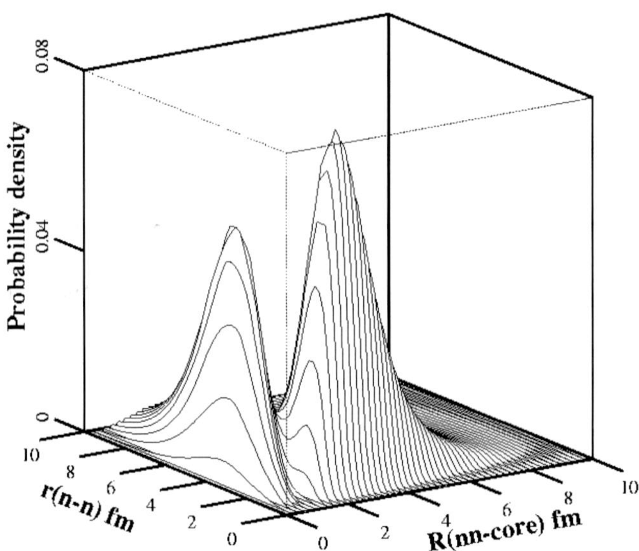

Fig. 1. – The cigar and dineutron configurations in the ^{6}He ground state.

The basic halo dynamics can be characterized as a coexistence of two subsystems: one which consists of core nucleons and the other of halo neutrons moving relative the core center of mass. With good accuracy the halo wave function Ψ can be written as a product of two functions

$$(1) \qquad \Psi(\mathbf{r}_1,\ldots,\mathbf{r}_A) = \varphi_C(\boldsymbol{\xi}_1,\ldots,\boldsymbol{\xi}_{A_C})\,\psi(\mathbf{x},\mathbf{y}).$$

The function $\varphi_C(\boldsymbol{\xi}_1,\ldots,\boldsymbol{\xi}_{A_C})$ describes the internal structure of the core, while $\psi(\mathbf{x},\mathbf{y})$ describes the relative motion of halo neutrons around the core CM. In (1) the coordinate $\boldsymbol{\xi}_j$ denotes the position of a core nucleon relative the core CM, $\mathbf{x} = \sqrt{\mu_x/\mathcal{M}}\,(\mathbf{r}_{n_1}-\mathbf{r}_{n_2})$ is the Jacobi scaled relative distance between halo neutrons and $\mathbf{y} = \sqrt{\mu_y/\mathcal{M}}\,(\mathbf{r}_C - (\mathbf{r}_{n_1}+\mathbf{r}_{n_2})/2)$ is the Jacobi scaled distance between the core and the center of mass of the two halo neutrons where μ_x and μ_y are corresponding reduced masses and $\mathcal{M}$ is a scaling mass (in our case it is convenient to choose it equal to the nucleon mass). The wave function $\psi(\mathbf{x},\mathbf{y})$ is solution of the Schrödinger three-body equation:

$$(2) \qquad (\hat{T}+\hat{V}-E)\psi(\mathbf{x},\mathbf{y}) = 0, \qquad \hat{V} = V_{Cn_1} + V_{Cn_2} + V_{n_1 n_2},$$

where $\hat{T}$ is the kinetic energy operator for relative motion of three particles, V_{Cn_i} and $V_{n_1 n_2}$ are the binary core-neutron and neutron-neutron interactions, respectively. Solutions of eq. (2) at a negative energy ($E < 0$) define the bound states of the halo nucleus and are completly characterized by the total angular momentum J and its projection M on a quantization axis. Within the method of the Hyperspherical Harmonics (HH) the bound-state wave function $\psi_{JM}(\mathbf{x}, \mathbf{y})$ can be decomposed in the following way:

$$(3) \qquad \psi_{JM}(\mathbf{x}, \mathbf{y}) = \frac{1}{\rho^{5/2}} \sum_{\gamma} \chi_{\gamma}^{J}(\rho) \left[\mathcal{Y}_{KL}^{l_x l_y}(\Omega_5^{\rho}) \otimes [\chi_{\frac{1}{2}}(\sigma_1) \otimes \chi_{\frac{1}{2}}(\sigma_2)]_S \right]_{JM}.$$

At positive energies ($E > 0$) solutions of the Schrödinger equation (2) describe $3 \to 3$ fragment scattering and can be characterized by momenta $(\mathbf{k}_x, \mathbf{k}_y)$ for the relative motion of the three fragments and their spin projections m_i. Accordingly, the decomposition of continuum wave functions $\psi_{m_1 m_2}^{(\pm)}(\mathbf{k}_x, \mathbf{k}_y; \mathbf{x}, \mathbf{y})$ within the HH method can be written as

$$(4) \quad \psi_{m_1 m_2}^{(\pm)}(\mathbf{k}_x, \mathbf{k}_y; \mathbf{x}, \mathbf{y}) = \sum_{J, \gamma, M, M_L, M_S} \imath^{K} \left(\frac{1}{2} m_1 \frac{1}{2} m_2 | S M_S \right)$$

$$\times (L M_L S M_S | J M) \, \mathcal{Y}_{K L M_L}^{l_x l_y *}(\Omega_5^{\kappa}) \, \psi_{\gamma J M}(\kappa, \mathbf{x}, \mathbf{y}),$$

$$\psi_{\gamma J M}(\kappa, \mathbf{x}, \mathbf{y}) = \frac{1}{(\kappa \rho)^{5/2}} \sum_{\gamma'} \chi_{\gamma \gamma'}^{J}(\kappa, \rho) \left[\mathcal{Y}_{K' L'}^{l_x' l_y'}(\Omega_5^{\rho}) \otimes [\chi_{\frac{1}{2}}(\sigma_1) \otimes \chi_{\frac{1}{2}}(\sigma_2)]_{S'} \right]_{JM}.$$

In the expressions above $(\rho, \Omega_5^{\rho}) = (\rho, \theta_{\rho}, \hat{\mathbf{x}}, \hat{\mathbf{y}})$ are hyperspherical coordinates in the 6-dimensional coordinate space of vectors $\mathbf{x}$ and $\mathbf{y}$ ($\rho = \sqrt{x^2 + y^2}$ means a hyperradius, $\theta_{\rho} = \arctan(x/y)$ is a hyperangle). Accordingly, $(\kappa, \Omega_5^{\kappa})$ are hyperspherical coordinates in the space of momenta $\mathbf{k}_x$ and $\mathbf{k}_y$, where $\kappa = \sqrt{(k_x^2/\mu_x + k_y^2/\mu_y)\mathcal{M}}$ is a hypermomentum and $\theta_{\kappa} = \arctan(\sqrt{\mu_y} k_x / \sqrt{\mu_x} k_y)$. The hyperspherical harmonic $\mathcal{Y}_{KLM}^{l_x l_y}(\Omega_5^{\rho})$ has the following analytical expression:

$$(5) \qquad \mathcal{Y}_{KLM}^{l_x l_y}(\Omega_5^{\rho}) = N_K^{l_x l_y} (\sin \theta_{\rho})^{l_x} (\cos \theta_{\rho})^{l_y} P_{(K - l_x - l_y)/2}^{l_x + 1/2, \, l_y + 1/2}(\cos 2\theta_{\rho})$$

$$\times \left[Y_{l_x}(\hat{\mathbf{x}}) \otimes Y_{l_y}(\hat{\mathbf{y}}) \right]_{LM},$$

where $P_n^{\alpha, \beta}$ are Jacobi polynomials and $N_K^{l_x l_y}$ is a normalization factor. $\gamma = (K, L, S, l_x, l_y)$ is an abbreviation for a set of quantum numbers, which characterizes the relative motion of the three fragments. Quantum number K has special structure, $K = l_x + l_y + 2n$, where n is a non-negative integer number. Inserting decompositions (3) and (4) into the Schrödinger equation (2) and projecting out the hyperharmonics $\mathcal{Y}_{KLM}^{l_x l_y}(\Omega_5^{\rho})$ we can get for hyperradial parts $\chi_{\gamma}^{J}(\rho)$ and $\chi_{\gamma \gamma'}^{J}(\kappa, \rho)$ of the bound and continuum wave functions the set of coupled K-harmonic equations.

$$(6) \quad \left(-\frac{\hbar^2}{2\mathcal{M}} \left[\frac{\mathrm{d}^2}{\mathrm{d}\rho^2} - \frac{\mathcal{L}(\mathcal{L}+1)}{\rho^2} \right] + V_{\gamma'\gamma'}^{J}(\rho) - E \right) \chi_{\gamma\gamma'}^{J}(\rho) = -\sum_{\gamma'' \neq \gamma'} V_{\gamma'\gamma''}^{J}(\rho) \, \chi_{\gamma\gamma''}^{J}(\rho),$$

where $\mathcal{L} = K + 3/2$ and matrix elements of interactions are defined the following way:

$$(7) \quad V_{\gamma'\gamma}^{J}(\rho) = \left\langle \left[\mathcal{Y}_{K'L'}^{l_x' l_y'}(\Omega_5^{\rho}) \otimes [\chi_{\frac{1}{2}} \otimes \chi_{\frac{1}{2}}]_{S'} \right]_{JM} \middle| \hat{V} \middle| \left[\mathcal{Y}_{KL}^{l_x l_y}(\Omega_5^{\rho}) \otimes [\chi_{\frac{1}{2}} \otimes \chi_{\frac{1}{2}}]_{S} \right]_{JM} \right\rangle.$$

The asymptotic behavior for a Borromean three-body halo system in the hyper radius, is similar to that of a two-body system in the binary separation, and for a bound state determined by the binding energy,

$$(8) \qquad \chi_\gamma^J(\rho \to 0) \sim \rho^{K+5/2}, \quad \chi_\gamma^J(\rho \to \infty) \sim \exp[-\kappa\rho].$$

For continuum wave functions $\chi_{\gamma\gamma'}^J(\rho)$ the boundary condition at the origin is the same as for bound states, while at $\rho \to \infty$ we have

$$(9) \qquad \chi_{\gamma\gamma'}^J(\rho \to \infty) \sim \rho^{1/2} \left[H_{K+2}^{(-)}(\kappa\rho)\, \delta_{\gamma,\gamma'} - S_{\gamma,\gamma'}^J\, H_{K'+2}^{(+)}(\kappa\rho) \right],$$

where $H_n^{(\pm)}$ are the Hankel functions of integer order ($n = K + 2$), that have asymptotic behaviour $\sim \rho^{-1/2}\exp[\pm\imath\kappa\rho]$ and describe the three-body outgoing and ingoing spherical waves. $S_{\gamma,\gamma'}^J$ is a S-matrix for $3 \to 3$ scattering when the 6-dimensional plane wave is an ingoing wave in the channel γ. Solutions of the system of eqs. (6) with boundary conditions that are appropriate for bound and continuum states describe the dynamics of the three-body halo systems.

Neglect of explicit considerations for internal core degrees of freedom is the main approximation in (1). These effects are treated approximately through the effective nucleon-core interaction V_{Cn_i}. Such factorization is a starting point for application of three-body models to description of halo structure [1]. Few-body models avoid the complicated and still partly open questions concerning the development of nuclear clustering, and instead calculate the halo wave function $\psi(\mathbf{x},\,\mathbf{y})$ directly. Within such models, extended for an approximate treatment of the Pauli principle, it is possible to give a consistent description of the main properties of both the ground state and the low-energy continuum wave functions of halo nuclei.

2˙1. *Spatial continuum correlations.* – The developments of the hyperspherical harmonics method have deepened and enriched the understanding of the Borromean three-body continuum. The method gives a unique formulation of the three-body in- and outgoing spherical waves, which was derived from Faddeev equations in the 70 s [9]. Using the example of ^{6}He as a testbench, we have analyzed several interesting continuum structures using both the Jacobian cluster $\mathbf{T}$-basis and the $\mathbf{Y}$-basis, and spatial correlated densities.

The 2_1^+ state of ^{6}He exhibits the characteristic features of a "true" three-body Borromean resonance: i) a narrow width, and ii) a strong concentration of the Wave Function (WF) in the internal region where all three particles are interacting close to each other. As a consequence, the hyperradial WF in the interaction region (see fig. 2) has a characteristic resonance amplification, for 2_1^+ three orders of magnitude larger than the Antisymmetrized three-body Plane Wave (APW) amplitude [10], and with an energy dependence which coincides with the energy dependence of the three-body scattering amplitude.

The 2_2^+ and 1_1^+ resonant states in ^{6}He (not shown) are wide above-barrier resonances, produced by the generally attractive averaged interactions, but with no pronounced

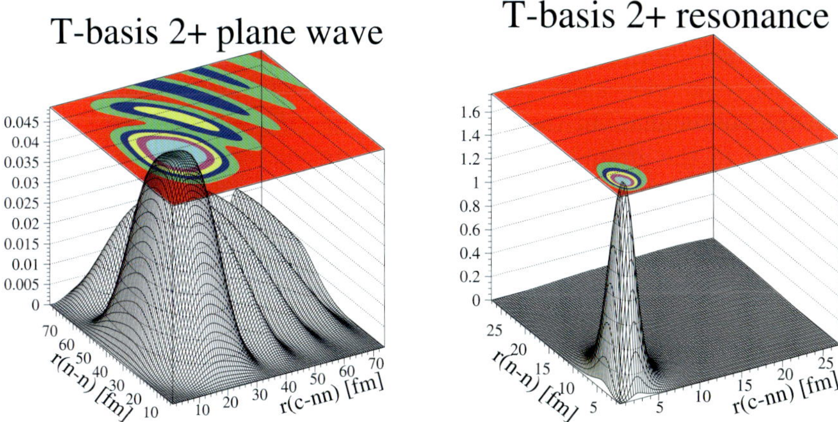

Fig. 2. – Spatial correlations for the 2^+ partial component of an APW (left) and the 2^+_1 resonance calculated at resonant energy in **T** coordinate system. Very similar correlations are found in the translation invariant "shell model" **Y** coordinate system.

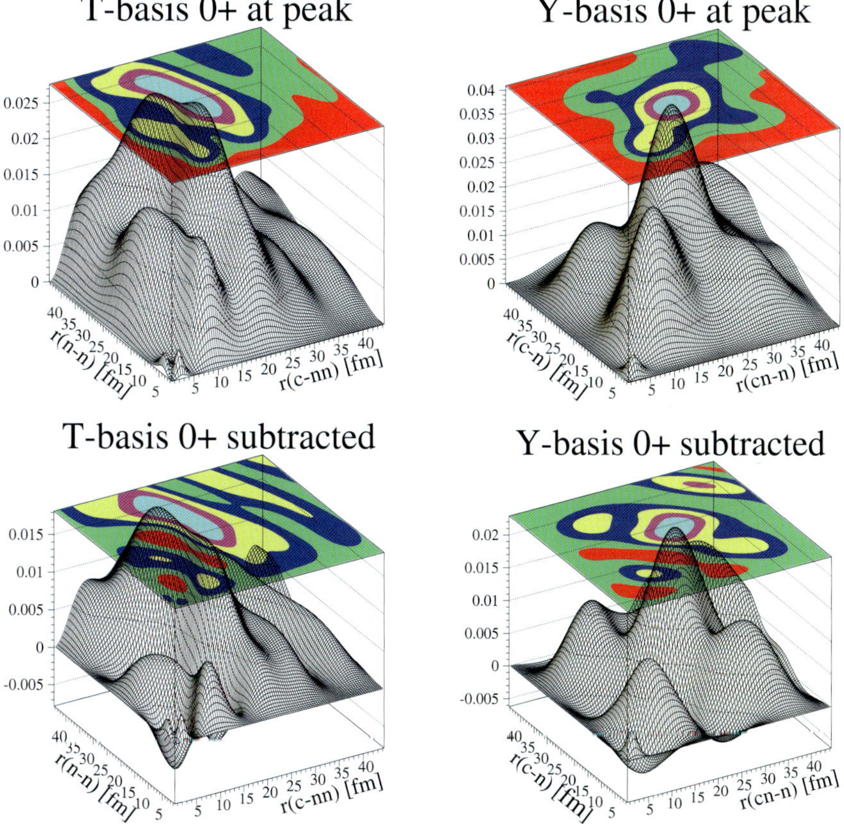

Fig. 3. – Spatial correlations for the energy peak position of the 0^+ soft monopole mode (upper row) and after subtraction of the APW density (lower row) in **T** and **Y** systems.

concentration of density in the interior region even after filtering out the non-resonant background coming from plane waves.

We also find some wide above-barrier resonance candidates which are completely spread into the continuum. In the 0^+ continuum of ^{6}He, at least two states should exist, orthogonal to the ground state: the soft monopole breathing and the spin-flip modes [11]. Our previous analysis in ref. [11] showed non-resonant behavior, although the response function exhibits a wide peak at 2.5 MeV. The correlated spatial density for 0^+ at this energy (see fig. 3) has a peaked structure in the region inside 30 fm, but after subtracting the density for plane waves, only half the density remains, revealing long-range correlations, but no concentration of density in the internal region. We can compare with the remnant at the origin of the halo ground state "di-neutron" and "cigar" configurations in the correlated spatial density for the ground-state structure (fig. 1). Perhaps this is a characteristic feature of any monopole three-body continuum, where there is a possibility for s-wave decay via the virtual s-wave state of the two halo neutrons. This important mode could give valuable information about the compression modulus of a dilute "neutron gas" which could be associated with the halo neutrons, and, in principle, could be extracted from the energy position of the monopole breathing mode.

There are also peaks in the response functions for soft modes, which have tentatively been related to large induced multipole moments. In ref. [11] we analyzed the 1^- soft dipole mode in ^{6}He and found no resonance but a virtual-state–like behavior of eigenphases, along with a bump in dipole strength functions. The corresponding 1^- correlated spatial density at the peak position shows, after subtraction of the plane wave density, that only a quarter of the dipole density remains, along with long-range correlations (at about 30 fm) analogous to the monopole continuum structure. A small peak in the interior region (at about 4 fm) resembles the above-barrier resonances (like in the 2_2^+ and 1_1^+ cases), but is not enough pronounced to justify classification as a resonant state. The monopole and dipole continuum of ^{6}He illustrate specific states which may be classified as Efimov-like continuum structures near the three-body threshold. In physical analogy with the bound-state Efimov effect, the origin of these states is the long-range effective three-body interactions with a range of the order of the sum of the scattering lengths in the binary subsystems. There is then a large correlation distance, and no concentration of the WF inside the region of interaction of all particles, only a long-range spreading of correlations. However, we need further exploration of these states in terms of energy correlations in the three-body continuum to distinguish between the decay through resonances in two-body subsystems and Efimov-like states.

$2{\cdot}2$. *Energy correlations*. – The uncertainty principle and halo structure are mostly responsible for the momenta (energy) correlations in the ground state and could be reflected in the specific conditions in transverse and longitudinal momenta distributions in high-energy fragmentation reactions (see fig. 4). Our analysis of ^{6}He in terms of Jacobi coordinates, displays all main sources of amplification of the continuum cross-sections: i) true three-body resonances, which are due to interaction of *all* three particles in the interior domain; ii) long-lived binary resonance in one of the pairs; iii) response of an

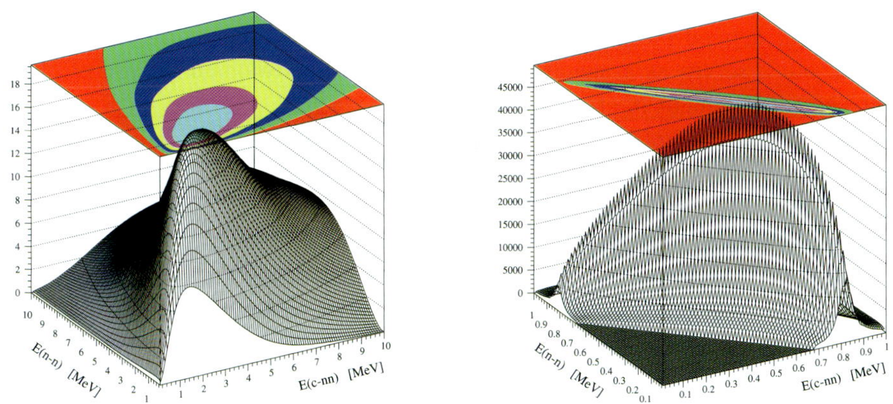

Fig. 4. – Energy correlation plot for 2^+ states in cluster **T** basis: left: NoFSI, right: with FSI (2_1^+ resonance).

extended system to the long-range transition operators.

There are three types of correlations: the intrinsic correlated structure of the three-body continuum itself, the correlations in the ground state only (or No final-state interactions), and the transition energy correlations with *all Final State Interactions* (FSI) included. The first would require $3 \rightarrow 3$ scattering. The second (NoFSI) is used in sudden approximation (or to some extent in the Serber break-up model), while the third (FSI) is probed by nuclear and electromagnetic responses which mix properties of the previous two and where the ground state still may play a very visible role. Only combined analysis of all three gives a complete understanding of the three-body continuum structure.

To characterize the relative motion of the three fragments, we need two relative Jacobi momenta: $\mathbf{k}_x$ between two constituents and $\mathbf{k}_y$ between the third constituent and the center of mass of the first pair. The excitation energy E_κ of the halo nucleus above threshold is equal to the sum of kinetic energies for the fragments' relative motion, $E_\kappa = \varepsilon_x + \varepsilon_y = \mathbf{k}_x^2/2\mu_x + \mathbf{k}_y^2/2\mu_y$. The energy E_κ and the three-body phase space $\sqrt{\varepsilon_x \varepsilon_y}\,\mathrm{d}\varepsilon_x \mathrm{d}\varepsilon_y$ are invariant, *i.e.* independent of the Jacobi system. We also introduce the variable $\epsilon = \varepsilon_x/E_\kappa$ ($\varepsilon_y/E_\kappa = 1 - \epsilon$) which describes the share of the relative kinetic energy residing within a pair of particles (or the third particle and the center mass of the pair), at three-body continuum energy E_κ.

For sharp three-body resonances the S-matrix poles coincide with peak energy in resonant amplification of the interior part of the wave function, and intrinsic energy (momentum) correlations for 3-3 scattering almost coincide with transition energy (momentum) correlations for reactions. The most remarkable feature of the "true" three-body resonance, exhibited by ^{6}He, is that it exists in lowest hypermoment configurations, which corresponds to the three particles interacting close to each other. It manifests itself as pockets in the diagonal hyperpotential terms and sharp crossing of $\pi/2$ in eigenphases.

The main criterion for existence and properties of any intrinsic resonant states, that they should *not* depend on the excitation mechanism (electromagnetic, strong or weak interaction, etc.), serves as a signature of the three-body resonant nature of observed resonant-like enhancement in cross-sections.

Resonance properties of transition amplitudes are revealed in the structure of the energy correlation functions, most pronounced for the 2_1^+ resonance in ^{6}He (see the maxima ridge in the right panel of fig. 4) and less for the wider 2_2^+ and 1_1^+ resonances. Besides "three-body virtual excitations" could exist, produced by low-lying resonances (or a virtual state) in the binary subsystems. Their pronounced inherent characteristic is a very long range of formation, involving an enormously large number of configurations in an HH basis, lack of pockets in the diagonal potential terms, and strong off-diagonal (coupling) terms. Our exploration of the binary energy correlations (n-n and α-n) in ^{6}He has shown complex interplay of the s-wave $n - n$ virtual state and the $p_{3/2}$ resonance in α-n (^{5}He) both for monopole and dipole soft modes. These correlations differ strongly from "free" binary correlations; we cannot suppose only binary FSI for decay into the three-body continuum; the presence of the third body in FSI is decisive. Three-body correlations for all configurations have demonstrated that, in spite of the presence of a soft dipole resonance-like peak in both nuclear and electromagnetic response functions at $\sim 2.5\,\mathrm{MeV}$, there is lack of noticeable true resonant behavior in the interior region. The same applies for the monopole case.

3. – Physics of the Borromean continuum; correlations of break-up fragments

Fragmentation reactions have complicated dynamics where nuclear structure and reaction mechanism are tightly intertwined. Our discussion will be confined by dissociation reactions with undestroyed core leading to the low-energy halo excitations. The exact transition matrix T_{fi} in the prior form can be written as

$$(10) \qquad T_{fi} = \langle \Psi^{(-)}(\boldsymbol{k}_x, \boldsymbol{k}_y, \boldsymbol{k}_f) \mid \sum_{p,t} V_{pt} - U_{aA} \mid \Psi_0,\ \Phi_A,\ \chi_i^{(+)}(\boldsymbol{k}_i) \rangle,$$

where $\Psi^{(-)}$ is the full scattering solution with ingoing wave boundary condition, Ψ_0 and Φ_A are ground-state wave functions of the halo projectile and the target, respectively. The distorted wave $\chi_i^{(+)}$ describing the relative motion of nuclei in the initial channel is a solution of the Schrödinger equation with optical potential U_{aA}. V_{pt} is the NN interaction between the projectile and target nucleons. Due to the translational invariance only relative momenta can characterize reaction dynamics. In expression (10) $\boldsymbol{k}_x$, $\boldsymbol{k}_y$ and $\boldsymbol{k}_{i,f}$ are momenta between a pair of fragments, between the center mass of a pair and the third fragment, between the center masses of halo and target nuclei in the initial and final channels, respectively. The exact T-matrix (10) cannot be calculated without approximations. Since our main goal is a study of the halo structure, the reaction is considered at conditions that allow a simplified treatment of the reaction mechanism making it both tractable and transparent. Hence, we study the collisions at high enough energy (large momenta $\boldsymbol{k}_i$ and $\boldsymbol{k}_f$) leading to the low-energy halo excitations when one-step processes dominate. The low-energy halo excitations correspond to the small values of relative momenta $\boldsymbol{k}_x$ and $\boldsymbol{k}_y$. Therefore the halo fragments spend some time together and interact between themselves. There are no spectators and all fragments are participants. At such conditions, a distorted wave treatment of reaction dynamics can be used and the exact scattering wave function $\Psi^{(-)}(\boldsymbol{k}_x, \boldsymbol{k}_y, \boldsymbol{k}_f)$ can be written as a product of wave functions: projectile, target and their relative motion. Then reaction amplitude (10) reduces to the next expression

$$(11) \qquad T_{fi} = \langle \chi_f^{(-)}(\boldsymbol{k}_f),\ \Phi_A,\ \Psi^{(-)}(\boldsymbol{k}_x, \boldsymbol{k}_y) \,|\, \sum_{p,t} V_{pt} |\, \Psi_0,\ \Phi_A,\ \chi_i^{(+)}(\boldsymbol{k}_i) \rangle,$$

where $\chi_f^{(-)}(\boldsymbol{k}_f)$ is the distorted wave describing relative motion of nuclei in the final channel, $\Psi^{(-)}(\boldsymbol{k}_x, \boldsymbol{k}_y)$ is the halo three-body continuum wave function. In calculation of $\Psi^{(-)}(\boldsymbol{k}_x, \boldsymbol{k}_y)$ all fragment pairwise interactions must be taken into account or, in other words, final-state interactions must be included in the full scale. The term with optical potential U_{aA} does not give a contribution to the T-matrix (11) due to the orthogonality between halo bound Ψ_0 and continuum $\Psi^{(-)}(\boldsymbol{k}_x, \boldsymbol{k}_y)$ wave functions. This low-energy region of nuclear excitations is the most sensitive to the three-body correlations and, consequently, the most useful for a study of halo structure. Such investigations require special selection of experimental data. It can be done by performing kinematically complete experiments when energies and momenta of all halo fragments are measured in coincidence. Then it is possible to restore an energy spectrum of the halo nucleus and select only events corresponding to the low-energy excitations. Simultaneously a variety of different energy and angular correlations become available and the possibility to describe them within the same model is a thorough test for our understanding of nuclear structure and reaction dynamics. The reaction model based on approximation (11) was applied successfully for description of many reactions with halo nuclei.

Recent experimental GSI data [12] on different angular and energy correlations of the three fragments from break-up of ^{6}He on lead target at collision energy 240 MeV/nucleon reveal a very interesting picture. The *low-energy spectrum* of ^{6}He shows a smooth behaviour, while with increasing excitation energy [13], the shape of some *correlations* changes dramatically along the spectrum.

Our theoretical analysis of the ^{6}He low-energy spectrum within the four-body microscopic distorted wave theory was reported in [14]. Figure 5 repeats the comparison of the calculated ^{6}He excitation spectrum (thick solid line) with experimental data. All necessary details about the nuclear structure and the reaction model used in the theoretical calculations can be found in [14]. The thin solid, dashed and dotted lines show the dipole 1^-, quadrupole 2^+, and monopole 0^+ contributions, respectively. The calculations reproduce both the shape and absolute value of the inclusive spectrum and display a small monopole contribution. The dipole dominates and the three-body 2^+ resonance at 1.8 MeV is strongly excited. Two energy regions have been singled out [12] in the ^{6}He excitation spectrum: $1 < E_\kappa < 3$ MeV and $3 < E_\kappa < 6$ MeV. To increase the

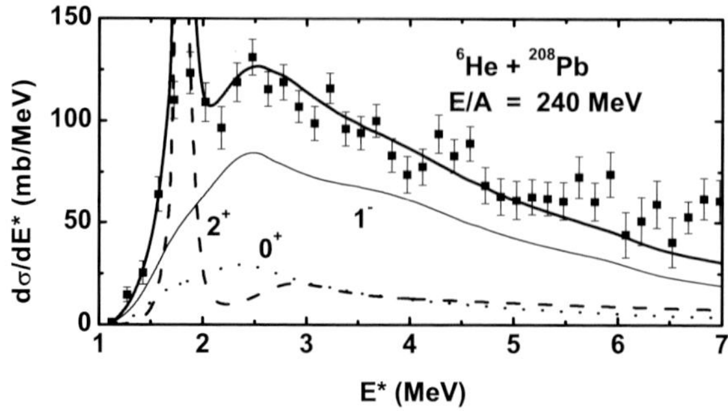

Fig. 5. – Comparison of the theoretical ^{6}He excitation spectrum (thick solid line) for ^{6}He$+^{208}$Pb break-up at 240 MeV/nucleon with experimental data. The thin solid, dashed and dotted lines show the dipole 1^{-}, quadrupole 2^{+}, and monopole 0^{+} contributions. Threshold is at 1 MeV.

sensitivity to the spectroscopic content of the continuum excitations, we calculate the modified energy correlations between fragments (denoted by W_{ph} in the figures) where the phase space energy factor $\sqrt{\epsilon(1-\epsilon)}$ has been divided out. The experimental data from [12] for energy correlations were scaled accordingly. Figure 6 for the energy interval $1 < E_{\kappa} < 3$ MeV suggests that our model assumptions on reaction mechanism and nuclear structure are correct. The situation in the next energy bin $3 < E_{\kappa} < 6$ MeV is, however, more perplexing. The angular correlation in the $\mathbf{T}$ system and for energy in $\mathbf{Y}$ now deviate strongly from experiment. We have not succeeded in attempts to improve the theoretical description of the experimental correlations by modifying the effective pair-wise interactions without destroying the good agreement obtained for excitations near threshold.

In the few-body cluster approach to inelastic excitation and inelastic break-up, transition densities are calculated with form factors in momentum space (as well as ordinary space), generated from a consistent set of hyper harmonic wave functions. Such quantities could in principle also be calculated in ab $initio$ formulations, at least estimates for transition strengths. For Borromean systems, these processes take us beyond the 3-body threshold to 3-body true resonances or more exotic continuum structures, followed by breakup of the halo structure and 4-body (or more) final reaction channels. For transfer processes involving Borromean systems alluded to above, a first step in ab $initio$ approaches would be to try to estimate spectroscopic factors; a more ambitious next step would be overlap functions where also the radial shapes of the transfer functions are calculated.

4. – Many-body ab $initio$ approaches

Reactions have hitherto exclusively been initiated as binary encounters between a halo projectile and a target. For Borromean projectiles, this may, as we have seen, lead to break-up into a 4-body final channel $A + C + n + n$ composed of the halo constituents and the target, where A and C also in general may become excited. Alternatively, halo

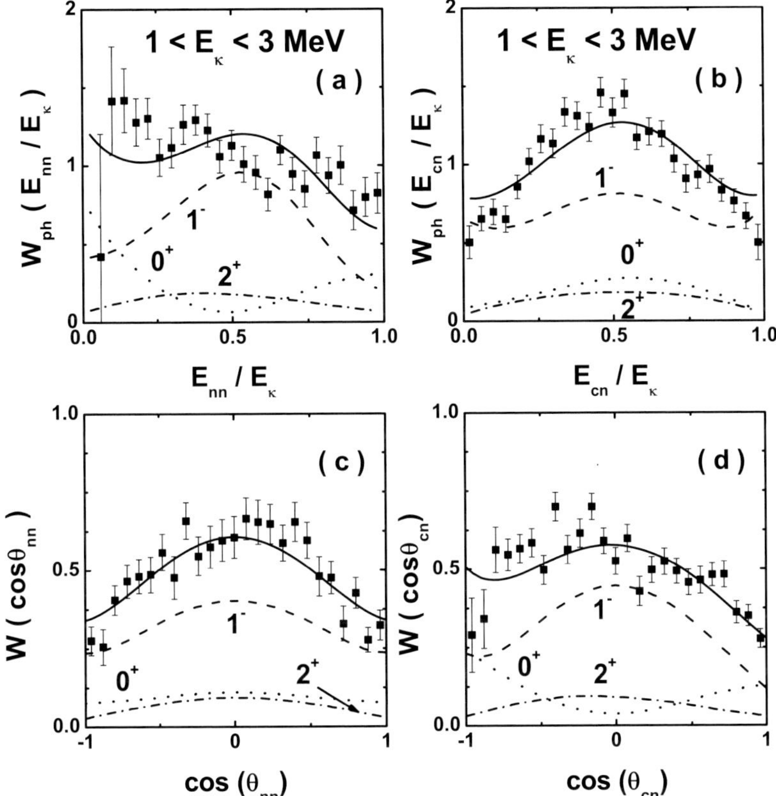

Fig. 6. – Energy ((a) and (b)) and angular ((c) and (d)) fragment correlations (solid line) in the ^{6}He + ^{208}Pb break-up at 240 MeV/nucleon for continuum energy region $1 < E_\kappa < 3$ MeV. (a) and (c) are shown in Jacobi configuration **T**, (b) and (d) in configuration **Y**. The dashed, dotted and dash-dotted lines show the dipole 1^-, monopole 0^+, and quadrupole 2^+ contributions, respectively.

constituents may be transferred to the target. A Borromean one-cluster transfer will imply a 3-body final channel $[A + n] + C + n$ or $[A + C] + n + n$ while a two-cluster transfer produces 2-body final channels. This picture is, however, deceptively simple. In general the transferred $[A + n]$ neutron may not be a halo neutron but may stem from the core C, depending on the collision energy. Hence, complete experiments are called for, where all exit fragments are recorded.

It is obvious that the few-body cluster picture is tailored to the reaction phenomena; this was historically also how the halo degrees of freedom were discovered. Here it has, however, to be kept in mind that the 3-body model, say $[\alpha + n + n]$ for ^{6}He is only "clean" up to the 2-body $[t + t]$ triton threshold at roughly 10 MeV.

We have demonstrated how the cluster picture is useful to get a handle on Borromean break-up via halo excitation. In this Fermi School we have also learned about attempts being made to address the reaction problem as an *ab initio* many-body problem in terms of a consistent use of realistic nucleon-nucleon interactions. Consistency was also attempted in the cluster formulations, but is not a straightforward task. We may ask about the challenges and hopes for a consistent many-body formulation.

The classical shell-model–type approach to structure description of a (sufficiently heavy) nucleus composed of a core surrounded by nucleons, is to employ expansions on antisymmetrized Slater-type wave functions, *i.e.* in terms of products of single-particle states corresponding to nucleons moving in the field of the core. This approach has also been tried for Borromean halo nuclei such as ^{6}He. A main challenge to this approach in this case is the Borromean nature implying no binary bound orbits between core (alpha) and neutron, only a few physical resonances/antibound states and a background continuum. The alpha particle is also not "sufficiently heavy" for core-recoil effects to be neglected. A way out are the so-called Berggren expansions, which also have been invoked in full *ab initio* approaches where, in principle, all nucleon degrees of freedom are included. It may be of interest to have a glimpse of how this has come about.

4˙1. *Historical perspective on the Berggren expansion.* – In the last decade progress has been made in understanding nuclear self-organization phenomena in terms of nucleonic degrees of freedom, in the spirit of Berggren completeness relations including binary resonances and background continuum. Borromean nuclei amplify the need for a complete approach. It is thus instructive to recall some of the historical development, following Berggren's work as an example.

In 1965 two pioneering papers on nuclear overlap functions and their role in direct nuclear reactions were published, one by Tore Berggren [15] and one by Pinkston and Satchler [16]. Berggren's paper "Overlap Integrals and Single-Particle Wave Functions in Direct Interaction Theories" was followed up two years later [17] by the paper "On the Use of Resonant States in Eigenfunction Expansions of Scattering and Reaction Amplitudes", opening up theoretical exploration of resonant final states, a field Berggren studied in depth in the next decade. The 1985 Physics Reports [18] "One- and Two-Nucleon Overlap Functions in Nuclear Physics" reviewed these vital quantities for nuclear reactions, and their basis in associated Berggren-Pinkston-Satchler (BPS) equations, with particular focus on asymptotic behaviour.

In 1965 direct light-ion–induced reactions such as (d, p), (p, d), (p, pn), and $(p, 2p)$ were in focus, and Berggren paid in particular attention to the latter, high-energy reactions $p+A \rightarrow B+p+p$, where a proton is knocked out by a beam with energy typically $100\,\mathrm{MeV}$, leaving B in a bound state. Ten years later the paper [19] concludes work to include final nuclei B which also may be in unbound states decaying into $C + n$ (C being an alpha-particle in the example), seeking a role for Gamow states to describe the exit channel.

The 1965 papers [15, 16] provided justification for the separation energy recipe (well depth method), used by practitioners to extract spectroscopic factors from direct transfer reactions. The argument was linked to the asymptotic tail behaviour of nuclear overlap functions $\langle x, B|A \rangle$, connecting bound states of nuclei A and $B = A - 1$, which for a neutron were shown to decay exponentially with just a rate determined by the separation energy $E(A, B) = E(B) - E(A)$ for extracting the neutron. The overlap functions do not fulfil any simple completeness relations, and are also in general not ortho-normal. Instead they satisfy a generalized completeness relation which includes center-of-mass (recoil) effects. Berggren also mentions cluster and multi-nucleon overlaps like $\langle x, x', C | A \rangle$, discussed in detail in [18]. Challenged by reactions to unbound final states, he moves on to this very important issue, which also involves overlaps, but with unbound final states. The 1978 paper puts Berggren's approach using Gamow states together, again with applications to the break-up reaction

$$(12) \qquad\qquad p + A \to B + p + p \mapsto C + n + p + p$$

treating $B = C + n$ as a Gamow resonance. Berggren's replacement of final B-states by complex energy Gamow states is an alternative to finding overlaps as solution to a non-Hermitian Hamilton operator in a BPS-like equation.

Berggren's work inspired a whole generation of reaction theorists, physicists who believed that reactions were far more than only tools for extracting spectroscopic bound-state information. Berggren's work inspires nowadays a new generation of physicists who believe that nuclei (organized hadronic matter) should be studied within one framework, structure and reactions—as potentially open systems. See ref. [20] as an example of application of Berggren expansion to a Borromean system.

4‎$\cdot$2. *Modern ab initio approaches.* – In *ab initio* many-body approaches to nuclear structure and reactions, one aims at a description of the nucleus as a whole starting from nucleon degrees of freedom and their mutual interactions. There are various methods that can provide virtually exact calculations of light nuclei. These are the Faddeev [21], Hyperspherical Harmonics [22,23], Greens-Function Monte Carlo [24] and No-Core-shell-model [25] approaches, to mention a few. Starting from any realistic interaction which fits two-nucleon phase shifts, these *ab initio* calculations give consistent underbinding for light nuclei, and have therefore revealed the need for a three-body force (3NF) in order to approach the experimental mass values. The appearance of 3NFs is not surprising since nucleons are not point particles, and by neglecting internal degrees of freedom three- and many-body forces are naturally induced.

One of the frontiers in nuclear structure today, therefore, concerns how one can consistently relate 3NFs to a given two-body interaction. Recently, highly accurate NN-interactions have been derived from Chiral perturbation theory [26-29]. Chiral Effective Field theory (EFT) starts from an effective Lagrangian consistent with the symmetries of QCD. Expanding the nuclear amplitude in powers of pions and/or a typical nucleon momentum over the chiral symmetry break-down scale, a perturbative series is obtained where NN forces, 3NFs and forces of higher rank appear systematically at a given order. This way, Chiral EFT gives a systematic and consistent link between nuclear structure and QCD, and further it also accounts for the natural hierarchy of forces, *i.e.* $V_{2N} > V_{3N} > V_{4N} \dots$.

So the modern understanding is that all realistic interactions are effective interactions. The effective Hamiltonian thus depends on a resolution scale Λ defining the energy scale at which properties are probed. Generally speaking, any two-body interaction induces three-body interactions when high-momentum degrees of freedom are integrated out. The dependence of observables on the resolution scale provides information on the physics which is missing, *i.e.* forces of higher rank induced by the renormalization [30]. In a fully renormalized theory, there will be a range of Λ in which observables are Λ independent. It is the hope that inclusion of effective 3NFs will be sufficient to approximately renormalize the many-body system.

The importance of 3NFs in spectra and properties of light nuclei is well established. But what is the role of 3NFs as one moves into the medium mass regime of the nuclear chart, and at the extremes of isospin? In order to answer this question, one needs a many-body method capable of an accurate description of medium mass nuclei and of nuclei located at the driplines, defining the limits of matter. A good candidate for such a method is the framework known as Coupled Cluster theory (CC).

Coupled Cluster theory is a method which has recently been revived in the nuclear-structure community. Since its development in the late 50 s with Coester and Kümmel [31, 32], it has mostly seen applications in the quantum chemistry community, and there it now defines the state-of-the-art theory for an accurate description of many-fermion systems. CC is a fully microscopic theory, and is an ideal compromise between accuracy on the one hand and computational cost on the other. Further, it is a theory which is *size-extensive*. Size-extensive theories ensure that the energy of the system scales correctly with an increasing number of particles. For non–size-extensive theories, like particle-hole truncated shell model approaches, one can not judge the quality of a calculation of large systems from the quality of a calculation of small systems. For these reasons, CC is very well suited for extending the *ab initio* program into the medium mass regime of the nuclear chart.

4'3. *Elements of Coupled Cluster theory*. – In CC one writes the exact many-body wave function as *the exponential ansatz*,

$$(13) \qquad |\Psi\rangle = e^T |\Phi_0\rangle .$$

$|\Phi_0\rangle$ is here an uncorrelated reference Slater determinant constructed from single-particle orbitals. It may be either the Hartree-Fock state or a naive filling of the lowest harmonic oscillator orbitals. The many-body correlations are introduced by the cluster operator T,

$$T = T_1 + T_2 + \ldots + T_A.$$

T is a linear combination of particle-hole excitation operators, where the n-particle n-hole excitation operator is given by

$$(14) \qquad T_n = \sum_{i_1 \ldots i_n a_1 \ldots a_n} t^{a_1 \ldots a_n}_{i_1 \ldots i_n} a^\dagger_{a_1} \ldots a^\dagger_{a_n} a_{i_n} \ldots a_{i_1} .$$

In the equation above, and in everything that follows, we employ the convention that indices $i, j, k, \ldots$ represent occupied orbitals in the reference determinant while $a, b, c, \ldots$ represent unoccupied states above the Fermi energy.

Inserting the exponential Ansatz in the time-independent Schrödinger equation, and multiplying with e^{-T}, we get

$$(15) \qquad e^{-T} H_N e^T \,|\Phi_0\rangle \equiv \bar{H}_N \,|\Phi_0\rangle = (E - E_0)\,|\Phi_0\rangle\,.$$

Here H_N denotes the normal-ordered Hamiltonian, while E_0 is the reference determinant energy. This is an eigenproblem for the *similarity transformed* Hamiltonian $\bar{H}_N$, or equivalently, we can interpret this as the conditions for a similarity transformation of H_N.

$\bar{H}_N$ can be rewritten with the Hausdorff expansion([1])

$$(16) \qquad \bar{H}_N = H_N + [H_N, T] + \frac{1}{2}\,[[H_N, T], T] + \frac{1}{3!}\,[[[H_N, T], T], T]$$
$$+ \frac{1}{4!}\,[[[[H_N, T], T], T], T]\,.$$

This form of the similarity transformation implies that the only terms which can contribute are those where the cluster operators are "connected" with the Hamiltonian. That is, if a term contains a cluster operator with no index in common with the Hamiltonian, the term vanishes. We may then write

$$(17) \qquad \bar{H}_N = \left(H_N e^T\right)_C,$$

where we have used the subscript C to indicate that only the connected terms should contribute. If the Hamiltonian is expressed with at most two- or three-body operators, the Hausdorff expansion will terminate naturally at the term with four or six commutators, respectively. The fact that only connected terms enter the CC equations ensures that the CC wave function is fully linked. This is what makes CC a *size extensive* theory.

By projecting eq. (15) on $\langle\Phi_0|$, we get the energy equation

$$(18) \qquad \langle\Phi_0|\bar{H}_N|\Phi_0\rangle = \langle\Phi_0|(H_N e^T)_C|\Phi_0\rangle = E - E_0.$$

To carry out the similarity transformation, we need to find the coefficients $t^{ab\ldots}_{ij\ldots}$. Equations for the coefficients are found by projecting on the excited states $\langle\Phi^{ab\ldots}_{ij\ldots}|$,

$$(19) \qquad \langle\Phi^{ab\ldots}_{ij\ldots}|\bar{H}|\Phi_0\rangle = \langle\Phi^{ab\ldots}_{ij\ldots}|\left(He^T\right)_C|\Phi_0\rangle = 0.$$

([1]) For a three-body Hamiltonian, the expansion would also include the terms

$$\frac{1}{5!}[[[[[H, T], T], T], T], T] + \frac{1}{6!}[[[[[[H, T], T], T], T], T], T].$$

The last term will, however, vanish in the CCSD approximation.

The further steps in a CC calculation involves diagrammatic techniques for expressing the left-hand sides of eqs. (18), (19). Once the diagrammatic expressions have been found and interpreted algebraically, the equations can be solved iteratively. A thorough description of the full procedure is available in [33].

What we have described so far is an exact calculation which would be just as expensive as a full diagonalization. The approximation employed in CC is to truncate T so that it only includes n-hole n-particle excitations up to a certain n. It should be pointed out that this is the *only* approximation needed in the CC formalism. Truncating to include only single and double particle-hole excitations, is known as Coupled-Cluster Singles and Doubles (CCSD). The inclusion of higher-order clusters is called CCSDT or even CCSDTQ (Triple and Quadruple excitations). The computational cost of CCSD scales as $\sim n_o^2 n_u^4$, while CCSDT scales as $\sim n_o^3 n_u^5$, (see below for n_o and n_u). Both of these are favourable compared with the combinatorial scaling of a standard shell model approach.

Even if we truncate the cluster operator, the exponential ansatz implies that the CC wave function includes not only singly and doubly excited states, but also higher order excitations from terms like $T_1 T_2, T_2 T_2, \ldots$. In fact, the wave function (13) contains contributions from all possible excitations of the reference determinant through repeated applications of T_1 and T_2. However, despite of having contributions from all possible excitations, the wave function is uniquely determined by a modest number of coefficients, t_i^a and t_{ij}^{ab}. For a model space with N orbits, filled with n_o particles (occupied orbits) and n_u unoccupied orbits, the number of coefficients is $N_C = n_o n_u + n_o(n_o - 1)n_u(n_u - 1)$. Symmetry relations will reduce the required storage even further, but the important point is that the number of coefficients in CCSD scales as

$$N_C \sim n_o^2 n_u^2.$$

We may compare this with the number of all possible excitations,

$$N_{tot} = \binom{N}{n_o} = \frac{N!}{n_u! n_o!} \, ,$$

and it is clear that the number of CCSD coefficients is indeed modest for a wave function including all possible excitations([2]).

But, what is the difference between for instance a quadruple excitation $T_4|\Phi_0\rangle$ and a double excitation applied twice, $T_2 T_2|\Phi_0\rangle$? After all, both procedures lead to the same excited Slater determinant, $|\Phi_{ijkl}^{abcd}\rangle$. The answer lies in how the excitations are treated in the Coupled Cluster equations. The operator $T_2 T_2$ forms two separate excitation clusters. In the equations, the $T_2 T_2$ operator takes part in terms describing the interaction between particles belonging to different clusters. The internal interaction in each cluster is taken

---

([2]) Three-body clusters has been shown to contribute significantly when CC is used in nuclear physics [34]. In CCSDT, the number of coefficients will scale as $\sim n_o^3 n_u^3$. This is still a modest number compared with the total number of possible excitations.

care of elsewhere by a term including only one T_2-operator. In contrast to this, the T_4 cluster operator represents a single excitation cluster of four particles. In the equations it would bring up all internal interactions in the 4-body cluster.

However, for a reasonably good choice of reference determinant, one would expect the physical ground state to have a rather small contribution from Slater determinants with simultaneous excitation of four particles. Moreover, visualizing the nucleons as particles switching randomly between predefined orbits, it would be much more likely that a fourfold excitation is reached through multiple steps of double or single excitations, rather than a synchronized "jump" of four nucleons. The probability of having a quadruply excited state should therefore somehow be linked to the probabilities of exciting clusters with one, two, or three particles. Operators like $T_2 T_2$ mirror these considerations, and this explains how N_C coefficients can be sufficient to parametrize a good wave function which includes all possible excitations.

In a recent paper [34], practically converged binding energies were reported for ^{40}Ca. This result alone clearly demonstrates the capabilities of CC as a method for *ab initio* calculations. The calculations in the paper were done with CCSD, but various schemes for including triple excitations as a perturbative correction were also employed. Most notably the relatively inexpensive CCSD(T) scheme which includes terms from CCSDT which appear up to 5-th order perturbation theory. It was also demonstrated that to achieve good agreement with analytical Faddeev-Yakubovsky calculations for ^{3}He and ^{4}He, triples corrections are necessary. Now, the question remains to be answered: how well does CC work on the drip line?

4'4. *Coupled Cluster approach to open quantum systems*. – By combining Coupled Cluster calculations with techniques for treating nuclei with strong couplings to the continuum structure, one has the possibility to extend the domain of both theoretical frameworks. Moving away from the valley of beta-stability, one reaches the limits for where stable matter exists. These limits, where neutrons/protons start to drip off the nuclei, are called the proton/neutron driplines. Exotic phenomena start to emerge as one moves away from the valley of stability, along various isotope chains. These are extreme matter clusterizations, halo-densities, melting and reorganizing of shell structures and ground states embedded in the continuum to mention a few. These exotic phenomena can not be described in terms of a closed quantum system description, rather the system should be treated as an open quantum system. In an open quantum system formulation the particles can be excited and correlate through continuum degrees of freedom, this is in contrast to a closed quantum system where the particles are trapped in an infinitely deep well, see fig. 7 for an illustration. The Fermi level might even be unbound/unstable as is the case for, *e.g.*, ^{5}He.

In [35] the first *ab initio* calculation of decay widths and lifetimes of a whole isotopic chain were reported. The calculations were done with Coupled Cluster, and it was argued that CC is an ideal approach to the *ab initio* description of loosely bound and unbound nuclei. The ground states of the helium chain $^{3-10}$He were calculated within the CCSD approximation, and the results were fully converged with respect to the model space. In

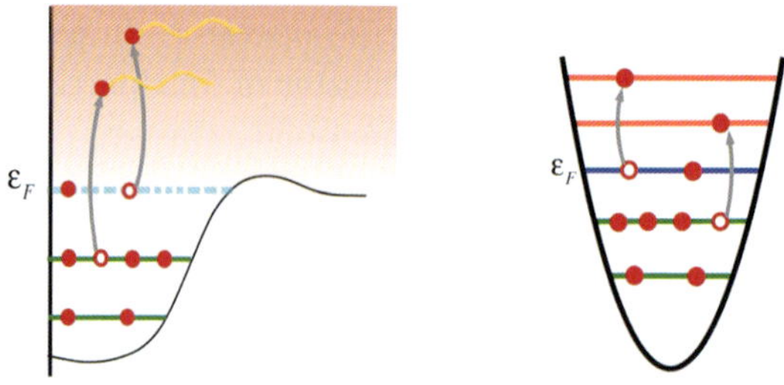

Fig. 7. – In an open quantum system (left) excitations to the continuum and the corresponding correlations should be treated correctly. For some nuclei, as depicted, the Fermi level is above the single particle threshold. In contrast, the particles in a closed quantum system (right) are trapped in a deep well.

order to correctly treat these nuclei as open-quantum systems, a Gamow-Hartree-Fock basis built from the N^3LO interaction model was used [36]. This basis is a Berggren basis, meaning that bound, resonant, and continuum states are treated on equal footing. To summarize, it was found that all calculated masses were consistently underbound with respect to the experimental mass values. However, the odd/even mass pattern was reproduced fairly well, and the calculated lifetimes were in semi-quantitative agreement

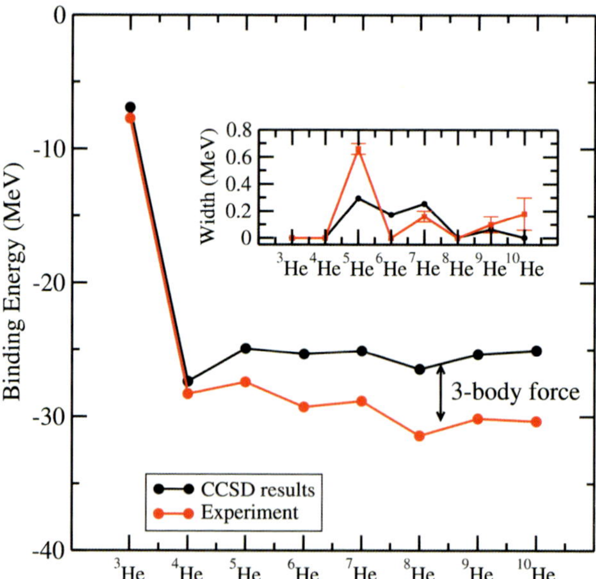

Fig. 8. – The binding energies for the helium chain reported in [35]. The nuclei are underbound, but the odd/even mass pattern is recognizable, and the lifetimes agree semi-quantitatively.

with experiment. The results are summarized in fig. 8, and, for example, it is seen that ^{5}He is unstable with respect to one-neutron decay, while ^{6}He is stable with respect to one-neutron decay but unstable with respect to two-neutron emission. ^{6}He was the only truly open-shell nucleus considered in [35], and therefore the most difficult to treat within single-reference Coupled-Cluster theory. It was argued that inclusion of full triples (CCSDT) and 3NFs in the calculation would significantly improve the results of ^{6}He.

The CC theory for a three-body Hamiltonian was addressed in another recent paper [37]. The presented results suggest that the main contribution from the three body forces can actually be expressed as density dependent zero-, one- and two-body terms. Those terms will only modify the corresponding two-body Hamiltonian, and will therefore not significantly increase the overall computational cost. This result, together with the promising results for the helium chain, could suggest that dripline *ab initio* CC calculations with a realistic, and consistent interaction might be just around the corner.

Full-scale CC calculations of drip-line nuclei will provide a solid fundament for exploring the possibility of an *ab initio* reaction formalism, and as a first step, we have started to investigate the calculation of spectroscopic factors within the CC formalism. The outcome of this is not yet clear, but the framework of CC seems to be flexible enough to be extended to such calculations.

5. – Future of many-body open quantum systems

The three-body break-up is an essentially richer but also more complicated process than binary break-up into two fragments. At fixed continuum energy, the relative motions of the three fragments have continuous distributions of kinetic energies, but it appears that at low excitation energies the leading physics for Borromean nuclei can be extracted from a few elementary modes characterized by a few orbital angular momenta. The task of continuum spectroscopy is to define the dominant excitation modes (multipolarities) and their quantum numbers (elementary modes). The way to achieve this is to explore the world of various correlations in fragment motions. This demands kinematically complete experiments and theoretical understanding of underlying reaction dynamics and nuclear structure. The first steps have now been taken in this direction.

Theoretical calculations reproduce quite well the low-lying excitation spectrum and fragment angular and energy correlations near break-up threshold. The fundamental fermionic nature of the halo neutrons is clearly exhibited by the correlations. While dipole dominates at most excitation energies other multipolarities can significantly distort the dipole correlation pictures and for a consistent analysis of experimental data all multipole excitations have to be taken into account. With increasing continuum energy ($3 < E_\kappa < 6\,\mathrm{MeV}$), some angular and energy correlations for ^{6}He are described within the theoretical model while some are not. Thus, also in the domain of high energies where the reaction mechanisms are simplest, we are presently left with some deficiency. The field is in its infancy and more experience has to be gained. The few-body study is being continued to ^{11}Li which exhibits strong s and p mixture.

The few-body approach also serves as a guide for *ab initio* formulations. We have briefly pointed to the promising Coupled Cluster method where the first results for reactions are expected in near future.

We have not addressed the low-energy reaction regime which is very challenging for loosely bound projectiles. Here reactions and structure are truly intertwined and coupled continuum channels required.

The way forward requires new ideas and new people, but the new generation of facilities such as FAIR open possibilities to bring the questions inside the laboratory, from theory to experiment.

$$* \ * \ *$$

One of the authors (JSV) thanks the organizer of this Fermi School for very instructive and pleasant days in Varenna. Two of the authors (BD and SE) are thankful to the University of Bergen for hospitality and also acknowledge support from RFBR Grants No. 05-02-17535 and 1885.2003.2, and Grant INTAS 03-54-6545. One of the authors (ØJ) thanks Oak Ridge National Laboratory for hospitality. This work was supported in part by the US Department of Energy under Contract Nos. DE-AC05-00OR22725 (Oak Ridge National Laboratory) and DE-FG02-96ER40963 (University of Tennessee).

REFERENCES

[1] Zhukov M. V., Danilin B. V., Fedorov D. V. et al., *Phys. Rep.*, **231** (1993) 151.

[2] Efimov V. M., *Phys. Lett. B*, **33** (1970) 563.

[3] Zhen Z. and Macek J., *Phys. Rev. A*, **38** (1988) 1193.

[4] Kraemer T. et al., *Nature*, **440** (2006) 315.

[5] Chichak K. S. et al., *Science*, **304** (2004) .

[6] Hiyama E., Kino Y. and Kamimura M., *Prog. Part. Nucl. Phys.*, **51** (2003) 223.

[7] Braaten E. and Hammer H.-W., *Phys. Rep.*, **428** (2006) 259.

[8] Oganessian Yu. Ts., Zagrebaev V. I. and Vaagen J. S., *Phys. Rev. Lett.*, **82** (1999) 4996.

[9] Merkuriev S. P., *Sov. J. Nucl. Phys.*, **19** (1974) 447.

[10] Danilin B. V., Rogde T., Vaagen J. S. et al., *Phys. Rev. C*, **69** (2004) 024609.

[11] Danilin B. V. and Zhukov M. V., *Yad. Fiz.*, **56** (1993) 67 (*Sov. J. Nucl. Phys.*, **56** (1993) 460); Danilin B. V., Thompson I. J., Vaagen J. S. and Zhukov M.V., *Nucl. Phys. A*, **632** (1998) 383.

[12] Chulkov L. V. et al., *Nucl. Phys. A*, **759** (2005) 23.

[13] Aumann T. et al., *Phys. Rev. C*, **59** (1999) 1252.

[14] Ershov S. N., Danilin B. V. and Vaagen J. S., *Phys. Rev. C*, **64** (2001) 064609.

[15] Berggren T., *Nucl. Phys.*, **72** (1965) 337.

[16] Pinkston W. T. and Satchler G. R., *Nucl. Phys.*, **72** (1965) 641.

[17] Berggren T., *Nucl. Phys. A*, **109** (1968) 265.

[18] Bang J. M., Gareev F. A., Pinkston W. T. and Vaagen J. S., *Phys. Rep.*, **125** (1985) 253.

[19] Ohlen G. and Berggren T., *Nucl. Phys. A*, **272** (1976) 21.

[20] Hagen G., Vaagen J. S. and Hjort-Jensen M., *J. Phys. A: Math. Gen.*, **37** (2004) 8991; Hagen G., Hjort-Jensen M. and Vaagen J. S., *Phys. Rev. C*, **71** (2005) 044314.

[21] Nogga A. et al., *Phys. Rev. C*, **65** (2002) 054003.

[22] Barnea N., Leidemann W. and Orlandini G., *Phys. Rev. C*, **61** (2000) 054001.

[23] Gazit D., Bacca S. *et al.*, *Phys. Rev. Lett.*, **96** (2006) 112301.

[24] Pieper S. C., *Nucl. Phys. A*, **751** (2005) 516.

[25] Navratil P., Vary J. P. and Barrett B. R., *Phys. Rev. Lett.*, **84** (2000) 5728.

[26] Bedaque P. F. and van Kolck U., *Annu. Rev. Nucl. Part. Sci.*, **52** (2002) 339.

[27] Epelbaum E., *Prog. Part. Nucl. Phys.*, **57** (2006) 654.

[28] Entem D. R. and Machleidt R., *Phys. Rev. C*, **68** (2003) 041001.

[29] Epelbaum E., Glockle W. and Meissner U.-G., *Nucl. Phys. A*, **747** (2005) 362.

[30] Bogner S. K., Schwenk A., Furnstahl R. J. and Nogga A., *Nucl. Phys. A*, **763** (2005) 59.

[31] Coester F., *Nucl. Phys.*, **7** (1958) 421.

[32] Coester F. and Kümmel H., *Nucl. Phys.*, **17** (1960) 477.

[33] Crawford T. D. and Schaefer H. F., *"An Introduction to Coupled Cluster Theory for Computational Chemists"*, *Rev. Comp. Chem.*, **14** (2000) 33.

[34] Hagen G., Dean D. J., Hjorth-Jensen M., Papenbrock T. and Schwenk A., *Phys. Rev. C*, **76** (2007) 044305, http://arxiv.org/abs/0707.1516v1.

[35] Hagen G., Dean D. J., Hjorth-Jensen M. and Papenbrock T., *Phys. Lett. B*, **656** (2007) 169.

[36] Entem D. R. and Machleidt R., *Phys. Lett. B*, **524** (2002) 93.

[37] Hagen G., Papenbrock T., Dean D. J., Schwenk A., Nogga A., Włoch M. and Piecuch P., *Phys. Rev. C*, **76** (2007) 034302.

DOI 10.3254/978-1-58603-885-4-261

Shell structure of exotic nuclei

TAKAHARU OTSUKA

Department of Physics, University of Tokyo - Hongo, Tokyo, 113-0033, Japan
Center for Nuclear Study, University of Tokyo - Hongo, Tokyo, 113-0033, Japan
RIKEN - Hirosawa, Wako-shi, Saitama
National Superconducting Cyclotron Laboratory, Michigan State University
East Lansing, MI, USA

Summary. — A basic introduction to the nuclear shell model is presented, with illustrative explanations of shell-model concepts and procedures. All details of many-body theories are avoided. First, we explain how magic numbers and shell structures appear from fundamental properties of nuclei such as the short-range attractive interaction and density saturation. Some concepts needed to understand the shell model are explained from scratch. The process of the shell model calculation is sketched also in a visual way. After the introduction, one of the recent topics, the evolution of shell structure in exotic nuclei, is briefly discussed, mentioning the importance of the tensor force.

1. – Basics of shell model

Nuclear theory has been developed in order to construct many-body systems from basic ingredients such as *nucleons* and *nuclear forces* (nucleon-nucleon interactions). The nuclear shell model has been an important part of nuclear theory, and should make crucial input to this end. We fbegin by asking three basic questions:

What is the shell model?

Why is it useful?

How do we perform calculations?

1‘1. *What is the shell model?* – We shall start with very basic points about nuclear shell structure and the shell model. Figure 1 shows somewhat schematically the nucleon-nucleon potential as a function of the distance between the two nucleons, for the spin-singlet (two interacting nucleons are coupled to total spin $S = 0$) and $L = 0$ (L: relative orbital angular momentum of the two nucleons) state. This state must have isospin $T = 1$ because of the antisymmetric coupling of the nucleons. This state is one of the most important states for the nucleon-nucleon potential, as the potential contains a strongly attractive part. Note that the nucleon-nucleon potential depends generally on S, L and J with $\vec{J} = \vec{L} + \vec{S}$. We find, in fig. 1, a hard-core repulsion inside a distance of $0.5\,\mathrm{fm}$, while it is strongly attractive around $1\,\mathrm{fm}$. These two features are found also in other important states, *e.g.*, of spin-triplet and even L states with $T = 0$. The potentials for such states are the origin of the proton-neutron binding in the deuteron.

From these two features, a hard repulsive core and a strong attraction around $1\,\mathrm{fm}$, we can easily expect that the nuclear potential is such that the balance between the attractive part around $1\,\mathrm{fm}$ and the inner repulsive part conspires to give a rather constant distance ($\sim 1\,\mathrm{fm}$) between nucleons, which are strongly bound together. The saturation of the density is thus realized at the same time. Although the actual mechanism contains more sophisticated dynamics —for instance, the density dependence of the potential— we will not go into such details. As the nucleon density should be rather constant, the surface can be defined clearly, despite the fact that the nucleus is such a complex quantum system with complicated interaction.

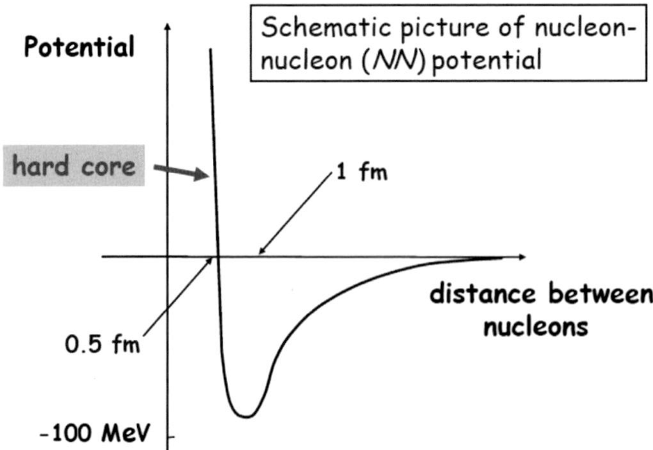

Fig. 1. – Schematic illustration of the nucleon-nucleon potential.

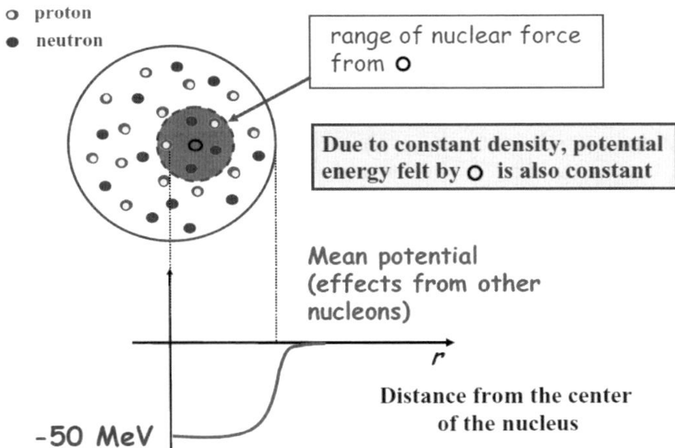

Fig. 2. – Schematic illustration of the mean potential for a nucleon open circle inside the nuclear surface.

The nucleon-nucleon interaction is very complicated, but can produce a simple mean potential. Figure 2 depicts this situation. A nucleon (open circle in fig. 2) well inside the nucleus feels the nucleon-nucleon interaction from the surrounding nucleons within reach (or *range*) of the interaction, which is about 1 fm. The sphere within this range is shown in shadow in fig. 2. As the density of the nucleon is constant inside the nucleus, the mean effect from surrounding nucleons should be almost constant, and the mean potential should be almost flat.

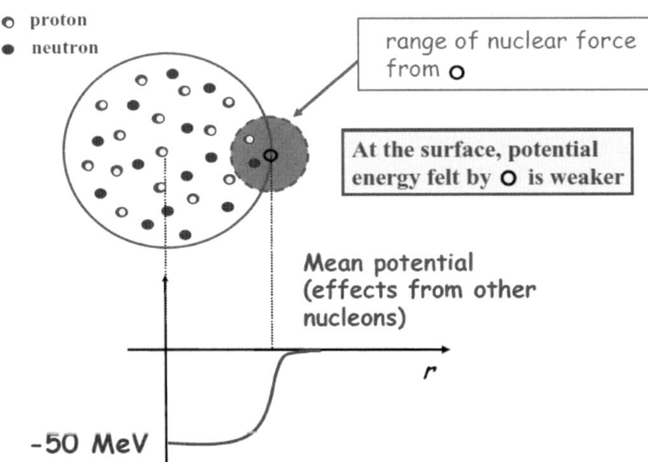

Fig. 3. – Schematic illustration of the mean potential for a nucleon open circle at the nuclear surface.

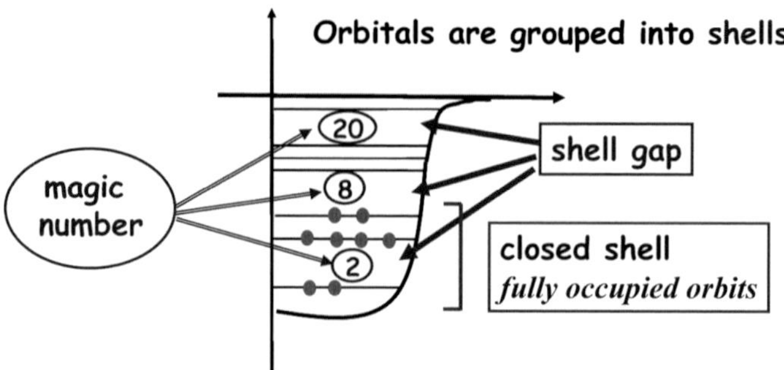

Fig. 4. – Single-particle motion in the mean potential. The horizontal lines indicate single-particle energies (SPEs), which stand for energy eigenvalues of the orbits. The orbits form shells, and gaps between shells define magic numbers. The number of particles below a shell gap corresponds to magic number, and the structure of single-particle orbits is called shell structure.

Figure 3 indicates the effect from surrounding nucleons for a nucleon at the surface (shown again by an open circle). Otherwise the legend of the figure is the same as fig. 2. The number of the surrounding nucleons becomes smaller, as this nucleon (open circle) moves out, resulting in less binding. Thus, the mean potential becomes shallower quickly at the surface.

Figure 4 displays schematically what the mean potential looks like. The single-particle motion inside this potential can be solved. This is just an eigenvalue problem with the eigenstates corresponding to various orbital motions, similar to electrons in a hydrogen-like atom. The eigenstate is referred to as an *orbit*, having its classical image in mind. Figure 4 shows energy eigenvalues of such orbits. These are usually called single-particle energies (SPEs). At this point, we assume that the nucleus is spherical, and the mean potential appears to be spherical too. The spherical potential gives us quantum numbers of these eigenstates such as the orbital angular momentum denoted by l, the total angular momentum denoted by j, and the number of nodes of the radial wave function denoted by n. Since the potential is spherical, orbits differing only by the z-component of $\vec{j}$, called j_z, are degenerate. Thus, there is a degeneracy of $(2j+1)$ magnetic substates for a given j. Because of this degeneracy, the SPE is referred to j (with l and n implicitly). If we discuss an "orbit" in a spherical potential, it means j, having $(2j+1)$ degenerate substates.

Each orbit has $(2j+1)$ substates with the same SPE. These substates are called single-particle states. Some orbits (with different j's) are grouped as shown in fig. 4. Such a group of orbits is called a *shell*. The energy spacing between two shells is called the *shell gap*. If all orbits below a given shell gap are occupied by protons, they form a proton *closed shell*. The number of protons in a closed shell is called the proton magic number. Figure 4 indicates that the magic numbers are actually $2, 8, 20, \ldots$. We shall discuss later why these are magic numbers. Likewise, neutron closed shells and magic numbers are defined. The closed shell is also called a *core*, and the pattern of the orbits stated above

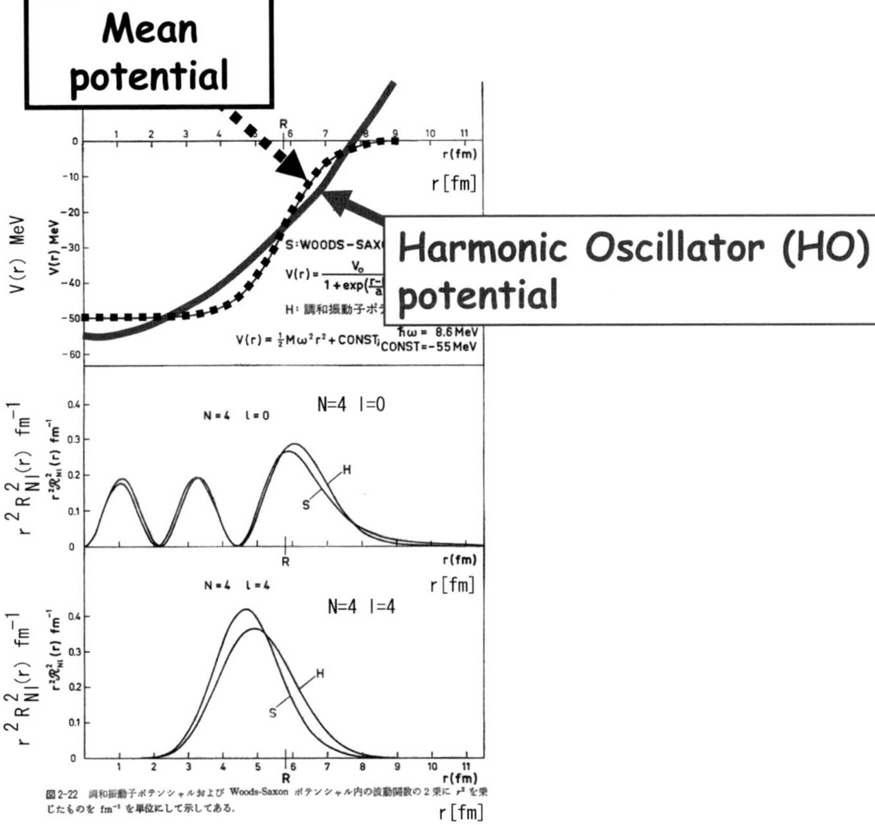

Fig. 5. – Comparison between harmonic-oscillator potential and Woods-Saxon potential.

is known as *shell structure*. We now turn to how to obtain the shell structure from simple arguments, without going into details.

The mean potential can be described, to a good approximation, in terms of the so-called Woods-Saxon (WS) potential. This is specified by three parameters: depth, radius and diffuseness. The eigenstates of the Woods-Saxon potential can be obtained only numerically. In order to make physics more transparent, we can introduce a harmonic-oscillator (HO) potential

$$(1) \qquad V_{HO} = m\omega^2 r^2/2,$$

where m is the mass of the nucleon, ω is the oscillator frequency, and r stands for the distance from the centre of the nucleus.

In fact, WS and HO potentials can be set to overlay each other as shown in fig. 5 if the bottom of the HO potential is adjusted and an appropriate value of ω is chosen. The properties of the eigenstates of a HO potential are much simpler than those of the Woods-

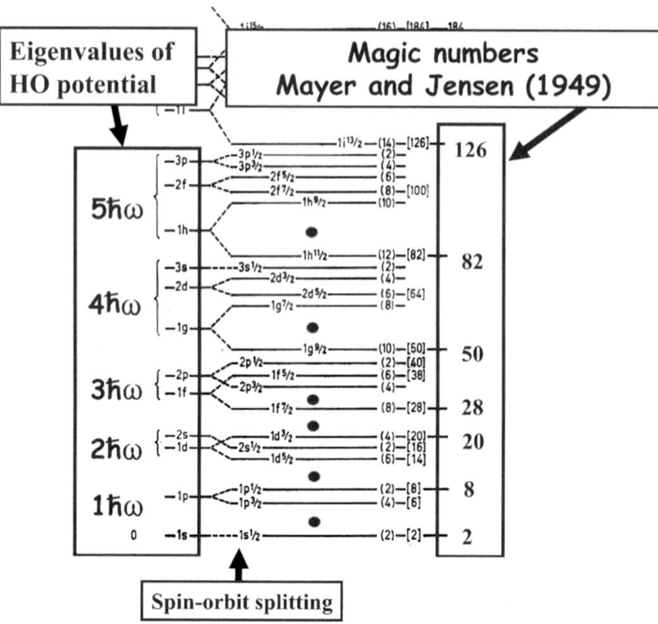

Fig. 6. – Energy eigenvalues of harmonic-oscillator potential with spin-orbit force, and Mayer-Jensen's magic numbers.

Saxon potential and can be described analytically. The eigenvalues are equally separated, as is well known, and are shown in fig. 6. Mayer and Jensen [1] proposed the shell structure of atomic nuclei by adding a *spin-orbit coupling* to the HO potential, explaining experimental data known at that time without a consistent theoretical description. The spin-orbit coupling is written as

$$(2) \qquad V_{ls}(r) = f(r)(\vec{l} \cdot \vec{s}),$$

where f is a function of r, and $\vec{l}$, $\vec{s}$ denote the orbital angular momentum and spin operators of a nucleon, respectively.

The function $f(r)$ is usually given by the derivative of the density divided by r with an appropriate strength. Naturally, $f(r)$ has a peak at the surface, because the density changes most rapidly at the surface.

The spin-orbit potential in eq. (2) can be included as a first-order perturbation and its effect is to lower the energy of the *j-upper* state

$$(3) \qquad j_{>} = l + 1/2$$

and raise the energy of the *j-lower* state

$$(4) \qquad j_{<} = l - 1/2.$$

These notations will be used later.

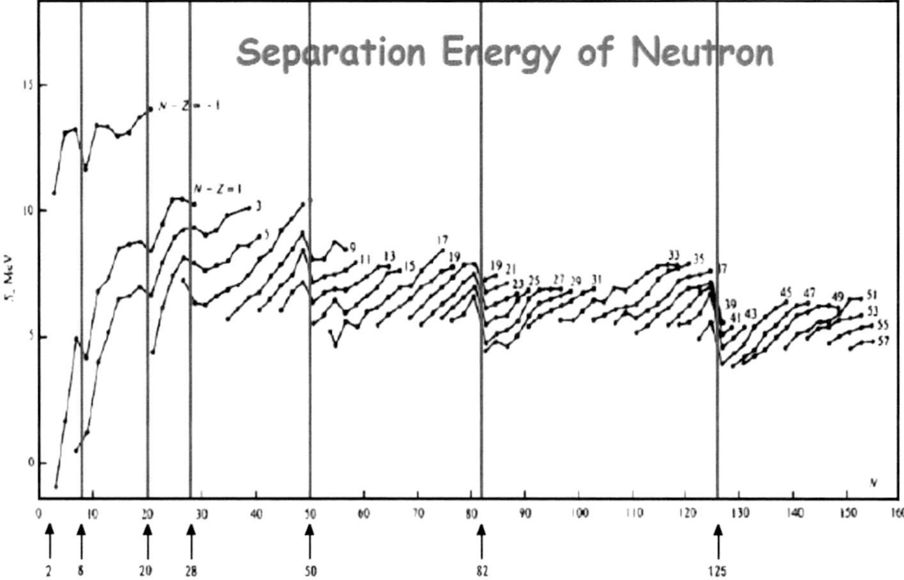

Fig. 7. – Observed neutron separation energy, the minimum energy to take a neutron out of the nucleus. Partly taken from [2].

After including the spin-orbit coupling, the degenerate orbits in the HO potential are split. The orbit is identified, for example, as $1f_{7/2}$ where the first number (integer) is the number of the node plus one, the second character is the usual notation of l, and the last part represents j.

The energy splitting due to the spin-orbit potential is approximately proportional to the value of l as expected from eq. (2). So it becomes more and more important for l larger. Mayer and Jensen included this effect, and predicted magic numbers as shown in fig. 6. The magic numbers 2, 8 and 20 are independent of the spin-orbit coupling. The orbits $1p_{3/2}$ and $1p_{1/2}$ are split, but the splitting is not large enough to break the magic numbers. Likewise, $1d_{5/2}$ and $1d_{3/2}$ are also split to a modest extent. However, the magic number 28 appears because the orbits $1f_{7/2}$ is pushed down significantly, whereas the next orbit $2p_{3/2}$ is lowered only moderately. So a gap is created between them, giving rise to a magic number 28. Similar situations occur for higher magic numbers 50, 82 and 126. These gaps are shown by full dots in fig. 6.

The success of Mayer-Jensen's magic number is tremendous. It really dominates the structure of nuclei at lower energies. There is much experimental evidence for magic numbers, one of which is the separation energy. Figure 7 shows the observed neutron separation energy S_n. Figure 8 shows how magic numbers are related to the separation energy. On the left, neutrons occupy orbits up to a shell gap, as the number of these neutrons is equal to a magic number. On the right, there is another neutron occupying an orbit above the shell gap. In order to take away one of the neutrons on the left, one needs

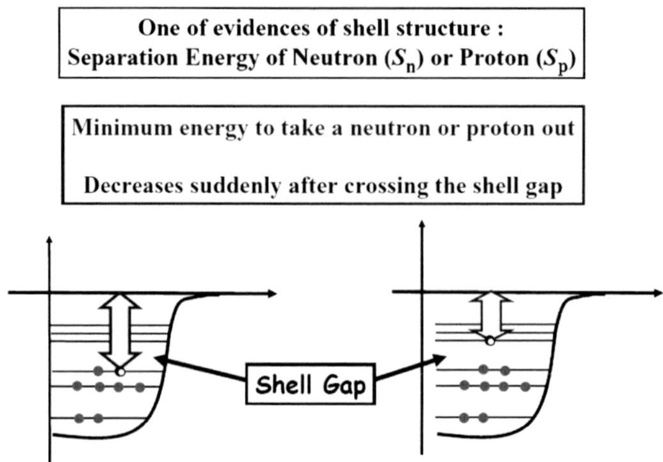

Fig. 8. – Relation of neutron magic numbers to neutron separation energy. Arrow indicates separation energy.

more energy as compared to the right part, where there is one neutron in a less bound orbit above the gap. The neutron separation energy means the minimum energy to take out one neutron from a given nucleus. So, it becomes smaller suddenly, by the amount of the shell gap, as the number of neutron goes beyond a magic number. This phenomenon can be seen in many places in fig. 7, where vertical lines indicate magic numbers and the separation energy decreases suddenly over these lines, particularly for neutron numbers $N = 50, 82, 126$. This is one of the pieces of evidence for magic numbers. However, if one looks at fig. 7 carefully, there are cases where the decrease in separation energy is not so large, or even an increase is seen. Thus, it can be expected that the magic numbers may not be perfect. However, such an idea has never been seriously discussed until recently. We will come back to possible changes of magic numbers later in this lecture.

At this point, we summarize this subsection. From its derivation, the magic numbers of Mayer and Jensen are direct and robust consequence of basic properties of the nuclear force and density, and there seems to be no way out. So, the magic numbers have been believed to remain the same for all (or most of) nuclei, stable or unstable.

1`2. *Why is the shell model useful?* – We now discuss how one can carry out shell model calculations. Through this, we would like to find an answer to the question as to why the shell model can be useful.

The shell model assumes that the orbits are already given as in fig. 6. The orbits of the proton (neutron) closed shell is completely occupied by protons (neutrons). In general, there can be some protons (or neutrons) occupying the next shell just above the closed shell. This shell is called the *valence shell,* and its nucleons are referred to as *valence nucleons.* The valence shell is, by definition, only partially occupied. If all orbits between two magic numbers is considered, this valence shell is called a *major shell.*

For instance, four orbits, $2p_{3/2}$, $2p_{1/2}$, $1f_{5/2}$ and $1g_{9/2}$, form a major shell between 28 and 50. (On the other hand, one might take a part of them as a valence shell. But the shell is not a major shell any more.) In the shell model calculation, the closed shells are treated as a vacuum, because the nucleons cannot change their single-particle states as long as they are in the closed shell. If one wants, one can make particle-hole excitations from the closed shell to valence shell. However, just for the sake of simplicity, we do not include such excitations for the time being. Namely, we look at degrees of freedom only in the valence shell. Because of the similarity between the closed-shell and the vacuum, the closed shell is often called the *(inert) core*. When we discuss dynamical properties, the (inert) core sounds more appropriate, but its meaning is the same as the closed shell.

We include two-body interaction between valence nucleons. Three-body interaction *etc.* are not included, however. Usually, it is supposed that effects of higher body (> 2 body) interactions are small enough in the energy scale of interest and/or their effects are renormalized into effective two-body interactions somehow. This is an approximation/assumption, but turns out to be reasonable from the viewpoint of comparison to experiment. The Hamiltonian then consists of the following terms:

$$(5) \qquad H = \sum_i \epsilon_i n_i + \sum_{i,j,k,l} v_{ij,kl} a_i^\dagger a_j^\dagger a_l a_k \, ,$$

where ϵ_i is the single-particle energy of the orbit i, n_i stands for the number operator of the orbit i, $v_{ij,kl}$ denotes two-body matrix element of the nucleon-nucleon (effective) interaction for orbits i, j, k, l, and $a^\dagger$ and a mean usual creation and annihilation operators.

In the single-particle picture of fig. 6, a nucleon stays in one of the orbits for ever. This is true in the closed shell, because all orbits are occupied, and a nucleon cannot move from one orbit to another within the closed shell. The situation differs in the valence shell. Figure 9 indicates how nucleons move via the nucleon-nucleon interaction. The occupancy pattern of nucleons over different orbits is called *configuration*. In fig. 9, the configuration on the left-hand side is changed to the one on the right-hand side due to the interaction between two valence nucleons. In other words, the interaction scatters two nucleons into a different pair of orbits compared to the pair of orbits before the scattering. Such scatterings occur an infinite number of times, and all possible configurations are mixed until a kind of equilibrium is achieved. The eigenstate of the shell-model Hamiltonian as in eq. (5) is thus obtained. It contains, in general, many components corresponding to different configurations. Such a mixing is called *configuration mixing*.

Figures 10 and 11 illustrate how this can be carried out computationally. As step 1 in fig. 10, we first calculate matrix elements of the Hamiltonian for the states of various configurations. These states can be represented by Slater determinants, $\phi_1, \phi_2, \phi_3, \dots$. Each Slater determinant is a product of single-particle states, $\alpha, \beta, \gamma, \dots$ for $\phi_1, \dots$ (see fig. 10). In step 2 in fig. 11, we construct the matrix of H and diagonalize it. This process is shown in the upper part of fig. 12, with eigenvalues being $\varepsilon_1, \varepsilon_2, \varepsilon_3, \dots$. We thus solve

A nucleon does not stay in an orbit for ever. The interaction between nucleons changes their occupations as a result of scattering.

Pattern of occupation : configuration

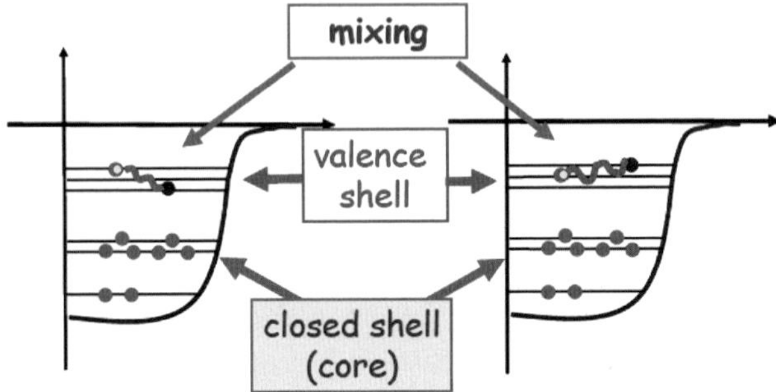

Fig. 9. – Mixing of different configurations due to the scattering between valence nucleons.

the eigenvalue problem,

$$(6) \qquad\qquad H\,\Psi = E\,\Psi,$$

where E is the energy eigenvalue and its eigenfunction is Ψ. The Ψ wave function is

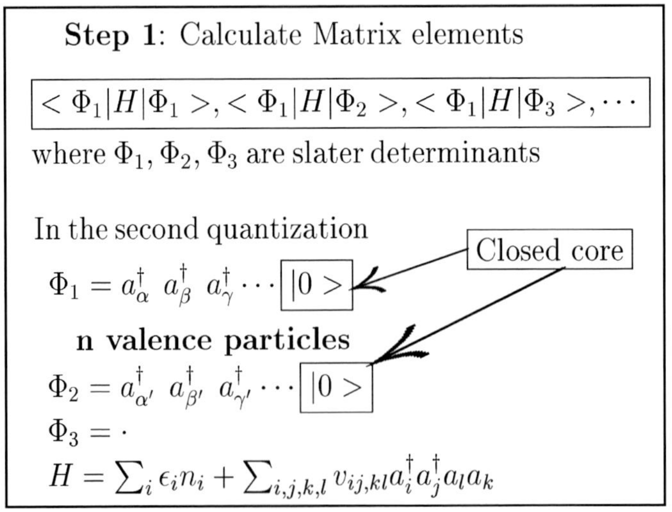

Fig. 10. – Calculation of shell-model Hamiltonian by Slater determinants.

> **Step 2** : Construct matrix of Hamiltonian, $\mathcal{H}$,
> and diagonalize it
>
> $$\mathcal{H} = \begin{bmatrix} <\phi_1|H|\phi_1> & <\phi_1|H|\phi_2> & <\phi_1|H|\phi_3> & \cdots \\ <\phi_2|H|\phi_1> & <\phi_2|H|\phi_2> & <\phi_2|H|\phi_3> & \cdots \\ <\phi_3|H|\phi_1> & <\phi_3|H|\phi_2> & <\phi_3|H|\phi_3> & \cdots \\ <\phi_4|H|\phi_1> & & \cdot & \cdot & \cdot \\ & \cdot & & \cdot & & \cdot & & \cdot \end{bmatrix}$$

Fig. 11. – Hamiltonian matrix.

expanded in terms of Slater determinants with probability amplitudes $c_1, c_2, c_3, \ldots$,

$$\text{(7)} \qquad \Psi = c_1\,\phi_1 + c_2\,\phi_2 + c_3\,\phi_3 + \ldots$$

Accordingly, the eigenvector of the matrix eigenvalue problem for eq. (6) is written as $(c_1, c_2, c_3, \ldots)$.

The precise evaluation of the energy eigenvalue and the configuration mixing is one of the major tasks of the shell model calculations. Once we obtain the wave function Ψ, we can calculate a variety of physical observables.

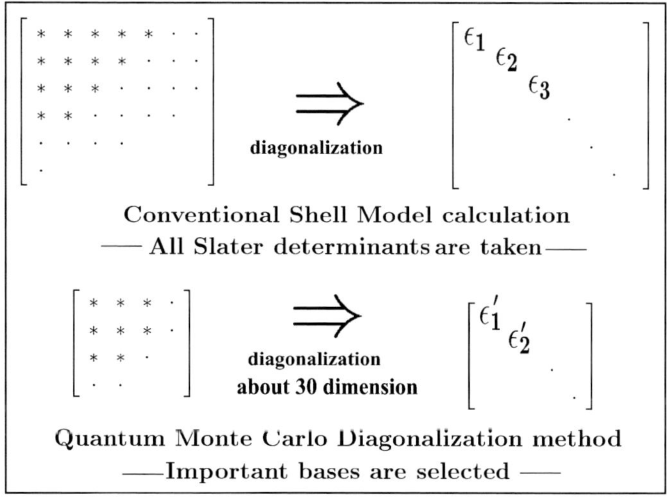

Fig. 12. – Signalization of Hamiltonian matrix.

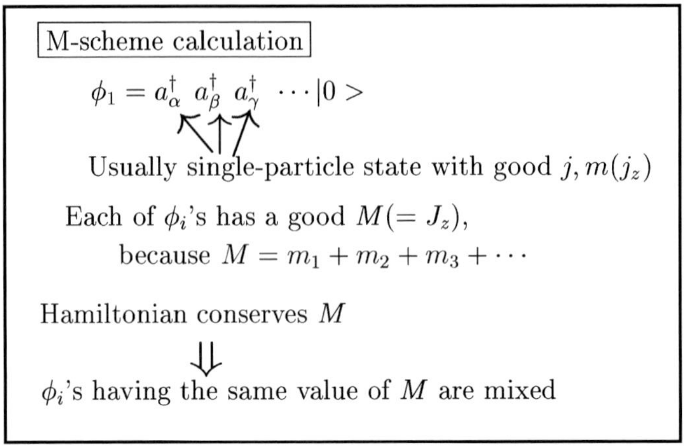

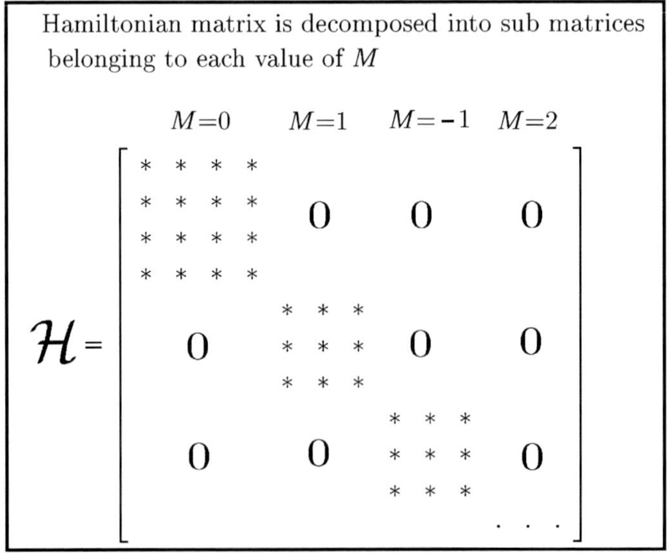

Fig. 13. – Multi-nucleon bases vectors with a good M $(= J_z)$.

1'3. *Some remarks on shell model calculations.* – The direct diagonalization of the matrix as in the upper part of fig. 12 is what the conventional shell model calculation does. We remark on some features of the shell model calculations so as to give some feeling for the actual calculation.

We first briefly discuss the *M-scheme*. Since the Hamiltonian in eq. (5) is rotationally invariant, the z-component of the total angular momentum, $\vec{J}$, is conserved for all eigenstates. The quantum number of J_z of Ψ in eq. (6) is denoted as M hereafter.

Fig. 14. – Decomposition of Hamiltonian matrix according to M.

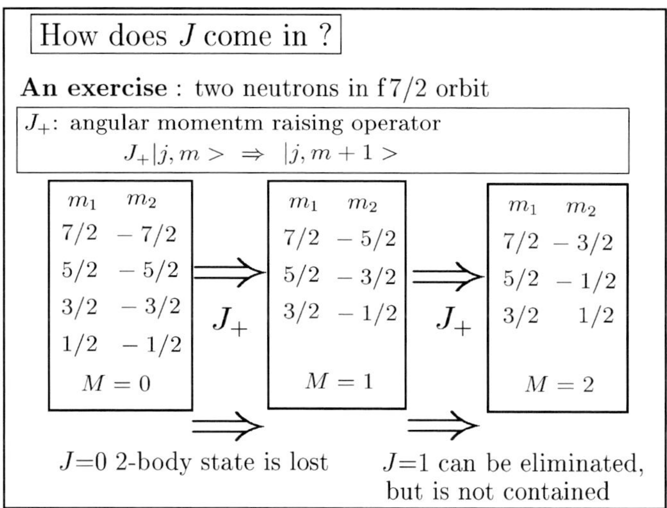

Fig. 15. – J quantum number of two-neutron system in $1f_{7/2}$.

Each single-particle state has a good quantum number j_z, the z-component of $\vec{j}$, because of the spherical mean potential. Each Slater determinant in eq. (7) can have a good J_z, if it is constructed from such single-particle states with good j_z's. Naturally, Slater determinants, $\phi_1, \phi_2, \phi_3, \ldots$ in eq. (7) should have the same value of J_z as the full Ψ. This situation is illustrated in fig. 13. As the Hamiltonian conserves J_z, the matrix elements in eq. (11) are finite only between the same values of J_z. Thus, one can construct the Hamiltonian matrix like fig. 14.

The Slater determinant is the product of single-particle states as seen in fig. 13. The next question is how the *conservation of* J can be achieved. This question can be answered by taking a simple example. Figure 15 shows the case of two neutrons in the $f_{7/2}$ orbit. The j_z values of the two neutrons are denoted as m_1 and m_2. There are four states in the $M = 0$ space as shown in the left column of the figure. Here, the space is a set of the states belonging to a given M ($M = 0$ in this case). The dimension of this space is four. By acting with the angular-momentum raising operator, J_+, we obtain an $M = 1$ space comprised of three states. They are shown in the middle column in fig. 15. The dimension is less by one than that of $M = 0$ space. One then knows that there is one $J = 0$ state in the $M = 0$ space, and it was eliminated by the J_+ operation (there is only an $M = 0$ state for $J = 0$).

By repeating the J_+ operation, one gets an $M = 2$ space. This space is shown in the right column of fig. 15. The dimension is still three, meaning that there is no $J = 1$ state in the spaces we are working on.

By doing similar analysis, one can see what J states are contained in each M space as shown in fig. 16.

The diagonalization of the Hamiltonian matrix is done for each M space separately.

	Dimension	Components of J values
$M = 0$	4	$J = 0, 2, 4, 6$
$M = 1$	3	$J = 2, 4, 6$
$M = 2$	3	$J = 2, 4, 6$
$M = 3$	2	$J = 4, 6$
$M = 4$	2	$J = 4, 6$
$M = 5$	1	$J = 6$
$M = 6$	1	$J = 6$

Fig. 16. – J contents of M spaces of two neutrons in $1f_{7/2}$.

By **diagonalizing** the matrix $\mathcal{H}$, we obtain wave functions of good J values by superposing Slater determinants.

In the case shown in fig. 16

$$M = 0$$

$$\mathcal{H} = \begin{bmatrix} * & * & * & * \\ * & * & * & * \\ * & * & * & * \\ * & * & * & * \end{bmatrix} \Rightarrow \begin{bmatrix} \mathbf{e_{J=0}} & 0 & 0 & 0 \\ 0 & \mathbf{e_{J=2}} & 0 & 0 \\ 0 & 0 & \mathbf{e_{J=4}} & 0 \\ 0 & 0 & 0 & \mathbf{e_{J=6}} \end{bmatrix}$$

e_J means the eigenvalue with the angular momentum, J.

Fig. 17. – J restoration by the diagonalization of Hamiltonian matrix for two neutrons in $1f_{7/2}$.

This property is a general one : valid for cases with more than 2 particles.

By **diagonalizing** the matrix $\mathcal{H}$, you get eigenvalues and wave functions of good J values
by superposing Slater determinants.

$$M$$

$$\mathcal{H} = \begin{bmatrix} * & * & * & * \\ * & * & * & * \\ * & * & * & * \\ * & * & * & * \end{bmatrix} \Rightarrow \begin{bmatrix} \mathbf{e_J} & 0 & 0 & 0 \\ 0 & \mathbf{e_{J'}} & 0 & 0 \\ 0 & 0 & \mathbf{e_{J''}} & 0 \\ 0 & 0 & 0 & \mathbf{e_{J'''}} \end{bmatrix}$$

Fig. 18. – J restoration by the diagonalization of Hamiltonian matrix for general cases.

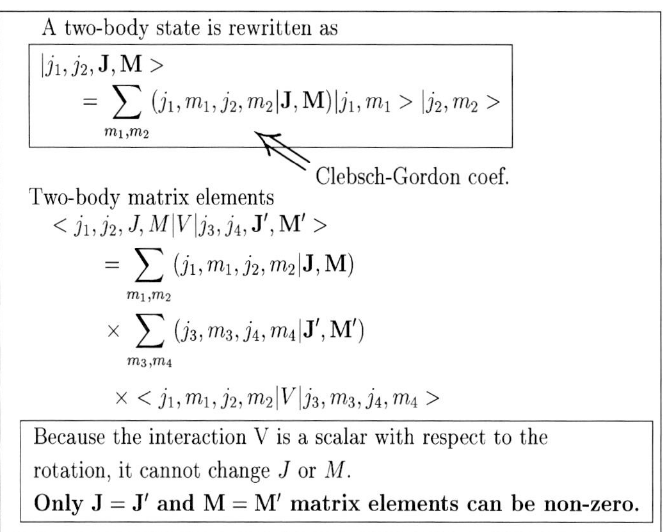

Fig. 19. – Two-body states and TBME.

Since the Hamiltonian conserves J in addition to M, the diagonalization of the Hamiltonian matrix produces eigenstates with *good J's*. This situation is depicted in fig. 17 for the $M = 0$ case of two neutrons in $f_{7/2}$ orbit. Such a restoration of J quantum number is a general one, and is achieved in general as a result of the diagonalization of the Hamiltonian, as shown in fig. 18. In other words, the eigenstate Ψ is given by a linear combination of Slater determinants, and the same linear combination is a simultaneous eigenstate of the J_z operator and the $(\vec{J} \cdot \vec{J})$ operator.

We now see some properties of *two-body matrix element (TBME)* of the nucleon-nucleon interaction. Two nucleons can have total angular momentum J and total isospin T. If nucleons are in single-particle states $|j_1 m_1\rangle$ and $|j_2 m_2\rangle$, they can be coupled to total angular momentum J and its z-projection M by Clebsch-Gordon coefficients, as shown in fig. 19. The same coupling occurs before and after the interaction (or scattering). We consider matrix elements between such coupled states. Since the interaction is rotationally invariant, it cannot change J or M, and consequently $J = J'$ and $M = M'$ are satisfied as in fig. 19. Moreover, the rotational invariance of the interaction makes all coupled matrix elements independent of M (see fig. 20). Thus, in this J-coupled scheme matrix elements of two-nucleon states are specified only by j_1, j_2, j_3, j_4 and J (see fig. 20). There is certainly another dependence on the isospin, T.

The two-body matrix elements are abbreviated as TBMEs *(Two-Body Matrix Elements)*. Of course, once the potential between two nucleons is given one can calculate all TBMEs. However, up to the present time, the potential between nucleons is not completely known. Moreover, there will be renormalization due to core-polarizations (as explained later), as well as many other different mechanisms that contribute to the TBMEs. It is true that no theory has succeeded in a perfect prediction of the interac-

Two-body matrix elements
$$< j_1, j_2, J, M|V|j_3, j_4, J, M >$$
are independent of M value, also because V is a scalar.

Two-body matrix elements are assigned by $j_{1,2}, j_3, j_4$ and J.

Jargon : Two-Body Matrix Element = **TBME**

Because of complexity of nuclear force, one can not express all TBME's by few empirical parameters.

Fig. 20. – TBME specification.

tion to be used in shell model calculations, and this situation will not be altered in the near future. Thus, for the practical use of shell model calculations, we need empirical corrections.

An example of successful calculations of TBMEs is by using the USD interaction. The USD interaction was proposed by Wildenthal and Brown [3] for the sd shell comprised of three orbits $1d_{5/2}$, $1d_{3/2}$ and $2s_{1/2}$. It consists of 63 TBMEs and three SPEs. The TBMEs are based on those given by the G-matrix interaction of Kuo [4]. Figure 21 shows a part of the USD TBMEs. The indices i, j, k and l stand, respectively, for j_1, j_2, j_3 and j_4 mentioned above. One finds the dependences on J and T. These TBMEs are calculated within the G-matrix formalism starting from meson exchange theory of the free nucleon-nucleon interaction. Such a calculation gives us reasonable numbers. But, once one performs shell model diagonalization with those TBMEs, the resultant energy levels are very different from experimental ones, and we do not learn much about the structure of the nucleus. We need to make empirical corrections. This is what was done by Wildenthal and Brown to obtain the USD interaction.

The nucleon-nucleon interactions used in shell model calculations are effective ones. They include effects of multiple scattering between nucleons with high-lying intermediate states in the sense of a second-order perturbation (Ladder diagram). Figure 22 indicates this contribution schematically as the effects of "higher shell".

Another contribution comes from the excitation of the core (closed shell), as referred to as the core polarization. In a first approximation, the core is completely occupied, but there are excitations in reality. The pairing interaction is enhanced by this mechanism.

Effects of these two types of outer shells are included, ending up with the so-called *effective interaction*.

The core polarization is also important for *effective charge* and the *effective g-factor*. Figure 23 shows how the electric quadrupole moment and magnetic moment are changed due to the excitation of the core. This phenomenon has been proposed by Arima and

USD interaction

i	j	k	l	J	T	V
d3/2	d3/2	d3/2	d3/2	0	1	-2.1845
d3/2	d3/2	d3/2	d3/2	1	0	-1.4151
d3/2	d3/2	d3/2	d3/2	2	1	-0.0665
d3/2	d3/2	d3/2	d3/2	3	0	-2.8842
d5/2	d3/2	d3/2	d3/2	1	0	0.5647
d5/2	d3/2	d3/2	d3/2	2	1	-0.6149
d5/2	d3/2	d3/2	d3/2	3	0	2.0337
d5/2	d3/2	d5/2	d3/2	1	0	-6.5058
d5/2	d3/2	d5/2	d3/2	1	1	1.0334
d5/2	d3/2	d5/2	d3/2	2	0	-3.8253
d5/2	d3/2	d5/2	d3/2	2	1	-0.3248
d5/2	d3/2	d5/2	d3/2	3	0	-0.5377
d5/2	d3/2	d5/2	d3/2	3	1	0.5894
d5/2	d3/2	d5/2	d3/2	4	0	-4.5062
d5/2	d3/2	d5/2	d3/2	4	1	-1.4497
d5/2	d3/2	s1/2	d3/2	1	0	-1.7080
d5/2	d3/2	s1/2	d3/2	1	1	0.1874
d5/2	d3/2	s1/2	d3/2	2	0	0.2832
d5/2	d3/2	s1/2	d3/2	2	1	-0.5247
d5/2	d3/2	s1/2	s1/2	1	0	2.1042
d5/2	d5/2	d3/2	d3/2	0	1	-3.1856
d5/2	d5/2	d3/2	d3/2	1	0	0.7221
d5/2	d5/2	d3/2	d3/2	2	1	-1.6221
d5/2	d5/2	d3/2	d3/2	3	0	1.8949
d5/2	d5/2	d5/2	d3/2	1	0	2.5435
d5/2	d5/2	d5/2	d3/2	2	1	-0.2828
d5/2	d5/2	d5/2	d3/2	3	0	2.2216
d5/2	d5/2	d5/2	d3/2	4	1	-1.2363
d5/2	d5/2	d5/2	d5/2	0	1	-2.8197
d5/2	d5/2	d5/2	d5/2	1	0	-1.6321
d5/2	d5/2	d5/2	d5/2	2	1	-1.0020
d5/2	d5/2	d5/2	d5/2	3	0	-1.5012

Fig. 21. – Part of USD TBMEs.

Horie [5,6], Blin-Stoyle and Perks [7] in 1954 and by Bohr and Mottelson [2].

We finally note that the single particle in the shell model is an "effective object" (or quasi-particle) with rather complicated correlations behind it like the core polarization. The shell model treats it as if it is a real particle, and includes those correlation effects in terms of the renormalization of effective interaction and operators. The same picture should be taken for mean-field models (or density functional theories) where the

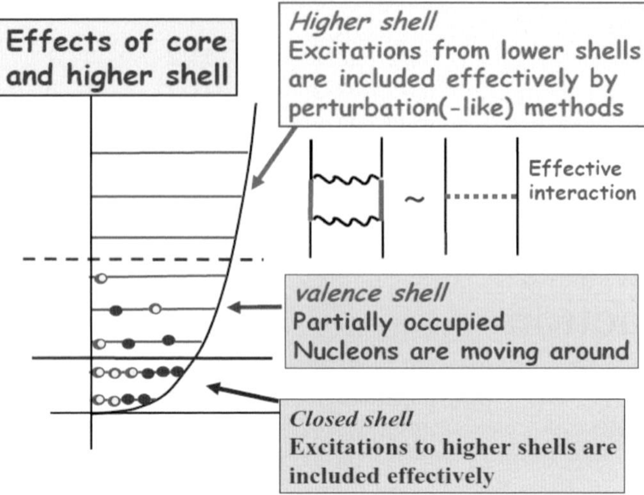

Fig. 22. – Corrections to effective interaction used in the shell model calculations. There are two types: one from higher shells, while the other from the closed shell.

renormalization is more severe due to more truncated effective interactions.

The coupling to various correlations reduces the "purity" of the single particle. Recent experiments show that about 60% of the "single-particle" probability remains in the shell model wave functions after the mixing of more complicated components [8]. However, we emphasize that this is not a catastrophe or anything like that, and is expected.

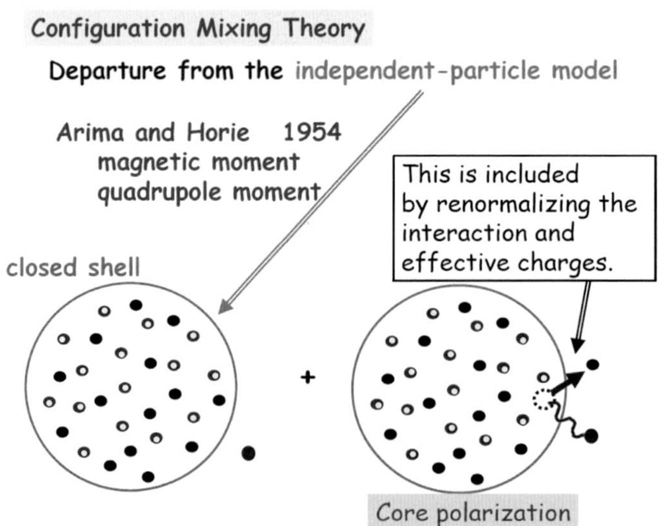

Fig. 23. – Schematic Illustration of Configuration Mixing Theory.

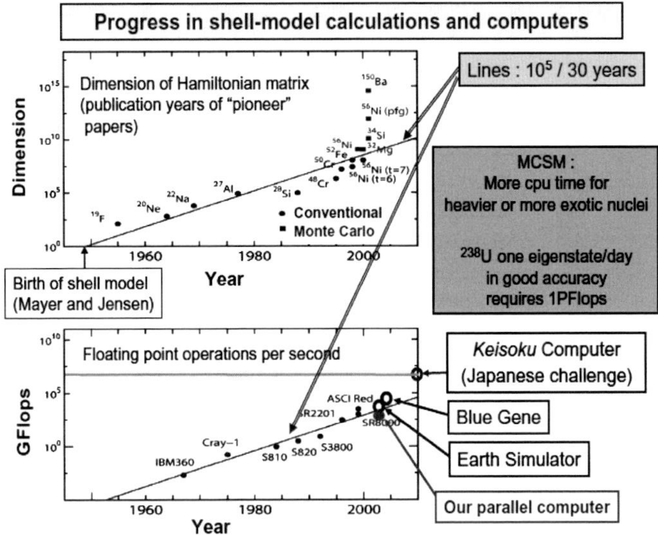

Fig. 24. – Upper panel: Maximum dimension of the Hamiltonian matrix feasible for the year of publication. Circular points are for conventional shell model, while box-shaped points are for Monte Carlo Shell Model. Lower panel: The computer capability (Flops) as a function of the year. The lines in the upper and lower panels indicate an increase of 10^5 times/30 years.

We have seen how the shell model is formulated in simple terms. We skipped many discussions as to how one can obtain good effective interactions from basic theories of the nucleon-nucleon interaction. This is rather complicated and still in progress.

The effective interaction and operators are essential parts of the shell model. These are also crucial in other approaches to nuclear structure. The shell model is constructed very carefully on nuclear forces. In other approaches of nuclear structure, forces are also models. In the shell model, although we use effective interactions, these interactions are built as realistic as possible, particularly in recent large scale-calculations. In this sense, the shell model can reflect various facets of nuclear forces into nuclear structure better than other models.

This is the merit of the shell model. Because of this, the shell model becomes more important in the region of exotica in the nuclear chart, as the predictive power is more needed than in the region of stable nuclei.

If the shell model is so useful, the next question is how to run a shell model calculation; or even whether it is always possible or not?

1˙4 *How do we perform shell model calculations?* – Figure 12 indicates the diagonalization of the Hamiltonian matrix. There are some computer programs for performing such shell model calculations. These include OXBASH by Brown, ANTOINE by Caurier *et al.*, MSHELL by Mizusaki, *etc.* The ANTOINE and MSHELL can be run on parallel computers, and can handle up to 1 billion dimension (as of the year 2007). One can

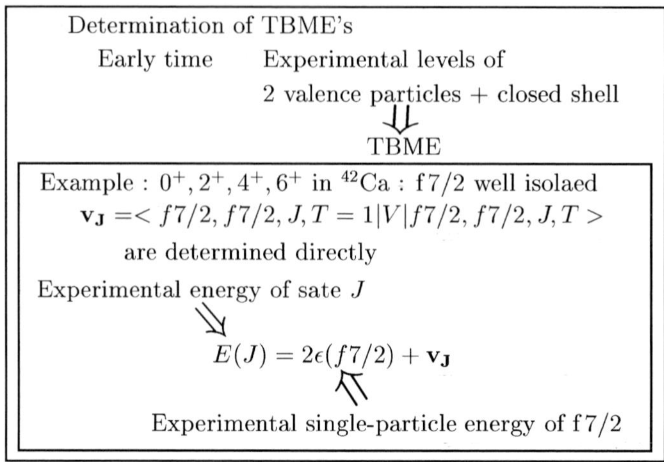

Determination of TBME's

Early time   Experimental levels of

2 valence particles + closed shell

$\Downarrow$

TBME

Example : $0^+, 2^+, 4^+, 6^+$ in ^{42}Ca : f7/2 well isolaed

$\mathbf{v_J} =< f7/2, f7/2, J, T = 1|V|f7/2, f7/2, J, T >$

are determined directly

Experimental energy of sate J

$E(J) = 2\epsilon(f7/2) + \mathbf{v_J}$

Experimental single-particle energy of f 7/2

Fig. 25. – Empirical determination of TBMEs for $1f_{7/2}$.

easily imagine that the practical difficulty should rise as the dimension becomes larger. The difficulties are in the computation and storage of so many matrix elements, and the diagonalization of such a huge matrix.

There have been steady and significant efforts with great success, as reviewed recently in [9]. Figure 24 shows how the maximum feasible dimension has been increased as a function of the year since 1949 (the birth year of the shell model). It is amazing that this growth is almost on a logarithmic scale. However, there exists a limit around 1 billion dimension, at least at present. In order to overcome this limit, the Monte Carlo Shell Model has been proposed [10-12]. Although the computation time becomes longer as the particle number and/or the number of valence orbits increases, this is not a very strong barrier for the Monte Carlo Shell Model. One could keep going to frontier cases with the dimension, for instance, 10^{15} [13]. In exotic nuclei, two conventional shells often merge, and the calculation becomes huge. The Monte Carlo Shell Model can still give us results.

2. – Construction of an effective interaction and an example in the pf shell

In this section, we show how one can determine TBMEs.

A simple example of experimental determination is the case of the $1f_{7/2}$ orbit. If this orbit is perfectly isolated, one can extract TBMEs from observed energy levels as shown in fig. 25.

However, the case of the $1f_{7/2}$ orbit is too simple. Other cases cannot be handled this way. For instance, Arima *et al.* have carried out a χ^2 fit of TBMEs for $1d_{5/2}$ and $2s_{1/2}$, as shown in fig. 26.

The idea of the χ^2 fit does not work for the full sd shell, however. In obtaining the USD interaction, Wildenthal and Brown carried out only a partial fit. One can choose some TBMEs whose linear combinations are sensitive to energies of low-lying states.

> The isolation of f7/2 is special. In other cases, several orbits must be taken into account.
>
> > In general, χ^2 fit is made
> > (i) TBME's are assumed,
> > (ii) energy eigenvalues are calculated,
> > (iii) χ^2 is calculated between theoretical and experimental energy levels,
> > (iv) TBME's are modified. Go to (i), and iterate the process until χ^2 becomes small enough.
>
> Example : $0^+, 2^+, 4^+$ in ^{18}O(oxyen) : d5/2 & s1/2
>
> $< d5/2, d5/2, J, T = 1|V|d5/2, d5/2, J, T >,$
>
> $< d5/2, s1/2, J, T = 1|V|d5/2, s1/2, J, T >,$ etc.
>
> Arima, Cohen, Lawson and McFarlane(Argonne group), 1968

Fig. 26. – Empirical determination of TBMEs for $1d_{5/2}$ and $2s_{1/2}$.

They adjusted 47 linear combinations out of 63 TBMEs and three SPEs. The rest were taken from G-matrix result of Kuo. The USD was thus created [3].

The same idea was taken for determining the GXPF1 interaction, which was created for the description of pf shell nuclei including the middle region [14, 15]. The GXPF1 is based on the G-matrix interaction by H.-Jensen et $al.$ [16]. There is also the KB3 interaction and its family for the pf shell, which are particularly good for the beginning of the pf shell [17, 18].

Figure 27 shows the correlations between the TBME of the G-matrix of H.-Jensen et $al.$ and the corresponding TBME of GXPF1. If the empirical fit does not change TBMEs, all the points should be lined up on the $y = x$ line in fig. 27. Indeed all points are near the $y = x$ line, but there are deviations. These deviations are results of the fit. The fit was made for 699 levels. One finds some general trends: $T = 0$ TBMEs are shifted to be more attractive as a whole, while $T = 1$ are made more repulsive. In fig. 27, the orbits, J and T are shown for some points. One sees that the $T = 0$ coupling between $1f_{7/2}$ and $1f_{5/2}$ is strong.

Using this GXPF1 interaction, many interesting results have been obtained [14, 19-56]. In particular, the issues like $N = 32$ and $N = 34$ new magic numbers have been extensively studied as well as deformation of Ti, Cr and Fe isotopes.

3. – The $N = 2$ problem: does the gap change?

One of the most prominent points around the so-called island of inversion has been the changing shell gap between the sd and pf shells. Figure 28 indicates how the gap depends on the proton number Z. This $N = 20$ gap is about 5-6 MeV for ^{40}Ca. The conventional idea leads us to a constant gap, as shown in fig. 28. The SDPF-M interaction, which was

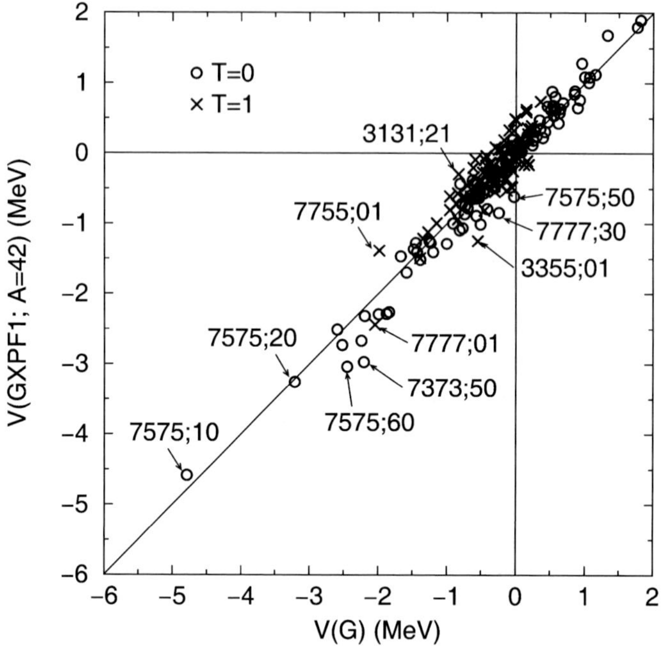

Fig. 27. – GXPF1 TBME *vs.* *G*-matrix TBME.

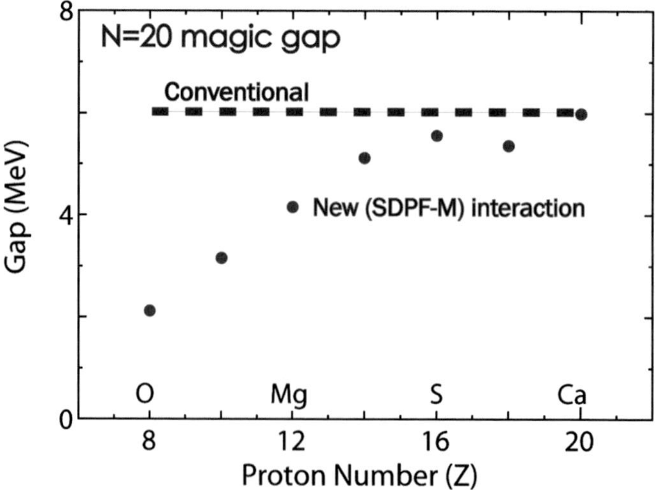

Fig. 28. – $N = 20$ gap obtained by the SDPF-M interaction (circlular points). Dashed line indicates a constant gap set by ^{40}Ca doubly magic nucleus.

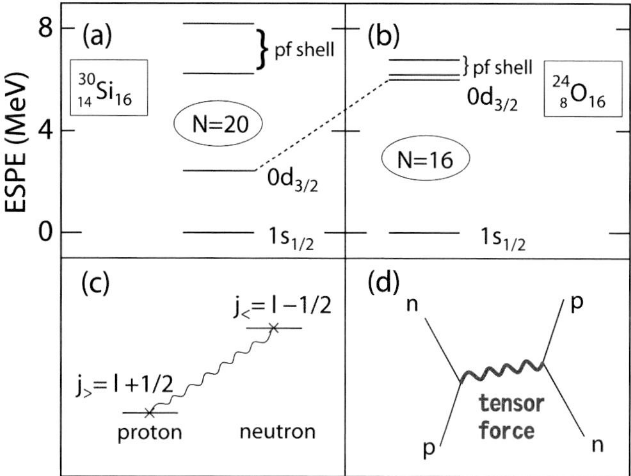

Fig. 29. – Mechanism of the shell evolution. Note that the gap shown is the $N = 20$ gap for $N = 16$, whereas the one shown in fig. 28 is the same gap but for $N = 20$.

obtained in [57] and was described in [58] and reproduces various peculiar phenomena in $N = 18$–22 isotopes of F, Ne, Na, Mg, Al and Si. This interaction produces a varying gap, as shown in fig. 28, which turned out to be essential for good agreement with many experimental data obtained in MSU, GANIL, GSI and RIKEN in recent years [57,59-72].

We discuss briefly below why the gap changes. Before moving there, we would like to point out that, once the gap becomes smaller, two shells tend to merge, and there will be many particle-hole excitations. This means that appropriate shell-model calculations can be done only with a huge Hamiltonian matrix, making conventional shell model calculations more difficult. Thus, the Monte Carlo Shell Model plays crucial roles in the studies of exotic nuclei.

The gap change, which is the most prominent consequence of the shell evolution, is now known to be primarily due to the tensor force. For this discussion, the reader is referred to the original papers [73-76]. Figure 29 was taken from [73] with an addition. This figure suggests how SPEs are changed due to a particular part the nucleon-nucleon interaction. When this paper was published, it was reported that the strong proton-neutron interaction between spin-flip partners ($j_>$ and $j_<$) is the origin of some shell evolution, but its fundamental origin was only speculation. Later, the tensor force has been shown to be the origin. So, we now put "tensor" into fig. 29 by a thick wavy line. The change of the $N = 20$ gap in fig. 28 is now understood as being due to the tensor force.

4. – Summary

The basic points of formulation and the properties of the shell model are presented without going into details or sophisticated theories. These can be found in standard textbooks.

We would like, however, to emphasize one point. Mayer-Jensen's magic numbers are a robust conclusion from the short-range attraction of the nuclear force and the density saturation together with the spin-orbit force. These are so sound, and cannot be thrown away. Therefore, if there is any deviation, one needs something new. The spin-isospin interactions have been looked over in the mean potential arguments in the past. At least, there have been no extensive studies, whereas these interactions have been known and been studied in other contexts. Once their effects were studied, there are robust and intuitively understood mechanism to change the shell structure. But, to see this, one needs to change the neutron (or proton) number considerably. It was done for Sb isotopes by Schiffer *et al.* [77], but requires rare isotope beam experiments in general. So, the shell evolution due to the tensor force, and maybe other unknown things, will open an new era of nuclear structure physics.

$$* \ * \ *$$

The author acknowledges Drs M. HONMA, Y. UTSUNO, T. MIZUSAKI, N. SHIMIZU, T. SUZUKI and N. AKAISHI for their help on theoretical parts of works related to this work. There are many experimentalists who helped this work. The author is grateful to Mr. K. TSUKIYAMA for his assistance in making the manuscript. This work has been supported in part by the JSPS Core-to-Core project "International Research Network for Exotic Femto Systems".

REFERENCES

[1] MAYER M. G., *Phys. Rev.*, **75** (1949) 1969; HAXEL O., JENSEN J. H. D. and SUESS H. E., *Phys. Rev.*, **75** (1949) 1766.

[2] BOHR A. and MOTTELSON B. R., *Nuclear Structure*, Vol. **1** (Benjamin, New York, 1969).

[3] BROWN B. A. and WILDENTHAL B. H., *Annu. Rev. Nucl. Part. Sci.*, **38** (1988) 29.

[4] KUO T. T. S., *Nucl. Phys. A*, **103** (1967) 71.

[5] ARIMA A. and HORIE H., *Prog. Theor. Phys.*, **11** (1954) 509.

[6] NOYA H., ARIMA A. and HORIE H., *Prog. Theor. Phys. Suppl.*, **8** (1958) 33.

[7] BLIN-STOYLE R. J. and PERKS M. A., *Proc. Phys. Soc. (London) Ser. A*, **67** (1954) 885.

[8] GADE A. *et al.*, *Phys. Rev. Lett.*, **93** (2004) 042501.

[9] CAURIER E., MARTINEZ-PINEDO G., NOWACKI F. *et al.*, *Rev. Mod. Phys.*, **77** (2005) 427.

[10] HONMA M., MIZUSAKI T. and OTSUKA T., *Phys. Rev. Lett.*, **75** (1995) 1284.

[11] OTSUKA T., HONMA M. and MIZUSAKI T., *Phys. Rev. Lett.*, **81** (1998) 1588.

[12] OTSUKA T., HONMA M., MIZUSAKI T., SHIMIZU N. and UTSUNO Y., *Prog. Part, Nucl. Phys.*, **47** (2001) 319.

[13] SHIMIZU N., OTSUKA T., MIZUSAKI T. and HONMA M., *Phys. Rev. Lett.*, **86** (2001) 1171.

[14] HONMA M., OTSUKA T., BROWN B. A. and MIZUSAKI T., *Phys. Rev. C*, **65** (2002) 061301.

[15] HONMA M., OTSUKA T., BROWN B. A. and MIZUSAKI T., *Phys. Rev. C*, **69** (2004) 034335.

[16] HJORTH-JENSEN M., KUO T. T. S. and OSNES E., *Phys. Rep.*, **261** (1995) 125.

[17] POVES A. and ZUKER A. P., *Phys. Rep.*, **70** (1981) 235.

[18] POVES A., SÁNCHEZ-SOLANO J., CAURIER E. and NOWACKI F., *Nucl. Phys. A*, **694** (2001) 157.

[19] JANSSENS R. V. F., FORNAL B., MANTICA P. F., BROWN B. A., BRODA R., BHATTACHARYYA P., CARPENTER M. P., CINAUSERO M., DALY P. J., DAVIES A. D., GLASMACHER T., GRABOWSKI Z. W., GROH D. E., HONMA M., KONDEV F. G., KRÓLAS W., LAURITSEN T., LIDDICK S. N., LUNARDI S., MARGINEAN N., MIZUSAKI T., MORRISSEY D. J., MORTON A. C., MUELLER W. F., OTSUKA T., PAWLAT T., SEWERYNIAK D., SCHATZ H., STOLZ A., TABOR S. L., UR C. A., VIESTI G., WIEDENHÖVER I. and WRZESIŃSKI J., *Phys. Lett. B*, **546** (2002) 55.

[20] MANTICA P. F., MORTON A. C., BROWN B. A., DAVIES A. D., GLASMACHER T., GROH D. E., LIDDICK S. N., MORRISSEY D. J., MUELLER W. F., SCHATZ H., STOLZ A., TABOR S. L., HONMA M., HOROI M. and OTSUKA T., *Phys. Rev. C*, **67** (2003) 014311.

[21] LISETSKIY A. F., PIETRALLA N., HONMA M., SCHMIDT A., SCHNEIDER I., GADE A., VON BRENTANO P., OTSUKA T., MIZUSAKI T. and BROWN B. A., *Phys. Rev. C*, **68** (2003) 034316.

[22] LIDDICK S. N., MANTICA P. F., JANSSENS R. V. F., BRODA R., BROWN B. A., CARPENTER M. P., FORNAL B., HONMA M., MIZUSAKI T., MORTON A. C., MUELLER W. F., OTSUKA T., PAVAN J., STOLZ A., TABOR S. L., TOMLIN B. E. and WIEDEKING M., *Phys. Rev. Lett.*, **92** (2004) 072502.

[23] HONMA MICHIO, OTSUKA TAKAHARU, ALEX BROWN B. and MIZUSAKI TAKAHIRO, *Phys. Rev. C*, **69** (2004) 034335.

[24] FREEMAN S. J., JANSSENS R. V. F., BROWN B. A., CARPENTER M. P., FISCHER S. M., HAMMOND N. J., HONMA M., LAURITSEN T., LISTER C. J., KHOO T. L., MUKHERJEE G., SEWERYNIAK D., SMITH J. F., VARLEY B. J., WHITEHEAD M. and ZHU S., *Phys. Rev. C*, **69** (2004) 064301.

[25] YURKEWICZ K. L., BAZIN D., BROWN B. A., CAMPBELL C. M., CHURCH J. A., DINCA D.-C., GADE A., GLASMACHER T., HONMA M., MIZUSAKI T., MUELLER W. F., OLLIVER H., OTSUKA T., RILEY L. A. and TERRY J. R., *Phys. Rev. C*, **70** (2004) 034301.

[26] YURKEWICZ K. L., BAZIN D., BROWN B. A., CAMPBELL C. M., CHURCH J. A., DINCA D.-C., GADE A., GLASMACHER T., HONMA M., MIZUSAKI T., MUELLER W. F., OLLIVER H., OTSUKA T., RILEY L. A. and TERRY J. R., *Phys. Rev. C*, **70** (2004) 054319.

[27] LIDDICK S. N., MANTICA P. F., BRODA R., BROWN B. A., CARPENTER M. P., DAVIES A. D., FORNAL B., GLASMACHER T., GROH D. E., HONMA M., HOROI M., JANSSENS R. V. F., MIZUSAKI T., MORRISSEY D. J., MORTON A. C., MUELLER W. F., OTSUKA T., PAVAN J., SCHATZ H., STOLTZ A., TABOR S. L., TOMLIN B. E. and WIEDEKING M., *Phys. Rev. C*, **70** (2004) 064303.

[28] FORNAL B., ZHU S., JANSSENS R. V. F., HONMA M., BRODA R., MANTICA P. F., BROWN B. A., CARPENTER M. P., DALY P. J., FREEMAN S. J., GRABOWSKI Z. W., HAMMOND N., KONDEV F. G., KROLAS W., LAURITSEN T., LIDDICK S. N., LISTER C. J., MOORE E. F., OTSUKA T., PAWLAT T., SEWERYNIAK D., TOMLIN B. E. and WRZESINSKI J., *Phys. Rev. C*, **70** (2004) 064304.

[29] YURKEWICZ K. L., BAZIN D., BROWN B. A., CAMPBELL C. M., CHURCH J. A., DINCA D.-C., GADE A., GLASMACHER T., HONMA M., MIZUSAKI T., MUELLER W. F., OLLIVER H., OTSUKA T., RILEY L. A. and TERRY J. R., *Phys. Rev. C*, **70** (2004) 064321.

[30] DINCA D.-C., JANSSENS R. V. F., GADE A., BAZIN D., BRODA R., BROWN B. A., CAMPBELL C. M., CARPENTER M. P., CHOWDHURY P., COOK J. M., DEACON A. N., FORNAL B., FREEMAN S. J., GLASMACHER T., HONMA M., KONDEV F. G., LECOUEY J.-L., LIDDICK S. N., MANTICA P. F., MUELLER W. F., OLLIVER H., OTSUKA T., TERRY J. R., TOMLIN B. A. and YONEDA K., *Phys. Rev. C*, **71** (2005) 041302.

[31] ANDERSSON L.-L., JOHANSSON E. K., EKMAN J., RUDOLPH D., DU RIETZ R., FAHLANDER C., GROSS C. J., HAUSLADEN P. A., RADFORD D. C. and HAMMOND G., *Phys. Rev. C*, **71** (2005) 011303.

[32] HAGEMANN M., BÄUMER C., VAN DEN BERG A. M., DE FRENNE D., FREKERS D., HANNEN V. M., HARAKEH M. N., HEYSE J., DE HUU M. A., JACOBS E., LANGANKE K., MARTÍNEZ-PINEDO G., NEGRET A., POPESCU L., RAKERS S., SCHMIDT R. and WÖRTCHE H. J., *Phys. Rev. C*, **71** (2005) 014606.

[33] BÜRGER A., SAITO T. R., GRAWE H., HÜBEL H., REITER P., GERL J., GÓRSKA M., WOLLERSHEIM H. J., AL-KHATIB A., BANU A., BECK T., BECKER F., BEDNARCZYK P., BENZONI G., BRACCO A., BRAMBILLA S., BRINGEL P., CAMERA F., CLÉMENT E., DOORNENBAL P., GEISSEL H., GÖRGEN A., GRĘBOSZ J., HAMMOND G., HELLSTRÖM M., HONMA M., KAVATSYUK M., KAVATSYUK O., KMIECIK M., KOJOUHAROV I., KORTEN W., KURZ N., LOZEVA R., MAJ A., MANDAL S., MILLION B., MURALITHAR S., NEUSSER A., NOWACKI F., OTSUKA T., PODOLYÁK ZS., SAITO N., SINGH A. K., WEICK H., WHELDON C., WIELAND O., WINKLER M. and RISING COLLABORATION, *Phys. Lett. B*, **622** (2005) 29.

[34] LESKE J., SPEIDEL K.-H., SCHIELKE S., KENN O., GERBER J., MAIER-KOMOR P., ROBINSON S. J., ESCUDEROS A., SHARON Y. Y. and ZAMICK L., *Phys. Rev. C*, **71** (2005) 044316.

[35] BRANDOLINI F. and UR C. A., *Phys. Rev. C*, **71** (2005) 054316.

[36] HONMA M., OTSUKA T., BROWN B. A. and MIZUSAKI T., *Eur. Phys. J. A*, **25** (2005) 499.

[37] HONMA M., OTSUKA T., MIZUSAKI T., HJORTH-JENSEN M. and BROWN B. A., *J. Phys. Conf.*, **20** (2005) 7.

[38] SILVEIRA M. A. G., MEDINA N. H., ALCÁNTARA-NÚÑEZ J. A., CYBULSKA E. W., DIAS H., OLIVEIRA J. R. B., RAO M. N., RIBAS R. V., SEALE W. A., WIEDEMANN K. T., BROWN B. A., HONMA M., MIZUSAKI T. and OTSUKA T., *J. Phys. G*, **31** (2005) s1577.

[39] DU RIETZ R., WILLIAMS S. J., RUDOLPH D., EKMAN J., FAHLANDER C., ANDREOIU C., AXIOTIS M., BENTLEY M. A., CARPENTER M. P., CHANDLER C., CHARITY R. J., CLARK R. M., CROMAZ M., DEWALD A., DE ANGELIS G., DELLA VEDOVA F., FALLON P., GADEA A., HAMMOND G., IDEGUCHI E., LENZI S. M., MACCHIAVELLI A. O., MARGINEAN N., MINEVA M. N., MÖLLER O., NAPOLI D. R., NESPOLO M., REVIOL W., RUSU C., SAHA B., SARANTITES D. G., SEWERYNIAK D., TONEV D. and UR C. A., *Phys. Rev. C*, **72** (2005) 014307.

[40] RILEY L. A., ABDELQADER M. A., BAZIN D., BOJAZI M. J., BROWN B. A., CAMPBELL C. M., CHURCH J. A., COTTLE P. D., DINCA D.-C., ENDERS J., GADE A., GLASMACHER T., HONMA M., HORIBE S., HU Z., KEMPER K. W., MUELLER W. F., OLLIVER H., OTSUKA T., PERRY B. C., ROEDER B. T., SHERRILL B. M., SPENCER T. P. and TERRY J. R., *Phys. Rev. C*, **72** (2005) 024311.

[41] FORNAL B., ZHU S., JANSSENS R. V. F., HONMA M., BRODA R., BROWN B. A., CARPENTER M. P., FREEMAN S. J., HAMMOND N., KONDEV F. G., KRÓLAS W., LAURITSEN T., LIDDICK S. N., LISTER C. J., LUNARDI S., MANTICA P. F., MARGINEAN N., MIZUSAKI T., MOORE E. F., OTSUKA T., PAWLAT T., SEWERYNIAK D., TOMLIN B. E., UR C. A., WIEDENHÖVER I. and WRZESINSKI J., *Phys. Rev. C*, **72** (2005) 044315.

[42] POVES A., NOWACKI F. and CAURIER E., *Phys. Rev. C*, **72** (2005) 047302.

[43] COSTIN A., PIETRALLA N., KOIKE T., VAMAN C., AHN T. and RAINOVSKI G., *Phys. Rev. C*, **72** (2005) 054305.

[44] LIDDICK S. N., MANTICA P. F., BRODA R., BROWN B. A., CARPENTER M. P., DAVIES A. D., FORNAL B., HOROI M., JANSSENS R. V. F., MORTON A. C., MUELLER W. F., PAVAN J., SCHATZ H., STOLZ A., TABOR S. L., TOMLIN B. E. and WIEDEKING M., *Phys. Rev. C*, **72** (2005) 054321.

[45] ADACHI T., FUJITA Y., VON BRENTANO P., LISETSKIY A. F., BERG G. P. A., FRANSEN C., DE FRENNE D., FUJITA H., FUJITA K., HATANAKA K., HONMA M., JACOBS

E., KAMIYA J., KAWASE K., MIZUSAKI T., NAKANISHI K., NEGRET A., OTSUKA T., PIETRALLA N., POPESCU L., SAKEMI Y., SHIMBARA Y., SHIMIZU Y., TAMESHIGE Y., TAMII A., UCHIDA M., WAKASA T., YOSOI M. and ZELL K. O., *Phys. Rev. C*, **73** (2006) 024311.

[46] GADE A., JANSSENS R. V. F., BAZIN D., BROWN B. A., CAMPBELL C. M., CARPENTER M. P., COOK J. M., DEACON A. N., DINCA D.-C., FREEMAN S. J., GLASMACHER T., KAY B. P., MANTICA P. F., MUELLER W. F., TERRY J. R. and ZHU S., *Phys. Rev. C*, **73** (2006) 037309.

[47] LIDDICK S. N., MANTICA P. F., BROWN B. A., CARPENTER M. P., DAVIES A. D., HOROI M., JANSSENS R. V. F., MORTON A. C., MUELLER W. F., PAVAN J., SCHATZ H., STOLZ A., TABOR S. L., TOMLIN B. E. and WIEDEKING M., *Phys. Rev. C*, **73** (2006) 044322.

[48] HOROI M., BROWN B. A., OTSUKA T., HONMA M. and MIZUSAKI T., *Phys. Rev. C*, **73** (2006) 061305.

[49] HONMA M., OTSUKA T., MIZUSAKI T. and HJORTH-JENSEN M., *J. Phys. Conf.*, **49** (2006) 45.

[50] LESKE J., SPEIDEL K.-H., SCHIELKE S., GERBER J., MAIER-KOMOR P., ROBINSON S. J. Q., ESCUDEROS A., SHARON Y. Y. and ZAMICK L., *Phys. Rev. C*, **74** (2006) 024315.

[51] ZHU S., DEACON A. N., FREEMAN S. J., JANSSENS R. V. F., FORNAL B., HONMA M., XU F. R., BRODA R., CALDERIN I. R., CARPENTER M. P., CHOWDHURY P., KONDEV F. G., KRÓLAS W., LAURITSEN T., LIDDICK S. N., LISTER C. J., MANTICA P. F., PAWLAT T., SEWERYNIAK D., SMITH J. F., TABOR S. L., TOMLIN B. E., VARLEY B. J. and WRZESINSKI J., *Phys. Rev. C*, **74** (2006) 064315.

[52] SILVEIRA M. A. G., MEDINA N. H., OLIVEIRA J. R. B., ALCÁNTARA-NÚÑEZ J. A., CYBULSKA E. W., DIAS H., RAO M. N., RIBAS R. V., SEALE W. A., WIEDEMANN K. T., BROWN B. A., HONMA M., MIZUSAKI T. and OTSUKA T., *Phys. Rev. C*, **74** (2006) 064312.

[53] ZHU S., JANSSENS R. V. F., FORNAL B., FREEMAN S. J., HONMA M., BRODA R., CARPENTER M. P., DEACON A. N., KAY B. P., KONDEV F. G., KRÓLAS W., KOZEMCZAK J., LARABEEE A., LAURITSEN T., LIDDICK S. N., LISTER C. J., MANTICA P. F., OTSUKA T., PAWLAT T., ROBINSON A., SEWERYNIAK D., SMITH J. F., STEPPENBECK D., TOMLIN B. E., WRZESINSKI J. and WANG X., *Phys. Lett. B*, **650** (2007) 135.

[54] HOROI M., STOICA S. and BROWN B. A., *Phys. Rev. C*, **75** (2007) 034303.

[55] DOSSAT C., ADIMI N., AKSOUH F., BECKER F., BEY A., BLANK B., BORCEA C., BORCEA R., BOSTON A., CAAMANO M., CANCHEL G., CHARTIER M., CORTINA D., CZAJKOWSKI S., DE FRANCE G., DE OLIVEIRA SANTOS F., FLEURY A., GEORGIEV G., GIOVINAZZO J., GRÉVY S., GRZYWACZ R., HELLSTRÖM M., HONMA M., JANAS Z., KARAMANIS D., KURCEWICZ J., LEWITOWICZ M., LÓPEZ JIMÉNEZ M. J., MAZZOCCHI C., MATEA I., MASLOV V., MAYET P., MOORE C., PFÜTZNER M., PRAVIKOFF M. S., STANOIU M., STEFAN I. and THOMAS J. C., *Nucl. Phys. A*, **792** (2007) 18.

[56] DEACON A. N., FREEMAN S. J., JANSSENS R. V., HONMA M., CARPENTER M. P., CHOWDHURY P., LAURITSEN T., LISTER C. J., SEWERYNIAK D., SMITH J. F., TABOR S. L., VARLEY B. J., XU F. R. and ZHU S., *Phys. Rev. C*, **76** (2007) 054303.

[57] UTSUNO Y., OTSUKA T., MIZUSAKI T. and HONMA M., *Phys. Rev. C*, **60** (1999) 054315.

[58] UTSUNO Y., OTSUKA T., GLASMACHER T., MIZUSAKI T. and HONMA M., *Phys. Rev. C*, **70** (2004) 044307.

[59] UTSUNO Y., OTSUKA T., MIZUSAKI T. and HONMA M., *Phys. Rev. C*, **64** (2001) 011301(R).

[60] UTSUNO Y., OTSUKA T., GLASMACHER T., MIZUSAKI T. and HONMA M., *Phys. Rev. C*, **70** (2004) 044307.

[61] ELEKES Z., DOMBRADI ZS., SAITO A., AOI N., BABA H., DEMICHI K., FULOP ZS., GIBELIN J., GOMI T., HASEGAWA H., IMAI N., ISHIHARA M., IWASAKI H., KANNO S., KAWAI S., KISHIDA T., KUBO T., KURITA K., MATSUYAMA Y., MICHIMASA S., MINEMURA T., MOTOBAYASHI T., NOTANI M., OHNISHI T., ONG H. J., OTA S., OZAWA A., SAKAI H. K., SAKURAI H., SHIMOURA S., TAKESHITA E., TAKEUCHI S., TAMAKI M., TOGANO Y., YAMADA K., YANAGISAWA Y. and YONEDA K., *Phys. Lett. B*, **599** (2004) 17.

[62] NEYENS G., KOWALSKA M., YORDANOV D., BLAUM K., HIMPE P., LIEVENS P., MALLION S., NEUGART R., UTSUNO Y. and OTSUKA T., *Phys. Rev. Lett.*, **94** (2005) 022501.

[63] TRIPATHI V., TABOR S. L., MANTICA P. F., HOFFMAN C. R., WIEDEKING M., DAVIES A. D., LIDDICK S. N., MUELLER W. F., OTSUKA T., STOLZ A., TOMLIN B. E., UTSUNO Y. and VOLYA A., *Phys. Rev. Lett.*, **94** (2005) 162501.

[64] MASON P., MARGINEAN N., LENZI S. M., IONESCU-BUJOR M., DELLA VEDOVA F., NAPOLI D. R., OTSUKA T., UTSUNO Y., NOWACKI F., AXIOTIS M., BAZZACCO D., BIZZETI P. G., BIZZETI-SONA A., BRANDOLINI F., CARDONA M., DE ANGELIS G., FARNEA E., GADEA A., HOJMAN D., IORDACHESCU A., KALFAS C., KROLL TH., LUNARDI S., MARTINEZ T., PETRACHE C. M., QUINTANA B., RIBAS R. V., ROSSI ALVAREZ C., UR C. A., VLASTOU R. and ZILIO S., *Phys. Rev. C*, **71** (2005) 014316.

[65] BELLEGUIC M., AZAIEZ F., DOMBRADI ZS., SOHLER D., LOPEZ-JIMENEZ M. J., OTSUKA T., SAINT-LAURENT M. G., SORLIN O., STANOIU M., UTSUNO Y., PENIONZHKEVICH YU.-E., ACHOURI N. L., ANGELIQUE J. C., BORCEA C., BOURGEOIS C., DAUGAS J. M., DE OLIVEIRA-SANTOS F., DLOUHY Z., DONZAUD C., DUPRAT J., ELEKES Z., GREVY S., GUILLEMAUD-MUELLER D., LEENHARDT S., LEWITOWICZ M., LUKYANOV S. M., MITTIG W., PORQUET M. G., POUGHEON F., ROUSSEL-CHOMAZ P., SAVAJOLS H., SOBOLEV Y., STODEL C. and TIMAR J., *Phys. Rev. C*, **72** (2005) 054316.

[66] IONESCU-BUJOR M., IORDACHESCU A., NAPOLI D. R., LENZI S. M., MARGINEAN N., OTSUKA T., UTSUNO Y., RIBAS R. V., AXIOTIS M., BAZZACCO D., BIZZETI-SONA A. M., BIZZETI P. G., BRANDOLINI F., BUCURESCU D., CARDONA M. A., DE ANGELIS G., DE POLI M., DELLA VEDOVA F., FARNEA E., GADEA A., HOJMAN D., KALFAS C. A., KROLL TH., LUNARDI S., MARTINEZ T., MASON P., PAVAN P., QUINTANA B., ROSSI ALVAREZ C., UR C. A., VLASTOU R. and ZILIO S., *Phys. Rev. C*, **73** (2006) 024310.

[67] TRIPATHI V., TABOR S. L., HOFFMAN C. R., WIEDEKING M., VOLYA A., MANTICA P. F., DAVIES A. D., LIDDICK S. N., MUELLER W. F., STOLZ A., TOMLIN B. E., OTSUKA T. and UTSUNO Y., *Phys. Rev. C*, **73** (2006) 054303.

[68] DOMBRÁDI ZS., ELEKES Z., SAITO A., AOI N., BABA H., DEMICHI K., FULOP ZS., GIBELIN J., GOMI T., HASEGAWA H., IMAI N., ISHIHARA M., IWASAKI H., KANNO S., KAWAI S., KISHIDA T., KUBO T., KURITA K., MATSUYAMA Y., MICHIMASA S., MINEMURA T., MOTOBAYASHI T., NOTANI M., OHNISHI T., ONG H. J., OTA S., OZAWA A., SAKAI H. K., SAKURAI H., SHIMOURA S., TAKESHITA E., TAKEUCHI S., TAMAKI M., TOGANO Y., YAMADA K., YANAGISAWA Y. and YONEDA K., *Phys. Rev. Lett.*, **96** (2006) 182501.

[69] TERRY J. R., BAZIN D., BROWN B. A., CAMPBELL C. M., CHURCH J. A., COOK J. M., DAVIES A. D., DINCA D.-C., ENDERS J., GADE A., GLASMACHER T., HANSEN P. G., LECOUEY J. L., OTSUKA T., PRITYCHENKO B., SHERRILL B. M., TOSTEVIN J. A., UTSUNO Y., YONEDA K. and ZWAHLEN H., *Phys. Lett. B*, **640** (2006) 86.

[70] HIMPE P., NEYENS G., BALABANSKI D. L., BELIER G., BORREMANS D., DAUGAS J. M., DE OLIVEIRA SANTOS F., DE RYDT M., FLANAGANA K., GEORGIEVD G., KOWALSKAE M., MALLION S., MATEA I., MOREL P., PENIONZHKEVICH YU. E., SMIRNOVAG. STODEL N. A., TURZOA K., VERMEULEN N. and YORDANOV D., *Phys. Lett. B*, **643** (2006) 257.

[71] GADE A., ADRICH P., BAZIN D., BOWEN M. D., BROWN B. A., CAMPBELL C. M., COOK J. M., ETTENAUER S., GLASMACHER T., KEMPER K. W., McDANIEL S., OBERTELLI A., OTSUKA T., RATKIEWICZ A., SIWEK K., TERRY J. R., TOSTEVIN J. A., UTSUNO Y. and WEISSHAAR D., *Phys. Rev. Lett.*, **99** (2007) 072502.

[72] TRIPATHI V., TABOR S. L., MANTICA P. F., UTSUNO Y., BENDER P., COOK J., HOFFMAN C. R., LEE S., OTSUKA T., PEREIRA J., PERRY M., PEPPER K., PINTER J., STOKER J., VOLYA A. and WEISSHAAR D., *Phys. Rev. C*, **76** (2007) 021301(R).

[73] OTSUKA T. *et al.*, *Phys. Rev. Lett.*, **87** (2001) 082502.

[74] OTSUKA T., SUZUKI T., FUJIMOTO R., GRAWE H. and AKAISHI Y., *Phys. Rev. Lett.*, **95** (2005) 232502.

[75] OTSUKA T., MATSUO T. and ABE D., *Phys. Rev. Lett.*, **97** (2006) 162501.

[76] OTSUKA T., HONMA M. and ABE D., *Nucl. Phys. A*, **788** (2007) 3c.

[77] SCHIFFER J. P. *et al.*, *Phys. Rev. Lett.*, **92** (2004) 162501.

DOI 10.3254/978-1-58603-885-4-291

Shell-model calculations with low-momentum realistic interactions

A. Covello

Dipartimento di Scienze Fisiche, Università di Napoli Federico II
INFN, Complesso Universitario di Monte S. Angelo
Via Cintia, I-80126 Napoli, Italy

Summary. — In this paper, I first give a brief survey of the theoretical framework for microscopic shell-model calculations starting from the free nucleon-nucleon (NN) potential. In this context, I discuss the use of the low-momentum NN potential $V_{\text{low-}k}$ in the derivation of the two-body effective interaction and emphasize its practical value as an alternative to the Brueckner G-matrix method. Then, I present some results of a recent shell-model study of exotic nuclei beyond ^{132}Sn, which we have obtained starting from the CD-Bonn potential renormalized by use of the $V_{\text{low-}k}$ approach. Comparison shows that they are in very good agreement with the available experimental data.

1. – Introduction

A main problem with the nuclear shell model, which is the basic framework for the study of complex nuclei in terms of nucleons, has long been the determination of the two-body effective interaction V_{eff}. Since the early 1950s through the mid 1990s, in the vast majority of shell-model calculations either empirical effective interactions containing adjustable parameters have been used or the two-body matrix elements themselves have been treated as free parameters.

It goes without saying that a fundamental goal of nuclear physics is to understand the properties of nuclei starting from the forces between nucleons. Within the framework of the shell model this implies the derivation of V_{eff} from the free nucleon-nucleon (NN) potential. Great progress in this direction has been achieved over the last decade, proving the ability of "realistic" effective interactions derived from modern NN potentials to provide, with no adjustable parameters, an accurate description of nuclear properties [1,2]. As a consequence, in the past few years the use of these interactions has been rapidly gaining ground, opening new perspectives to nuclear structure theory.

A main difficulty encountered in the derivation of V_{eff} from the free NN potential is the existence of a strong repulsive core. As is well known, the traditional way to overcome this difficulty is the Brueckner G-matrix method. A few years ago, a new approach was proposed [3,4] which consists in deriving from V_{NN} a renormalized low-momentum potential, $V_{\text{low-}k}$, that preserves the physics of the original potential up to a certain cutoff momentum Λ. This is a smooth potential which can be used directly to derive V_{eff}. As we shall discuss in more detail in sect. **4**, we have shown [2-5] that this approach provides an advantageous alternative to the use of the G matrix.

Recently, we have studied [6-9] several neutron-rich nuclei around ^{132}Sn in terms of the shell model, employing as input for the calculation of V_{eff} a low-momentum potential $V_{\text{low-}k}$ obtained by renormalizing the CD-Bonn NN potential [10]. The study of exotic nuclei around ^{132}Sn is currently drawing much attention, as it offers the opportunity to explore for possible modifications of the shell structure when approaching the neutron drip line. In this context, of special interest are exotic nuclei beyond the $N = 82$ shell closure. The experimental study of these nuclei is very difficult, but in recent years new facilities and techniques have made some of them accessible to spectroscopic studies. Today, the development of radioactive ion beams opens up the prospect of gaining substantial information on far-from-stability nuclei in the ^{132}Sn region. This makes it very interesting and timely not only to try to explain the available data but also to make predictions which may stimulate, and be helpful to, future experiments.

The purpose of this paper is twofold. On the one hand, it gives a brief survey of the theoretical framework for realistic shell-model calculations employing low-momentum realistic interactions. On the other hand, it presents some results of our recent calculations on exotic nuclei beyond ^{132}Sn. More precisely, I shall focus attention on ^{134}Sn and on the two odd-odd Sb isotopes ^{134}Sb and ^{136}Sb. The first and last nucleus with an N/Z ratio of 1.68 and 1.67, respectively, are at present the most exotic nuclei beyond ^{132}Sn for which information exists on excited states.

The outline of the paper is as follows. Following this introduction, a brief review of the NN interaction is given in sect. **2**, which is mainly aimed at sketching the present state of the art in the field. The derivation of the shell-model effective interaction is discussed in sect. **3**, while the low-momentum NN potential $V_{\text{low-}k}$ is introduced in sect. **4**. The results of our calculations for the nuclei mentioned above are reported and compared with experiment in sect. **5**. The last section, sect. **6**, provides a brief summary and concluding remarks.

2. – High-quality nucleon-nucleon potentials

While the NN interaction has been extensively studied since the discovery of the neutron, it is only from the mid 1990s on that high-quality NN potentials have come into play which fit the about 6000 proton-proton and neutron-proton scattering data with a $\chi^2/\text{datum} \approx 1$. These are the potentials constructed by the Nijmegen group, Nijm-I and NijmII [11], the Argonne V_{18} potential [12], and the CD-Bonn potential [10]. While Nijm-II is a totally local potential, the Nijm-I contains momentum-dependent terms which in configuration space give rise to nonlocalities in the central force component. Except for the One-Pion-Exchange (OPE) tail, these potentials are purely phenomenological with a total of 41 and 47 parameters for Nijm-I and Nijm-II, respectively. The CD-Bonn potential is a charge-dependent, nonlocal one-boson-exchange potential. It includes the π, ρ, and ω mesons plus two effective scalar-isoscalar σ bosons, the parameters of which are partial-wave dependent; the total number of parameters is 43. The Argonne V_{18} model, so named for its operator content, is a purely phenomenological (except for the correct OPE tail) nonrelativistic NN potential with a local operator structure, having in total 40 adjustable parameters.

All the high-precision NN potentials mentioned above have a large number of free parameters, say about 45, which is the price one has to pay to achieve a very accurate fit of the world NN data. This makes it clear that, to date, high-quality potentials with an excellent $\chi^2/\text{datum} \approx 1$ can only be obtained within the framework of a substantially phenomenological approach. Since these potentials fit almost equally well the NN data up to the inelastic threshold, their on-shell properties are essentially identical, namely they are "phase-shift equivalent". In addition, they all predict almost identical deuteron observables. While they have also in common the inclusion of the OPE contribution, their off-shell behavior may be quite different. A detailed comparison between their predictions is given in the review paper [13]. I only mention here that the predicted D-state probability of the deuteron ranges from 4.85% for CD-Bonn to 5.76% for V_{18}. In this context, the question arises of how much nuclear structure results may depend on the NN potential one starts with. We shall consider this important point in sect. **4**.

In recent years, efforts have been made to construct NN potentials within the framework of the chiral effective theory. The literature on this subject, which is still actively pursued, is by now very extensive and even a brief summary is outside the limits of this paper. Thus, I shall only give here a bare outline of the essentials of this important current issue in the theory of the NN interaction. A comprehensive review may be found in ref. [14].

The approach to the NN interaction based upon chiral effective field theory was started by Weinberg [15,16] some fifteen years ago and then developed by several authors. The basic idea [15] is to derive the NN potential starting from the most general chiral Lagrangian for low-energy pions and nucleons, consistent with the symmetries of quantum chromodynamics, in particular the spontaneously broken chiral symmetry. The chiral Lagrangian provides a perturbative framework for the derivation of the nucleon-nucleon potential. In fact, it was shown by Weinberg [16] that a systematic expansion of the nuclear potential exists in powers of the small parameter Q/Λ_χ, where Q denotes a generic low-momentum and $\Lambda_\chi \approx 1\,\mathrm{GeV}$ is the chiral symmetry-breaking scale. This perturbative low-energy theory is called chiral perturbation theory (χPT).

Very recently, chiral potentials at the next-to-next-to-next-to-leading order (N^{3}LO, fourth order) have been constructed [17, 18]. The potential developed in [17], dubbed Idaho N^{3}LO, with 29 parameters in all, gives a χ^2/datum for the reproduction of the np and pp data below 290 MeV of 1.10 and 1.50, respectively. A brief survey of the current status of the chiral potentials as well as a list of references to recent nuclear structure studies employing the Idaho N^{3}LO potential may be found in ref. [19]. We only mention here that the use of this potential by our own group [20-22] has produced very promising results.

The foregoing discussion has all been focused on the two-nucleon force. As is well known, the role of three-nucleon ($3N$) interactions in light nuclei has been, and is currently, actively investigated. However, as regards the derivation of a realistic shell-model effective interaction, the $3N$ forces have not been taken into account up to now.

3. – The shell-model effective interaction

The shell-model effective interaction V_{eff} is defined, as usual, in the following way. In principle, one should solve a nuclear many-body Schrödinger equation of the form

$$H\Psi_i = E_i\Psi_i, \tag{1}$$

with $H = T + V_{NN}$, where T denotes the kinetic energy. This full-space many-body problem is reduced to a smaller model-space problem of the form

$$PH_{\mathrm{eff}}P\Psi_i = P(H_0 + V_{\mathrm{eff}})P\Psi_i = E_iP\Psi_i. \tag{2}$$

Here $H_0 = T + U$ is the unperturbed Hamiltonian, U being an auxiliary potential introduced to define a convenient single-particle basis, and P denotes the projection operator onto the chosen model space,

$$P = \sum_{i=1}^{d} |\psi_i\rangle\langle\psi_i|, \tag{3}$$

d being the dimension of the model space and $|\psi_i\rangle$ the eigenfunctions of H_0. The effective interaction V_{eff} operates only within the model space P. In operator form it can be schematically written [23] as

$$V_{\mathrm{eff}} = \hat{Q} - \hat{Q}'\int\hat{Q} + \hat{Q}'\int\hat{Q}\int\hat{Q} - \hat{Q}'\int\hat{Q}\int\hat{Q}\int\hat{Q} + \ldots, \tag{4}$$

where $\hat{Q}$, usually referred to as the $\hat{Q}$-box, is a vertex function composed of irreducible linked diagrams, and the integral sign represents a generalized folding operation. $\hat{Q}'$ is obtained from $\hat{Q}$ by removing terms of first order in the interaction. Once the $\hat{Q}$-box is calculated, the folded-diagram series of eq. (4) can be summed up to all orders by iteration methods.

As already pointed out in the Introduction, a main difficulty with the derivation of V_{eff} from any modern NN potential is the existence of a repulsive core which prevents its direct use in nuclear structure calculations. This difficulty is usually overcome by resorting to the well known Brueckner G-matrix method. The G-matrix is obtained from the bare NN potential V_{NN} by solving the Bethe-Goldstone equation

$$(5) \qquad G(\omega) = V_{NN} + V_{NN} Q_2 \frac{1}{\omega - Q_2 T Q_2} Q_2 G(\omega),$$

where T is the two-nucleon kinetic energy and ω is an energy variable, commonly referred to as starting energy. The operator Q_2 is the Pauli exclusion operator for two interacting nucleons, to make sure that the intermediate states of G must not only be above the filled Fermi sea but also outside the model space within which eq. (5) is to be solved.

The use of the G matrix has long proved to be a valuable tool to overcome the difficulty posed by the strong short-range repulsion contained in the free NN potential. However, the G matrix is model-space dependent as well as energy dependent; these dependences make its actual calculation rather involved. Recently, a new approach that bypasses the use of the G-matrix has been proposed [3,4]. In the next section, I shall only give a brief outline of it, while a detailed description can be found in ref. [4].

4. – The low-momentum nucleon-nucleon potential $V_{\text{low-}k}$

As pointed out in the Introduction, we "smooth out" the strong repulsive core contained in the bare NN potential V_{NN} by constructing a low-momentum potential $V_{\text{low-}k}$. This is achieved by integrating out the high-momentum modes of V_{NN} down to a cut-off momentum Λ. This integration is carried out with the requirement that the deuteron binding energy and low-energy phase shifts of V_{NN} are preserved by $V_{\text{low-}k}$. This requirement may be satisfied by the following T-matrix equivalence approach. We start from the half-on-shell T matrix for V_{NN}

$$(6) \qquad T(k',k,k^2) = V_{NN}(k',k) + \wp \int_0^\infty q^2 dq\, V_{NN}(k',q) \frac{1}{k^2 - q^2} T(q,k,k^2),$$

where $\wp$ denotes the principal value and k, k', and q stand for the relative momenta. The effective low-momentum T matrix is then defined by

$$(7) \qquad T_{\text{low-}k}(p',p,p^2) = V_{\text{low-}k}(p',p)$$

$$+ \wp \int_0^\Lambda q^2 dq\, V_{\text{low-}k}(p',q) \frac{1}{p^2 - q^2} T_{\text{low-}k}(q,p,p^2),$$

where the intermediate state momentum q is integrated from 0 to the momentum space cut-off Λ and $(p',p) \leq \Lambda$. The above T matrices are required to satisfy the condition

$$(8) \qquad T(p',p,p^2) = T_{\text{low-}k}(p',p,p^2); \quad (p',p) \leq \Lambda .$$

The above equations define the effective low-momentum interaction $V_{\text{low-}k}$, and it has been shown [4] that they are satisfied when $V_{\text{low-}k}$ is given by the Kuo-Lee-Ratcliff (KLR) folded-diagram expansion [23, 24], originally designed for constructing shell-model effective interactions, see eq. (4). In addition to the preservation of the half-on-shell T matrix, which implies preservation of the phase shifts, this $V_{\text{low-}k}$ preserves the deuteron binding energy, since eigenvalues are preserved by the KLR effective interaction.

For any value of Λ, $V_{\text{low-}k}$ can be calculated very accurately using iteration methods. Our calculation is performed by employing the iterative implementation of the Lee-Suzuki method [25] proposed in ref. [26]. The $V_{\text{low-}k}$ given by the T-matrix equivalence approach mentioned above is not Hermitian. Therefore, an additional transformation is needed to make it Hermitian. To this end, we resort to the Hermitization procedure suggested in [26], which makes use of the Cholesky decomposition of symmetric positive definite matrices.

Once the $V_{\text{low-}k}$ is obtained, we use it, plus the Coulomb force for protons, as input interaction in eq. (4) for the calculation of V_{eff}. This folded-diagram method was previously applied to many nuclei [13] using G-matrix interactions. Since $V_{\text{low-}k}$ is already a smooth potential, it is no longer necessary to calculate the G-matrix. More precisely, to obtain the shell-model effective interaction we first calculate the $\hat{Q}$-box including diagrams up to second order in $V_{\text{low-}k}$ and then sum up the folded-diagram series of eq. (4) using the Lee-Suzuki iteration method [11].

As mentioned in the Introduction, we have assessed the merit of the $V_{\text{low-}k}$ approach in practical applications. To this end, we have compared the results of shell-model calculations performed by starting from the CD-Bonn potential and deriving V_{eff} through both the $V_{\text{low-}k}$ and G-matrix approaches. In particular, results for ^{18}O are presented in [4] while the calculations of ref. [5] concern ^{132}Sn neighbors. A comparison between the G matrix and $V_{\text{low-}k}$ spectra for the heavy-mass nucleus ^{210}Po can be found in [2]. In all these calculations the cut-off parameter Λ has been chosen around $2 \, \text{fm}^{-1}$, in accord with the criterion given in [4]. It has been a remarkable finding of these studies that the $V_{\text{low-}k}$ results are as good as, or even better than, the G-matrix ones.

As we have discussed in sect. **2**, there are several high-quality potentials which fit equally well the NN scattering data. The results of our realistic shell-model calculations reported in the next section have all been obtained using as input the CD-Bonn potential. This may raise the question of how much they depend on this choice of the NN potential. We have verified that shell-model effective interactions derived from phase-shift equivalent NN potentials through the $V_{\text{low-}k}$ approach do not lead to significantly different results. Here, by way of illustration, we present the results obtained for the nucleus ^{134}Te. This nucleus has only two valence protons and thus offers the opportunity to test directly the matrix elements of the various effective interactions. In fig. 1 we show,

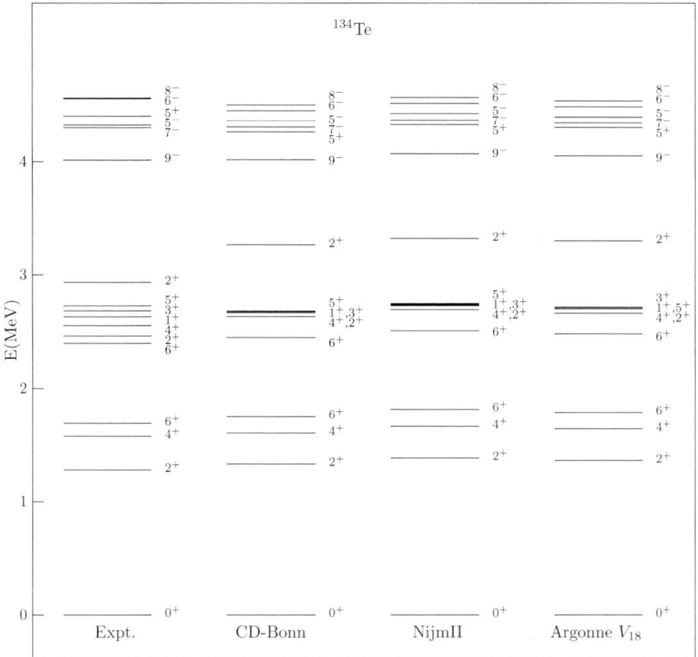

Fig. 1. – Spectrum of ^{134}Te. Predictions by various NN potentials are compared with experiment.

together with the experimental spectrum, the spectra obtained by using the CD-Bonn, NijmII, and Argonne V_{18} potentials, all renormalized through the $V_{\text{low-}k}$ procedure with a cutoff momentum $\Lambda = 2.2\,\text{fm}^{-1}$. From fig. 1 we see that the calculated spectra are very similar, the differences between the level energies not exceeding $80\,\text{keV}$. It is also seen that the agreement with experiment is very good for all the three potentials.

5. – Exotic nuclei beyond ^{132}Sn: comparison with available results and predictions

In our calculations we assume that ^{132}Sn is a closed core and let the valence neutrons occupy the six levels $0h_{9/2}$, $1f_{7/2}$, $1f_{5/2}$, $2p_{3/2}$, $2p_{1/2}$, and $0i_{13/2}$ of the 82-126 shell, while for the odd proton in ^{134}Sb and ^{136}Sb the model space includes the five levels $0g_{7/2}$, $1d_{5/2}$, $1d_{3/2}$, $2s_{1/2}$, and $0h_{11/2}$ of the 50-82 shell. As mentioned in the previous section, the two-body matrix elements of the effective interaction are derived from the CD-Bonn NN potential renormalized through the $V_{\text{low-}k}$ procedure with a cut-off momentum $\Lambda = 2.2\,\text{fm}^{-1}$. The computation of the diagrams included in the $\hat{Q}$-box is performed within the harmonic-oscillator basis using intermediate states composed of all possible hole states and particle states restricted to the five shells above the Fermi surface. The oscillator

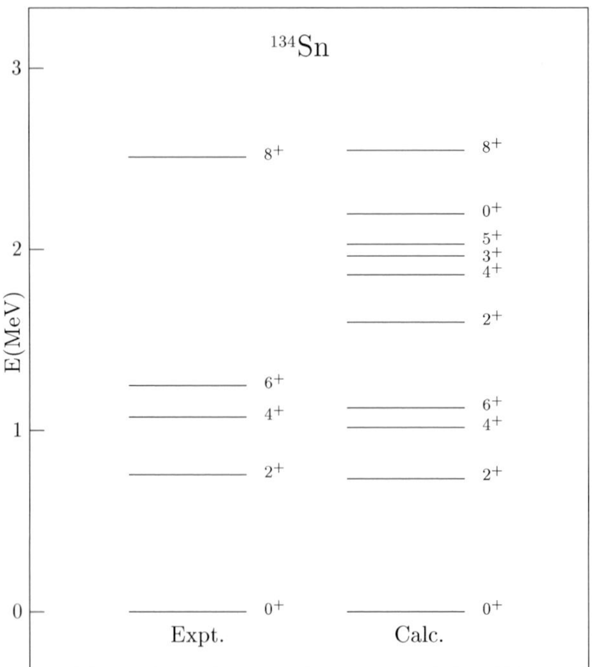

Fig. 2. – Experimental and calculated spectrum of ^{134}Sn.

parameter used is $\hbar\omega = 7.88\,\text{MeV}$. As regards the single-particle energies, they have been taken from experiment. All the adopted values are reported in ref. [6].

Let us start with ^{134}Sn. From fig. 2 we see that while the theory reproduces very well all the observed levels it also predicts, in between the 6^+ and 8^+ states, the existence of five states with spin ≤ 5. Clearly, the latter could not be seen from the γ-decay of the 8^+ state populated in the spontaneous fission experiment of ref. [27].

Recently, the $B(E2; 0^+ \to 2_1^+)$ value in ^{134}Sn has been measured [28] using Coulomb excitation of neutron-rich radioactive ion beams. We have calculated this $B(E2)$ with an effective neutron charge of $0.70\,e$. We obtain $B(E2; 0^+ \to 2_1^+) = 0.033\ e^2\text{b}^2$, in excellent agreement with the experimental value $0.029(4)\ e^2\text{b}^2$.

Let us now come to ^{134}Sb. In fig. 3 we show the energies of the first eight calculated states, which are multiplet, and compare them with the eight lowest-lying experimental states. The wave functions of these states are characterized by very little configuration mixing, the percentage of the leading component having a minimum value of 88% for the $J^\pi = 2^-$ state while ranging from 94% to 100% for all other states.

We see that the agreement between theory and experiment is very good, the discrepancies being in the order of a few tens of keV for most of the states. The largest discrepancy occurs for the 7^- state, which lies at about $130\,\text{keV}$ above its experimental counterpart. It is an important outcome of our calculation that we predict almost the

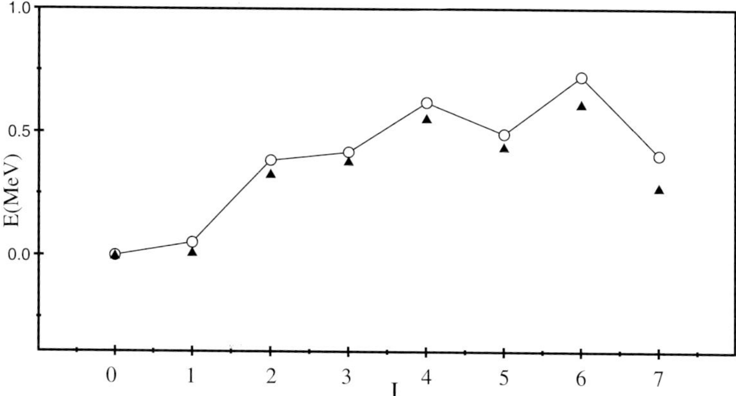

Fig. 3. – Proton-neutron $\pi g_{7/2}\nu f_{7/2}$ multiplet in ^{134}Sb. The theoretical results are represented by open circles while the experimental data by solid triangles.

right spacing between the 0^- ground state and first excited 1^- state. In fact, the latter has been observed at $13\,\mathrm{keV}$ excitation energy, our value being $53\,\mathrm{keV}$.

As regards ^{136}Sb, with two more valence neutrons, its ground state was identified as 1^- in the early β-decay study of ref. [29]. A few years ago, the spectroscopic study of ref. [30] led to the observation of a μs isomeric state, which was tentatively assigned a spin and parity of 6^-. Very recently [9], new experimental information has been obtained on the μs isomeric cascade, leading to the identification of two more excited states.

This achievement is at the origin of our realistic shell-model calculation for this nucleus [9], whose results we are now going to present. The four observed levels are all identified with states which are dominated by the configuration $\pi g_{7/2}\nu(f_{7/2})^3$. In fig. 4, we show these levels together with the calculated yrast states having angular momentum from 0^- to 7^-, which all arise from this configuration. These states may be viewed as the evolution of the $\pi g_{7/2}\nu f_{7/2}$ multiplet in ^{134}Sb. From fig. 4 we see that the agreement between theory and experiment is very good, the largest discrepancy being about $70\,\mathrm{keV}$.

An important piece of information is provided by the measured half life of the 6^- state, from which a $B(E2; 6^- \to 4^-)$ value of $170(40)\,e^2\,\mathrm{fm}^4$ is extracted. Using effective proton and neutron charges of $1.55\,e$ and $0.70\,e$, respectively, we obtain the value $131\,e^2\,\mathrm{fm}^4$, which compares very well with experiment. It is worth mentioning that these values of the effective charges have been consistently used in our previous calculations for nuclei in the ^{132}Sb region [31].

It is now interesting to compare the pattern of the calculated multiplet in ^{136}Sb with that of the $\pi g_{7/2}\nu f_{7/2}$ multiplet in ^{134}Sb. From fig. 5 we see that the two patterns are quite different. Both exhibit staggering starting from $J = 3^-$, but with different magnitude and opposite phase. In ^{136}Sb, the ground state is predicted to have $J^\pi = 1^-$ while the 0^- is raised at about $380\,\mathrm{keV}$, in contrast to the very small spacing between the 0^- ground state and the first excited 1^- state in ^{134}Sb. It is worth mentioning that

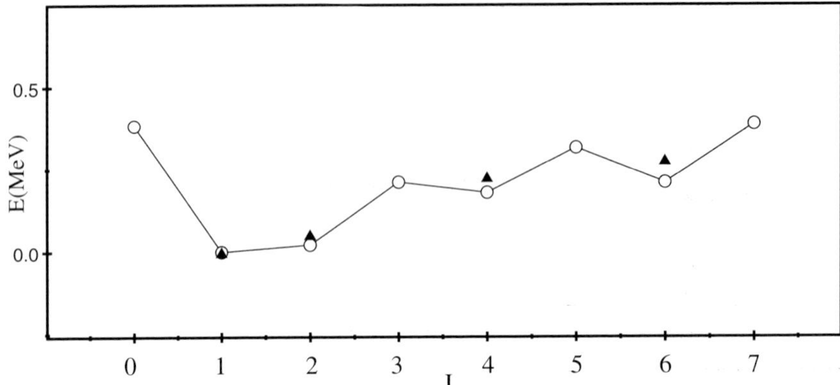

Fig. 4. – Low-lying levels in ^{136}Sb. The theoretical results are represented by open circles and the experimental data by solid triangles.

the percentage of configurations other than the dominant one in the considered states of ^{136}Sb is rather large, ranging from 24% to 32%. The configuration mixing, however, does not affect significantly the behavior of the spectrum of ^{136}Sb shown in fig. 5.

One may be surprised at the fact that an additional pair of neutrons can cause such rapid a change from the particle-particle spectrum of ^{134}Sb to a spectrum which is similar to the particle-hole one. Actually, the overall action of the neutron-proton interaction is significantly influenced by the presence of the two additional neutrons. We have verified that while the shift in the excitation energy of the 0^- state may be explained even by considering, as usual, the two additional neutrons coupled to zero

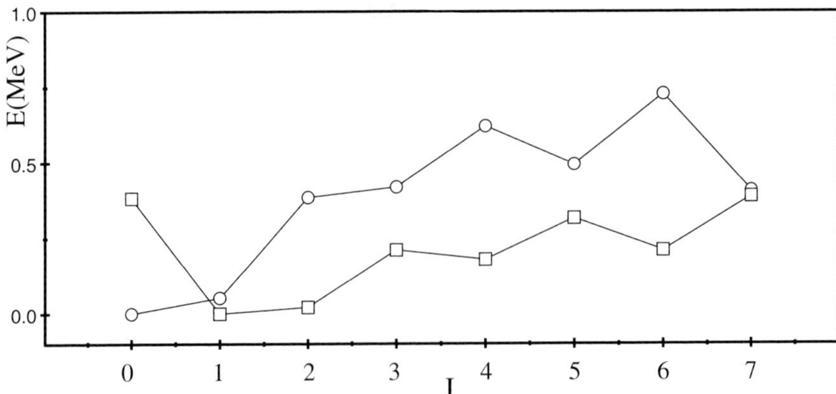

Fig. 5. – Low-lying calculated levels of ^{134}Sb (open circles) and ^{136}Sb (open squares).

angular momentum, the role of components with a non-zero coupled pair is crucial in determining the calculated staggering.

The observation of the missing states of the "$\pi g_{7/2} \nu f_{7/2}$ multiplet" in ^{136}Sb is certainly needed to verify the soundness of the above findings. However, the experimental level scheme resulting from the study of ref. [9] shows that there is no state with $J^\pi = 3^-$, 5^-, or 7^- below the isomeric 6^- state, which seems to confirm the theoretical predictions.

6. – Concluding remarks

This paper may be regarded as composed of two parts. In the first part, I have tried to give a self-contained yet brief survey of modern shell-model calculations employing two-body effective interactions derived from the free nucleon-nucleon potential. A main feature of these calculations is that no adjustable parameter appears in the determination of the effective interaction. This removes the uncertainty inherent in the traditional use of empirical interactions, making the shell model a truly microscopic theory of nuclear structure.

I have also shown how the $V_{\text{low-}k}$ approach to the renormalization of the strong short-range repulsion contained in all modern NN potentials is a valuable tool for nuclear structure calculations. The main merits of $V_{\text{low-}k}$ may be summarized as follows: i) $V_{\text{low-}k}$, since it is a smooth NN potential, can be used directly in nuclear structure calculations within a perturbative approach; ii) it does not depend either on the starting energy or on the model space, as is instead the case of the G matrix, which is defined in the nuclear medium; iii) the $V_{\text{low-}k}$'s extracted from various phase-shift equivalent potentials are very similar to each other, thus suggesting the realization of a nearly unique low-momentum NN potential.

In the second part of the paper, I have presented some results of our recent realistic shell-model calculations for nuclei beyond doubly magic ^{132}Sn. These neutron-rich nuclei offer the opportunity for a stringent test of the matrix elements of the effective interaction. The very good agreement with the available experimental data shown in sect. **5** supports confidence in the predictive power of our calculations in this region of exotic nuclei, which is of great current interest.

* * *

This paper is dedicated to my long-standing friend R. A. RICCI on the occasion of his 80th birthday. It has been my great pleasure to be one of the organizers of the CLXIX Varenna Course *Nuclear Structure far from Stability: new Physics and new Technology*, during which we celebrated this joyful event. Dear Renato, I wish you many more years of health and fruitful activity. This work was supported in part by the Italian Ministero dell'Istruzione, dell'Università e della Ricerca (MIUR).

REFERENCES

[1] COVELLO A., CORAGGIO L., GARGANO A. and ITACO N., *Acta Phys. Pol. B*, **32** (2001) 871, and references therein.

[2] COVELLO A., in *Proceedings of the International School of Physics "Enrico Fermi" Course CLIII*, edited by MOLINARI A., RICCATI L., ALBERICO W. M. and MORANDO M. (IOS Press, Amsterdam) 2003, pp. 79-91.

[3] KUO T. T. S., BOGNER S. K., CORAGGIO L., COVELLO A. and ITACO N., in *Challenges of Nuclear Stucture, Proceedings of the 7th International Spring Seminar on Nuclear Physics, Maiori, 2001*, edited by COVELLO A. (World Scientific, Singapore) 2002, pp. 129-138.

[4] BOGNER S. K., KUO T. T. S., CORAGGIO L., COVELLO A. and ITACO N., *Phys. Rev. C*, **65** (2002) 051301(R).

[5] COVELLO A., CORAGGIO L., GARGANO A. and ITACO N., in *Challenges of Nuclear Stucture, Proceedings of the 7th International Spring Seminar on Nuclear Physics, Maiori, 2001*, edited by COVELLO A. (World Scientific, Singapore) 2002, pp. 139-146.

[6] CORAGGIO L., COVELLO A., GARGANO A. and ITACO N., *Phys. Rev. C*, **72** (2005) 057302.

[7] CORAGGIO L., COVELLO A., GARGANO A. and ITACO N., *Phys. Rev. C*, **73** (2006) 031302(R).

[8] COVELLO A., CORAGGIO L., GARGANO A. and ITACO N., *Eur. Phys. J. ST*, **149** (2007) 93.

[9] SIMPSON G. S., ANGELIQUE J. C., GENEVEY J., PINSTON J. A., COVELLO A., GARGANO A., KÖSTER U., ORLANDI R. and SCHERILLO A., *Phys. Rev. C*, **76** (2007) 041303(R).

[10] MACHLEIDT R., *Phys. Rev. C*, **63** (2001) 031302(R).

[11] STOKS V. G. J., KLOMP A. M., TERHEGGEN C. P. F. and DE SWART J., *Phys. Rev. C*, **49** (1994) 2950.

[12] WIRINGA R. B., STOKS V. G. J. and SCHIAVILLA R., *Phys. Rev. C*, **51** (1995) 38.

[13] MACHLEIDT R. and SLAUS I., *J. Phys. G*, **27** (2001) R69.

[14] BEDAQUE P. F. and VAN KOLCK U., *Annu. Rev. Nucl. Part. Sci.*, **52** (2002) 339.

[15] WEINBERG S., *Phys. Lett. B*, **251** (1990) 288.

[16] WEINBERG S., *Nucl. Phys. B*, **363** (1991) 3.

[17] ENTEM D. R. and MACHLEIDT R., *Phys. Rev. C*, **68** (2003) 041001(R).

[18] EPELBAUM E., GLÖCKLE W. and MEISSNER U.-G., *Nucl. Phys. A*, **747** (2005) 362.

[19] MACHLEIDT R., *Nucl. Phys. A*, **790** (2007) 17c.

[20] CORAGGIO L., COVELLO A., GARGANO A., ITACO N. and KUO T. T. S., *Phys. Rev. C*, **68** (2003) 034320.

[21] CORAGGIO L., COVELLO A., GARGANO A., ITACO N., KUO T. T. S. and MACHLEIDT R., *Phys. Rev. C*, **71** (2005) 014307.

[22] CORAGGIO L., COVELLO A., GARGANO A., ITACO N. and KUO T. T. S., *Phys. Rev. C*, **73** (2006) 014304.

[23] KUO T. T. S., LEE S. Y. and RATCLIFF K. F., *Nucl. Phys. A*, **176** (1971) 65.

[24] KUO T. T. S. and OSNES E., *Lect. Notes Phys.*, Vol. **364** (Springer-Verlag, Berlin) 1990.

[25] SUZUKY K. and LEE S. Y., *Prog. Theor. Phys.*, **64** (1980) 2091.

[26] ANDREOZZI F., *Phys. Rev. C*, **54** (1996) 684.

[27] KORGUL A. *et al.*, *Eur. Phys. J. A*, **7** (2000) 167.

[28] BEENE J. R. *et al.*, *Nucl. Phys. A*, **746** (2004) 471c.

[29] HOFF P., OMTVED J. P., FOGELBERG B., MACH H. and HELLSTRÖM M., *Phys. Rev. C*, **56** (1997) 2865.

[30] MINEVA M. N. *et al.*, *Eur. Phys. K. A*, **11** (2001) 9.

[31] COVELLO A., CORAGGIO L., GARGANO A. and ITACO N., *Prog. Part. Nucl. Phys.*, **59** (2007) 401.

DOI 10.3254/978-1-58603-885-4-303

Studying nuclear structure by means of Coulomb energy differences

S. M. LENZI

Dipartimento di Fisica dell'Università and INFN - I-35131 Padova, Italy

Summary. — The study of differences in excitation energy between analogue states in isobaric multiplets allows to verify the validity of isospin symmetry and independence as a function of the angular momentum. These differences are of the order of tens of keV and can be well reproduced by state-of-the-art shell model calculations. Several nuclear-structure properties can be deduced from these data, such as the alignment of nucleons along rotational bands, the evolution of the nuclear radius and the identification of pure single-particle excitations across two main shells. In addition, the isospin breaking of the nuclear interaction is suggested by the systematic comparison with data. The different ingredients that enter the calculation of the Coulomb energy differences between mirror nuclei are introduced and shown in different examples of mirror nuclei in the $f_{7/2}$ and the sd shell.

1. – Introduction

Symmetries in physics are related to conserved quantities that in quantum mechanics give rise to good quantum numbers. Among the variety of symmetries we encounter in nuclear physics, the isospin symmetry is one of the fundamental ones: it is based on the charge symmetry and independence of the nuclear interaction. This allows to treat the proton and the neutron as the same particle, the nucleon, characterised by the isospin quantum number $t = 1/2$ and projection $t_z = -1/2$ and $t_z = 1/2$, respectively

(Heisemberg, 1932). This can be generalised to the whole nucleus (Wigner, 1937), where the projection of the isospin results from the contribution of the A nucleons

$$(1) \qquad T_z = \sum_{i=1}^{A} t_{z,i} = \frac{N-Z}{2},$$

and the total isospin T can get the values

$$(2) \qquad \left| \frac{N-Z}{2} \right| \leq T \leq \frac{N+Z}{2}.$$

Whilst T_z is always unique in a given nucleus, T can take different values, following eq. (2). In general (but there are many exceptions), the isospin of the ground and low-lying states is $T = |T_z| = |\frac{N-Z}{2}|$, the larger values of T lying higher in energy.

The simplest example is the deuteron. It is composed by a proton and a neutron, therefore $T_z = 0$ and $T = 0, 1$. The $J = 1, T = 0$ state is the ground state while the state $J = 0, T = 1$ is not bound. ($J + T$ has to be odd as the wave function of the two nucleons is antisymmetric.) This is telling us that the nuclear force is stronger in the $T = 0$ state. A system formed by two protons or two neutrons can be coupled to $J = 0, T = 1$, but not to $J = 1, T = 0$ as it is forbidden by the Pauli principle. The three states $J = 0, T = 1$ in the proton-proton, neutron-neutron and proton-neutron pairs form a triplet of isobaric analogue states. (It is well known that such two-nucleon states are not bound.)

A group of nuclei with the same mass is called an *isobaric multiplet*. The simplest example is a pair of mirror nuclei, which have exchanged the number of protons and neutrons. Such a pair should have identical structure in the hypothesis of the charge symmetry of the strong interaction. Another example of isobaric multiplets is the quadruplet with $A = 21$, $^{21}_{9}\mathrm{F}_{12}$, $^{21}_{10}\mathrm{Na}_{11}$, $^{21}_{11}\mathrm{Ne}_{10}$ and $^{21}_{12}\mathrm{Mg}_{9}$. In these nuclei we can identify analogue states, *i.e.* those states that have the same J and T and the same configuration in the isospin formalism. In this quadruplet, the mirror pair $^{21}_{10}\mathrm{Na}_{11}$, $^{21}_{11}\mathrm{Ne}_{10}$ has $|T_z| = 1/2$ and the pair $^{21}_{9}\mathrm{F}_{12}$, $^{21}_{12}\mathrm{Mg}_{9}$ has $|T_z| = 3/2$. There will be analogue states with $T = 1/2$ only in the first pair, while analogue states with $T = 3/2$ can be found in all the four nuclei.

Very difficult to reach experimentally are isobaric analogue states in multiplets with $T \geq 1$, where, generally, available data are limited to low masses and spins. This is due to the fact that, while nuclei with neutron excess are easily reachable, the proton-rich members of the multiplets are farther from the stability line and the cross section for their population via fusion-evaporation or transfer reactions using stable beams and targets is very low. Recently, excited states up to spin $J \sim 12$–13 have been observed in $T = 1$ isobaric triplets thanks to important improvements in gamma-ray spectroscopic techniques [1-4]. Using radioactive beams, more recently, the $J^{\pi} = 2^{+}$ $T = 2$ state in $^{36}\mathrm{Ca}$ has been observed and compared to the analogue state in its mirror $^{36}\mathrm{S}$ both in GSI and GANIL [5,6]. It is thus clear why the most studied mirror pairs are those with $|N - Z| = 1$, $T_z = \pm 1/2$.

Why is it interesting to study isobaric multiplets and what can we learn from these investigations? The answer comes from the breakdown of the isospin symmetry. In fact, in a nucleus the interaction between nucleons has two terms, a nuclear term and a Coulomb term. The latter acts only between protons and therefore it is not isospin symmetric. Therefore, when we compare two mirror nuclei, for example, we will find differences between analogue states induced by the Coulomb interaction. These differences are important if we compare the absolute energy of the states as the Coulomb contribution to the bulk energy is of the order of hundreds of MeV. This can be easily verified by a simple estimate. Consider the nucleus as a sphere of radius R_C. The electrostatic energy is given by

$$(3) \qquad E_C = \frac{3Z(Z-1)e^2}{5R_C} .$$

For a nucleus of mass $A = 49$, $Z = 25$ the (bulk) electrostatic energy amounts to $E \sim 120\,\text{MeV}$. Taking the difference in energy with the analogue ground state in its mirror with $A = 49$, $Z = 24$ ($T = 1/2$ states), we obtain $\Delta E \sim 10\,\text{MeV}$. These differences are called Coulomb Displacement Energies (CDE). If we now consider *excited* states in both nuclei and calculate the differences between the excitation energies (referred to the ground state in each partner), the differences we measure are generally of the order of tens of keV (three orders of magnitude smaller that the CDE) and in some few cases they can reach few hundreds of keV. These (small) differences have been given the name Coulomb Energy Differences (CED).

In this lecture we will concentrate on the CED to see how we can get from their measurement many interesting features of the *nuclear*-structure dynamics as a function of the angular momentum. These studies have become experimentally and theoretically feasible in the last decade when large arrays of Ge detectors were constructed and shell model codes that could deal with large matrices were developed.

In the laboratory, these nuclei are populated in general using fusion-evaporation reactions and the excited states are determined by measuring the gamma-rays emitted in the decay along the yrast line. In such reactions many channels open and detection systems with high sensitivity and resolving power are needed to get information on the weak proton-rich channels. Among the different experimental set-ups adopted for these studies using stable beams and targets, the most reliable ones use a gamma-ray array coupled to ancillary devices such as charged-particle and neutron detectors or mass spectrometers.

In the last decade, medium-light nuclei have been studied at high angular momentum along the $N = Z$ line. In particular, nuclei in the middle of the $f_{7/2}$ shell that present rotational behaviour near the ground state have been extensively investigated from both experimental and theoretical sides. It is important to note that it has been this collaboration between experimentalists and theoreticians which allowed to have a complete picture of the physics in this mass region. In particular, the CED have revealed very interesting features such as how the nucleons align along a rotational band, how the radius of the nucleus changes as a function of the angular momentum and how pure single-particle configurations in excitations across two main shells can be identified. Evidence of possible charge-asymmetric terms in the nuclear interaction has been also reported. These investigations have been recently extended to other mass regions.

An extensive description of the issues discussed in this work can be found in the recent review article of ref. [7]. In the following sections I will concentrate in some few examples to illustrate the different aspects of the nuclear structure that are evidenced when studying CED.

2. – The mirror pair ^{50}Fe-^{50}Cr and the nucleon alignment

The mirror nuclei $^{50}_{26}$Fe$_{24}$ and $^{50}_{24}$Cr$_{26}$ are deformed in the ground state ($\beta \sim 0.26$) and present a rotational-like band [1, 8, 9]. As these systems are rather small, above some value of the nuclear spin, it results energetically favourable to the nucleus to increase the angular momentum by aligning the spins of their valence nucleons to generate angular momentum. This causes a discontinuity in the regular sequence of transition energies between consecutive levels. In a plot of the angular momentum $vs.$ rotational frequency or transition energy, this translates in an upbending or even a backbending at the aligning frequency. This can be seen in fig. 1 for the case of ^{50}Cr: the regular "rotor" behaviour interrupts at $J = 8$ to give rise first to an upbending and later to a backbending.

Two identical nucleons in the $f_{7/2}$ orbital that are coupled to $J = 0$, can recouple gradually to higher values of J until they reach the maximum allowed value $J = 6$, or can directly align from $J = 0$ to $J = 6$. Which type of nucleons are aligning at the backbending is not easy to determine, mainly when we are dealing with nuclei near the

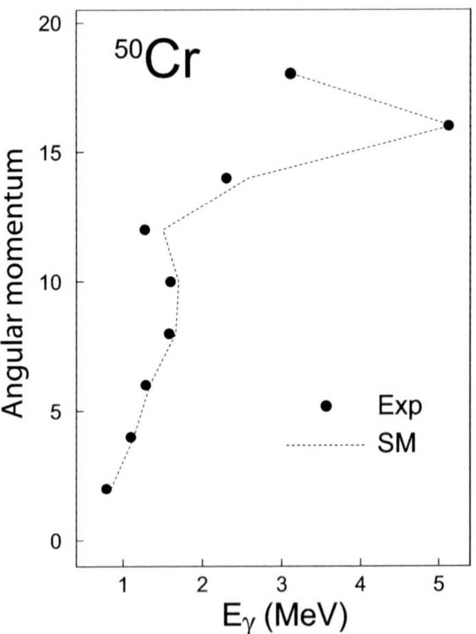

Fig. 1. – Angular momentum $vs.$ transition energy for the yrast band in ^{50}Cr [8].

$N = Z$ line and both the number of protons and neutrons are even. (In odd-mass nuclei, the blocking effect due to the unpaired nucleon favours the alignment of a pair of the other nucleon type.) The analysis of the CED can give the answer to this question. This is due to the fact that when a pair of protons align or recouple there is, in addition to a nuclear effect (that is the same for the neutrons), another effect due to a change of the Coulomb energy (that acts only on protons). This can be seen as follows: in the $J = 0$ coupled configuration, the overlap of the wave functions of the two protons is maximum, and so is the Coulomb energy. As they are recoupled to larger values of J, the interaction decreases and in the maximum aligned $J = 6$ configuration, the Coulomb energy is minimum. As the Coulomb interaction is repulsive, this translates into a more bound state in the nucleus where the protons align, with respect to the mirror partner where a neutron pair aligns. This effect was suggested for the first time by Cameron and collaborators [10] in the case of the $T = 1/2$ $A = 49$ mirror pair.

By comparing the level schemes of two mirror nuclei, the CED is obtained as

$$(4) \qquad\qquad CED_{J,T} = E^*_{J,T,T_z} - E^*_{J,T,T_z+n},$$

where n is the difference between the number of neutrons and protons, $n = N - Z$, and J is the spin of the state. (Remember that $N + Z = A$ is the same in both mirrors.) An alignment along a rotational band will show as a jump in the CED: a positive value will indicate that a pair of protons have aligned in the $T_z + n$ ($Z_<$) partner, while a negative value will correspond to the alignment of a proton pair in the T_z ($Z_>$) one. Simultaneously, of course, a pair of neutrons will align in the corresponding mirror partners. Therefore, by just looking at the experimental trend of the CED as a function of the angular momentum, one can deduce which type of nucleons aligns at the backbending.

In the left-hand side of fig. 2 the level schemes of the mirror pair ^{50}Fe-^{50}Cr are reported. Only those states in ^{50}Cr relevant for the comparison are shown, as just states up to 11^+ have been identified in the proton-rich ^{50}Fe. The similarity between them is evident. In the right-hand side of fig. 2, the experimental CED, called for the case of mirror pairs Mirror Energy Differences (MED $= E^*(^{50}$Fe$) - E^*(^{50}$Cr$))$, are shown in the upper panel. The MED will tell us many different things as we will see in the following sections, in particular, the decrease of the MED at low spins will be due to changes of the nuclear shape (see sect. **3**), but this needs some calculations. From the figure, however, we can deduce what happens at the backbending following the arguments given above. The smooth rise of the MED up to $J = 8$ indicates the recoupling of protons in ^{50}Cr (neutrons in ^{50}Fe), followed by a sudden neutron alignment in ^{50}Cr (protons in ^{50}Fe) at $J = 10$.

In the shell model framework, the alignment can be computed by *counting* the number of proton pairs in a j orbital coupled to the maximum spin ($J = 2j - 1$). In the $f_{\frac{7}{2}}$ shell, this can be performed by calculating the expectation value of the operator (that acts only on protons) $A = [(a^+ a^+)^{J=6} (a\, a)^{J-6}]^0$ for each excited state as a function of the angular momentum. Doing this for both nuclei, one can then calculate the difference $\Delta A = A(Z_>) - A(Z_<)$ for the mirror pair, as a function of the angular momentum. If the alignment of a pair of protons in the nucleus with charge $Z_>$ – and, consequently,

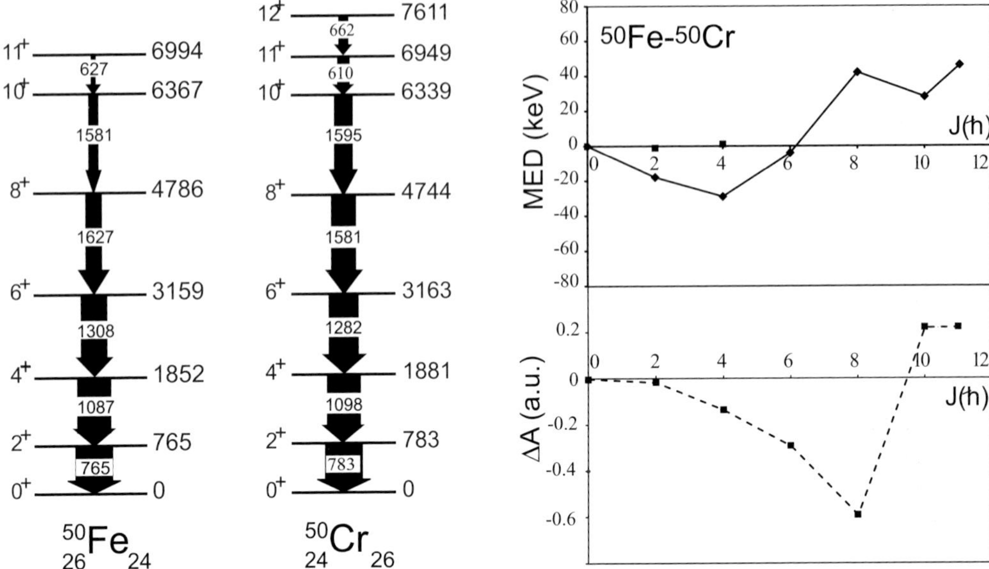

Fig. 2. – On the left side: MED and alignment for the mirror pair ^{50}Fe $-^{50}$Cr as a function of the angular momentum [1]. On the right side: The experimental MED (upper panel); the calculated "alignment" in the shell-model framework (lower panel)—see text for details.

the alignment of a pair of neutrons in the $Z_<$ – occurs first, ΔA will increase, whilst it will decrease if the opposite happens. This was introduced by Poves and Sánchez-Solano in ref. [11] for the case of $A = 51$. This is computed in the right-lower panel of fig. 2 for the mirror pair ^{50}Fe-^{50}Cr. From this figure we can observed that the conclusions regarding the recoupling and alignment we derived from the MED in the upper panel are confirmed by the shell model calculations in the lower panel. It is thus clear how we can get information of the nuclear-structure properties by just looking at the MED in mirror nuclei. Other features necessitate a more deep understanding of the wave functions and a good shell model description of the spectroscopy of the nuclei of interest is mandatory.

In this mass region, exact shell model calculations in the full fp space are feasible. For the specific case of $A = 50$, both the excitation energy and transition probabilities can be reproduced with good accuracy (see fig. 1 and refs. [8, 9]). This encourages us to perform the MED with such calculations, although these energy differences are small and it is not *a priori* obvious that this goal could be achieved. In fact first attempts of reproducing the MED in nuclei of the $f_{7/2}$ shell failed, as some important effects were not taken into account. It was in ref. [12] where Zuker *et al.* gave a complete and successful description of all the CED so far measured in this mass region. In the following section the different contributions to the MED deduced from ref. [12] are described.

3. – The mirror pair ^{48}Mn-^{48}V and the Monopole Coulomb radial term

By adding the Coulomb potential to the nuclear effective interaction and performing the diagonalization of the corresponding shell model matrix in the valence space, only the multipole contribution of the Coulomb interaction (V_{CM}) is considered. Earlier tentatives of reproducing the MED as a function of the angular momentum using this approach failed as monopole contributions were not taken into account [13]. In ref. [1] a renormalization of the multipole Coulomb interactions was suggested together with the introduction of a monopole term called *radial term*. This term takes into account the changes of shape (radius) of the nucleus along the yrast sequence.

In $f_{\frac{7}{2}}$-shell nuclei, the occupation of orbits different from the $f_{\frac{7}{2}}$, important to create collectivity near the ground state, decreases along the rotational bands, due to the alignment process, producing changes of the nuclear radius. In particular, at low spin, the occupation of the $p_{3/2}$ orbital is important but decreases gradually with increasing angular momentum and becomes negligible at the band-terminating state. In the fp shell, p orbits have larger radius than f orbits. The nucleus, therefore decreases its radius as a function of the angular momentum. This affects the MED as valence protons in orbitals with smaller radii are *nearer* to the charged core and have *more* Coulomb energy.

Following refs. [1,12], the monopole Coulomb contribution to the MED can be deduced by considering the Coulomb energy of a uniformly charged sphere of radius R_C (eq. (3)). When calculating the mirror energy differences for each state of spin J as a function of the angular momentum, we refer the MED(J) values to the ground state. On doing so, the monopole effect of E_C almost vanishes. A small contribution remains, however, due to the change in charge radius with the angular momentum, as discussed above. The monopole Coulomb radial contribution to the MED can be written:

$$(5) \qquad MED^J(V_{Cr}) = \Delta E_C(J) - \Delta E_C(\text{g.s.})$$
$$= \frac{3}{5}n(2Z-n)e^2\left(\frac{1}{R_C(J)} - \frac{1}{R_C(\text{g.s.})}\right)$$
$$= -\frac{3}{5}n(2Z-n)e^2\frac{\Delta R(J)}{R_C^2} \, ,$$

where $\Delta R(J) = R_C(J) - R_C(\text{g.s.})$ and we assume, following refs. [1,12,14], that it is *the same in both mirror nuclei.*

Changes in the charge radius can be accounted for, within the shell model, by considering the evolution of the occupation numbers of the different orbits as a function of the angular momentum along the yrast bands [1]. As stated above, for nuclei in the $f_{\frac{7}{2}}$ shell, it is the relative occupation of the $p_{\frac{3}{2}}$ orbit which determines the main changes of radii. The assumption of equal radii of the mirror partners means that a calculation of the *average* of proton and neutron occupation numbers of the $p_{\frac{3}{2}}$, m_π and m_ν, respectively, is required: $\frac{m_\pi + m_\nu}{2}$. The contribution of the monopole Coulomb radial term to the MED

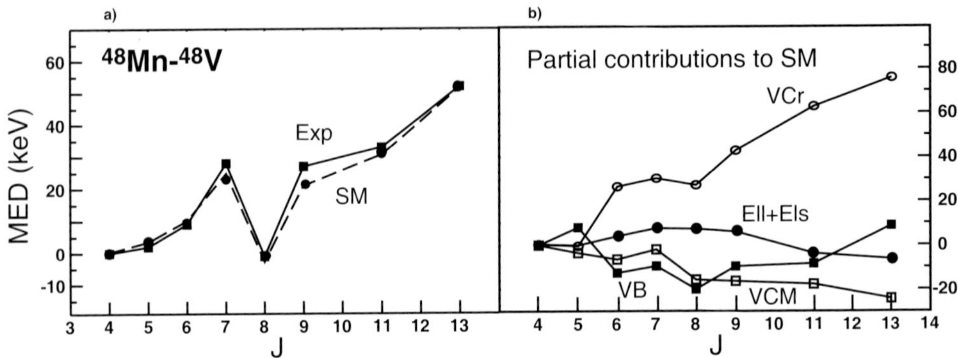

Fig. 3. – MED for the mirror pair ^{48}Mn-^{48}V as a function of the angular momentum [3]. (a) The experimental MED in comparison with shell model calculations (SM). (b) The different contributions to the shell model MED (see text for details).

of mirror nuclei with $|T_z| = n/2$ can be parameterised as

$$(6) \qquad MED^J(V_{Cr}) = n\, \alpha_r \left(\frac{m_\pi(\text{g.s.}) + m_\nu(\text{g.s.})}{2} - \frac{m_\pi(J) + m_\nu(J)}{2} \right),$$

where the constant α_r can be deduced from the single-particle relative energies in mass $A = 41$, as discussed in ref. [12]. It is important to note, however, that the radial term in the calculation of the MED is not a single-particle effect and therefore it cannot be accounted for by setting different single-particle energies for protons and neutrons in the shell-model calculation.

A good example of mirror nuclei where this term gives a significant contribution, and where multipole effects almost cancel due to the symmetry between particles and holes in the $f_{7/2}$ shell for these nuclei, is the odd-odd pair $A = 48$, $T = 1$ [3]. They are described in detail in ref. [3]. Here, the MED are reported in fig. 3(a) and compared to the shell model calculations, the agreement is excellent. In fig. 3(b), the different terms entering the calculation are displayed. In particular, the radial term (V_{Cr}) gives the main contribution. The other (multipole) terms are V_{CM}, the multipole Coulomb, and V_B, the isospin symmetry breaking term, that will be discussed in the next section. Finally, monopole corrections to the single particle energies are also shown in the figure. These latter do not have a sizable contribution to the MED of $f_{7/2}$-shell nuclei but are very important in sd-shell nuclei (see sect. **5**).

4. – The mirror pair ^{54}Ni-^{54}Fe and the ISB term

Although the identification of the different electromagnetic terms that enter in the calculation of the MED allows to improve the description of the data, a satisfactory solution was only reached when A.P. Zuker [12] introduced the V_B contribution to the

MED. This is a schematic term, deduced from the data of the $A = 42$ isobaric triplet. By considering the yrast states $J = 0, 2, 4, 6; T = 1$ in the mirror nuclei ^{42}Ti and ^{42}Ca and assuming that they have essentially $f_{7/2}^2$ configurations, the Coulomb contribution to the MED can be calculated. In the hypothesis of isospin symmetry of the nuclear interaction, the experimental MED values and those obtained using the Coulomb interaction should be similar. This is, however, not the case for $A = 42$. The result of subtracting to the experimental MED the Coulomb contribution is of the same order of magnitude of the latter. The maximum value is obtained at $J = 2$.

This means that Isospin-Symmetry Breaking (ISB) effects cannot be ignored in the calculation of the excitation energy differences. This effect was already noted in early studies performed by Brown and Sherr in ref. [15] but the origin of this charge-dependent interaction is still an open question. This was considered of nuclear nature in ref. [12], but contributions from the renormalisation of the Coulomb interaction could also contribute. In ref. [12], an extremely simple ansatz was proposed which consists on constructing an ISB Hamiltonian in the fp shell by just taking the $f_{\frac{7}{2}}^2$ matrix element for $J = 2$ with a strength of $\sim 100\,\mathrm{keV}$ for protons and zero for neutrons, as suggested from the data of $A = 42$.

$$(7) \qquad\qquad V_B = V_{B,f_{\frac{7}{2}}}(J = 2),$$

where $V_{B,f_{7/2}}(J)$ is a matrix element of value $\sim$100 keV for protons. The results for the MED obtained using this schematic term together with those introduced above (monopole and multiple Coulomb) are very successful. Using the same parametrization of eq. (7), the data on MED for all the $f_{7/2}$-shell isobaric multiplets measured so far can be reproduced with very good accuracy [7, 12]. The fact that the ISB is maximum at $J = 2$ and that the same schematic term is absolutely needed to reproduce the data along the whole $f_{7/2}$ shell is very significant. In fig. 4 the MED values for the $T = 1$ mirror pair ^{54}Ni-^{54}Fe are reported. The role of the V_B term is clearly evidenced and results of the same order of magnitude than the V_{CM} contribution. The monopole contribution of V_{Cr} is small, which can be understood because these nuclei, that are near closed shell, are not deformed near the ground state. In fig. 4(b), the experimental MED for $A = 54$ are compared to those for $A = 42$. These nuclei are symmetric in the number of particles/holes in the $f_{7/2}$ shell and are therefore called cross-conjugate. (Actually, for $A = 42$ the values reported in the figure correspond to $-$MED to reflect this symmetry between particles and holes.) The similarity between the two MED curves is impressive. This clearly shows that the ISB term is important all across the $f_{7/2}$ shell. Moreover, analysing the wave functions for $A = 42$ and $A = 54$ it is clear that they are different: while in $A = 42$ the $f_{7/2}^2$ configuration is dominating, only 60% of the wave functions in $A = 54$ are pure $f_{7/2}^{-2}$. As the MED reflect the changes of configuration as a function of the angular momentum, the similarity of both MED, in spite of the fact that the wave functions are quite different, indicates that the configurations must be very much independent of J. In other words, the pure $f_{7/2}$ description works well because the terms that may break cross-conjugation symmetry do exist but they have J-independent effects that vanish for the MED.

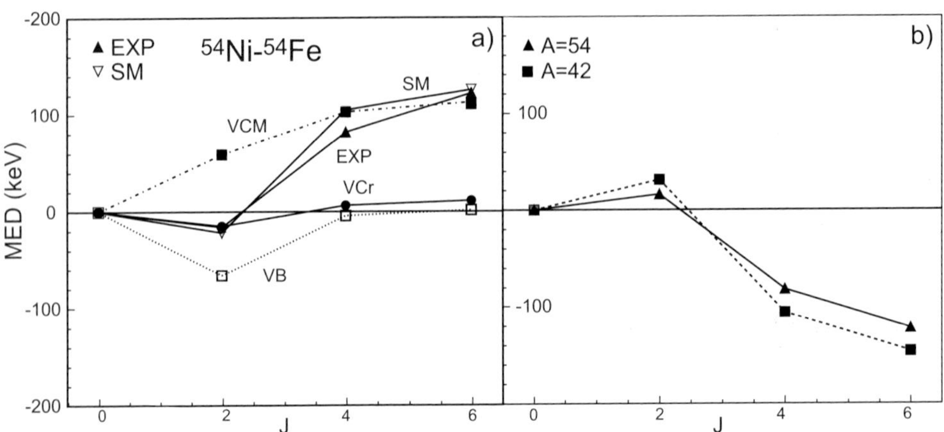

Fig. 4. – MED for the mirror pair ^{54}Ni-^{54}Fe as a function of the angular momentum [4]. (a) The experimental MED in comparison with shell model calculations (SM) and the different contributions to the latter (see text for details). (b) Comparison between the experimental MED for the cross-conjugate mirror pairs $A = 54$ and $A = 42$ (in this latter case the plot corresponds to $-$MED to reflect the cross-conjugate symmetry).

5. – The mirror pair ^{39}Ca-^{39}K and the electromagnetic spin-orbit term

Single-particle energies are modified by monopole effects due to the Coulomb potential. A first correction has been introduced by Duflo and Zuker [14] in a work devoted to the study of Coulomb displacement energies. This term is proportional to the square of the orbital angular momentum and results from shell corrections to the Coulomb potential generated by a uniformly charged sphere. The expression of the energy correction to the single-particle energy of the proton above a core results [14]:

$$(8) \qquad E_{ll} = \frac{-4.5 Z_{cs}^{13/12}[2l(l+1) - N(N+3)]}{A^{1/3}\left(N + \frac{3}{2}\right)} \ \text{keV}.$$

where Z_{cs} is the proton number of the core and N is the principal quantum number of the shell. The effect is sizable. As an example, the relative energy between the proton orbitals $f_{\frac{7}{2}}$ and $p_{\frac{3}{2}}$ is increased by $\sim 150\,\text{keV}$ with respect to the neutron energy difference.

A second correction term to the single-particle energies arise from the electromeagnetic spin-orbit interaction (EMSO) [16, 17]. This interaction, analogous to the atomic case, results from the Larmor precession of the nucleons in the nuclear electric field due to their intrinsic magnetic moments and to the Thomas precession experienced by the protons because of their charge. The effect of the *nuclear* spin-orbit hamiltonian in the single-particle spectrum is well known. It amounts to several MeV and acts on both protons and neutrons. The EMSO effect is almost 50 times smaller than the nuclear spin-orbit potential. However, as it acts differently on neutrons than on protons, its effect can become very important for some particular states when calculating the MED [18, 19]. The general expression of the electromagnetic spin-orbit potential [16] is

$$(9) \qquad V_{ls} = (g_s - g_l) \frac{1}{2m_N^2 c^2} \left(\frac{1}{r} \frac{dV_C}{dr} \right) \vec{l} \cdot \vec{s},$$

where g_s and g_l are the gyromagnetic factors and m_N is the nucleon mass. The term proportional to g_s is the Larmor term; the Thomas term, proportional to g_l, vanishes in the neutron case.

A rough estimate of the energy shift produced by the relativistic electromagnetic spin-orbit term has been given by Nolen and Schiffer [16] assuming that V_C is generated by a uniformly charged sphere of radius R_C

$$(10) \qquad E_{ls} \simeq (g_s - g_l) \frac{1}{2m_N^2 c^2} \left(-\frac{Ze^2}{R_C^3} \right) \langle \vec{l} \cdot \vec{s} \rangle$$

E_{ls} has opposite sign for neutrons and protons and for orbits with parallel or anti-parallel spin-orbit coupling. For the nuclei of interest, which excite nucleons from the sd to the fp shell, the effect of the EMSO is to reduce the energy gap between the $d_{3/2}$ and $f_{7/2}$ orbitals for protons by $\sim 120\,\mathrm{keV}$ and to increase it for neutrons by a similar amount.

The influence of this term will be important for states whose configuration involves *pure single-particle* excitations from the $d_{3/2}$ to the $f_{7/2}$. The net effect on the MED will be of the order of $\approx 240\,\mathrm{keV}$ if the configuration of the state corresponds to a proton excitation across the $Z = 20$ gap in one nucleus and, of course, due to the mirror symmetry, a neutron excitation in the mirror partner. On the contrary, small effects on the MED will be obtained whenever the configuration of the state involves the excitation of one proton *or* one neutron with similar probabilities in the same nucleus (idem in its mirror), as the effects are compensated.

In fig. 5 the MED for the mirror pair ^{39}Ca-^{39}K [20] are reported in panel (a). The experimental values are compared to the shell model results. These calculations have been performed using the $sdfp$ interaction [21] with the code ANTOINE [22]. To calculate the MED we have considered, in addition to the contribution of the multipole Coulomb interaction, the single-particle effects. The three contributions are reported in fig. 5(b). An interesting result of the calculations is that the multipole Coulomb term (V_{CM}) assumes large values and follows the trend of the experimental curves. This is contrary to what was expected from the systematics in the $f_{7/2}$ shell, where the calculated values did not exceed $\sim 100\,\mathrm{keV}$, as discussed in ref. [19]. The contributions arising from the single-particle terms are proportional to the difference between the occupation numbers of protons and neutrons of the different orbits. As shown in fig. 5(b) the EMSO interaction gives a contribution to the MED similar to that of the multipole Coulomb. Smaller contributions are obtained from the E_{ll} term.

The MED provide also a useful tool to recognise the configuration of the states. In ^{35}Ar there are two close states with $J^\pi = 11/2^-$, which lie at 5384 and 5614 keV excitation energy, respectively; in the mirror nucleus ^{35}Cl, the isobaric analogue states

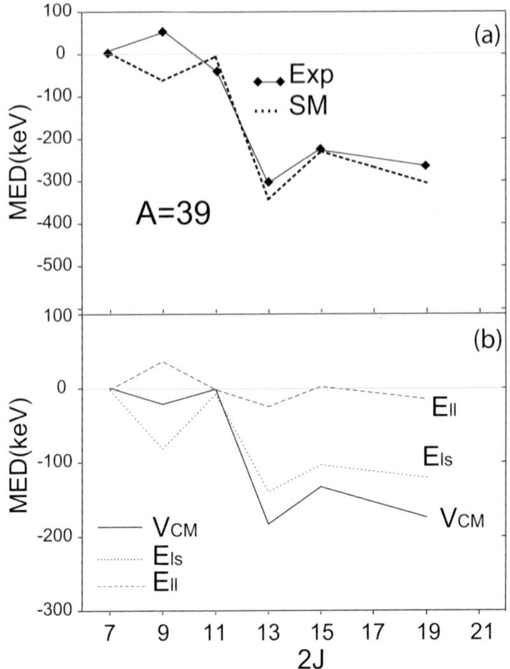

Fig. 5. – MED for the mirror pair ^{39}Ca-^{39}K as a function of the angular momentum [20]. (a) The experimental MED in comparison with shell model calculations (SM); (b) the different contributions to the latter (see text for details).

lie at 5407 and 5927 keV. The MED values are, respectively, $\mathrm{MED}(11/2_1^-) = -23$ and $\mathrm{MED}(11/2_2^-) = -313$. Quite a large difference between the two MED values. Following the arguments given above, one could immediately deduce that the first excited state has a mixed configuration where with the same probability a proton or a neutron are excited from the $d_{3/2}$ to the fp shell, while for the second case one proton undergoes the excitation in ^{35}Ar and a neutron in ^{35}Cl. This agrees with the wave functions calculated by the shell model [23].

In reproducing the MED in this mass region, the V_B and V_{Cr} terms have not been included. The latter contribution should not be important as the composition of the wave functions does not change along the sequence of levels considered. The treatment of the V_B term, however, is complex, as it has to be determined to take into account effects in the sd shell. Although the quality of the fit is very good in the case of $A = 39$, it is not so good for the $A = 35$ mirror pair ^{35}Ar-^{35}Cl, where the trend of the MED is very well reproduced but the absolute values are overestimated [23]. Further developments are needed in this mass region from the theoretical side, in particular, in the development of effective interactions, that consider both the sd and fp shells together, to render the shell model calculations as good as in the $f_{7/2}$ shell.

The importance of single-particle effects has been recently shown in fp shell nuclei [24, 25] and in non-natural parity states in $A = 46$ [26] and $A = 48$ [3], where the MED have been very well reproduced both qualitatively and quantitatively.

6. – Conclusions

In this lecture, the main features that can be put in evidence by analysing the energy differences between excited analogue states in mirror nuclei have been briefly discussed. Although not treated in here, also the energy differences among analogue states in $T = 1$ triplets (which include also the $N = Z$ members of the multiplets) can be reproduced very well with the shell model [4, 12]. A more complete and extensive description of both theoretical and experimental methods to study CED can be found in the review article of ref. [7]. What makes this subject fascinating is the fact that such small energy differences, caused mainly by the Coulomb potential, can reveal so many details on the *nuclear* structure that cannot be evidenced by other, more standard, methods. The bonus of the possible evidence of ISB nuclear interaction makes it still more interesting and merits further investigation.

The extension of these studies to other mass regions such as the sd or fp shells, which have been recently addressed [7, 19, 20, 25], has put in evidence the importance of another monopole term: the electromagnetic spin-orbit interaction. This changes the single-particle energy of both protons and neutrons and has sizable effects on the MED whenever single-particle excitations are dominant in the wave functions.

Exploration of the isospin degree of freedom is certainly one of the key nuclear-structure goals of the present and future radioactive beam facilities. Exciting opportunities will open that will allow to test the present methods in more extreme situations and will help understanding the nuclear structure in great detail.

$$* \quad * \quad *$$

This work has been performed with the collaboration of several colleagues. In particular I would like to thank A. P. ZUKER, M. A. BENTLEY, F. DELLA VEDOVA, A. GADEA and N. MARGINEAN.

REFERENCES

[1] LENZI S. M. *et al.*, *Phys. Rev. Lett.*, **87** (2001) 122501.
[2] GARRETT P. E. *et al.*, *Phys. Rev. Lett.*, **87** (2001) 122502.
[3] BENTLEY M. A. *et al.*, *Phys. Rev. Lett.*, **97** (2006) 132501.
[4] GADEA A. *et al.*, *Phys. Rev. Lett.*, **97** (2006) 152501.
[5] DOORNENBAL P. *et al.*, *Phys. Lett. B*, **647** (2007) 237.
[6] BÜRGER A. *et al.*, *Acta Phys. Pol. B*, **38** (2007) 1353.
[7] BENTLEY M. A. and LENZI S. M., *Prog. Part. Nucl. Phys.*, **59** (2007) 497.
[8] LENZI S. M. *et al.*, *Phys. Rev. C*, **56** (1997) 1313.
[9] BRANDOLINI F. *et al.*, *Nucl. Phys. A*, **642** (1998) 387.
[10] CAMERON J. A. *et al.*, *Phys. Rev. C*, **44** (1991) 1882.

[11] Bentley M. A. *et al.*, *Phys. Rev. C*, **62** (2000) 051303.

[12] Zuker A. P., Lenzi S. M., Martinez-Pinedo G. and Poves A., *Phys. Rev. Lett.*, **89** (2002) 142502.

[13] Bentley M. A. *et al.*, *Phys. Lett. B*, **437** (1998) 243.

[14] Duflo J. and Zuker A. P., *Phys. Rev. C*, **66** (2002) 051304(R).

[15] Brown B. A. and Sherr R., *Nucl. Phys. A*, **322** (1979) 61.

[16] Nolen J. A. and Schiffer J. P., *Annu. Rev. Nucl. Sci.*, **19** (1969) 471.

[17] Inglis D. R., *Phys. Rev.*, **82** (1951) 181.

[18] Trache L. *et al.*, *Phys. Rev. C*, **54** (1996) 2361.

[19] Ekman J. *et al.*, *Phys. Rev. Lett.*, **92** (2004) 132502.

[20] Andersson Th. *et al.*, *Eur. Phys. J. A*, **6** (1999) 5.

[21] Caurier E. *et al.*, *Phys. Lett. B*, **522** (2001) 240.

[22] Caurier E., Code ANTOINE, Strasbourg, 1989; Caurier E. and Nowacki F., *Acta Phys. Pol.*, **30** (1999) 705.

[23] Della Vedova F. *et al.*, *Phys. Rev. C*, **75** (2007) 034317.

[24] Andersson L.-L., *Phys. Rev. C*, **71** (2005) 011303.

[25] Ekman J., Fahlander C. and Rudolph D., *Mod. Phys. Lett.*, **20** (2005) 2977.

[26] Garrett P. E. *et al.*, *Phys. Rev. C*, **75** (2007) 014307.

DOI 10.3254/978-1-58603-885-4-317

Selected topics in nuclear astrophysics

K. Langanke

Gesellschaft für Schwerionenforschung and Technische Universität Darmstadt
D-64291 Darmstadt, Germany

Summary. — These lectures discuss selected topics in nuclear astrophysics. They include hydrogen burning, solar neutrinos, advanced stellar burning stages, the final fate of massive stars as core-collapse supernovae and the associated explosive nucleosynthesis. The emphasis is on the nuclear ingredients, which determine the evolution and dynamics of the astrophysical objects.

1. – Introduction

Nuclear astrophysics aims at describing the origin of the chemical elements in the Universe as well as of the various nuclear processes, occurring in and powering astrophysical objects, which lead to the elemental production. It is impossible to cover the large width of this truely interdisciplinary field in three lectures. Hence I have made a biased choice and concentrate only on three topics: i) the nuclear reactions occurring during the long lifetime of stars when these generate their energy in a sequence of nuclear burning stages in hydrostatic equilibrium; ii) the nuclear aspect of the collapse of the inner core of massive stars once it has run out of its nuclear energy source and of the star's explosion as a type II supernova; iii) the explosive nucleosynthesis occurring during this explosion which leads to the production of heavy elements by the rapid neutron capture process and potentially also by the recently discovered νp process.

Due to this selection I will omit several other fascinating and timely topics of nuclear astrophysics. These include the slow neutron capture process, which produces about half of the nuclei heavier than iron, the p process, which makes the proton rich, heavy nuclei, explosive hydrogen burning which occurs in novae and X-ray bursts, nucleosynthesis during the Big Bang, in cosmic rays and by neutrinos during a core-collapse supernova, gamma-ray bursts, and also thermonuclear (type Ia) supernova. Recent developments on these subjects are reviewed in a special issue on nuclear astrophysics: K. Langanke, F. K. Thielemann, M. Wiescher (Editors), *Nucl. Phys. A*, **777** (2006). The interested reader is also referred to excellent recent reviews on aspects of nuclear astrophysics:

- General nucleosynthesis: G. Wallerstein *et al.*, *Rev. Mod. Phys.*, **69** (1997) 795; M. Arnould and K. Takahashi, *Rep. Prog. Phys.*, **62** (1999) 395; F. Käppeler, F.-K. Thielemann and M. Wiescher, *Annu. Rev. Nucl. Part. Sci.*, **48** (1998) 175.

- Core-collapse supernovae: H. Th. Janka, K. Kifonidis and M. Rampp, in *Physics of Neutron Star Interiors*; edited by D. Blaschke, N. K. Glendenning and A. Sedrakian, *Lecture Notes in Physics*, Vol. **578** (Springer, Berlin), p. 333; A. Burrows, *Prog. Part. Nucl. Phys.*, **46** (2001) 59; H. A. Bethe, *Rev. Mod. Phys.*, **62** (1990) 801.

- Type-Ia supernovae: W. Hillebrandt and J. C. Niemeyer, *Annu. Rev. Astron. Astrophys.*, **38** (2000) 191.

- S-process: F. Käppeler, *Prog. Part. Nucl. Phys.*, **43** (1999) 419; M. Busso, R. Gallino and G. J. Wasserburg, *Annu. Rev. Astron. Astrophys.*, **37** (1999) 239.

- R-process: J. J. Cowan, F.-K. Thielemann and J. W. Truran, *Phys. Rep.*, **208** (1991) 267; Y.-Z. Qian, *Prog. Part. Nucl. Phys.*, **50** (2003) 153.

Of course, it is still very much recommended to read the two pioneering papers: E. M. Burbidge, G. R. Burbidge, W. A. Fowler and F. Hoyle, *Rev. Mod. Phys.*, **29** (1957) 547 and A. G. W. Cameron, *Stellar Evolution, Nuclear Astrophysics, and Nucleogenesis*, Report CRL-41, Chalk River, Ontario.

2. – Astrophysical nuclear reaction rates

During the hydrostatic burning stages of a star, charged-particle reactions most frequently occur at energies far below the Coulomb barrier, and are possible only via the *tunnel effect*, the quantum-mechanical possibility of penetrating through a barrier at a classically forbidden energy. At these low energies, the cross-section $\sigma(E)$ is dominated by the penetration factor, which for point-like particles and in the absence of a centrifugal barrier (s-wave) is well approximated by [1]

$$(1) \qquad P(E) = \exp\left[-\frac{2\pi Z_1 Z_2 e^2}{\hbar v}\right] \equiv \exp\left[-2\pi\eta(E)\right],$$

where $\eta(E)$ is often called the Sommerfeld parameter. In numerical units,

$$(2) \qquad 2\pi\eta(E) = 31.29 Z_1 Z_2 \sqrt{\frac{\mu}{E}},$$

where the energy E is defined in keV and μ is the reduced mass number.

It is convenient and customary to redefine the cross-section in terms of the astrophysical S-factor by factoring out the known energy dependences of the penetration factor and the de Broglie factor, in the *model-independent* way,

$$(3) \qquad S(E) = \sigma(E)(E)\exp[2\pi\eta(E)].$$

For low-energy, nonresonant reactions, the astrophysical S-factor should have only a weak energy dependence that reflects effects arising from the strong interaction, as from antisymmetrization, and from small contributions from partial waves with $l > 0$ and for the finite size of the nuclei. The validity of this approach has been justified in numerous (nonresonant) nuclear reactions for which the experimentally determined S-factors show only weak E-dependences at low energies.

The reaction rate of two nuclear species x, y with number densities N_x and N_y, respectively, is given by

$$(4) \qquad R_{xy} = \frac{1}{1+\delta_{xy}} N_x N_y \langle \sigma(v)v \rangle,$$

where one has to account for the velocity distributions of target and projectile in the astrophysical environment. As these are given by Maxwell-Boltzmann distributions, one has ($E = \mu/2\,v^2$, where here μ is the reduced mass)

$$(5) \qquad \langle \sigma(v)v \rangle = \left(\frac{8}{\pi\mu}\right)^{1/2} \left(\frac{1}{kT}\right)^{3/2} \int_0^\infty \sigma(E)E\ \exp\left[-\frac{E}{kT}\right]\,\mathrm{d}E,$$

$$(6) \qquad = \left(\frac{8}{\pi\mu}\right)^{1/2} \left(\frac{1}{kT}\right)^{3/2} \int_0^\infty S(E)\ \exp\left[-\frac{E}{kT} - 2\pi\eta(E)\right]\,\mathrm{d}E.$$

For typical applications in hydrostatic stellar burning, the product of the two exponentials forms a peak ("Gamow-peak") which may be well approximated by a Gaussian,

$$(7) \qquad \exp\left[-\frac{E}{kT} - 2\pi\eta(E)\right] \cong I_{\mathrm{max}} \exp\left[-\left(\frac{E - E_0}{\Delta/2}\right)^2\right]$$

Table I. – *Values for E_0, Δ, and $I_{\max}$ at solar core temperature ($T_6 = 15.6$) [1].*

Reaction	E_0 (keV)	$\Delta/2$ (keV)	$I_{\max}$
$p + p$	5.9	3.2	1.1×10^{-6}
$^3\mathrm{He} + {}^3\mathrm{He}$	22.0	6.3	4.5×10^{-23}
$^3\mathrm{He} + {}^4\mathrm{He}$	23.0	6.4	5.5×10^{-23}
$p + {}^7\mathrm{Be}$	18.4	5.8	1.6×10^{-18}
$p + {}^{14}\mathrm{N}$	26.5	6.8	1.8×10^{-27}
$\alpha + {}^{12}\mathrm{C}$	56.0	9.8	3.0×10^{-57}
$^{16}\mathrm{O} + {}^{16}\mathrm{O}$	237.0	20.2	6.2×10^{-239}

with [1]

$$E_0 = 1.22\left(Z_1^2 Z_2^2 \mu T_6^2\right)^{1/3} (\mathrm{keV}), \tag{8}$$

$$\Delta = \frac{4}{\sqrt{3}}\sqrt{E_0 kT} \tag{9}$$

$$= 0.749\left(Z_1^2 Z_2^2 \mu T_6^5\right)^{1/6} (\mathrm{keV}),$$

$$I_{\max} = \exp\left[-\frac{3E_0}{kT}\right]. \tag{10}$$

T_6 measures the temperature in units of 10^6 K. Examples of E_0, Δ and $I_{\max}$, evaluated for some nuclear reactions at the solar core temperature ($T_6 \approx 15.6$), are summarized in table I.

We can draw a few interesting conclusions from table I. At first, reactions operate in relatively narrow energy windows around the astrophysically most effective energy E_0. Secondly, the penetration factor dominates the reaction rates at astrophysical energies which is also true if one accounts for the different strengths of the fundamental interactions. Hence stellar burning will start with the fusion of those nuclei with the smallest product of charges, *i.e.* the smallest Coulomb barrier. As neutrons are not available when stars form this will be hydrogen burning, as we will discuss in the next section. It becomes also clear from inspecting the different $I_{\max}$ values that reactions of nuclei with large charge numbers effectively cannot occur in the Sun as, for these reactions, even the solar core is far too cold.

However, it usually turns out that the astrophysically most effective energy E_0 is smaller than the energies at which the reaction cross-section can be measured directly in the laboratory. Thus for astrophysical applications, an extrapolation of the measured cross-section to stellar energies is usually necessary, often over many orders of magnitude. In the case of nonresonant reactions, the extrapolation can be safely performed in terms of the astrophysical S-factor, because of its rather weak energy dependence.

We finally have to mention that the nuclear reaction rates are enhanced in stellar environments as the repulsive Coulomb barrier is somewhat shielded by other plasma particles [2]. Such shielding effects are also encountered in laboratory measurements of nuclear cross-sections at low energies as the electrons present in the target (and perhaps also in the projectile) effectively reduce the Coulomb barrier [3]. We stress, however, that laboratory and plasma screening are different. Hence, the first has to be removed from the data to determine the cross-sections for bare nuclei which then, in astrophysical applications, has to be modified due to plasma effects.

3. – Hydrostatic burning stages

When the temperature and density in a star's interior rises as a result of gravitational contraction, it will be the lightest charged species that can react first and supply the energy and pressure to stop the gravitational collapse of the gaseous cloud. Thus it is hydrogen burning (the fusion of four protons into a ^{4}He nucleus) in the stellar core that stabilizes the star first and for the longest time. However, because of the larger charge ($Z = 2$), helium, the ashes of hydrogen burning, cannot effectively react at the temperature and pressure present during hydrogen burning in the stellar core. After exhaustion of the core hydrogen, the resulting helium core will gravitationally contract, thereby raising the temperature and density in the core until the temperature and density are sufficient to ignite helium burning, starting with the triple-alpha reaction, the fusion of three ^{4}He nuclei to ^{12}C. In massive stars, this sequence of contraction of the core nuclear ashes continues until ignition of these nuclei in the next burning stage repeats itself several times. After helium burning, the massive star goes through periods of carbon, neon, oxygen, and silicon burning in its central core. As the binding energy per nucleon is a maximum near iron (the end-product of silicon burning), freeing nucleons from nuclei in and above the iron peak, to build still heavier nuclei, requires more energy than is released when these nucleons are captured by the nuclei present. Therefore, the procession of nuclear burning stages ceases. This results in a collapse of the stellar core and an explosion of the star as a type II supernova. As an example, table II shows the timescales and conditions for the various hydrostatic burning stages of a $25 M_\odot$ star. One observes that stars spend most of their lifetime ($\sim 90\%$) during hydrogen burning (then the stars will be found on the main sequence in the Hertzsprung-Russell diagram). The rest is basically spent during core helium burning. During this evolutionary stage, the star expands dramatically and becomes a Red Giant.

Stellar evolution depends very strongly on the mass of the star. On general grounds, the more massive a star the higher the temperatures in the core at which the various burning stages are ignited. Moreover, as the nuclear reactions depend very sensitively on temperature, the nuclear fuel is faster exhausted the larger the mass of the star (or the core temperature). This is quantitatively demonstrated in table III which shows the timescales for core hydrogen burning as a function of the main-sequence mass of the star. One observes that stars with masses less than $\sim 0.5 M_\odot$ burn hydrogen for times which are significantly longer than the age of the Universe (about 13 billion years). Thus such low-

TABLE II. – *Evolutionary stages of a $25M_\odot$ star (from [4]).*

Evolutionary stage	Temperature (keV)	Density (g/cm^3)	Time scale
Hydrogen burning	5	5	7×10^6 y
Helium burning	20	700	5×10^5 y
Carbon burning	80	2×10^5	600 y
Neon burning	150	4×10^6	1 y
Oxygen burning	200	10^7	6 mo
Silicon burning	350	3×10^7	1 d
Core collapse	700	3×10^9	seconds
Core bounce	15000	4×10^{14}	10 ms
Explosion	100–500	10^5-10^9	0.01 to 0.1 s

mass stars have not completed one lifecycle and did also not contribute to the elemental abundances in the Universe. A star like our Sun has a life expectation due to core hydrogen burning of about 10^{10} y, which is about double its current age ($\sim 4.55 \times 10^9$ y).

As the temperatures and densities required for the higher burning stages increase successively, stars need a minimum mass to ignite such burning phases. For example, a core mass slightly larger than $1M_\odot$ is required to ignite carbon burning. One also has to consider that stars, mainly during core helium burning, have mass losses due to flashes or stellar winds. In summary, as a rule-of-thumb, stars with main-sequence masses $\leq 8M_\odot$ end their lifes as White Dwarfs. These are stars which are dense enough so that their electrons are highly degenerate and which are stabilized by the electron degeneracy pressure. There exists an upper limit for the mass of a star which can be stabilized by electron degeneracy. This is the Chandrasekhar mass of $\sim 1.4M_\odot$. Stars with masses of $\geq 13M_\odot$ go through the full cycle of hydrostatic burning stages and end with a collapse

TABLE III. – *Hydrogen burning timescales τ_H as a function of stellar mass (from [4]).*

M ($M_\odot$)	τ_H (y)
0.40	2×10^{11}
0.80	1.4×10^{10}
1.00	1×10^{10}
1.10	9×10^9
1.70	2.7×10^9
2.25	5×10^8
3.00	2.2×10^8
5.00	6×10^7
9.00	2×10^7
16.00	1×10^7
25.00	7×10^6
40.00	1×10^6

of their internal iron core. If the mass of the star is less than a certain limit, $M \sim 30 M_\odot$, the star becomes a supernova leaving a compact remnant behind after the explosion; this is a neutron star. More massive stars might collapse directly into black holes.

3`1. *Hydrogen burning.* – In low-mass stars, like our Sun, hydrogen burning proceeds mainly via the *pp* chains (see fig. 1), with small contributions from the CNO cycle. The later becomes the dominant energy source in hydrogen-burning stars, if the temperature in the stellar core exceeds about 20 million degrees (the temperature in the solar core is about 15.6×10^6 K).

The pp chains start with the fusion of two proton nuclei to the only bound state of the two-nucleon system, the deuteron. This reaction is mediated by the weak interaction, as a proton has to be converted into a neutron. Correspondingly, the cross-section for this reaction is very small and no experimental data at low proton energies exist. Although the reaction rate at solar energies is based purely on theory, the calculations are generally considered to be under control and the uncertainty of the solar p + p rate is estimated to be better than a few percent [5]. The next reaction (p + d → ^{3}He) is mediated by the electromagnetic interaction. It is therefore much faster than the p + p fusion with the consequence that deuterons in the solar core are immediately transformed to ^{3}He nuclides and no significant abundance of deuterons is present in the core (there is about 1 deuteron per 10^{18} protons in equilibrium under solar core conditions). As no deuterons are available and ^{4}Li (the endproduct of p + ^{3}He) is not stable, the reaction chain has to continue with the fusion of two ^{3}He nuclei via ^{3}He + ^{3}He → 2p + ^{4}He. This reaction terminates the ppI chain where in summary four protons are fused to one ^{4}He nucleus with an energy gain of 26.2 MeV; the rest of the mass difference is spent to produce two positrons and is radiated away by the neutrinos produced in the initial p + p fusion reaction. Once ^{4}He is produced in sufficient abundance, the pp chain can be completed

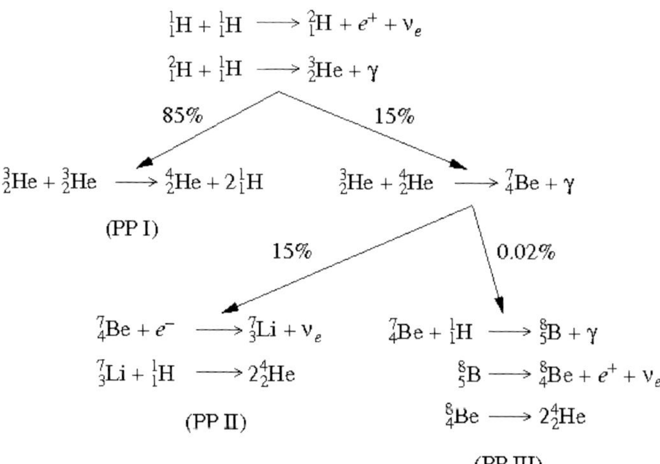

Fig. 1. – The reactions of the solar pp chains.

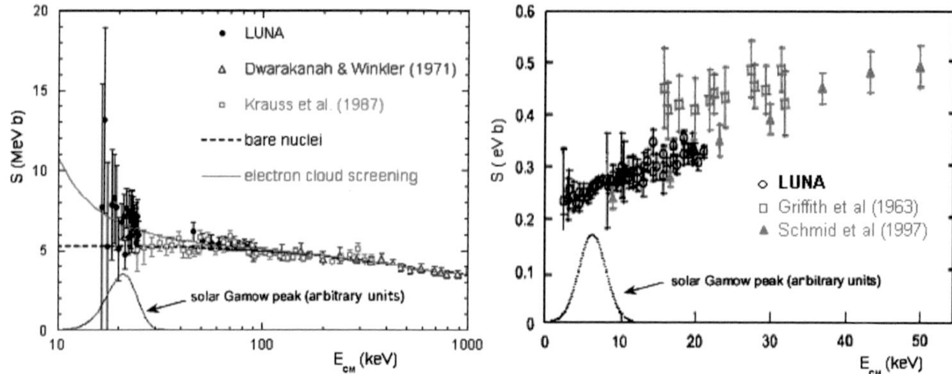

Fig. 2. – S-factors for the low-energy ^{3}He + ^{3}He and d + p fusion reactions. The data have been taken at the Gran Sasso Underground Laboratory by the LUNA collaboration. For the first time, it has been possible to measure nuclear cross-sections at the astrophysically most effective energies covering the regime of the Gamow peak. The ^{3}He + ^{3}He data include electron screening effects which have been removed to obtain the cross-section for bare nuclei (from [7]).

by two other routes. At first ^{3}He and ^{4}He fuse to ^{7}Be which then either captures a proton (producing ^{8}B) or, more likely, an electron (producing ^{7}Li). The two chains are then terminated by the weak decay of ^{8}B to an excited state in ^{8}Be, which subsequently decays into two ^{4}He nuclei, and by the ^{7}Li(p, ^{4}He)^{4}He reaction. In summary, both routes fuse 4 protons into one ^{4}He nucleus. The energy gain of these two branches of the pp chains is slightly smaller than in the dominating ppI chain, as neutrinos with somewhat larger energies are produced en route [6].

All of the pp chain reactions are non-resonant at low energies such that the extrapolation of data taken in the laboratory to stellar (solar) energies is quite reliable [5]. For two reactions (^{3}He + ^{3}He $\to$ ^{4}He + 2p and p + d $\to$ ^{3}He) the cross-sections have been measured at the solar Gamow energies, thus making extrapolations of data unnecessary (see fig. 2). These important measurements have been performed by the LUNA group at the Gran Sasso Laboratory [7] far underground to effectively remove background events due to the shielding by the rocks above the experimental hall. For many years the p + ^{7}Be fusion reaction has been considered the most uncertain nuclear cross-sections in solar models. However, impressive progress has been achieved in determining this reaction rate in recent years and it appears that the goal of knowing the solar reaction rate for this reaction better than the limit of 5%, as desired for solar models, has been reached [8].

Also in the CNO cycle four protons are fused to one ^{4}He nucleus. However, this reaction chain requires the presence of ^{12}C as catalyst. It then proceeds through the following sequence of reactions:

$$^{12}\text{C}(\text{p}, \gamma)^{13}\text{N}(\beta)^{13}\text{C}(\text{p}, \gamma)^{14}\text{N}(\text{p}, \gamma)^{15}\text{O}(\beta)^{15}\text{N}(\text{p}, \alpha)^{12}\text{C}.$$

The slowest step is the p + ^{14}N reaction. There has been decisive experimental progress in determining this reaction rate in the last years by the LUNA and LENA collaborations which both showed that the reaction rate is actually smaller by a factor 2 than previously believed [9, 10]. As a consequence the CNO cycle contributes less than 1% of the total energy generation in the SUN. The new p + ^{14}N reaction rate has also interesting consequences for the age determination of globular clusters [11].

$3^{.}2$. *Solar neutrinos*. – Every second more than 100 billion neutrinos, which are produced by hydrogen burning in the interior of the Sun, pass through every square-centimeter of the earth surface (fig. 3). As the neutrino-matter interaction is extremely small, most of these neutrinos go unnoticed. However, a few of them are detected by large, dedicated detectors and indeed have proven that the Sun generates its energy by nuclear burning. Nevertheless, the event rate of detected solar neutrinos turned out be less than predicted by the solar models (fig. 4); this constituted the famous "solar neutrino problem" [6] which could only be solved very recently. The solution to the problem are neutrino oscillations. Within the solar pp chains only ν_e neutrinos are produced. Enhanced by matter effects some of these neutrinos are transformed into ν_μ or ν_τ neutrinos on their way out of the Sun. As the original solar neutrino detectors can only observe ν_e neutrinos (with energies larger than a certain threshold which depends on the mass difference of the target nucleus and its daughter, see fig. 4), the observed neutrino flux in these detectors is smaller than the total neutrino flux generated in the Sun. This picture was confirmed by the SNO collaboration in the recent years [12]. The SNO detector has the capability to observe neutral current events induced by neutrinos (the neutrino dissociation of deuterons into protons and neutrons in heavy water). As the neutral current events can be induced by all neutrino families such signal determines the total solar neutrino flux. It is found that it is larger than the ν_e flux and agrees nicely with the predictions of the solar models. It is also important to stress that the Superkamiokande detector, with its sensitivity to the direction of the incoming neutrino, has unequivocally proven that the detected neutrinos are indeed coming from the Sun!

The important SNO result is shown in fig. 5. At first, charged-current reactions can only be mediated by ν_e neutrinos via $\nu_e + d \rightarrow p + p + e^-$; a potential flux of other neutrino species (ν_μ and ν_τ produced by oscillations) cannot be detected. Hence such charged-current events are presented by a band with fixed ν_e, but unspecified $\nu_\mu + \nu_\tau$ flux in the neutrino flux diagram shown in fig. 4. On the other hand, the neutral-current reactions $\nu + d \rightarrow p + n + \nu$ can be induced by all three neutrino species. Thus, its observed event rate constrains the sum of ν_e and $\nu_\mu + \nu_\tau$ fluxes and corresponds to the "diagonal" band in the flux diagram. The diagram contains additionally results from the Superkamiokande detector where solar neutrinos can be detected by scattering off electrons. Such a process is possible for all 3 neutrino species; however, the cross-section is larger by about a factor 6 for ν_e neutrinos than for the other two due to a charged-current contribution which is only possible for ν_e neutrinos. The results from electron scattering thus correspond to the slightly tilted band shown in the flux diagram. Importantly the three independent experimental constraints have a common overlap corresponding to a non-vanishing $\nu_\mu + \nu_\tau$ flux. As ν_μ, ν_τ neutrinos are not generated by solar hydrogen

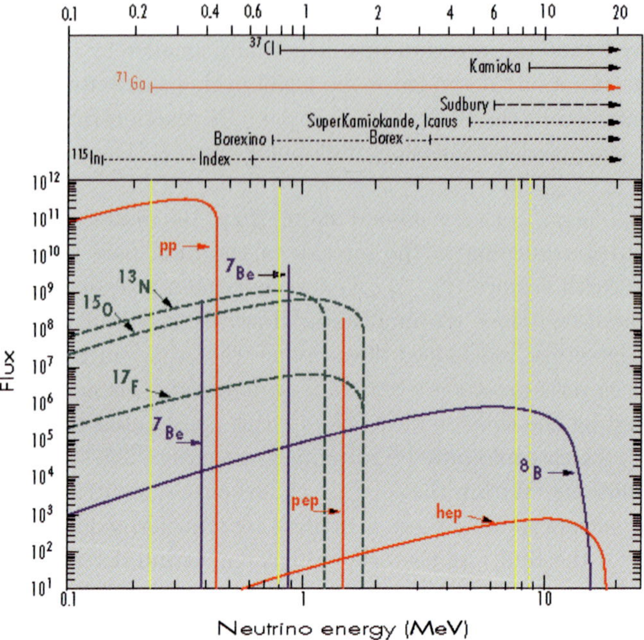

Fig. 3. – Solar neutrino flux per second and cm^2 on the Earth surface, as expected from the various hydrogen burning reactions. The arrows show the energy range for which various detectors can observe solar neutrinos (courtesy J. N. Bahcall).

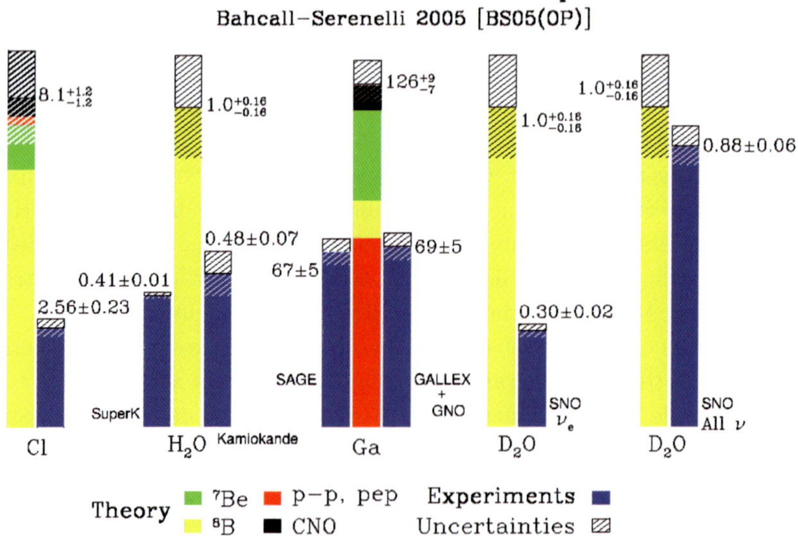

Fig. 4. – Comparison of detected neutrino rates (or fluxes) with the predictions of the solar model (from [13]).

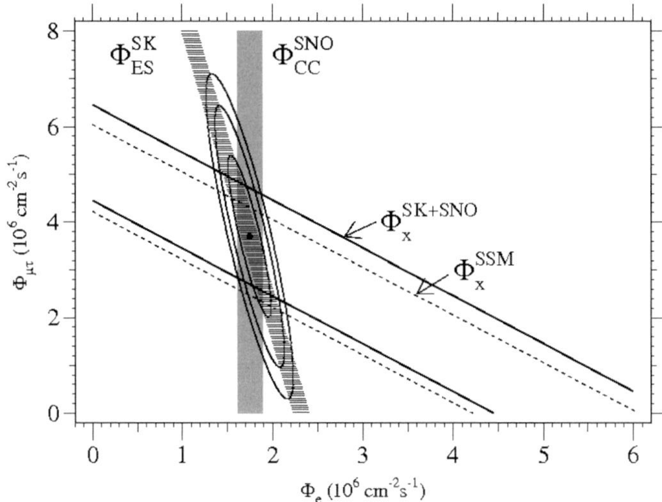

Fig. 5. – Constraints of the solar ν_e and $\nu_\mu + \nu_\tau$ fluxes from the observations of charged-current and neutral-current events at SNO and of electron-scattering events at Superkamiokande. The observed total solar neutrino flux agrees with the predictions from the solar model (from [12]).

burning, they must be produced by oscillations of solar ν_e neutrinos. As such oscillations are only possible if not all neutrinos are massless, the solution of the solar neutrino problem also opens the door to physics beyond the standard model of particle physics.

3˙3. *Helium burning*. – No stable nuclei with mass numbers $A = 5$ and $A = 8$ exist. Thus, fusion reactions of $p + {}^4\mathrm{He}$ and ${}^4\mathrm{He} + {}^4\mathrm{He}$ lead to unstable resonant states in ${}^5\mathrm{Li}$ and ${}^8\mathrm{Be}$ which decay extremely fast. However, the lifetime of the ${}^8\mathrm{Be}$ ground-state resonance ($\sim 10^{-16}$ s) is long enough to establish a small ${}^8\mathrm{Be}$ equilibrium abundance under helium burning conditions ($T \sim 10^8$ K, $\rho \sim 10^5$ g/cm^3), which amounts to about $\sim 5 \times 10^{-10}$ of the equilibrium ${}^4\mathrm{He}$ abundance. As pointed out by Salpeter [14], this small ${}^8\mathrm{Be}$ equilibrium abundance allows the capture of another ${}^4\mathrm{He}$ nucleus to form the stable ${}^{12}\mathrm{C}$ nucleus. The second step of the triple-alpha fusion reaction is highly enlarged by the presence of an s-wave resonance in ${}^{12}\mathrm{C}$ at a resonance energy of 287 keV (the Hoyle state). To derive the triple-alpha reaction rate under helium burning conditions, it is sufficient to know the properties of this resonance (its γ-width and its α-width). These quantities have been determined experimentally and it is generally assumed that the triple-alpha rate is known with an accuracy of about 15% for helium burning in Red Giants. A current estimate for the uncertainty of the triple-α rate is given in ref. [15] which discusses also the influence of the rate on some aspects of subsequent stellar evolution. An improved triple-α rate for temperatures higher than in Red Giant helium burning is given in [16].

The second step in helium burning, the ${}^{12}\mathrm{C}(\alpha, \gamma){}^{16}\mathrm{O}$ reaction, is the crucial reaction in stellar models of massive stars. Its rate determines the relative importance of the subsequent carbon and oxygen burning stages, including the abundances of the elements produced in these stages. The ${}^{12}\mathrm{C}(\alpha, \gamma){}^{16}\mathrm{O}$ reaction determines also the relative abundance of ^{12}C and ^{16}O, the two bricks for the formation of life, in the Universe. Stellar models are very sensitive to this rate and its determination at the most effective energy in helium burning ($E = 300\,\text{keV}$) with an accuracy of better than 20% is asked for. Despite enormous experimental efforts in the last 3 decades, this goal has not been achieved yet, as the low-energy ^{12}C$(\alpha, \gamma)^{16}$O reaction cross-section is tremendously tricky. Data have been taken down to energies of about $E = 1\,\text{MeV}$, requiring an extrapolation of the S-factor to $E = 300\,\text{keV}$. The data are dominated by a $J = 1^-$ resonance at $E = 2.418\,\text{MeV}$. Unfortunately there is another $J = 1^-$ level at $E = -45\,\text{keV}$, just below the $\alpha + {}^{12}$C threshold. These two states interfere. In the data the broad resonance at $E = 2.418\,\text{MeV}$ dominates, while at stellar burning energies it is likely to be the other way round. It turns out to be quite difficult to determine the properties of the particle-bound states, although a major step forward has been achieved using indirect means by studying the β-decay of the ^{16}N ground state to states in ^{16}O above the α-threshold and their subsequent decay into the $\alpha + {}^{12}$C channel [17]. The γ-decay of the $J = 1^-$ states to the ^{16}O ground state is of dipole ($E1$) nature. If isospin were a good quantum number, this transition would be exactly forbidden, as all involved nuclei (^{4}He, ^{12}C, ^{16}O) have isospin quantum numbers $T = 0$. The observed dipole transition must then come from small isospin-symmetry breaking admixtures. The data, indeed, suggest that the transitions are suppressed by about 4 orders of magnitude compared to "normal" $E1$ transitions. Such a large suppression makes it possible that $E2$ (quadrupole) transitions can compete with the dipole contributions. This is confirmed by measurements of the ^{12}C$(\alpha, \gamma)^{16}$O angular distributions at low energies which are mixtures of dipole and quadrupole contributions. While both, dipole and quadrupole, cross-sections can thus be determined from the data (although the measurement of angular distributions is much more challenging than the determination of the total cross-section), the extrapolation of the $E2$ data to stellar energies at $E = 300\,\text{keV}$ is strongly hampered by the fact that the stellar cross-section is dominated by the tail of a particle-bound $J = 2^+$ state at $E = -245\,\text{keV}$, which, however, is much weaker in the data taken at higher energies. Recent results for the ^{12}C$(\alpha, \gamma)^{16}$O S-factor are presented in [18], [19] and in [20].

The ^{16}O$(\alpha, \gamma)^{20}$Ne is non-resonant at stellar energies and hence very slow, compared to the $\alpha + {}^{12}$C reaction. Thus, helium burning finishes with the ^{12}C$(\alpha, \gamma)^{16}$O reaction.

3'4. *Carbon, neon, oxygen, silicon burning.* – In the fusion of two ^{12}C nuclei, the α and proton-channels have positive Q-values ($Q = 4.62\,\text{MeV}$ and $Q = 2.24\,\text{MeV}$). Thus, the fusion produces nuclides with smaller charge numbers, which can then interact with other carbon nuclei or produced elements. The main reactions in carbon burning are: ^{12}C$(^{12}$C$, \alpha)^{20}$Ne, ^{12}C$(^{12}$C$, \text{p})^{23}$Na, ^{23}Na$(\text{p}, \alpha)^{20}$Ne, ^{23}Na$(\text{p}, \gamma)^{24}$Mg, ^{12}C$(\alpha, \gamma)^{16}$O, which determine the basic energy generation. However, many other reactions can occur, even producing elements beyond ^{24}Mg like ^{26}Mg and ^{27}Al.

Neon burning occurs at temperatures just above $T = 10^9\,\text{K}$. At these conditions the presence of high-energy photons is sufficient to photodissociate ^{20}Ne via the ^{20}Ne$(\gamma, \alpha)^{16}$O reaction which has a Q-value of $-4.73\,\text{MeV}$. This reaction liberates α-particles which react then very fast with other ^{20}Ne nuclei leading to the production of

^{28}Si via the chain ^{20}Ne$(\alpha, \gamma)^{24}$Mg$(\alpha, \gamma)^{28}$Si. Again, many other reactions induced by protons, ^{4}He and also neutrons, which are produced within the occuring reaction chains, occur.

In the fusion of two ^{16}O nuclei, the α and proton-channels have positive Q-values ($Q = 9.59$ MeV and $Q = 7.68$ MeV). Like in carbon burning, the liberated protons and ^{4}He nuclei react with other ^{16}O nuclei. Among the many nuclides produced during oxygen burning are ^{33}Si and ^{35}Cl. These have quite low Q-values against electron captures making it energetically favorable to capture electrons from the degenerate electron sea (the Fermi energies of electrons during core oxygen burning is of order 1 MeV) via the ^{33}Si$(e^-\nu)^{33}$P and ^{35}Cl$(e^-\nu)^{35}$S reactions. The emitted neutrinos carry energy away, thus cooling the star. As we will see in the next section, electron captures play a decisive role in core-collapse supernovae.

The nuclear reaction network during silicon burning is initiated by the photodissociation of ^{28}Si, producing protons, neutrons and α-particles. These particles react again with ^{28}Si or the nuclides produced. During silicon burning, the temperature is already quite high ($T \sim 3$–4×10^9 K). This makes the nuclear reactions mediated by the strong and electromagnetic force quite fast and a chemical equilibrium between reactions and their inverse processes establishes. Under such conditions, the abundance distributions of the nuclides present in the network becomes independent of the reaction rates, establishing the Nuclear Statistical Equilibrium (NSE) (see [21, 22]). Then the abundance of a nuclide with proton and neutron numbers Z and N can be expressed in terms of the abundance of free protons and neutrons by

$$(11) \qquad Y_{Z,N} = G_{Z,N}(\rho N_A)^{(A-1)} \frac{A^{3/2}}{2^A} \left(\frac{2\pi\hbar^2}{m_u kT} \right)^{3/2(A-1)} \exp\left[\frac{B_{Z,N}}{kT} \right] Y_n^N Y_p^Z,$$

where $G_{Z,N}$ is the nuclear partition function (at temperature T), $B_{Z,N}$ the binding excess of the nucleus, m_u the unit nuclear mass and Y_p, Y_n are the abundances of free protons and neutrons. The NSE distribution is subject to the mass and charge conservation, which can be formulated as

$$(12) \qquad \sum_i A_i Y_i = 1; \qquad \sum_i Z_i Y_i = Y_e,$$

where Y_e is the electron-to-nucleon ratio and the sum runs over all nuclides. Until the onset of oxygen burning, one has $Y_e = 0.5$ (nuclei like ^{12}C have identical proton and neutron numbers, while the number of electrons equals the proton number). Once electron capture processes start, the Y_e value is reduced (protons are changed into neutrons, while the total number of nucleons is preserved).

We finally note that, during silicon burning, not a full NSE is established. However, the nuclear chart breaks into several regions in which NSE equilibrium is established. The different regions are not yet in equilibrium, as the nuclear reactions connecting them are yet not fast enough. One uses the term "Quasi-NSE" for these conditions.

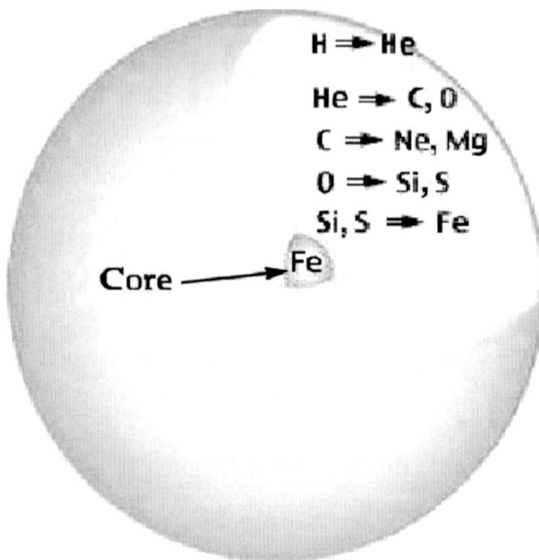

Fig. 6. – At the end of hydrostatic burning a star has an onion-like structure with nuclear burning occurring in various shells.

4. – Core collapse supernovae

Massive stars end their lifes as type II supernovae, triggered by a collapse of their central iron core with a mass of more than $1M_\odot$. The general picture of a core-collapse supernova is probably well understood and has been confirmed by various observations from supernova 1987A. It can be briefly summarized as follows.

At the end of its hydrostatic burning stages (fig. 6), a massive star has an onion-like structure with various shells where nuclear burning still proceeds (hydrogen, helium, carbon, neon, oxygen and silicon shell burning). As nuclei in the iron/nickel range have the highest binding energy per nucleon, the iron core in the star's center has no nuclear energy source to support itself against gravitational collapse. As mass is added to the core, its density and temperature raises, finally enabling the core to reduce its free energy by electron captures of the protons in the nuclei [23]. This reduces the electron degeneracy pressure and the core temperature as the neutrinos produced by the capture can initially leave the star unhindered. Both effects accelerate the collapse of the star. With increasing density, neutrino interactions with matter become decisively important and neutrinos have to be treated by Boltzmann transport. Nevertheless the collapse proceeds until the core composition is transformed into neutron-rich nuclear matter. Its finite compressibilty brings the collapse to a halt, a shock wave is created which traverses outwards through the infalling matter of the core'e envelope. This matter is strongly heated and dissociated into free nucleons. Due to current models the shock has not sufficient energy to explode the star directly. It stalls, but is shortly after revived by energy transfer from the neutrinos which are produced by the cooling of the neutron star

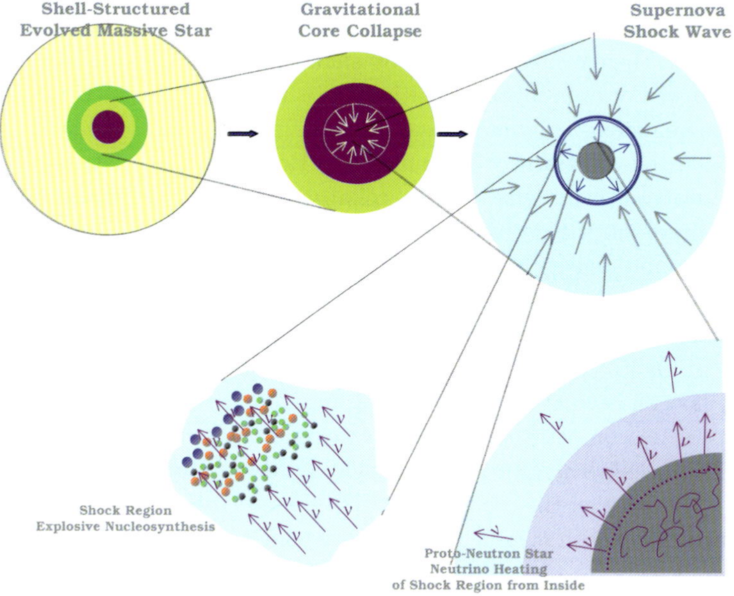

Fig. 7. – Schematic stages of the collapse of a massive star and a supernova explosion. (Courtesy of Roland Diehl.)

born in the center of the core. The neutrinos carry away most of the energy generated by the gravitational collapse and a fraction of the neutrinos are absorbed by the free nucleons behind the stalled shock. The revived shock can then explode the star and the stellar matter outside of a certain mass cut is ejected into the Interstellar Medium. A brief sketch of the various supernova phases is given in fig. 7. Due to the high temperatures associated with the shock's passage, nuclear reactions can proceed rather fast giving rize to explosive nucleosynthesis which is particularly important in the deepest layers of the ejected matter. Reviews on core-collapse supernovae can be found in [24, 25].

Nevertheless, the most sophisticated spherical supernova simulations, including detailed neutrino transport [26-28], currently fail to explode indicating that improved input or numerical treatment is required. Among these microscopic inputs are nuclear processes mediated by the weak interaction, where recent progress has been made possible by improved many-body models and better computational facilities, as is summarized in [29]. Here we focus on the electron capture on nuclei, which strongly influences the dynamics of the collapse and produces the neutrinos present during the collapse, and on recent developments in explosive nucleosynthesis, where again neutrino-induced reactions are essential. Recent two-dimensional simulations stress also the importance of plasma instabilities, which in fact initiated successful numerical explosions [30, 31].

4'1. *Electron captures in core-collapse supernovae—the general picture*. – Late-stage stellar evolution is described in two steps. In the presupernova models the evolution is studied through the various hydrostatic core and shell burning phases until the central core density reaches values up to 10^{10} g/cm^3. The models consider a large nuclear reaction network. However, the densities involved are small enough to treat neutrinos solely as an energy loss source. For even higher densities this is no longer true as neutrino-matter interactions become increasingly important. In modern core-collapse codes neutrino transport is described self-consistently by multigroup Boltzmann simulations [26-28]. While this is computationally very challenging, collapse models have the advantage that the matter composition can be derived from Nuclear Statistical Equilibrium (NSE) as the core temperature and density are high enough to keep reactions mediated by the strong and electromagnetic interactions in equilibrium. This means that for sufficiently low entropies, the matter composition is dominated by the nuclei with the highest binding energies for a given Y_e. The presupernova models are the input for the collapse simulations which follow the evolution through trapping, bounce and hopefully explosion.

The collapse is a competition of the two weakest forces in nature: gravity *vs.* weak interaction, where electron captures on nuclei and protons and, during a period of silicon burning, also β-decay play the crucial roles [32]. The weak-interaction processes become important when nuclei with masses $A \sim 55$–60 (pf-shell nuclei) are most abundant in the core (although capture on sd shell nuclei has to be considered as well). As weak interactions changes Y_e and electron capture dominates, the Y_e value is successively reduced from its initial value ~ 0.5. As a consequence, the abundant nuclei become more neutron-rich and heavier, as nuclei with decreasing Z/A ratios are more bound in heavier nuclei. Two further general remarks are useful. There are many nuclei with appreciable abundances in the cores of massive stars during their final evolution. Neither the nucleus with the largest capture rate nor the most abundant one are necessarily the most relevant for the dynamical evolution: what makes a nucleus relevant is the product of rate times abundance.

For densities $\rho < 10^{11}$ g/cm^3, stellar weak-interaction processes are dominated by Gamow-Teller (GT) and, if applicable, by Fermi transitions. At higher densities forbidden transitions have to be included as well. To understand the requirements for the nuclear models to describe these processes (mainly electron capture), it is quite useful to recognize that electron capture is governed by two energy scales: the electron chemical potential μ_e, which grows like $\rho^{1/3}$, and the nuclear Q-value. Further, μ_e grows much faster than the Q values of the abundant nuclei. We can conclude that at low densities, where one has $\mu_e \sim Q$ (*i.e.* at presupernova conditions), the capture rates will be very sensitive to the phase space and require an accurate as possible description of the detailed GT$_+$ distribution of the nuclei involved. Furthermore, the finite temperature in the star requires the implicit consideration of capture on excited nuclear states, for which the GT distribution can be different than for the ground state. It has been demonstrated [33,34] that modern shell model calculations are capable to describe nuclear properties relevant to derive stellar electron capture rates (spectra and GT$_+$ distributions) rather well (an example is shown in fig. 8) and are therefore the appropriate tool to calculate the weak-interaction rates for those nuclei ($A \sim 50$–65) which are relevant at such densities. At higher densities, when μ_e is sufficiently larger than the respective nuclear Q values, the capture rate becomes less sensitive to the detailed GT$_+$ distribution and is mainly only dependent on the total GT strength. Thus, less sophisticated nuclear models might be sufficient. However, one is facing a nuclear structure problem which has been over-come only very recently. Once the matter has become sufficiently neutron-rich, nuclei with proton numbers $Z < 40$ and neutron numbers $N > 40$ will be quite abundant in the core. For such nuclei, Gamow-Teller transitions would be Pauli forbidden (GT_+ transitions change a proton into a neutron in the same harmonic-oscillator shell) were it not for nuclear correlation and finite-temperature effects which move nucleons from the pf shell into the gds shell (see fig. 9). To describe such effects in an appropriately large model space (*e.g.*, the complete $fpgds$ shell) is currently only possible by means of the Shell Model Monte Carlo approach (SMMC) [35-37]. In [38] SMMC-based electron capture rates have been calculated for many nuclei which are present during the collapse phase.

$4`2$. *Weak-interaction rates and presupernova evolution*. – Up to densities of a few 10^{10} g/cm^3 electron capture is still dominated by capture on nuclei in the $A \sim 45$–65 mass range, for which capture rates have been derived on the basis of large-scale shell model diagonalization studies. Importantly, the shell model rates are noticeably smaller than those derived previously on the basis of the independent particle model [34]. To study the influence of these slower shell model rates on presupernova models Heger *et al.* [32,39] have repeated the calculations of Weaver and Woosley [40] keeping the stellar physics, except for the weak rates, as close to the original studies as possible. Figure 10 examplifies

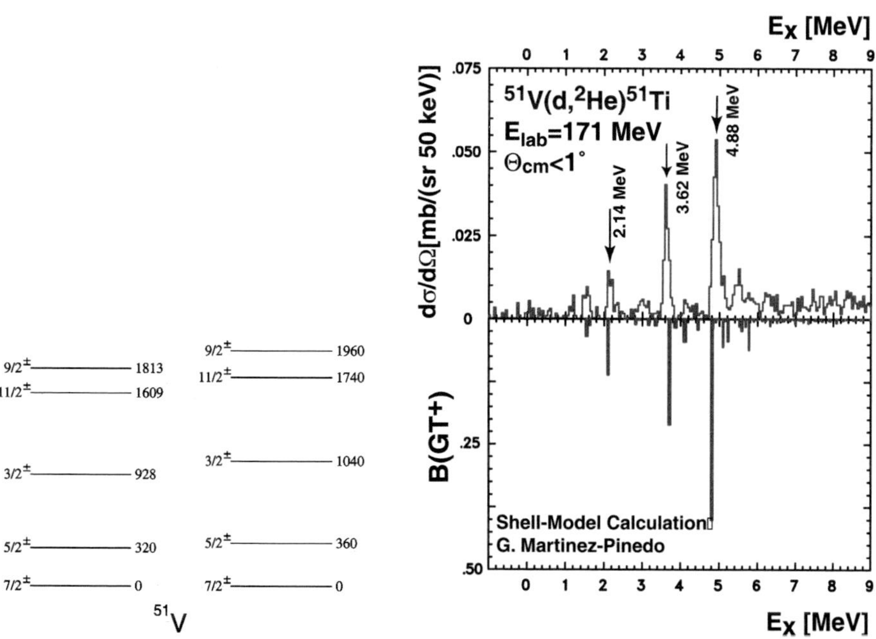

Fig. 8. – Comparison of the measured spectrum (left) and ^{51}V(d, ^{2}He)^{51}Ti cross-section (right) at forward angles (which is proportional to the GT_+ strength) with the shell model spectrum and GT distribution in ^{51}V (partially from [41]).

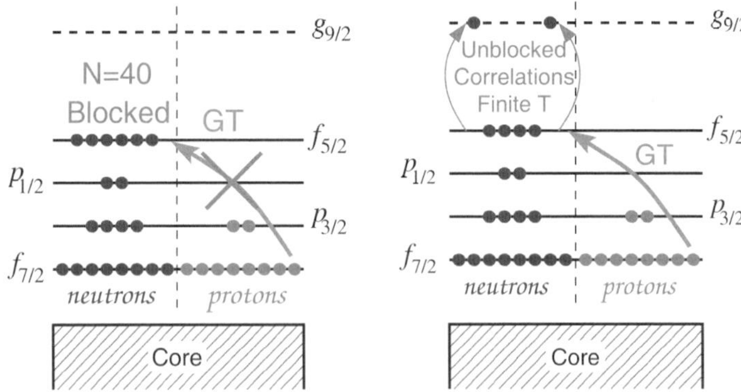

Fig. 9. – In the independent particle model GT transitions are blocked at neutron number $N = 40$ (left). The blocking is overcome by correlations and finite temperature effects (right).

the consequences of the shell model weak-interaction rates for presupernova models in terms of the three decisive quantities: the central Y_e value and entropy and the iron core mass. The central values of Y_e at the onset of core collapse increased by 0.01–0.015 for the new rates. This is a significant effect. We note that the new models also result in lower core entropies for stars with $M \leq 20 M_\odot$, while for $M \geq 20 M_\odot$, the new models actually have a slightly larger entropy. The iron core masses are generally smaller in the new models where the effect is larger for more massive stars ($M \geq 20 M_\odot$), while for the most common supernovae ($M \leq 20 M_\odot$) the reduction is by about $0.05 M_\odot$.

Electron capture dominates the weak-interaction processes during presupernova evolution. However, during silicon burning, β decay (which increases Y_e) can compete and

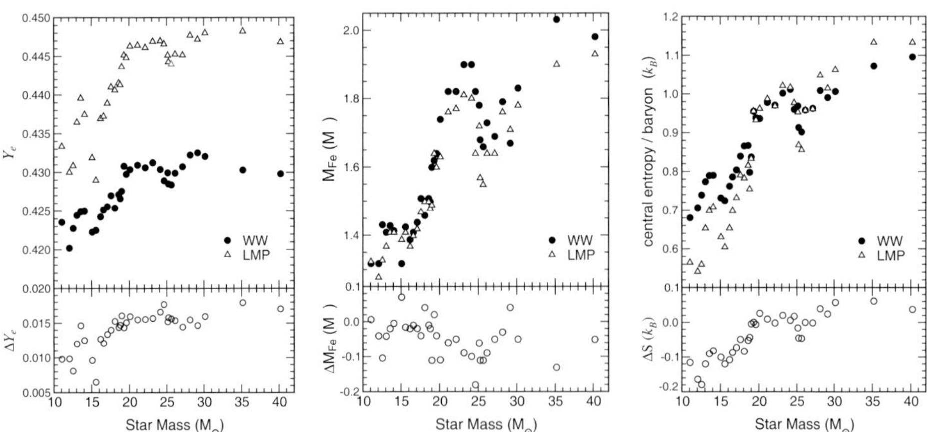

Fig. 10. – Comparison of the center values of Y_e (left), the iron core sizes (middle) and the central entropy (right) for 11–40$M_\odot$ stars between the WW models, which used the FFN rates, and the ones using the shell model weak-interaction rates (LMP) (from [39]).

adds to the further cooling of the star. With increasing densities, β-decays are hindered as the increasing Fermi energy of the electrons blocks the available phase space for the decay. Thus, during collapse β-decays can be neglected.

We note that the shell model weak interaction rates predict the presupernova evolution to proceed along a temperature-density-Y_e trajectory where the weak processes are dominated by nuclei rather close to stability. Thus it will be possible, after radioactive ion-beam facilities become operational, to further constrain the shell model calculations by measuring relevant beta decays and GT distributions for unstable nuclei. References [32,39] identify those nuclei which dominate (defined by the product of abundance times rate) the electron capture and beta decay during various stages of the final evolution of a $15M_\odot$, $25M_\odot$ and $40M_\odot$ star.

4'3. *The role of electron capture during collapse.* – Until recently core-collapse simulations assumed that electron capture on nuclei are prohibited by the Pauli blocking mechanism [42]. However, based on the SMMC calculations it has been shown in [38] that capture on nuclei dominates over capture on free protons (the later was evaluated in [43] and has been included in the simulations). This is demonstrated in fig. 11 which compares the two capture rates. The figure also shows that neutrinos from captures on nuclei have a mean energy 40–60% less than those produced by capture on protons. Although capture on nuclei under stellar conditions involves excited states in the parent and daughter nuclei, it is mainly the larger $|Q|$-value which significantly shifts the energies of the emitted neutrinos to smaller values. These differences in the neutrino spectra strongly influence neutrino-matter interactions, which scale with the square of

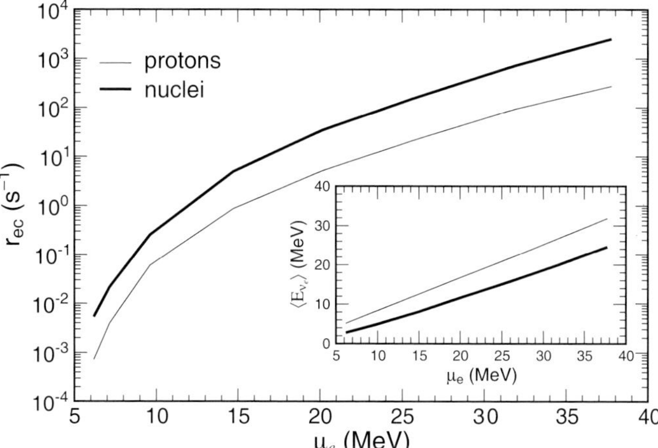

Fig. 11. – The reaction rates for electron capture on protons (thin line) and nuclei (thick line) are compared as a function of electron chemical potential along a stellar collapse trajectory. The insert shows the related average energy of the neutrinos emitted by capture on nuclei and protons. The results for nuclei are averaged over the full nuclear composition (see text, from [38]).

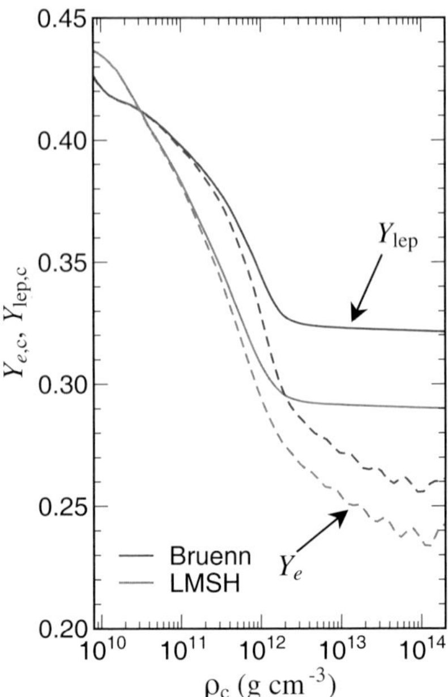

Fig. 12. – The electron and lepton fraction in the center during the collapse phase. The thin line is a simulation using the Bruenn parameterization [43], while the thick line is for a simulation using the shell model rates. (Courtesy of Hans-Thomas Janka.)

the neutrino energy and are essential for collapse simulations [28, 27].

The effects of this more realistic implementation of electron capture on heavy nuclei have been evaluated in independent self-consistent neutrino radiation hydrodynamics simulations by the Oak Ridge and Garching collaborations [38, 44]. The changes compared to the previous simulations, which basically ignored electron capture on nuclei, are significant: in denser regions, the additional electron capture on heavy nuclei results in more electron capture in the new models. In lower density regions, where nuclei with $A < 65$ dominate, the shell model rates [34] result in less electron capture. The results of these competing effects can be seen in fig. 12, which shows the center value of Y_e and Y_{lep} (the lepton-to-baryon ratio) during the collapse. At densities above $10^{12}\,\mathrm{g/cm^3}$, Y_{lep} is (nearly) constant indicating that the neutrinos are trapped and a Fermi sea of neutrinos is being built up. Weak-interaction processes are now in equilibrium with their inverse. The changes in electron captures strongly reduce the temperatures and entropies in the inner core and hence affect its composition. They also strongly influence the shock. The combination of increased electron capture in the interior with reduced electron capture in the outer regions causes the shock to form with 16% less mass interior to it and a 10% smaller velocity difference across the shock (see fig. 13). In spite of this mass reduction, the radius from which the shock is launched is actually displaced slightly outwards to $15.7\,\mathrm{km}$ from $14.8\,\mathrm{km}$ in the old models. Also the altered gradients in density and lepton

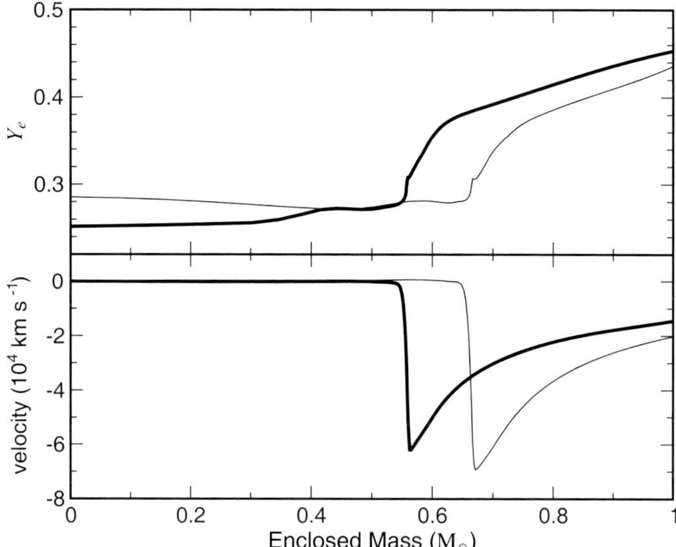

Fig. 13. – The electron fraction and velocity as functions of the enclosed mass at bounce for a 15 $M_\odot$ model [32]. The thin line is a simulation using the Bruenn parameterization [43], while the thick line is for a simulation using the shell model rates.

fraction play an important role in the behavior of the shock. Though also the new models fail to produce explosions in the spherically symmetric limit, the altered gradients allow the shock in the case with improved capture rates to reach 205 km, which is about 10 km further out than in the old models.

Astrophysics simulations have demonstrated that electron capture rates on nuclei have a strong impact on the core collapse trajectory and the properties of the core at bounce. The evaluation of the rates has to rely on theory as a direct experimental determination of the rates for the relevant stellar conditions (*i.e.* rather high temperatures) is currently impossible. Nevertheless it is important to experimentally explore the configuration mixing between *pf* and *sdg* shell in extremely neutron-rich nuclei as such understanding will guide and severely constrain nuclear models. Such guidance is expected from future radioactive ion-beam facilities like FAIR.

5. – Making heavy elements in explosive nucleosynthesis

When in a successful explosion the shock passes through the outer shells, its high temperature induces an explosive nuclear burning on short time scales. This explosive nucleosynthesis can alter the elemental abundance distributions in the inner (silicon, oxygen) shells. More interestingly the shock has sufficiently high temperatures that nuclear binding cannot withstay and matter is ejected as free protons and neutrons from the neutron star surface. As the outflow is adiabatic, the ejected matter reaches cooler regions at larger distances and nuclei can be assembled from free nucleons. The abundance outcome obviously depends strongly on the ratio of neutrons and protons which is set by the competition of neutrino and anti-neutrino absorption on free nucleons.

Recently explosive nucleosynthesis has been investigated consistently within supernova simulations, where a successful explosion has been enforced by slightly enlargening the neutrino absorption cross-section on nucleons or the neutrino mean-free path, which both increase the efficiency of the energy transport to the stalled shock. The results presented in [45, 46] showed that in an early phase after the bounce the ejected matter is actually proton rich as the ν_e neutrinos have sufficiently large energies to drive the matter proton rich. In later stages, *i.e.* a few seconds after bounce, the neutrino opacities in the neutron-rich matter ensure that $\bar{\nu}_e$ have larger average energies than ν_e and $\bar{\nu}_e$ absorption on protons dominates driving the matter neutron rich (allowing for the r-process to occur).

5`1. *The νp process.* – The studies presented in [45-47] show that matter with Y_e larger than 0.5 will always be present in core-collapse supernovae explosions with ejected matter irradiated by a strong neutrino flux, independently of the details of the explosion. As this proton-rich matter expands and cools, nuclei can form resulting in a composition dominated by $N = Z$ nuclei, mainly ^{56}Ni and ^{4}He, and protons. Without the further inclusion of neutrino and antineutrino reactions the composition of this matter will finally consist of protons, alpha-particles, and heavy (Fe-group) nuclei (in nucleosynthesis terms a proton- and alpha-rich freeze-out), with enhanced abundances of ^{45}Sc, ^{49}Ti, and ^{64}Zn [45, 46]. In these calculations the matter flow stops at ^{64}Ge (fig. 14) with a small proton capture probability and a beta-decay half-life (64 s) that is much longer than the expansion time scale (~ 10 s) [45].

As noted by Martinez-Pinedo and explored in [48] the synthesis of nuclei with $A > 64$ can be obtained, if one explores the previously neglected effect of neutrino interactions on the nucleosynthesis of heavy nuclei. $N \sim Z$ nuclei are practically inert to neutrino capture (converting a neutron in a proton), because such reactions are endoergic for neutron-deficient nuclei located away from the valley of stability. The situation is different for antineutrinos that are captured in a typical time of a few seconds, both on protons and nuclei, at the distances at which nuclei form (~ 1000 km). This time scale is much shorter than the beta-decay half-life of the most abundant heavy nuclei reached without neutrino interactions (*e.g.*, ^{56}Ni, ^{64}Ge). As protons are more abundant than heavy nuclei, antineutrino capture occurs predominantly on protons, causing a residual density of free neutrons of 10^{14}–10^{15} cm^{-3} for several seconds, when the temperatures are in the range 1–3 GK. The neutrons produced via antineutrino absorption on protons can easily be captured by neutron-deficient $N \sim Z$ nuclei (for example ^{64}Ge), which have large neutron capture cross-sections. The amount of nuclei with $A > 64$ produced is then directly proportional to the number of antineutrinos captured. While proton capture, (p, γ), on ^{64}Ge takes too long, the (n, p) reaction dominates (with a lifetime of 0.25 s at a temperature of 2 GK), permitting the matter flow to continue to nuclei heavier than ^{64}Ge via subsequent proton captures (see fig. 14).

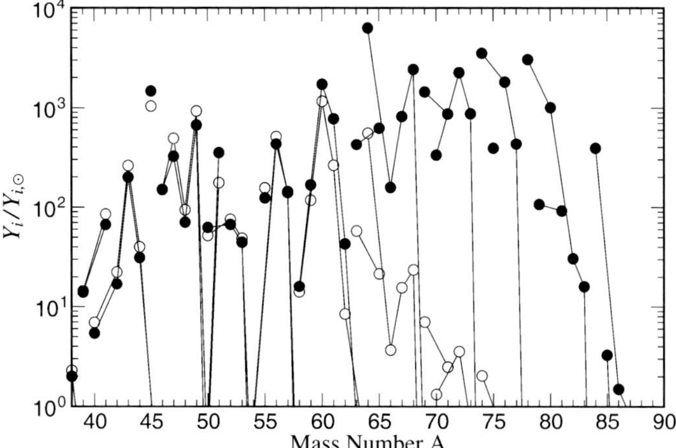

Fig. 14. – Elemental abundance yields (normalized to solar) for elements produced in the proton-rich environment shortly after the supernova shock formation. The matter flow stops at nuclei like ^{56}Ni and ^{64}Ge (open circles), but can proceed to heavier elements if neutrino reactions are included during the network (full circles) (from [48]).

Fröhlich *et al.* argue that all core-collapse supernovae will eject hot, explosively processed matter subject to neutrino irradiation and that this novel nucleosynthesis process (called νp process) will operate in the innermost ejected layers producing neutron-deficient nuclei above $A > 64$. However, how far the massflow within the νp process can proceed, strongly depends on the environment conditions, most noteably on the Y_e value of the matter. Obviously the larger Y_e, the larger the abundance of free protons

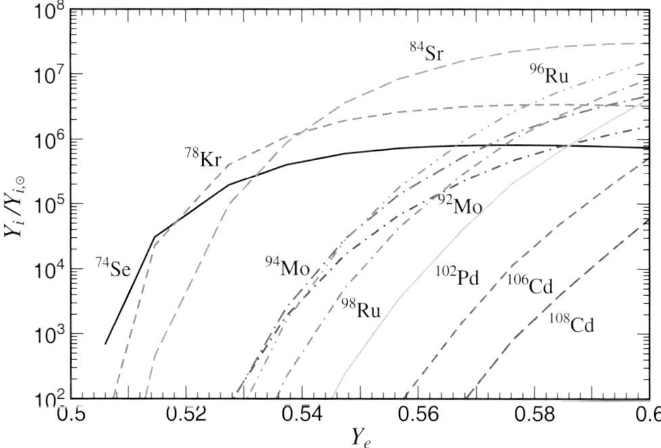

Fig. 15. – Light p-nuclei abundances in comparison to solar abundances as a function of Y_e. The Y_e-values given are the ones obtained at a temperature of $3\,$GK that corresponds to the moment when nuclei are just formed and the νp-process starts to act. (From [48].)

which can be transformed into neutrons by antineutrino absorption. Figure 15 shows the dependence of the νp-process abundances as a function of the Y_e value of the ejected matter. Nuclei heavier than $A = 64$ are only produced for $Y_e > 0.5$, showing a very strong dependence on Y_e in the range 0.5–0.6. A clear increase in the production of the light p-nuclei, ^{92,94}Mo and ^{96,98}Ru, is observed as Y_e gets larger. Thus the νp process offers the explanation for the production of these light p-nuclei, which was yet unknown. It might also explain the presence of strontium in the extremely metalpoor, and hence very old, star HE 1327-2326 [49]. The νp process results of Fröhlich *et al.* have been recently confirmed by Pruet *et al.* [50], in a study of the nucleosynthesis that occurs in the early proton-rich neutrino wind.

5˙2. *The r-process.* – About half of the elements heavier than mass number $A \sim 60$ are made within the *r-process*, a sequence of rapid neutron captures and β decays [51,52]. The process is thought to occur in environments with extremely high neutron densities. Then neutron captures are much faster than the competing decays and the r-process path runs through very neutron-rich, unstable nuclei. Once the neutron source ceases, the process stops and the produced nuclides decay towards stability producing the neutron-rich heavier elements. Many parameter studies of the reaction network, aimed at reproducing the observed r-process abundances, have progressed our general understanding. So it is generally accepted that the r-process operates in an environment with temperatures of order 10^9 K and with neutron densities in excess of 10^{20}/cm^3 [53]. Simulations of the r-process reaction network indicate that under such conditions for temperature and neutron densities the reaction path runs through extremely neutron rich nuclei with separation energies, $S_n < 2 - 3$ MeV [53], for which most of the required properties (*i.e.* mass, lifetime and neutron capture cross-sections) are experimentally unknown. Thus, these quantities have to be modelled, based on experimental guidance. As we will discuss in the next subsection some progress has been achieved recently, but the real breakthrough is expected from future rare isotope facilities. Finally we mention that the observed peaks in the r-process abundance distribution are related to the magic neutron numbers, $N = 50, 82, 126$, where the r-process runs through several, rather long β decays at $N = 50, 82$ and 126 (the respective nuclei are called r-process waiting points) and a relatively large amount of matter is built up.

Despite many promising attempts the actual site of the r-process has not been identified yet. However, parameter studies have given clear evidence that the observed r-process abundances cannot be reproduced at one site with constant temperature and neutron density [54]. Thus the abundances require a superposition of several (at least three) r-process components. This likely implies a dynamical r-process in an environment in which the conditions change during the duration of the process. The currently favored r-process sites (type II supernovae [55] and neutron-star mergers [56]) offer such dynamical scenarios. However, recent meteoritic clues might even point to more than one distinct site for our solar r-process abundance [57]. The same conclusion can be derived from the observation of r-process abundances in low-metallicity stars (*e.g.*, [58]), a milestone of r-process research.

5‘2.1. Nuclear r-process input. Arguably the most important nuclear ingredient in r-process simulations are the nuclear masses as they determine the flow-path. Unfortunately nearly all of them are experimentally unknown and have to be theoretically estimated. Traditionally this is done on the basis of parametrizations to the known masses. Although these empirical mass formulae achieve rather remarkable fits to the data (the standard deviation is of order 700 keV), extrapolation to unknown masses appears less certain and different mass formulae can predict quite different trends for the very neutron-rich nuclei of relevance to the r-process. The most commonly used parametrizations are based on the Finite Range Droplet Model (FRDM), developed by Móller and collaborators [59], and on the ETFSI (Extended Thomas-Fermi with Strutinski integral) model of Pearson [60]. A new era has been opened very recently, as for the first time, nuclear mass tables have been derived on the basis of nuclear many-body theory (Hartree-Fock-Bogoliobov model) [61-63] rather than by parameter fit to data.

The nuclear halflives strongly influence the relative r-process abundances. In a simple β-flow equilibrium picture the elemental abundance is proportional to the halflife, with some corrections for β-delayed neutron emission [64]. As r-process halflives are longest for the magic nuclei, these waiting point nuclei determine the minimal r-process duration time; *i.e.* the time needed to build up the r-process peak around $A \sim 200$. We note, however, that this time depends also crucially on the r-process path and can be as short as a few 100 milliseconds if the r-process path runs close to the neutron dripline.

There are a few milestone halflife measurements including the $N = 50$ waiting point nuclei ^{80}Zn and ^{79}Cu and the $N = 82$ waiting point nuclei ^{130}Cd and ^{129}Ag [65, 66]. Although no halflives for $N = 126$ waiting points have yet been determined, there has been decisive progress towards this goal recently [67].

These data play crucial roles in constraining and testing nuclear models which are still necessary to predict the bulk of halflives required in r-process simulations. It is generally assumed that the halflives are dominated by allowed Gamow-Teller (GT) transitions, with forbidden transitions contributing noticeably for the heavier r-process nuclei [68]. The β decays only probe the weak low-energy tail of the GT distributions and provide quite a challenge to nuclear modeling as they are not constrained by sumrules. Traditionally the estimate of the halflives are based on the quasi-particle random phase approximation on top of the global FRDM or ETFSI models. Recently halflives for selected (spherical) nuclei have been presented using the QRPA approach based on the microscopic Hartree-Fock-Bogoliubov method [69] or a global density functional [70, 71]; in particular the later approach achieved quite good agreement with data for spherical nuclei in different ranges of the nuclear chart. Applications of the interacting shell model [72-74] have yet been restricted to waiting point nuclei with magic neutron numbers. Here, however, this model which accounts for correlations beyond the QRPA approach, obtains quite good results (for an example see fig. 16). We remark that, besides a good description of the allowed (and forbidden) transition matrix elements, an accurate reproduction of the Q_{n} values is required from the models.

Further r-process simulations require rates for neutron capture, and for the various fission processes (neutron-induced, beta-delayed, spontaneous, perhaps neutrino induced)

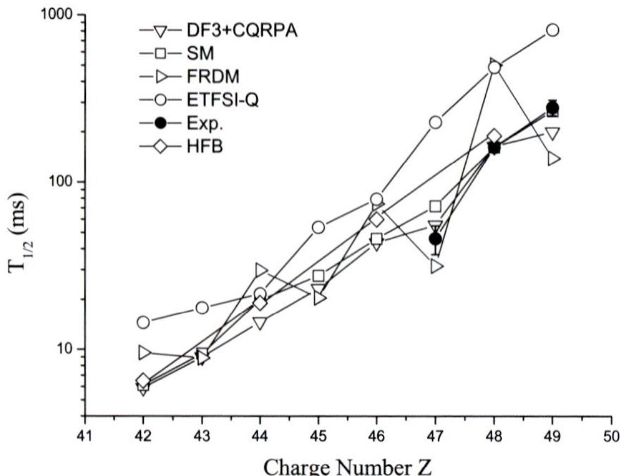

Fig. 16. – Comparison of various theoretical halflife predictions with data for the $N = 82$ r-process waiting points (from [74]).

together with the corresponding yield distributions. If the r-process occurs in strong neutrino fluences, different neutrino-induced charged-current (*e.g.* (ν_e, e^-)) and neutral-current (*e.g.* (ν, ν')) reactions, which are often accompanied by the emission of one or several neutrons [75-77], have to be modelled and included as well. We mention that the occurence of low-lying dipole strength in neutron-rich nuclei, as predicted by nuclear models [78], can have noticeable effects on neutron capture cross-sections [79].

Recently r-process simulations, which for the first time included detailed fission yield distributions derived on the basis of the abration-ablation model code ABLA [80], have been performed [81,82]. The studies have been performed assuming the neutrino-driven wind scenario (*i.e.* matter ejected from the neutron star in a core-collapse supernova)

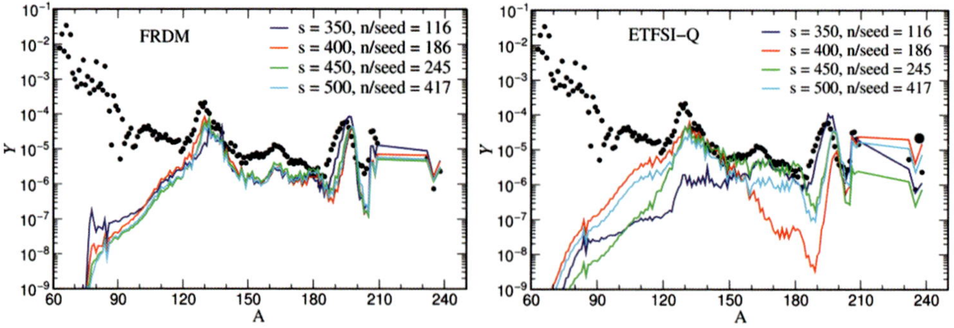

Fig. 17. – r-process abundances obtained in simulations which explore the potential impact of fission. The calculations are performed for different neutron-to-seed ratios and two different mass tabulations. The solid line is the solar r-process abundance distribution (from [82]).

as the r-process site. The entropy of the ejected matter, or equivalently the neutron-to-seed (n/s) ratio, has been treated as a free parameter. To explore the potential effect of fission on the r-process, this ratio has to be relatively large ($n/s > 250$). If this condition is fulfilled, then the r-process runs quite fast along a path close to the dripline reaching heavy nuclei with neutron numbers close to the magic number $N = 184$ and above. Once the neutron source is used, *i.e.* the ejected matter has reached radii where the neutron number density is too low to support fast neutron captures, the produced nuclides decay. For the heavy nuclei the dominant decay mode is neutron-induced fission which via their fission decay products produce peaks around mass numbers $A \sim 130$ and 195, corresponding to the second and third peaks in the observed r-process abundance distributions (fig. 17). These simulations clearly stress the importance of the nuclear masses. While the studies, which use the FRDM masses, give quite similar abundance distributions for different n/s ratios once a threshold value is overcome, the calculations based on the ETFSI masses yield abundance results which vary strongly with the assumed n/s value. This strikingly different behavior can be traced back to differences in the predicted masses for very neutron rich nuclei. The FRDM mass tabulation predicts that certain nuclei just above the magic neutron number $N = 82$ act as obstacles for the r-process matter flow holding back material and ensuring that more free neutrons are available when part of the matter reaches heavy nuclei with $N = 184$ to guarantee production of nuclides even beyond these waiting points. The ETFSI mass tabulation does not identify such obstacles on the r-process path and, as no matter is kept back at

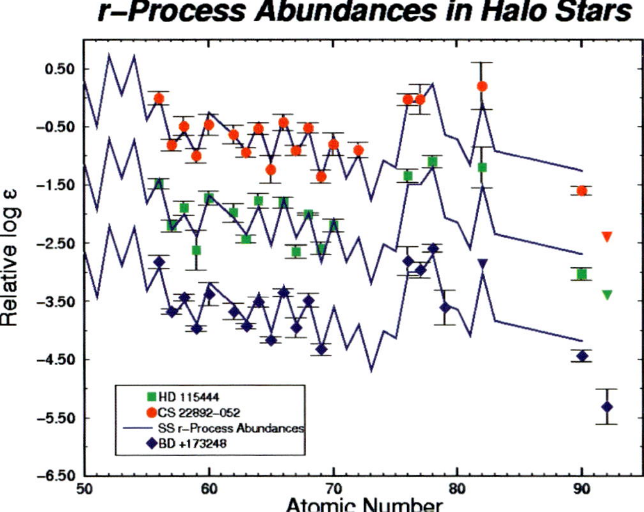

Fig. 18. – Observed r-process abundances in three old stars in the galactical halo, compared to the solar r-process abundance (blue line) which reflects the chemical history of the galaxy. For better viewing the different curves are off-set by arbitrary numbers. (Courtesy John Cowan.)

relatively small mass numbers, less neutrons are available once the matter flow reaches $N = 184$ suppressing matter flow beyond this waiting point [82]. We will know which of the two mass models is more realistic once masses of very neutron rich nuclei will be measured; this is one of the important aims at radioactive ion-beam facilities like FAIR. It should be stressed that observations of r-process abundances in old halo stars in our galaxy show always the same pattern between mass numbers $A = 130$ and 195 (fig. 18), while the patterns differ for $A < 130$. It is intriguing that these old stars in completely different locations of the Milky Way, which have witnessed only a few, and importantly not the same, supernova outbursts, show identical r-process patterns between the second and third peaks as obtained when averaged over the chemical history of the galaxy (solar r-process abundance). The fact that r-process simulations including fission and based on the FRDM mass model yield just such a behavior is certainly interesting. However, the nuclear and astrophysical input in these simulations is yet too uncertain to draw any conclusions.

$$* \ * \ *$$

The work presented here has benefitted from a close and intensive collaboration with GABRIEL MARTINEZ-PINEDO during the last decade. I also would like to thank I. BORZOV, DAVID DEAN, CARLA FRÖHLICH, ALEXANDER HEGER, RAPHAEL HIX, THOMAS JANKA, ANDRIUS JUODAGALVIS, ALEKSANDRA KELIC, TONY MEZZACAPPA, BRONSON MESSER, MATTHIAS LIEBENDÖRFER, PETER VON NEUMANN-COSEL, ACHIM RICHTER, JORGE SAMPAIO, KARL-HEINZ SCHMIDT, F.-K. THIELEMANN and STAN WOOSLEY for many discussions and fruitful collaborations.

REFERENCES

[1] ROLFS C. E. and RODNEY W. S., *Cauldrons in the Cosmos* (Chicago Press, Chicago) 1988.
[2] SALPETER E. E., *Austr. J. Phys.*, **7** (1954) 373.
[3] ASSENBAUM H. J., LANGANKE K. and ROLFS C. E., *Z. Phys. A*, **327** (1987) 461.
[4] WOOSLEY S. E., in *Proceedings of the Accelerated Radioactive Beam Workshop*, edited by BUCHMANN L. and D'AURIA J. M. (TRIUMF, Canada) 1985.
[5] ADELBERGER E. *et al.*, *Rev. Mod. Phys.*, **70** (1998) 1265.
[6] BAHCALL J. N., *Neutrino Astrophysics* (Cambridge University Press, Cambridge) 1989.
[7] BONETTI R. *et al.*, *Phys. Rev. Lett.*, **82** (1999) 5205.
[8] JUNGHANS A. R. *et al.*, *Phys. Rev. C*, **68** (2003) 065803.
[9] IMBRIANI G. *et al.*, *Eur. Phys. J. A*, **25** (2005) 455.
[10] RUNKLE R. C. *et al.*, *Phys. Rev. Lett.*, **94** (2005) 082503.
[11] DEGL'INNOCENTI S. *et al.*, *Phys. Lett. B*, **590** (2004) 13.
[12] AHMED S. N. *et al.*, *Phys. Rev. Lett.*, **87** (2001) 071301; **89** (2002) 011301; preprint nucl-ex/0309004.
[13] BAHCALL J. N., BASU S. and SERENELLI A. M., *Astrophys. J.*, **631** (2005) 1281.
[14] SALPETER E. E., *Phys. Rev.*, **88** (1952) 547; **107** (1957) 516.
[15] AUSTIN S. M., *Nucl. Phys. A*, **758** (2005) 375.
[16] FYNBO H. *et al.*, *Nature*, **433** (2005) 135.

[17] BUCHMANN L. *et al.*, *Phys. Rev. Lett.*, **70** (1993) 726.

[18] KUNZ R., JAEGER M., MAYER A., HAMMER J. W., STAUDT G., HARISSOPULOS S. and PARADELLIS T., *Phys. Rev. Lett.*, **86** (2001) 3244.

[19] TISCHHAUSER P. *et al.*, *Phys. Rev. Lett.*, **88** (2002) 072501.

[20] BUCHMANN L. and BARNES C. A., *Nucl. Phys. A*, **777** (2006) 254.

[21] SHAPIRO S. L. and TEUKOLSKY S. A., *Black Holes, White Dwarfs and Neutron Stars* (Wiley-Interscience, New York) 1983.

[22] MEYER B. S. and WALSH J. H., *Nuclear Physics in the Universe*, edited by GUIDRY M. W. and STRAYER M. R. (Institute of Physics, Bristol) 1993.

[23] BETHE H. A., BROWN G. E., APPLEGATE J. and LATTIMER J. M., *Nucl. Phys. A*, **324** (1979) 487.

[24] BETHE H. A., *Rev. Mod. Phys.*, **62** (1990) 801.

[25] JANKA H.-TH., KIFONIDIS K. and RAMPP M., *Lecture Notes in Physics*, Vol. **578** (Springer) 2001, p. 333.

[26] BURAS R., RAMPP M., JANKA H.-TH. and KIFONIDIS K., *Phys. Rev. Lett.*, **90** (2003) 241101.

[27] JANKA H. TH. and RAMPP M., *Astrophys. J. Lett.*, **539** (2000) L33.

[28] MEZZACAPPA A. *et al.*, *Phys. Rev. Lett.*, **86** (2001) 1935.

[29] LANGANKE K. and MARTINEZ-PINEDO G., *Rev. Mod. Phys.*, **75** (2003) 819.

[30] BURROWS A. *et al.*, *Astrophys. J.*, **640** (2006) 878.

[31] MAREK A., Ph.D. Thesis, TU Munich (2006); MAREK A. and JANKA H.-TH., submitted to *Astrophys. J.*

[32] HEGER A., LANGANKE K., MARTINEZ-PINEDO G. and WOOSLEY S. E., *Phys. Rev. Lett.*, **86** (2001) 1678.

[33] CAURIER E., LANGANKE K., MARTÍNEZ-PINEDO G. and NOWACKI F., *Nucl. Phys. A*, **653** (1999) 439.

[34] LANGANKE K. and MARTÍNEZ-PINEDO G., *Nucl. Phys. A*, **673** (2000) 481.

[35] JOHNSON C. W., KOONIN S. E., LANG G. H. and ORMAND W. E., *Phys. Rev. Lett.*, **69** (1992) 3157.

[36] LANGANKE K. *et al.*, *Phys. Rev. C*, **52** (1995) 718.

[37] KOONIN S. E., DEAN D. J. and LANGANKE K., *Phys. Rep.*, **278** (1997) 2.

[38] LANGANKE K. *et al.*, *Phys. Rev. Lett.*, **90** (2003) 241102.

[39] HEGER A., WOOSLEY S. E., MARTINEZ-PINEDO G. and LANGANKE K., *Astrophys. J.*, **560** (2001) 307.

[40] WOOSLEY S. E. and WEAVER T. A., *Astrophys. J. Suppl.*, **101** (1995) 181.

[41] BÄUMER C. *et al.*, *Phys. Rev. C*, **68** (2003) 031303(R).

[42] FULLER G. M., *Astrophys. J.*, **252** (1982) 741.

[43] BRUENN S. W., *Astrophys. J. Suppl.*, **58** (1985) 771; MEZZACAPPA A. and BRUENN S. W., *Astrophys. J.*, **405** (1993) 637; **410** (1993) 740.

[44] HIX R. W. *et al.*, *Phys. Rev. Lett.*, **91** (2003) 210102.

[45] PRUET J. *et al.*, *Astrophys. J.*, **644** (2006) 1028.

[46] FRÖHLICH C. *et al.*, *Astrophys. J.*, **637** (2006) 415.

[47] WANAJO S., *Astrophys. J.*, **647** (2006) 1323.

[48] FRÖHLICH C. *et al.*, *Phys. Rev. Lett.*, **96** (2006) 142502.

[49] FREBEL A. *et al.*, *Nature (London)*, **434** (2005) 871.

[50] PRUET J., HOFFMAN R., WOOSLEY S. E., JANKA H. T. and BURAS R., *Astrophys. J.*, **644** (2006) 1028.

[51] BURBIDGE E. M., BURBIDGE G. R., FOWLER W. A. and HOYLE F., *Rev. Mod. Phys.*, **29** (1957) 547.

[52] CAMERON A. G. W., Chalk River Report CRL-41 (1957).

[53] COWAN J. J., THIELEMANN F.-K. and TRURAN J. W., *Phys. Rep.*, **208** (1991) 267.

[54] KRATZ K. L. *et al.*, *Astrophys. J.*, **402** (1993) 216.

[55] WOOSLEY S. E. *et al.*, *Astrophys. J.*, **399** (1994) 229.

[56] FREIBURGHAUS C., ROSSWOG S. and THIELEMANN F.-K., *Astrophys. J. Lett.*, **525** (1999) L121.

[57] WASSERBURG G. J., BUSSO M., GALLINO R. and NOLLETT K. M., *Nucl. Phys. A*, **777** (2006) 5.

[58] SNEDEN C. *et al.*, *Astrophys. J.*, **591** (2003) 936.

[59] MÖLLER P., NIX J. R. and KRATZ K. L., *At. Data Nucl. Data Tables*, **66** (1997) 131.

[60] BORSOV I. N., GORIELY S. and PEARSON J. M., *Nucl. Phys. A*, **621** (1997) 307c.

[61] SAMYN M., GORIELY S. and PEARSON J. M., *Nucl. Phys. A*, **725** (2003) 69.

[62] GORIELY S., SAMYN M., PEARSON J. M. and ONSI M., *Nucl. Phys. A*, **750** (2005) 425.

[63] STONE J. R., *J. Phys. G*, **31** (2005) 211.

[64] KRATZ K. L. *et al.*, *J. Phys. G*, **24** (1988) S331.

[65] KRATZ K. L. *et al.*, *Z. Phys. A*, **325** (1986) 489.

[66] LETTRY J. *et al.*, *Rev. Sci. Instrum.*, **69** (1998) 761.

[67] KURTUKIAN-NIETO T. *et al.*, submitted to *Phys. Rev. Lett.*

[68] BORZOV I. N., *Phys. Rev. C*, **67** (2003) 025802.

[69] ENGEL J. *et al.*, *Phys. Rev. C*, **60** (1999) 014302.

[70] BORZOV I. N. *et al.*, *Z. Phys. A*, **355** (1996) 117.

[71] BORZOV I. N. *et al.*, submitted to *Nucl. Phys. A*.

[72] MARTINEZ-PINEDO G. and LANGANKE K., *Phys. Rev. Lett.*, **83** (1999) 4502.

[73] MARTINEZ-PINEDO G., *Nucl. Phys. A*, **688** (2001) 57c.

[74] CUENCA-GARCIA J. J. *et al.*, *Eur. Phys. J. A*, **34** (2007) 99.

[75] FULLER G. M. and MEYER B. S., *Astrophys. J.*, **453** (1995) 792.

[76] QIAN Y.-Z., HAXTON W. C., LANGANKE K. and VOGEL P., *Phys. Rev. C*, **55** (1997) 1532.

[77] HAXTON W. C., LANGANKE K., QIAN Y.-Z. and VOGEL P., *Phys. Rev. Lett.*, **78** (1997) 2694.

[78] CATARA F., LANZO E. G., NAGARAJAN M. A. and VITTURI A., *Nucl. Phys. A*, **624** (1997) 449.

[79] GORIELY S., *Phys. Lett.*, **436** (1998) 10.

[80] BENLLIURE J. *et al.*, *Nucl. Phys. A*, **628** (1998) 458.

[81] MARTINEZ-PINEDO G. *et al.*, *Prog. Part. Nucl. Phys.*, **59** (2007) 199.

[82] MARTINEZ-PINEDO G. *et al.*, in preparation.

DOI 10.3254/978-1-58603-885-4-347

The interacting boson model for exotic nuclei

P. Van Isacker

Grand Accélérateur National d'Ions Lourds, CEA/DSM–CNRS/IN2P3
BP 55027, F-14076 Caen Cedex 5, France

Summary. — These lectures notes give an introduction to the use of algebraic techniques for obtaining analytic eigensolutions of quantum-mechanical systems consisting of many particles in interaction. The notions of symmetry and dynamical symmetry in quantum physics are introduced and subsequently illustrated with the example of the interacting boson model of atomic nuclei. Some recent applications of this model to exotic nuclei are discussed.

1. – Introduction

The Interacting Boson Model (IBM) has developed steadily since its inception more than three decades ago [1] and consists by now of a family of nuclear-structure models linked by common underlying assumptions concerning their microscopic foundation and by the algebraic origin of their formalism. In a wider context, the IBM represents one of the three principal approaches for the theoretical description of the atomic nucleus. The first is the independent-particle shell model which describes the nucleus in terms of independently moving particles. The second approach is the geometric model which pictures the nucleus as a dense liquid drop exhibiting vibrations around an equilibrium shape which can be spherical or deformed. The reconciliation of these two different views of the nucleus (which both have strong empirical backing) has been one of the major endeavours of theoretical nuclear physics since the 1950s. The IBM can be viewed as a third, alternative way to understand nuclei, where the focus is on the search for symmetries in a quantal system of many interacting particles which have a geometric interpretation in the limit of large particle number. This defines two important aspects of the IBM: i) the use of symmetry methods to obtain analytical eigensolutions of quantal many-body systems; ii) the link between the liquid-drop and shell-model views of the nucleus.

Although the algebraic methods to be explained below have formed the basis of the IBM and its offshoots, it should be stressed that the techniques are of general interest and can be applied to any quantum-mechanical many-body system. To emphasize this aspect, the review starts with a brief reminder of the role of symmetry in quantum mechanics (sect. **2**) and of the application of group theory in quantal many-body systems (sect. **3**). As an illustration of these generic methods the example of the IBM is discussed in sect. **4**. Obviously, no attempt can be made at a complete review of the model but an outline of its symmetry structure is given, its connection with the underlying fermionic degrees of freedom is briefly outlined and its geometric interpretation in terms of the classical limit is discussed. The final two sections present examples of applications of the IBM to exotic nuclei. The first, presented in sect. **5**, gives a detailed account on how three-body interactions between the bosons can improve the description of the spectroscopic properties of certain nuclei. The second application (sect. **6**) outlines a method whereby the full IBM Hamiltonian is used to obtain a simultaneous description of the binding energies and excitation spectra of a large number of nuclei in a single major shell.

2. – Symmetry in quantum mechanics

In this section it is shown how group theory can be applied to quantum mechanics. First a reminder of the concepts of symmetry and dynamical symmetry, and their consequences is given.

2$^{.}$1. *Symmetry*. – A Hamiltonian $\hat{H}$ which commutes with the generators $\hat{g}_i$ that form a Lie algebra $\mathcal{G}$, that is

$$(1) \qquad \forall \hat{g}_i \in \mathcal{G} : [\hat{H}, \hat{g}_i] = 0,$$

is said to have a *symmetry* $\mathcal{G}$ or, alternatively, to be *invariant under* $\mathcal{G}$.

The determination of operators $\hat{g}_i$ that leave invariant the Hamiltonian of a given physical system is central to any quantum-mechanical description. The reasons for this are profound and can be understood from the correspondence between geometrical and quantum-mechanical transformations. It can be shown that the transformations $\hat{g}_i$ with the symmetry property (1) are induced by geometrical transformations that leave unchanged the corresponding classical Hamiltonian. So it is that the classical notion of a conserved quantity is transcribed in quantum mechanics in the form of the symmetry property (1) of the Hamiltonian.

2˙2. *Degeneracy and state labelling.* – A well-known consequence of a symmetry is the occurrence of degeneracies in the eigenspectrum of $\hat{H}$. Given an eigenstate $|\gamma\rangle$ of $\hat{H}$ with energy E, condition (1) implies that the states $\hat{g}_i|\gamma\rangle$ all have the same energy:

$$\hat{H}\hat{g}_i|\gamma\rangle = \hat{g}_i\hat{H}|\gamma\rangle = E\hat{g}_i|\gamma\rangle.$$

An arbitrary eigenstate of $\hat{H}$ shall be written as $|\Gamma\gamma\rangle$, where the first quantum number Γ is different for states with different energies and the second quantum number γ is needed to label degenerate eigenstates. The eigenvalues of a Hamiltonian that satisfies condition (1) depend on Γ only,

$$(2) \qquad\qquad\qquad \hat{H}|\Gamma\gamma\rangle = E(\Gamma)|\Gamma\gamma\rangle,$$

and, furthermore, the transformations $\hat{g}_i$ do not admix states with different Γ:

$$(3) \qquad\qquad\qquad \hat{g}_i|\Gamma\gamma\rangle = \sum_{\gamma'} a^{\Gamma}_{\gamma'\gamma}(i)|\Gamma\gamma'\rangle.$$

This simple discussion of the consequences of a Hamiltonian symmetry illustrates at once the relevance of group theory in quantum mechanics. Symmetry implies degeneracy and eigenstates that are degenerate in energy provide a Hilbert space in which a matrix representation of the symmetry group can be constructed. Consequently, the representations of a given group directly determine the degeneracy structure of a Hamiltonian with that symmetry.

A sufficient condition for a Hamiltonian to have the symmetry property (1) is that it can be expressed in terms of Casimir operators of various orders. The eigenequation (2) then becomes

$$(4) \qquad\qquad \left(\sum_m \kappa_m \hat{C}_m[\mathcal{G}]\right)|\Gamma\gamma\rangle = \left(\sum_m \kappa_m E_m(\Gamma)\right)|\Gamma\gamma\rangle.$$

In fact, the following discussion is valid for any analytic function of the various Casimir operators but mostly a linear combination is taken, as in eq. (4). The energy eigenvalues $E_m(\Gamma)$ are functions of the labels that specify the irreducible representation Γ, and are known for all classical Lie algebras.

2˙3. *Dynamical symmetry breaking.* – The concept of a dynamical symmetry can now be introduced for which (at least) two algebras $\mathcal{G}_1$ and $\mathcal{G}_2$ with $\mathcal{G}_1 \supset \mathcal{G}_2$ are needed. The eigenstates of a Hamiltonian $\hat{H}$ with symmetry $\mathcal{G}_1$ are labelled as $|\Gamma_1\gamma_1\rangle$. But, since $\mathcal{G}_1 \supset \mathcal{G}_2$, a Hamiltonian with $\mathcal{G}_1$ symmetry necessarily must also have a symmetry $\mathcal{G}_2$ and, consequently, its eigenstates can also be labelled as $|\Gamma_2\gamma_2\rangle$. Combination of the two properties leads to the eigenequation

$$(5) \qquad\qquad \hat{H}|\Gamma_1\eta_{12}\Gamma_2\gamma_2\rangle = E(\Gamma_1)|\Gamma_1\eta_{12}\Gamma_2\gamma_2\rangle,$$

where the role of γ_1 is played by $\eta_{12}\Gamma_2\gamma_2$. In eq. (5) the irreducible representation $[\Gamma_2]$ may occur more than once in $[\Gamma_1]$, and hence an additional quantum number η_{12} is needed to uniquely label the states. Because of $\mathcal{G}_1$ symmetry, eigenvalues of $\hat{H}$ depend on Γ_1 only.

In many examples in physics, the condition of $\mathcal{G}_1$ symmetry is too strong and a *possible* breaking of the $\mathcal{G}_1$ symmetry can be imposed via the Hamiltonian

$$(6) \qquad \hat{H}' = \sum_{m_1} \kappa_{m_1} \hat{C}_{m_1}[\mathcal{G}_1] + \sum_{m_2} \kappa_{m_2} \hat{C}_{m_2}[\mathcal{G}_2],$$

which consists of a combination of Casimir operators of $\mathcal{G}_1$ *and* $\mathcal{G}_2$. The symmetry properties of the Hamiltonian $\hat{H}'$ are now as follows. Since $[\hat{H}', \hat{g}_i] = 0$ for $\hat{g}_i \in \mathcal{G}_2$, $\hat{H}'$ is invariant under $\mathcal{G}_2$. The Hamiltonian $\hat{H}'$, since it contains $\hat{C}_{m_2}[\mathcal{G}_2]$, does not commute, in general, with all elements of $\mathcal{G}_1$ and for this reason the $\mathcal{G}_1$ symmetry is broken. Nevertheless, because $\hat{H}'$ is a combination of Casimir operators of $\mathcal{G}_1$ and $\mathcal{G}_2$, its eigenvalues can be obtained in closed form:

$$(7) \qquad \left(\sum_{m_1} \kappa_{m_1} \hat{C}_{m_1}[\mathcal{G}_1] + \sum_{m_2} \kappa_{m_2} \hat{C}_{m_2}[\mathcal{G}_2] \right) |\Gamma_1 \eta_{12} \Gamma_2 \gamma_2\rangle$$

$$= \left(\sum_{m_1} \kappa_{m_1} E_{m_1}(\Gamma_1) + \sum_{m_2} \kappa_{m_2} E_{m_2}(\Gamma_2) \right) |\Gamma_1 \eta_{12} \Gamma_2 \gamma_2\rangle.$$

The conclusion is thus that, although $\hat{H}'$ is not invariant under $\mathcal{G}_1$, its eigenstates are the same as those of $\hat{H}$ in eq. (5). The Hamiltonian $\hat{H}'$ is said to have $\mathcal{G}_1$ as a *dynamical symmetry*. The essential feature is that, although the eigenvalues of $\hat{H}'$ depend on Γ_1 *and* Γ_2 (and hence $\mathcal{G}_1$ is not a symmetry), the eigenstates do not change during the breaking of the $\mathcal{G}_1$ symmetry: the dynamical symmetry breaking splits but does not admix the eigenstates. A convenient way of summarizing the symmetry character of $\hat{H}'$ and the ensuing classification of its eigenstates is as follows:

$$(8) \qquad \begin{array}{ccc} \mathcal{G}_1 & \supset & \mathcal{G}_2 \\ \downarrow & & \downarrow \\ \Gamma_1 & & \eta_{12}\Gamma_2 \end{array} \ .$$

This equation indicates the larger algebra $\mathcal{G}_1$ (sometimes referred to as the *dynamical algebra* or *spectrum generating algebra*) and the symmetry algebra $\mathcal{G}_2$, together with their associated labels with possible multiplicities.

3. – Dynamical symmetries in quantal many-body systems

So far the discussion of symmetries has been couched in general terms leading to results that are applicable to any quantum-mechanical system. We shall now be somewhat more specific and show how the concept of dynamical symmetry can be applied systematically to find analytical eigensolutions for a system of interacting bosons and/or fermions. As the results are most conveniently discussed in a second-quantization formalism, first a brief reminder of some essential formulas is given.

3˙1. *Many-particle states in second quantization*. – In general, particle creation and annihilation operators shall be denoted as $c_i^\dagger$ and c_i, respectively. (Note that, as these are operators, consistency with earlier notational conventions would require the notation $\hat{c}_i^\dagger$ and $\hat{c}_i$; the hats are omitted for notational simplicity.) The index i comprises the complete quantum-mechanical labelling of a single-particle state. In many applications i coincides with the labels of a stationary quantum state for a single particle in which case $c_i^\dagger$ creates a particle in that stationary state. The index i may include intrinsic quantum numbers such as spin, isospin, colour etc.

The particles are either fermions or bosons, for which the notations $c \equiv a$ and $c \equiv b$, respectively, shall be reserved. They obey different statistics, of Fermi-Dirac and of Bose-Einstein, respectively, which in second quantization is imposed through the (anti)commutation properties of creation and annihilation operators:

$$(9) \qquad \{a_i, a_j^\dagger\} = \delta_{ij}, \qquad \{a_i^\dagger, a_j^\dagger\} = \{a_i, a_j\} = 0,$$

and

$$(10) \qquad [b_i, b_j^\dagger] = \delta_{ij}, \qquad [b_i^\dagger, b_j^\dagger] = [b_i, b_j] = 0.$$

Introducing the notation

$$(11) \qquad [\hat{u}, \hat{v}]_q \equiv \hat{u}\hat{v} - (-)^q \hat{v}\hat{u},$$

with $q = 0$ for bosons and $q = 1$ for fermions, one can express these relations as

$$(12) \qquad [c_i, c_j^\dagger]_q = \delta_{ij}, \qquad [c_i^\dagger, c_j^\dagger]_q = [c_i, c_j]_q = 0.$$

A many-particle state can be written as

$$|\bar{n}\rangle \equiv \prod_i \frac{(c_i^\dagger)^{n_i}}{\sqrt{n_i!}} |\mathrm{o}\rangle,$$

where $|\mathrm{o}\rangle$ is the vacuum state which satisfies

$$(13) \qquad \forall i : c_i|\mathrm{o}\rangle = 0.$$

A many-particle state is thus completely determined by specifying the number of particles n_i in each quantum state i, and the (possibly infinite) set of numbers n_i is collectively denoted as $\bar{n}$. For fermions only $n_i = 0$ and $n_i = 1$ are allowed (since $a_i^\dagger a_i^\dagger |o\rangle = -a_i^\dagger a_i^\dagger |o\rangle = 0$) but for bosons no restrictions on n_i exist.

As an example of the application of the second-quantization formalism, we prove that $c_j^\dagger c_j$ is an operator that counts the number of particles in state j. For bosons, its action on a many-particle state can be worked out from the commutation relations (10) and the vacuum property (13):

$$b_j^\dagger b_j |\bar{n}\rangle = \prod_{i \neq j} \frac{(b_i^\dagger)^{n_i}}{\sqrt{n_i!}} b_j^\dagger b_j \frac{(b_j^\dagger)^{n_j}}{\sqrt{n_j!}} |o\rangle = \prod_{i \neq j} \frac{(b_i^\dagger)^{n_i}}{\sqrt{n_i!}} b_j^\dagger \frac{1}{\sqrt{n_j!}} [b_j, (b_j^\dagger)^{n_j}] |o\rangle.$$

Since, for arbitrary operators $\hat{u}$, $\hat{v}$ and $\hat{w}$,

$$[\hat{u}, \hat{v}\hat{w}] = \hat{v}[\hat{u}, \hat{w}] + [\hat{u}, \hat{v}]\hat{w},$$

one finds

$$[b_j, (b_j^\dagger)^{n_j}] = (b_j^\dagger)^{n_j - 1} + [b_j, (b_j^\dagger)^{n_j - 1}]b_j^\dagger = \cdots = n_j (b_j^\dagger)^{n_j - 1},$$

and thus

$$b_j^\dagger b_j |\bar{n}\rangle = n_j |\bar{n}\rangle.$$

The equivalent property for fermions can be shown simply by noting that either $n_j = 0$ or $n_j = 1$, in which case $a_j^\dagger a_j$ gives zero or one, respectively. Hence, in summary we have

$$c_j^\dagger c_j |\bar{n}\rangle = n_j |\bar{n}\rangle.$$

and, because of this property, $c_j^\dagger c_j$ is called a number operator.

3'2. *Particle-number conserving spectrum generating algebras.* – The determination of the properties of a quantal system of N interacting particles requires the solution of the eigenvalue equation associated with the Hamiltonian

$$\text{(14)} \qquad \hat{H} = \sum_i \epsilon_i c_i^\dagger c_i + \sum_{ijkl} v_{ijkl} c_i^\dagger c_j^\dagger c_k c_l + \cdots,$$

containing one-body terms ϵ_i, two-body interactions v_{ijkl} and so on; higher-order interactions can be included in the expansion, if needed. Hamiltonian (14) satisfies the requirement of particle-number conservation; the case that does not conserve particle number is discussed in the next subsection.

Note that the assumed diagonality of the one-body term does not make Hamiltonian (14) less general. In fact, a non-diagonal one-body term,

$$\sum_{ij} \epsilon_{ij} c_i^\dagger c_j,$$

can always be brought in diagonal form through a unitary transformation of the form

$$c_j = \sum_s u_{js} c'_s, \qquad c_i^\dagger = \sum_r u_{ir}^* c_r'^\dagger,$$

which yields

$$\sum_{ij} \epsilon_{ij} c_i^\dagger c_j = \sum_{rs} \left(\sum_{ij} u_{ir}^* \epsilon_{ij} u_{js} \right) c_r'^\dagger c'_s \equiv \sum_r \epsilon'_r c_r'^\dagger c'_r,$$

where the last equality is obtained by choosing the unitary transformation such that it diagonalizes ϵ_{ij}:

$$\sum_{ij} u_{ir}^* \epsilon_{ij} u_{js} = \epsilon'_r \delta_{rs}.$$

Since the transformation is unitary, its inverse is also and the new single-particle states still obey the same (anti)commutation relations (12):

$$[c'_r, c_s'^\dagger\}_q = \sum_{ij} u_{ri}^{-1} u_{sj}^{-1*} [c_i, c_j^\dagger\}_q = \sum_i u_{ri}^{-1} u_{si}^{-1*} = \delta_{rs}.$$

With use of the property (see eqs. (11), (12))

$$c_j^\dagger c_k = (-)^q c_k c_j^\dagger - (-)^q \delta_{jk},$$

Hamiltonian (14) can be written in a different form as

(15) $$\hat{H} = \sum_{il} \left(\epsilon_i \delta_{il} - (-)^q \sum_j v_{ijkl} \right) \hat{u}_{il} + (-)^q \sum_{ijkl} v_{ijkl} \hat{u}_{ik} \hat{u}_{jl} + \cdots,$$

where the notation $\hat{u}_{ij} \equiv c_i^\dagger c_j$ is introduced. The reason for doing so becomes clear when the commutator of the $\hat{u}_{ij}$ operators is considered:

$$[\hat{u}_{ij}, \hat{u}_{kl}] = c_i^\dagger c_k^\dagger [c_j, c_l] + c_i^\dagger [c_j, c_k^\dagger] c_l - c_k^\dagger [c_l, c_i^\dagger] c_j + [c_i^\dagger, c_k^\dagger] c_l c_j,$$

which, because of the identity

$$[\hat{u}, \hat{v}] = [\hat{u}, \hat{v}\} - (1 - (-)^q)\, \hat{v}\hat{u},$$

can be brought into the form

$$[\hat{u}_{ij}, \hat{u}_{kl}] = \hat{u}_{il}\delta_{jk} - \hat{u}_{kj}\delta_{il} - (1 - (-)^q) \left[c_i^\dagger c_k^\dagger c_l c_j + c_i^\dagger c_k^\dagger c_j c_l \right].$$

The last term on the right-hand-side of this equation is zero for bosons (when $q = 0$) as it is for fermions since in that case the expression between square brackets vanishes. This shows that

$$[\hat{u}_{ij}, \hat{u}_{kl}] = \hat{u}_{il}\delta_{jk} - \hat{u}_{kj}\delta_{il} \tag{16}$$

is valid for a boson as well as a fermion realization of the $\hat{u}_{ij}$ and that these operators in both cases satisfy the commutation relations of the unitary algebra $U(n)$ where n is the dimensionality of the single-particle space.

The equivalent form (15) shows that the solution of the eigenvalue problem for N particles associated with the Hamiltonian (14) requires the diagonalization of $\hat{H}$ in the symmetric representation $[N]$ of $U(n)$ in case of bosons or in its antisymmetric representation $[1^N]$ in case of fermions. This, for a general Hamiltonian, is a numerical problem which quickly becomes intractable with increasing numbers of particles N or increasing single-particle space n. A strategy for solving *particular classes* of the many-body Hamiltonian (14) can be obtained by considering the algebra $U(n)$ as a spectrum generating or dynamical algebra $\mathcal{G}_{\mathrm{dyn}}$ on which a dynamical symmetry breaking is applied. The generalization of the procedure of subsect. **2·3** is straightforward and starts from a chain of nested algebras

$$\mathcal{G}_1 \equiv \mathcal{G}_{\mathrm{dyn}} \supset \mathcal{G}_2 \supset \cdots \supset \mathcal{G}_s \equiv \mathcal{G}_{\mathrm{sym}}, \tag{17}$$

where the last algebra $\mathcal{G}_s$ in the chain is the symmetry algebra of the problem. To appreciate the relevance of this classification in connection with the many-body problem (14), one associates with the chain (17) the Hamiltonian

$$\hat{H} = \sum_{r=1}^{s}\sum_{m} \kappa_{rm}\hat{C}_m[\mathcal{G}_r], \tag{18}$$

which represents a direct generalization of eq. (6) and where κ_{rm} are arbitrary coefficients. The operators in Hamiltonian (18) satisfy

$$\forall m, m', r, r' : [\hat{C}_m[\mathcal{G}_r], \hat{C}_{m'}[\mathcal{G}_{r'}]] = 0.$$

This property is evident from the fact that all elements of $\mathcal{G}_r$ are in $\mathcal{G}_{r'}$ or *vice versa*. Hence, Hamiltonian (18) is written as a sum of commuting operators and as a result its eigenstates are labelled by the quantum numbers associated with these operators. Note that the condition of the *nesting* of the algebras is crucial for constructing a set of commuting operators and hence for obtaining an analytical solution. Since the Casimir operators can be expressed in terms of the operators $\hat{u}_{ij}$, expansion (18) can, in principle, be rewritten in the form (15) with the order of the interactions determined by the maximal order m of the invariants.

To summarize these results, Hamiltonian (18)—which can be obtained from the general Hamiltonian (14) for specific coefficients ϵ_i, v_{ijkl}...—can be solved analytically. Its eigenstates do not depend on the coefficients κ_{rm} and are labelled by

$$
(19) \qquad
\begin{array}{ccccccc}
\mathcal{G}_1 & \supset & \mathcal{G}_2 & \supset & \cdots & \supset & \mathcal{G}_s \\
\downarrow & & \downarrow & & & & \downarrow \\
\Gamma_1 & & \eta_{12}\Gamma_2 & & & & \eta_{s-1,s}\Gamma_s
\end{array} \;.
$$

Its eigenvalues are given in closed form as

$$
(20) \qquad \hat{H}|\Gamma_1\eta_{12}\Gamma_2\ldots\eta_{s-1,s}\Gamma_s\rangle = \sum_{r=1}^{s}\sum_{m}\kappa_{rm}E_m(\Gamma_r)|\Gamma_1\eta_{12}\Gamma_2\ldots\eta_{s-1,s}\Gamma_s\rangle,
$$

where $E_m(\Gamma_r)$ are known functions introduced in subsect. $2^{\cdot}2$.

Thus a generic scheme is established for finding analytically solvable Hamiltonians (14): it requires an enumeration of all nested chains of the type (17) which is a purely algebraic problem. The symmetry of the dynamical algebra $\mathcal{G}_{\mathrm{dyn}}$ is broken dynamically and the only remaining symmetry is $\mathcal{G}_{\mathrm{sym}}$ which is the true symmetry of the problem.

This approach to find analytical eigensolutions for a system of interacting bosons and/or fermions has received prominence with the work of Arima and Iachello [1,2] where the dynamical algebra $\mathcal{G}_{\mathrm{dyn}}$ is $U(6)$ and the symmetry algebra $\mathcal{G}_{\mathrm{sym}}$ is the rotational algebra $SO(3)$, as will be described in detail in sect. **4**. It should not be forgotten, however, that these symmetry methods had been used before in different models in physics. The Isobaric Multiplet Mass Equation [3] as well as the Gell-Mann–Okubo mass formula [4,5] can be viewed as examples of symmetry breaking of the type (19). A beautiful example in nuclear physics is Elliott's rotational $SU(3)$ model [6] in which Wigner's supermultiplet [7] degeneracy associated with $SU(4)$ is lifted dynamically by the quadrupole interaction. The technique has also been applied to the nuclear shell model. Based on the methods developed for the IBM, a systematic procedure for constructing analytically solvable Hamiltonians was devised by Ginocchio [8], drawing on earlier ideas by Hecht *et al.* [9] and using a method which resembles that of pseudo spin [10,11]. The theory was later developed under the name of fermion dynamical symmetry model [12].

$\mathbf{3^{\cdot}3}$. *Particle-number non-conserving dynamical algebras.* – The Hamiltonians constructed from the unitary generators u_{ij} necessarily conserve particle number since that is so for the generators themselves. In many cases (involving, *e.g.*, virtual particles, effective phonon-like excitations...) no particle-number conservation can be imposed and a more general formalism is required. Another justification for such generalizations is that the strategy outlined in subsect. $3^{\cdot}2$ has the drawback that the dynamical algebra $\mathcal{G}_{\mathrm{dyn}}$ can become very large (due a large single-particle space combined with possible intrinsic quantum numbers such as spin and isospin) which makes the analysis of the group-theoretical reduction (17), and the associated labelling (19) in particular, too difficult to be of practical use. In some cases the following, more economical, procedure is called for.

In addition to the unitary generators u_{ij}, also the operators $\hat{s}_{ij} \equiv c_i c_j$ and $\hat{s}_{ij}^\dagger \equiv c_i^\dagger c_j^\dagger$ are considered. (Note that this notation implies $\hat{s}_{ij}^\dagger = (\hat{s}_{ji})^\dagger$.) We now show that the set of operators $\hat{s}_{ij}$ and $\hat{s}_{ij}^\dagger$, added to $\hat{u}_{ij}' \equiv \hat{u}_{ij} + \frac{1}{2}(-)^q \delta_{ij}$, forms a closed algebra. Since the operators $\hat{u}_{ij}$ and $\hat{u}_{ij}'$ differ by a constant only, they satisfy the same commutation relations (16), and in the same way it can be shown that

$$[\hat{u}_{ij}', \hat{s}_{kl}] = -\hat{s}_{kj}\delta_{il} - \hat{s}_{jl}\delta_{ik}, \qquad [\hat{u}_{ij}', \hat{s}_{kl}^\dagger] = \hat{s}_{il}^\dagger \delta_{jk} + \hat{s}_{ki}^\dagger \delta_{jl}.$$

The commutator of $\hat{s}_{ij}$ with $\hat{s}_{kl}^\dagger$ deserves a more detailed consideration:

$$[\hat{s}_{ij}, \hat{s}_{kl}^\dagger] = c_i c_k^\dagger \delta_{jl} + c_i c_l^\dagger \delta_{jk} + c_k^\dagger c_j \delta_{il} + c_l^\dagger c_j \delta_{ik}$$
$$- (1 - (-)^q) \left[c_i c_k^\dagger c_l^\dagger c_j + c_i c_k^\dagger c_j c_l^\dagger + c_k^\dagger c_l^\dagger c_i c_j + c_k^\dagger c_i c_l^\dagger c_j \right].$$

The term between square brackets can be worked out by assuming that the c's are fermions since for bosons the factor $(1 - (-)^q)$ in front is zero. This gives

$$[\hat{s}_{ij}, \hat{s}_{kl}^\dagger] = (-)^q c_k^\dagger c_i \delta_{jl} + (-)^q c_l^\dagger c_i \delta_{jk} + c_k^\dagger c_j \delta_{il} + c_l^\dagger c_j \delta_{ik} + \delta_{il}\delta_{jk} + \delta_{ik}\delta_{jl}$$
$$- (1 - (-)^q) \left[-c_k^\dagger c_i \delta_{jl} + c_k^\dagger c_j \delta_{il} + \delta_{ik}\delta_{jl} \right]$$
$$= (-)^q \left(\hat{u}_{li}\delta_{jk} + \hat{u}_{kj}\delta_{il} + \delta_{ik}\delta_{jl} \right) + \left(\hat{u}_{ki}\delta_{jl} + \hat{u}_{lj}\delta_{ik} + \delta_{il}\delta_{jk} \right),$$

and leads to the following result, valid for both fermions and bosons:

$$[\hat{s}_{ij}, \hat{s}_{kl}^\dagger] = (-)^q \hat{u}_{li}'\delta_{jk} + (-)^q \hat{u}_{kj}'\delta_{il} + \hat{u}_{ki}'\delta_{jl} + \hat{u}_{lj}'\delta_{ik}.$$

The modification $\hat{u}_{ij} \rightarrow \hat{u}_{ij}'$ is thus necessary to ensure closure of the commutator. The set $\{\hat{u}_{ij}', \hat{s}_{ij}, \hat{s}_{ij}^\dagger\}$ contains $n(2n+1)$ or $n(2n-1)$ independent generators for bosons or fermions, respectively. From dimensionality (but also from the commutation relations) it can be inferred that the respective Lie algebras are $Sp(2n)$ and $SO(2n)$.

It is clear that these algebras can be used to construct number non-conserving Hamiltonians. However, the addition of the pair creation and annihilation operators enlarges rather than diminishes the dimension of the dynamical algebra and does not lead to a simplification of the algebraic structure of the problem. The latter can be achieved by considering specific linear combinations

$$\hat{U}(\bar{\alpha}) \equiv \sum_{ij} \alpha_{ij}\hat{u}_{ij}, \quad \hat{S}_+(\bar{\beta}) \equiv \sum_{ij} \beta_{ij}\hat{s}_{ij}^\dagger, \quad \hat{S}_-(\bar{\beta}) \equiv \left(\hat{S}_+(\bar{\beta}) \right)^\dagger,$$

where the coefficients α_{ij} and β_{ij} are chosen to ensure closure to a *subalgebra* of either $Sp(2n)$ or $SO(2n)$.

This procedure will not be formally developed further here. We note that, among the nuclear models, several examples are encountered that illustrate the approach, such as the $SU(2)$ quasi-spin algebra [13] or the $SO(8)$ algebra of neutron-proton pairing [14].

4. – The interacting boson model

In this section an introduction to the IBM is given with particular emphasis on the version of the model which includes higher-order interactions between the bosons. A full account of the IBM is given by Iachello and Arima [2].

Before turning to a detailed discussion of the IBM, it is worthwhile to summarize the philosophy of the model as well as the most important results obtained with it. In the IBM the nucleus is described in terms of interacting s and d bosons (more about its justification in subsect. **4**`4). For such a system three different classes of analytical solutions or *limits* exist: the vibrational $U(5)$ limit [15], the rotational $SU(3)$ limit [16] and the γ-unstable $SO(6)$ limit [17]. While at the time of arrival of the IBM (1975), the vibrational and rotational limits were well-recognized features in the nuclear landscape, this was not the case for the third limit. The $SO(6)$ limit still stands out as an excellent example of the value and power of symmetry methods; its origins were purely algebraic, but its structure was later found to resemble that of a γ-unstable rotor. Its predictions were found to correspond closely to the empirical structure of some Pt nuclei [18]. Since that early work, it has become increasingly evident that the $SO(6)$ symmetry in fact represents the third commonly occurring class of nuclei, which have been identified in several regions, most notably around $A = 130$ [19].

A second landmark contribution of the IBM to nuclear physics is *supersymmetry*. Its starting point is the extension of the IBM to odd-mass nuclei, achieved by considering, in addition to the bosons, a single fermion [20]. The resulting Interacting Boson-Fermion Model (IBFM) lends itself equally well as the IBM to a study based on symmetry considerations whereby certain classes of model Hamiltonians can be solved analytically [21]. A particularly attractive feature is the conceptual similarity in the description of even-even and odd-mass nuclei. While the spectrum generating algebra of the IBM is $U^B(6)$, the one of the IBFM is $U^B(6) \otimes U^F(\Omega)$, where Ω is the size of the single-particle space available to the fermion and the superscripts B and F are added to indicate the boson or fermion realization of the Lie algebra. The Lie algebra $U^B(6) \otimes U^F(\Omega)$ provides a *separate* description of even-even and odd-mass nuclei: although the treatment is similar in both cases, no operator exists that *connects* even-even and odd-mass states. An extension, proposed by Iachello [22], considers in addition operators that transform a boson into a fermion or *vice versa*. The resulting set of operators does not any longer form a classical Lie algebra which is defined in terms of commutation relations. Instead, to define a closed algebraic structure, one needs to introduce an internal operation that corresponds to a mixture of commutation and anticommutation and the resulting algebra is called a *graded* or *super*algebra, denoted by $U(6/\Omega)$. The supersymmetric generators thus induce a connection between even-even and odd-mass nuclei and lead to a *simultaneous* treatment of such pairs of nuclei. At the basis of this unified treatment is the enlargement of the dynamical algebra. This process of enlargement can be continued and from it there result unified descriptions of ever higher numbers of nuclei. A further example of this mechanism is obtained if a distinction is made between neutrons and protons, both for fermions and for bosons. It then follows that a quartet of nuclei (even-even, even-odd, odd-even and odd-odd) is connected by the supersymmetric operators and that this quartet can be described simultaneously with a single Hamiltonian [23]. This approach continues to inspire the study of odd-odd nuclei to the present day, see for example [24].

A third landmark contribution of the IBM is the prediction of neutron-proton *non-symmetric* states. Given the microscopic interpretation of the bosons as correlated pairs of nucleons, a natural extension of the IBM-1 (the simplest version of the IBM) is to assume two different types of bosons, neutron and proton, giving rise to the neutron–proton interacting boson model or IBM-2 [25, 26]. The algebraic structure of IBM-2 is a product of $U(6)$ algebras, $U_\nu(6) \otimes U_\pi(6)$, consisting of neutron (ν) and proton (π) generators, respectively. The most important aspect of IBM-2 is that it predicts states which are additional to those found in IBM-1 [27]. The lowest states in energy are symmetric in $U(6)$ and are the analogues of those in IBM-1. The next class of states no longer is symmetric in $U(6)$; they are observed experimentally [28] and seem to be a persistent feature of nuclei [29]. From a geometric analysis of non-symmetric states emerges that they correspond to linear or angular displacement oscillations in which the neutrons and protons are out of phase, in contrast to the symmetric IBM-2 states for which such oscillations are in phase. The occurrence of such states was first predicted in the context of geometric two-fluid models in vibrational [30] and deformed [31] nuclei in which they appear as neutron-proton counter oscillations. The IBM-2 thus confirms these geometric descriptions but at the same time generalizes them to *all* nuclei, not only spherical and deformed, but γ unstable and transitional as well. It is precisely in the latter case that the example of ^{94}Mo has been shown to agree to a remarkable extent with the predictions of the $SO(6)$ limit of the IBM-2 [32].

4˙1. *Hamiltonian.* – The building blocks of the IBM are s and d bosons with angular momenta $\ell = 0$ and $\ell = 2$. A nucleus is characterized by a constant total number of bosons N which equals half the number of valence nucleons (particles or holes, whichever is smaller). In these notes no distinction is made between neutron and proton bosons, an approximation which is known as IBM-1.

Since the Hamiltonian of the IBM-1 conserves the total number of bosons, it can be written in terms of the 36 operators $b^\dagger_{\ell m} b_{\ell' m'}$ where $b^\dagger_{\ell m}$ ($b_{\ell m}$) creates (annihilates) a boson with angular momentum ℓ and z projection m. According to eq. (16) this set of 36 operators generates the Lie algebra $U(6)$. A Hamiltonian that conserves the total number of bosons is of the generic form

$$(21) \qquad\qquad \hat{H} = E_0 + \hat{H}_1 + \hat{H}_2 + \hat{H}_3 + \cdots,$$

where the index refers to the order of the interaction in the generators of $U(6)$. The first term E_0 is a constant which represents the binding energy of the core. The second term is the one-body part

$$(22) \qquad \hat{H}_1 = \epsilon_s [s^\dagger \times \tilde{s}]^{(0)} + \epsilon_d \sqrt{5} [d^\dagger \times \tilde{d}]^{(0)} \equiv \epsilon_s \hat{n}_s + \epsilon_d \hat{n}_d,$$

where $\times$ refers to coupling in angular momentum (shown as an upperscript in round brackets), $\tilde{b}_{\ell m} \equiv (-)^{\ell-m} b_{\ell,-m}$ and the coefficients ϵ_s and ϵ_d are the energies of the s and d bosons. The third term in the Hamiltonian (21) represents the two-body interaction

$$(23) \qquad \hat{H}_2 = \sum_{\ell_1 \le \ell_2, \ell_1' \le \ell_2', L} \tilde{v}^L_{\ell_1 \ell_2 \ell_1' \ell_2'} [[b^\dagger_{\ell_1} \times b^\dagger_{\ell_2}]^{(L)} \times [\tilde{b}_{\ell_2'} \times \tilde{b}_{\ell_1'}]^{(L)}]^{(0)}_0,$$

where the coefficients $\tilde{v}$ are related to the interaction matrix elements between normalized two-boson states,

$$\langle \ell_1 \ell_2; LM | \hat{H}_2 | \ell_1' \ell_2'; LM \rangle = \sqrt{\frac{(1 + \delta_{\ell_1 \ell_2})(1 + \delta_{\ell_1' \ell_2'})}{2L+1}} \, \tilde{v}^L_{\ell_1 \ell_2 \ell_1' \ell_2'}.$$

Since the bosons are necessarily symmetrically coupled, allowed two-boson states are s^2 ($L = 0$), sd ($L = 2$) and d^2 ($L = 0, 2, 4$). Since for n states with a given angular momentum one has $n(n+1)/2$ interactions, seven independent two-body interactions v are found: three for $L = 0$, three for $L = 2$ and one for $L = 4$.

This analysis can be extended to higher-order interactions. One may consider, for example, the three-body interactions $\langle \ell_1 \ell_2 \ell_3; LM | \hat{H}_3 | \ell_1' \ell_2' \ell_3'; LM \rangle$. The allowed three-boson states are s^3 ($L = 0$), $s^2 d$ ($L = 2$), sd^2 ($L = 0, 2, 4$) and d^3 ($L = 0, 2, 3, 4, 6$), leading to $6 + 6 + 1 + 3 + 1 = 17$ independent three-body interactions for $L = 0, 2, 3, 4, 6$, respectively. Note that any three-boson state $s^i d^{3-i}$ is fully characterized by its angular momentum L; this is no longer the case for higher boson numbers when additional labels must be introduced.

The number of possible interactions at each order n is summarized in table I for up to $n = 3$. Some of these interactions contribute to the binding energy but do not influence

TABLE I. – *Enumeration of n-body interactions in IBM-1 for $n \le 3$.*

Order	Number of interactions		
	Total	Type I[a]	Type II[b]
$n = 0$	1	1	0
$n = 1$	2	1	1
$n = 2$	7	2	5
$n = 3$	17	7	10

[a] Interaction energy is constant for all states with the same N.
[b] Interaction energy varies from state to state.

the excitation spectrum of a nucleus. To determine the number of such interactions, one notes that the Hamiltonian $\hat{N}\hat{H}_{n-1}$ for constant boson number (*i.e.* a single nucleus) essentially reduces to the $(n-1)$-body Hamiltonian $\hat{H}_{n-1}$. Consequently, of the $\mathcal{N}_n$ independent interactions of order n contained in $\hat{H}_n$, $\mathcal{N}_{n-1}$ terms of the type $\hat{N}\hat{H}_{n-1}$ must be discarded if one wishes to retain only those that influence the excitation energies. For example, given that there is one term of order zero (*i.e.* a constant), one of the two first-order terms (*i.e.* the combination $\hat{N}$) does not influence the excitation spectrum. Likewise, there are two first-order terms (*i.e.* $\hat{n}_s$ and $\hat{n}_d$) and hence two of the seven two-body interactions do not influence the excitation spectrum. This argument leads to the numbers quoted in table I.

We conclude that for fits of excitation spectra there is a single one-boson energy of relevance, as well as five two-body and ten three-body interactions. If also binding energies are included in the analysis, an additional one-boson energy can be considered as well as two two-body and seven three-body interactions. These numbers of parameters are rather high for practical applications and simplifications must be sought on the basis of physical, empirical or symmetry arguments. To the latter we now turn.

4'2. *Dynamical symmetries.* – The characteristics of the most general IBM Hamiltonian which includes up to two-body interactions and its group-theoretical properties are by now well understood [33]. Numerical procedures exist to obtain its eigensolutions but the problem can be solved analytically for particular choices of boson energies and boson-boson interactions. For an IBM Hamiltonian with up to two-body interactions between the bosons, three different analytical solutions or limits exist: the *vibrational* $U(5)$ [15], the *rotational* $SU(3)$ [16] and the *γ-unstable* $SO(6)$ limit [17]. They are associated with the algebraic reductions

$$
(24) \qquad U(6) \supset \left\{ \begin{array}{c} U(5) \supset SO(5) \\ SU(3) \\ SO(6) \supset SO(5) \end{array} \right\} \supset SO(3).
$$

The algebras appearing in the lattice (24) are subalgebras of $U(6)$ generated by operators of the type $b^{\dagger}_{lm} b_{l'm'}$, the explicit form of which is listed, for example, in ref. [2]. With the subalgebras $U(5)$, $SU(3)$, $SO(6)$, $SO(5)$ and $SO(3)$ there are associated one linear (of $U(5)$) and five quadratic Casimir operators. This matches the number of one- and two-body interactions quoted in the last column of table I. The total of all one- and two-body interactions can be represented by including in addition the operators $\hat{C}_1[U(6)]$, $\hat{C}_2[U(6)]$ and $\hat{C}_1[U(6)]\hat{C}_1[U(5)]$. The most general IBM Hamiltonian with up to two-body interactions can thus be written in an *exactly* equivalent way with Casimir operators. Specifically, the Hamiltonian reads

$$
(25) \qquad \begin{aligned} \hat{H}_{1+2} = {} & \kappa_1 \hat{C}_1[U(5)] + \kappa'_1 \hat{C}_2[U(5)] + \kappa_2 \hat{C}_2[SU(3)] + \\ & + \kappa_3 \hat{C}_2[SO(6)] + \kappa_4 \hat{C}_2[SO(5)] + \kappa_5 \hat{C}_2[SO(3)], \end{aligned}
$$

which is just an alternative way of writing $\hat{H}_1 + \hat{H}_2$ of eqs. (22), (23) if interactions are omitted that contribute to the binding energy only.

The representation (25) is much more telling when it comes to the symmetry properties of the IBM Hamiltonian. If some of the coefficients κ_i vanish such that $\hat{H}_{1+2}$ contains Casimir operators of subalgebras belonging to a *single* reduction in the lattice (24), then, according to the discussion of subsect. **2·3**, the eigenvalue problem can be solved analytically. Three classes of spectrum generating Hamiltonians can thus be constructed of the form

$$(26) \qquad U(5) : \hat{H}_{1+2} = \kappa_1 \hat{C}_1[U(5)] + \kappa_1' \hat{C}_2[U(5)] + \kappa_4 \hat{C}_2[SO(5)] + \kappa_5 \hat{C}_2[SO(3)],$$
$$SU(3) : \hat{H}_{1+2} = \kappa_2 \hat{C}_2[SU(3)] + \kappa_5 \hat{C}_2[SO(3)],$$
$$SO(6) : \hat{H}_{1+2} = \kappa_3 \hat{C}_2[SO(6)] + \kappa_4 \hat{C}_2[SO(5)] + \kappa_5 \hat{C}_2[SO(3)].$$

In each of these limits the Hamiltonian is written as a sum of commuting operators and, as a consequence, the quantum numbers associated with the different Casimir operators are conserved. They can be summarized as follows:

$$(27) \qquad
\begin{array}{ccccccccc}
U(6) & \supset & U(5) & \supset & SO(5) & \supset & SO(3) & \supset & SO(2) \\
\downarrow & & \downarrow & & \downarrow & & \downarrow & & \downarrow \\
[N] & & n_d & & \tau & & \nu_\Delta L & & M_L
\end{array} ,$$

$$
\begin{array}{ccccccc}
U(6) & \supset & SU(3) & \supset & SO(3) & \supset & SO(2) \\
\downarrow & & \downarrow & & \downarrow & & \downarrow \\
[N] & & (\lambda, \mu) & & K_L L & & M_L
\end{array} ,$$

$$
\begin{array}{ccccccccc}
U(6) & \supset & SO(6) & \supset & SO(5) & \supset & SO(3) & \supset & SO(2) \\
\downarrow & & \downarrow & & \downarrow & & \downarrow & & \downarrow \\
[N] & & \sigma & & \tau & & \nu_\Delta L & & M_L
\end{array} .$$

Furthermore, for each of the three Hamiltonians in eq. (26) an analytic eigenvalue expression is available,

$$(28) \qquad U(5) : E(n_d, v, L) = \kappa_1 n_d + \kappa_1' n_d(n_d + 4) + \kappa_4 \tau(\tau + 3) + \kappa_5 L(L + 1),$$
$$SU(3) : E(\lambda, \mu, L) = \kappa_2(\lambda^2 + \mu^2 + \lambda\mu + 3\lambda + 3\mu) + \kappa_5 L(L + 1),$$
$$SO(6) : E(\sigma, \tau, L) = \kappa_3 \sigma(\sigma + 4) + \kappa_4 \tau(\tau + 3) + \kappa_5 L(L + 1).$$

One can add Casimir operators of $U(6)$ to the Hamiltonians in eq. (25) without breaking any of the symmetries. For a given nucleus they reduce to a constant contribution. They can be omitted if one is only interested in the spectrum of a single nucleus but they should be introduced if one calculates binding energies. Note that none of the Hamiltonians in eq. (26) contains a Casimir operator of $SO(2)$. This interaction breaks the $SO(3)$ symmetry (lifts the M_L degeneracy) and would only be appropriate if the nucleus is placed in an external electric or magnetic field.

The dynamical symmetries of the IBM arise if combinations of certain coefficients κ_i in Hamiltonian (25) vanish. The converse, however, cannot be said: even if all parameters κ_i are non-zero, the Hamiltonian $\hat{H}_{1+2}$ still may exhibit a dynamical symmetry and be analytically solvable. This is a consequence of the existence of unitary transformations which preserve the eigenspectrum of the Hamiltonian $\hat{H}_{1+2}$ (and hence its analyticity properties) and which can be represented as transformations in the parameter space $\{\kappa_i\}$. A *systematic* procedure exists for finding such transformations or parameter symmetries [34] which can, in fact, be applied to any Hamiltonian describing a system of interacting bosons and/or fermions.

While a numerical solution of the shell-model eigenvalue problem in general rapidly becomes impossible with increasing particle number, the corresponding problem in the IBM with s and d bosons remains tractable at all times, requiring the diagonalization of matrices with dimension of the order of $\sim 10^2$. One of the main reasons for the success of the IBM is that it provides a workable, albeit approximate, scheme which allows a description of transitional nuclei with a few relevant parameters. Numerous papers have been published on such transitional calculations. We limit ourselves here to citing those that first treated the transitions between the three limits of the IBM: from $U(5)$ to $SU(3)$ [35], from $SO(6)$ to $SU(3)$ [36] and from $U(5)$ to $SO(6)$ [37].

4'3. *Partial dynamical symmetries.* – As argued in sect. **2**, a dynamical symmetry can be viewed as a generalization and refinement of the concept of symmetry. Its basic paradigm is to write a Hamiltonian in terms of Casimir operators of a set of nested algebras. Its hallmarks are i) solvability of the complete spectrum, ii) existence of exact quantum numbers for all eigenstates and iii) pre-determined structure of the eigenfunctions, independent of the parameters in the Hamiltonian. A further enlargement of these ideas is obtained by means of the concept of *partial dynamical symmetry*. The idea is to relax the conditions of *complete* solvability and this can be done in essentially two different ways:

1) *Some of the eigenstates keep all of the quantum numbers.* In this case the properties of solvability, good quantum numbers, and symmetry-dictated structure are fulfilled exactly, but only by a subset of eigenstates [38, 39].

2) *All eigenstates keep some of the quantum numbers.* In this case none of the eigenstates is solvable, yet some quantum numbers (of the conserved symmetries) are retained. In general, this type of partial dynamical symmetry arises if the Hamiltonian preserves some of the quantum numbers in a dynamical-symmetry classification while breaking others [40, 41].

Combinations of 1) and 2) are possible as well, for example, if some of the eigenstates keep some of the quantum numbers [42].

We emphasize that dynamical symmetry, be it partial or not, is a notion that is not restricted to a specific model but can be applied to any quantal system consisting of interacting particles. Quantum Hamiltonians with a partial dynamical symmetry can be constructed with general techniques and their existence is closely related to the order of the interaction among the particles. Applications of these concepts continue to be explored in all fields of physics.

4'4. *Microscopy.* – The connection with the shell model arises by identifying the s and d bosons with correlated (or Cooper) pairs formed by two nucleons in the valence shell coupled to angular momentum $J = 0$ and $J = 2$. There exists a rich and varied literature on general procedures to carry out boson mappings in which pairs of fermions are represented as bosons. They fall into two distinct classes. In the first one establishes a correspondence between boson and fermion operators by requiring them to have the same algebraic structure, that is, the same commutation relations. In the second class the correspondence is established rather between state vectors in both spaces. In each case further subclasses exist that differ in their technicalities (*e.g.*, the nature of the operator expansion or the hierarchy in the state correspondence). In the specific example at hand, namely the mapping between the IBM and the shell model, the most successful procedure arguably has been the so-called OAI mapping [43] which associates vectors based on a seniority ($U(5)$) hierarchy in fermion (boson) space. It has been used in highly complex situations that go well beyond the simple version of IBM-1 with just identical s and d bosons and which include, for example, neutron-proton $T = 1$ and $T = 0$ pairs [44, 45]. In a similar vein a miscroscopic foundation has been given to the IBFM; examples of various mapping techniques for odd-mass nuclei can be found in refs. [46-50].

4'5. *The classical limit.* – The coherent-state formalism [51-53] represents a bridge between algebraic and geometric nuclear models. The central outcome of the formalism is that for any IBM-1 Hamiltonian a corresponding potential $V(\beta, \gamma)$ can be constructed where β and γ parametrize the intrinsic quadrupole deformation of the nucleus [54]. This procedure is known as the classical limit of the IBM.

The coherent states used for obtaining the classical limit of the IBM are of the form

$$(29) \qquad |N; \alpha_\mu\rangle \propto \left(s^\dagger + \sum_\mu \alpha_\mu d_\mu^\dagger \right)^N |o\rangle,$$

where $|o\rangle$ is the boson vacuum and α_μ are five complex variables. These have the interpretation of (quadrupole) shape variables and their associated conjugate momenta. If one limits oneself to static problems, the α_μ can be taken as real; they specify a shape and are analogous to the shape variables of the droplet model of the nucleus [54]. The α_μ can be related to three Euler angles which define the orientation of an intrinsic frame of reference, and two intrinsic shape variables, β and γ, that parametrize quadrupole vibrations of the nuclear surface around an equilibrium shape. In terms of the latter variables, the coherent state (29) is rewritten as

$$(30) \qquad |N; \beta\gamma\rangle \propto \left(s^\dagger + \beta \left[\cos\gamma d_0^\dagger + \sqrt{\frac{1}{2}} \sin\gamma (d_{-2}^\dagger + d_{+2}^\dagger) \right] \right)^N |o\rangle.$$

The expectation value of Hamiltonian (21) in this state can be determined by elementary methods [55] and yields a functional expression in β and γ which is identified with a potential $V(\beta, \gamma)$, familiar from the geometric model. In this way the following classical limit of Hamiltonian (21) is found:

$$(31) \qquad V(\beta, \gamma) = E_0 + \sum_{n \geq 1} \frac{N(N-1) \cdots (N-n+1)}{(1+\beta^2)^n} \sum_{kl} a_{kl}^{(n)} \beta^{2k+3l} \cos^l 3\gamma,$$

where the non-zero coefficients $a_{kl}^{(n)}$ of order $n = 1, 2$ and 3 are given by

$$(32) \qquad a_{00}^{(1)} = \epsilon_s, \quad a_{10}^{(1)} = \epsilon_d,$$

$$a_{00}^{(2)} = \frac{1}{2} v_{ssss}^0, \quad a_{10}^{(2)} = \sqrt{\frac{1}{5}} v_{ssdd}^0 + v_{sdsd}^2, \quad a_{01}^{(2)} = -\frac{2}{\sqrt{7}} v_{sddd}^2,$$

$$a_{20}^{(2)} = \frac{1}{10} v_{dddd}^0 + \frac{1}{7} v_{dddd}^2 + \frac{9}{35} v_{dddd}^4,$$

$$a_{00}^{(3)} = \frac{1}{6} v_{ssssss}^0, \quad a_{10}^{(3)} = \sqrt{\frac{1}{15}} v_{ssssdd}^0 + \frac{1}{2} v_{ssdssd}^2,$$

$$a_{01}^{(3)} = -\frac{1}{3} \sqrt{\frac{2}{35}} v_{sssddd}^0 - \sqrt{\frac{2}{7}} v_{ssdsdd}^2,$$

$$a_{20}^{(3)} = \frac{1}{10} v_{sddsdd}^0 + \sqrt{\frac{1}{7}} v_{ssdddd}^2 + \frac{1}{7} v_{sddsdd}^2 + \frac{9}{35} v_{sddsdd}^4,$$

$$a_{11}^{(3)} = -\frac{1}{5} \sqrt{\frac{2}{21}} v_{sddddd}^0 - \frac{\sqrt{2}}{7} v_{sddddd}^2 - \frac{18}{35} \sqrt{\frac{2}{11}} v_{sddddd}^4,$$

$$a_{30}^{(3)} = \frac{1}{14} v_{dddddd}^2 + \frac{1}{30} v_{dddddd}^3 + \frac{3}{154} v_{dddddd}^4 + \frac{7}{165} v_{dddddd}^6,$$

$$a_{02}^{(3)} = \frac{1}{105} v_{dddddd}^0 - \frac{1}{30} v_{dddddd}^3 + \frac{3}{110} v_{dddddd}^4 - \frac{4}{1155} v_{dddddd}^6,$$

in terms of the single boson energies ϵ_s and ϵ_d, and the matrix elements between normalized two- and three-body states,

$$v_{\ell_1 \ell_2 \ell_1' \ell_2'}^L = \langle \ell_1 \ell_2; LM | \hat{H}_2 | \ell_1' \ell_2'; LM \rangle,$$

$$v_{\ell_1 \ell_2 \ell_3 \ell_1' \ell_2' \ell_3'}^L = \langle \ell_1 \ell_2 \ell_3; LM | \hat{H}_3 | \ell_1' \ell_2' \ell_3'; LM \rangle.$$

The expressions (31), (32) are useful for choosing between the many possible three-body interactions.

A *catastrophe* analysis [56] of the potential surfaces in (β, γ) as a function of the Hamiltonian parameters determines the stability properties of these shapes. This analysis was carried out for the general IBM Hamiltonian with up to two-body interactions by López-Moreno and Castaños [57]. The results of this study are confirmed if a simplified IBM Hamiltonian is considered of the form [58]

$$(33) \qquad \hat{H}_{1+2}^{\text{ecqf}} = \epsilon \hat{n}_d + \kappa \hat{Q} \cdot \hat{Q}.$$

This Hamiltonian provides a simple parametrization of the essential features of nuclear structural evolution in terms of a vibrational term $\hat{n}_d$ (the number of d bosons) and a quadrupole interaction $\hat{Q} \cdot \hat{Q}$ with

$$(34) \qquad \hat{Q}_\mu = [s^\dagger \times \tilde{d} + d^\dagger \times s]^{(2)}_\mu + \chi [d^\dagger \times \tilde{d}]^{(2)}_\mu.$$

Besides an overall energy scale, the spectrum of the Hamiltonian (33) is determined by two parameters: the ratio ϵ/κ and χ. The three limits of the IBM are obtained with an appropriate choice of parameters: $U(5)$ if $\kappa = 0$, $SU_\pm(3)$ if $\epsilon = 0$ and $\chi = \pm\sqrt{7}/2$, and $SO(6)$ if $\epsilon = 0$ and $\chi = 0$. One may thus represent the parameter space of the simplified IBM Hamiltonian (33) on a triangle with vertices that correspond to the three limits $U(5)$, $SU(3)$ and $SO(6)$, and where arbitrary points correspond to specific values of ϵ/κ and χ. Since there are two possible choices for $SU(3)$, $\chi = -\sqrt{7}/2$ and $\chi = +\sqrt{7}/2$, the triangle can be extended to cover both cases by allowing χ to take negative as well as positive values.

The geometric interpretation of any IBM Hamiltonian on the triangle can now be found from its expectation value in the coherent state (30) which for the particular Hamiltonian (33) gives

$$(35) \qquad V(\beta, \gamma) = \frac{N\epsilon\beta^2}{1 + \beta^2} + \kappa \left[\frac{N(5 + (1 + \chi^2)\beta^2)}{1 + \beta^2} \right.$$
$$\left. + \frac{N(N-1)}{(1 + \beta^2)^2} \left(\frac{2}{7}\chi^2\beta^4 - 4\sqrt{\frac{2}{7}}\chi\beta^3 \cos 3\gamma + 4\beta^2 \right) \right].$$

The catastrophe analysis of this surface is summarized with the phase diagram shown in fig. 1. Analytically solvable limits are indicated by the black dots. Two different $SU(3)$ limits occur corresponding to two possible choices of the quadrupole operator, $\chi = \pm\sqrt{7}/2$. Close to the $U(5)$ vertex, the IBM Hamiltonian has a vibrational-like spectrum. Towards the $SU(3)$ and $SO(6)$ vertices, it acquires rotational-like characteristics. This is confirmed by a study of the character of the potential surface in β and γ associated with each point of the triangle. In the region around $U(5)$, corresponding to large ϵ/κ ratios, the minimum of the potential is at $\beta = 0$. On the other hand, close to the $SU_+(3)$–$SO(6)$–$SU_-(3)$ axis the IBM Hamiltonian corresponds to a potential with a deformed minimum between $\beta = 0$ and $\beta = \sqrt{2}$. Furthermore, in the region around prolate $SU_-(3)$ ($\chi < 0$) the minimum occurs for $\gamma = 0°$ while around oblate $SU_+(3)$ ($\chi > 0$) it does for $\gamma = 60°$. In this way the picture emerges that the IBM parameter space can be divided into three regions according to the character of the associated potential having (I) a spherical minimum, (II) a prolate deformed minimum or (III) an oblate deformed minimum. The boundaries between the different regions (the so-called Maxwell set) are indicated by the dashed lines in fig. 1 and meet in a triple point. The spherical-deformed border region displays another interesting phenomenon. Since the *absolute* minimum of the potential must be either spherical, or prolate or oblate deformed, its character uniquely determines the three regions and the dividing Maxwell lines. Nevertheless,

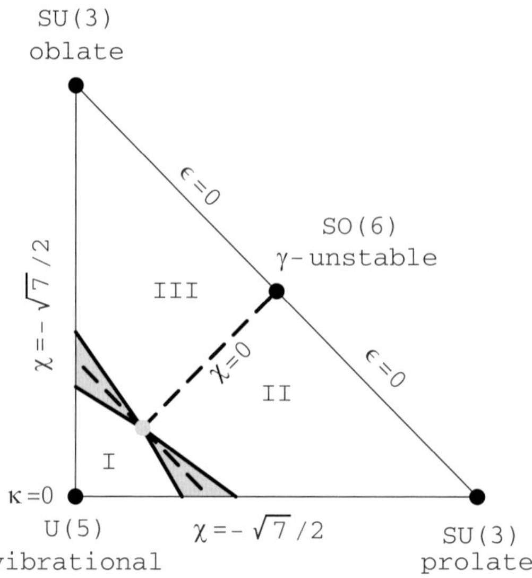

Fig. 1. – Phase diagram of the Hamiltonian (33) and the associated geometric interpretation. The parameter space is divided into three regions depending on whether the corresponding potential has (I) a spherical, (II) a prolate deformed or (III) an oblate deformed absolute minimum. These regions are separated by dashed lines and meet in a triple point (grey dot). The shaded area corresponds to a region of coexistence of a spherical and a deformed minimum. Also indicated are the points on the triangle (black dots) which correspond to the dynamical-symmetry limits of the Hamiltonian (33) and the choice of parameters ϵ, κ and χ for specific points or lines of the diagram.

this does not exclude the possibility that, in passing from one region to another, the potential may display a second *local* minimum. This indeed happens for the $U(5)$-$SU(3)$ transition [59] where there is a narrow region of coexistence of a spherical and a deformed minimum, indicated by the shaded area in fig. 1. Since, at the borders of this region of coexistence, the potential undergoes a *qualitative* change of character, the boundaries are genuine critical lines of the potential surface [56].

Although these geometric results have been obtained with reference to the simplified Hamiltonian (33) and its associated "triangular" parameter space, it must be emphasized that they remain valid for the general IBM Hamiltonian with up to two-body interactions [57].

5. – Triaxiality in the interacting boson model

In this section a first application of the IBM is discussed, namely the use of higher-order interactions between the d bosons and its relation to triaxiality. First, a simplified IBM Hamiltonian with up to two-body interactions is described which has been used in the systematic analysis of the collective properties of many nuclei. Although generally yielding satisfactory results when compared to available spectroscopic data, systematic deviations are observed for properties that are related to (rigid or soft) triaxial nuclear behaviour. It is then argued that such observed deficiencies call for the introduction of three-body interactions between the bosons. Such interactions can be applied to a variety of $SO(6)$-like nuclei and the example of neutron-rich ruthenium isotopes is presented here. The study described in this section is based on the paper by Stefanescu *et al.* [60] but employs a modified numerical procedure as introduced in ref. [61].

5'1. *A specific two-body Hamiltonian.* – From a great number of standard IBM-1 studies [2] one has a good idea of a workable Hamiltonian with up to two-body interactions which is of the form

$$(36) \qquad \hat{H}'^{\mathrm{ecqf}}_{1+2} = \epsilon\,\hat{n}_d + \kappa\,\hat{Q}\cdot\hat{Q} + \kappa'\hat{L}\cdot\hat{L} + \lambda\,\hat{n}_d^2,$$

where $\hat{Q}$ is the quadrupole operator (34) and $\hat{L}$ is the angular-momentum operator, $\hat{L}_\mu = \sqrt{10}\,[d^\dagger \times \tilde{d}]^{(1)}_\mu$. The $\hat{Q}^2$ and $\hat{L}^2$ terms in eq. (36) constitute the Hamiltonian of the so-called Consistent-Q Formalism (CQF) [62]. Its eigenfunctions are fully determined by χ which for $\chi = \pm\sqrt{7}/2$ gives rise to the deformed or $SU(3)$ limit and for $\chi = 0$ to the γ-unstable or $SO(6)$ limit. In an Extended Consistent-Q Formalism (ECQF) [63] a further term $\epsilon\,\hat{n}_d$ is added with which the third, vibrational or $U(5)$ limit of the IBM-1 can be obtained. The ECQF Hamiltonian thus allows one to reach all three limits of the model with four parameters. In some nuclei an additional term $\lambda\,\hat{n}_d^2$ further improves the description of the excitation spectrum. The effect of this term with $\lambda < 0$ is an increase of the moment of inertia with increasing angular momentum (or d-boson seniority τ). This so-called "τ-compression" has been used for the first time in ref. [64].

For the calculation of electric quadrupole properties an $E2$ transition operator is needed. In the IBM-1 it is defined as $\hat{T}_\mu(E2) = e_\mathrm{b}\hat{Q}_\mu$, where e_b is an effective charge for the bosons. In CQF the quadrupole operators in the $E2$ operator and in the Hamiltonian are the same [62], that is, they contain the same χ.

5'2. *A specific three-body Hamiltonian.* – Many nuclear properties can be correctly described by the simple Hamiltonian (36) but some cannot. A notable example is the even-odd staggering in the quasi-γ band of nuclei that are close to the $SO(6)$ limit. A characteristic feature of the γ-unstable limit of IBM-1 is a bunching of quasi-γ-band states according to 2^+, $(3^+, 4^+)$, $(5^+, 6^+),\dots$, that is, 3^+ and 4^+ are close in energy, *etc.* This even-odd staggering is observed in certain $SO(6)$ nuclei but not in all and in some it is, in fact, replaced by the opposite bunching $(2^+, 3^+)$, $(4^+, 5^+)$, $\dots$, which is typical of a rigid triaxial rotor [65]. From these qualitative observations it is clear that the even-odd quasi-γ-band staggering is governed by the γ degree of freedom (*i.e.* triaxiality) as it changes character in the transition from a γ-soft vibrator to a rigid triaxial rotor.

A proper description of triaxiality in the IBM-1 must necessarily involve higher-order interactions as can be shown from the expressions given in subsect. **4'5**. The minimum of the potential $V(\beta, \gamma)$ in eq. (31) (which can be thought of as the equilibrium shape of the nucleus) of an IBM-1 Hamiltonian with up to two-body interactions is either spherical

TABLE II. – *Normalization coefficients $N_{\lambda L}$ for three-d-boson states.*

L	0	2	3	4	6
$\lambda = 0$	—	$\sqrt{\frac{5}{14}}$	—	—	—
$\lambda = 2$	$\sqrt{\frac{1}{6}}$	$\sqrt{\frac{7}{8}}$	$\sqrt{\frac{7}{30}}$	$\sqrt{\frac{7}{22}}$	—
$\lambda = 4$	—	$\sqrt{\frac{35}{72}}$	$-\sqrt{\frac{7}{12}}$	$\sqrt{\frac{7}{20}}$	$\sqrt{\frac{1}{6}}$

($\beta = 0$), prolate deformed ($\beta > 0, \gamma = 0°$) or oblate deformed ($\beta > 0, \gamma = 60°$). The lowest term in eq. (31) with a triaxial extremum is quadratic in $\cos 3\gamma$ ($l = 2$) and this requires a non-zero $a_{02}^{(3)}$ coefficient. From the explicit expressions given in eqs. (32) it is seen that the lowest-order interactions possibly leading to a triaxial minimum in $V(\beta, \gamma)$ are thus necessarily of the form

$$(37) \qquad \hat{H}_3^d = \sum_L \tilde{v}_{dddddd}^L [[d^\dagger \times d^\dagger]^{(\lambda)} \times d^\dagger]^{(L)} \cdot [[\tilde{d} \times \tilde{d}]^{(\lambda')} \times \tilde{d}]^{(L)},$$

where the allowed angular momenta are $L = 0, 2, 3, 4, 6$. For several L more than one combination of intermediate angular momenta λ and λ' is possible; these do not give rise to independent terms but differ by a scale factor. To avoid the confusion caused by this scale factor, we rewrite the Hamiltonian (37) as

$$(38) \qquad \hat{H}_3^d = \sum_L v_{dddddd}^L \hat{B}_L^\dagger \cdot \tilde{B}_L, \qquad \hat{B}_{LM}^\dagger = N_{\lambda L}[[d^\dagger \times d^\dagger]^{(\lambda)} \times d^\dagger]_M^{(L)}.$$

For simplicity's sake the coefficients v_{dddddd}^L shall be denoted as v_L in the following. The normalization coefficient $N_{\lambda L}$ is defined such that $B_{LM}|d^3; LM\rangle$ yields the vacuum state $|o\rangle$, where $|d^3; LM\rangle$ is a normalized, symmetric state of three bosons coupled to total angular momentum L and z projection M. The normalization coefficients $N_{\lambda L}$ are given in table II for the different combinations of λ and L. Results are independent of λ provided the appropriate coefficient $N_{\lambda L}$ is used.

While there are good arguments for choosing any of the three-body terms $\hat{B}_L^\dagger \cdot \tilde{B}_L$, it is more difficult to distinguish *a priori* between these five different interactions. From the expression for $a_{02}^{(3)}$ given in eqs. (32) it is seen that the three-body term $\hat{B}_L^\dagger \cdot \tilde{B}_L$ with $L = 3$ is proportional to $\sin^2 3\gamma$. It is therefore the interaction which is most effective to create a triaxial minimum in the potential $V(\beta, \gamma)$ and for this reason it has been studied in most detail. The effect of $\hat{B}_3^\dagger \cdot \tilde{B}_3$ on even-odd staggering in the quasi-γ band was demonstrated with numerical calculations [66]. Applications of the $L = 3$ three-body term were proposed in ref. [67] for $SO(6)$-like Xe and Ba isotopes in the mass region around $A = 130$, as well as for ^{196}Pt.

Most of the results presented below are obtained with the d-boson cubic interaction with $L = 3$ which in general reproduces best the quasi-γ-band properties. The terms with $L \neq 3$ nevertheless have been systematically investigated and those results will occasionally be referred to in the following.

5'3. *Numerical procedure.* – To test the effectiveness of the various cubic interactions in reproducing the data in near-$SO(6)$ nuclei, the following fitting procedure has been used. The nuclei considered should have enough known states in the ground-state and quasi-γ bands—preferably up to angular momentum $J^\pi = 10^+$—for the procedure to be meaningful. The first step is to determine the parameters in the standard IBM-1 Hamiltonian (36). For an initial choice of χ, the parameters κ and κ' are first determined while keeping ϵ_d and λ_d zero. With (κ, κ') thus found as starting values, a new fit is performed setting ϵ_d free as well, leading to the best values $(\kappa, \kappa', \epsilon_d)$. Finally, this process is repeated by letting also λ_d free, leading to a final set $(\kappa, \kappa', \epsilon_d, \lambda_d)$ for a given χ. The parameter χ cannot be reliably determined from energies but is fixed from $E2$ transition rates which are calculated in the CQF. If not enough $E2$ data are available, we take χ from a neighbouring isotope. The entire procedure is repeated for different χ, retaining the value that gives the best agreement with the $E2$ data. In a last step the importance of the $\hat{B}_L^\dagger \cdot \tilde{B}_L$ terms is tested in a similar way by allowing the variation of all five parameters $(\kappa, \kappa', \epsilon_d, \lambda_d, v_L)$ while keeping χ constant. Since one is particularly interested in the influence of v_L on the even-odd staggering, in this final step this parameter is adjusted to the members of the quasi-γ band only. For reasons of numerical stability the weight given to the ground-state-band members is not exactly zero but small.

The accuracy of the fits can be tested by plotting the signature splitting $S(J)$ of the quasi-γ band given by [68]

$$(39) \qquad S(J) = \frac{E(J) - E(J-1)}{E(J) - E(J-2)} \cdot \frac{J(J+1) - (J-1)(J-2)}{J(J+1) - J(J-1)} - 1,$$

which vanishes if there is no even-odd staggering.

5'4. *Results for the neutron-rich ruthenium isotopes.* – In recent years gamma-ray spectroscopy of fission fragments has significantly improved our knowledge of the structure of medium-mass neutron-rich nuclei. In particular, for the heavy fission products ^{108,110,112}Ru, produced by a ^{252}Cf source and studied with the Gammasphere array, new data have become available [69, 70]. As a result of these studies, many more members of the ground-state and quasi-γ bands are now known and this yields important information on the triaxiality character of these nuclei, as will be shown in this section.

The isotopes ^{108,110,112}Ru were already considered in ref. [60] in the context of the IBM-1 with cubic interactions. The advantage of the method presented in this section is that a consistent one- and two-body IBM-1 Hamiltonian is taken to which a three-body term is added without changing the value of χ. In this way any improvement of the description of the quasi-γ-band staggering can be unambiguously attributed to the three-body term. Also, a least-squares fit is performed to the parameters in the Hamiltonian according to the procedure outlined in subsect. **5'3**. Although all results are obtained with a parameter-search routine, there is no guarantee that the absolute value of the

TABLE III. – *Parameters and r.m.s. deviation for ruthenium isotopes in units of keV.*

Nucleus	ϵ_d	κ	κ'	λ_d	v_3	$\chi^{(a)}$	σ
^{108}Ru	1078	-57.6	12.1	-144.9	—	-0.10	23
	852	-66.8	8.3	-130.7	-13.1	-0.10	45
	732	-74.6	14.0	-157.8	$30.5^{(b)}$	-0.10	19
^{110}Ru	1053	-46.1	15.5	-123.7	—	-0.10	39
	873	-56.9	9.9	-108.5	-28.1	-0.10	20
^{112}Ru	837	-45.3	15.2	-116.8	—	-0.10	55
	424	-57.8	7.7	-73.7	-46.8	-0.10	38

$^{(a)}$ Dimensionless.

$^{(b)}$ Value of the coefficient v_2.

root-mean-square (r.m.s.) deviation is obtained but the parameters shown in table III define at least a local minimum.

In spite of the differences in fitting procedure the results obtained here are globally in agreement with those of Stefanescu *et al.* [60]. The main conclusion is that, while the staggering pattern of the quasi-γ band is much improved with the $\hat{B}_3^\dagger \cdot \tilde{B}_3$ interaction in ^{110,112}Ru, this is not the case for ^{108}Ru (see fig. 2). This is also evident from the parameters shown in table III: the r.m.s. deviation σ actually increases for ^{108}Ru when the three-body interaction is added to the Hamiltonian. This is a consequence of our fit procedure which gives (almost) exclusive weight to the quasi-γ-band members when also v_3 is fitted. This increase in σ illustrates that the quasi-γ-band energies in ^{108}Ru cannot be reproduced by adding $\hat{B}_3^\dagger \cdot \tilde{B}_3$ without destroying the agreement for the ground-state band.

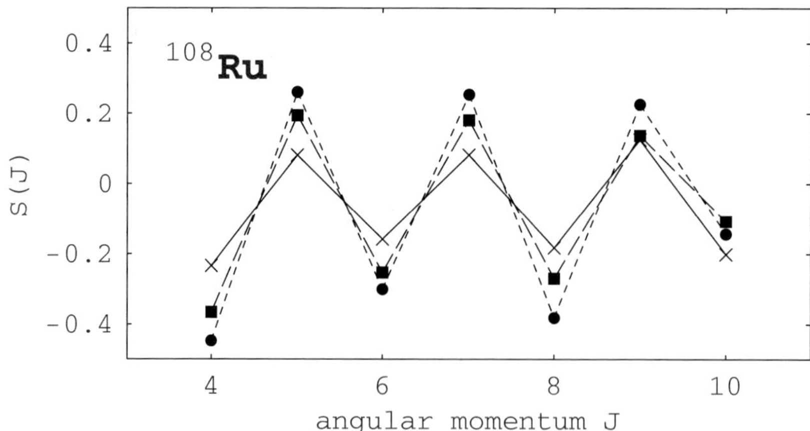

Fig. 2. – Observed and calculated signature splitting for the quasi-γ band in ^{108}Ru. The data are indicated by crosses and the results of the IBM-1 with and without the three-body interaction $\hat{B}_3^\dagger \cdot \tilde{B}_3$ by squares and dots, respectively.

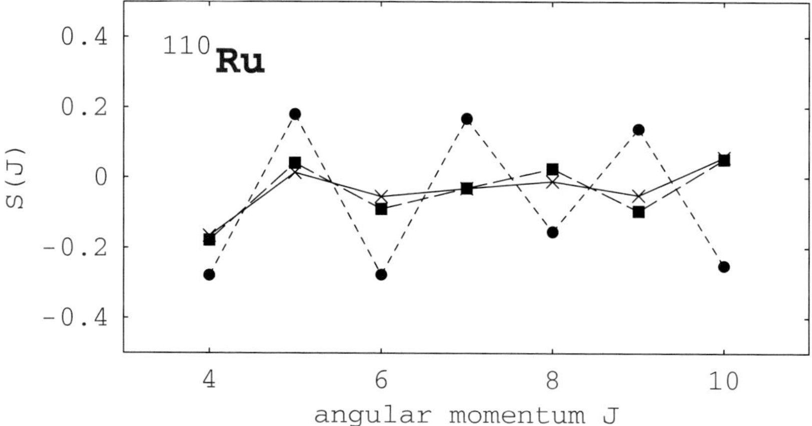

Fig. 3. – Same caption as fig. 2 for ^{110}Ru.

As one goes to the heavier ruthenium isotopes, one notices a distinct evolution of the odd-even staggering pattern (see figs. 3 and 4). Whereas the staggering pattern is essentially consistent with the IBM-1 calculation without cubic interactions in ^{108}Ru, this is no longer the case in the two heavier isotopes. In ^{110}Ru there is very little staggering at all, $S(J) \approx 0$, and in ^{112}Ru the staggering pattern in the data is in fact the reverse of what is obtained without cubic interactions, especially at higher angular momenta. The $\hat{B}_3^\dagger \cdot \tilde{B}_3$ interaction shifts levels with even (odd) angular momentum upwards (downwards) in energy and it does so increasingly with increasing spin. This is exactly what can be observed from the data in ^{110}Ru and ^{112}Ru and this provides a strong phenomenological argument for the use of the $\hat{B}_3^\dagger \cdot \tilde{B}_3$ interaction.

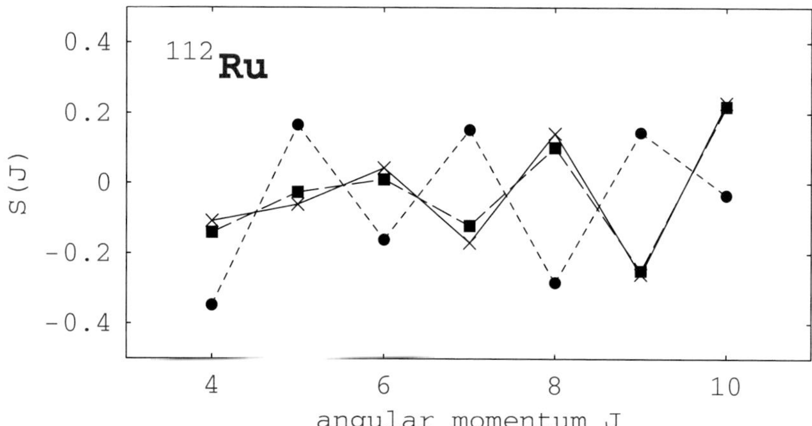

Fig. 4. – Same caption as fig. 2 for ^{112}Ru.

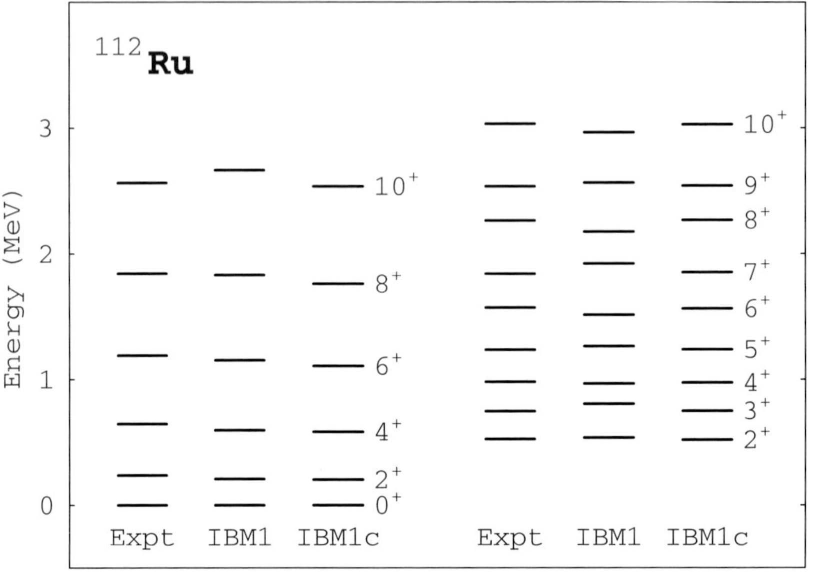

Fig. 5. – Energies of the levels of the ground-state and the γ band up to angular momentum $J^\pi = 10^+$ in the nucleus ^{112}Ru. The three columns correspond to the experimental energies, and the IBM-1 calculation without ("IBM1") and with ("IBM1c") a cubic interaction.

One should appreciate the sensitivity of the signature splitting $S(J)$ to the energies of the γ-band levels. This is illustrated with fig. 5 where the energy spectrum of the nucleus ^{112}Ru is shown. It is indeed visible from the figure that energies of the levels of the γ band are better reproduced if cubic interactions are considered but the improvement is much more tellingly illustrated by the plot of $S(J)$ in fig. 4.

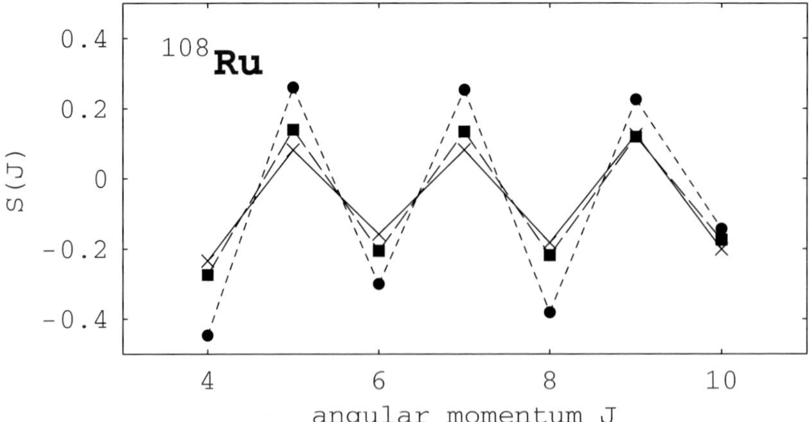

Fig. 6. – Same caption as fig. 2 for the interaction $\hat{B}_2^\dagger \cdot \tilde{B}_2$.

TABLE IV. – *Experimental and calculated E2 branching ratios for* ^{108,110,112}Ru.

Ratio	^{108}Ru		^{110}Ru		^{112}Ru	
	Expt	IBM1c	Expt	IBM1c	Expt	IBM1c
$\dfrac{2^+_2 \to 2^+_1}{2^+_2 \to 0^+_1}$	8.6(20)	4.6	14.9(2)	10.1	22.2(3)	12.3
$\dfrac{3^+_1 \to 2^+_2}{3^+_1 \to 2^+_1}$	15.9(14)	11.8	20.4(3)	15.0	21.7(4)	16.2
$\dfrac{4^+_2 \to 2^+_2}{4^+_2 \to 2^+_1}$	100(5)	53.8	100(6)	373	318(26)	∞
$\dfrac{4^+_2 \to 2^+_2}{4^+_2 \to 4^+_1}$	2.1(1)	2.6	1.1(1)	1.6	0.94(4)	1.4
$\dfrac{5^+_1 \to 3^+_1}{5^+_1 \to 4^+_1}$	10(1)	17.6	25(1)	29.3	37(2)	40.8

From the plot of the signature splitting we can also "understand" why the $\hat{B}^\dagger_3 \cdot \tilde{B}_3$ interaction fails in ^{108}Ru: the deviations in staggering between the data and the IBM-1 calculation without cubic interactions actually decrease rather than increase with angular momentum. This feature is incompatible with the $L = 3$ term in Hamiltonian (38) but is exactly what is obtained with the $L = 2$ term as shown in fig. 6.

It is important to check that the cubic Hamiltonian thus obtained gives reasonable results as regards electric quadrupole transitions. In the initial two-body Hamiltonian the E2 transition rates are essentially determined by the value of χ in the quadrupole operator. It is expected that this is still the case when cubic terms are added as long as these do not substantially alter the eigenstates of the Hamiltonian. A number of $E2$ branching ratios from γ-band states are known from recent γ-spectroscopy work on fission products [60] and are compared in table IV with the results of a cubic Hamiltonian with the fitted values of v_2 in ^{108}Ru and of v_3 in ^{110,112}Ru in table III. There is a satisfactory overall agreement with the data. The fitted cubic Hamiltonian gives a very large $B(E2; 4^+_2 \to 2^+_2)/B(E2; 4^+_2 \to 2^+_1)$ ratio in ^{112}Ru which is due to the accidental vanishing of the $4^+_2 \to 2^+_1$ transition and which "agrees" with the large value found experimentally. The largest discrepancy between theory and experiment is the $B(\mathrm{E2}; 2^+_2 \to 2^+_1)/B(E2; 2^+_2 \to 0^+_1)$ ratio which is systematically underpredicted but, on the other hand, the trend of increasing ratio as the mass number increases is correctly obtained in the calculation.

Once the parameters of the Hamiltonian have been fitted to the energy spectrum and $E2$ transition rates, its classical limit yields a potential energy surface $V(\beta, \gamma)$ as obtained from expression (31). In this way it can be verified to what extent triaxial features are introduced by the cubic interactions. Figure 7 provides an illustration by showing the potential energy surfaces $V(\beta, \gamma)$ for ^{112}Ru obtained in the classical limit of the IBM-1 Hamiltonian without and with the $\hat{B}^\dagger_3 \cdot \tilde{B}_3$ interaction. The surface on the left-hand side is obtained from the two-body Hamiltonian and has a minimum at $\beta = 0$ (spherical). The Hamiltonian which includes the $\hat{B}^\dagger_3 \cdot \tilde{B}_3$ interaction yields the surface on

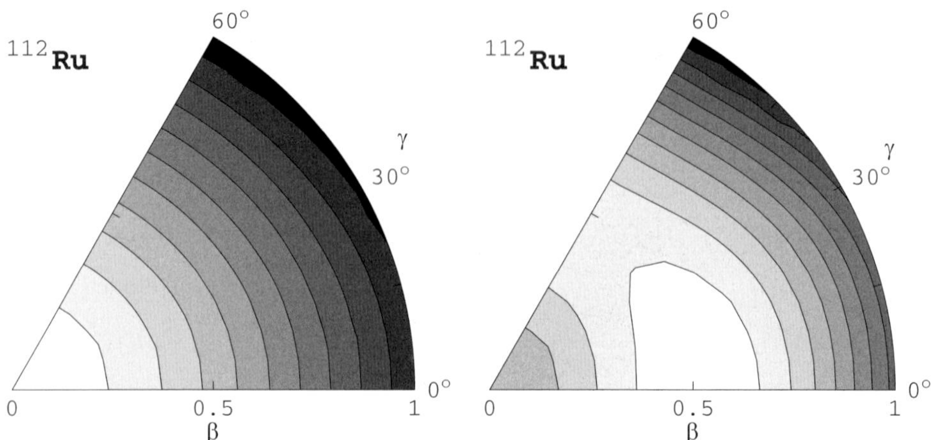

Fig. 7. – Potential energy surfaces $V(\beta,\gamma)$ for ^{112}Ru. The plot on the left-hand side shows the classical limit of the IBM-1 Hamiltonian with only two-body interactions while on the right-hand side the effect of $\hat{B}_3^\dagger \cdot \tilde{B}_3$ is included.

the right-hand side which exhibits a (shallow) minimum at prolate deformation ($\beta \neq 0$ and $\gamma = 0^\circ$). In this example a noticeable change of the potential $V(\beta,\gamma)$ is found as a result of including cubic interactions which perhaps is not surprising since parameter variations are rather important between IBM-1 and IBM-1c in ^{112}Ru (see table III). However, even in this extreme example no triaxial minimum is obtained.

These examples illustrate how a careful analysis of available data may guide the selection of the different interactions in the IBM-1 Hamiltonian. While we have currently a good working Hamiltonian which includes up to two-body interactions and which describes nuclei throughout the nuclear chart, little is known of the overall trends for three-body interactions. A systematic study of three-body interactions in several series of isotopes where sufficient data are available is currently under way [61]. Although in many nuclei cubic interactions considerably improve the staggering properties of the γ band, in none of the nuclei studied so far a triaxial minimum is obtained and the changes in the potential $V(\beta,\gamma)$ induced by the cubic interactions usually are minor.

6. – Global calculations for spectra and binding energies

If one limits the Hamiltonian of the IBM-1 to interactions that are at most of two-body nature between the bosons, the total number of parameters is ten. As shown in table I, six of the parameters determine the energy spectrum of individual nuclei while the four remaining ones exclusively contribute to the binding energy. The parameter systematics of the former is by now well established through phenomenological studies with input from microscopic theory (for references, see [2]). Surprisingly little has been done with IBM concerning absolute binding energies and in most cases only two-nucleon separation energies have been considered, such as in the recent detailed studies of García-Ramos *et al.* [71] and Fossion *et al.* [72]. The work reported here is most closely related to that of Davis *et al.* [73].

In this section a method based on IBM is proposed that combines spectroscopic information with mass data. For this purpose it is essential to keep in mind that the IBM is a valence-nucleon model which lumps all information on the core of the nucleus into a single constant E_0, its binding energy. Hamiltonian (21) by itself, therefore, cannot provide an adequate description of the *total* binding energy of the nucleus. The method proposed here consists of subtracting a global liquid-drop contribution (which does not include shell or deformation effects) from the nuclear binding energy and modeling the remainder with an IBM-1 Hamiltonian. An outline of the method is given as well as a first application of it in the rare-earth region.

Recall that the binding energy $B(N, Z)$ of a nucleus with N neutrons and Z protons is defined through

$$(40) \qquad M(N, Z)c^2 = N m_{\mathrm{n}} c^2 + Z m_{\mathrm{p}} c^2 - B(N, Z),$$

where $M(N, Z)$ is the mass of the nucleus and m_{n} (m_{p}) the mass of the neutron (proton). The binding energy $B(N, Z)$ thus represents the energy needed to pull a nucleus into its $N + Z$ separate nucleons. The binding energy $B(N, Z)$ is positive if the nucleus is bound and energy has to be supplied to pull it apart. Note also that $M(N, Z)$ here refers to the mass of the nucleus only and *not* to that of the atom; so the binding energy $B(N, Z)$ is that of the neutrons and the protons and does not include contributions from the electrons.

A simple, yet surprisingly accurate formula for the binding energy of an atomic nucleus is given by

$$(41) \qquad B(N, Z) = a_{\mathrm{v}} A - a_{\mathrm{s}} A^{2/3} - a_{\mathrm{c}} \frac{Z(Z-1)}{A^{1/3}} - \frac{S_{\mathrm{v}}}{1 + y_{\mathrm{s}} A^{-1/3}} \frac{4T(T+r)}{A}$$
$$+ a_{\mathrm{p}} \frac{\Delta(N, Z)}{A^{1/2}},$$

where $A = N + Z$ is the total number of nucleons. Equation (41) is known as the liquid-drop mass formula. The first three terms appearing in the mass formula are referred to as volume, surface and Coulomb, and have a macroscopic origin that can be understood intuitively by viewing the nucleus as a dense, charged liquid drop. The fourth so-called symmetry term is a consequence of the Pauli principle: nuclear matter prefers to be symmetric ($N = Z$) because, at constant A, such configuration maximizes availability of the lowest quantum states. Formula (41) uses a somewhat sophisticated form of the symmetry energy where surface and the so-called Wigner effects are considered via the inclusion of y_{s} and r, respectively. The last term represents a simple parametrization of the most important correlation in nuclei, pairing, by assuming $\Delta(N, Z) = +1$, 0 and -1 in even-even, odd-mass and odd-odd nuclei, respectively. In the convention of

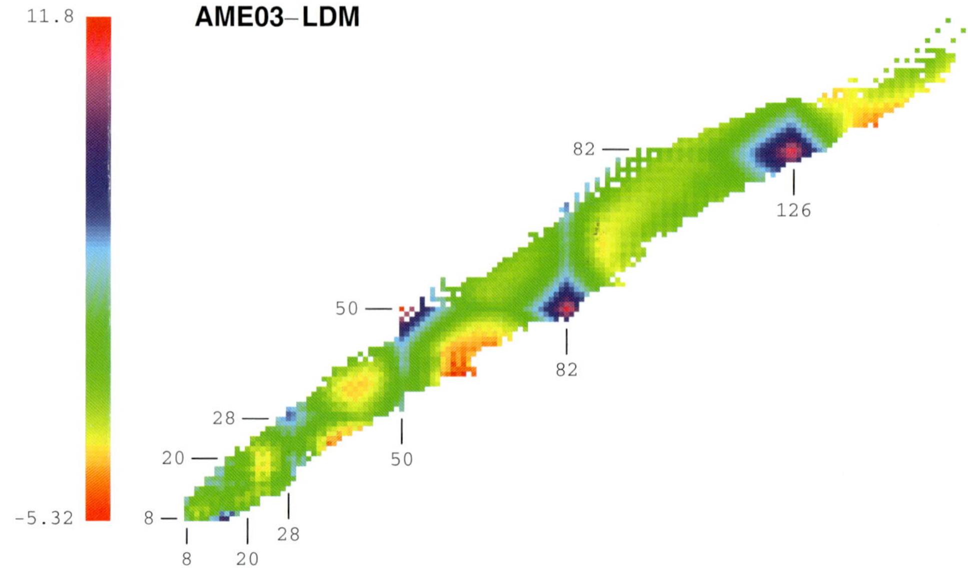

Fig. 8. – Differences between measured and calculated binding energies for nuclei with $N, Z \geq 8$. The binding energies are calculated with the mass formula (41).

positive binding energies, the volume and pairing contributions are positive while others are negative; as a result all a coefficients in formula (41) are positive. Experimental and theoretical progress over the last years has given rise to much more sophisticated mass formulae [74] but the version (41) is sufficient for the present purpose.

In fig. 8 are shown the differences between formula (41) with $r = 1$ and the measured nuclear binding energies taken from the 2003 atomic mass evaluation [75]. Immediately obvious from the figure are the large deviations that occur for doubly magic nuclei such as ^{100}Sn, ^{132}Sn or ^{208}Pb which have a diamond-like appearance. This suggests the use of a term linear in $N_\nu + N_\pi$ with N_ρ the number of valence neutron $(\rho = \nu)$ or proton $(\rho = \pi)$ bosons. Furthermore, the ellipse-like deviations in mid-shell regions suggest another term which is quadratic in $N_\nu + N_\pi$. This simple visual inspection of the deviations thus suggests to add to the liquid-drop mass formula (41) the two-parameter term [76]

$$(42) \qquad B_{\text{shell}}(N, Z) = a_1(N_\nu + N_\pi) + a_2(N_\nu + N_\pi)^2.$$

This prescription is equivalent to counting valence particles or holes from the nearest closed shell and requires pre-defined magic numbers in the nuclear shell model which are here taken to be $N, Z = 8$, 20, 28, 50, 82, 126 and 184. Note that $N_\nu + N_\pi$ coincides with the total number of bosons which was introduced in sect. **4**; the notation N for this number is not used here in order to avoid confusion with neutron number. The corrections (42) can be considered as a basic version of the successful mass formula of Duflo and Zuker [77].

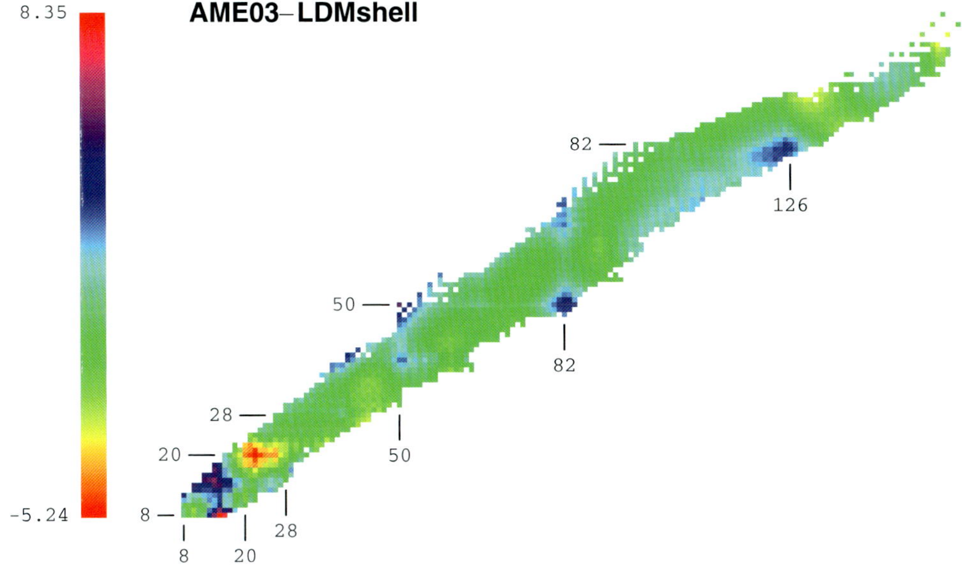

Fig. 9. – Differences between measured and calculated binding energies for nuclei with $N, Z \geq 8$. The binding energies are calculated with the mass formula (41) to which the two-parameter term (42) is added.

The use of these two simple corrections reduces the r.m.s. deviation for more than 2000 nuclear masses from 2.48 to 1.41 MeV while the values of the macroscopic coefficients remain stable [76]. The shell-corrected plot shown in fig. 9 has much reduced deviations for the doubly-magic nuclei and in the mid-shell regions of the heavier nuclei. A large fraction of the remaining r.m.s. deviation of 1.41 MeV is due to nuclei lighter than ^{56}Ni where shell effects are large and cannot be so easily parametrized.

The significant reduction of the r.m.s. deviation with just two terms in the mass formula shows that valence effects are a crucial element in the calculation of nuclear binding energies. This suggests the use of the IBM for mass calculations precisely because it is a valence-nucleon model. Another way of writing the shell correction (42) is in terms of the linear and quadratic Casimir operators of $U(6)$,

$$(43) \qquad B_{\text{shell}}(N, Z) = a_1' \hat{C}_1[U(6)] + a_2' \hat{C}_2[U(6)],$$

with $a_1' = a_1 - 5a_2$ and $a_2' = a_2$.

Given the success of Hamiltonian (25) in describing the spectral properties of separate nuclei and of the combination (43) in reducing deviations from a liquid-drop mass formula for many nuclei, one may attempt a description of both properties simultaneously with the following Hamiltonian:

$$(44) \quad \hat{H}_{1+2}^{\text{full}} = -B(N,Z) + E_0 + \kappa_0 \hat{C}_1[U(6)] + \kappa_0' \hat{C}_2[U(6)] + \kappa_0'' \hat{C}_1[U(5)]\hat{C}_1[U(6)]$$
$$+\kappa_1 \hat{C}_1[U(5)] + \kappa_1' \hat{C}_2[U(5)] + \kappa_2 \hat{C}_2[SU(3)] + \kappa_3 \hat{C}_2[SO(6)]$$
$$+\kappa_4 \hat{C}_2[SO(5)] + \kappa_5 \hat{C}_2[SO(3)].$$

This expression coincides with the full IBM-1 Hamiltonian with up to two-body interactions between the bosons. Note the minus sign in front of $B(N,Z)$ which is needed to convert from positive binding energies to negative absolute energies. The third and fourth terms in eq. (44) are exactly those needed to reduce the r.m.s. deviation in the nuclear-mass calculation. The term in κ_0'' represents the product $(N_\nu + N_\pi)\hat{n}_d$ and allows for a d-boson energy which changes with boson number, a feature which is suggested by microscopic theory [43]. The remaining terms coincide with the IBM-1 Hamiltonian (25). In summary, Hamiltonian (44) contains all terms up to second order, including a contribution from the core inspired by the liquid-drop model and a one-body term ϵ_d which varies linearly with $N_\nu + N_\pi$. All two-body interactions between the bosons are assumed constant throughout the entire shell; only three-body interactions can represent $(N_\nu + N_\pi)$-dependent two-body interactions.

Hamiltonian (44) can be applied to a set of nuclei belonging to a single major shell which, by way of example, is chosen here to be all even-even nuclei with $82 < N < 126$ and $50 < Z < 82$. Semi-magic nuclei are excluded because they are known to exhibit a seniority spectrum which does not allow an interpretation in terms of IBM. Since a simultaneous fit of many nuclei is attempted with spectra that vary from vibrational to rotational, there exists no obvious *ansatz* for the correct parameter set and an efficient fitting procedure is needed. The method followed here is based on the diagonalization of the error matrix which establishes a hierarchy of the most relevant parameter combinations. The approach is identical to that of the determination of shell-model matrix elements in the sd shell [78] and is summarized here for completeness.

Hamiltonian (44) is first written in a simplified notation as

$$\hat{H} = \sum_{i=1}^{P} \kappa_i \hat{O}_i,$$

where κ_i are the P parameters that need to be determined and $\hat{O}_i$ are the P Casimir operators. In the present application the parameters a_i in $B(N,Z)$ have been determined first from a fit to all masses of nuclei with $N, Z \geq 8$. These parameters are kept fixed in the subsequent adjustment of the κ_i to the data set in the shell with $82 < N < 126$ and $50 < Z < 82$. More sophisticated procedures can be envisaged involving iterative or even simultaneous adjustments of a_i and κ_i. If both pieces of the Hamiltonian are treated consistently, it will then probably be possible to absorb the constant E_0 into the liquid-drop expression for $B(N,Z)$.

The parameters κ_i are fitted to a data set consisting of M experimental energies E^k_{expt}, $k = 1, \ldots, M$. In the shell with $82 < N < 126$ and $50 < Z < 82$, the available data set comprises 128 ground-state and 1019 excited-state energies. One of the main difficulties in carrying out the present analysis is the selection of relevant data. As the IBM is a model of collective behaviour of nuclei, only excited states of such character should be included, and this selection is far from obvious in many cases. Nevertheless, a selection of this kind has to be carried out and for each selected level a theoretical counterpart is proposed with an energy

$$\lambda_k \equiv \langle \Phi_k | \hat{H} | \Phi_k \rangle = \sum_{i=1}^{P} \kappa_i \langle \Phi_k | \hat{O}_i | \Phi_k \rangle \equiv \sum_{i=1}^{P} \kappa_i \beta_i^k.$$

The wave functions $|\Phi_k\rangle$ are obtained by diagonalizing $\hat{H}$ for an initial choice of parameters $\{\kappa_i^0\}$ and are iteratively improved in the manner explained below.

The optimal set of parameters is obtained by minimization of the r.m.s. deviation

$$\chi^2 = \sum_{k=1}^{M} \left(\frac{E^k_{\text{expt}} - \lambda_k}{\sigma^k_{\text{expt}}} \right)^2,$$

where σ^k_{expt} is the error on the experimental energy. Minimization with respect to $\{\kappa_i\}$, under the assumption of κ_i-independent matrix elements β_i^k, leads to a set of linear equations of the form

$$\sum_{i=1}^{P} G_{ij}\kappa_i = e_j, \qquad \text{or} \qquad \kappa_i = \sum_{j=1}^{P}(G^{-1})_{ji}e_j,$$

where $\mathbf{G}$ and $\mathbf{e}$ are $P \times P$ and $P \times 1$ matrices, respectively, defined as

$$G_{ij} = \sum_{k=1}^{M} \frac{\beta_i^k \beta_j^k}{\left(\sigma^k_{\text{expt}}\right)^2}, \qquad e_i = \sum_{k=1}^{M} \frac{E^k_{\text{expt}} \beta_i^k}{\left(\sigma^k_{\text{expt}}\right)^2}.$$

The inverse matrix $\mathbf{G}^{-1}$ is known as the error matrix and contains all information on correlations between parameters. In particular, diagonalization of $\mathbf{G}$ (or $\mathbf{G}^{-1}$) yields a hierarchy of parameters. The diagonalization of $\mathbf{G}$ amounts to finding a unitary transformation $\mathbf{A}$ such that $\mathbf{D} = \mathbf{A}\mathbf{G}\mathbf{A}^{\mathrm{T}}$ is diagonal, $D_{ij} = D_i\delta_{ij}$, or, equivalently, $\mathbf{D}^{-1} = \mathbf{A}\mathbf{G}^{-1}\mathbf{A}^{\mathrm{T}}$ with $(D^{-1})_{ij} = d_i\delta_{ij} = (1/D_i)\delta_{ij}$. The transformation $\mathbf{A}$ defines a set of *uncorrelated* parameters $\nu_i = \sum_j A_{ij}\kappa_j$ with associated errors given by d_i. Consequently, the parameter ν_i can be considered as well determined if the corresponding eigenvalue d_i is small; the ordering of d_i in increasing size thus provides a hierarchy of parameters ν_i. This enables one to use the full Hamiltonian (44) with all P Casimir

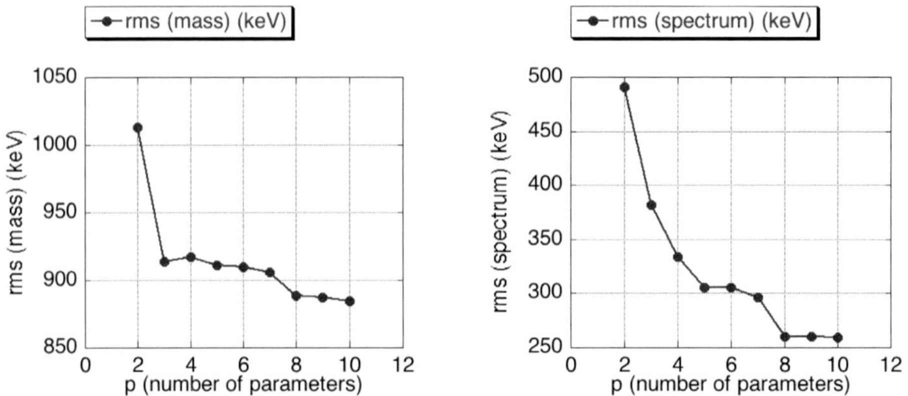

Fig. 10. – The r.m.s. deviation σ in units of keV for masses (left) and for excitation energies (right) as a function of the number of parameters p.

operators but to fit only $p \leq P$ parameter combinations ν_i. For a given number of parameters $p \leq P$ the following fitting procedure can therefore be defined [78]. From an initial choice of parameters $\{\nu_i^0\}$ a subsequent set is defined according to

$$
\nu_i^1 = \begin{cases} \sum_{j=1}^{P} A_{ij}\kappa_j = \sum_{j,j'=1}^{P} A_{ij}(G^{-1})_{j'j}e_{j'}, & \text{if} \quad i \leq p, \\ \nu_i^0, & \text{if} \quad i > p, \end{cases}
$$

where it is assumed that $\mathbf{A}$ is the unitary matrix which diagonalizes $\mathbf{D}^{-1}$ into eigenvalues d_i that are ordered in increasing value. With this set of parameters $\{\nu_i^1\}$ new wave functions $|\Phi_k\rangle$, matrix elements β_i^k, and matrices $\mathbf{G}$ and $\mathbf{e}$ are obtained with which the next set of parameters $\{\nu_i^2\}$ can be calculated, and so on, until convergence is reached.

Two additional points should be mentioned. The first is that, although ultimately one would like to treat ground and excited states on the same footing, this is not done at present. The *absolute* energies of the 128 ground states are fitted while for excited states the fitted quantity is the excitation energy, that is, the energy *relative* to the ground state. The second point is that the use of the experimental error σ_{expt}^k on its own is unsatisfactory since in many cases (*e.g.*, most excitation energies) this error is negligible compared to the r.m.s. deviation σ. The proper way to deal with this issue is to consider instead for each experimental data point the error

$$
\sqrt{\left(\sigma_{\text{expt}}^k\right)^2 + \sigma^2}, \quad \text{with} \quad \sigma^2 = \sum_{k=1}^{M} \left(E_{\text{expt}}^k - \lambda_k\right)^2,
$$

where σ should be determined iteratively. It is clear that the consideration of the experimental error becomes important only when it is larger than or of the same order as

σ. In the present calculation no experimental errors have been taken into account since σ is still relatively large. Strategies for improving the precision of the calculation, as mentioned in the concluding paragraph of this section, might require the consideration of the experimental errors in the future.

Figure 10 shows the r.m.s. deviation σ for masses and for excitation energies as a function of the number of parameters up to $p = 10$. In spite of the sophisticated fitting procedure explained in the preceding discussion, convergence towards the optimal parameter set is not guaranteed. In fact, the final parameters, obtained by gradually increasing p starting from $p = 2$, may depend on the choice of the initial set $\{\kappa_i^0\}$. The r.m.s. deviations shown in fig. 10, $\sigma_{\mathrm{masses}} = 884\,\mathrm{keV}$ and $\sigma_{\mathrm{spectra}} = 259\,\mathrm{keV}$, are those found after a preliminary exploration of the parameter space but they are not necessarily the lowest that can be obtained with the full one- plus two-body IBM-1 Hamiltonian.

To summarize this section, a strategy has been outlined for merging the calculations of ground- and excited-state energies and preliminary results have been presented for even-even nuclei in the major shell with $82 < N < 126$ and $50 < Z < 82$. No definitive results for the one- and two-body parameter space are available yet. An further obvious improvement is to include three-body interactions between the bosons which would allow for boson-number–dependent two-body interactions. The overall purpose of the present approach is that once a reliable parameter set can be determined from known nuclei, it might be of use for the *prediction* of spectral properties of nuclei further removed from the line of stability.

* * *

The interacting boson model has been my favoured research topic for many years. Over those years I had the pleasure to work with many collaborators and it would be difficult to name all of them here. Nevertheless, I would like to reserve a special mention of thanks to F. IACHELLO for his initial and continuing support, and to A. FRANK and J. JOLIE who helped to shape sects. **2** and **3** on symmetries and dynamical symmetries and sect. **4** on the interaction boson model. The work on triaxiality and cubic interactions in the IBM was initiated together with I. STEFANESCU and A. GELBERG. A special thanks goes also to B. SORGUNLU who studied the problem of triaxiality in further detail in many other nuclei and was of invaluable help in compiling the data for carrying out the global calculations reported in sect. **6**. Finally, I would like to thank L. DIEPERINK with whom I developed some of the ideas related to nuclear masses.

REFERENCES

[1] ARIMA A. and IACHELLO F., *Phys. Rev. Lett.*, **35** (1975) 1069.
[2] IACHELLO F. and ARIMA A., *The Interacting Boson Model* (Cambridge University Press, Cambridge) 1987.
[3] WIGNER E. P., *Proceedings of Robert A Welch Foundation Conference on Chemical Research: I. The Structure of the Nucleus* (Welch Foundation, Houston) 1958, p. 88.
[4] GELL-MANN M., *Phys. Rev.*, **125** (1962) 1067.

[5] Okubo S., *Prog. Theor. Phys.*, **27** (1962) 949.

[6] Elliott J. P., *Proc. R. Soc. (London), Ser. A*, **245** (1958) 128; 562.

[7] Wigner E. P., *Phys. Rev.*, **51** (1937) 106.

[8] Ginocchio J. N., *Ann. Phys. (N.Y.)*, **126** (1980) 234.

[9] Hecht K. T., McGrory J. B. and Draayer J. P., *Nucl. Phys. A*, **197** (1972) 369.

[10] Hecht K. T. and Adler A., *Nucl. Phys. A*, **137** (1969) 129.

[11] Arima A., Harvey M. and Shimizu K., *Phys. Lett. B*, **30** (1969) 517.

[12] Chen J.-Q., Chen X.-G., Feng D.-H., Wu C.-L., Ginocchio J. N. and Guidry M. W., *Phys. Rev. C*, **40** (1989) 2844.

[13] Kerman A. K., *Ann. Phys. (N.Y.)*, **12** (1961) 300.

[14] Flowers B. H. and Szpikowski S., *Proc. Phys. Soc.*, **84** (1964) 673.

[15] Arima A. and Iachello F., *Ann. Phys. (N.Y.)*, **99** (1976) 253.

[16] Arima A. and Iachello F., *Ann. Phys. (N.Y.)*, **111** (1978) 201.

[17] Arima A. and Iachello F., *Ann. Phys. (N.Y.)*, **123** (1979) 468.

[18] Cizewski J. A., Casten R. F., Smith G. J., Stelts M. L., Kane W. R., Börner H. G. and Davidson W. F., *Phys. Rev. Lett.*, **40** (1978) 167.

[19] Casten R. F., von Brentano P. and Haque A. M. I., *Phys. Rev. C*, **31** (1985) 1991.

[20] Iachello F. and Scholten O., *Phys. Rev. Lett.*, **43** (1979) 679.

[21] Iachello F. and Van Isacker P., *The Interacting Boson-Fermion Model* (Cambridge University Press, Cambridge) 1991.

[22] Iachello F., *Phys. Rev. Lett.*, **44** (1980) 772.

[23] Van Isacker P., Jolie J., Heyde K. and Frank A., *Phys. Rev. Lett.*, **54** (1985) 653.

[24] Metz A., Jolie J., Graw G., Hertenberger R., Gröger J., Günther C., Warr N. and Eisermann Y., *Phys. Rev. Lett.*, **83** (1999) 1542.

[25] Arima A., Otsuka T., Iachello F. and Talmi I., *Phys. Lett. B*, **66** (1977) 205.

[26] Otsuka T., Arima A., Iachello F. and Talmi I., *Phys. Lett. B*, **76** (1978) 139.

[27] Iachello F., *Phys. Rev. Lett.*, **53** (1984) 1427.

[28] Bohle D., Richter A., Steffen W., Dieperink A. E. L., Lo Iudice N., Palumbo F. and Scholten O., *Phys. Lett. B*, **137** (1984) 27.

[29] Richter A., *Prog. Part. Nucl. Phys.*, **34** (1995) 261.

[30] Faessler A., *Nucl. Phys. A*, **85** (1966) 653.

[31] Lo Iudice N. and Palumbo F., *Phys. Rev. Lett.*, **53** (1978) 1532.

[32] Pietralla N. *et al.*, *Phys. Rev. Lett.*, **83** (1999) 1303.

[33] Castaños O., Chacón E., Frank A. and Moshinsky M., *J. Math. Phys.*, **20** (1979) 35.

[34] Shirokov A. M., Smirnova N. A. and Smirnov Yu. F., *Phys. Lett. B*, **434** (1998) 237.

[35] Scholten O., Iachello F. and Arima A., *Ann. Phys. (N.Y.)*, **115** (1978) 324.

[36] Casten R. F. and Cizewski J. A., *Nucl. Phys. A*, **309** (1978) 477.

[37] Stachel J., Van Isacker P. and Heyde K., *Phys. Rev. C*, **25** (1982) 650.

[38] Alhassid Y. and Leviatan A., *J. Phys. A*, **25** (1992) L1265.

[39] Leviatan A., *Phys. Rev. Lett.*, **77** (1996) 818.

[40] Leviatan A., Novoselski A. and Talmi I., *Phys. Lett. B*, **172** (1986) 144.

[41] Van Isacker P., *Phys. Rev. Lett.*, **83** (1999) 4269.

[42] Leviatan A. and Van Isacker P., *Phys. Rev. Lett.*, **89** (2002) 222501.

[43] Otsuka T., Arima A. and Iachello F., *Nucl. Phys. A*, **309** (1978) 1.

[44] Thompson M. J., Elliott J. P. and Evans J. A., *Phys. Lett. B*, **195** (1987) 511.

[45] Juillet O., Van Isacker P. and Warner D. D., *Phys. Rev. C*, **63** (2001) 054312.

[46] Talmi I., *Interacting Bose-Fermi Systems in Nuclei*, edited by Iachello F. (Plenum Press, New York) 1981, p. 329.

[47] Scholten O. and Dieperink A. E. L., *Interacting Bose-Fermi Systems in Nuclei*, edited by Iachello F. (Plenum Press, New York) 1981, p. 343.

[48] VITTURI A., *Interacting Bose-Fermi Systems in Nuclei*, edited by IACHELLO F. (Plenum Press, New York) 1981, p. 355.

[49] OTSUKA T., YOSHIDA N., VAN ISACKER P., ARIMA A. and SCHOLTEN O., *Phys. Rev. C*, **35** (1987) 328.

[50] KLEIN A. and CHEN J.-Q., *Phys. Rev. C*, **47** (1993) 612.

[51] GINOCCHIO J. N. and KIRSON M. W., *Phys. Rev. Lett.*, **44** (1980) 1744.

[52] DIEPERINK A. E. L., SCHOLTEN S. and IACHELLO F., *Phys. Rev. Lett.*, **44** (1980) 1747.

[53] BOHR A. and MOTTELSON B. R., *Phys. Scripta*, **22** (1980) 468.

[54] BOHR A. and MOTTELSON B. R., *Nuclear Structure. II Nuclear Deformations* (Benjamin, Reading, Massachusetts) 1975.

[55] VAN ISACKER P. and CHEN J.-Q., *Phys. Rev. C*, **24** (1981) 684.

[56] GILMORE R., *Catastrophe Theory for Scientists and Engineers* (Wiley, New York) 1981.

[57] LÓPEZ-MORENO E. and CASTAÑOS O., *Phys. Rev. C*, **54** (1996) 2374

[58] JOLIE J., CASTEN R. F., VON BRENTANO P. and WERNER V., *Phys. Rev. Lett.*, **87** (2001) 162501

[59] IACHELLO F., ZAMFIR N. V. and CASTEN R. F., *Phys. Rev. Lett.*, **81** (1998) 1191.

[60] STEFANESCU I. *et al.*, *Nucl. Phys. A*, **789** (2007) 125.

[61] VAN ISACKER P. and SORGUNLU B., submitted to *Nucl. Phys. A*.

[62] WARNER D. D. and CASTEN R. F., *Phys. Rev. Lett.*, **48** (1982) 1385.

[63] LIPAS P. O., TOIVONEN P. and WARNER D. D., *Phys. Lett. B*, **155** (1985) 295.

[64] PAN X.-W., OTSUKA T., CHEN J.-Q. and ARIMA A., *Phys. Lett. B*, **287** (1992) 1.

[65] DAVYDOV A. S. and FILIPPOV G. F., *Nucl. Phys.*, **8** (1958) 237.

[66] HEYDE K., VAN ISACKER P., WAROQUIER M. and MOREAU J., *Phys. Rev. C*, **29** (1984) 1420.

[67] CASTEN R. F., VON BRENTANO P., HEYDE K., VAN ISACKER P. and JOLIE J., *Nucl. Phys. A*, **439** (1985) 289.

[68] ZAMFIR N. V. and CASTEN R. F., *Phys. Lett. B*, **260** (1991) 265.

[69] CHE X.-L. *et al.*, *China Phys. Lett.*, **21** (2003) 1904.

[70] HUA H. *et al.*, *China Phys. Lett.*, **20** (2003) 350.

[71] GARCÍA-RAMOS J. E., DE COSTER C., FOSSION R. and HEYDE K., *Nucl. Phys. A*, **688** (2001) 735.

[72] FOSSION R., DE COSTER C., GARCÌA-RAMOS J. E., WERNER T. and HEYDE K., *Nucl. Phys. A*, **697** (2002) 703.

[73] DAVIS E. D., DIALLO A. F., BARRETT B. R. and BALANTEKIN A. B., *Phys. Rev. C*, **44** (1991) 1655.

[74] MÖLLER P. and NIX R., *Nucl. Phys. A*, **536** (1992) 20.

[75] AUDI G., WAPSTRA A. H. and THIBAULT C., *Nucl. Phys. A*, **729** (2003) 337.

[76] DIEPERINK A. E. L. and VAN ISACKER P., *Eur. Phys. J. A*, **32** (2007) 11.

[77] DUFLO J. and ZUKER A. P., *Phys. Rev. C*, **52** (1995) R23.

[78] BROWN B. A. and RICHTER W. A., *Phys. Rev. C*, **74** (2006) 034315.

DOI 10.3254/978-1-58603-885-4-385

The Interacting Boson Approximation model

R. F. CASTEN

A.W. Wright Nuclear Structure Laboratory, Yale University - New Haven, CT, 06520, USA

Summary. — The Interacting Boson Approximation (IBA) model is discussed in the context of microscopic and macroscopic approaches to nuclear collectivity. The Hamiltonian, group theoretical structure, dynamical symmetries, the Consistent Q Formalism (CQF and ECQF), the technique of Orthogonal Crossing Contours (OCC), and practical calculations with the model are discussed.

1. – Introduction

One of the most pervasive features of atomic nuclei is the manifestation of collectivity, and correlations in the behavior of the constituent nucleons, which occurs in a stunning variety of forms. Large amplitude examples are giant resonance modes of excitation and the process of fission. Small amplitude collective motions, that occur at low energy, and have been extensively studied, show up as, for example, vibrations of spherical nuclei, and in rotational and vibrational modes in deformed nuclei.

The study of nuclei, and our understanding of them, can be looked at from two complementary perspectives, which can (slightly inaccurately) be labeled the microscopic and macroscopic. In the former, the emphasis is on the motions of the nucleons, and their interactions. The degrees of freedom are those of the nucleons themselves. The latter views the nucleus from a conceptual "distance" of a few fermis, focusing on the overall shape of the nucleus and oscillations in that shape. Here the relevant degrees of freedom

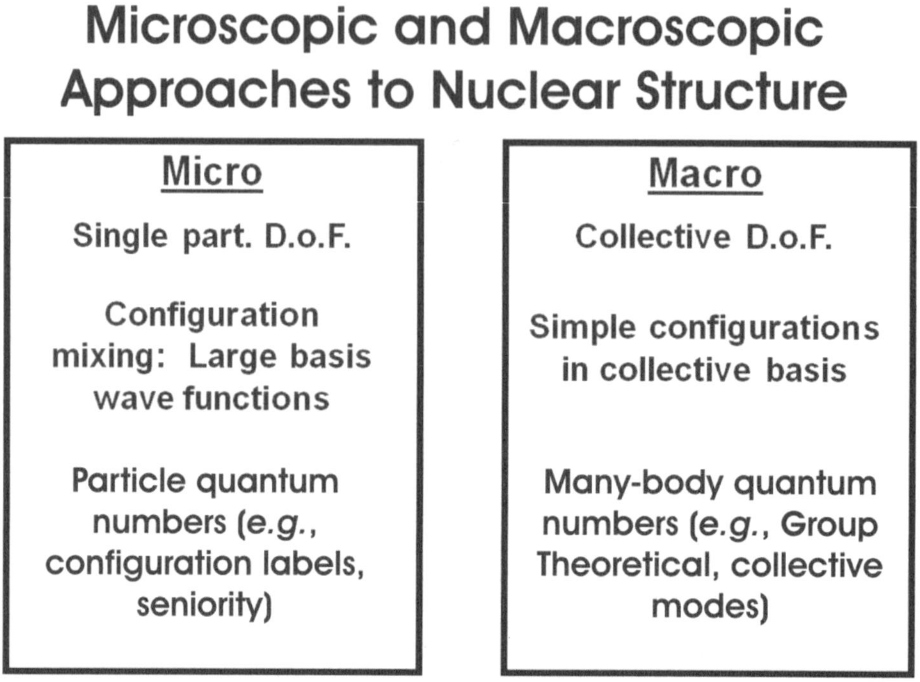

Microscopic and Macroscopic Approaches to Nuclear Structure

<table>
<tr><td>

<u>Micro</u>

Single part. D.o.F.

Configuration mixing: Large basis wave functions

Particle quantum numbers (*e.g.,* configuration labels, seniority)

</td><td>

<u>Macro</u>

Collective D.o.F.

Simple configurations in collective basis

Many-body quantum numbers (*e.g.,* Group Theoretical, collective modes)

</td></tr>
</table>

Fig. 1. – Complementarity of nucleonic and many-body approaches to nuclear structure.

are those of the many-body system taken as a whole and descriptions of the structure are couched in terms of shape variables and symmetries. These ideas are sketched in fig. 1.

Of course, the two perspectives must be consistent and much work was done in the early 1960's, with techniques such as the Random Phase Approximation (RPA), to deduce and predict collective structures from a microscopic perspective. That said, both perspectives have an independent value. For example, an excellent shell model description of a heavy nucleus with many valence nucleons may accurately reproduce a number of observables but it is unlikely to easily give an insight into the overall structure: it might reproduce, for example, the yrast energies, but would not see that these are close to those of a quantum-mechanical symmetric spinning top. Conversely, a description in terms of collective motion might give an excellent understanding of the low-lying excitations in a simple picture but would not inform as to the key nucleonic orbits and configurations underlying such motions. Thus both perspectives are needed and complementary.

These lectures will focus on the collective approach. There have been many successful models of collective behavior. These range from extreme idealizations such as the perfect, axial rotor or harmonic vibrational motion, to complex, multi-parameter treatments that account for fine details in the data. However, both approaches are somewhat unsatisfactory—the former because few nuclei exhibit the idealized behavior envisioned

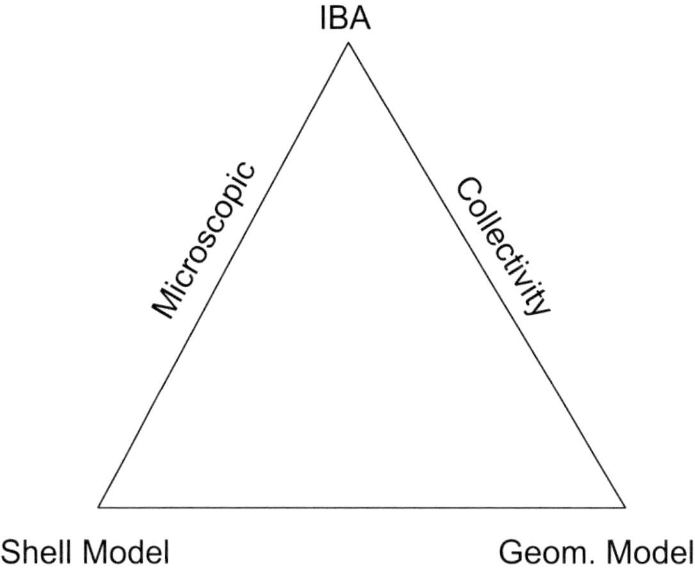

Fig. 2. – Illustration of the links between the IBA and the shell and geometrical models.

(partly because actual nuclei are finite systems which limits the degree to which they can be modeled by constructs with smooth shapes), and the latter because multi-parameter fitting is risky and complex at best with no guarantees of a unique solution.

Needed and desirable would be a model that is based on simple ideas of collective shapes and excitation modes, but flexible enough to incorporate a wide variety of structures, and relying on a minimum of parameters. It would be an additional advantage if it were to be grounded ultimately in a microscopic foundation.

Such a model exists—called the Interacting Boson Approximation (IBA) model [1,2]—and it is the subject of these lectures. Before continuing, there is one caveat that is important to understand, regarding the so-called "predictive power" of models. This is often misunderstood.

Microscopic models have a certain predictive power in the sense that one chooses a set of single-particle energies and two-body interactions, and then diagonalizes a Hamiltonian matrix to get energies and other observables. Once the initial choices are made, predictions for many nuclei can be obtained and the model predicts the relationships of observables as a function of N and Z. This, however, is limited in two ways: a "mean field" (one-body single particle) potential is assumed and no account is taken of sub-nucleonic degrees of freedom, and, secondly, in practice the single-particle energies are themselves N and Z dependent (for example, due to the monopole proton-neutron interaction [3]) such that even the magic numbers themselves may change as has long been known [3-5] and recently highlighted [6] far from stability. Likewise, the interaction matrix elements are not constant. Therefore, calculations even for neighboring nuclei must be parameterized. Nevertheless, the microscopic basis provides a degree of predictive power. Macroscopic or phenomenological models, on the other hand, are not predictive in this sense. Generally, one needs a couple of known observables in a given nucleus to pin down the basic structure. Then such models, like the IBA, provide a wealth of predictions of other observables that can be tested. That is not to say that the predictions do not change in generally smooth and somewhat predictable ways, but, in principle, such models are fit to each nucleus.

One last introductory comment that will be important in understanding the IBA model: shell structure is dominated by the concept of magic numbers and shell gaps such that, to good approximation, it is often sufficient to consider only the valence nucleons, those outside the nearest closed shells. This tremendously simplifies microscopic calculations and provides a key ingredient in the IBA which is specifically a valence space model. The IBA model therefore has foundations in the shell model but predicts collective behavior. It therefore "looks" in both directions—toward microscopy and towards macroscopic collectivity. This "Janus"-like feature is schematically illustrated in fig. 2, and will be discussed below.

2. – Foundations of the IBA

To understand the IBA and its motivation, it is useful to start from a microscopic perspective. Consider two identical valence nucleons in an $h_{9/2}$ orbit. They can couple their angular momenta to 0,2,4,6,8. Thus there is only a single 2^+ state. Now consider two identical nucleons that can occupy freely either an $h_{9/2}$ orbit or an $f_{7/2}$ orbit. How many 2^+ states can be made subject to the Pauli Principle? There are 3 such states comprised as follows using standard shell model configuration notation: $(h_{9/2})^2$, $(f_{7/2})^2$, $(h_{9/2}, f_{7/2})$. As a third example, consider four identical nucleons that can occupy these same orbits. Now one can form 24 2^+ states as for example: $|(h_{9/2})^2 j = 2, (f_{7/2})^2 j = 0; \ J = 2\rangle$, $|(h_{9/2})^2 j = 0, (f_{7/2})^2 j = 2; \ J = 2\rangle$, $|(h_{9/2})^2 j = 2, (f_{7/2})^2 j = 2; \ J = 2\rangle$, $|(h_{9/2})^2 j = 2, (f_{7/2})^2 j = 4; \ J = 2\rangle, \ldots$

Clearly, comparing these examples, it should be clear that, in nuclei far from shell closures, a huge number of states can be made. One well-known example of this is ^{154}Sm, which has 12 valence protons in the 50-82 shell and 10 valence neutrons in the 82-126 shell. With these available orbits, and subject to Pauli principle limitations, it is nevertheless possible to make $\sim 3 \times 10^{14}$ 2^+ states! Clearly this is an impossible problem to tackle by diagonalization! Even with advances in computing, if it became feasible, it would be virtually impossible to visually see from the wave functions (each of which would have $\sim 3 \times 10^{14}$ components) any simplicities and, in particular to see that this nucleus is an excellent example of an axially symmetric deformed rotor. Thus, the call for models that somehow simplify the problem.

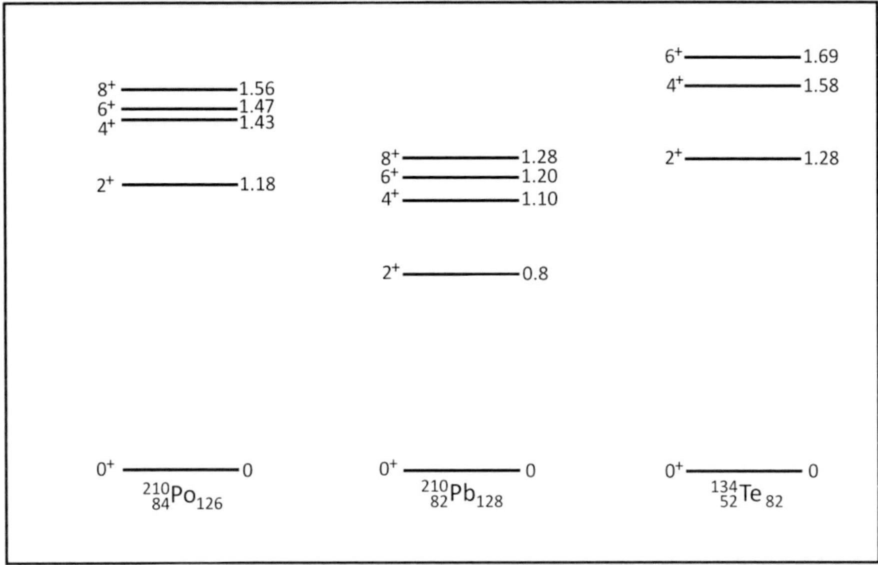

Fig. 3. – Yrast spectra of nuclei with two nucleons outside a doubly-closed shell. Energies are in MeV.

There are various possible approaches to this. One is to truncate the space, either by eliminating certain orbits, or limiting allowed seniority values or other techniques. One then has to assess if such truncations keep the main part of the physics or throw away too much. Random truncations surely will. Therefore, some physics insight into the problem should dictate the truncation.

Though the IBA is a collective model it is founded microscopically precisely in such a truncation, and a monumental one at that. In the case of ^{154}Sm, for example, the IBA ansatz results in a space of only 26 2^+ states! That such a truncation is able to select out the most important physical collective properties of the nucleus (at least up to the pairing gap at about 2 MeV) is incredible to imagine and was audacious to suggest. Nevertheless, as we shall see, the model is extremely successful, probably the most successful collective model there is.

Therefore it is worthwhile beginning our study of the IBA by discussing the shell model truncation scheme it embodies. To do so, it is useful to look at empirical spectra of certain simple nuclei, namely those with two identical nucleons outside closed shells. Examples of these are shown in fig. 3.

These spectra all have a characteristic behavior that results from any short range attractive residual interaction. The spectra always have a 0^+ ground state, followed by a large spacing to the other states, which appear in order of increasing spin, with successively smaller spacings up to the maximum spin $(2j - 1)$. Note the important point that the ratio of the energies of the 4^+ and 2^+ states, $R_{4/2} = E(4_1^+)/E(2_1^+)$, is < 2. For all known *collective* motions, $R_{4/2} \geq 2.0$ so this is an extremely useful, simple, first signature of structure for any even-even nucleus. It does not determine the structure unambiguously, as we shall see, but it gives the first indication and determines the "class" of nucleus at hand.

The immediate relevance of this to the IBA is that the lowest two states are 0^+ and 2^+. The coupling of two identical nucleons in the same orbit under the influence of a short-range attractive interaction always favors these two angular momenta. In fact, this is true for a much broader class of interactions as well, such as a quadrupole force.

Inspired by this, and the concept of Cooper pairs from condensed matter systems, Iachello and Arima proposed that a suitable truncation of the shell model to accommodate low-lying collective states was to consider only configurations consisting of pairs of valence nucleons whose spins are coupled to angular momenta 0 or 2. Pairs of fermions act like bosons and these two configurations are called s bosons and d bosons, respectively. We cannot stress enough how utterly drastic this truncation of the shell model space is. That it works at all is remarkable. That it works as well as it does is astonishing. Of course, it cannot account for all the features, even of collective nuclei, and various extensions of the model have been discussed over the years, such as including g bosons (corresponding to pairs of nucleons coupled to spin 4). But these are generally needed either to fine tune predictions or to account for specific states in specific nuclei. We will not consider such extensions further here but there is an extensive literature on them. (For the foundations of the model, see refs. [1, 2]. For a valuable resource on the IBA literature, see ref. [7]. For an early review of the model, see ref. [8]; for a discussion on which much of the present treatment is based, see ref. [9]. Further references are given for specific situations below.)

The next step in developing the IBA is to decide on the Hamiltonian. The IBA model uses an extremely simple form. First, however, we need to decide how many bosons a given nucleus has. The IBA is a valence space model and the bosons are each considered as a pair of fermions. Therefore the total number of bosons, N_B, that is, s bosons plus d bosons, is simply half the number of valence protons plus half the number of valence neutrons, each counted to the *nearest* closed shell. In the so-called IBA-1 model, where no distinction is made between proton and neutron degrees of freedom, one simply takes half the total number of valence nucleons. A few examples illustrate the counting procedure. Thus ^{154}Sm has 11 bosons while ^{172}Hf has 14 and so does ^{180}Hf, and ^{196}Pt has 6.

Of course, one can imagine that ignoring the distinction between proton and neutron degrees of freedom may sometimes be inadequate. While most of the low-lying states in even-even nuclei are proton-neutron symmetric to a good approximation, there is also a wealth of so-called "mixed-symmetry states" occurring higher in energy where the proton-neutron degrees of freedom need to be taken into account. This has led to an extension of the original IBA model to the IBA-2 model. Though containing quite a few more parameters, and though not used nearly as much as the IBA-1, it is often essential and is a valuable complement to the IBA-1. There are also models, called IBA-3 and IBA-4, useful especially for light nuclei, that take isospin into account, and likewise models for odd-A and odd-odd nuclei, called the IBFM and IBFFM, where "F" stands for fermion, in which either one or two fermions are coupled to a boson core. Again, these models have a significantly greater number of parameters and have had limited use but can sometimes be very valuable. We will here restrict our attention to the original and simple IBA-1.

3. – The IBA Hamiltonian and group theoretical concepts

We are now in a position to discuss the IBA Hamiltonian. It will be written in terms of creation and destruction operators for s and d bosons. For those readers unfamiliar with the use of second quantized operators, we summarize here the simple rules for them in terms of an operator b that creates or destroys an excitation of type b.

$$\tilde{b}|n_b\rangle = \sqrt{n_b}|n_b - 1\rangle, \tag{1}$$

$$b^\dagger|n_b\rangle = \sqrt{(n_b + 1)}|n_b + 1\rangle, \tag{2}$$

$$b^\dagger \tilde{b}|n_b\rangle = b^\dagger \sqrt{n_b}|n_b - 1\rangle = \sqrt{n_b}\sqrt{(n_b - 1) + 1}|n_b\rangle = n_b|n_b\rangle. \tag{3}$$

Note that the operator combination $b^\dagger \tilde{b}$ gives back the initial wave function and therefore both conserves and counts the number of b-phonons.

As noted, the bosons of the IBA can be of either s or d character and, in fact, different states in a given nucleus will be characterized by different numbers and angular momentum couplings of s and d bosons, as we shall see. Of course, since s and d bosons correspond to different fermion angular momentum couplings they should occur at quite different energies (see fig. 3). In the language of second quantization, the operators $s^\dagger s$ and $(d^\dagger \tilde{d})^0$ give, respectively, the number of s and d bosons, n_s and n_d. Thus we can write a Hamiltonian for the IBA as

$$\begin{aligned} H_{\text{IBA}} &= H_s + H_d + H_{\text{int}} \\ &= \epsilon_s n_s + \epsilon_d n_d + H_{\text{int}}(\text{with terms containing up to four operators}). \end{aligned} \tag{4}$$

Here, ϵ_s and ϵ_d are the energies of an s boson and a d boson, respectively. Hence, the first two terms just give the total energies of the s bosons and d bosons, respectively, and H_{int} gives the interactions between the bosons. Since the number of bosons is fixed for a given nucleus, the operators allowed in eq. (4) always occur in the bi-linear combinations $s^\dagger s$, $d^\dagger \tilde{d}$, $s^\dagger \tilde{d}$, and $d^\dagger s$. Since, in this paper we will only consider excitation energies, we have an arbitrary overall energy zero and we can therefore set $\epsilon_s = 0$.

The most general form of this Hamiltonian with up to four operators in each term is then given by

$$\begin{aligned} H = \epsilon_d n_d &+ \frac{1}{2}\sum_J C_J \left(d^\dagger d^\dagger\right)^{(J)} \cdot \left(\tilde{d}\tilde{d}\right)^{(J)} \\ &+ \frac{v_2}{\sqrt{10}}\left[(d^\dagger d^\dagger)^{(2)} \cdot \tilde{d}s + \text{h.c.}\right] + \frac{v_0}{\sqrt{5}}\left(d^{\dagger 2}s^2 + \text{h.c.}\right), \end{aligned} \tag{5}$$

where h.c. stands for the Hermitian conjugate.

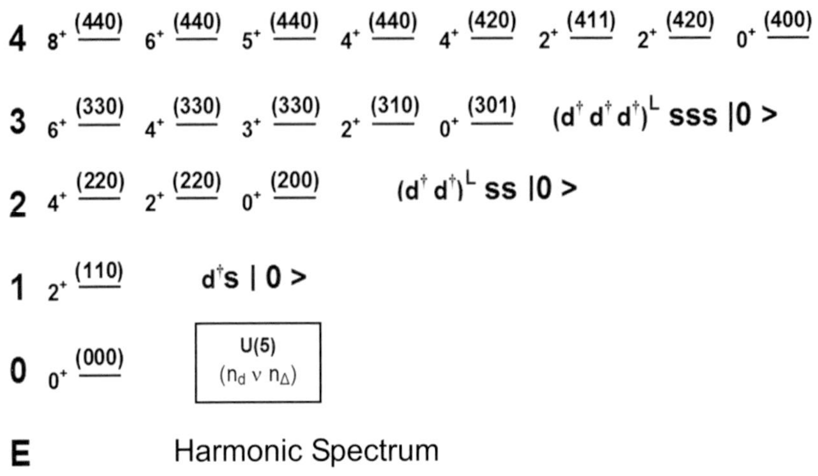

4 8^+ (440) 6^+ (440) 5^+ (440) 4^+ (440) 4^+ (420) 2^+ (411) 2^+ (420) 0^+ (400)

3 6^+ (330) 4^+ (330) 3^+ (330) 2^+ (310) 0^+ (301) $(d^\dagger d^\dagger d^\dagger)^L$ sss $|0>$

2 4^+ (220) 2^+ (220) 0^+ (200) $(d^\dagger d^\dagger)^L$ ss $|0>$

1 2^+ (110) $d^\dagger$s $|0>$

0 0^+ (000) $\boxed{\begin{array}{c} U(5) \\ (n_d \vee n_\Delta) \end{array}}$

E   Harmonic Spectrum

Fig. 4. – Spectrum of low-lying levels in $U(5)$ for the case of a harmonic vibrator with degenerate multiplets. The quantum number labels are the number of d-bosons, the numbers that are not coupled in pairs or triplets to $J = 0$, and the number of triplets of d-bosons coupled to $J = 0$.

It is useful to already take a quick look at the structure of this Hamiltonian. As noted, the first term gives the total energy of all the d bosons. If we ignore for now the subsequent terms, two conclusions are immediately clear: first, the lowest energy state will have the minimum number of d bosons, namely, 0 and $n_s = N_B$, where N_B is the total boson number. Thus the ground state must have spin 0^+. The first excited state—the lowest possible excitation given by the first term in eq. (5)—will have one d boson (and hence N_B - 1 s bosons) and occur at an excitation energy of ϵ_d. It is created from the ground state, $|n_s = N_B, n_d = 0\rangle$, by the operator $d^\dagger s$ such that $d^\dagger s|N_B, 0\rangle = \sqrt{N_B}|N_B - 1, 1\rangle$. The second excited "state" will correspond to a 2-d boson configuration. But 2 spin-2 bosons can couple to total spins 0^+, 2^+ and 4^+. These states will be degenerate at an energy of 2 ϵ_d. The next states will be a degenerate quintuplet with spins 0^+, 2^+, 3^+, 4^+ and 6^+ at $E = 3$ ϵ_d, and so on. Figure 4 illustrates this spectrum and it should be instantly obvious that it corresponds to the geometrical model of a spherical harmonic quadrupole phonon vibrator.

The second conclusion from eq. (5) is that the s and d boson numbers are good quantum numbers for a Hamiltonian given by the first term in eq. (5). We will return to this simple limiting case shortly. For now, let us briefly look at the effects of the other terms in eq. (5). The second term also conserves the number of d-bosons, therefore n_s and n_d remain good quantum numbers. The following terms conserve N_B (as they must) but do not conserve n_s or n_d individually. Note that the states given by the first term of eq. (5) are a complete orthonormal basis for any IBA-1 Hamiltonian. Therefore, the eigenstates of the full Hamiltonian in eq. (5) can be expanded in the basis provided by the first term. In this case, if the last two terms have non-zero coefficients, each wave function in general will consist of a mixture (a linear combination) of these basis states. That is, n_s and n_d will no longer be good quantum numbers. We will return to this point.

The Hamiltonian of eq. (5) was indeed the original Hamiltonian but, in general, does not admit of easy physical intuition. Therefore a completely equivalent alternate version was almost immediately introduced. Called the "multipole" form of the IBA Hamiltonian it is nothing more than a convenient rearrangement of terms. It is given by

$$(6) \qquad H = \epsilon n_d + a_0 P^\dagger P + a_1 J^2 + \kappa Q^2 + a_3 T_3^2 + a_4 T_4^2,$$

where

$$(7) \qquad Q = \left(d^\dagger s + s^\dagger \tilde{d}\right) - \frac{\sqrt{7}}{2} \left(d^\dagger \tilde{d}\right)^{(2)},$$

$$(8) \qquad P = \frac{1}{2} \left(\tilde{d}^{(2)} - s^2\right),$$

$$(9) \qquad T_J = \left(d^\dagger \tilde{d}\right)^{(J)}, \quad J = 0, 1, 2, 3, 4,$$

$$n_d = \sqrt{5} T_0, \quad J = \sqrt{10} T_1.$$

Here, ϵ has the same meaning as ϵ_d before (where, since $\epsilon_s = 0$, we have simply dropped the subscript to simplify the notation), and the other terms are defined above. In a qualitative way the term in $P^\dagger P$ is like a kind of pairing interaction between bosons, while the operator Q is the most general operator up to second order in boson operators that has angular momentum 2. Thus $Q \cdot Q$ has the form of a quadrupole interaction between bosons. The term in J^2 gives contributions to rotational energies—its eigenvalues are proportional to $J(J+1)$—but is seldom used except for the most extreme fine tuning of calculations. The other terms are various combinations of the elementary s and d operators with less obvious physical sense.

Before proceeding further, we need to take a brief excursion into group theory. The IBA has deep roots in group theory and at least a rudimentary understanding of group theory greatly facilitates understanding the model. Here, however, we will skirt this issue for the most part and only introduce some basic terminology and a few concepts that will provide a convenient framework later on.

Imagine a set of operators (such as the bilinear combinations of s and d operators). If they close on commutation, they are said to be generators of a group. The group name is characteristic of the number and character of these operators. Taking account of magnetic substates, there are a total of 36 bilinear IBA operators ($s^\dagger s$ (1 of them), $s^\dagger \tilde{d}$ (5), $d^\dagger s$ (5), and $d^\dagger \tilde{d}$ (25)). These are the generators of the group $U(6)$. An absolutely essential concept is that a set of generators defines and conserves a quantum number. In our example, some of these 36 operators do not, for example, conserve n_d: $s^\dagger \tilde{d}$ is an example. However, clearly, all of them conserve the total boson number N_B.

Groups can have subgroups. In the IBA, the 25 operators of the form $d^\dagger \tilde{d}$ are the generators of the group $U(5)$. Here, the quantum number defined by this subset of operators, which is also conserved by them, is clearly the d-boson number. Other subsets of operators will define further quantum numbers.

An operator that commutes with all the generators of a group is called a Casimir operator of that group. For example, the operator $N_B = n_s + n_d$ is a Casimir operator of the group $U(6)$. Owing to the commutation relations, that operator must conserve the characteristic quantum number of the group—in our case, N_B.

The impact of this is the following: Suppose some Hamiltonian can be written so that successive terms are written solely in terms of Casimir operators of a group and its subgroups. Then each term has a characteristic quantum number. Moreover, that quantum number is a good quantum number for each state which is an eigenvalue of that Hamiltonian. An outcome of the group theory is that the eigenvalues of the system can be written *analytically* in terms of those quantum numbers. This is an extremely powerful concept because it implies that in such a Hamiltonian the excitation energies, as well as the wave functions and all transition rates, are completely analytic! Even though the most general IBA Hamiltonian is not written in terms of a specific set of Casimir operators of a group and its sub-groups, these ideas are particularly valuable because there do exist three independent ways of writing the IBA Hamiltonian in terms of such "group chains". These three ways correspond to group structures labeled $U(5)$, $SU(3)$ and $O(6)$. These will be discussed extensively below. Here our purpose is to point out the existence of these three analytic solutions. They are called *dynamical symmetries*.

There is one other important consequence of this discussion. Consider a group and its subgroups (the detailed physics of the groups is not important until later):

$$U(6) \supset U(5) \supset O(3),$$

where the characteristic quantum number for $O(3)$ is the total angular momentum J. Since each group has a characteristic quantum number and since the energies of states corresponding to that group are given (analytically) in terms of that quantum number alone, it follows, in this case, that the energies of the states corresponding to ("spanning") $U(6)$—though they will have a variety of values of the quantum numbers n_d and J—will not depend on n_d or J, but only on N_B. If the Hamiltonian has only the Casimir operator of $U(6)$, then all states of a given N_B, that is, all the states for a given nucleus, will be degenerate! This is obviously not a very useful situation. However, if the Hamiltonian now consists of Casimir operators of $U(6)$ and $U(5)$, the energies will depend on both N_B and n_d. That is, the degeneracies of $U(6)$ will now be broken. The magnitude of the degeneracy breaking will be proportional to the coefficient in front of the $U(5)$ Casimir operator in the Hamiltonian. Still, of course, a state with given n_d can have several angular momenta: for example, for $n_d = 2$, angular momenta 0^+, 2^+, and 4^+ are allowed (see fig. 4). However, those states will be degenerate. If we now add in to the Hamiltonian the Casimir operator of $O(3)$, the states will be labeled by J and their energies will now depend on J as well (in this case, according to $J(J+1)$). Thus

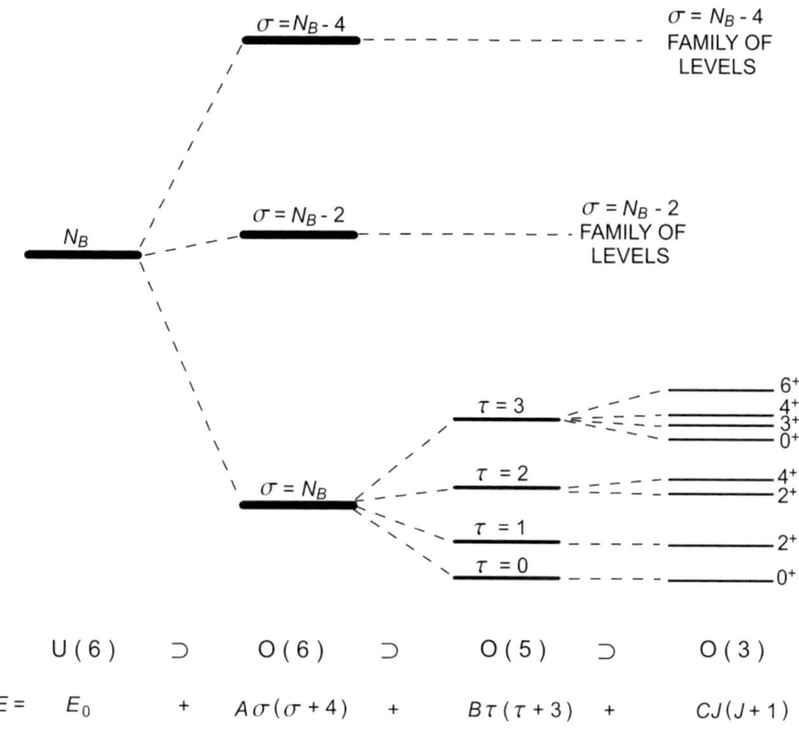

Fig. 5. – Illustration of the quantum number labeling of states in successive sub-groups of $U(6)$ and of the degeneracy breaking in such a group chain. Based on ref. [9].

we have the important results that, in a group chain, each step defines a new quantum number(s), and each step breaks a previous degeneracy such that the eigenvalues now depend in some analytic form on the values of that new quantum number.

This process is illustrated schematically for the $O(6)$ group chain in fig. 5. Here the parent group is $U(6)$ in which the representations are labeled by N_B, the total boson number. Thus each representation corresponds to an entire nucleus, in which all the levels are degenerate. Different representations correspond to different nuclei. The spectrum of $U(6)$ would be useful, for example, in a study of the relative masses of different nuclei. However, it offers little for the study of low-lying excitations of nuclei. Hence the next step, called $O(6)$, labels and splits the levels according to the quantum number σ. The next step involves the group $O(5)$ and a quantum number τ, followed by the group $O(3)$, with representations defined by the total angular momentum quantum number, J. These two sub-groups together break the degeneracy of states of a given σ.

With this background we now consider each of the three dynamical symmetries of the IBA, namely $U(5)$, $SU(3)$, and $O(6)$, with group chains:

$$\text{I.} \quad \begin{array}{ccccccc} U(6) & \supset & U(5) & \supset & O(5) & \supset & O(3) \\ N & & n_d & & \nu & & n_\Delta J \end{array} \quad U(5), \tag{10}$$

$$\text{II.} \quad \begin{array}{ccccc} U(6) & \supset & SU(3) & \supset & O(3) \\ N & & (\lambda, \mu) & & K J \end{array} \quad SU(3), \tag{11}$$

$$\text{III.} \quad \begin{array}{ccccccc} U(6) & \supset & O(6) & \supset & O(5) & \supset & O(3) \\ N & & \sigma & & \tau & & \nu_\Delta J \end{array} \quad O(6), \tag{12}$$

where the characteristic quantum numbers for each sub-group are indicated.

4. – Discussion of the IBA and its predictions

We first discuss these three dynamical symmetries and their characteristic properties and then go on to discuss the nature of the IBA for more general Hamiltonians that give structures intermediate between these symmetries. We start with the simplest case, $U(5)$.

$4\dot{}1.$ $U(5)$. – (See ref. [10].) The $U(5)$ symmetry is, in some ways, the simplest and we start with that. The simplest imaginable IBA Hamiltonian is

$$H = \epsilon_s n_s + \epsilon_d n_d, \tag{13}$$

where the coefficients of n_s and n_d are the energies of each boson type. As before, we are only interested in excitation energies, that is, relative energies, so we can set one of these to zero. Since pairs of nucleons favor coupling to angular momentum zero, it is natural to set $\epsilon_s = 0$. Then

$$H = \epsilon n_d, \tag{14}$$

where we have again dropped the unneeded subscript on the d-boson energy.

Clearly, this Hamiltonian gives a spectrum of energies proportional to n_d and is directly analogous to the geometric concept of a harmonic vibrator with $R_{4/2} = 2$. This spectrum was shown in fig. 4. Of course, realistic vibrational nuclei exhibit broken degeneracies. The concept of successive degeneracy breaking in a dynamical symmetry is most useful when successive scales of breaking become smaller and smaller so that the overall structure is visually clear. This is beautifully illustrated in the nucleus ^{110}Cd whose experimental level scheme [11] is shown in fig. 6. The different multiplets are quite distinct, even though one level of each (corresponding to successive members of an excited 0^+ quasi-band) is pushed up a bit in energy.

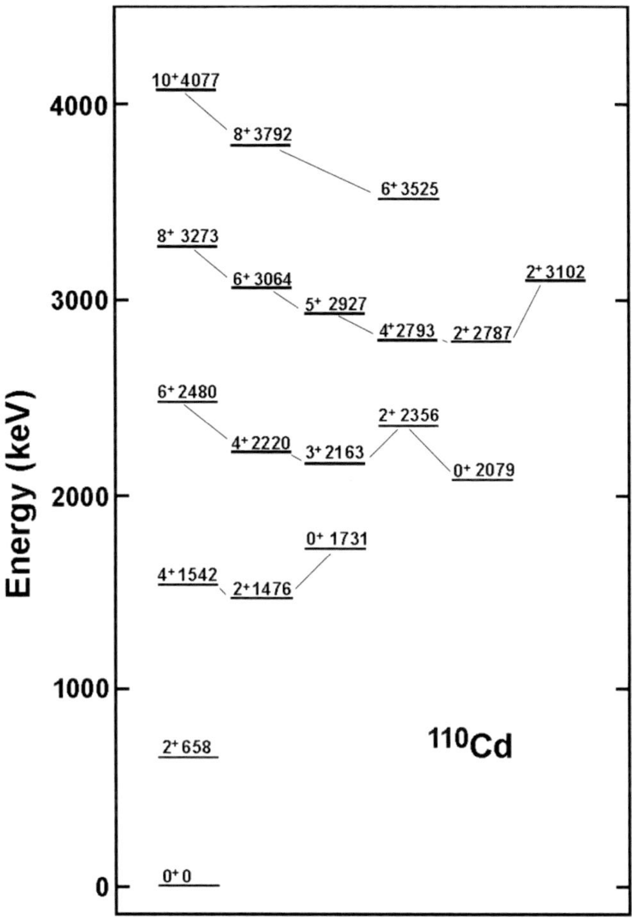

Fig. 6. – Low-lying levels of ^{110}Cd arranged so as to display the vibrational multiplets. For each of the higher multiplets, one state appears higher than the others. Based on ref. [11].

These broken degeneracies can be obtained by including more terms in the Hamiltonian which still conserve n_d. This is achieved with the Hamiltonian

$$(15) \qquad H = \epsilon n_d + \frac{1}{2} \Sigma_J C_J \left(d^\dagger d^\dagger \right)^J \cdot \left(\tilde{d}\tilde{d} \right)^J ,$$

where the C_J terms create and destroy pairs of d-bosons coupled to different angular momenta and therefore split the members of the phonon multiplets according to spin. Note that n_d is still a good quantum number so there is no mixing of the harmonic $U(5)$ eigenstates. The eigenvalues are defined in terms of the quantum numbers of the $U(5)$

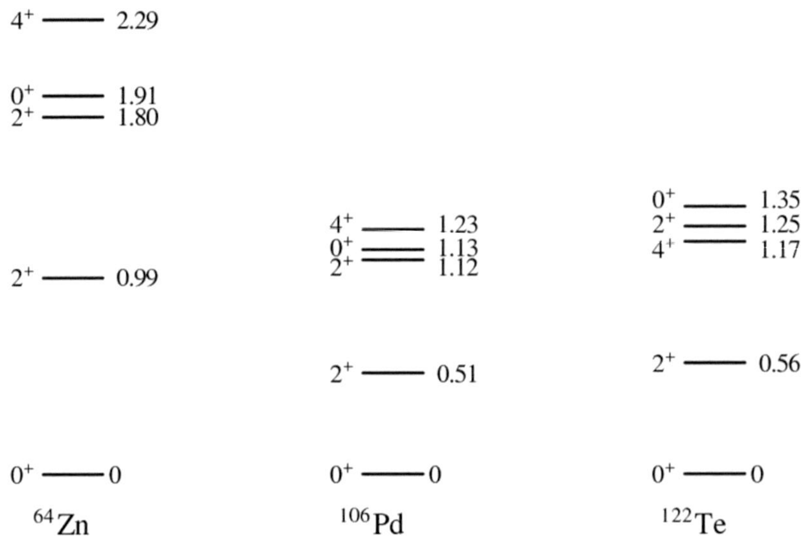

4^+ —— 2.29

0^+ —— 1.91
2^+ —— 1.80

4^+ —— 1.23
0^+ —— 1.13
2^+ —— 1.12

0^+ —— 1.35
2^+ —— 1.25
4^+ —— 1.17

2^+ —— 0.99

2^+ —— 0.51

2^+ —— 0.56

0^+ ——0

0^+ ——0

0^+ ——0

^{64}Zn

^{106}Pd

^{122}Te

Fig. 7. – Typical examples of vibrational spectra. See also the spectrum of ^{110}Cd in fig. 6.

group chain:

$$(16) \qquad E(n_d, v, J) = \alpha n_d + \beta n_d(n_d + 4) + 2\gamma v(v + 4) + 2\delta J(J + 1),$$

where the Greek coefficients α, β, γ, and δ are parameters. Clearly, if only α is non-zero, one has a harmonic spectrum. Figure 7 gives some empirical examples of the lowest levels of some typical candidates for $U(5)$ symmetry showing different ways of breaking the degeneracies.

Each of the symmetries has characteristic, analytic predictions for various other quantities besides energies. The most important of these are $E2$ transitions. The most general $E2$ operator in up to bilinear terms in the s and d boson operators is given by

$$(17) \qquad Q = \left(d^\dagger s + s^\dagger \tilde{d}\right) + \chi \left(d^\dagger \tilde{d}\right)^{(2)}.$$

The first two terms convert an s boson into a d boson or vice versa. The last conserves n_d. In order for this operator Q to reproduce the characteristic vibrator selection rules that the quadrupole moment is zero and $E2$ transitions can create or destroy only a single vibrational phonon, requires that $\chi = 0$. This operator Q, acting on the wave functions defined by the Hamiltonians of eqs. (14) or (15), gives a number of characteristic predictions. An important result is

$$(18) \qquad \sum_{J'} B\left(E2 : J, n_d + 1 \rightarrow J', n_d\right) = e_B^2(n_d + 1)(N_B - n_d).$$

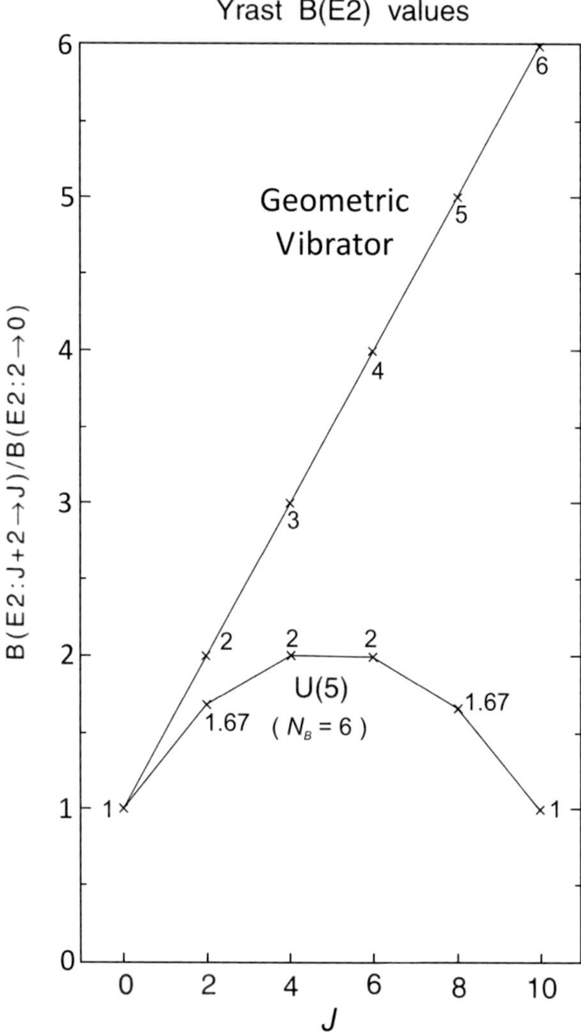

Fig. 8. – Yrast $B(E2)$ values in the geometrical harmonic vibrator and in $U(5)$ for $N_B = 6$.

This describes the sum of $B(E2)$ values from a state of a given n_d value to states with n_d reduced by one—that is, in the language of the vibrator, the de-excitations from an n-phonon state to the $(n\text{-}1)$ phonon states. The structure of this equation is very revealing. Think in terms of the operator $(s^\dagger \tilde{d})$. The first factor on the right in eq. (18) (other than the scale factor, c_B^2) is the square of the square root of the number of d-bosons in the initial state: this arises from the $\tilde{d}$ operator (in analogy with eq. (1)). This term gives the same predictions as the geometric vibrator model where the $E2$ operator is the single-phonon destruction operator, b. In *that* model, the number of phonons is decreased by

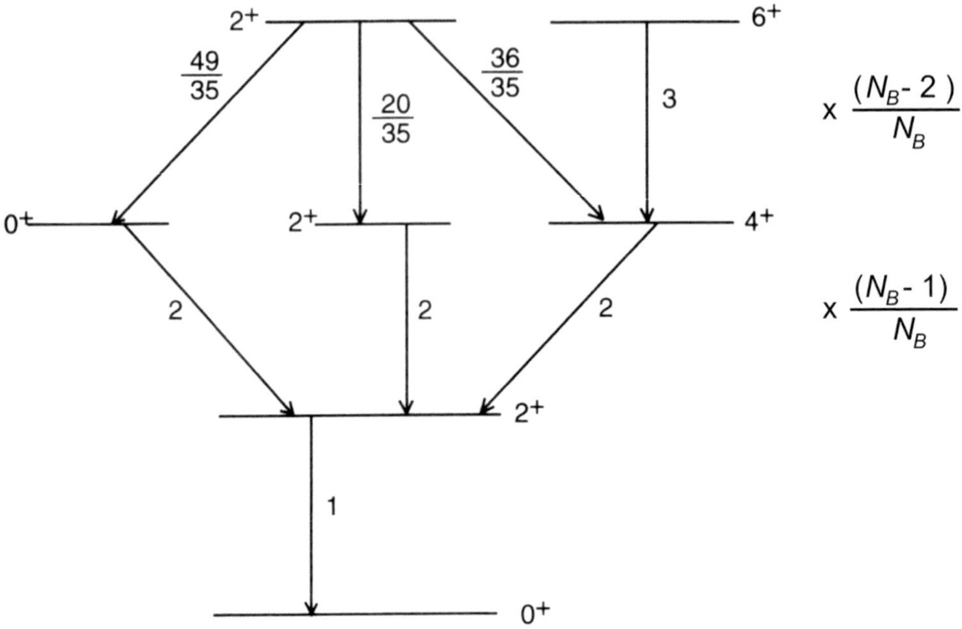

Fig. 9. – Some $B(E2)$ values normalized to $B(E2 : 2^+_1 \to 0^+_1)$ up through the 3-phonon multiplet of a vibrator. The numbers on the arrows are for the geometric model. In $U(5)$ they are multiplied by the finite boson number correction factors on the right. Since these correction factors depend only on $N_B - n_d$, they are the same for all the states in a given multiplet. Therefore, *relative* $B(E2)$ values from a given initial state are the same as in the vibrator model.

the transition—phonons are thought of as particle-hole excitations relative to the Fermi surface. In the IBA, on the other hand, the total number of bosons is conserved—they represent the valence nucleons. Therefore the $E2$ operator must have a product of two operators (creation and destruction) *each* giving a square root factor (that gets squared in the $B(E2)$ value). This gives rise to the second factor on the right in eq. (18) which is proportional to $n_s = N_B - n_d$. As the number of d-bosons in a given nucleus increases, the number of s-bosons decreases. Thus, while the first factor gives $B(E2)$ values that are linear in the d-boson number, the second term produces a modification of this increase due to the finite and conserved total number of bosons. Note that the second factor in eq. (18) depends on N_B. This result is typical of many in the IBA. The model shows a finite N_B dependence but it goes over to the geometrical model for infinite N_B. Figure 8 compares the IBA and vibrator model predictions for the yrast $B(E2)$ values for $N_B = 6$, a value typical of vibrational nuclei (These predictions would be nearly indistinguishable if, for example, we had used $N_B = 1000$.).

Note that the sum in eq. (18) means that, for multi-phonon states that can decay to more than one final state, it is the *sum* of the $B(E2)$ values that is proportional to

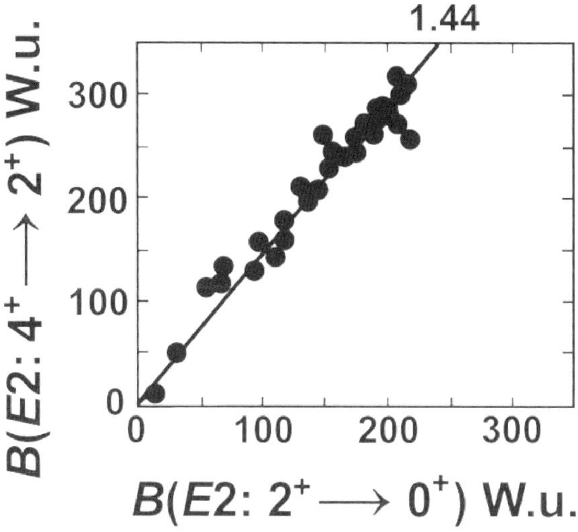

Fig. 10. – Empirical values for $B(E2 : 4_1^+ \to 2_1^+)$ values for collective nuclei (with $Z > 40$ and $R_{4/2} > 2$) plotted against $B(E2 : 2_1^+ \to 0_1^+)$. The least-squares fitted slope of the line through the data points, 1.44, is indicated.

the number of phonons in the initial state. The relative $B(E2)$ values originate in the $O(5)$ sub-group, which also characterizes the geometric vibrator and therefore they are the same as in the vibrator model. Some are illustrated in fig. 9.

Finally, eq. (18) gives the yrast $B(E2)$ ratio

$$(19) \qquad R = \left[\frac{B\left(E2 : 4_1^+ \to 2_1^+\right)}{B\left(E2 : 2_1^+ \to 0_1^+\right)} \right]_{U(5)} = \left[\frac{N_B - 1}{N_B} \right] \left[\frac{B\left(E2 : 4_1^+ \to 2_1^+\right)}{B\left(E2 : 2_1^+ \to 0_1^+\right)} \right]_{phonon} .$$

Since vibrational nuclei typically occur near closed shells, N_B is usually small, typically, say, $N_B \sim 4$, so that the $B(E2)$ ratio in eq. (19) is around $2 \times \frac{3}{4} = 1.5$. This reduction was seen explicitly for $N_B = 6$ in fig. 8. We will see later that the rotor model gives almost exactly the same result as the Alaga ratio, namely $10/7 = 1.43$, so that there is almost no difference in this ratio as a function of structure. It is interesting that this is fully consistent with the data which are summarized for collective nuclei in fig. 10.

To repeat one additional remark on $U(5)$: as we shall see, the other dynamical symmetries, and, indeed, all calculations deviating from $U(5)$, involve terms in the Hamiltonian that do not conserve n_d. That is, they mix states of good n_d. In carrying out such calculations numerically the resulting wave functions must be expressed in some basis. Since $U(5)$ states have good n_d, it is almost universal practice to use $U(5)$ as the set of basis states for the IBA.

4`2. $SU(3)$. – (See ref. [12].) We now turn to the $SU(3)$ symmetry.

$SU(3)$ corresponds to a deformed rotor but to a very specific type of rotor with characteristic signatures. Indeed, most rotational nuclei are described in the IBA with wave functions deviating substantially from $SU(3)$. Nevertheless, $SU(3)$ provides an excellent and useful paradigm. To discuss this limit, it is useful to exploit the multipole version of the full IBA Hamiltonian in eq. (6). The Hamiltonian for $SU(3)$ is given by

$$(20) \qquad\qquad H = \kappa Q \cdot Q + a_1 J^2$$

with Q given by eq. (7)

The group chain of eq. (11) has quantum numbers (λ, μ) K and J. The first pair defines the representations of $SU(3)$, and λ ranges from $2N_B$ down to zero according to certain rules (see ref. [12]), while the latter two define the levels within a given representation by their K projection number and total angular momentum, J. The allowed values of K are $0, 2, 4, \ldots \mu$.

The eigenvalues are given in terms of these quantum numbers by

$$(21) \qquad E(\lambda, \mu, J) = A \left[\lambda^2 + \mu^2 + \lambda\mu + 3(\lambda + \mu) \right] + BJ(J + 1).$$

Since A is invariably negative (the $Q \cdot Q$ interaction is attractive), simple trial calculations for various allowed values of (λ, μ) show that this equation gives that the lowest representation is labeled $(2\,N_B, 0)$ and therefore has only a single sequence of states with $K = 0$. This corresponds to the ground-state rotational band of a deformed nucleus, with energies proportional to $J(J + 1)$. The next lowest representation is found to be $(2N_B - 4, 2)$. This will have sequences of states with $K = 0$ and $K = 2$. These corresponds to a band sometimes (inaccurately) associated with the geometric concept of a β-vibration, an oscillation about the equilibrium value of the quadrupole deformation β, and a γ-band excitation that is an oscillation away from axial symmetry. A typical $SU(3)$ level scheme is shown in fig. 11.

There are several interesting and characteristic features of this level scheme. First, since the energies depend only on (λ, μ) and J (but not K), states of the same spin in different bands within a given representation are degenerate. In particular, this refers to the γ and $K = 0$ bands of the first-excited representation. In the vast majority of deformed nuclei these two bands are far from degenerate: usually (but not always) the γ-band is lower. Thus $SU(3)$ must be broken in these nuclei. We will see below which terms in the general Hamiltonian accomplish this.

Secondly, the $E2$ selection rule with the operator Q as defined in eq. (7) is $\Delta(\lambda, \mu) = 0$. That is, $E2$ transitions cannot change representation. This prediction is exactly contrary to the geometrical model in which the lowest intrinsic states have collective transitions to the ground state band whereas, in $SU(3)$, those transitions are forbidden. Perhaps even more strangely, the same $SU(3)$ selection rules implies collective transitions *between* the γ and $K = 0$ bands. Moreover, such "$\beta \to \gamma$" transitions persist in broken $SU(3)$ Hamiltonians suitable for typical well-deformed nuclei (see below, fig. 17). In

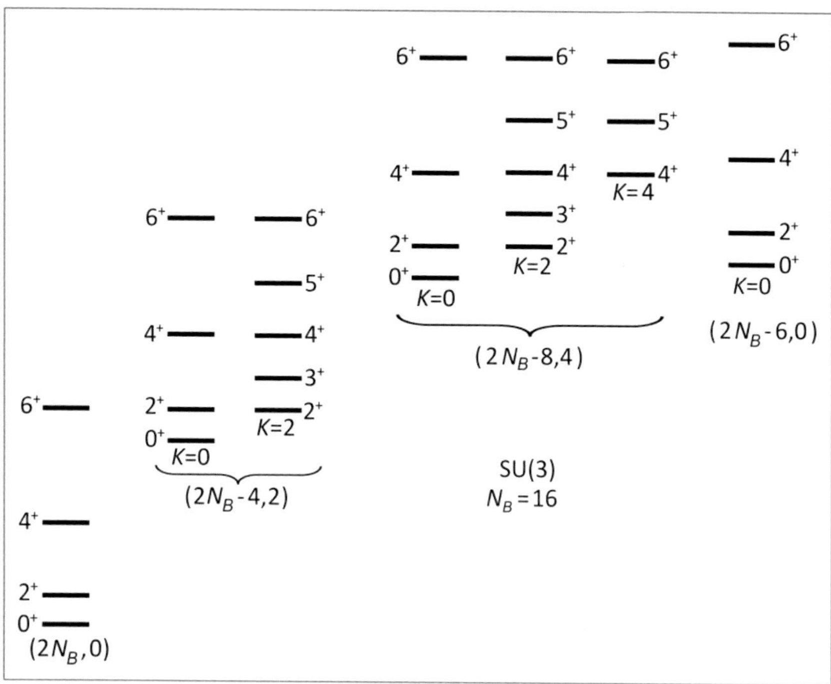

Fig. 11. – Low-lying levels of $SU(3)$ for $N_B = 16$. The parentheses show the (λ, μ) values. Based on ref. [9].

the geometric model these are forbidden because they would require the destruction of one type of vibrational excitation and the creation of another (and a single operator, b, cannot do this). When this prediction was originally recognized, it was thought to be a serious problem for the IBA since such γ-band to $K = 0$ band transitions were not known—they are low-energy transitions, hindered by a factor E_γ^5, and had never been seen. However, soon thereafter, highly sensitive experiments [13] with the GAMS spectrometers at the ILL in Grenoble in ^{168}Er revealed that such transitions do exist and are collective. Lastly, we note that, in the IBA, these intra-representation, inter-band transitions vanish in the large N_B limit, again showing that the IBA goes over to the geometric model.

A third characteristic of $SU(3)$ is that there is no mixing of the γ- and ground-bands. Such mixing is well known throughout deformed nuclei but nuclei well described by $SU(3)$ should have little or no mixing. A final prediction of $SU(3)$ is that the ratio of $K = 0$ to ground $B(E2)$ values to γ-band to ground-band transitions, though both are forbidden, is finite, namely 1/6 for $2^+ \to 0^+$ transitions.

In addition, though, in principle, the parameters of the IBA Hamiltonian can change with Z and N, eq. (21) shows that the γ-bandhead energies in $SU(3)$ have an N_B dependence as $(2N_B - 1)$. Therefore, in a region of $SU(3)$-like nuclei, one would at least expect that the γ and $K = 0$ excitations would increase in energy with N_B.

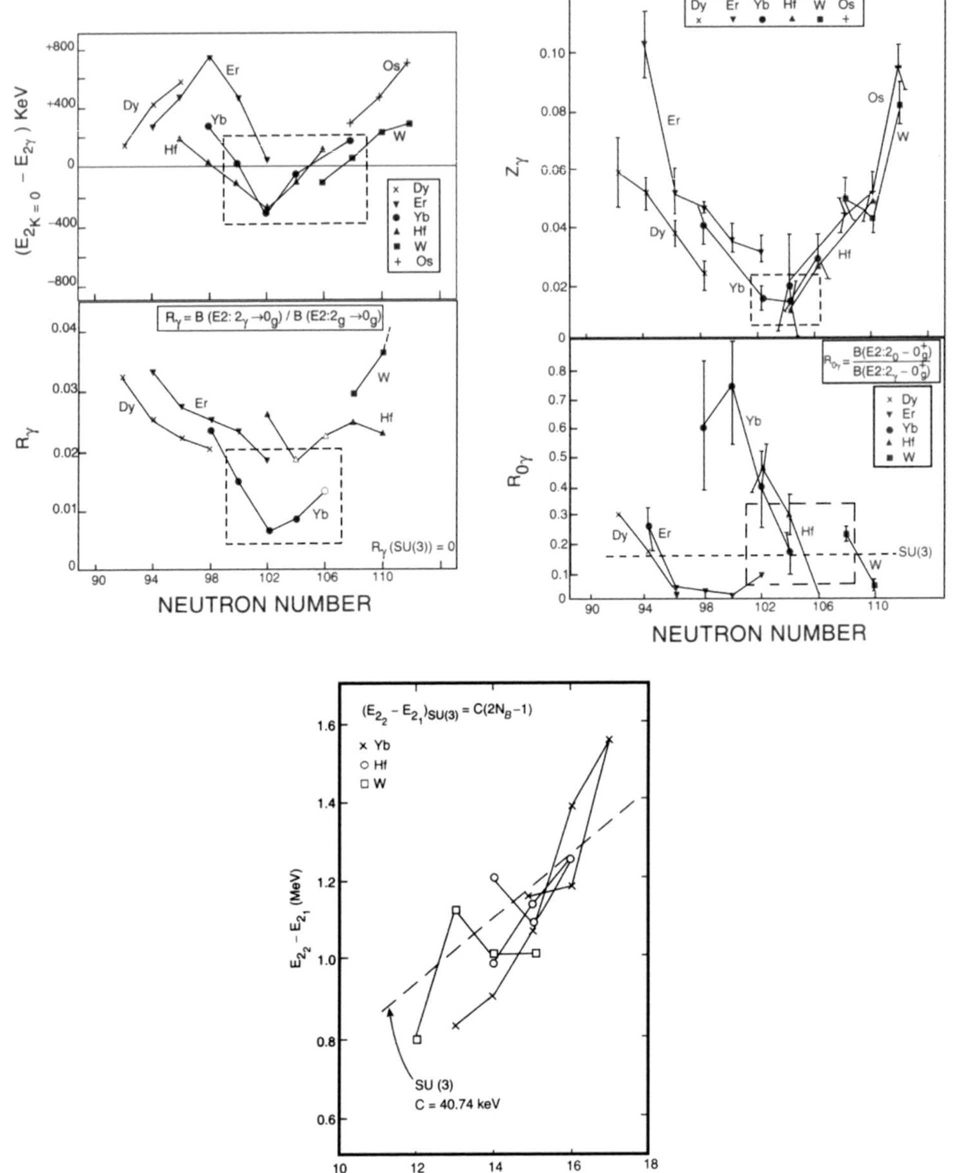

Fig. 12. – Comparison of data in the rare-earth region with characteristic signatures of $SU(3)$. Based on ref. [9].

Very few nuclei exhibit all of these signatures. ^{156}Gd has some of them (but in that region $E(2^+_\gamma)$ *decreases* with N_B). A few nuclei near $N = 102$ do seem to exhibit all as shown in fig. 12.

The yrast $B(E2)$ values in $SU(3)$ are given by

$$(22) \qquad B\left(E2 : J+2 \to J\right) = e_B^2 \frac{3}{4} \left[\frac{(J+2)(J+1)}{(2J+3)(2J+5)} \right] (2N_B - J)(2N_B + J + 3).$$

Note again that this has a factor that depends only on J and one that depends on both J and N_B. The first term is exactly the same as in the geometrical model where is it known as an Alaga rule, while the second reflects the finite particle number microscopic aspect of the IBA. Using this equation, we can calculate the $B(E2)$ ratio

$$(23) \qquad \frac{B\left(E2 : 4^+_1 \to 2^+_1\right)}{B\left(E2 : 2^+_1 \to 0^+_1\right)} = \frac{10}{7} \left[\frac{(2N_B - 2)(2N_B + 5)}{(2N_B)(2N_B + 3)} \right],$$

where the first factor is precisely the Alaga ratio and the second is a finite N_B factor that goes to unity for large N_B, recovering, as usual, the geometric model.

$4\dot{}3.$ $O(6)$. – (See refs. [14-17].) The final dynamical symmetry of the IBA-1 is $O(6)$. This symmetry has a particularly important historical significance. Though now recognized [16] to correspond to the geometrical limit of the γ-flat Wilets Jean model, when first proposed in 1978, it appeared as a new symmetry, experimentally unknown. The immediate discovery [17] that ^{196}Pt is an excellent manifestation of $O(6)$ helped establish the viability of the IBA and the usefulness of its algebraic foundations.

To describe $O(6)$ we deviate from the historical route and turn to a special formalism for the IBA called the Consistent Q Formalism (CQF) [18]. The idea behind it is seen by comparing two Hamiltonians that produce the $O(6)$ symmetry. The original $O(6)$ Hamiltonian used parts of eq. (6):

$$(24) \qquad H = a_0 P^\dagger P + a_1 J^2 + a_3 T_3^2.$$

This gives rise to the eigenvalue equation

$$(25) \qquad E(\sigma, \tau, J) = A(N - \sigma)(N + \sigma + 4) + B\tau(\tau + 3) + CJ(J + 1)$$

which is written in terms of the three quantum numbers of the $O(6)$ group chain of eq. (12), where major families of states are labeled by a quantum number σ, further specified by a phonon-like quantum number τ, and finally fully specified by their angular momenta J.

The key term defining the wave functions in the Hamiltonian in eq. (24) is $P^\dagger P$ where P is defined in eq. (8). Consider the effect of this term on the $U(5)$ basis states. It mixes states with $\Delta n_d = 0$ and 2. Now consider the $Q \cdot Q$ term in eq. (6). It is easily seen using eq. (7) for Q that this will mix $U(5)$ states by $\Delta n_d = 0, 1, 2$. However, if we drop

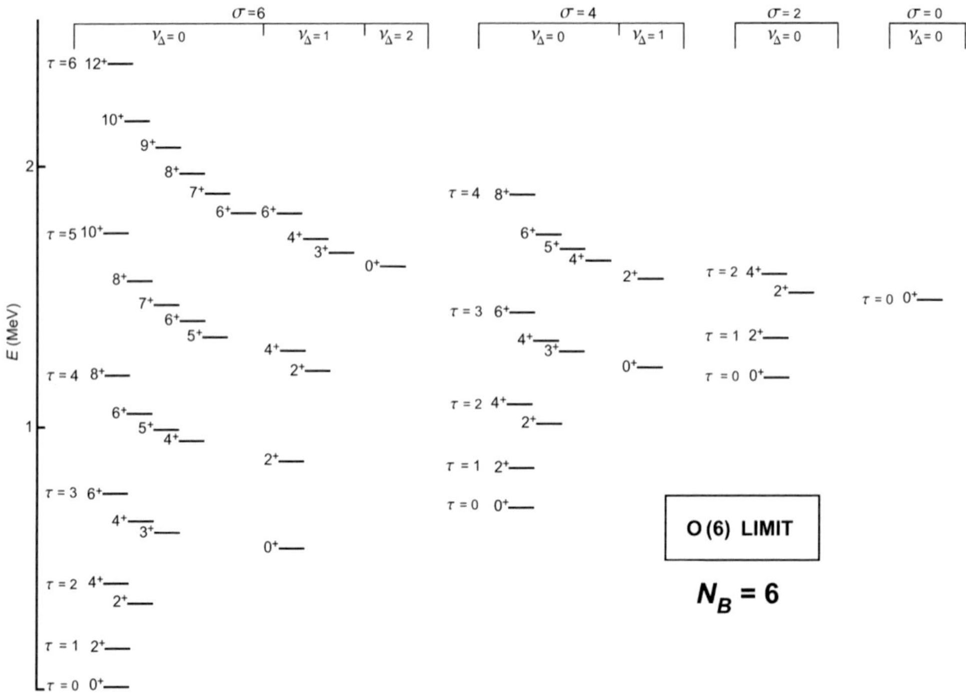

Fig. 13. – Levels of $O(6)$ for $N_B = 6$ displayed according to their σ, τ, and J quantum numbers. Based on ref. [17].

the term in $(d^\dagger \tilde{d})^{(2)}$ in Q, then the mixing is $\Delta n_d = 0, 2$ only. Hence, the operator $Q \cdot Q$, with no $(d^\dagger \tilde{d})^{(2)}$ term, has the same effect on the $U(5)$ wave function as the $P^\dagger P$ term. Hence, if we write a more general form of Q, namely

$$(26) \qquad Q = \left(s^\dagger \tilde{d} + d^\dagger s \right) + \chi \left(d^\dagger \tilde{d} \right)^{(2)},$$

where χ is a parameter, it might be possible to reproduce $O(6)$ with the simpler Hamiltonian

$$(27) \qquad H = \kappa Q^2 + a_1 J^2 \qquad \text{(CQF)}$$

by setting $\chi = 0$.

It turns out that this Hamiltonian does indeed give an $O(6)$ spectrum. However, it is a special case of it: since eq. (27) has only two terms it can only give an eigenvalue expression with two independent parameters. In fact, it gives the same eigenvalues as eq. (25) with $A = B$. Remarkably, the known empirical examples of $O(6)$ seem to display this special form showing the usefulness of the CQF.

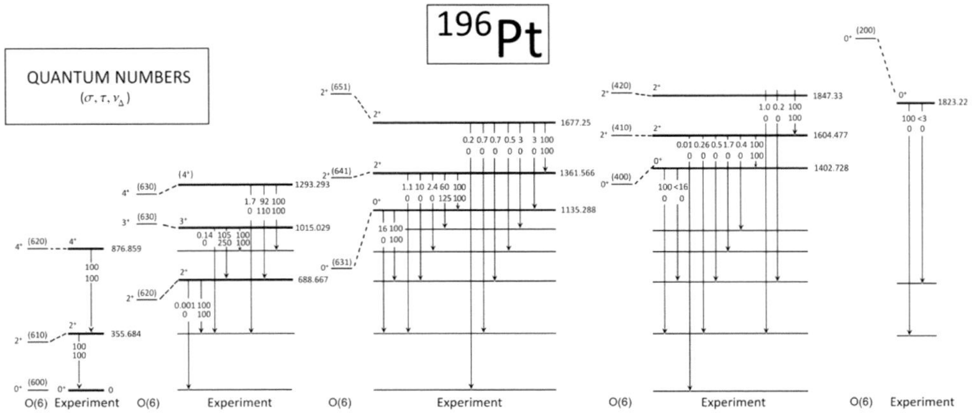

Fig. 14. – Comparison of the low-lying levels and relative $B(E2)$ values with the predictions of $O(6)$. Relative experimental $B(E2)$ values are given as the upper numbers on the transition arrows while the $O(6)$ predictions are the lower numbers. Based on ref. [17].

(The name CQF and the main use of this approach actually has its origin, not in this application, but in a more general one, in which the structure deviates from any of the symmetries. Then, χ is taken as a free parameter in the Hamiltonian but is kept the same in the $E2$ transition quadrupole operator of eq. (17) as well. The CQF is nowadays used in nearly all IBA calculations since it is simpler than using the full Hamiltonian of eqs. (5) or (6), is simpler than the original $O(6)$ Hamiltonian, but works equally well and with one fewer parameter.)

A typical $O(6)$ spectrum is given in fig. 13.

Here the roles of the σ, τ and J quantum numbers in the degeneracy breaking is clear. At first glance $O(6)$ may look like a vibrator spectrum but more careful inspection shows important differences. First of all, the multiplets do not have the same spin content. The $\tau = 2$ multiplet, for example, has $J = 4$ and 2, whereas the vibrator has an additional 0^+ state. Likewise, for $\tau = 3$, $O(6)$ lacks the 2^+ state that appears in the three phonon multiplet of $U(5)$. These two states have, in a sense, evolved in $O(6)$ to form the first two states of a different excited 0^+ band.

Secondly, in the vibrator, the energies are proportional to the d-boson number and the spectrum is "linear" (see fig. 4) and $R_{4/2} = 2$. In $O(6)$, they go as $\tau(\tau+3)$ and $R_{4/2} = 2.5$.

The $E2$ selection rules are $\Delta\sigma = 0$ and $\Delta\tau = 1$. These lead to phonon-like predictions (albeit with different numerical $B(E2)$ values than in $U(5)$) and to the particular result that there is no allowed decay of the 0^+ state with $\sigma = N_B - 2$, $\tau = 0$. Another characteristic prediction of $O(6)$, arising from the τ selection rule, is that the first excited 0^+ state should decay to the 2_2^+ state not the 2_1^+ state as in typical vibrator spectra. Also typical of $O(6)$ is the appearance of repeated sequences of states 0^+-2^+-2^+.

The first nucleus observed [17] to reflect $O(6)$, and still the best example, is ^{196}Pt whose level scheme is shown in fig. 14. Given that there are only 2 or 3 parameters for

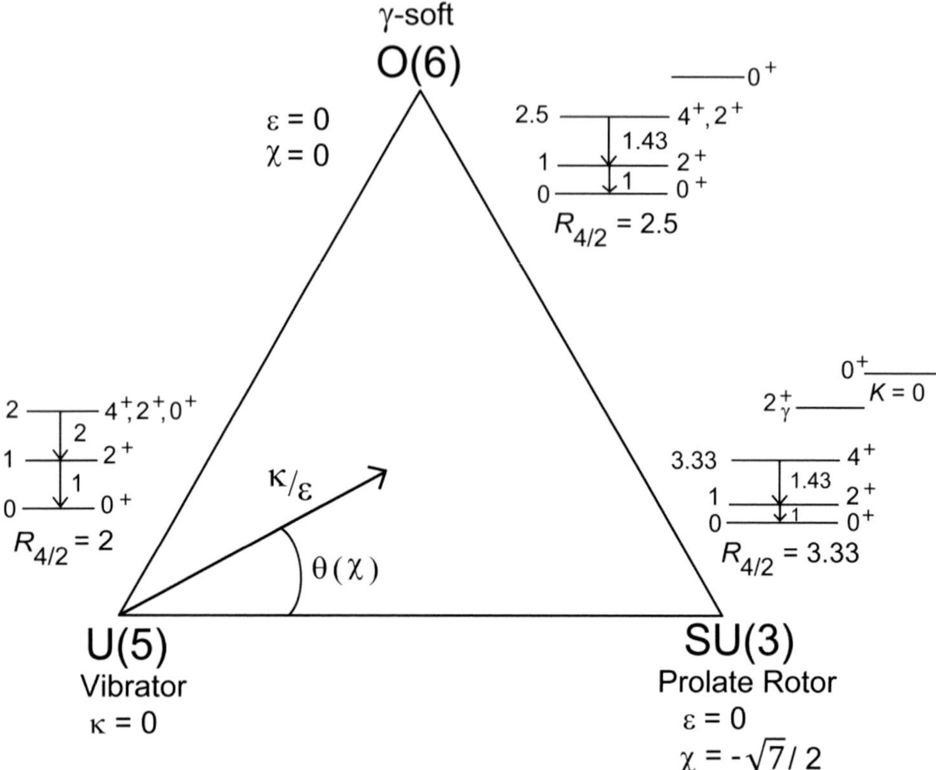

Fig. 15. – The symmetry triangle of the IBA [21] showing the three dynamical symmetries along with their geometrical analogues and parameters of eq. (28) that give each symmetry. The parameters κ/ϵ and χ are discussed below. χ is quantitatively expressed in terms of an angle θ off the bottom axis (see later discussion).

both the energies and $B(E2)$ values, the agreement is remarkable. Note the characteristic 0^+-2^+-2^+ sequences as well as the agreement with the τ selection rule. The 2_2^+ state, for example, has a branching ratio of 10^5. In general the selection rule is satisfied to about an order of magnitude—forbidden transitions are typically an order of magnitude weaker than allowed ones. The σ selection rule, the forbiddeness of any decay for the 0^+ state at 1402 keV, can only be tested by the measurement of absolute $B(E2)$ values. When $O(6)$ was first proposed, the $B(E2 : 0_3^+ \rightarrow 2_1^+)$ was not known but subsequent measurements showed [19] that it is indeed about an order of magnitude weaker than the $B(E2 : 2_1^+ \rightarrow 0_1^+)$ value. Other nuclei, especially in the Xe-Ba region, have also been found [20] to mirror the $O(6)$ dynamical symmetry.

More details of all these symmetries may be found in the literature, in particular in refs. [1, 2, 7-12, 15-18].

The three symmetries emerging from the IBA suggest a depiction in the form of a triangle [21] with each vertex denoting one of the symmetries. This is shown in fig. 15. The figure shows mini-level schemes of the first few levels and transitions for each symmetry.

The remainder of our discussion will be in terms of this symmetry triangle.

5. – More general properties of the IBA: Calculations throughout the triangle

Of course, very few nuclei manifest any of the symmetries but rather have structures (and therefore spectra) differing from them. It is in treating this vast bulk of all collective nuclei where both the CQF and the symmetry triangle are particularly useful, as we shall see. Simply put, most nuclei occupy interior positions in the triangle. We will shortly discuss simple methods to locate any given nucleus in the triangle as well as to determine trajectories of structural evolution. First, however, it is useful to discuss how one proceeds along one such trajectory. Consider first the three symmetries and the three connecting legs of the triangle. For this discussion, and indeed, for most IBA calculations, we can drop the J^2 term in eq. (27) and use an extension of the CQF to include an ϵn_d term

$$\text{(28)} \qquad\qquad\qquad H = \epsilon n_d + \kappa Q \cdot Q.$$

$U(5)$ is defined by finite ϵ and $\kappa = 0$, $SU(3)$ by $\epsilon = 0$, finite κ and $\chi = -\sqrt{7}/2$, while $O(6)$ is defined by $\epsilon = 0$, finite κ and $\chi = 0$. χ has no meaning in $U(5)$ so, in effect, $U(5)$ contains all χ values. Therefore, to go from $U(5)$ to $SU(3)$ one simply keeps $\chi = -\sqrt{7}/2$ and varies κ/ϵ from zero to infinity. (We shall see shortly a way to avoid such infinities.) To go from $U(5)$ to $O(6)$, again involves the same variation of κ/ϵ but with $\chi = 0$. Finally, the simplest of the transition legs is $O(6)$ to $SU(3)$, in which ϵ is kept at zero and only χ is varied, from zero to $-\sqrt{7}/2$. One easy way to parameterize IBA calculations, then, is to treat κ/ϵ as a radius vector from the $U(5)$ origin and χ as an angle off the bottom leg. Since the bottom leg of the triangle corresponds to nuclei that are rather stiff in the γ degree of freedom while the entire $U(5)$ to $O(6)$ leg is rigorously γ-soft, χ in effect specifies the γ-dependence of the effective geometric potential corresponding to the IBA Hamiltonian: $\chi = -\sqrt{7}/2$ implies an essentially axially symmetric nucleus (albeit with zero point fluctuations in γ) while $\chi = 0$ is γ-soft.

As noted, the simplest leg is $SU(3)$ to $O(6)$ in which the only parameter that affects the wave functions and transition rates is χ (Energies can depend on terms in the Hamiltonian of eq. (6) such as $a_3 T_3^2$.). Since many predictions of energy ratios and transition rates therefore depend only on χ and on the boson number N_B, one can make [18] universal contour plots (universal for this leg) of any observable against N_B and χ. We now discuss predictions along this trajectory. Examples of the results are shown in figs. 16 and 17.

These show a number of fascinating features that agree well with extensive data. First consider the general magnitudes of the predicted values. Note that there is nothing in the input to the model calculations that specifies these values—they come out of the model assumptions and the Hamiltonian. Figure 16 shows a feature that often appears—namely observables that vanish in two or three limits (dynamical symmetries) but are finite in

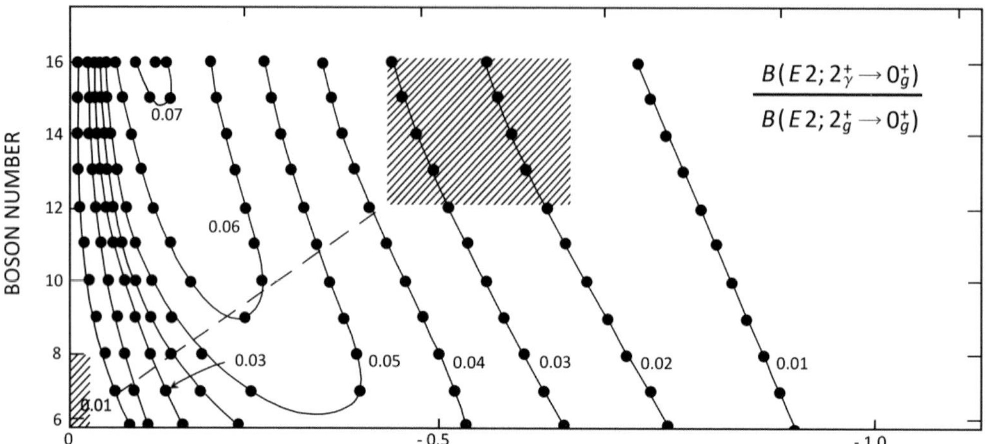

Fig. 16. – Contour plot of the indicated $B(E2)$ ratio against χ and boson number. The cross-hatched areas denote the regions typical of $O(6)$ and typical well-deformed (but not $SU(3)$) nuclei. Based on ref. [18].

between. Thus, by definition, measured finite values imply intermediate structure. Here the $B(E2)$ ratio is zero in $O(6)$ where the numerator violates the τ selection rule, and in $SU(3)$ where it violates the (λ, μ) selection rule. Interestingly, the maximum values are about 0.06 or 0.07. This is an automatic and unavoidable prediction of the IBA and yet it is found experimentally as is seen in fig. 18, where in the rare earth region, the Os isotopes show values maximizing just a little higher at about 0.09.

Of course, a figure such as fig. 16 does not in itself specify *how* structure varies, that is, along a series of nuclei, what the trajectory in χ is as a function of, say, neutron number. Nevertheless, it is instructive to choose a trajectory and inspect the predictions. To do this, we note that it has been found that typical deformed nuclei have $\chi \sim -0.5$ (and boson numbers from 14-16 or so). Thus the upper cross hatched area in the figure gives the best results (along this leg) for the majority of deformed rare-earth or actinide nuclei. The simplest path from $O(6)$ to rotor nuclei is therefore a straight line. Taking that trajectory gives the solid curve in fig. 18. Remarkably, this mirrors the experimental trends remarkably well given how schematic the calculation is.

Figure 17 illustrates other key collective observables. The bottom panel reflects the idea that, in $SU(3)$, $E2$ transitions between the γ- and first-excited $K = 0$ bands are allowed and collective (comparable to γ- to ground-band transition strengths). For deformed nuclei, the ratio is, in fact, near unity. The upper two panels show other properties of the γ and $K = 0$ excitations. The top panel shows that the energy of the $K = 0$ band, along the $SU(3)$ to $O(6)$ leg of the symmetry triangle, is predicted to occur in deformed nuclei at an energy roughly 50% higher than the γ-band. Experimentally the ratio is almost always between 1.2 and 1.8.

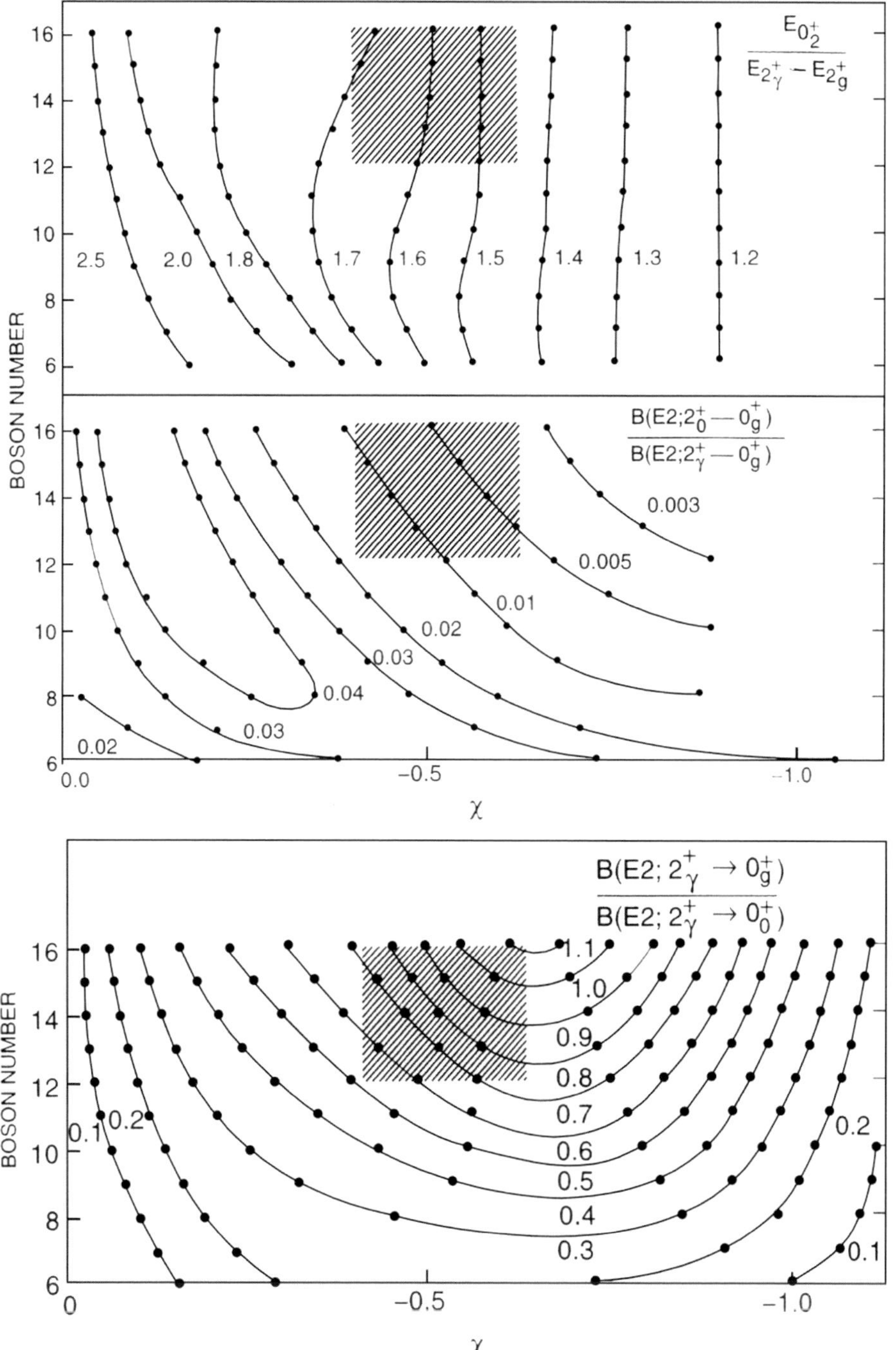

Fig. 17. – Same as fig. 16, for three additional observables. Based on ref. [18].

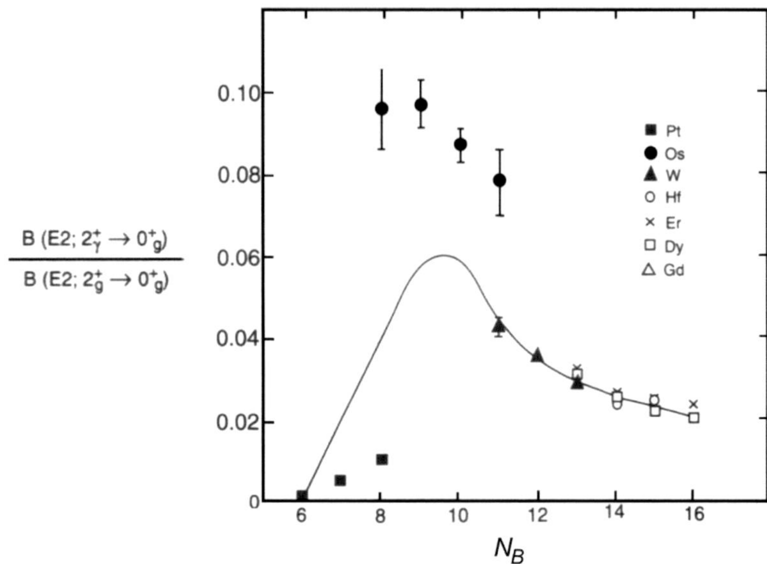

Fig. 18. – Predictions of the decay of the 2^+_γ level for the linear trajectory in fig. 16, compared to the data for nuclei from Gd to Pt. Based on ref. [18].

Finally, the middle panel shows that $B(E2 : K = 0\text{-ground}) \ll B(E2 : \gamma\text{-ground})$. Again, there is nothing in the input to these calculations that specifies this—it is a result that emerges and it is exactly what is seen experimentally: $K = 0$ to ground-band transitions are invariably much weaker than those from the γ-band to the ground-band.

Figure 19 illustrates specific calculations [22] of ground-and γ-bands of two Os nuclei, ^{192}Os and ^{186}Os. These were actually done prior to the invention of the CQF and used the Hamiltonian of eq. (6) with the $P^\dagger P$, Q^2, T^2_3 and J^2 terms. However, they give results for these $B(E2)$ values that are similar to those in the CQF obtained by changing χ alone. The nucleus ^{192}Os is very γ-soft and at the beginning of the transition from $O(6)$ in ^{196}Pt towards a rotor in the lighter Os isotopes, The nucleus ^{186}Os is further along this trajectory. The changes in γ-ground-band $B(E2)$ branching ratios from those close to $O(6)$ towards a deformed structure are clear. For example, the ratio $B(E2 : 2^+_\gamma \to 0^+_1)/B(E2 : 2^+_\gamma \to 2^+_1)$, which is zero ($\tau$ forbidden) in $O(6)$ and 0.7 in the rotor (Alaga ratio), changes from 0.11 in ^{192}Os to 0.43 in ^{186}Os. The calculations reproduce these trends extremely well. Furthermore, the ratios of intraband to interband $B(E2)$ values, such as $B(E2 : 4^+_\gamma \to 2^+_\gamma)/B(E2 : 4^+_\gamma \to 2^+_g)$, are also exceptionally well reproduced.

This comparison of IBA calculations along the $SU(3)$ to $O(6)$ leg with experimental results dramatically shows the power of the IBA and the success of its truncation of the shell model and the choice of the IBA Hamiltonian. However, the symmetry triangle is a two-dimensional surface that spans a much richer set of IBA Hamiltonians. Up until about five years ago, it was indeed thought that most deformed nuclei lay along

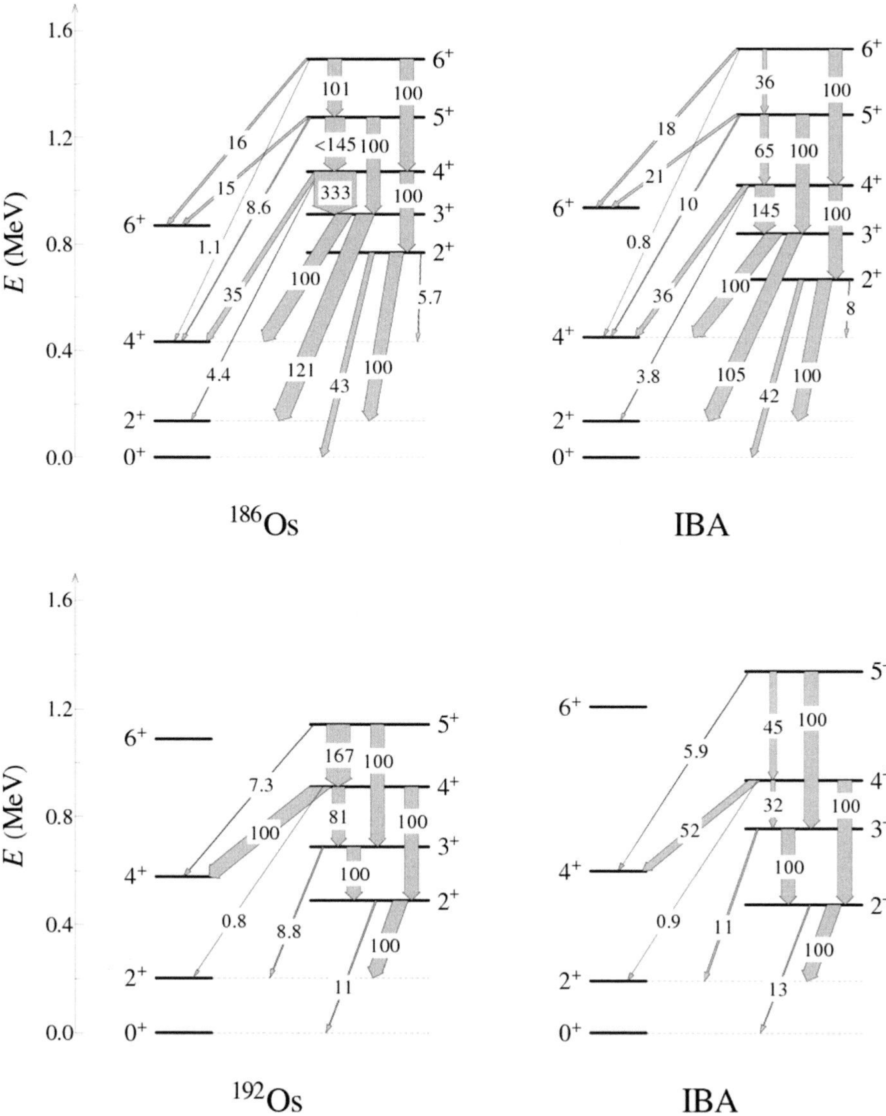

Fig. 19. – Comparison of the experimental level schemes of the γ- and ground-bands of two Os nuclei with IBA calculations along the $O(6)$ to $SU(3)$ leg of the symmetry triangle. These pre-CQF results are similar to those obtained by varying χ in the $E2$ operator. Based on the calculations of ref. [22].

the $SU(3)$ to $O(6)$ leg. Nevertheless, most of the observables tested related to energies and transitions involving the ground- and γ-bands. Predictions of excited 0^+ states and the bands built on them had traditionally been notably poorer [23]. However, this situation has recently changed due to the influence of a development that is seemingly unrelated. Descriptions of the rapid shape changes that occur in certain regions, couched in terms of quantum phase transitions [24, 25], along with geometric models of critical point symmetries such as $X(5)$, have successfully described the properties of the lowest 0^+ states and the levels built on them [26, 27]. This has encouraged a further study of these excitations with other models such as the IBA and has led to an extensive mapping (see below) of structural evolution in the rare earth region that preserves the agreement obtained in earlier CQF calculations but gives improved results for the 0^+ sequences. To discuss this, we consider Hamiltonians which correspond to interior portions in the symmetry triangle. We now describe how to obtain such Hamiltonians and then describe a simple technique of orthogonal crossing contours to determine the IBA parameters for virtually any nucleus.

To see this, consider the ϵn_d term in eqs. (5) or (6). Since this term dominates in the $U(5)$ symmetry but vanishes along the $O(6)$ to $SU(3)$ leg (see eq. (28)), it is reasonable to expect that it decreases as a function of distance from the $U(5)$ vertex.

The Hamiltonian has 2 parameters (except for scale), namely κ/ϵ and χ, the former controlling the spherical–deformed character of the solutions and the latter the γ-softness. Their meaning in the context of the triangle is illustrated in fig. 15, where χ is defined by $\theta = (\pi/3)[1 + (2/\sqrt{7})\chi]$ [28]. Use of this form with Q defined as in eq. (26) with the parameter χ is called the Extended CQF or ECQF [29], and with it, all points in the triangle can be parameterized easily. The use of both terms in eq. (28) gives much improved agreement with the data.

It is therefore important to understand how the model predictions behave throughout the triangle. To study this, we introduce the concept of contours of constant values and of the technique of orthogonal crossing contours.

Let us start with the simple and extremely useful structural signature $R_{4/2}$. $R_{4/2}$ is 2.0 for a harmonic vibrator and 3.33 for $SU(3)$. Therefore, presumably, somewhere along the bottom leg of the triangle, it will pass through a value such as, say, 2.9. (Where this happens is boson-number dependent since the $Q \cdot Q$ term in the Hamiltonian tends to scale roughly as N_B^2 while the ϵn_d term goes as N_B—the calculations below are for $N_B = 10$.) This is shown in fig. 20.

Now consider the $O(6)$ to $SU(3)$ leg. $R_{4/2}$ goes from 2.5 in $O(6)$ to, again, 3.33 in $SU(3)$. Once more, there must be a point along this leg where $R_{4/2} = 2.9$. Hence we immediately see that $R_{4/2}$ alone does not specify the structure uniquely. Indeed, looking at fig. 20, we see that there is a locus, cutting through the triangle, where $R_{4/2} = 2.9$. Similar contours define other $R_{4/2}$ values as indicated. Thus it is clear that we need another observable to pinpoint the structure. Figure 20 shows that, for the most part, the contours of constant $R_{4/2}$ run vertically through the triangle. Thus, to fix the position of a nucleus in the triangle we need a set of contours that tend to run horizontally. Interestingly, these are not easy to find as fig. 21 shows.

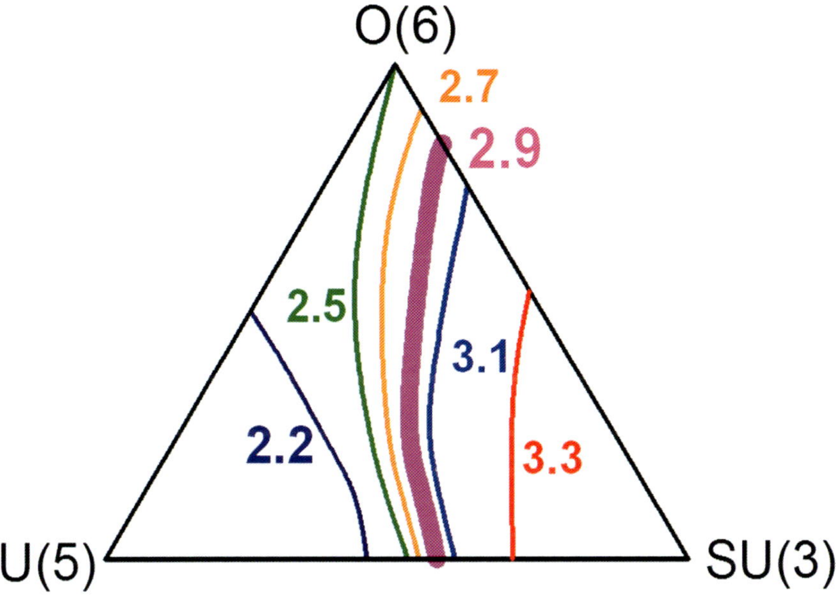

Fig. 20. – Contours of constant $R_{4/2}$ in the symmetry triangle for $N_B = 10$. The contour for $R_{4/2} = 2.9$ is highlighted.

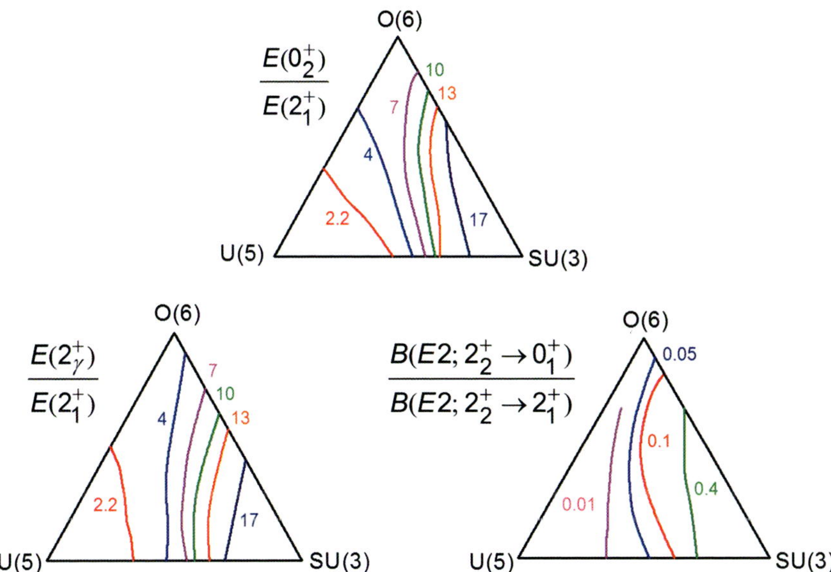

Fig. 21. – Same as fig. 20 for other observables.

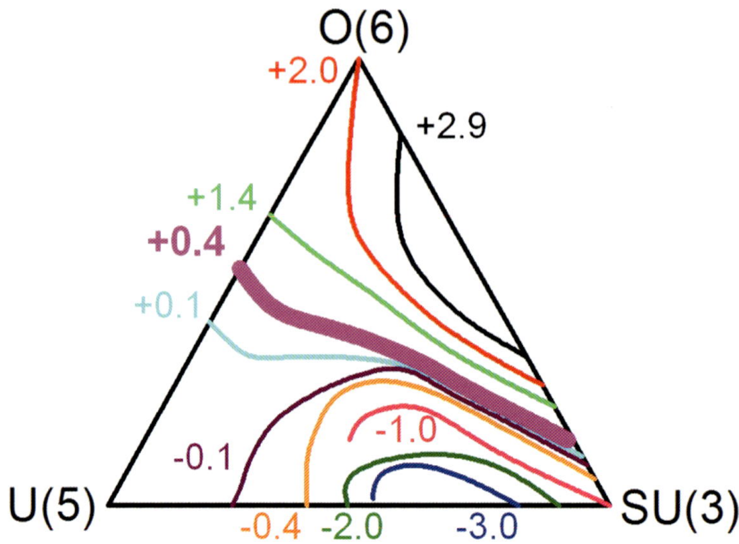

Fig. 22. – Same as figs. 20, 21 for the energy difference $R_{0\gamma} = [E(0_2^+) - E(2_2^+)]/E2(2_1^+)$. The contour for $R_{0\gamma} = +0.4$ is highlighted.

All these cases correspond to a class of observables involving ratios of energy levels or transition rates. However, another class, involving *differences* in energies or $B(E2)$ values does provide sets of orthogonal contours. The most useful of these, in terms of ease of measurement, is shown in fig. 22.

The technique of Orthogonal Crossing Contours (OCC) [23, 30, 31] exploits these distinct trajectories. To illustrate this, let us suppose that we have a nucleus with $R_{4/2} = 2.9$ and with $R_{0\gamma} = +0.4$ (and $N_B = 10$). These are the contour values that are highlighted in figs. 20 and 22. We show their crossing point in fig. 23 and mark it with a pin.

This point is defined by a specific ratio of κ/ϵ and χ and gives a good starting set of parameters. This is the essence of the OCC technique. Of course, some fine tuning around this point in the triangle will often improve the overall agreement.

It is possible to carry out such fits for large regions of nuclei rather simply given the 2-parameter nature of the problem. This has been done [28, 32, 33] for the rare-earth region for nuclei with $N \leq 104$. In figs. 24 and 25 we give examples of experimental level schemes and the IBA fits to them where the parameters have been obtained using the OCC technique. Clearly the agreement is excellent even for such complex level schemes of nuclei quite far from any of the three dynamical symmetries. An overview of structural evolution in this region is provided by the set of trajectories in the symmetry triangle in fig. 26.

There are three aspects of these results worth comment. First, contrary to what had previously been thought, they do not all lie close to the $SU(3)$ to $O(6)$ leg. Here we stress strongly that structure does not change linearly within the triangle. See, for example, the $R_{4/2}$ contours in fig. 20, where, quite far into the triangle from $SU(3)$, $R_{4/2}$ remains

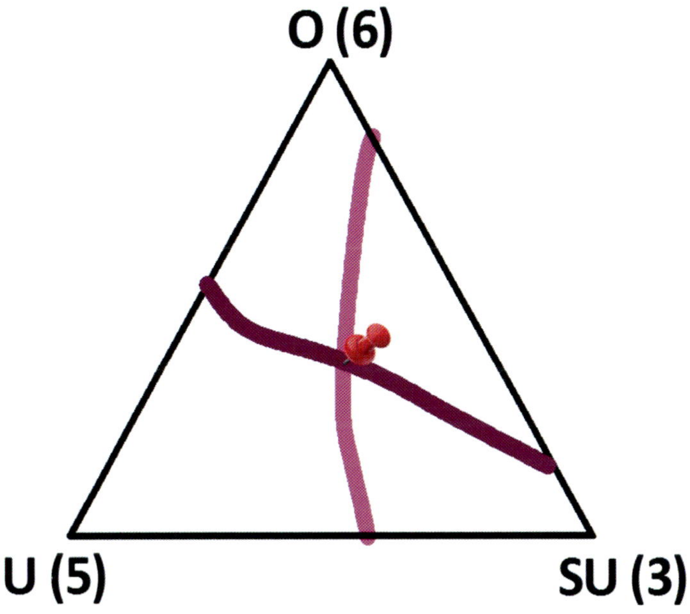

Fig. 23. – Crossing contours for $R_{4/2} = 2.9$ and $R_{0\gamma} = +0.4$ for $N_B = 10$.

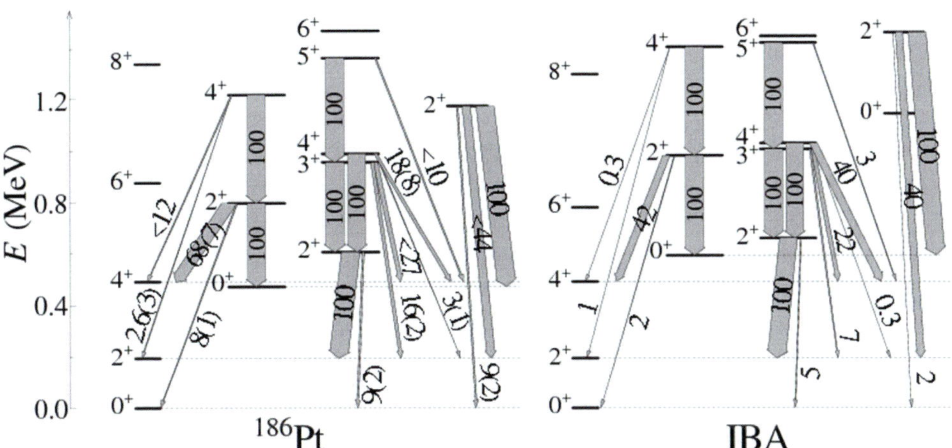

Fig. 24. – Comparison of predictions for ^{186}Pt obtained with the OCC technique using the ECQF. Based on refs. [32, 33].

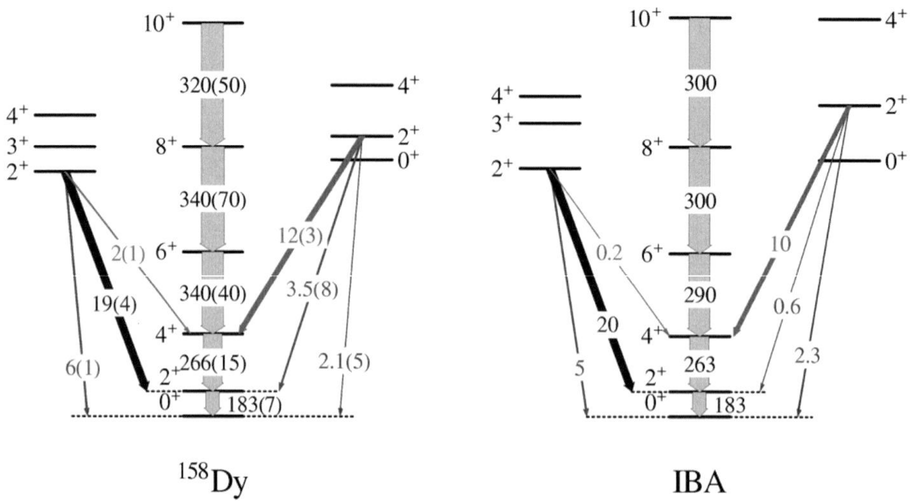

Fig. 25. – Same as for fig. 24, for ^{158}Dy. Based on ref. [28].

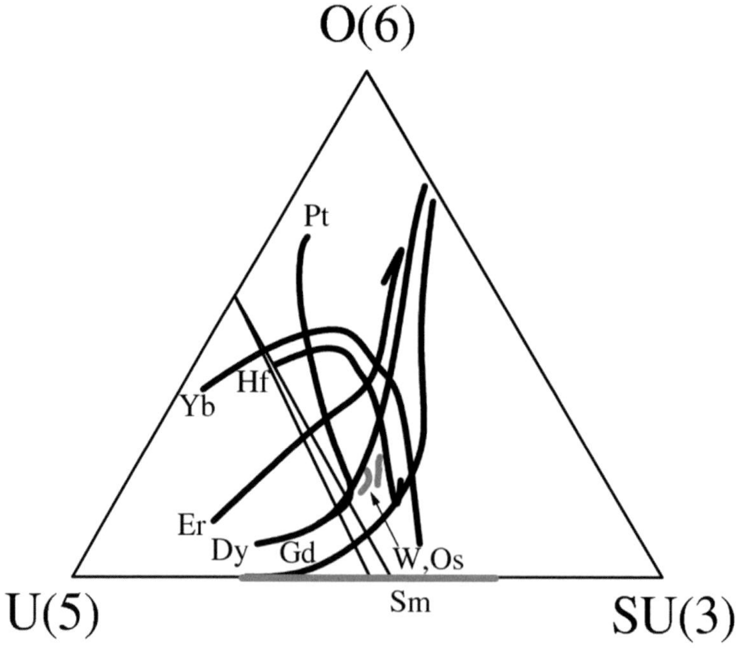

Fig. 26. – Trajectories of structural evolution for $N \leq 104$ in the rare-earth region. Note that, for Pt, to show the evolution towards $O(6)$ in ^{196}Pt, we include nuclei with $N > 104$. Based on ref. [28], but augmented with newer calculations from refs. [32, 33].

> 3.3, and quite far from $U(5)$, $R_{4/2}$ remains < 2.3, and the main changes in $R_{4/2}$ from vibrator-like values to rotor-like values occur in a very narrow region. (This is a region of quantum phase transition which is beyond the scope of the present treatment. Note that the rapidity of the phase transition increases with N_B.) Thus the structural deviations of the rare-earth nuclei from the $SU(3)$ to $O(6)$ leg, while significant, are perhaps not as large as might at first appear.

Secondly, the differences are largely due to properly accounting for the observed 0_2^+ states and the improved agreement for them does not come at the expense of agreement for the ground- and γ-bands.

Thirdly, there is a feature of structural evolution in the triangle that we have not discussed but which is perhaps of deep significance although its meaning is still being interpreted. In $O(6)$ and along the $SU(3)$ to $O(6)$ leg (see the Os isotopes in fig. 19), the γ-band comes quite low in energy, well below the lowest 0^+ band. Along the $U(5)$ to $SU(3)$ leg, the opposite occurs—the 0^+ band is lower. This can be easily seen in fig. 22.

Thus, there is a region cutting through the interior of the triangle corresponding to $R_{0\gamma} = 0$, that is, to nearly degenerate 0^+ and 2^+ intrinsic excitation modes. This contour seems to separate the triangle into two distinct regions. A number of years ago in a study [34] of order and chaos in IBA spectra, it was found that the degree of chaotic behavior generally increases as one deviates from the dynamical symmetries at the vertices. However, surprisingly, it was found that there was a unique narrow region, in the interior, where ordered spectra reappear. This region is called the arc of regularity and it was recently found [35] that its locus corresponds closely to that of $R_{0\gamma} = 0$, and that several nuclei in the rare earth region indeed lie close to it. This "Arc of Regularity" seems to play the role of an interior boundary or internal "frontier", dividing nuclei into two classes, and to point to a region where perhaps some as yet unidentified symmetry or partial symmetry exists. Extensive calculations [36], including those with large boson numbers, have mapped the spectra in detail throughout the triangle, elucidating many fascinating results beyond the scope of this work.

Lastly, we note that, in practice, the Hamiltonian of eq. (28) is often written in a more transparent form that avoids parameter infinities, namely as

$$(29) \qquad H_{\mathrm{ECQF}} = a\left((1 - \zeta)n_d - \frac{\zeta}{4N_B} Q \cdot Q \right).$$

In this Hamiltonian, $\zeta = 0$ gives a $U(5)$ spectrum while $\zeta = 1$ gives a deformed spectrum, either $SU(3)$ if $\chi = -\sqrt{7}/2$, $O(6)$ if $\chi = 0$, or something in between for intermediate χ values. Thus, ζ specifies the length of a radius vector from the $U(5)$ vertex of the triangle to any given point. Comparing eqs. (28) and (29), ζ is given by

$$(30) \qquad \zeta = \frac{4N_B}{4N_B - \epsilon/\kappa},$$
$$a = \kappa\left(\epsilon/\kappa - 4N_B\right).$$

Other forms of the Hamiltonian in eq. (28) have also been used. The most common alternate version [37] is

$$(31) \qquad H_{\mathrm{ECQF}} = c \left(\eta\, n_d - \frac{1-\eta}{N_B}\, Q \cdot Q \right),$$

where $\eta = 1$ for $U(5)$ and 0 for deformed nuclei along the $SU(3)$ to $O(6)$ leg, and

$$(32) \qquad \begin{aligned} \eta &= \frac{-\epsilon/\kappa}{N_B - \epsilon/\kappa}\,, \\ c &= \kappa\,(\epsilon/\kappa - N_B)\,. \end{aligned}$$

Finally, we note an important point relative to all IBA calculations that returns us to the discussion at the beginning, namely the relation of the IBA calculations to an underlying microscopic understanding. Such calculations, or, for example, the trajectories of fig. 26, tell us *what* these nuclei are doing but they do not address the question of *why*. This latter question can only be answered in the framework of a truly microscopic approach and the present results are a challenge to further work in this direction. With recent developments in theoretical techniques such as density functional theory [38, 39], there is reason for optimism here.

$$* \quad * \quad *$$

I would like to thank F. Iachello, D. Warner, V. Zamfir, L. McCutchan, J. Jolie, P. von Brentano, S. Pittel, K. Heyde, P. Van Isacker, R. Burcu Cakirli, A. Aprahamian, J. Cizewski and H. Börner, among many others. Work supported by the US DOE under Grant No. DE-FG02-91ER-40609.

REFERENCES

[1] Arima A. and Iachello F., *Phys. Rev. Lett.*, **35** (1975) 1069.
[2] Iachello F. and Arima A., *The Interacting Boson Model* (Cambridge University Press, Cambridge, England) 1987.
[3] Heyde K., Van Isacker P., Casten R. F. and Wood J. L., *Phys. Lett. B*, **155** (1985) 303.
[4] Casten R. F., Warner D. D., Brenner D. S. and Gill R. L., *Phys. Rev. Lett.*, **47** (1981) 1433.
[5] Federman P. and Pittel S., *Phys. Lett. B*, **69** (1977) 385.
[6] Janssens R. V. F., *Nature*, **435** (2005) 897.
[7] Bonatsos D., *Interacting Boson Models of Nuclear Structure* (Oxford University Press, Clarendon) 1988.
[8] Casten R. F. and Warner D. D., *Rev. Mod. Phys.*, **60** (1988) 389.
[9] Casten R. F., *Nuclear Structure from a Simple Perspective*, 2nd edition (Oxford University Press, New York) 2000, pp. 173-288.
[10] Arima A. and Iachello F., *Ann. Phys. (N.Y.)*, **99** (1976) 253.
[11] Kern J., Garrett P. E., Jolie J. and Lehmann H., *Nucl. Phys. A*, **593** (1995) 21.

[12] ARIMA A. and IACHELLO F., *Ann. Phys. (N.Y.)*, **111** (1978) 201.

[13] WARNER D. D., CASTEN R. F. and DAVIDSON W. F., *Phys. Rev. Lett.*, **45** (1980) 1761.

[14] ARIMA A. and IACHELLO F., *Ann. Phys. (N.Y.)*, **123** (1979) 468.

[15] ARIMA A. and IACHELLO F., *Phys. Rev. Lett.*, **40** (1978) 385.

[16] CIZEWSKI J. A., CASTEN R. F., SMITH G. J., MACPHAIL M. R., STELTS M. L., KANE W. R., BÖRNER H. G. and DAVIDSON W. F., *Nucl. Phys. A*, **323** (1979) 349.

[17] CIZEWSKI J. A., CASTEN R. F., SMITH G. J., STELTS M. L., KANE W. R., BÖRNER H. G. and DAVIDSON W. F., *Phys. Rev. Lett.*, **40** (1978) 167.

[18] WARNER D. D. and CASTEN R. F., *Phys. Rev. Lett.*, **48** (1982) 1385; *Phys. Rev. C*, **28** (1983) 1798.

[19] BÖRNER H. G., JOLIE J., ROBINSON S. J., KRUSCHE B., PIEPENBRING R., CASTEN R. F., APRAHAMIAN A. and DRAAYER J. P., *Phys. Rev. Lett.*, **66** (1991) 691; 837 (Erratum).

[20] CASTEN R. F. and VON BRENTANO P., *Phys. Lett. B*, **152** (1985) 22.

[21] CASTEN R. F., *Interacting Bose-Fermi Systems in Nuclei*, edited by IACHELLO F. (Plenum, New York) 1981, p. 1.

[22] CASTEN R. F. and CIZEWSKI J. A., *Nucl. Phys. A*, **309** (1978) 477.

[23] CHOU W.-T., CASTEN R. F and VON BRENTANO P., *Phys. Rev. C*, **45** (1992) R9.

[24] IACHELLO F., *Phys. Rev. Lett.*, **85** (2000) 3580.

[25] IACHELLO F., *Phys. Rev. Lett.*, **87** (2001) 052502.

[26] CASTEN R. F. and ZAMFIR N. V., *Phys. Rev. Lett.*, **85** (2000) 3584.

[27] CASTEN R. F. and ZAMFIR N. V., *Phys. Rev. Lett.*, **87** (2001) 052503.

[28] McCUTCHAN E. A., ZAMFIR N. V. and CASTEN R. F., *Phys. Rev. C*, **69** (2004) 064306.

[29] LIPAS P. O., TOIVONEN P. and WARNER D. D., *Phys. Lett. B*, **155** (1985) 295.

[30] HARDER M. K. and TANG K. T., *Phys. Lett. B*, **369** (1996) 1.

[31] McCUTCHAN E. A. and CASTEN R. F., *Phys. Rev. C*, **74** (2006) 057302.

[32] McCUTCHAN E. A. and ZAMFIR N. V., *Phys. Rev. C*, **71** (2005) 054306.

[33] McCUTCHAN E. A., CASTEN R. F. and ZAMFIR N. V., *Phys. Rev. C*, **71** (2005) 061301(R).

[34] ALHASSID Y. and WHELAN N., *Phys. Rev. Lett.*, **67** (1975) 816.

[35] JOLIE J., CASTEN R. F., CEJNAR P., HEINZE S., McCUTCHAN E. A. and ZAMFIR N. V., *Phys. Rev. Lett.*, **93** (2004) 132501.

[36] CEJNAR P. and JOLIE J., *Phys. Rev. E*, **61** (2000) 6237.

[37] WERNER V., VON BRENTANO P., CASTEN R. F. and JOLIE J., *Phys. Lett. B*, **527** (2002) 55.

[38] STOITSOV M. V., DOBACZEWSKI J., NAZAREWICZ W. and BORYCKI P., *Int. J. Mass Spectrom.*, **251** (2006) 243.

[39] STOITSOV M. V., CAKIRLI R. B., CASTEN R. F., NAZAREWICZ W. and SATULA W., *Phys. Rev. Lett.*, **98** (2007) 132502.

DOI 10.3254/978-1-58603-885-4-423

Symmetry and supersymmetry in nuclear physics

A. B. BALANTEKIN

Department of Physics, University of Wisconsin - Madison, WI 53706, USA

Summary. — A survey of algebraic approaches to various problems in nuclear physics is given. Examples are chosen from pairing of many-nucleon systems, nuclear structure, fusion reactions below the Coulomb barrier, and supernova neutrino physics to illustrate the utility of group-theoretical and related algebraic methods in nuclear physics.

1. – Introduction

Symmetry concepts play a very important role in all of physics. Originally symmetries utilized in physics did not change the particle statistics: such symmetries either transform bosons into bosons or fermions into fermions. Natural mathematical tools to explore symmetries are Lie algebras, *i.e.* the set of operators closing under commutation relations, schematically shown as

$$[G_{\mathrm{B}}, G_{\mathrm{B}}] = G_{\mathrm{B}}. \tag{1}$$

Lie algebras are associated with Lie groups. One could think of Lie groups as the exponentiation of Lie algebras.

On the other hand, supersymmetries transform bosons into bosons, fermions into fermions, *and* bosons into fermions and vice versa. Natural mathematical tools to explore them are superalgebras and supergroups: Superalgebras, sets containing bosonic (G_B) as well as fermionic (G_F) operators, close under commutation *and* anticommutation relations as shown below:

$$[G_B, G_B] = G_B,$$
$$[G_B, G_F] = G_F,$$
$$(2) \qquad \{G_F, G_F\} = G_B$$

The simplest superalgebra can be easily worked out. Consider three-dimensional harmonic-oscillator creation and annihilation operators and define

$$(3) \qquad K_0 = \frac{1}{2}\left(b_i^\dagger b_i + \frac{3}{2}\right) \quad K_+ = \frac{1}{2}b_i^\dagger b_i^\dagger = (K_-)^\dagger.$$

It is easy to show that the operators defined in eq. (3) satisfy

$$(4) \qquad [K_0, K_\pm] = \pm K_\pm, \quad [K_+, K_-] = -2K_0.$$

The algebra depicted in eq. (4) is the $SU(1,1)$ algebra [1]. Casimir operators are operators that satisfy the condition

$$(5) \qquad [C(G), K_0] = 0 = [C(G), K_\pm].$$

Casimir operators obtained by multiplying one, two, three elements of the algebra are called linear, quadratic, cubic Casimir operators. For $SU(1,1)$ the quadratic Casimir operator is

$$(6) \qquad C_2 = K_0^2 - \frac{1}{2}\left(K_+K_- + K_-K_+\right).$$

The concept of dynamical symmetries found many applications in nuclear physics (see, *e.g.*, [2]). Consider a chain of algebras (or associated groups):

$$(7) \qquad G_1 \supset G_2 \supset \cdots \supset G_n.$$

If a given Hamiltonian can be written in terms of the Casimir operators of the algebras in this chain, then such a Hamiltonian is said to possess a dynamical symmetry:

$$(8) \qquad H = \sum_{i=1}^{n} \left[\alpha_i C_1(G_i) + \beta_i C_2(G_i)\right].$$

Obviously all these Casimir operators commute with each other. Consequently, finding the energy eigenvalues of this Hamiltonian reduces to the problem of reading the eigenvalues of the Casimir operators off the existing tabulations.

Introducing spin (*i.e.* fermionic) degrees of freedom represented by the Pauli matrices as well as the bosonic (harmonic-oscillator) ones and defining

$$(9) \qquad F_+ = \frac{1}{2} \sum_i \sigma_i b_i^\dagger, \quad F_- = \frac{1}{2} \sum_i \sigma_i b_i,$$

one can write the following additional commutation and anticommutation relations:

$$\begin{aligned}
[K_0, F_\pm] &= \pm \frac{1}{2} F_\pm, \\
[K_+, F_+] &= 0 = [K_-, F_-], \\
[K_\pm, F_\mp] &= \mp F_\pm, \\
\{F_\pm, F_\pm\} &= K_\pm, \\
\{F_+, F_-\} &= K_0.
\end{aligned}$$

The operators K_+, K_-, K_0, F_+, and F_- generate the Osp(1/2) superalgebra, which is non-compact [1,3]. Since the operators K_+, K_-, K_0 alone generate the $SP(2) \sim SU(1,1)$ subalgebra we can write the group chain

$$(10) \qquad \mathrm{Osp}(1/2) \supset SU(1,1) \supset SO(2).$$

(Note that the operator K_0 can be viewed as the Casimir operator of the $SO(2)$ subalgebra of $SU(1,1)$.) The Casimir operators of Osp(1/2) and Sp(2) $\sim SU(1,1)$ are given by

$$(11) \qquad C_2\left(\mathrm{Osp}(1/2)\right) = \frac{1}{4}\left(\mathbf{L} + \frac{\boldsymbol{\sigma}}{2}\right)^2 = \frac{1}{4}\mathbf{J}^2,$$

$$(12) \qquad C_2\left(\mathrm{Sp}(2)\right) = \frac{1}{2}\mathbf{L}^2 - \frac{3}{16},$$

where $\mathbf{L}$ is the angular momentum carried by the oscillator, *i.e.* $\mathbf{L}_i = \epsilon_{ijk}\mathbf{r}_j\mathbf{p}_k = i\epsilon_{ijk}b_k^\dagger b_j$. It can easily be shown that a harmonic-oscillator Hamiltonian with a constant spin-orbit coupling

$$(13) \qquad H = \frac{1}{2}\left(\mathbf{p}^2 + \mathbf{r}^2\right) + \lambda\left(\boldsymbol{\sigma} \cdot \mathbf{L} + \frac{3}{2}\right)$$

can be written in terms of the Casimir operators of the group chain given in eq. (10):

$$(14) \qquad H = 4\lambda C_2\left(\mathrm{Osp}(1/2)\right) - 4\lambda C_2\left(\mathrm{Sp}(2)\right) + 2K_0.$$

This provides perhaps the simplest example of a dynamical supersymmetry.

2. – Fermion pairing

Pairing is a salient property of multi-fermion systems and as such it has a long history in nuclear physics (for a recent review see [4]). Already in 1950 Meyer suggested that short-range attractive nucleon-nucleon interaction yields nuclear ground states with angular momentum zero [5]. Mean-field calculations with effective interactions describe many nuclear properties, however they cannot provide a complete solution of the underlying complex many-body problem. For example, after many years of investigations, we know that the structure of low-lying collective states in medium-heavy to heavy nuclei are determined by pairing correlations with $L = 0$ and $L = 2$. This was exploited by many successful models of nuclear structure such as the interacting boson model of Arima and Iachello [6-8].

Pairing plays a significant role not only in finite nuclei, but also in nuclear matter and can directly effect related observables. For example, we know that neutron superfluidity is present in the crust and the inner part of a neutron star. Pairing could significantly effect the thermal evolution of the neutron star by suppressing neutrino (and possibly exotics such as axions) emission [9].

Charge symmetry implies that interactions between two protons and two neutrons are very similar; hence proton-proton and neutron-neutron pairs play a similar role in nuclei. In addition isospin symmetry implies that proton-neutron interaction is also very similar to the proton-proton and neutron-neutron interactions. Currently there is very little experimental information about neutron-proton pairing in heavier nuclei as one needs to study proton rich nuclei to achieve this goal. But one expects that data from the current and future radiative beam facilities will change this picture.

First microscopic theory of pairing was the Bardeen, Cooper, Schriffer (BCS) theory [10]. Soon after its introduction, the BCS theory was applied to nuclear structure [11-13]. Application of the BCS theory to nuclear structure has a main drawback: BCS wave function is *not* an eigenstate of the number operator. Several solutions were offered to remedy this shortcoming such as adding Random-Phase Approximation (RPA) to the BCS theory [14], projection of the particle number after variation [15], or projection of the particle number before the variation. The last technique was recently much utilized in nuclei (see, *e.g.*, [16]). Pairing correlations may also play an interesting role in halo nuclei [17]. It should also be noted that the theory of pairing in nuclear physics has many parallels with the theory of ultrasmall metallic grains in condensed-matter physics (see, *e.g.*, [18]).

$2^{\cdot}1$. *Quasi-spin algebra*. – The concept of seniority was introduced by Racah to aid the classification of atomic spectra [19]. Seniority quantum number is basically the number of unpaired particles in the j^n configuration. Seniority-conserving pairing interactions are a very limited class, however such interactions make an interesting case study. Kerman introduced the quasi-spin scheme to treat such cases [20]. In this scheme nucleons are placed at time-reversed states $|j\,m\rangle$ and $(-1)^{(j-m)}|j\,-m\rangle$. Introducing the creation and annihilation operators for nucleons at level j, $a^{\dagger}_{j\,m}$ and $a_{j\,m}$, the quasi-spin operators are written as

$$\hat{S}_j^+ = \sum_{m>0} (-1)^{(j-m)} a_{j\,m}^\dagger a_{j\,-m}^\dagger, \tag{15}$$

$$\hat{S}_j^- = \sum_{m>0} (-1)^{(j-m)} a_{j\,-m} a_{j\,m} \tag{16}$$

and

$$\hat{S}_j^0 = \frac{1}{2} \sum_{m>0} \left(a_{j\,m}^\dagger a_{j\,m} + a_{j\,-m}^\dagger a_{j\,-m} - 1 \right). \tag{17}$$

These operators form a set of mutually commuting $SU(2)$ algebras:

$$\left[\hat{S}_i^+, \hat{S}_j^-\right] = 2\delta_{ij}\hat{S}_j^0, \qquad \left[\hat{S}_i^0, \hat{S}_j^\pm\right] = \pm\delta_{ij}\hat{S}_j^\pm. \tag{18}$$

The operator $\hat{S}_j^0$ can be related to the number operator

$$\hat{S}_j^0 = \hat{N}_j - \frac{1}{2}\Omega_j, \tag{19}$$

where $\Omega_j = j + 1/2$ is the maximum number of pairs that can occupy the level j and the number operator is

$$\hat{N}_j = \frac{1}{2} \sum_{m>0} \left(a_{j\,m}^\dagger a_{j\,m} + a_{j\,-m}^\dagger a_{j\,-m} \right). \tag{20}$$

Since $0 < \hat{N}_j < \Omega_j$ these $SU(2)$ algebras are realized in the representation with the total angular-momentum quantum number $\frac{1}{2}\Omega_j$.

The most general Hamiltonian for nucleons interacting with a pairing force can be written as

$$\hat{H} = \sum_{jm} \epsilon_j a_{j\,m}^\dagger a_{j\,m} - |G| \sum_{jj'} c_{jj'} \hat{S}_j^+ \hat{S}_{j'}^-, \tag{21}$$

where ϵ_j are the single-particle energy levels, $|G|$ is the pairing interaction strength with dimensions of energy, and $c_{jj'}$ are dimensionless parameters describing the distribution of this strength between different orbitals. When the latter are separable ($c_{jj'} = c_j^* c_{j'}$) we get

$$\hat{H} = \sum_{jm} \epsilon_j a_{j\,m}^\dagger a_{j\,m} - |G| \sum_{jj'} c_j^* c_{j'} \hat{S}_j^+ \hat{S}_{j'}^-. \tag{22}$$

There are a number of approximations one can make to simplify the Hamiltonians in eqs. (21) or (22). If we assume that the NN interaction is determined by a single parameter (usually chosen to be the scattering length), all c_j's are the same and we get

$$(23) \qquad \hat{H} = \sum_{jm} \epsilon_j a^\dagger_{j\,m} a_{j\,m} - |G| \sum_{jj'} \hat{S}^+_j \hat{S}^-_{j'}.$$

This case was solved by Richardson [21].

If we assume that the energy levels are degenerate, then the first term is a constant for a given fixed number of pairs. This case can be solved by using the quasi-spin algebra since $H \propto S^+ S^-$. A list of the exactly solvable cases can then be given as follows:

– Quasi-spin limit [20]:

$$(24) \qquad \hat{H} = -|G| \sum_{jj'} \hat{S}^+_j \hat{S}^-_{j'}.$$

– Richardson's limit [21]:

$$(25) \qquad \hat{H} = \sum_{jm} \epsilon_j a^\dagger_{j\,m} a_{j\,m} - |G| \sum_{jj'} \hat{S}^+_j \hat{S}^-_{j'}.$$

– An algebraic solution developed by Gaudin [22] which is sketched out in the next section (subsect. **2**'2). Gaudin's model is closely related to Richardson's limit.

– The limit in which the energy levels are degenerate (when the first term becomes a constant for a given number of pairs):

$$(26) \qquad \hat{H} = -|G| \sum_{jj'} c^*_j c_{j'} \hat{S}^+_j \hat{S}^-_{j'}.$$

Different aspects of this solution were worked out in refs. [23-25].

– Most general separable case with two shells [26].

Clearly we can solve the pairing problem numerically in the quasispin basis [27]. Note that it is also possible to utilize the quasi-spin concept for mixed systems of bosons and fermions in a supersymmetric framework [28].

2'2. *Gaudin algebra*. – To study the problem of interacting spins on a lattice Gaudin introduced a method [22] closely related to Richardson's solution. Our presentation of Gaudin's method here follows that given in ref. [29]. This method starts with three operators $J^{\pm}(\lambda)$, $J^0(\lambda)$, parametrized by one parameter λ, satisfying the commutation relations

$$(27) \qquad \left[J^+(\lambda), J^-(\mu)\right] = 2\frac{J^0(\lambda) - J^0(\mu)}{\lambda - \mu},$$

$$(28) \qquad \left[J^0(\lambda), J^{\pm}(\mu)\right] = \pm\frac{J^{\pm}(\lambda) - J^{\pm}(\mu)}{\lambda - \mu},$$

and

$$(29) \qquad \left[J^0(\lambda), J^0(\mu)\right] = \left[J^{\pm}(\lambda), J^{\pm}(\mu)\right] = 0.$$

The Lie algebra depicted above is referred to as the rational Gaudin algebra. A possible realization of this algebra is given by

$$(30) \qquad J^0(\lambda) = \sum_{i=1}^{N} \frac{\hat{S}_i^0}{\epsilon_i - \lambda} \quad \text{and} \quad J^{\pm}(\lambda) = \sum_{i=1}^{N} \frac{\hat{S}_i^{\pm}}{\epsilon_i - \lambda},$$

where ϵ_i are, in general, arbitrary parameters. In eq. (30), instead of the quasi-spin algebra one can obviously use any mutually-commuting N $SU(2)$ algebras (in fact, copies of any other algebra). The operator

$$(31) \qquad H(\lambda) = J^0(\lambda)J^0(\lambda) + \frac{1}{2}J^+(\lambda)J^-(\lambda) + \frac{1}{2}J^-(\lambda)J^+(\lambda)$$

is not the Casimir operator of the Gaudin algebra, but such operators commute for different values of the parameter:

$$(32) \qquad [H(\lambda), H(\mu)] = 0.$$

Lowest weight vector, $|0\rangle$, is chosen to satisfy the conditions

$$(33) \qquad J^-(\lambda)|0\rangle = 0, \quad \text{and} \quad J^0(\lambda)|0\rangle = W(\lambda)|0\rangle.$$

Hence it is an eigenstate of the Hamiltonian given in eq. (31):

$$(34) \qquad H(\lambda)|0\rangle = \left[W(\lambda)^2 - W'(\lambda)\right]|0\rangle.$$

To find other eigenstates consider the state $|\xi\rangle = J^+(\xi)|0\rangle$ for an arbitrary complex number ξ. Since

$$(35) \qquad \left[H(\lambda), J^+(\xi)\right] = \frac{2}{\lambda - \xi}\left(J^+(\lambda)J^0(\xi) - J^+(\xi)J^0(\lambda)\right),$$

we conclude that if $W(\xi) = 0$, then $J^+(\xi)|0\rangle$ is an eigenstate of $H(\lambda)$ with the eigenvalue

$$E_1(\lambda) = \left[W(\lambda)^2 - W'(\lambda)\right] - 2\frac{W(\lambda)}{\lambda - \xi} . \tag{36}$$

Gaudin showed that this approach can be generalized and a state of the form

$$|\xi_1, \xi_2, \ldots, \xi_n\rangle \equiv J^+(\xi_1)J^+(\xi_2)\ldots J^+(\xi_n)|0\rangle \tag{37}$$

is an eigenvector of $H(\lambda)$ if the numbers $\xi_1, \xi_2, \ldots, \xi_n \in \mathbf{C}$ satisfy the so-called Bethe Ansatz equations:

$$W(\xi_\alpha) = \sum_{\substack{\beta=1 \\ (\beta \neq \alpha)}}^{n} \frac{1}{\xi_\alpha - \xi_\beta} \quad \text{for} \quad \alpha = 1, 2, \ldots, n. \tag{38}$$

The corresponding eigenvalue is

$$E_n(\lambda) = \left[W(\lambda)^2 - W'(\lambda)\right] - 2\sum_{\alpha=1}^{n} \frac{W(\lambda) - W(\xi_\alpha)}{\lambda - \xi_\alpha} . \tag{39}$$

To make a connection to Richardson's solution we define the so-called $\mathcal{R}$-operators as

$$\lim_{\lambda \to \epsilon_k} \left(\lambda - \epsilon_k\right) H(\lambda) = \mathcal{R}_k. \tag{40}$$

In the realization of eq. (30), these operators take the form

$$\mathcal{R}_k = -2\sum_{j \neq k} \frac{\mathbf{S}_k \cdot \mathbf{S}_j}{\epsilon_k - \epsilon_j} . \tag{41}$$

Taking the limits $\mu \to \epsilon_k$ first and $\lambda \to \epsilon_j$ second in eq. (32), one easily obtains

$$[H(\lambda), \mathcal{R}_k] = 0, \quad [\mathcal{R}_j, \mathcal{R}_k] = 0. \tag{42}$$

One can also prove the equalities

$$\sum_i \mathcal{R}_i = 0, \tag{43}$$

and

$$\sum_i \epsilon_i \mathcal{R}_i = -2\sum_{i \neq j} \mathbf{S}_i \cdot \mathbf{S}_j. \tag{44}$$

A careful examination of the Gaudin algebra in eqs. (27), (28), and (29) indicates that not only the operators $\mathbf{J}(\lambda)$, but also the operators $\mathbf{J}(\lambda) + \mathbf{c}$ satisfy this algebra for a constant $\mathbf{c}$. In this case the conserved quantity is replaced by

$$(45) \qquad H(\lambda) = \mathbf{J}(\lambda) \cdot \mathbf{J}(\lambda) \Rightarrow H(\lambda) + 2\mathbf{c} \cdot \mathbf{J}(\lambda) + \mathbf{c}^2$$

which has the same eigenstates. One can define the Richardson operators, R_k, in an analogous way to the $\mathcal{R}_k$ defined in eq. (40):

$$(46) \qquad \lim_{\lambda \to \epsilon_k} (\lambda - \epsilon_k)\,(H(\lambda) + 2\mathbf{c} \cdot \mathbf{S}) = R_k,$$

which implies

$$(47) \qquad R_k = -2\mathbf{c} \cdot \mathbf{S}_k - 2\sum_{j \neq k} \frac{\mathbf{S}_k \cdot \mathbf{S}_j}{\epsilon_k - \epsilon_j}\,.$$

Equation (42) is then replaced by

$$(48) \qquad [H(\lambda) + 2\mathbf{c} \cdot \mathbf{S}, R_k] = 0 \quad [R_j, R_k] = 0,$$

with the conditions

$$(49) \qquad \sum_i R_i = -2\mathbf{c} \cdot \sum_k \mathbf{S}_k$$

and

$$(50) \qquad \sum_i \epsilon_i R_i = -2\sum_i \epsilon_i \mathbf{c} \cdot \mathbf{S}_i - 2\sum_{i \neq j} \mathbf{S}_i \cdot \mathbf{S}_j.$$

Rewriting the Hamiltonian of eq. (25) in the form

$$(51) \qquad \Rightarrow H = \sum_j \epsilon_j S_j^0 - |G| \left(\left(\sum_i \mathbf{S}_i\right) \cdot \left(\sum_i \mathbf{S}_i\right) - \left(\sum_i S_i^0\right)^2 + \left(\sum_i S_i^0\right) \right)$$
$$+ \text{ constant terms,}$$

and choosing the constant vector of eq. (45) to be

$$(52) \qquad \mathbf{c} = (0, 0, -1/2|G|)$$

one immediately obtains

$$(53) \qquad \frac{H}{|G|} = \sum_i \epsilon_i R_i + |G|^2 \left(\sum_i R_i\right)^2 - |G| \sum_i R_i + \cdots$$

Since all R_k and $H(\lambda)+2\mathbf{c}\cdot\mathbf{S}$ mutually commute (cf. eq. (48)), they have the same eigenvalues. Equation (53) then tells us that they are also the eigenvalues of the Richardson Hamiltonian, eq. (25).

2'3. *Exact solution for degenerate spectra.* – If the single-particle spectrum is degenerate, then it is possible to find the eigenvalues and eigenstates of the Hamiltonian in eq. (22). If all the single-particle energies are the same, the first term in eq. (22) is a constant and can be ignored. Defining the operators

$$(54) \qquad \hat{S}^+(0) = \sum_j c_j^* \hat{S}_j^+ \qquad \text{and} \qquad \hat{S}^-(0) = \sum_j c_j \hat{S}_j^-,$$

the Hamiltonian of eq. (22) can be rewritten as

$$(55) \qquad \hat{H} = -|G|\hat{S}^+(0)\hat{S}^-(0) + \text{constant}.$$

(The operators in eq. (54) are defined with argument 0 for reasons explained in the following discussion). In the 1970's Talmi showed that, under certain assumptions, a state of the form

$$(56) \qquad \hat{S}^+(0)|0\rangle = \sum_j c_j^* \hat{S}_j^+ |0\rangle,$$

with $|0\rangle$ being the particle vacuum, is an eigenstate of a class of Hamiltonians including the one above [30]. A direct calculation yields the eigenvalue equation

$$(57) \qquad \hat{H}\hat{S}^+(0)|0\rangle = \left(-|G|\sum_j \Omega_j |c_j|^2\right) \hat{S}^+(0)|0\rangle.$$

However, there are other one-pair states besides the one in eq. (57). For example for two levels j_1 and j_2, the orthogonal state,

$$(58) \qquad \left(\frac{c_{j_2}}{\Omega_{j_1}}\hat{S}_{j_1}^+ - \frac{c_{j_1}}{\Omega_{j_2}}\hat{S}_{j_2}^+\right)|0\rangle,$$

is also an eigenstate with $E = 0$. In ref. [23] it was shown that there is a systematic way to derive these states. To see their solution we define the operators

$$(59) \qquad \hat{S}^+(x) = \sum_j \frac{c_j^*}{1 - |c_j|^2 x}\hat{S}_j^+ \qquad \text{and} \qquad \hat{S}^-(x) = \sum_j \frac{c_j}{1 - |c_j|^2 x}\hat{S}_j^-.$$

Note that if one substitutes $x = 0$ in the operators of eq. (59), one obtains the operators in eq. (54). Further defining the operator

$$(60) \qquad \hat{K}^0(x) = \sum_j \frac{1}{1/|c_j|^2 - x}\hat{S}_j^0,$$

one can prove the following commutation relations:

$$(61) \qquad \left[\hat{S}^+(x), \hat{S}^-(0)\right] = \left[\hat{S}^+(0), \hat{S}^-(x)\right] = 2K^0(x),$$

$$(62) \qquad \left[\hat{K}^0(x), \hat{S}^\pm(y)\right] = \pm \frac{\hat{S}^\pm(x) - \hat{S}^\pm(y)}{x - y}.$$

These commutators are very similar to, but not the same as, those of the Gaudin algebra described in eqs. (27), (28), and (29). Using the commutators in eqs. (61) and (62) one can easily show that

$$(63) \qquad \hat{S}^+(0)\hat{S}^+\left(z_1^{(N)}\right) \ldots \hat{S}^+\left(z_{N-1}^{(N)}\right)|0\rangle$$

is an eigenstate of the Hamiltonian in eq. (55) if the following Bethe ansatz equations are satisfied:

$$(64) \qquad \sum_j \frac{-\Omega_j/2}{1/|c_j|^2 - z_m^{(N)}} = \frac{1}{z_m^{(N)}} + \sum_{k=1(k\neq m)}^{N-1} \frac{1}{z_m^{(N)} - z_k^{(N)}}, \qquad m = 1, 2, \ldots N-1.$$

The energy of the state in eq. (63) is

$$(65) \qquad E_N = -|G| \left(\sum_j \Omega_j |c_j|^2 - \sum_{k=1}^{N-1} \frac{2}{z_k^{(N)}} \right).$$

The authors of ref. [23] used a Laurent expansion of the operators given in eq. (59) around $x = 0$ followed by an analytic continuation argument to show the validity of their results in the entire complex plane except some singular points. The derivation sketched above instead utilizes the algebra depicted in eqs. (61) and (62); it is significantly simpler. The state in eq. (63) is an eigenstate if the shell is at most half full. Similarly the state

$$(66) \qquad \hat{S}^+\left(x_1^{(N)}\right)\hat{S}^+\left(x_2^{(N)}\right) \ldots \hat{S}^+\left(x_N^{(N)}\right)|0\rangle$$

is an eigenstate with zero energy if the following Bethe ansatz equations are satisfied:

$$(67) \qquad \sum_j \frac{-\Omega_j/2}{1/|c_j|^2 - x_m^{(N)}} = \sum_{k=1(k\neq m)}^{N} \frac{1}{x_m^{(N)} - x_k^{(N)}} \qquad \text{for every} \quad m = 1, 2, \ldots, N.$$

Again this is an eigenstate if the shell is at most half full.

To figure out what happens if the available states are more than half full we note that there are degeneracies in the spectra. Let us denote the state where all levels are completely filled by $|\bar{0}\rangle$. It is easy to show that the state $|\bar{0}\rangle$ and the particle vacuum, $|0\rangle$ are both eigenstates of the Hamiltonian of eq. (55) with the same energy,

Table I. – *Particle-hole degeneracy—two different states with the same energy.*

No. of pairs	State
N	$\hat{S}^+(0)\hat{S}^+(z_1^{(N)})\ldots\hat{S}^+(z_{N-1}^{(N)})|0\rangle$
$N_{\max}+1-N$	$\hat{S}^-(z_1^{(N)})\hat{S}^-(z_2^{(N)})\ldots\hat{S}^-(z_{N-1}^{(N)})|\bar{0}\rangle$

$E = -|G|\sum_j \Omega_j |c_j|^2$. This suggests that if the shells are more than half full, then one should start with a state of the form

$$(68) \qquad \hat{S}^-\left(z_1^{(N)}\right)\hat{S}^-\left(z_2^{(N)}\right)\ldots\hat{S}^-\left(z_{N-1}^{(N)}\right)|\bar{0}\rangle.$$

Indeed the state in eq. (68) is an eigenstate of the Hamiltonian in eq. (55) with the energy

$$(69) \qquad E = -G\left(\sum_j \Omega_j |c_j|^2 - \sum_{k=1}^{N-1}\frac{2}{z_k^{(N)}}\right),$$

if the following Bethe ansatz equations are satisfied [24, 25]:

$$(70) \qquad \sum_j \frac{-\Omega_j/2}{1/|c_j|^2 - z_m^{(N)}} = \frac{1}{z_m^{(N)}} + \sum_{k=1(k\neq m)}^{N-1}\frac{1}{z_m^{(N)} - z_k^{(N)}}.$$

In eq. (70) $N_{\max}+1-N$ is the number of particle pairs. If the available particle states are more than half full, then there are no zero energy states. Note that the Bethe ansatz conditions in eqs. (70) and (64) as well the energies in eqs. (69) and (65) are identical. This particle-hole degeneracy is due to a hidden supersymmetry [25] and is illustrated in table I. This supersymmetry is described in subsect. 3˙2.

2˙4. *Exact solutions with two shells.* – Consider the most general pairing Hamiltonian with only two shells:

$$(71) \qquad \frac{\hat{H}}{|G|} = \sum_j 2\varepsilon_j \hat{S}_j^0 - \sum_{jj'} c_j^* c_{j'}\hat{S}_j^+ \hat{S}_{j'}^- + \sum_j \varepsilon_j \Omega_j,$$

with $\varepsilon_j = \epsilon_j/|G|$. It turns out that this problem can be solved using techniques illustrated in the previous sections. Eigenstates of the Hamiltonian in eq. (71) can be written using the new step operators [26]

$$(72) \qquad \mathcal{J}^+(x) = \sum_j \frac{c_j^*}{2\varepsilon_j - |c_j|^2 x}S_j^+$$

as

$$(73) \qquad \mathcal{J}^+(x_1)\mathcal{J}^+(x_2)\cdots\mathcal{J}^+(x_N)|0\rangle.$$

Introducing the definitions

$$(74) \qquad \beta = 2\frac{\varepsilon_{j_1} - \varepsilon_{j_2}}{|c_{j_1}|^2 - |c_{j_2}|^2}, \qquad \delta = 2\frac{\varepsilon_{j_2}|c_{j_1}|^2 - \varepsilon_{j_1}|c_{j_2}|^2}{|c_{j_1}|^2 - |c_{j_2}|^2},$$

we obtain the energy eigenvalue to be

$$(75) \qquad E_N = -\sum_{n=1}^{N}\frac{\delta x_n}{\beta - x_n},$$

if the parameters x_k satisfy the Bethe ansatz equations

$$(76) \qquad \sum_j \frac{\Omega_j |c_j|^2}{2\varepsilon_j - |c_j|^2 x_k} = \frac{\beta}{\beta - x_k} + \sum_{n=1(\neq k)}^{N}\frac{2}{x_n - x_k}.$$

2˙5. *Solutions of Bethe ansatz equations.* – Various solutions of the pairing problem discussed in the previous sections are typically considered as semi-analytical solutions since one still needs to find the solutions of the Bethe ansatz equations. This a task which, quite often, needs to be tackled numerically. However, in certain limits it is possible to find solutions of the Bethe ansatz equations analytically. The method outlined here was first presented in ref. [31].

Consider the Bethe ansatz equations for the degenerate single-particle levels and zero-energy eigenstate given in eq. (67). Introducing new variables, η_i,

$$(77) \qquad x_i^{(N)} = \frac{1}{|c_{j_2}|^2} + \eta_i^{(N)}\left(\frac{1}{|c_{j_1}|^2} - \frac{1}{|c_{j_2}|^2}\right).$$

Equation (67) can be rewritten as

$$(78) \qquad \sum_{k=1(k\neq i)}^{N}\frac{1}{\eta_i^{(N)} - \eta_k^{(N)}} - \frac{\Omega_{j_2}/2}{\eta_i^{(N)}} + \frac{\Omega_{j_1}/2}{1 - \eta_i^{(N)}} = 0.$$

It can be easily shown that the polynomial admitting the solutions of eq. (78) as zeros

$$(79) \qquad p_N(z) = \prod_{i=1}^{N}\left(z - \eta_i^{(N)}\right)$$

satisfies the hypergeometric equation [32]

$$(80) \qquad z(1-z)p_N'' + \left[-\Omega_{j_2} + (\Omega_{j_1}\Omega_{j_2})z\right]p_N' + N(N - \Omega_{j_1} - \Omega_{j_2} - 1)p_N = 0.$$

Consequently the problem of finding the solutions of the Bethe ansatz equation (67) reduces to calculating the roots of hypergeometric functions. For analytical expressions of the energy eigenvalues obtained in this manner the reader is referred to ref. [24].

3. – Supersymmetric quantum mechanics in nuclear physics

Consider two Hamiltonians

$$(81) \qquad H_1 = G^\dagger G, \ H_2 = GG^\dagger,$$

where G is an arbitrary operator. The eigenvalues of these two Hamiltonians

$$(82) \qquad \begin{aligned} G^\dagger G|1,n\rangle &= E_n^{(1)}|1,n\rangle, \\ GG^\dagger|2,n\rangle &= E_n^{(2)}|2,n\rangle \end{aligned}$$

are the same:

$$(83) \qquad E_n^{(1)} = E_n^{(2)} = E_n$$

and the eigenvectors are related:

$$(84) \qquad |2,n\rangle = G\left[G^\dagger G\right]^{-1/2}|1,n\rangle.$$

This works for all cases except when $G|1,n\rangle = 0$, which should be the ground state energy of the positive-definite Hamiltonian H_1.

To see why this is called supersymmetry we define

$$(85) \qquad Q^\dagger = \begin{pmatrix} 0 & 0 \\ G^\dagger & 0 \end{pmatrix}, \quad Q = \begin{pmatrix} 0 & G \\ 0 & 0 \end{pmatrix},$$

Then

$$(86) \qquad H = \{Q, Q^\dagger\} = \begin{pmatrix} H_2 & 0 \\ 0 & H_1 \end{pmatrix}.$$

with

$$(87) \qquad [H, Q] = 0 = [H, Q^\dagger].$$

Clearly the operators H, Q, and $Q^\dagger$ close under the commutation and anticommutation relations of eqs. (86) and (87), forming a very simple superalgebra. The two Hamiltonians depicted in eq. (81) are said to form a system of supersymmetric quantum mechanics [33]. Many aspects of the supersymmetric quantum mechanics has been investigated in detail [34-36]. In the following sections two applications of supersymmetric quantum mechanics to nuclear physics are summarized.

3˙1. *Application of supersymmetric quantum mechanics to pseudo-orbital angular momentum and pseudo-spin.* – The nuclear shell model is a mean-field theory where the single-particle levels can be taken as those of a three-dimensional harmonic oscillator (labeled with $SU(3)$ quantum numbers) for the lowest ($A \leq 20$) levels. For heavier nuclei with more than 20 protons or neutrons, different parity orbitals mix. The Nilsson Hamiltonian of the spherical shell model is [37]

$$(88) \qquad H = \omega b_i^\dagger b_i - 2k\mathbf{L} \cdot \mathbf{S} - k\mu \mathbf{L}^2,$$

where the second term mixes opposite parity orbitals and the last term mocks up the deeper potential felt by the nucleons as L increases.

Fits to data indicate that $\mu \approx 0.5$, hence there exists degeneracies in the single-particle spectra. In the 50–82 shell (the $SU(3)$ label or the principal harmonic-oscillator quantum number of which is $N = 4$) the $s_{1/2}$ and $d_{3/2}$ orbitals and further $d_{5/2}$ and $g_{7/2}$ orbitals are almost degenerate. It is possible to give a phenomenological account of this degeneracy by introducing a second $SU(3)$ algebra called the pseudo-$SU(3)$ [38,39]. Assuming that those orbitals belong to the $N = 3$ (with $\ell = 1, 3$) representation of the latter $SU(3)$ algebra, the quantum numbers of the $SO(3)$ algebra included in this new $SU(3)$ are called pseudo-orbital-angular momentum ($\ell = 1, 3$ in this case). We can also introduce a "pseudo-spin" ($s = 1/2$). One can easily show that $j = 1/2$ and $3/2$ orbitals (and also $j = 5/2$ and $7/2$ orbitals) are degenerate if pseudo-orbital angular momentum and pseudo-spin coupling vanishes. This degeneracy follows from the supersymmetric quantum-mechanical nature of the problem.

It can be shown that two Hamiltonians written in the $SU(3)$ and the pseudo-$SU(3)$ bases are supersymmetric partners of each other [40]. The operator that transforms these two bases into one another is [41]

$$(89) \qquad U = G \left[G^\dagger G\right]^{-1/2} = \sqrt{2}F_- \left(K_0 + [F_+, F_-]\right)^{-1/2}$$
$$= \left(\sigma_i b_i^\dagger\right)\left(b_i^\dagger b_i - \sigma_i L_i\right)^{-1/2}$$

yielding the supersymmetry transformation between the pseudo-$SU(3)$ Hamiltonian H' and the $SU(3)$ Hamiltonian H:

$$(90) \qquad H' = U H U^\dagger = b_i^\dagger b_i - 2k\left(2\mu - 1\right)\mathbf{L} \cdot \mathbf{S} - k\mu \mathbf{L}^2 + \left[1 - 2k(\mu - 1)\right].$$

The transformation depicted in eq. (90) can also be expressed in terms of the generators of the orthosymplectic superalgebra Osp(1/2) [40].

3˙2. *Supersymmetric quantum mechanics and pairing in nuclei.* – Let us consider the separable pairing Hamiltonian with degenerate single-particle spectra given in eq. (55):

$$(91) \qquad \hat{H}_{\text{SC}} \sim -|G|\hat{S}^+(0)\hat{S}^-(0),$$

and introduce the operator

$$(92) \qquad \hat{T} = \exp\left[-i\frac{\pi}{2}\sum_i \left(\hat{S}_i^+ + \hat{S}_i^-\right)\right].$$

This operator transforms the empty shell, $|0\rangle$, to the fully occupied shell, $|\bar{0}\rangle$:

$$(93) \qquad \hat{T}|0\rangle = |\bar{0}\rangle.$$

To establish the connection to the supersymmetric quantum mechanics we define the operators

$$(94) \qquad \hat{B}^- = \hat{T}^\dagger \hat{S}^-(0), \qquad \hat{B}^+ = \hat{S}^+(0)\hat{T}.$$

Supersymmetric quantum mechanics tells us that the partner Hamiltonians $\hat{H}_1 = \hat{B}^+\hat{B}^-$ and $\hat{H}_2 = \hat{B}^-\hat{B}^+$ have identical spectra except for the ground state of $\hat{H}_1$. It can easily be shown that in this case two Hamiltonians $\hat{H}_1$ and $\hat{H}_2$ are actually identical and equal to the pairing Hamiltonian (eq. (91)). Hence the role of the supersymmetry is to connect the states $|1, n\rangle$ and $|2, n\rangle$. These are the "particle" and "hole" states. Hence if the shell is less than half-full ("particles") there is a zero-energy state (note that the Hamiltonian in eq. (55) negative definite, hence this is the highest energy state). If the shell is more than half-full ("holes") this state disappears. Otherwise the spectra for particle and hole states are the same as the rules of the supersymmetric quantum mechanics implies [25].

4. – Dynamical supersymmetries in nuclear physics

Dynamical supersymmetries in nuclear physics start with the algebraic model of nuclear collectivity called interacting boson model [6-8]. In this model, low-lying quadrupole collective states of even-even nuclei are generated as states of a system of bosons occupying two levels, one with angular momentum zero (s-boson) and one with angular momentum two (d-boson). It is discussed elsewhere in great detail in these proceedings [42, 43]. There are three exactly solvable limits of the simplest form of the interacting boson model where neutron and proton bosons are not distinguished:

– Vibrational limit: $SU(6) \supset SU(5) \supset SO(5) \supset SO(3)$.

– Rotational limit: $SU(6) \supset SU(3) \supset SO(3)$.

– Gamma-unstable limit: $SU(6) \supset SO(6) \supset SO(5) \supset SO(3)$.

It is possible to extend the interacting boson model to describe odd-even and odd-odd nuclei [44]. Dynamical supersymmetries arise out of certain exactly solvable limits of this interacting boson-fermion model [45, 46].

In an odd-even nucleus, in addition to the correlated nucleon pairs (s and d bosons) we need the degrees of freedom of the unpaired fermions. If those unpaired fermions are in the $j_1, j_2, j_3, \cdots$ orbitals, then the fermionic sector of the theory is represented by the fermionic algebra $SU_{\mathrm{F}}(\sum_i(2j_i+1))$ and the resulting $SU(6)_{\mathrm{B}} \times SU_{\mathrm{F}}(\sum_i(2j_i+1))$ algebra is embedded in the superalgebra $SU(6/\sum_i(2j_i+1))$. In the first example of dynamical supersymmetry worked out the unpaired fermion was in a $j = 3/2$ $(d_{3/2})$ orbital coupled to the nuclei described by the gamma-unstable $(SO(6))$ limit of the interacting boson model. The resulting $SU(6/4)$ supersymmetry was used to describe many properties of nuclei in the Os-Pt region [47]. Immediately afterwards this supersymmetry was extended to the $SU(6/12)$ superalgebra including fermions in $s_{1/2}$, $d_{3/2}$, and $d_{5/2}$ orbitals [48]. Theoretical implications of nuclear supersymmetries were extensively investigated by many authors [49-54]. The existence of dynamical supersymmetries in nuclei is experimentally established [55-60].

There exist other symmetries of nuclei describing phase transitions between different dynamical symmetry limits. For these so-called critical-point symmetries zeros of wavefunctions in confining potentials of the geometric model of nuclei give rise to point groups symmetries [2]. This scheme is generally applicable to the spectra of systems undergoing a second-order phase transition between the dynamical symmetry limits $SU(n-1)$ and $SO(n)$. The resulting symmetries are either named after the discrete subgroups of the Euclidean group $E(5)$, or $X(5)$. A review of the experimental searches for critical-point symmetries at nuclei in the $A = 130$ and $A = 150$ regions is given in ref. [61]. It is also possible to couple Bohr Hamiltonian of the geometric model with a five-dimensional square well potential to a fermion using the five-dimensional generalization of the spin-orbit interaction. In doing so the $E(5)$ symmetry of the even-even nuclei goes into the $E(5/4)$ supersymmetry for odd-even nuclei near the critical point [62]. Initial experimental tests of this latter supersymmetry are encouraging [63].

5. – Application of symmetry techniques to subbarrier fusion

In the study of nuclear reactions one needs to describe translational motion coupled with internal degrees of freedom representing the structure of colliding nuclei. The Hamiltonian for such a multidimensional quantum problem is taken to be

$$(95) \qquad H = -\frac{\hbar^2}{2\mu}\nabla^2 + V(\mathbf{r}) + H_0(q) + H_{\mathrm{int}}(\mathbf{r}, q),$$

where $\mathbf{r}$ is the relative coordinate of the target-projectile pair and q represents any internal degrees of freedom of the system. To study the effect of the structure of the target nuclei on fusion cross-sections near and below the Coulomb barrier, one can take $V(r)$ to be the potential barrier and the third term in eq. (95) represents the internal structure of the target nuclei. All the dynamical information about the system can be obtained by solving the evolution equation

$$(96) \qquad i\hbar\frac{\partial \hat{U}}{\partial t} = H\hat{U}$$

with the initial condition $\hat{U}(t_i) = 1$. If part of the Hamiltonian in eq. (95), say $H_0(q) + H_{\mathrm{int}}(\mathbf{r}, q)$, is an element of a particular Lie algebra, then symmetry methods can significantly simplify the solution of the problem. In such a case, if one considers a single trajectory $\mathbf{r}(t)$, the q-dependent part of the evolution operator becomes an element of the Lie group associated with the Lie algebra mentioned above. (It is possible to make this statement rigorous in the context of path-integral formalism [64]).

It is by now well-established that heavy-ion fusion cross-sections below the Coulomb barrier are several orders of magnitude larger than one would expect from a one dimensional barrier penetration picture, an enhancement which is attributed to the coupling of the translational motion to additional degrees of freedom such as nuclear and Coulomb excitation, nucleon transfer, or neck formation [65, 66]. The multidimensional barrier penetration problem inherent in subbarrier fusion can be addressed in the coupled-channels formalism and state-of-the-art coupled-channel codes are currently available (see, *e.g.*, [67]). Although several puzzles remain (such as the large values of the surface diffuseness parameter in the nuclear potential required to fit the data [68,69]; very steep fall-off of the fusion data for some systems at extreme subbarrier energies [70]; or inadequacy of the standard nuclear potentials to simultaneously reproduce fusion and elastic-scattering measurements [71]) there is overall good agreement between coupled-channels calculations and the experimental data.

An alternative approach is to formulate the problem algebraically by using a model amenable to such an approach for describing the nuclear structure, such as the interacting boson model [72]. Not only this approach fits the data well, it can also be used for transitional nuclei, the treatment of which could be more complicated for coupled-channels calculations [72-75].

6. – Application of algebraic techniques in nuclear astrophysics

Nuclear astrophysics has been very successful exploring the origin of elements. As our understanding of the heavens evolved, it was realized that nuclear data was needed for astrophysics calculations, such as nucleosynthesis, stellar evolution, the Big-Bang cosmology, X-ray bursts, and supernova dynamics. Nuclear astrophysics initially started with reaction rate measurements, but with the recent rapid growth of the observational data, expanding computational capabilities, and availability of exotic nuclear beams, interest in all aspects of nuclear physics relevant to astrophysical phenomena has significantly increased [76]. In the rest of this section an application of algebraic techniques to a nuclear astrophysics problem is presented.

Light nuclei are formed during the Big-Bang nucleosynthesis era and nuclei up to the iron, nickel and cobalt group are formed during the stellar evolution. A good fraction of the nuclei heavier than iron were formed in the rapid neutron capture (r-process) nucleosynthesis. The astrophysical location of the r-process is expected to be where explosive phenomena are present since a large number of interactions are required to take place at this location during a rather short time interval. Core-collapse supernovae which occur following the stages of nuclear burning during stellar evolution after the formation of an iron core is such a site. Neutrino interactions play a very important role in the evolution of core-collapse supernovae [77]. Almost all (99%) of the gravitational binding energy (10^{53} ergs) of the progenitor star is released in the neutrino cooling of the neutron star formed after the collapse. It was suggested that neutrino-neutrino interactions could play a potentially very significant role in core-collapse supernovae [78-82]. A key quantity for determining the r-process yields is the neutron to seed-nucleus ratio (or equivalently neutron-to-proton ratio). Interactions of the neutrinos and antineutrinos streaming out of the core both with nucleons and seed nuclei determine the neutron-to-proton ratio. Before these neutrinos reach the r-process region they undergo matter-enhanced neutrino oscillations as well as coherently scatter over other neutrinos. Many-body behavior of this neutrino gas is not completely understood, but may have a significant impact on r-process nucleosynthesis.

It was shown that neutrinos moving in matter gain an effective potential due to forward scattering from the background particles such as the electrons [83-85]. One can write the Hamiltonian describing neutrino transport in dense matter for two neutrino flavors as

$$(97) \qquad H_\nu = \int \mathrm{d}p \left(\frac{\delta m^2}{2p} \cos 2\theta - \sqrt{2} G_\mathrm{F} N_\mathrm{e} \right) J_0(p)$$
$$+ \frac{1}{2} \int \mathrm{d}p \, \frac{\delta m^2}{2p} \sin 2\theta \left(J_+(p) + J_-(p) \right),$$

where N_e is the electron density of the medium (assumed to be charge neutral and unpolarized), θ is the mixing angle between neutrino flavors, $\delta m^2 = m_2^2 - m_1^2$ is the difference between squares of the neutrino masses, and G_F is the Fermi coupling strength of the weak interactions. In the above equation the operators $J^\pm$, J^0 has been written in terms of the neutrino creation and annihilation operators:

$$(98) \qquad J_+(p) = a_\mathrm{x}^\dagger(p) a_\mathrm{e}(p), \quad J_-(p) = a_\mathrm{e}^\dagger(p) a_\mathrm{x}(p),$$
$$J_0(p) = \frac{1}{2} \left(a_\mathrm{x}^\dagger(p) a_\mathrm{x}(p) - a_\mathrm{e}^\dagger(p) a_\mathrm{e}(p) \right).$$

Here $a_\mathrm{e}^\dagger(p)$ and $a_\mathrm{x}(p)$ are the creation and annihilation operators for the electron neutrino with momentum p and either muon or tau neutrino with momentum p, respectively. The operators in eq. (98) form as many mutually commuting $SU(2)$ algebras as the number of allowed values of neutrino momenta:

$$(99) \qquad [J_+(p), J_-(q)] = 2\delta^3(p-q) J_0(p), \quad [J_0(p), J_\pm(p)] = \pm \delta^3(p-q) J_\pm(p)$$

If, instead of two neutrino flavors, one considers all three active neutrino flavors then one should use copies of $SU(3)$ algebras. The contribution of neutrino-neutrino forward scattering terms to the neutrino Hamiltonian is given by

$$(100) \qquad H_{\nu\nu} = \sqrt{2} G_F \int \mathrm{d}p \, \mathrm{d}q \left(1 - \cos \vartheta_{pq} \right) \mathbf{J}(p) \cdot \mathbf{J}(q),$$

where ϑ_{pq} is the angle between neutrino momenta $\mathbf{p}$ and $\mathbf{q}$. It is easy to include neutrino-antineutrino and antineutrino-antineutrino scattering terms in this Hamiltonian [81], but we ignore them here to keep the presentation simple.

Since a large number of neutrinos (10^{58}) are emitted during a typical core-collapse it is very difficult to evaluate neutrino evolution exactly using the two-body Hamiltonian of eq. (100). Instead one uses a mean field approximation where the product of two commuting arbitrary operators $\hat{\mathcal{O}}_1$ and $\hat{\mathcal{O}}_2$ can be approximated as

$$(101) \qquad \hat{\mathcal{O}}_1\hat{\mathcal{O}}_2 \sim \hat{\mathcal{O}}_1\langle\xi|\hat{\mathcal{O}}_2|\xi\rangle + \langle\xi|\hat{\mathcal{O}}_1|\xi\rangle\hat{\mathcal{O}}_2 - \langle\xi|\hat{\mathcal{O}}_1|\xi\rangle\langle\xi|\hat{\mathcal{O}}_2|\xi\rangle,$$

provided that the condition

$$(102) \qquad \langle\xi|\hat{\mathcal{O}}_1\hat{\mathcal{O}}_2|\xi\rangle = \langle\xi|\hat{\mathcal{O}}_1|\xi\rangle\langle\xi|\hat{\mathcal{O}}_2|\xi\rangle$$

is satisfied. In eqs. (101) and (102) One can then replace the exact Hamiltonian of eq. (100) with the approximate expression

$$H_{\nu\nu} \sim 2\frac{\sqrt{2}G_F}{V} \int \mathrm{d}p\mathrm{d}q\, R_{pq} \left(J_0(p)\langle J_0(q)\rangle + \frac{1}{2}J_+(p)\langle J_-(q)\rangle + \frac{1}{2}J_-(p)\langle J_+(q)\rangle \right),$$

where $R_{pq} = (1 - \cos\vartheta_{pq})$ and the averages are calculated over the entire ensemble of neutrinos. Systematic corrections to this expression are explored in [81]. Calculations using this approximation do not yield conditions (*i.e.* large enough neutron to seed nucleus ratio) favorable to r-process nucleosynthesis [78]. There is encouraging progress in numerical calculations using the exact Hamiltonian of eq. (100) [79, 80]. However an algebraic solution to this problem is currently lacking.

7. – Conclusions

In many-fermion physics, mean-field approaches can describe many properties; but are inadequate to describe the whole picture; pairing correlations play a crucial role. In fact, pairing correlations are not only necessary to understand the structure of rare-earths and actinides, but, since they are essential for the description of the neutrino gas in a core-collapse supernova where many nuclei are produced, also necessary to understand the existence of such nuclei in the first place. Models exploiting symmetry properties and pairing correlations have been very successful. These also gave rise to dynamical supersymmetries. We showed that using algebraic techniques it is possible to solve the s-wave pairing problem almost exactly (*i.e.* reducing it to Bethe ansatz equations) at least for a number of simplified cases.

* * *

This work was supported in part by the US National Science Foundation Grant No. PHY-0555231 and in part by the University of Wisconsin Research Committee with funds granted by the Wisconsin Alumni Research Foundation.

REFERENCES

[1] Balantekin A. B., *Ann. Phys. (N.Y.)*, **164** (1985) 277.

[2] Iachello F., *Phys. Rev. Lett.*, **85** (2000) 3580; **44** (1980) 772.

[3] Balantekin A. B., Schmitt H. A. and Barrett B. R., *J. Math. Phys.*, **29** (1988) 1634.

[4] Dean D. J. and Hjorth-Jensen M., *Rev. Mod. Phys.*, **75** (2003) 607 [arXiv:nucl-th/0210033].

[5] Mayer M. G., *Phys. Rev.*, **78** (1950) 22.

[6] Arima A. and Iachello F., *Ann. Phys. (N.Y.)*, **99** (1976) 253.

[7] Arima A. and Iachello F., *Ann. Phys. (N.Y.)*, **111** (1978) 201.

[8] Arima A. and Iachello F., *Ann. Phys. (N.Y.)*, **123** (1979) 468.

[9] Page D., Prakash M., Lattimer J. M. and Steiner A., *Phys. Rev. Lett.*, **85** (2000) 2048 [arXiv:hep-ph/0005094].

[10] Bardeen J., Cooper L. N. and Schrieffer J. R., *Phys. Rev.*, **108** (1957) 1175.

[11] Bohr A., Mottelson B. and Pines D., *Phys. Rev.*, **110** (1958) 936.

[12] Belyaev S., *Mat. Fys. Medd. K. Dan. Vidensk. Selsk.*, **31** (1959) 641.

[13] Migdal A., *Nucl. Phys.*, **13** (1959) 655.

[14] Unna I. and Weneser J., *Phys. Rev.*, **137** (1965) B1455.

[15] Kerman A. K., Lawson R. D. and MacFarlane M. H., *Phys. Rev.*, **124** (1961) 162.

[16] Hagino K. and Bertsch G. F., *Nucl. Phys. A*, **679** (2000) 163 [arXiv:nucl-th/0003016];

[17] Hagino K., Sagawa H., Carbonell J. and Schuck P., *Phys. Rev. Lett.*, **99** (2007) 022506 [arXiv:nucl-th/0611064].

[18] von Delft J. and Braun F., arXiv:cond-mat/9911058; von Delft J., Zaikin A. D., Golubev D. S. and Tichy W., *Phys. Rev. Lett.*, **77** (1996) 3189.

[19] Racah G., *Phys. Rev.*, **63** (1943) 367.

[20] Kerman A. K., *Ann. Phys. (N.Y.)*, **12** (1961) 300.

[21] Richardson R. W., *Phys. Lett*, **3** (1963) 277; **14** (1965) 325; *J. Math. Phys.*, **6** (1965) 1034; **18** (1967) 1802; *Phys. Rev.*, **141** (1966) 949; **144** (1966) 874.

[22] Gaudin M., *J. Physique*, **37** (1976) 1087; *La Fonction d'onde de Bethe, Collection du Commissariat a l'énergie atomique* (Masson, Paris) 1983.

[23] Pan F., Draayer J. P. and Ormand W. E., *Phys. Lett. B*, **422** (1998) 1 [arXiv:nucl-th/9709036].

[24] Balantekin A. B., de Jesus J. H. and Pehlivan Y., *Phys. Rev. C*, **75** (2007) 064304 [arXiv:nucl-th/0702059].

[25] Balantekin A. B. and Pehlivan Y., *J. Phys. G*, **34** (2007) 1783 [arXiv:0705.1318 [nucl-th]].

[26] Balantekin A. B. and Pehlivan Y., *Phys. Rev. C*, **76** (2007) 050019(R).

[27] Volya A., Brown B. A. and Zelevinsky V., *Phys. Lett. B*, **509** (2001) 37 [arXiv:nucl-th/0011079].

[28] Schmitt H. A., Halse P., Barrett B. R. and Balantekin A. B., *Phys. Lett. B*, **210** (1988) 1.

[29] Balantekin A. B., Dereli T. and Pehlivan Y., *J. Phys. A*, **38** (2005) 5697 [arXiv:math-ph/0505071].

[30] Talmi T., *Nucl. Phys.*, **A172** (1971) 1.

[31] Balantekin A. B., Dereli T. and Pehlivan Y., *J. Phys. G*, **30** (2004) 1225 [arXiv:nucl-th/0407006]; *Int. J. Mod. Phys. E*, **14** (2005) 47 [arXiv:nucl-th/0505023].

[32] Stieltjes T. J., *Sur Quelques Theoremes d'Algebre, Oeuvres Completes*, Vol. **11** (Noordhoff, Groningen) 1914.

[33] Witten E., *Nucl. Phys. B*, **188** (1981) 513.

[34] COOPER F., KHARE A. and SUKHATME U., *Phys. Rep.*, **251** (1995) 267 [arXiv:hep-th/9405029].

[35] COOPER F., GINOCCHIO J. N. and KHARE A., *Phys. Rev. D*, **36** (1987) 2458.

[36] FRICKE S. H., BALANTEKIN A. B., HATCHELL P. J. and UZER T., *Phys. Rev. A*, **37** (1988) 2797.

[37] NILSSON S. G., *Mat. Fys. Medd. K. Dan. Viden. Selsk.*, **29** (1955) 16.

[38] RATNA RAJU R. D., DRAAYER J. P. and HECHT K. T., *Nucl. Phys. A*, **202** (1973) 433.

[39] ARIMA A., HARVEY M. and SHIMIZU K., *Phys. Lett. B*, **30** (1969) 517.

[40] BALANTEKIN A. B., CASTANOS O. and MOSHINSKY M., *Phys. Lett. B*, **284** (1992) 1.

[41] CASTANOS O., MOSHINSKY M. and QUESNE C., *Phys. Lett. B*, **277** (1992) 238.

[42] CASTEN R., this volume, p. 385.

[43] VAN ISACKER P., this volume, p. 347.

[44] ARIMA A. and IACHELLO F., *Phys. Rev. C*, **14** (1976) 761.

[45] IACHELLO F., *Phys. Rev. Lett.*, **44** (1980) 772.

[46] IACHELLO F. and SCHOLTEN O., *Phys. Rev. Lett.*, **43** (1979) 679.

[47] BALANTEKIN A. B., BARS I. and IACHELLO F., *Phys. Rev. Lett*, **47** (1981) 19; *Nucl. Phys. A*, **370** (1981) 284.

[48] BALANTEKIN A. B., BARS I., BIJKER R. and IACHELLO F., *Phys. Rev. C*, **27** (1983) 1761.

[49] BALANTEKIN A. B. and PAAR V., *Phys. Rev. C*, **34** (1986) 1917.

[50] SCHMITT H. A., HALSE P., BALANTEKIN A. B. and BARRETT B. R., *Phys. Rev. C*, **39** (1989) 2419.

[51] NAVRATIL P., GEYER H. B. and DOBACZEWSKI J., *Nucl. Phys. A*, **607** (1996) 23 [arXiv:nucl-th/9606043]; CEJNAR P. and GEYER H. B., *Phys. Rev. C*, **68** (2003) 054324 [arXiv:nucl-th/0306058];

[52] BAREA J., BIJKER R., FRANK A. and LOYOLA G., *Phys. Rev. C*, **64** (2001) 064313 [arXiv:nucl-th/0107048].

[53] LEVIATAN A., *Phys. Rev. Lett.*, **92** (2004) 202501 (**92** (2004) 219902 (Erratum)) [arXiv:nucl-th/0312018]; *Int. J. Mod. Phys. E*, **14** (2005) 111 [arXiv:nucl-th/0407107].

[54] BAREA J., ALONSO C. E., ARIAS J. M. and JOLIE J., *Phys. Rev. C*, **71** (2005) 014314.

[55] MAUTHOFER A. *et al.*, *Phys. Rev. C*, **34** (1986) 1958; **39** (1989) 1111.

[56] ROTBARD G. *et al.*, *Phys. Rev. C*, **55** (1997) 1200.

[57] METZ A., EISERMANN Y., GOLLWITZER A., HERTENBERGER R., VALNION B. D., GRAW G. and JOLIE J., *Phys. Rev. C*, **61** (2000) 064313 (**67** (2003) 049901 (Erratum)).

[58] ALGORA A., JOLIE J., DOMBRADI Z., SOHLER D., PODOLYAK Z. and FENYES T., *Phys. Rev. C*, **67** (2003) 044303.

[59] BAREA J., BIJKER R. and FRANK A., *Phys. Rev. Lett.*, **94** (2005) 152501 [arXiv:nucl-th/0412090].

[60] WIRTH H. F. *et al.*, *Phys. Rev. C*, **70** (2004) 014610.

[61] MCCUTCHAN E. A., ZAMFIR N. V. and CASTEN R. F., *J. Phys. G*, **31** (2005) S1485.

[62] CAPRIO M. A. and IACHELLO F., *Nucl. Phys. A*, **781** (2007) 26 [arXiv:nucl-th/0610026].

[63] FETEA M. S. *et al.*, *Phys. Rev. C*, **73** (2006) 051301.

[64] BALANTEKIN A. B. and TAKIGAWA N., *Ann. Phys. (N.Y.)*, **160** (1985) 441.

[65] BALANTEKIN A. B. and TAKIGAWA N., *Rev. Mod. Phys.*, **70** (1998) 77 [arXiv:nucl-th/9708036].

[66] DASGUPTA M., HINDE D. J., ROWLEY N. and STEFANINI A. M., *Annu. Rev. Nucl. Part. Sci.*, **48** (1998) 401.

[67] HAGINO K., ROWLEY N. and KRUPPA A. T., *Comput. Phys. Commun.*, **123** (1999) 143 [arXiv:nucl-th/9903074].

[68] HAGINO K., ROWLEY N. and DASGUPTA M., *Phys. Rev. C*, **67** (2003) 054603 [arXiv:nucl-th/0302025].

[69] HAGINO K., TAKEHI T., BALANTEKIN A. B. and TAKIGAWA N., *Phys. Rev. C*, **71** (2005) 044612 [arXiv:nucl-th/0412044].

[70] JIANG C. L. *et al.*, *Phys. Rev. C*, **69** (2004) 014604; *Phys. Rev. Lett.*, **89** (1992) 052701.

[71] MUKHERJEE A., HINDE D. J., DASGUPTA M., HAGINO K, NEWTON J. O. and BUTT R. D., *Phys. Rev. C*, **75** (2007) 044608.

[72] BALANTEKIN A. B., BENNETT J. R. and TAKIGAWA N., *Phys. Rev. C*, **44** (1991) 145.

[73] BALANTEKIN A. B, BENNETT J. R., DEWEERD A. J. and KUYUCAK S., *Phys. Rev. C*, **46** (1992) 2019.

[74] BALANTEKIN A. B., BENNETT J. R. and KUYUCAK S., *Phys. Rev. C*, **48** (1993) 1269; **49** (1994) 1079; 1294; *Phys. Lett. B*, **335** (1994) 295 [arXiv:nucl-th/9407037].

[75] BALANTEKIN A. B. and KUYUCAK S., *J. Phys. G*, **23** (1997) 1159 [arXiv:nucl-th/9706068].

[76] LANGANKE K., this volume, p. 317.

[77] BALANTEKIN A. B. and FULLER G. M., *J. Phys. G*, **29** (2003) 2513 [arXiv:astro-ph/0309519].

[78] BALANTEKIN A. B. and YUKSEL H., *New J. Phys.*, **7** (2005) 51 [arXiv:astro-ph/0411159].

[79] DUAN H., FULLER G. M. and QIAN Y. Z., *Phys. Rev. D*, **74** (2006) 123004 [arXiv:astro-ph/0511275]; **76** (2007) 085013 [arXiv:0706.4293 [astro-ph]].

[80] DUAN H., FULLER G. M., CARLSON J. and QIAN Y. Z., *Rev. Lett.*, **97** (2006) 241101 [arXiv:astro-ph/0608050]; *Phys. Rev. D*, **74** (2006) 105014 [arXiv:astro-ph/0606616]; **75** (2007) 125005 [arXiv:astro-ph/0703776].

[81] BALANTEKIN A. B. and PEHLIVAN Y., *J. Phys. G*, **34** (2007) 47 [arXiv:astro-ph/0607527].

[82] HANNESTAD S., RAFFELT G. G., SIGL G. and WONG Y. Y., *Phys. Rev. D*, **74** (2006) 105010 (**76** (2007) 029901 (Erratum)) [arXiv:astro-ph/0608695]; RAFFELT G. G. and SMIRNOV A. Y., *Phys. Rev. D*, **76** (2007) 081301 [arXiv:0705.1830 [hep-ph]].

[83] WOLFENSTEIN L., *Phys. Rev. D*, **17** (1978) 2369.

[84] MIKHEEV S. P. and SMIRNOV A. Y., *Sov. J. Nucl. Phys.*, **42** (1985) 913 (*Yad. Fiz.*, **42** (1985) 1441).

[85] MIKHEEV S. P. and SMIRNOV A. Y., *Nuovo Cimento C*, **9** (1986) 17.

DOI 10.3254/978-1-58603-885-4-447

Dynamical symmetries and regular *vs.* chaotic quantum motion in realistic models of nuclear structure

G. MAINO(*)

ENEA - via Martiri di Montesole 1, 40129, Bologna, Italy
Università di Bologna, Sede di Ravenna - via degli Ariani 1, 48100 Ravenna, Italy

Summary. — A quantum-statistical analysis of the regular and chaotic dynamics of medium-mass even-even nuclei within the framework of the interacting boson model-2 (IBM-2) has been carried out. In particular, in the $SU_{\pi+\nu}(3) \to U_{\pi+\nu}(5)$ transition a broad nearly regular region has been observed, while the $SU_{\pi+\nu}(3)^* \to U_{\pi+\nu}(5)$ one shows an unexpected completely regular behavior. These results confirm and extend the observations previously made by other authors in the frame of IBM-1, strengthening the hypothesis of the basic role played by the partial dynamical symmetries in preserving regular motion patterns even far from the usual dynamical limits of IBM-2. A new important feature of the $SU_{\pi+\nu}(3)^*$ limit has also been found, confirming the importance of the $SU(3)$ symmetry of IBM in studying nuclear-structure properties.

(*) E-mail: giuseppe.maino@bologna.enea.it

1. – Introduction

The algebraic models of nuclear structure have both boson and fermion degrees of freedom. Hence, on the one hand they can describe both collective and single-particle nuclear states, on the other hand, they lend themselves to a microscopical interpretation within the framework of the shell model [1], since the so-called bosons are comparable to correlated pairs of nucleons. Moreover, if the topologies joined to every algebraic structure are considered, it is possible to define a univoque correspondence between the algebraic model and the usual geometric models (collective model, asymmetric rotor model (ARM), and so on). Thus algebraic models, such as the interacting boson and fermion model originally introduced by Arima and Iachello [2], are able to establish a link between the semiclassical traditional description of the nucleus and the shell model, supplying a generalization of the results concerned with the collective rotational modes of light nuclei ($A < 40$), which in 1958, for the first time, Elliott obtained in the frame of the $SU(3)$ scheme [3].

The Interacting Boson Model (IBM) [4] represents a suitable and efficient way to truncate the complete space of the shell model states, leading to a simple numerical calculation of the eigenvalues for a many-body non-relativistic quantum system, such as a medium- or heavy-mass nucleus. Moreover, there are particular situations in which the algebraic structure of the model allows to obtain analytical solutions of the eigenvalue problem, in that case the IBM Hamiltonian is said to have an exact dynamical symmetry. The interacting boson model is therefore a realistic model of nuclear structure, which has two fundamental features:

– it is able to give expectations as regards the nuclei far from the beta stability valley. It is therefore interesting from an astrophysics point of view.

– It displays a remarkable computational simplicity and permits to gain analytical solutions which, in the classical limit, correspond to completely integrable systems showing a regular dynamical behavior.

In this paper I present and discuss a quantum investigation of the regular and chaotic dynamics of medium-mass ($A > 40$) even-even nuclei within the framework of the interacting boson model version 2 (IBM-2) [4], which is a realistic model of nuclear structure where the isospin degree of freedom is explicitly introduced. This analysis has been performed studying the fluctuation properties (departures from spectral uniformity) of nuclear levels by means of the Random Matrix Theory (RMT) [5]. In particular, use has been made of the Gaussian Orthogonal Ensemble (GOE) since chaotic many-body systems with time reversal symmetry (like nuclei) are associated with it. A detailed description of the behavior of the corresponding semiclassical systems by using the usual tools of the classical investigation (namely, analysis of Poincaré maps, determination of Lyapunov exponents, and so on) that confirms the present results is worth mentioning and will be the object of further investigation.

The results I obtained show a persistence of regular motion patterns even far from dynamical symmetry limits of the IBM-2 model ($U_{\pi+\nu}(5)$ vibrational nucleus, $SU_{\pi+\nu}(3)$ rotational nucleus, $O_{\pi+\nu}(6)$ γ-unstable nucleus, and $SU_{\pi+\nu}(3)^*$ triaxial nucleus). An

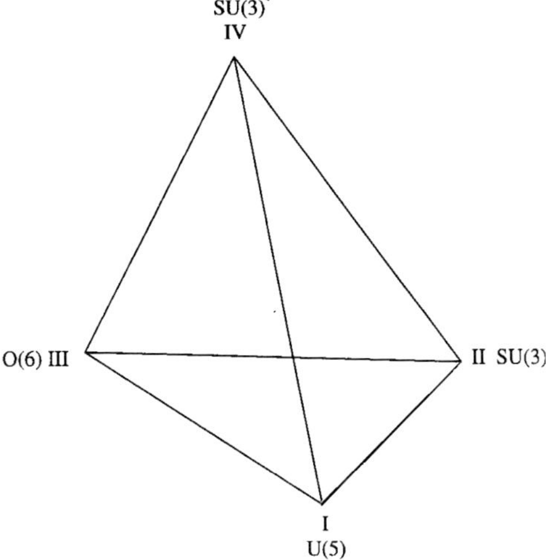

Fig. 1. – Dieperink's tetrahedron.

explanation of the persistence of regular, rather than chaotic, features even when strong violations of the usual dynamical symmetries have occurred, can be connected with the existence—investigated in the last years [6-9]—of the so-called *partial dynamical symmetries*, since they may cause suppression (namely, reduction) of the chaotic behavior, as shown in ref. [7]. In this connection extremely interesting is the $SU_{\pi+\nu}(3) \to U_{\pi+\nu}(5)$ transition which displays a broad *nearly regular region* whose presence can be explained in terms of partial dynamical $SU_{\pi+\nu}(3)$ symmetry, like previously made by Alhassid, Leviatan, and coworkers [6-12] in the frame of the original version of the interacting boson model (IBM-1) [2]. In particular, the observation of a broad nearly regular region in the $SU_{\pi+\nu}(3) \to U_{\pi+\nu}(5)$ shape-phase transition, represents a significant extension of the results formerly obtained by Whelan and Alhassid [10] within the framework of IBM-1. Indeed, they found that the nearly regular region connecting the $SU_{\pi+\nu}(3)$ and $U_{\pi+\nu}(5)$ vertices inside the Casten's triangle was a narrow band (see figs. 4 and 5 in ref. [10]), while our results clearly show that it is a larger region. The fundamental role played by $SU_{\pi+\nu}(3)$ partial dynamical symmetry as well as critical point and phase transitions symmetries, introduced some years ago by Iachello [11], in maintaining regular motion patterns even far from the usual dynamical limits of the IBM model [12] is further confirmed by the unexpected completely regular behavior observed in the $SU_{\pi+\nu}(3)^* \to U_{\pi+\nu}(5)$ transition. The latter result, with other ones that will be presented later, induced us to suppose, for the first time, that the $SU_{\pi+\nu}(3)^*$ limit has a deeper attraction basin than that of the $SU_{\pi+\nu}(3)$ dynamical symmetry. Our results allowed also us to get a further confirmation of the importance of the $SU(3)$ limit of IBM in analyzing many nuclear structure properties, such as energy levels, electromagnetic transitions and moments. For example, let us remember Elliott's $SU(3)$ model of collective rotational motions in light-mass nuclei [4], and the liquid-drop model.

Section **2** is devoted to a very brief picture of the IBM-2 algebraic tools used to evaluate the shape-phase transitions along Dieperink's tetrahedron (see fig. 1). In sect. **3** the main features of the statistical distributions, employed to carry out the quantum investigation of the regular and chaotic behavior of nuclear spectra, are given and in the following section **4** the main results of our theoretical analysis of IBM-2 spectra are presented and discussed. Finally, conclusions and perspectives for future developments of these studies are drawn in sect. **5**.

2. – Algebraic features of the IBM-2

The IBM-2 Hamiltonian is the so-called Talmi Hamiltonian [13]

$$(1) \qquad H_{\text{TALMI}} = E_0 + \varepsilon_\pi \, \widehat{n}_{d_\pi} + \varepsilon_\nu \, \widehat{n}_{d_\nu} + k \, \widehat{Q}_\pi^\chi \cdot \widehat{Q}_\nu^\chi + \lambda' \widehat{M}_{\pi\nu} + \widehat{V}_{\pi\pi} + \widehat{V}_{\nu\nu} \,,$$

where, quite often, the assumption $\varepsilon_\pi = \varepsilon_\nu = \varepsilon$ is made. In eq. (1), E_0 is a function which does not contribute to the excitation energy, while $\varepsilon_\pi \, \widehat{n}_{d_\pi} + \varepsilon_\nu \, \widehat{n}_{d_\nu}$ is the one-body term originating from the pairing between identical nucleons. The next term in eq. (1) is the quadrupole operator $\widehat{Q}_\rho^\chi$ ($\rho = \pi, \nu$), expressing the quadrupole-quadrupole interaction between different nucleons, whose explicit expression is given by

$$(2) \qquad \widehat{Q}_\rho^\chi = \left[d_\rho^\dagger \times \widetilde{s}_\rho + s_\rho^\dagger \times \widetilde{d}_\rho \right]^{(2)} + \chi_\rho \left[d_\rho^\dagger \times \widetilde{d}_\rho \right]^{(2)} \qquad (\rho = \pi, \nu).$$

In eq. (2), $\chi_\pi(\chi_\nu)$ represents the proton (neutron) quadrupole deformation parameter. The form assumed by the nucleus essentially depends on the values taken by χ_π and χ_ν, since a strong connection exists between the parameter χ_ρ ($\rho = \pi, \nu$) and the IBM-2 dynamical symmetries, $U_{\pi+\nu}(5)$, $SU_{\pi+\nu}(3)$, $O_{\pi+\nu}(6)$, and $SU_{\pi+\nu}(3)^*$. In particular, the $U_{\pi+\nu}(5)$ dynamical symmetry (vibrational nucleus) arises when $\chi_\rho = 0$ ($\rho = \pi, \nu$). In the $SU_{\pi+\nu}(3)$ limit (rotational nucleus), one has $\chi_\rho = \pm\frac{\sqrt{7}}{2}$ ($\rho = \pi, \nu$)[1], while the $SU_{\pi+\nu}(3)^*$ case (triaxial nucleus) corresponds to $\chi_\pi = -\frac{\sqrt{7}}{2}$ and $\chi_\nu = +\frac{\sqrt{7}}{2}$, or vice versa. Finally, in the $O_{\pi+\nu}(6)$ dynamical symmetry (γ-unstable nucleus), the χ_π and χ_ν parameters have opposite sign ($-\frac{\sqrt{7}}{2} < \chi_\pi < 0$ and $0 < \chi_\nu < +\frac{\sqrt{7}}{2}$) [14]. In eq. (1), $\widehat{M}_{\pi\nu}$ represents the so-called Majorana operator, which is expressed as follows:

$$(3) \qquad \widehat{M}_{\pi\nu} = -2 \sum_{L=1,3} \alpha_L \left[d_\nu^\dagger \times d_\pi^\dagger \right]^{(L)} \cdot \left[\widetilde{d}_\nu \times \widetilde{d}_\pi \right]^{(L)}$$

$$+ \alpha_2 \left[s_\pi^\dagger \times d_\nu^\dagger - s_\nu^\dagger \times d_\pi^\dagger \right]^{(2)} \cdot \left[\widetilde{s}_\pi \times \widetilde{d}_\nu - \widetilde{s}_\nu \times \widetilde{d}_\pi \right]^{(2)} .$$

The $\widehat{M}_{\pi\nu}$ operator introduces the symmetry energy favoring states where the protons and neutrons move in phase (symmetric states). Indeed, $\widehat{M}_{\pi\nu}$ shifts the mixed-symmetry

---

[1] Usually, when the IBM calculations are carried out the decimal value $\chi_\rho = \pm 1.30$ is used, since $\pm\sqrt{7}/2 = \pm 1.30$.

states (also called no-symmetric states)[2] to energies higher than those where the symmetric states are lying. The last two terms of the Talmi Hamiltonian (eq. (1)) are given by

$$\widehat{V}_{\rho\rho} = \frac{1}{2} \sum_{L=0,2,4} (-1)^L \sqrt{2L+1}\, c_L^\rho \left[\left[d_\rho^\dagger \times d_\rho^\dagger \right]^{(L)} \times \left[\tilde{d}_\rho \times \tilde{d}_\rho \right]^{(L)} \right]^{(0)} \qquad (\rho = \pi, \nu).$$

The $\widehat{V}_{\rho\rho}$ ($\rho = \pi, \nu$) operator describes the residual interaction between like bosons.

The IBM-2 exhibits four dynamical symmetries[3] associated with two different types of group chains, the F-spin[4] symmetry limit (I_1, II_1, III_1, and IV) and the coupled chains (I_2, II_2, III_2, and IV) [4]. The symmetries I, II, III, and IV can be placed at the vertices of a tetrahedron (fig. 1), as suggested by Dieperink (1983) [16].

In this work, use has been made of the former limit whose group chains are the following:

$$(4) \quad U_\pi(6) \otimes U_\nu(6) \supset U_{\pi+\nu}(6) \supset U_{\pi+\nu}(5) \supset O_{\pi+\nu}(5) \supset O_{\pi+\nu}(3) \supset O_{\pi+\nu}(2), \qquad (I_1)$$

$$U_\pi(6) \otimes U_\nu(6) \supset U_{\pi+\nu}(6) \supset SU_{\pi+\nu}(3) \supset O_{\pi+\nu}(3) \supset O_{\pi+\nu}(2), \qquad (II_1)$$

$$U_\pi(6) \otimes U_\nu(6) \supset U_{\pi+\nu}(6) \supset O_{\pi+\nu}(6) \supset O_{\pi+\nu}(5) \supset O_{\pi+\nu}(3) \supset O_{\pi+\nu}(2), \qquad (III_1)$$

$$U_\pi(6) \otimes U_\nu(6) \supset SU_\pi(3) \otimes \overline{SU_\nu(3)} \supset SU_{\pi+\nu}(3)^* \supset O_{\pi+\nu}(3) \supset O_{\pi+\nu}(2). \qquad (IV)$$

The IV chain is a distinctive feature of the IBM-2, and the bar over $SU_\nu(3)$ denotes a particular choice in the sign of the second $SU(3)$ generator.

The powerful algebraic structure of IBM-2 provides simplified expressions for the whole Hamiltonian when the considered nucleus is close to one of the limiting dynamical symmetries of the model ($U_{\pi+\nu}(5)$, $SU_{\pi+\nu}(3)$, $O_{\pi+\nu}(6)$, and $SU_{\pi+\nu}(3)^*$) [4,17]. In these cases, the Hamiltonian of the interacting boson model-2 can be written in terms only of the Casimir (or invariant) operator of a particular chain of groups, beginning from a general algebra which in IBM-2 is $U_{\pi+\nu}(6)$. When that occurs, the IBM-2 Hamiltonian is said to have a dynamical symmetry and the classical eigenvalue problem of diagonalizing the nuclear Hamiltonian is greatly reduced in numerical complexity, since analytical expressions can be carried out. In this situation, the semiclassical limit is completely integrable. Hence, the fact that the IBM Hamiltonian has a dynamical symmetry provides

---

[2] The mixed-symmetry states are levels not fully symmetric in the neutron and proton components and the capability of predicting them is an important feature of the IBM-2.

[3] For an exhaustive review of the role played by the dynamical symmetries in the study of the nuclear structure, the reader is referred to ref. [15].

[4] The F-spin is, by definition, the isotopic spin for the bosons, namely a quantum number which allows to distinguish between proton bosons ($F = -\frac{1}{2}$) and neutron bosons ($F = +\frac{1}{2}$).

a number of advantages, first of all the possibility to work out analytical expressions for the spectroscopic properties. Moreover, the solutions of the eigenvalue problem can be simply analyzed in order to extract valuable information on the nuclear dynamics, allowing for a clear identification of fully symmetric, mixed-symmetry or intruder states.

3. – Statistical analysis of regular and chaotic behavior of nuclear spectra

The atomic nuclei are quantum systems where both regular and chaotic states co-exist [18]. In general, in order to analyze the chaotic dynamics of a quantum system use is made of the Random Matrix Theory (RMT) [5], which gives an excellent and very elegant theoretical structure for studying the fluctuations of the energy levels. In particular, since the nuclei have both time reversal and rotational symmetries, the statistical properties of the spectra are analyzed by means of the GOE (Gaussian Orthogonal Ensemble).

A very useful tool to describe the spectral statistics of a large set of quantum systems is the *nearest-neighbor level-spacing distribution* $P(s)$ [5, 18-25]. This distribution allows one to know the probability with which two adjacent eigenvalues are at the distance s. If the classical analog of a quantum system is integrable, then the $P(s)$ distribution is that of a Poisson ensemble [23],

$$(5) \qquad P(s) = \exp(-s).$$

When this situation occurs (eq. (5)), one has a high probability to have close degeneracies. Moreover, neither repulsions nor correlations between levels are observed. Instead, if the quantum systems are similar to chaotic classical systems (non-integrable) the $P(s)$ statistics is approximated by the Wigner distribution (GOE),

$$(6) \qquad P(s) = \frac{\pi}{2}\, s\, \exp\left[-\frac{\pi}{4}\, s^2\right].$$

Equation (6) shows that the levels have the tendency to repel each other ($P(s) \to 0$ as $s \to 0$).

It is a usual practice to quantify the chaotic behavior in terms of the *Brody parameter* ω, interpolating the distributions (5) and (6) by means of the Brody distribution [23]

$$(7) \qquad P(s,\omega) = \alpha\,(\omega + 1)\, s^{\omega}\, \exp\left[-\alpha\, s^{\omega+1}\right],$$

where

$$(8) \qquad \alpha = \left[\Gamma\left(\frac{\omega + 2}{\omega + 1}\right)\right]^{\omega+1}.$$

It is easy to see that, if $\omega = 0$, eq. (5) is recovered, while when $\omega = 1$ one gets the Wigner distribution (6).

3'1. *Phenomenological descriptions of the nuclear level densities.* – In the literature a number of phenomenological nuclear level density models exists, the most widely
used being the *Gilbert-Cameron* (GC) model [26], the *Back-Shifted Fermi Gas* (BSFG)
model [27], that proposed by *Ignatyuk* [28, 29], and an extension of the *interacting boson
model* to the description of collective features at finite temperature [30]. This is mainly
so since they are rather simple models, they parametrize the principal features of nuclear
level density and are easily included in statistical model calculations. In 1965 Gilbert
and Cameron [26] proposed a composite nuclear level density formula, which assumes
two different forms depending on the excitation energy E less or greater, respectively,
than a matching energy E_x (in MeV), whose systematic is

$$ (9) \qquad\qquad E_x = 2.5 + 150/A, $$

with A the mass number. E_x is the matching point where the level density at high energy
ρ_{BF} (eqs. (10) and (16) below), is equal to the density at low energy ρ_{CTF} (eqs. (20)
and (22) below), and both them have the same slope.

 – At *high energies* $(E > E_x)$, the level density (in MeV^{-1}) of all spins J, is given
by the *Bethe Formula* (BF) [31] for the Fermi gas model([5]) [26, 27, 32-36] (total level
density),

$$ (10) \qquad\qquad \rho_{\mathrm{BF}}(E) = \frac{\exp[2\sqrt{aU}]}{12\sqrt{2}\,\sigma\,a^{1/4}U^{5/4}}\, , $$

where $U = E - U_0 = E - \Delta$ is measured from the fully degenerate state occurring at U_0.
Δ is the pairing energy([6]) (in MeV^{-1}), which has to be subtracted from the excitation
energy E, in order to remove the systematic differences in the values of the a parameter
for neighboring even-even, odd-A, and odd-odd nuclei (even-odd effects). In eq. (10),
a is the level density parameter (in MeV^{-1}), which is proportional to the density of
single-particle states g, near the Fermi level [26],

$$ (11) \qquad\qquad a = \frac{\pi^2}{6}\,g. $$

Moreover, σ is the spin cut-off parameter describing the spin distribution,

$$ (12) \qquad\qquad \sigma^2 = g\,\langle m^2\rangle t. $$

([5]) In the Fermi gas model, the nucleus is considered to be an ideal gas of non-interacting neutrons and protons bound in a potential well. This assumption makes the use of thermodynamic
methods possible.
([6]) The pairing energy Δ is calculated from the proton and neutron pairing energy Δ_p and Δ_n,
respectively. In particular, one has $\Delta = \Delta_p + \Delta_n$ for even-even nuclei, $\Delta = \Delta_p$ for even-odd
nuclei, $\Delta = \Delta_n$ for odd-even nuclei, and $\Delta = 0$ for odd-odd nuclei.

Here, $\langle m^2 \rangle$ is the r.m.s. value of the projection of the angular momentum J, of single-particle states at the Fermi level, and t is the nuclear temperature (in MeV), $t \equiv \sqrt{U/a}$. Gilbert and Cameron [26] found the systematic behavior of a over a range of mass number A, to be linearly correlated to shell corrections([7]) $S = S_p + S_n$ (sum of the shell corrections for neutron and protons),

$$(13) \qquad a/A = 0.00917S + C,$$

where C is a constant equal to 0.14 for spherical nuclei and to 0.12 for the deformed ones. The GC model employs an energy dependent spin cut-off parameter σ (eq. (12)), whose systematic behavior is the following:

$$(14) \qquad \sigma^2 = 0.0888 \sqrt{aU}\, A^{2/3}\,.$$

Equation (14) shows the weak dependence of σ on the excitation energy E $[\sigma \sim (E - \Delta)^{1/4}]$. The parameterization (14) has been modified by Reffo [37] obtaining

$$(15) \qquad \sigma^2 = 0.1446 \sqrt{aU}\, A^{2/3}\,.$$

The density of levels with spin J (both parities included) and energy E is provided by [26, 27, 32-36, 38-42] (partial level density),

$$(16) \qquad \rho_{\mathrm{BF}}(E, J) = f(J)\, \rho_{\mathrm{BF}}(E) =$$
$$= \frac{2J + 1}{24\sqrt{2}\,\sigma^3\, a^{1/4} U^{5/4}} \cdot \exp\left[2\sqrt{aU} - \frac{(J + \frac{1}{2})^{1/2}}{2\sigma^2}\right],$$

where $\rho_{\mathrm{BF}}(E)$ is the total level density (10) and $f(J)$ has the form

$$(17) \qquad f(J) = \exp\left[-\frac{J^2}{2\sigma^2}\right] - \exp\left[-\frac{(J + 1)^2}{2\sigma^2}\right]$$
$$\approx \frac{2J + 1}{2\sigma^2} \exp\left[-\frac{(J + \frac{1}{2})^{1/2}}{2\sigma^2}\right].$$

From the concepts of the classical thermodynamics, the nuclear temperature T can be defined by the nuclear level density $\rho(E)$ [43],

$$(18) \qquad \frac{1}{T} \equiv \frac{\mathrm{d}}{\mathrm{d}E} \ln \rho(E).$$

([7]) The shell corrections allow to remove the differences between nuclei near and far from closed shells.

Hence, in the GC model at high energies the energy-dependent temperature $T_{\mathrm{BF}}(E)$ is

$$(19) \qquad T_{\mathrm{BF}}(E) \equiv \left(\frac{\mathrm{d}}{\mathrm{d}E} \ln \rho_{\mathrm{BF}}(E) \right)^{-1}$$

$$= \left(\frac{2U\sqrt{a} - 3\sqrt{U}}{2U^{3/2}} \right)^{-1}.$$

It is also possible to define a nuclear temperature for levels with spin J, which, at high energies, tends towards to $T = \sqrt{U/a}$.

– At *low energies* ($E < E_x$), the total level density is provided by the *Constant Temperature Formula* (CTF) [24-26, 32-35, 38, 44],

$$(20) \qquad \rho_{\mathrm{CTF}}(E) = \frac{1}{T_{\mathrm{CTF}}} \exp\left[\frac{E - E_0}{T_{\mathrm{CTF}}} \right].$$

The T_{CTF} and E_0 parameters are determined by fitting the densities ρ_{BF} (eq. (10)) and ρ_{CTF} (eq. (20)) at $E = E_x$, with $E_x = U_x + \Delta$ [26]

$$(21) \qquad T_{\mathrm{CTF}} = T_{\mathrm{BF}}(U_x),$$

$$E_0 = E_x - T_{\mathrm{CTF}} \ln[T_{\mathrm{CTF}}\, \rho_{\mathrm{BF}}(U_x)].$$

The partial level density is given by

$$(22) \quad \rho_{\mathrm{CTF}}(E, J) = f(J)\, \rho_{\mathrm{CTF}}(E) = \frac{2J + 1}{2\sigma^2\, T_{\mathrm{CTF}}} \exp\left[\frac{E - E_0}{T_{\mathrm{CTF}}} - \frac{(J + \frac{1}{2})^{1/2}}{2\sigma^2} \right],$$

where $f(J)$ has the form (17). Since the dependence of the spin cut-off parameter σ on the energy E, is weak (eq. (14)), at low energies a constant σ value may be used.

As is well known, the $N(E)$ integral of the level density $\rho(E)$ is the number of levels with energy less than or equal to E. At low energies, the cumulative number $N_{\mathrm{CTF}}(E)$ of all levels up to an excitation energy E, is

$$(23) \qquad N_{\mathrm{CTF}}(E) = \int_0^E \rho_{\mathrm{CTF}}(E')\mathrm{d}E'$$

$$= \exp\left[\frac{E - E_0}{T_{\mathrm{CTF}}} \right] - \exp\left[\frac{-E_0}{T_{\mathrm{CTF}}} \right] + N_0.$$

If the partial level density $\rho_{\mathrm{CTF}}(E, J)$ (eq. (22)) is considered, the expression for the staircase function $N_{\mathrm{CTF}}(E, J)$ is

$$(24) \qquad N_{\mathrm{CTF}}(E, J) = \int_0^E \rho_{\mathrm{CTF}}(E', J)\mathrm{d}E'$$

$$= \frac{2J + 1}{2\sigma^2} \left\{ \exp\left[\frac{E - E_0}{T_{\mathrm{CTF}}} - \frac{(J + \frac{1}{2})^{1/2}}{2\sigma^2} \right] \right.$$

$$\left. - \exp\left[\frac{-E_0}{T_{\mathrm{CTF}}} - \frac{(J + \frac{1}{2})^{1/2}}{2\sigma^2} \right] \right\} + N_0 \, .$$

In eqs. (23) and (24), the integration constant N_0 is a fictive number of levels below the ground state, which allows us to reduce the discrepancies between the experimental and fitted cumulative numbers of levels [24, 33, 34, 42]. The introduction of N_0 is necessary since the occurrence of the very low-lying levels in nuclei depends on specific structures and not on statistical rules.

At high energies, the cumulative number $N_{\mathrm{BF}}(E, J)$ of all the levels with spin J (eq. (16)) up to an excitation energy E, is

$$(25) \qquad N_{\mathrm{BF}}(E, J) = \int_0^E \rho_{\mathrm{BF}}(E', J) \, \mathrm{d}E'$$

$$= f(J) \int_0^E \rho_{\mathrm{BF}}(E') \, \mathrm{d}E' \, .$$

If the total level density (10) is considered, the $N_{\mathrm{BF}}(E, J)$ staircase function is given by

$$(26) \qquad N_{\mathrm{BF}}(E) = \int_0^E \rho_{\mathrm{BF}}(E') \, \mathrm{d}E' \, .$$

Unlike the low-energy case (eqs. (23) and (24)), the number of levels calculated with the Bethe formula (eqs. (25) and (26)) has to be obtained by numerical integration due to the more complicated dependence on the excitation energy.

Differently from the GC model, the BSFG formalism [27, 33-36, 38, 40, 42, 45, 46] employs a single Fermi gas formula [31] in order to describe the level density over the whole excitation energy range. The main feature of this model is that both the level density parameter a, and the pairing energy Δ, are adjusted to obtain the best fit to experimental results. In particular, Dilg $et\ al.$ [27] provided systematics for both a and Δ parameters, for $A < 65$,

$$(27a) \qquad\qquad a = 2.40 + 0.067A,$$

$$(27b) \qquad\qquad \Delta = -130A^{-1} + P,$$

where P is a pairing energy correction based on the even-odd characteristics of the nucleus.

The total level density then reads as

$$(28) \qquad \rho(E) = \frac{\exp[2\sqrt{a(E-\Delta)}]}{12\sqrt{2}\sigma\, a^{1/4}(E-\Delta+T)^{5/4}},$$

where the definition of nuclear temperature, $E - \Delta = aT^2 - T$, given by Lang and LeCouteur [47], is adopted. Δ is the back shift of the ground state[8], accounting for both pairing and shell effects.

The partial level density is

$$(29) \qquad \rho(E,J) = \frac{2J+1}{24\sqrt{2}\,\sigma^3\, a^{1/4}(E-\Delta+t)^{5/4}} \exp\left[2\sqrt{a(E-\Delta)} - \frac{(J+\frac{1}{2})^{1/2}}{2\sigma^2}\right].$$

In the BSFG model, the spin cut-off parameter σ is related to the moment of inertia for a rigid body[9] and to the nuclear temperature T [27, 43],

$$(30) \qquad \sigma^2 = \frac{I_{\text{rigid}}\, T}{\hbar^2} \approx 0.015 A^{5/3} T.$$

Usually, the expression exploited for the spin cutoff parameter σ, is that computed from Gilbert and Cameron (eq. (14)). As a consequence, for the Bethe formula (28) the energy dependent nuclear temperature $T(E)$ is [33, 34]

$$(31) \qquad T(E) = \frac{E-\Delta}{\sqrt{a\,(E-\Delta)} - 3/2}.$$

As far as the calculation of the cumulative number of levels is concerned, equations like those of GC model for both total and partial level densities (eqs. (23) and (24), respectively) are gained.

Both the GC and BSFG models are based upon the Fermi gas model (non-interacting particles), where the level density parameter a does not depend on energy (eq. (13)). It follows that neither the GC nor the BSFG model takes into account a number of important features of real nuclei such as shell inhomogeneities in the single-particle spectrum, pairing correlations of nucleons of the superconducting type and collective effects due to coherent superposition of individual nucleon excitations. These peculiarities of real nuclei are quite significant at low excitation energies, but disappear as the energy increases. Ignatyuk *et al.* [29, 48] developed a method of calculation of the level density, in which the collective and shell effects are included. These new formulas are based on the generalized superfluid model (GSM) [39, 49, 50].

---

[8] In order to improve the fit to experimental data, the ground-state position has to be back-shifted, namely lowered with respect to the actual ground state [26, 27].
[9] The moment of inertia for a rigid body is $I_{\text{rigid}} = \frac{2}{5}MR^2$, where M is the nuclear mass and $R = 1.25 \times A^{1/3}$ fm is the nuclear radius.

In the GSM model, there exists a critical temperature T_c, at the phase transition point from the superfluid state to the normal (Fermi gas) one [29, 48, 51, 52]

$$T_c = 0.567\Delta_0 \,, \tag{32}$$

where Δ_0 is the correlation function of the ground state $(T = 0)$. The critical energy is

$$U_c = 0.472 \, a_c \, \Delta_0^2 \,, \tag{33}$$

with a_c value of the parameter $a(U)$ at the critical point [48, 51]. For $U \geq U_c$, the equations of state of the superfluid model differ from those of the Fermi gas model by only a shift of the excitation energy [48],

$$U = aT^2 + E_{\text{cond}} \,. \tag{34}$$

Here, E_{cond} is the ground-state condensation energy which characterizes the decrease of the ground state energy of the Fermi gas due to correlation interaction. The main feature of the formalism proposed by Ignatyuk is that here, unlike the GC and BSFG models, one has an energy dependence of the level density parameter [28, 53]

$$a(Z, A, U) = \widetilde{a}(A) \left\{ 1 - f(U) \frac{\delta W(Z, A)}{U - E_{\text{cond}}} \right\}, \tag{35}$$

where $\widetilde{a}(A)$ is the asymptotic value of the level density parameter at high energies[10] and $\delta W(Z, A)$ represents the shell correction in the nuclear mass formula [54]. In eq. (35), $f(U)$ is a dimensionless *universal* function (see fig. 2 in ref. [53]), which determines the energy trend of the level density parameter a at lower excitation energies[11] [28, 53]

$$f(U) = 1 - \exp[-\gamma(U - E_{\text{cond}})]. \tag{36}$$

Here, $\gamma \, (= 0.054\,\text{MeV}^{-1})$ is the parameter characterizing the rate of rearrangement of the shells with energy.

In order to take into account the collective excitations, the level density for a given angular momentum J, assumes the form [55]

$$\rho(U, J) = \rho_{\text{int}}(U, J) \, K_{\text{coll}}(U), \tag{37}$$

---

[10] The shell effects disappear as the energy increases, thus $a \to \widetilde{a}$ for U large and the Fermi gas approximation $a = \text{const}$, is valid.
[11] In particular, one has $f(U) \to 1$ as $U \to \infty$, and $f(U)/U \to \text{const}$ as $U \to 0$.

where $\rho_{\text{int}}(U, J)$ is the density of internal (non-collective) states and $K_{\text{coll}}(U)$ is the collective enhancement coefficient [43,55]. The density of internal nuclear levels is [29,48,56]

$$(38a) \qquad \rho_{\text{int}}(U, J) = \frac{(2J + 1)\pi}{24\sqrt{2}\,\sigma_{\text{eff}}^3\,\sqrt{a^3\,T^5}} \exp\left[2at - \frac{(J + \frac{1}{2})^2}{2\,\sigma_{\text{eff}}^2}\right],$$

$$(38b) \qquad \sigma_{\text{eff}}^2 = \begin{cases} I_\perp^{2/3}\,I_\parallel^{1/3}\,T, & \text{for deformed nuclei,} \\ I_\parallel\,T, & \text{for spherical nuclei.} \end{cases}$$

In eqs. (38a), (38b), $I_\perp$ and $I_\parallel$ are the perpendicular and parallel moments of inertia, respectively. The collective enhancement coefficient $K_{\text{coll}}(U)$ is [29, 48]

$$(39) \qquad K_{\text{coll}}(U) = K_{\text{rot}}(U)\,K_{\text{vib}}(U),$$

with $K_{\text{rot}}(U)$ and $K_{\text{vib}}(U)$ the coefficients of the rotational and vibrational enhancement of the level density, respectively. The adiabatic estimate for these coefficients is usually exploited [29, 48, 55, 56]

$$(40a) \qquad K_{\text{rot}}(U) = \begin{cases} I_\perp\,T, & \text{for deformed nuclei,} \\ 1, & \text{for spherical nuclei,} \end{cases}$$

$$(40b) \qquad K_{\text{vib}}(U) \approx \exp\left[1.69\left(\frac{3m_0 A}{4\pi\sigma_{\text{drop}}}\frac{C_{\text{drop}}}{C}\right)^{2/3} T^{4/3}\right].$$

In the expression for the $K_{\text{vib}}(U)$ factor (eq. (40b)), σ_{drop} ($= 1.2\,\text{MeV} \cdot \text{fm}^{-2}$) is the surface tension in the liquid-drop model, which corresponds to the analog phenomenological parameter of the mass formula. The ratio C_{drop}/C characterizes the difference of the rigidity coefficient of the excited nucleus from the corresponding coefficients of the liquid-drop model.

   3'2. *IBM collective enhancement factors for level densities.* – In 1990, I [30] applied the interacting boson model-1 [4] at finite temperature in order to evaluate the collective enhancement factor in nuclear level densities in the three dynamical symmetries of the IBM-1 Hamiltonian. As usual, when collective and single-particle degrees of freedom are decoupled, the total level density can be written as the intrinsic (non-collective) level density times a collective factor [55]

$$(41) \qquad \rho(U, J) = \rho_{\text{int}}(U, J)Z_{\text{coll}}(U).$$

In eq. (41), $\rho_{\text{int}}(U, J)$ is the density of internal states (eq. (38a)) and $Z_{\text{coll}}(U)$ is the collective enhancement factor. The $Z_{\text{coll}}(U)$ coefficient can be approximated by the canonical partition function of a collective Hamiltonian, and in the three symmetry limits [30] one obtains:

– chain I: $U(6) \supset U(5) \supset O(5) \supset O(3)$

$$(42a) \qquad Z_{\text{coll}}^{(I)}(\beta(U), J) = \sum_{n_d} \sum_{\nu, n_\Delta} \sum_J (2J + 1)$$

$$\times \exp\left[- \beta\big[E_0 + \varepsilon_0 n_d + \alpha_0 n_d(n_d + 4) + 2\beta_0\nu(\nu + 3) + 2\gamma_0 J(J + 1)\big]\right],$$

– chain II: $U(6) \supset SU(3) \supset O(3)$

$$(42b) \qquad Z_{\text{coll}}^{(II)}(\beta(U), J) = \sum_{\lambda, \mu, \widetilde{\chi}} \sum_J (2J + 1)$$

$$\times \exp\left[- \beta\left[E_0 + \frac{2}{3}\delta_0(\lambda^2 + \mu^2 + \lambda\mu + 3\lambda + 3\mu) + 2\gamma_0 J(J + 1)\right]\right],$$

– chain III: $U(6) \supset O(6) \supset O(5) \supset O(3)$

$$(42c) \qquad Z_{\text{coll}}^{(III)}(\beta(U), J) = \sum_{\sigma} \sum_{\tau, \widetilde{\nu}_\Delta} \sum_J (2J + 1)$$

$$\times \exp[-\beta[E_0 + 2\beta_0\tau(\tau + 3) + 2\eta_0\sigma(\sigma + 4) + \gamma_0 J(J + 1)]],$$

where $\beta = 1/T$ is the inverse of the nuclear temperature. In eqs. $(42a)$-$(42c)$, the quantum numbers labeling the collective states associated with each subgroup in the group chains I, II, and III are defined as in ref. [4]. It is worth noting that, since the summations over the quantum numbers are not mutually independent, $Z_{\text{coll}}(\beta(U), J)$ cannot be written as a product of vibrational and rotational partition functions, $Z_{\text{rot}}Z_{\text{vib}}$, as is usually made in the traditional approaches to the calculation of collective enhancement factor (eq. (39)). As a consequence, rotations and vibrations are coupled in the irreducible representations of the reduction chains I, II, and III.

4. – Results and discussion

I made use of the Brody distribution $P(s, \omega)$, given by eq. (7), in order to analyze the regular and chaotic behavior of theoretical IBM-2 nuclear spectra, which are obtained by means of a modified version of the NPBOS code [57,58].

The procedure adopted to carry out the statistical analysis of these spectra is described step-by-step in the following discussion.

– In order to consider only the energy levels which can significantly contribute to the spectral analysis, it is necessary to fit the staircase function $N(E)$, by means of a suitable expression for the integral of the level density $\rho(E)$. Since I worked within the framework of the neutron-proton interacting boson model, I exploited eqs. $(42a)$-$(42c)$ appropriately extended to the IBM-2. As an example, fig. 2 illustrates the fit to the theoretical energy levels in the $SU_{\pi+\nu}(3) \to U_{\pi+\nu}(5)$ transition case for $\chi_\pi = -1.25$ and $\chi_\nu = -1.05$. The fluctuation of the first levels here observed is due to their non-statistical behavior [33,42]. To verify this trend of the low-lying level density, we repeated

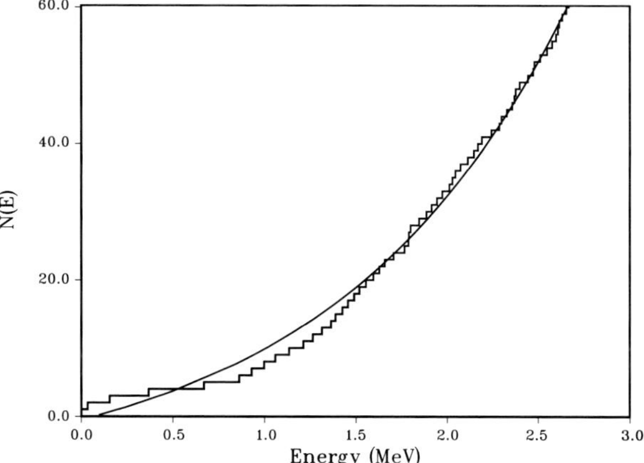

Fig. 2. – The best fit to the staircase function $N(E)$, in the $SU_{\pi+\nu}(3) \to U_{\pi+\nu}(5)$ transition for $\chi_\pi = -1.25$ and $\chi_\nu = -1.05$.

the calculations by using both the Gilbert-Cameron expression (eq. (16)) and the back-shifted Fermi gas formula (29). An analog behavior to that shown in fig. 2 was obtained.

– The staircase function $N(E)$, can be written as follows [59]:

$$(43) \qquad N(E) = N_{\mathrm{SM}}(E) + N_{\mathrm{FLUCT}}(E),$$

where $N_{\mathrm{SM}}(E)$ represents the smoothed behavior (eq. (41) in the present case) and $N_{\mathrm{FLUCT}}(E)$ expresses the fluctuations of the spectrum. Due to the non-universal character of the smoothed behavior of the energy spectrum, this spectral property cannot be treated by means of the RMT and therefore must be removed. In order to perform this task, we unfolded the $\{E_1, E_2, \ldots, E_N\}$ theoretical spectrum obtaining the following unfolded spectrum $\{\widetilde{E}_1, \widetilde{E}_2, \ldots, \widetilde{E}_N\}$ [25, 44]

$$(44) \qquad \widetilde{E}_i = E_{\mathrm{MIN}} + \frac{F(E_i) - F(E_{\mathrm{MIN}})}{F(E_{\mathrm{MAX}}) - F(E_{\mathrm{MIN}})}(E_{\mathrm{MAX}} - E_{\mathrm{MIN}}).$$

In eq. (44), $F(E)$ is the best fit, previously obtained, to the cumulative number of levels $N(E)$, and E_{MIN} and E_{MAX} represent the first and last energies of the transitional nucleus spectrum considered, respectively. As a consequence of the transformation (44), the unfolded level density does not show an exponential character, like the original levels (see eqs. (41) and (42a)-(42c)), but it is constant. This is why the level density is now denoted by ρ_C.

– Finally, I calculate $P(s)$. The actual distances s_i $(i = 1, \ldots, N)$, between adjacent energy levels are computed in the following way [25]:

$$(45) \qquad s_i = \left(\widetilde{E}_{i+1} - \widetilde{E}_i\right)\rho_C.$$

TABLE I. – $SU_{\pi+\nu}(3) \to U_{\pi+\nu}(5)$ transition. Values assumed by Talmi Hamiltonian parameters (in MeV) and the Brody parameter ω. The uncertainty (in %) in the chaoticity parameter ω is given within parentheses; $N_\pi = 4$ and $N_\nu = 5$.

Dynamical symmetry	ε	k	χ_ν	χ_π	α_1	α_2	α_3	$c_0^\nu = c_2^\nu = c_4^\nu$	ω
$SU_{\pi+\nu}(3)$	0.300	-0.080	-1.30	-1.25	1.000	0.050	-0.070	0.0	-0.12 (16)
:	0.300	-0.080	-1.10		0.700	0.050	-0.070		$+0.50$ (16)
:	0.300	-0.080	-1.05		0.700	0.200	-0.090		$+0.52$ (15)
:	0.300	-0.080	-0.95		0.700	0.180	-0.090		$+0.40$ (15)
:	0.300	-0.080	-0.80		0.700	0.180	-0.090		$+0.06$ (15)
:	0.600	-0.050	-0.40		1.100	0.110	-0.050		-0.06 (15)
$U_{\pi+\nu}(5)$	0.540	-0.030	0.00		0.900	0.100	-0.050		-0.15 (15)
$SU_{\pi+\nu}(3)$	0.300	-0.080	-1.30	-1.20	1.000	0.050	-0.070	0.0	-0.38 (15)
:	0.300	-0.080	-1.10		0.700	0.050	-0.070		-0.19 (16)
:	0.300	-0.080	-1.05		0.700	0.200	-0.090		$+0.51$ (16)
:	0.300	-0.080	-0.80		0.700	0.200	-0.090		-0.03 (16)
:	0.600	-0.070	-0.60		1.000	0.110	-0.070		$+0.66$ (16)
:	0.600	-0.050	-0.20		1.000	0.120	-0.070		-0.17 (15)
$U_{\pi+\nu}(5)$	0.600	-0.020	0.00		1.000	0.150	-0.070		-0.23 (15)
$SU_{\pi+\nu}(3)$	0.300	-0.080	-1.30	-1.10	0.500	0.050	-0.070	0.0	-0.40 (15)
:	0.300	-0.080	-1.10		0.500	0.050	-0.070		$+0.05$ (15)
:	0.300	-0.080	-1.05		0.700	0.200	-0.090		$+0.56$ (16)
:	0.300	-0.080	-0.80		0.700	0.180	-0.090		$+0.05$ (15)
:	0.600	-0.050	-0.40		1.000	0.120	-0.050		-0.23 (15)
$U_{\pi+\nu}(5)$	0.600	-0.050	0.00		0.900	0.150	-0.080		-0.35 (16)
$SU_{\pi+\nu}(3)$	0.200	-0.085	-1.30	-0.80	0.900	0.090	-0.090	0.0	-0.19 (16)
:	0.200	-0.077	-1.10		0.900	0.090	-0.050		-0.17 (15)
:	0.500	-0.050	-0.95		0.900	0.070	-0.010		$+0.06$ (14)
:	0.570	-0.030	-0.60		0.900	0.070	-0.010		-0.40 (15)
:	0.570	-0.020	-0.20		0.900	0.070	-0.010		-0.46 (15)
$U_{\pi+\nu}(5)$	0.515	-0.020	0.00		0.900	0.090	-0.010		-0.29 (16)

The $P(s)$ distribution is the histogram of the number of distances s_i, included in the $(s, s + \Delta s)$ range where Δs is equal to one half of the average distance between the levels [44]

$$(46) \qquad \Delta s = \frac{1}{2} \bar{s}.$$

$P(s)$ must then be normalized to unity

$$(47) \qquad \int_0^\infty P(s) \, \mathrm{d}s = 1.$$

TABLE II. – $SU_{\pi+\nu}(3)^* \to U_{\pi+\nu}(5)$ transition. Values taken by Talmi Hamiltonian parameters (in MeV) and the Brody parameter ω. The error (in %) in the ω parameter is supplied within parentheses; $N_\pi = 4$ and $N_\nu = 5$.

Dynamical symmetry	ε	k	χ_ν	χ_π	α_1	α_2	α_3	$c_0^\nu = c_2^\nu = c_4^\nu$	ω
$SU_{\pi+\nu}(3)^*$	0.350	−0.050	+1.30	−1.25	0.200	0.040	−0.010	0.0	−0.35 (15)
:	0.340	−0.050	+1.10		0.200	0.035	−0.010		−0.17 (14)
:	0.340	−0.050	+0.90		0.200	0.040	−0.010		−0.23 (14)
:	0.400	−0.060	+0.50		0.300	0.090	−0.030		+0.12 (16)
:	0.500	−0.060	+0.30		0.700	0.100	−0.040		+0.06 (14)
$U_{\pi+\nu}(5)$	0.570	−0.060	0.00		1.000	0.100	−0.050		−0.12 (15)
$SU_{\pi+\nu}(3)^*$	0.395	−0.040	+1.30	−0.80	0.300	0.060	−0.030	0.0	−0.23 (15)
:	0.395	−0.050	+1.10		0.300	0.040	−0.030		−0.15 (15)
:	0.350	−0.050	+0.90		0.300	0.040	−0.030		−0.06 (14)
:	0.450	−0.060	+0.50		0.500	0.060	−0.020		+0.03 (15)
:	0.550	−0.060	+0.20		0.700	0.050	−0.020		−0.05 (14)
$U_{\pi+\nu}(5)$	0.550	−0.070	0.00		1.000	0.040	−0.020		−0.12 (15)

TABLE III. – $O_{\pi+\nu}(6) \to SU_{\pi+\nu}(3)^*$ transition. Values assumed by Talmi Hamiltonian parameters (in MeV) and the Brody parameter ω. The uncertainty (in %) in the chaoticity parameter ω is given within parentheses; $N_\pi = 4$ and $N_\nu = 5$.

Dynamical symmetry	ε	k	χ_ν	χ_π	α_1	α_2	α_3	$c_0^\nu = c_2^\nu = c_4^\nu$	ω
$O_{\pi+\nu}(6)$	0.430	−0.080	0.00	−1.25	0.900	0.090	−0.010	0.0	+0.05 (15)
:	0.470	−0.055	+0.30		0.900	0.090	−0.010		+0.05 (14)
:	0.480	−0.055	+0.50		0.900	0.090	−0.010		+0.09 (14)
:	0.480	−0.055	+0.90		0.900	0.090	−0.010		−0.17 (15)
:	0.340	−0.050	+1.20		0.200	0.040	−0.010		−0.29 (14)
$SU_{\pi+\nu}(3)^*$	0.350	−0.050	+1.30		0.200	0.040	−0.010		−0.31 (15)
$O_{\pi+\nu}(6)$	0.400	−0.050	+0.10	−0.80	0.900	0.150	−0.050	0.0	−0.17 (16)
:	0.450	−0.050	+0.30		0.900	0.150	−0.050		−0.15 (15)
:	0.450	−0.070	+0.50		0.900	0.050	−0.050		−0.20 (14)
:	0.445	−0.070	+0.70		0.900	0.040	−0.050		−0.23 (16)
·	0.455	−0.070	+0.90		0.900	0.040	−0.050		−0.14 (15)
$SU_{\pi+\nu}(3)^*$	0.395	−0.040	+1.30		0.300	0.060	−0.030		−0.08 (16)

TABLE IV. – $O_{\pi+\nu}(6) \to SU_{\pi+\nu}(3)$ *transition. Experimental IBM-2 Hamiltonian parameters taken from ref. [10].* $N_\pi = 2(3)$ *for* Pt(Os) *isotopes, while* N_ν *ranges from 4* (^{194}Os, ^{196}Pt) *to 9* (^{184}Os, ^{186}Pt). *The error (in %) in the* ω *parameter is given within parentheses.*

Isotope	ε	k	χ_ν	χ_π	$\alpha_1 = \alpha_3$	α_2	c_0^ν	c_2^ν	c_4^ν	ω
^{194}Os	0.450	−0.150	+1.05	−1.30	−0.100	+0.040	+0.60	+0.02	0.0	+0.14 (15)
^{192}Os	0.450	−0.150	+0.95				+0.55	+0.04		−0.15 (15)
^{190}Os	0.450	−0.150	+0.80				+0.45	0.00		+0.06 (15)
^{188}Os	0.450	−0.150	+0.45				0.00	−0.09		+0.07 (14)
^{186}Os	0.450	−0.140	0.00				−0.25	−0.13		−0.15 (16)
^{184}Os	0.500	−0.135	−0.50				−0.25	−0.16		−0.06 (16)
^{196}Pt	0.580	−0.180	+1.05	−0.80	−0.100	+0.040	+0.60	+0.02	0.0	+0.05 (15)
^{194}Pt	0.580	−0.180	+0.95				+0.55	+0.04		+0.28 (14)
^{192}Pt	0.580	−0.180	+0.80				+0.45	0.00		+0.22 (15)
^{190}Pt	0.580	−0.180	+0.45				0.00	−0.09		+0.20 (15)
^{188}Pt	0.580	−0.160	0.00				−0.25	−0.13		+0.64 (16)
^{186}Pt	0.620	−0.145	−0.50				−0.25	−0.16		+0.05 (14)

TABLE V. – $U_{\pi+\nu}(5) \to O_{\pi+\nu}(6)$ *transition. Values taken by Talmi Hamiltonian parameters (in MeV) and the Brody parameter* ω. *The uncertainty (in %) in the chaoticity parameter* ω *is supplied within parentheses;* $N_\pi = 4$ *and* $N_\nu = 5$.

Dynamical symmetry	ε	k	χ_ν	χ_π	α_1	α_2	α_3	$c_0^\nu = c_2^\nu = c_4^\nu$	ω
$U_{\pi+\nu}(5)$	0.530	−0.020	+0.10	−1.25	0.900	0.090	−0.070	0.0	−0.29 (14)
:	0.520	−0.020	+0.30		0.900	0.090	−0.050		+0.06 (16)
:	0.510	−0.020	+0.40		0.900	0.090	−0.090		−0.06 (16)
:	0.510	−0.020	+0.50		0.900	0.090	−0.090		−0.27 (16)
:	0.480	−0.055	+0.90		0.900	0.090	−0.050		−0.21 (14)
$O_{\pi+\nu}(6)$	0.470	−0.050	+1.30		0.900	0.090	−0.050		−0.23 (15)
$U_{\pi+\nu}(5)$	0.515	−0.020	+0.10	−0.80	0.900	0.090	−0.010	0.00	−0.46 (15)
:	0.550	−0.020	+0.30		0.900	0.090	−0.010		−0.52 (15)
:	0.550	−0.020	+0.60		0.900	0.070	−0.010		−0.46 (16)
:	0.480	−0.100	+0.80		0.900	0.070	−0.050		−0.17 (15)
:	0.490	−0.080	+1.10		0.900	0.080	−0.050		+0.23 (15)
$O_{\pi+\nu}(6)$	0.490	−0.070	+1.30		0.900	0.090	−0.050		+0.15 (15)

TABLE VI. – $SU_{\pi+\nu}(3)^* \to SU_{\pi+\nu}(3)$ transition. Values assumed by Talmi Hamiltonian parameters (in MeV) and the Brody parameter ω. The uncertainty (in %) in the ω parameter is provided within parentheses; $N_\pi = 4$ and $N_\nu = 5$.

Dynamical symmetry	ε	k	χ_ν	χ_π	α_1	α_2	α_3	$c_0^\nu = c_2^\nu = c_4^\nu$	ω
$SU_{\pi+\nu}(3)^*$	0.350	-0.050	$+1.30$	-1.25	0.200	0.040	-0.010	0.0	-0.46 (15)
:	0.340	-0.060	$+0.40$		0.200	0.035	-0.010		-0.44 (15)
:	0.340	-0.060	$+0.10$		0.100	0.030	-0.010		$+0.17$ (14)
:	0.400	-0.077	-0.50		0.900	0.150	-0.010		-0.25 (14)
:	0.300	-0.073	-1.10		1.000	0.170	-0.010		-0.41 (15)
$SU_{\pi+\nu}(3)$	0.300	-0.073	-1.30		1.000	0.180	-0.020		-0.46 (16)
$SU_{\pi+\nu}(3)^*$	0.395	-0.040	$+1.30$	-0.80	0.300	0.030	-0.030	0.0	-0.28 (15)
:	0.395	-0.050	$+1.10$		0.300	0.040	-0.030		-0.15 (15)
:	0.350	-0.050	$+0.50$		0.010	0.040	-0.030		-0.46 (14)
:	0.300	-0.090	-0.30		1.000	0.100	-0.010		-0.15 (16)
:	0.150	-0.080	-0.95		0.900	0.090	-0.040		$+0.06$ (15)
$SU_{\pi+\nu}(3)$	0.200	-0.080	-1.30		0.900	0.090	-0.090		-0.29 (15)

As already mentioned, I obtained the IBM-2 energy spectrum by means of a suitably adapted version of the NPBOS code [57,58], which calculates the excitation energies and the eigenvectors of the Talmi Hamiltonian (eq. (1)). By this way, fourteen shape-phase transitions along the Dieperink's tetrahedron (fig. 1) have been evaluated. Six of them have been approximated by keeping constant the χ_π parameter, $\chi_\pi = -1.25$, and varying the χ_ν parameter in the $(-1.30, +1.30)$ range. The other six transitions have been carried out by putting χ_π equal to -0.80 and ranging χ_ν from -1.30 to $+1.30$[12]. Finally, the last two transitions are concerned with the $SU_{\pi+\nu}(3) \to U_{\pi+\nu}(5)$ transition and have been calculated by fixing χ_π first to -1.20 and then to -1.10 and varying again the χ_ν parameter in the $(-1.30, +1.30)$ range. Hence, the calculated shape-phase transitions are the following:

$$SU_{\pi+\nu}(3) \to U_{\pi+\nu}(5); \qquad U_{\pi+\nu}(5) \to O_{\pi+\nu}(6),$$
$$O_{\pi+\nu}(6) \to SU_{\pi+\nu}(3)^*; \qquad SU_{\pi+\nu}(3)^* \to SU_{\pi+\nu}(3),$$
$$SU_{\pi+\nu}(3)^* \to U_{\pi+\nu}(5); \qquad O_{\pi+\nu}(6) \to SU_{\pi+\nu}(3).$$

In every transition, the values 4(5) for the proton (neutron) boson numbers $N_\pi(N_\nu)$ have been assumed, since the quantum fluctuations are independent of $N_\pi(N_\nu)$. The only exception is the $O_{\pi+\nu}(6) \to SU_{\pi+\nu}(3)$ transition since for it realistic calculations for the platinum and osmium isotopes have been carried out (the highest Pt isotopes

---

[12] The fact of keeping constant χ_π, instead of χ_ν, is a purely convenient choice and has no physical relevance in the present context.

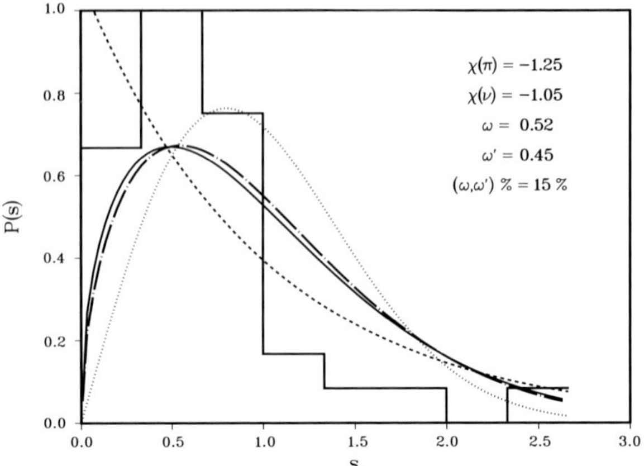

Fig. 3. – $P(s)$ distribution for the $SU_{\pi+\nu}(3) \to U_{\pi+\nu}(5)$ transition with $\chi_\pi = -1.25$ and $\chi_\nu = -1.05$. The dashed, dotted, and dot-dashed curves correspond with Poisson (eq. (6)), GOE (eq. (7)), and Brody (eq. (8), with $\omega = 0.52$) distributions, respectively. The solid curve refers to the Brody distribution calculated within the uncertainty (15%) in the ω parameter (eq. (8), with $\omega' = 0.45$).

have a γ-unstable feature, while the lightest Pt and Os isotopes exhibit an axially symmetric deformed character). In particular, use has been made of the IBM-2 Hamiltonian parameters given by Bijker *et al.* [14] by putting $N_\pi = 2(3)$ for Pt(Os) isotopes, while N_ν ranges from 4 (^{194}Os, ^{196}Pt) to 9 (^{184}Os, ^{186}Pt). Tables I–IV provide both the values assumed by the Talmi Hamiltonian parameters ε, k, χ_ν, χ_π, α_L ($L = 1, 2, 3$), and c_L^ρ ($L = 0, 2, 4$; $\rho = \pi, \nu$), in the different shape-phase transitions and those taken by the Brody parameter ω. Negative values of the ω parameter are often found which indicate a larger number of small level spacings (namely additional degeneracies) than in the Poisson distribution (eq. (5)). When this situation occurs the corresponding dynamical symmetry limit is said to be "overintegrable" [10].

Extremely interesting is table I where the $SU_{\pi+\nu}(3) \to U_{\pi+\nu}(5)$ shape-phase transition is shown. From it, the presence of a nearly regular region is clearly seen in agreement with the results previously obtained by Whelan and Alhassid (1993) [10] within the framework of IBM-1. The chaoticity parameter ω assumes values larger than or equal to $+0.50$ when the χ_π parameter ranges from -1.25 to -1.10. The uncertainty in the ω parameter is less than or equal to 16%: if a new value ω' of the Brody parameter is calculated within this error, the two curves $P(s, \omega)$ and $P(s, \omega')$ are quite near each other (figs. 3, 4, and 5).

For $\chi_\pi < -1.10$, the dynamical behavior of the nuclear spectra becomes completely regular: the maximum value of the ω parameter is $\omega = +0.06$, with an error less than 15% (fig. 6). It is interesting to note an anomaly which comes into sight when χ_π is equal to -1.20: a unexpected $\omega = +0.66$ value (with an uncertainty of about 16%) appears in correspondence with $\chi_\nu = -0.60$ (see table I and fig. 7). I suppose that this non-regular behavior is due to an irregularity in the energy spectrum since when one is in the

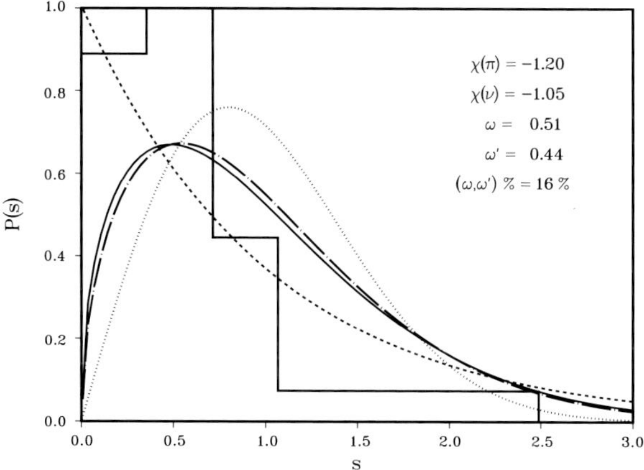

Fig. 4. – $P(s)$ distribution for the $SU_{\pi+\nu}(3) \rightarrow U_{\pi+\nu}(5)$ transition with $\chi_\pi = -1.20$ and $\chi_\nu = -1.05$. The dashed, dotted, and dot-dashed curves refer to Poisson (eq. (6)), GOE (eq. (7)), and Brody (eq. (8), with $\omega = 0.51$) distributions, respectively. The solid curve stands for the Brody distribution computed within the error (16%) in the ω parameter (eq. (8), with $\omega' = 0.44$).

partial dynamical symmetry domain (as in the $SU_{\pi+\nu}(3) \rightarrow U_{\pi+\nu}(5)$ transition) a very unstable situation occurs. Hence, also the variation of a single Hamiltonian parameter is enough to originate some irregularities in nuclear spectra. In spite of this, the fact that $\omega \geq +0.50$ when χ_π ranges from -1.25 to -1.10 clearly proves that the nearly regular

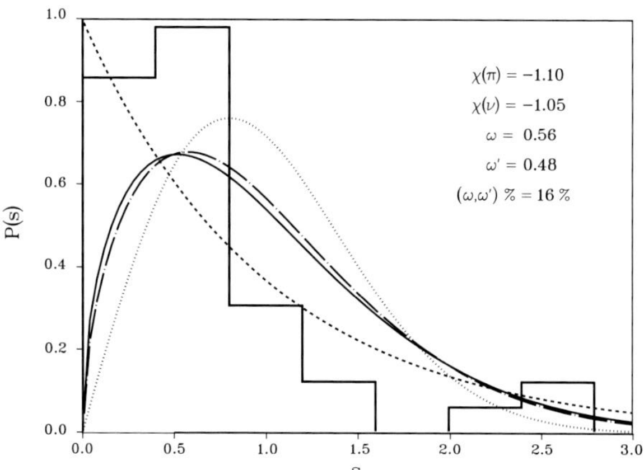

Fig. 5. – $P(s)$ distribution for the $SU_{\pi+\nu}(3) \rightarrow U_{\pi+\nu}(5)$ transition with $\chi_\pi = -1.10$ and $\chi_\nu = 1.05$. A not completely regular feature of the nuclear dynamics clearly appears. The dashed, dotted, and dot-dashed curves refer to Poisson (eq. (6)), GOE (eq. (7)) and Brody (eq. (8), with $\omega = 0.56$) distributions, respectively. The solid curve corresponds to the Brody distribution evaluated within the uncertainty (16%) in the ω parameter (eq. (8), with $\omega' = 0.48$).

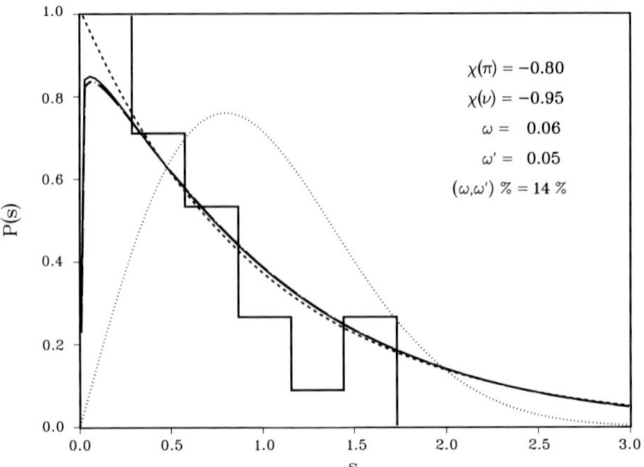

Fig. 6. – $P(s)$ distribution for the $SU_{\pi+\nu}(3) \rightarrow U_{\pi+\nu}(5)$ transition with $\chi_\pi = -0.80$ and $\chi_\nu = -0.95$. The dashed, dotted, and dot-dashed curves correspond with Poisson (eq. (6)), GOE (eq. (7)), and Brody (eq. (8), with $\omega = 0.06$) distributions, respectively. The solid curve refers to the Brody distribution calculated within the uncertainty (14%) in the ω parameter (eq. (8), with $\omega' = 0.05$).

region connecting the $SU_{\pi+\nu}(3)$ and $U_{\pi+\nu}(5)$ dynamical limits is not a narrow band, as Whelan and Alhassid supposed [10], but is a broad region. It has been the rich algebraic structure of the IBM-2 model which allowed to obtain this important result confirming and extending the previous observations [10].

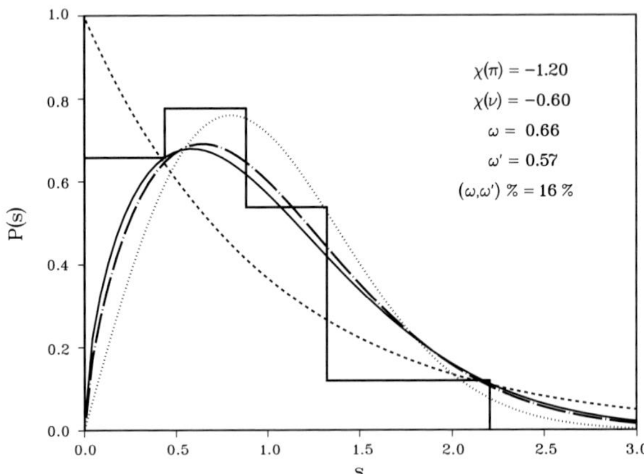

Fig. 7. – $P(s)$ distribution for the $SU_{\pi+\nu}(3) \rightarrow U_{\pi+\nu}(5)$ transition with $\chi_\pi = -1.20$ and $\chi_\nu = -0.60$. We think that this non-regular behavior is due to an irregularity in the IBM-2 nuclear spectrum. The dashed, dotted, and dot-dashed curves stand for Poisson (eq. (6)), GOE (eq. (7)) and Brody (eq. (8), with $\omega = 0.66$) distributions, respectively. The solid curve refers to the Brody distribution evaluated within the error (16%) in the ω parameter (eq. (8), with $\omega' = 0.57$).

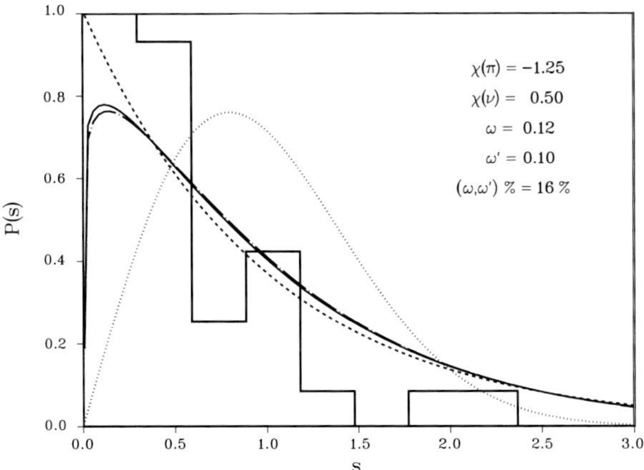

Fig. 8. – $P(s)$ distribution for the $SU_{\pi+\nu}(3)^* \to U_{\pi+\nu}(5)$ transition with $\chi_\pi = -1.25$ and $\chi_\nu = 0.50$. An unexpected completely regular feature appears even for an intermediate value of the χ_ν parameter. The dashed, dotted, and dot-dashed curves correspond to Poisson (eq. (6)), GOE (eq. (7)) and Brody (eq. (8), with $\omega = 0.12$) distributions, respectively. The solid curve refers to the Brody distribution computed within the uncertainty (16%) in the ω parameter (eq. (8), with $\omega' = 0.10$).

Indeed, Whelan and Alhassid carried out the $SU(3) \to U(5)$ transition by varying the χ parameter (they did not distinguish between χ_π and χ_ν parameters for proton and neutron bosons, respectively) which can be considered as the weighted average of χ_π and χ_ν. As a consequence, it was impossible for them to look through the $SU(3) \to U(5)$ transition from a microscopical point of view as I did. An explanation of the presence of a nearly regular region even far from the $SU_{\pi+\nu}(3)$ and $U_{\pi+\nu}(5)$ dynamical limits can be connected to the existence of partial dynamical $SU_{\pi+\nu}(3)$ symmetry and to phase-transition symmetries. In general, a partial dynamical symmetry [8, 12] arises when a dynamical symmetry of the model is broken in such a way that some of the eigenstates of the Hamiltonian (but not all) still exhibit the properties of that dynamical symmetry. Therefore, even if a dynamical symmetry is broken (as in shape-transitional regions) it is still possible to observe a regular feature of the nuclear dynamics. A further confirmation of the basic role played by partial dynamical $SU_{\pi+\nu}(3)$ symmetry in keeping regular motion patterns in nuclear spectra is supplied by the completely regular behavior of the $SU_{\pi+\nu}(3)^* \to U_{\pi+\nu}(5)$ transition (table II). Since $SU_{\pi+\nu}(3)$ and $SU_{\pi+\nu}(3)^*$ dynamical limits have the same generators up to a sign, the presence of a nearly regular region like that observed in the $SU_{\pi+\nu}(3) \to U_{\pi+\nu}(5)$ case (table I) was expected.

Instead, table II clearly shows a completely regular behavior as indicated by the maximum value assumed by the Brody parameter, $\omega = +0.12$, with an error of about 16% (fig. 8). I have interpreted the full regularity observed in the $SU_{\pi+\nu}(3)^* \to U_{\pi+\nu}(5)$ shape-phase transition as a clear indication that the $SU_{\pi+\nu}(3)^*$ limit has an attraction basin deeper than that of the $SU_{\pi+\nu}(3)$ symmetry. This fact is strengthened by the $O_{\pi+\nu}(6) \to SU_{\pi+\nu}(3)^*$ (table III) and $O_{\pi+\nu}(6) \to SU_{\pi+\nu}(3)$ (table IV, in particular the platinum chain) transitions.

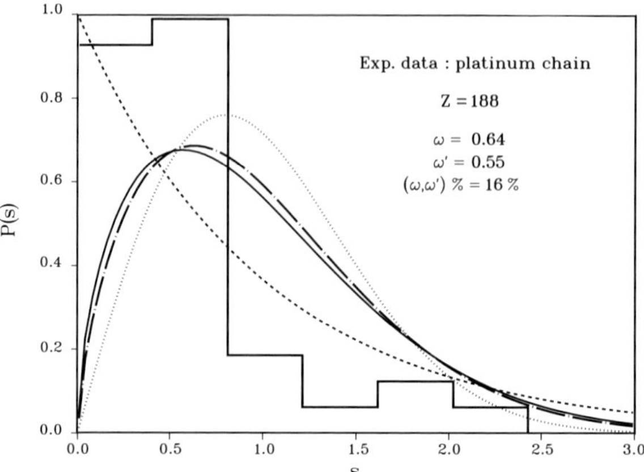

Fig. 9. – $P(s)$ distribution for the $O_{\pi+\nu}(6) \rightarrow SU_{\pi+\nu}(3)$ transition with $\chi_\pi = -0.80$ and $\chi_\nu = 0.0$. They are experimental data concerning the platinum chain (taken from ref. [10]). The presence of a nearly regular region is clearly shown. The dashed, dotted, and dot-dashed curves correspond to Poisson (eq. (6)), GOE (eq. (7)) and Brody (eq. (8), with $\omega = 0.64$) distributions, respectively. The solid curve refers to the Brody distribution calculated within the error (16%) in the ω parameter (eq. (8), with $\omega' = 0.55$).

Table IV shows the presence of a nearly regular region, we have $\omega = +0.64$ as maximum value with an uncertainty of about 16% (fig. 9), while table III exhibits a completely regular behavior (maximum absolute value $\omega = 0.23$ with an error of about 16% (fig. 10)).

The appearing of $\omega = +0.64$ in correspondence with $\chi_\nu = 0.0$ in the $O_{\pi+\nu}(6) \rightarrow SU_{\pi+\nu}(3)$ case is acceptable since Bijker *et al.* [14] stopped their investigations at $\chi_\nu = -0.50$ (remember that in the $SU_{\pi+\nu}(3)$ limit $\chi_\rho = -1.30$ ($\rho = \pi, \nu$)). Hence, $\omega = +0.64$ comes into sight at intermediate values of the χ_ν parameter as it is usually expected. The presence of the nearly regular region in the $O_{\pi+\nu}(6) \rightarrow SU_{\pi+\nu}(3)$ transition is a partial confirmation of the results previously obtained by Alhassid *et al.* [60], since it does not indicate a chaotic feature, as observed by these authors (see fig. 1 in the first paper of ref. [60]), but it only expresses a not completely regular motion pattern. As in the $SU_{\pi+\nu}(3) \rightarrow U_{\pi+\nu}(5)$ case (see table I), the difference existing between the present results and those gained by Alhassid *et al.* in the frame of IBM-1, is due to the more complex algebraic structure of the IBM-2 model. In particular, since in IBM-1 there is no distinction between χ_π and χ_ν, Alhassid *et al.* carried out the $O(6) \rightarrow SU(3)$ transition considering $\chi = 0$ for γ-unstable nuclei ($O(6)$ limit) and $\chi = -1.30$ for rotational nuclei ($SU(3)$ limit).

However, the IBM-2 model shows that from a microscopical point of view in the $O_{\pi+\nu}(6)$ dynamical symmetry the χ_π and χ_ν parameters have opposite sign ($\chi_\pi < 0$ and $\chi_\nu > 0$) and range from $|1.30|$ to 0. It follows that the guess made by Alhassid *et al.* [60] ($\chi = 0$ in the $O(6)$ limit) is only an approximation. This is why, working in IBM-2, the presence of a nearly regular region has been observed (table IV), while a

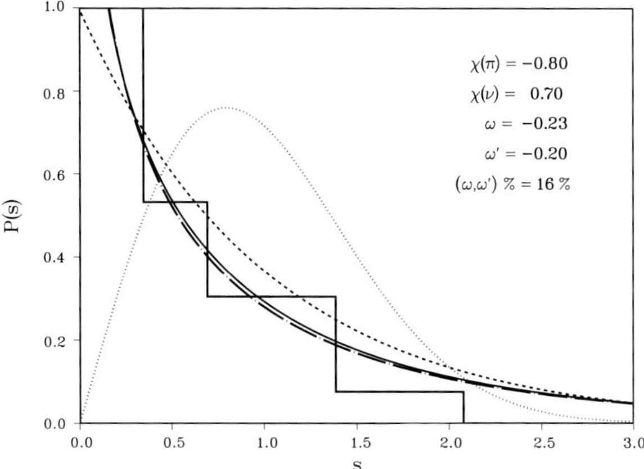

Fig. 10. – $P(s)$ distribution for the $O_{\pi+\nu}(6) \to SU_{\pi+\nu}(3)^*$ transition with $\chi_\pi = -0.80$ and $\chi_\nu = 0.70$. A completely regular dynamics comes into sight even far from the $O_{\pi+\nu}(6)$ and $SU_{\pi+\nu}(3)^*$ dynamical limits. In particular, the ω parameter assumes a negative value indicating a larger number of small level spacings than in the Poisson distribution. Such situation is termed overintegrable. The dashed, dotted, and dot-dashed curves stand for Poisson (eq. (6)), GOE (eq. (7)) and Brody (eq. (8), with $\omega = -0.23$) distributions, respectively. The solid curve corresponds to the Brody distribution evaluated within the uncertainty (16%) in the ω parameter (eq. (8), with $\omega' = -0.20$).

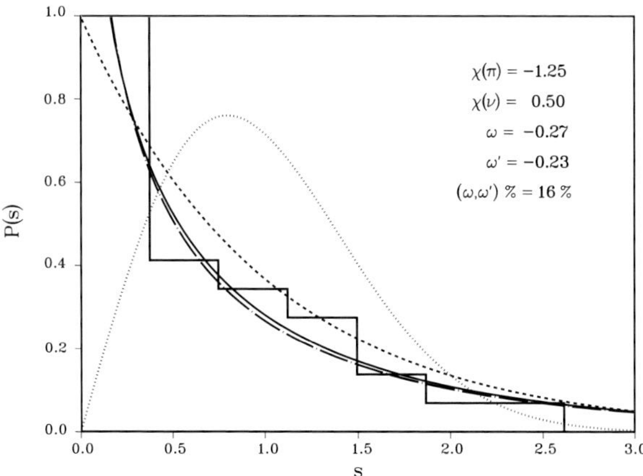

Fig. 11. – $P(s)$ distribution for the $U_{\pi+\nu}(5) \to O_{\pi+\nu}(6)$ transition with $\chi_\pi = -1.25$ and $\chi_\nu = 0.50$. The dashed, dotted, and dot-dashed curves stand for Poisson (eq. (6)), GOE (eq. (7)) and Brody (eq. (8), with $\omega = -0.27$) distributions, respectively. The solid curve corresponds to the Brody distribution calculated within the error (16%) in the ω parameter (eq. (8), with $\omega' = -0.23$).

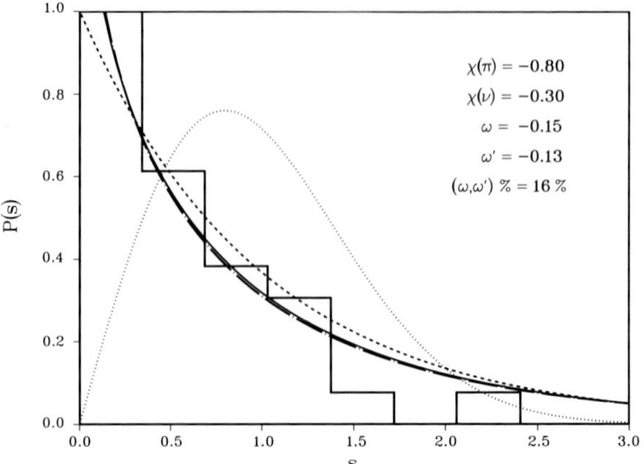

Fig. 12. – $P(s)$ distribution for the $SU_{\pi+\nu}(3)^* \to SU_{\pi+\nu}(3)$ transition with $\chi_\pi = -0.80$ and $\chi_\nu = -0.30$. An expected completely regular behavior is shown. The dashed, dotted, and dot-dashed curves stand for Poisson (eq. (6)), GOE (eq. (7)) and Brody (eq. (8), with $\omega = -0.15$) distributions, respectively. The solid curve corresponds to the Brody distribution computed within the error (16%) in the ω parameter (eq. (8), with $\omega' = -0.13$).

chaotic behavior appears if the statistical analysis is performed within the framework of IBM-1 [60]. Note that the use of realistic calculations [14] makes our observations about the $O_{\pi+\nu}(6) \to SU_{\pi+\nu}(3)$ transition very reliable. As far as the $U_{\pi+\nu}(5) \to O_{\pi+\nu}(6)$ transition (table V) is concerned, it displays a completely regular motion pattern (fig. 11).

This is due to the fact that the $U_{\pi+\nu}(5)$ and $O_{\pi+\nu}(6)$ limits have $O_{\pi+\nu}(5)$ as common subalgebra (see eqs. (4)), thus two of their generators are the same. This result confirms the observations previously made by Whelan and Alhassid (1993) [10]. Finally, the $SU_{\pi+\nu}(3)^* \to SU_{\pi+\nu}(3)$ shape-phase transition (table VI) shows an expected full regularity (fig. 12), since the $SU_{\pi+\nu}(3)^*$ and $SU_{\pi+\nu}(3)$ dynamical symmetries have the same generators up to a sign.

These results can be summarized by saying that the upper pyramid in the Dieperink's tetrahedron (fig. 1) exhibits a complete regularity and only in the lower one a non-regular motion appears (we refer to the $SU_{\pi+\nu}(3) \to U_{\pi+\nu}(5)$ and $O_{\pi+\nu}(6) \to SU_{\pi+\nu}(3)$ transitions).

5. – Conclusions

The rich algebraic structure of the IBM-2 model allowed us to carry out a careful quantum investigation of the regular and chaotic dynamics of medium mass even-even nuclei. This analysis permitted to gain a better knowledge of the nuclear dynamics in shape-phase transition regions examining thoroughly and extending the investigations previously made by other authors [7,9,10,12,60] within the framework of IBM-1. In particular, the observations of an unexpected complete regularity in the upper pyramid in Dieperink's tetrahedron (fig. 1) led to deem that the $SU_{\pi+\nu}(3)^*$ dynamical limit has an attraction basin deeper than that of the $SU_{\pi+\nu}(3)$ symmetry. A careful investigation of this new feature of the $SU_{\pi+\nu}(3)^*$ limit needs a further classical analysis (study of Poincaré maps, determination of Lyapunov exponents, and so on) which will be presented in a forthcoming paper. The supposed high depth of the attraction basin of the $SU_{\pi+\nu}(3)^*$ limit along with the presence of a broad nearly regular region in the $SU_{\pi+\nu}(3) \to U_{\pi+\nu}(5)$ transition, and of a narrower nearly regular region in the $O_{\pi+\nu}(6) \to SU_{\pi+\nu}(3)$ transition is evidence for the fundamental role played by the partial dynamical symmetries (in particular, the partial dynamical $SU_{\pi+\nu}(3)$ symmetry) and, more generally, the critical point and phase transitions symmetries in conserving regular motion patterns even when strong violations of the usual dynamical symmetries of IBM-2 occur. Moreover, these results further confirm the importance of the $SU(3)$ limit of IBM in studying many nuclear properties.

REFERENCES

[1] MAYER M. G. and JENSEN J. H. D., *Elementary Theory of Nuclear Shell Structure* (Wiley, New York) 1955.
[2] ARIMA A. and IACHELLO F., *Phys. Rev. Lett.*, **35** (1975) 1069.
[3] ELLIOTT J. P., *Proc. R. Soc. London Ser. A*, **245** (1959) 128.
[4] IACHELLO F. and ARIMA A., *The Interacting Boson Model* (Cambridge University Press, Cambridge) 1987.
[5] WIGNER E., *SIAM Rev.*, **9** (1967) 1.
[6] ALHASSID Y. and LEVIATAN A., *J. Phys. A*, **25** (1992) L1265.
[7] WHELAN N., ALHASSID Y. and LEVIATAN A., *Phys. Rev. Lett.*, **71** (1993) 2208.
[8] LEVIATAN A., in *Symmetries in Science VII*, edited by GRUBER B. and OTSUKA T. (Plenum Press, New York) 1994, p. 383.
[9] LEVIATAN A. and WHELAN N., *Phys. Rev. Lett.*, **77** (1996) 5202.
[10] WHELAN N. and ALHASSID Y., *Nucl. Phys. A*, **556** (1993) 42.
[11] IACHELLO F., *Phys. Rev. Lett.*, **85** (2000) 3580; **87** (2001) 052502-1; CASTEN R. F. and ZAMFIR N. V., *Phys. Rev. Lett.*, **87** (2001) 052503-1; IACHELLO F., *Phys. Rev. Lett.*, **91** (2003) 132502-1; CAPRIO M. A. and IACHELLO F., *Phys. Rev. Lett.*, **93** (2004) 242502-1; BONATSOS D., LENIS D., PETRELLIS D. and TERZIEV P. A., *Phys. Lett. B*, **588** (2004) 172; FRANK A., VAN ISACKER P. and IACHELLO F., *Phys. Rev. C*, **73** (2006) 061302(R).
[12] LEVIATAN A., in *Perspectives for the Interacting Boson Model*, edited by CASTEN R. F. *et al.* (Word Scientific Publ., Singapore) 1994, p. 129.
[13] OTSUKA T., ARIMA A., IACHELLO F. and TALMI I., *Phys. Lett. B*, **76** (1978) 139.
[14] BIJKER R., DIEPERINK A. E. L., SCHOLTEN O. and SPANHOFF R., *Nucl. Phys. A*, **344** (1980) 207.
[15] VAN ISACKER P., *Rep. Prog. Phys.*, **62** (1999) 1661.
[16] DIEPERINK A. E. L., in *Proceedings of the International School of Nuclear Physics, Erice, 1982*, Vol. **9**, edited by WILKINSON D. (Pergamon Press, Oxford) 1983, p. 121.
[17] TALMI I., *Simple Models of Complex Nuclei* (Harwood Academic Publ., Chur, Switzerland) 1993, Chapt. 39.
[18] LOPEZ-ARIAS M. T., MANFREDI V. R. and SALASNICH L., *Nuovo Cimento*, **17** (1994) 1.
[19] DYSON F. J., *J. Math. Phys.*, **3** (1962) 140.
[20] DYSON F. J. and MEHTA M. L., *J. Math. Phys.*, **4** (1963) 701.
[21] BOHIGAS O., GIANNONI M. J. and SCHMIT C., *Phys. Rev. Lett.*, **52** (1984) 1.

[22] BOHIGAS O. and WEIDENMÜLLER H. A., *Annu. Rev. Nucl. Part. Sci.*, **38** (1988) 421.

[23] TABOR M., *Chaos and Integrability in Nonlinear Dynamics* (John Wiley & Sons, New York) 1989.

[24] CAURIER E., GÓMEZ J. M. G., MANFREDI V. R. and SALASNICH L., *Phys. Lett. B*, **365** (1996) 7.

[25] GÓMEZ J. M. G., MANFREDI V. R. and SALASNICH L., in *New Perspective in Nuclear Structure*, edited by COVELLO A. (World Scientific Publ., Singapore) 1996, p. 225.

[26] GILBERT A. and CAMERON A. G. W., *Can. J. Phys.*, **43** (1965) 1446.

[27] DILG W., SCHANTL W., VONACH H. and UHL M., *Nucl. Phys. A*, **217** (1973) 269.

[28] IGNATYUK A. V., *Yad. Fiz.*, **21** (1975) 20 (*Sov. J. Nucl. Phys.*, **21** (1975) 10).

[29] IGNATYUK A. V., ISTEKOV K. K. and SMIRENKIN G. N., *Yad. Fiz.*, **30** (1979) 1205 (*Sov. J. Nucl. Phys.*, **29** (1979) 450).

[30] MAINO G., MENGONI A. and VENTURA A., *Phys. Rev. C*, **42** (1990) 988.

[31] BETHE H. A., *Phys. Rev.*, **50** (1936) 332; *Rev. Mod. Phys.*, **9** (1937) 69.

[32] REFFO G., in *Theory and Applications of Moment Methods in Many-Fermion Systems*, edited by DALTON B. J. *et al.* (Plenum Publ. Corp., New York) 1980, p. 167.

[33] VON EGIDY T., BEHKAMI A. N. and SCHMIDT H. H., *Nucl. Phys. A*, **454** (1986) 109.

[34] KRUSCHE B. and LIEB K. P., *Phys. Rev. C*, **34** (1986) 2103.

[35] ARTHUR E. D., GUENTHER P. T., SMITH A. B. and SMITH D. L., in *Proceedings of the OECD Meeting Nuclear Level Density, Bologna 1989*, edited by REFFO G. *et al.* (World Scientific, Singapore) 1992, p. 177.

[36] MUSTAFA M. G., BLANN M., IGNATYUK A. V. and GRIMES S. M., *Phys. Rev. C*, **45** (1992) 1078.

[37] REFFO G., *Parameter Systematics for Statistical Theory Calculations of Neutron Reaction Cross Sections* in Nuclear Theory for Applications, IAEA-SMR-43 report (1979), p. 205.

[38] HUIZENGA J. R., VONACH H. K., KATSANOS A. A., GORSKI A. J. and STEPHAN C. J., *Phys. Rev.*, **182** (1969) 1149.

[39] IGNATYUK A. V. and STAVINSKIĬ V. S., *Yad. Fiz.*, **11** (1970) 1213 (*Sov. J. Nucl. Phys.*, **11** (1970) 674).

[40] HUIZENGA J. R. and MORETTO L. G., *Annu. Rev. Nucl. Sci.*, **22** (1972) 427.

[41] SOLOVIEV V. G., STOYANOV CH. and VDOVIN A. I., *Nucl. Phys. A*, **224** (1974) 411.

[42] VON EGIDY T., SCHMIDT H. H. and BEHKAMI A. N., *Nucl. Phys. A*, **481** (1988) 189.

[43] ERICSON T., *Nucl. Phys.*, **11** (1959) 481; *Adv. Phys.*, **9** (1960) 425.

[44] PAAR V. and VORKAPIĆ D., *Phys. Lett. B*, **205** (1988) 7.

[45] GADIOLI E. and ZETTA L., *Phys. Rev.*, **167** (1968) 1016.

[46] STELTS M. L., CHRIEN R. E. and MARTEL M. K., *Phys. Rev. C*, **24** (1981) 1419.

[47] LANG J. M. B. and LECOUTEUR K. J., *Proc. R. Soc. London, Ser. A*, **67** (1954) 586.

[48] IGNATYUK A. V., ISTEKOV K. K. and SMIRENKIN G. N., *Yad. Fiz.*, **30** (1979) 1205 (*Sov. J. Nucl. Phys.*, **30** (1979) 626).

[49] IGNATYUK A. V. and SHUBIN YU. N., *Yad. Fiz.*, **8** (1968) 1135 (*Sov. J. Nucl. Phys.*, **8** (1969) 660).

[50] IGNATYUK A. V., *Yad. Fiz.*, **17** (1973) 502 (*Sov. J. Nucl. Phys.*, **17** (1973) 258).

[51] KUDAYEV G. A., OSTAPENKO YU. B., SVIRIN M. I. and SMIRENKIN G. N., *Yad. Fiz.*, **47** (1988) 341 (*Sov. J. Nucl. Phys.*, **47** (1988) 215).

[52] SVIRIN M. I. and SMIRENKIN G. N., *Yad. Fiz.*, **48** (1988) 682 (*Sov. J. Nucl. Phys.*, **48** (1988) 437).

[53] IGNATYUK A. V., SMIRENKIN G. N. and TISHIN A. S., *Yad. Fiz.*, **21** (1975) 485 (*Sov. J. Nucl. Phys.*, **21** (1975) 255).

[54] MEYERS W. D. and SWIATECKI W. S., *Ark. Fysik.*, **36** (1967) 593.

[55] BJØRNHOLM S., BOHR A. and MOTTELSON B. R., in *Proceedings of the Third International Symposium on Physics and Chemistry of Fission, Rochester, 1973*, Vol. **1** (IAEA, Vienna) 1974, p. 367.

[56] PLATONOV S. YU., FOTINA O. V. and YUMINOV O. A., *Nucl. Phys. A*, **503** (1989) 461.

[57] OTSUKA T. and YOSHIDA N., *User's manual of program NPBOS*, Japan Atomic Energy Research Institute Report JAERI-M/85-094, 1985.

[58] MAINO G., *Int. J. Mod. Phys. E*, **6** (1997) 287.

[59] MEREDITH D. C., KOONIN S. E. and ZIRNBAUER M. R., *Phys. Rev. A*, **37** (1988) 3499.

[60] ALHASSID Y., NOVOSELSKY A. and WHELAN N., *Phys. Rev. Lett.*, **65** (1990) 2971; ALHASSID Y. and WHELAN N., *Phys. Rev. C*, **43** (1991) 2637; ALHASSID Y. and NOVOSELSKY A., *Phys. Rev. C*, **45** (1992) 1677.

DOI 10.3254/978-1-58603-885-4-477

Weak interaction in nuclei

E. FIORINI(*)

Dipartimento di Fisica G.P.S. Occhialini, Università di Milano-Bicocca and INFN
Sezione di Milano-Bicocca - Piazza della Scienza 3, 20126 Milano, Italy

Summary. — A few examples will be given of the essential role played by low-energy nuclear physics in the fundaments of elementary particles and in particle astrophysics. The crucial impact in weak-interaction physics by the discovery of parity violation, which is now fifty years old, and the corresponding experiments will be summarized. A brief discussion will be devoted to the recent experiments on neutrino oscillations which prove that the difference between the square masses of two neutrinos of different flavour is different from zero. As a consequence the mass of at least one neutrino has to be finite, but oscillations cannot provide a direct indication of its value. Stimulated by these exciting results a vast series of experiments aiming to determine directly the neutrino mass has been carried out and is running or planned.

1. – Introduction

A fundamental step in the history of weak interactions dates back to the '30s of the last century and is closely bound to the suggested properties of the then still hypothetical particle, named neutron by Wolfang Pauli and later neutrino (in Italian) by Enrico Fermi.

(*) E-mail: Ettore.Fiorini@mib.infn.it

In fact it was only shortly after the suggestion by Pauli that Enrico Fermi constructed the beautiful theory where beta decay was studied as a *local* process with charged parity-conserving currents. I would like to stress that it was impossible in those times to think to a mediator as massive as the W and Z particles, and thus we can consider the *locality* of Fermi as a more than reasonable assumption.

A second fundamental step about twenty years after was the discovery of parity non-conservation in weak interactions to which I will devote here some attention. As a consequence of this discovery the neutrino was assumed as a massless particle, totally different from its antineutrino with full conservation of the total lepton number. Conservation was later assumed also for a new quantum number: the flavour which distinguishes the three types (electronic, muonic and tauonic) of the neutrino.

A third step which occurred about thirty years ago, was the discovery of the existence of neutral currents in weak interactions and of the presence of the above-mentioned heavy mediating particles, the W and Z bosons. As a consequence many processes and in particular many neutrino interactions, which were considered as totally forbidden before, where actually found, but neutrino was still considered as a massless lepton and flavour-conserving particle.

The last step dates to the beginning of this millennium and is the consequence of a series of experiments initiated decades before on neutrino oscillations. This process, suggested by Bruno Pontecorvo about fifty years ago, consists in the spontaneous transformation of neutrinos in neutrinos of a different flavour. Its existence implies that the difference of the squared masses of neutrinos of different flavours is different from zero and that as a consequence at least one neutrino has a finite mass. A massive neutrino indicates that the lepton number is not conserved and that therefore there is a certain equality between the neutrino and its antiparticle as suggested by Ettore Majorana in 1937, only one year before his tragic disappearance. This fact bring us back again to the discovery of parity violation just at the beginning of these last exciting fifty years of weak interactions.

2. – Parity violation in weak interactions

The suggestion of parity violation in weak interactions was introduced in 1956 in a beautiful theoretical paper [1] aiming to cure the so-called θ-t puzzle. The θ-particle, which we now call K^+, decays into $\pi^+ + \pi^0$. It was however found to have exactly the same mass as the τ which decays into $\pi^+ + \pi^0 + \pi^+$, namely in a different parity state. Lee and Yang [1] noticed that no evidence existed for parity conservation in β decays and in hyperon or meson decays, namely in all weak-interaction processes known at that time. As a consequence, they brilliantly suggested a series of possibile experiments to test parity conservation in various weak-interaction processes. I believe that one should particularly appreciate a theoretical paper like this, which is not based on existing results, but indeed suggests *a priori* experiments on a process not yet found.

The suggestion of Lee and Yang stimulated experiments which started immediately [2,3] and whose results were all published within one year in 1957. The first of these has

Fig. 1. – C. S. Wu.

been carried out by a group led by the great physicist Wu (fig. 1) with the experimental set-up shown in fig. 2. The decay investigated is

$$(1) \qquad\qquad {}^{60}\mathrm{Co} \rightarrow {}^{60}\mathrm{Fe}^* + \mathrm{e}^- + \bar{\nu}_\mathrm{e}$$

followed by the electromagnetic decay of ${}^{60}\mathrm{Fe}^*$ into two gamma-rays.

The angular distribution of the gamma-rays at temperatures above $110\,\mathrm{mK}$ is isotropic, but at lower temperature, where the cobalt crystal is polarized, it becomes more and more anisotropic. This is incidentally an excellent way to measure the temperature of the crystal. The angular distribution is however symmetric with respect to the polarization direction, since the γ decay is a parity-conserving electromagnetic process. The angular distribution of the electrons was on the contrary not only anisotropic, but also asymmetric: this is a clear prove of parity non-conservation.

Three other experiments almost immediately confirmed this result, but it is interesting to note that all authors thank Ms. Wu for informing them of her result prior to publication.

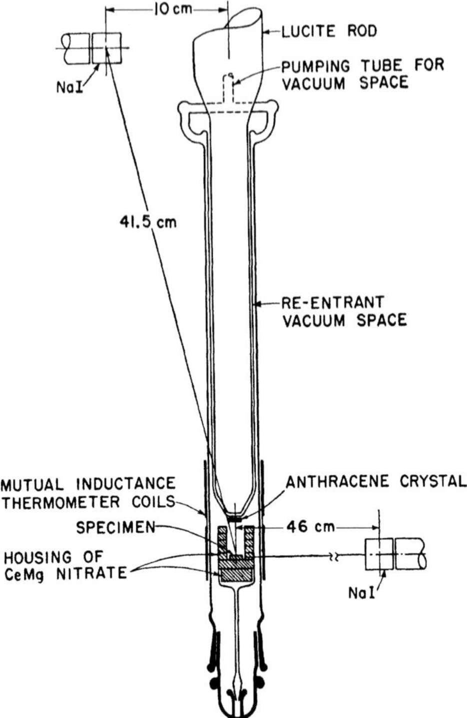

Fig. 2. – The experiment by C. S. Wu *et al.*

In the decay

$$\pi^+ \Rightarrow \nu_\mu + \mu^+ \qquad \text{followed by} \quad \mu^+ \to e^+ + \nu_e + \bar{\nu}_\mu \tag{2}$$

the angular distribution of the electron is found to be asymmetric with respect to the spin of the muon.

Asymmetry was also found in nuclear emulsion in the chain

$$\pi^+ \to \nu_\mu + \mu^+ \to e^+ + \nu_e + \nu_\mu \tag{3}$$

and in the decay

$$\Lambda \to \pi^- + p, \tag{4}$$

where a polarized Λ particle is produced in the reaction

$$\pi^- + p \to \Lambda + K^0. \tag{5}$$

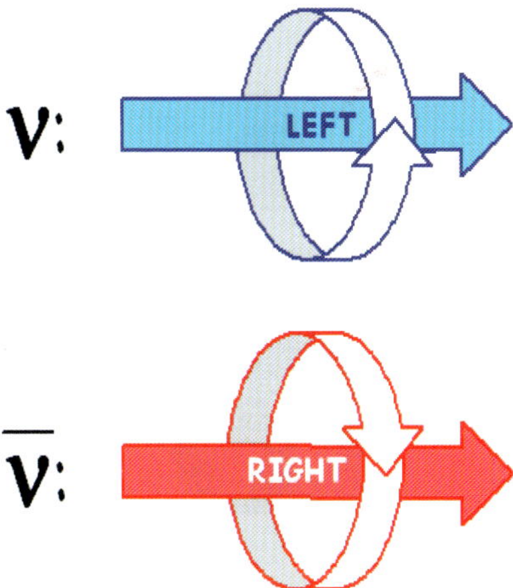

Fig. 3. – Chirality, namely the alignment of neutrino spin towards the sense of motion.

Of considerable interest are experiments carried out, also recently, searching for a *parity-violating impurity* in strong and electromagnetic interactions. One has in fact to note that a strong interaction like, for instance, an α decay contains a weak component, proportional to the parity-violating term $\varphi\varphi^*$. This contribution is generally of the order of 10^{-13}–10^{-14} with respect to the strong cross-section. When looking for an angular distribution or for a polarization, one expects a weak impurity which is on the contrary of the order of φ. Parity-violating effects were found in experiments on the angular distribution of γ-rays from polarized nuclear states or weak and unexpected circular polarization of γ-rays from decays of excited nuclear states [3,4].

Of considerable interest in nuclear physics are experiments searching for *parity forbidden* α decays [5]. An interesting example is the search for α decay into carbon of the 2^- excited state of ^{16}O at 8.87 MeV. Since both ^{12}C and ^{4}He are 0^+ states, this decay is forbidden by parity conservation and can only be produced by an *impurity* of weak interactions.

Very delicate optical experiments are also carried out [6] to search for an anomalous presence in circular polarization on an effect due to the contribution of weak interactions in a parity-conserving electromagnetic interaction process.

The discovery of parity non-conservation led to the above-mentioned two-neutrino theory (fig. 3).

3. – Neutrino oscillations and the problem of the neutrino mass

As mentioned before, the three neutrino families are identified by a quantum number named *flavour* (*electronic, muonic or tauonic*). In the Standard Model of weak interactions this number is conserved. A very important event in fundamental physics has been in the last years the discovery of neutrino oscillations [7,8] which were predicted almost fifty years ago by the great physicist Bruno Pontecorvo. Let us consider as an example

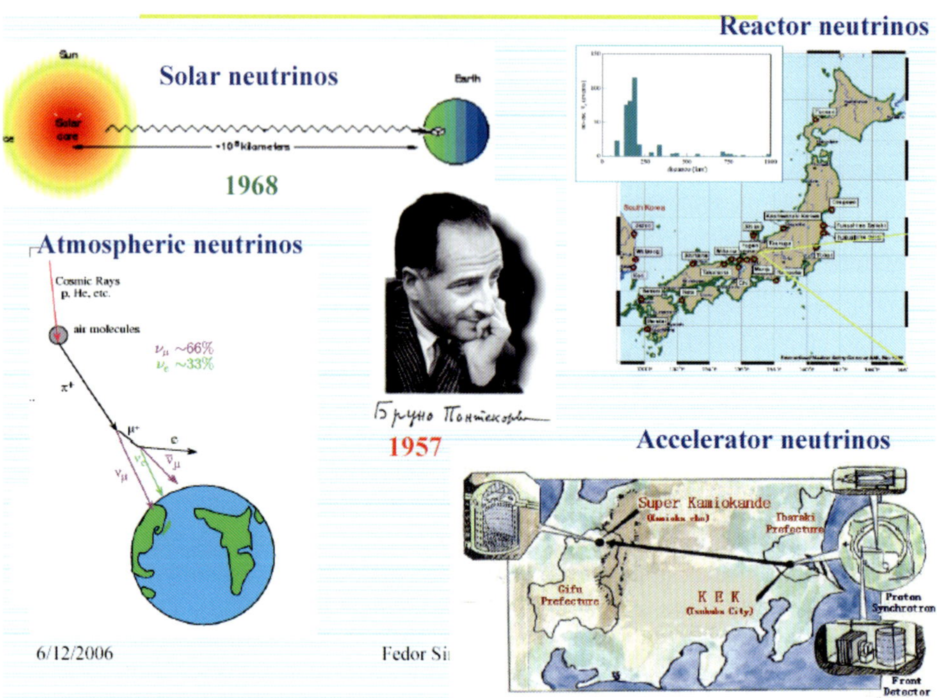

Fig. 4. – The various experiments showing neutrino oscillations, as suggested by B. Pontecorvo (center).

the neutrinos produced by the fusion processes which take place in the central region of the Sun and which are the source of the great energy produced in this and in all other stars. The copious flux of these neutrinos, which are of the electronic type, is such that their interactions, even if indeed rare, can be revealed in a very massive detector placed underground to avoid the "noise" due to cosmic rays. A pioneering experiment carried out in the United States and further searches performed in Japan, Russia, in the Gran Sasso Laboratory in Italy and more recently in Canada have clearly shown the presence of these neutrinos, but with a flux definitely lower than the expected one. This is due to the fact that solar neutrinos *oscillate* inside the Sun and in their long path toward the Earth transforming themselves into neutrinos of muonic or tauonic flavours. As a consequence the flux of electronic neutrinos on the Earth is lower than predicted by the so-called *Solar Model*.

Neutrino oscillations have been confirmed with neutrinos produced by cosmic rays in the atmosphere, and artificially by particle accelerators and nuclear reactors (fig. 4). These oscillations, which obviously violate the conservation of the flavour number, can only occur if the difference of the squared masses of two neutrinos of different flavours is finite. This obviously means that at least one neutrino has a mass different from zero, but neutrino oscillations are unable to determine its absolute value.

The problem of the neutrino mass is crucial in fundamental physics: if it is finite the neutrino can propagate with a velocity lower than the velocity of light and the alignment of its spin (fig. 3) with respect to the direction of motion would be less than 100%, etc. Another consequence would be that the total lepton number is likely violated and that there is not an absolute distinction between a neutrino and an antineutrino, as suggested since 1937 by Ettore Majorana.

4. – Direct and indirect ways to determine the neutrino mass

Various experimental and cosmological approaches were and are considered for the direct and indirect measurement of the neutrino mass.

4ʹ1. *Single beta decay.* – The most direct method to determine the mass of the neutrino [7,8] is the study of the deformation at the end point of the spectrum of the electron in single beta decay (fig. 5). No evidence for a finite neutrino mass has been obtained, but the present upper limits of about $2\,\mathrm{eV}$ are still far from what is suggested by neutrino oscillations. A new experiment, KATRIN, to be carried out, as most of the previous ones, on the decay of tritium is being designed in Germany (fig. 6) and aims at reaching a sensitivity of $0.2\,\mathrm{eV}$.

4ʹ2. *Measurements on the Cosmic Ray Background.* – A more powerful, but model-dependent, method to determine the mass of the neutrino comes from cosmology. Our Universe is presently embedded in a "sea" of photons decoupled from matter about $400\,000$ years after the big bang. It represents the so-called cosmic microwave background (CMB). We are also *swimming* in a sea of relic neutrinos decoupled much before, about a second after the big bang. The mass of these neutrinos would modify the distribution in space of CMB. Recent measurements on this CMB background have set a model-dependent upper limit on the neutrino masses slightly lower than that obtained

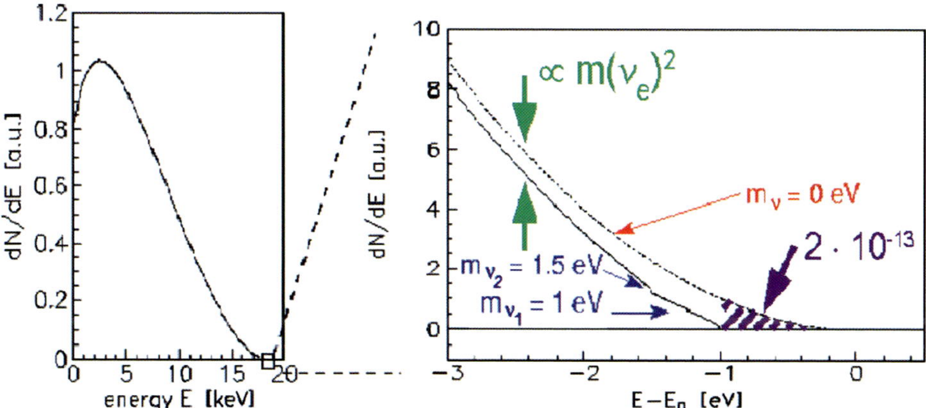

Fig. 5. – Deformation of the beta decay spectrum due to the neutrino mass.

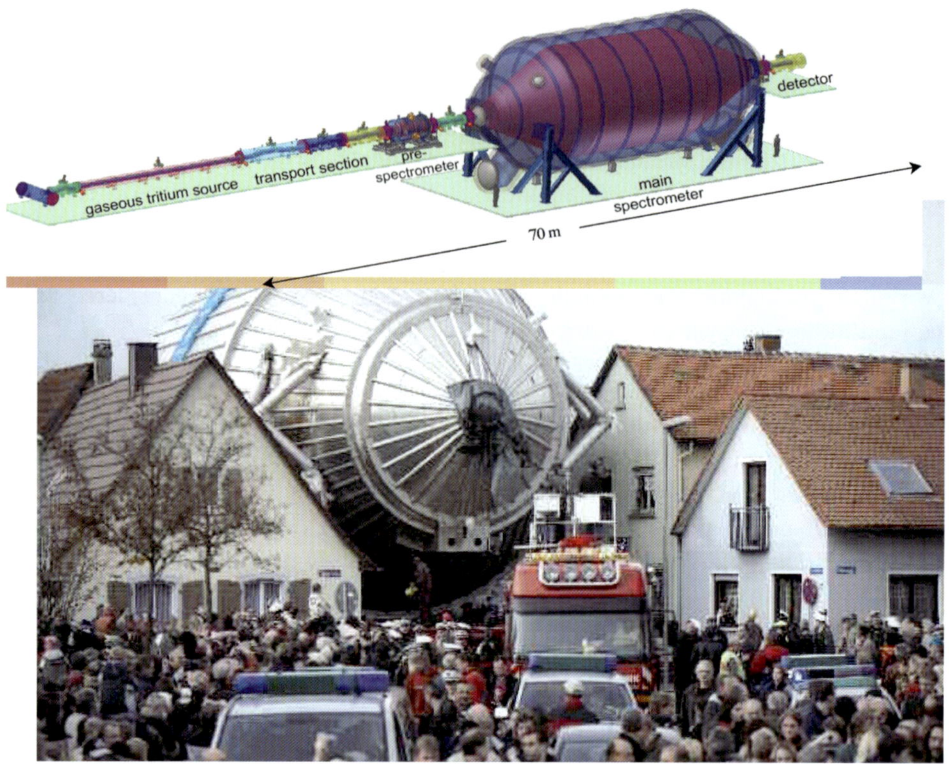

Fig. 6. – Arrival of the KATRIN structure in Karlsruhe.

in the direct measurements mentioned before. They are however still far from the values predicted by oscillations.

4'3. *Double beta decay*. – A third method to determine the effective neutrino mass is connected to a fundamental puzzle in neutrino physics: is neutrino a Dirac or a Majorana

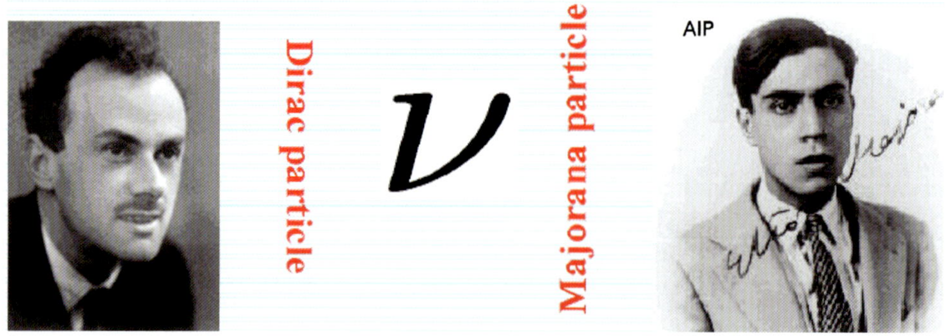

Fig. 7. – Dirac and Majorana.

Fig. 8. – Maria Goeppert Mayer and Enrico Fermi.

particle? (fig. 7). In the former hypothesis neutrino would be totally different from the antineutrino, his chirality, namely the property reported in fig. 3, would be 100% and its mass most likely null. In the latter, based on a brilliant theory suggested in 1937 by Ettore Majorana, the neutrino would not be distinct from its antiparticle, his mass would be finite and the lepton number would be violated. The most powerful method to investigate lepton number conservation is double beta decay (DBD), a rare nuclear process suggested by Maria Goeppert Mayer [9] in 1935, only one year after the Fermi weak-interaction theory (fig. 8). This process (fig. 9) consists in the direct transition from a nucleus (A, Z) to its isobar $(A, Z + 2)$ and can be investigated when the single beta decay of (A, Z) to $(A, Z+1)$ is energetically forbidden or at least strongly hindered.

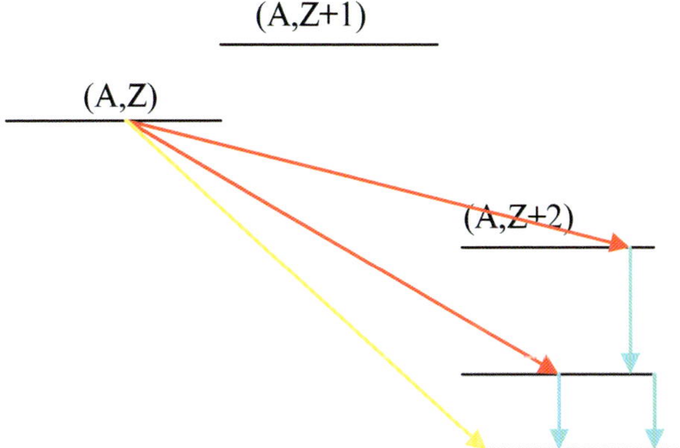

Fig. 9. – Scheme of two neutrino and neutrinoless DBD.

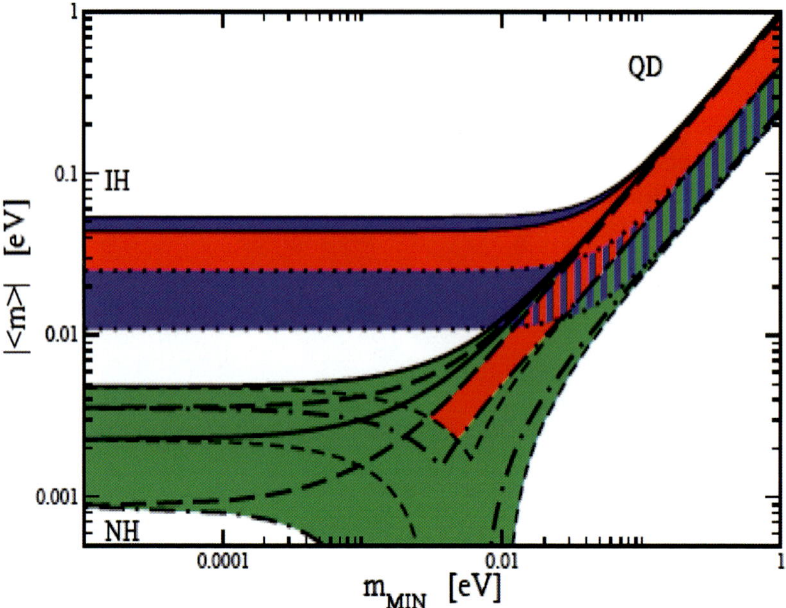

Fig. 10. – Effective neutrino mass expected in DBD experiments from neutrino oscillations. The upper and lower curves refer to the so-called *inverted* and *normal* hierarchies.

The decay can occur in three channels

$$(6) \qquad (A, Z + 2) \rightarrow (A, Z + 2) + 2e^- + 2\,\bar{\nu}_{e},$$

$$(7) \qquad (A, Z + 2) \rightarrow (A, Z + 2) + 2e^- + (1, 2 \ldots, \chi),$$

$$(8) \qquad (A, Z + 2) \rightarrow (A, Z + 2) + 2e^-.$$

In the first channel two antineutrinos are emitted. This process does not violate the lepton number, it is allowed by the Standard Model, and has been found in ten nuclei. We will not consider the second channel which violates the lepton number with the emission of one or more massless Goldstone particles named "Majorons". Our interest will be devoted to the third process which is normally called *neutrinoless* DBD, even if also in process (7) no neutrino is emitted. This process would strongly dominate on the two-neutrino channel if lepton number is violated. From the experimental point of view, in neutrinoless DBD the two electrons would share the total transition energy since the energy of the nuclear recoil is negligible. A peak would therefore appear in the spectrum of the sum of the two electron energies in contrast with the wide bump expected, and already found, for the two-neutrino DBD. The presence of neutrinoless DBD almost naturally implies that a term $\langle m_\nu \rangle$ called the "effective neutrino mass" is different from zero.

DBD is a very rare process both in the case of the two-neutrino and of the neutrinoless channel. In the latter its rate would be proportional to a phase space term, to the square of the nuclear matrix element and to the square of the above-mentioned term $\langle m_\nu \rangle$. While the phase space term can be easily calculated, this is not true for the nuclear matrix element whose evaluation is a source of sometime excited debates. The calculated values could vary by factors up to two. As a consequence the discovery of neutrinoless DBD should be made on two or more different nuclei. From the experimental point of view there is an even more compelling reason to do that. In a common background spectrum many peaks appear due to radioactive contaminations and many of them can hardly be attributed to a clear origin. It is not possible therefore to exclude that a peak in the region of neutrinoless DBD could be mimicked by some unknown radioactive event. Investigation of spectra obtained from different nuclear candidates where the neutrinoless DBD peak is expected in different regions would definitely prove the existence of this important phenomenon.

The value of $\langle m_\nu \rangle$ and therefore the rate of neutrinoless DBD is correlated to properties of oscillations. As shown in fig. 10 values of a few tens or units of meV are expected in the case of the two different orderings of neutrino masses named "inverted" and "normal" hierarchy, respectively.

4'3.1. *Experimental approach.* Two different experimental approaches can be adopted to search for DBD [10-14]: the indirect and the direct one.

Indirect experiments

The most common indirect approach is the *geochemical* one. It consists in the isotopic analysis of a rock containing a relevant percentage of the nucleus (A, Z) to search for an abnormal isotopic abundance of the nucleus $(A, Z + 2)$ produced by DBD. This method was very successful in the first searches for DBD and led to its discovery in various nuclei, but could not discriminate among the various DBD modes (two neutrino or neutrinoless decay, decays to excited levels, *etc.*). The same is true for the *radiochemical* methods consisting in storing for long time large masses of DBD candidates (*e.g.*, ^{238}U) and in searching later the presence of a radioactive product (*e.g.*, ^{238}Th) due to DBD.

Direct experiments

Direct experiments are based on two different approaches (fig. 11). In the *calorimetric* or *source = detector* one the detector itself is made by a material containing the DBD candidate nucleus (*e.g.*, ^{76}Ge in a germanium semiconductor detector or ^{136}Xe in a xenon TPC, scintillation or ionization detector). In the *source $\neq$ detector* approach sheets of the DBD source are interleaved with suitable detectors of ionizing particles. A weak magnetic field could be present to eliminate various sources of background. Thin sheets have to be used to optimize the resolution in the measurement of the sum of the two electron energies.

Thermal detectors

A new approach [15-18] based on the calorimetric detection of DBD is the use of thermal or cryogenic detectors, amply adopted also in searches on dark-matter particles and for direct measurements of the neutrino mass in single beta decay. An *absorber* is

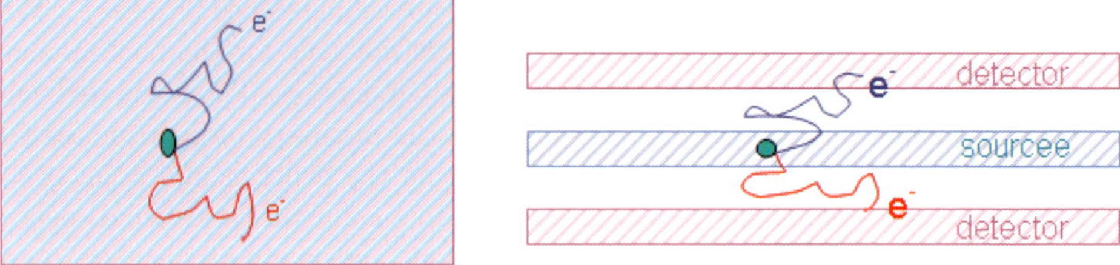

Fig. 11. – The two different approaches to direct search for DBD.

made by a crystal, possibly of diamagnetic and dielectric type, kept at low temperature where its heat capacity is proportional to the cube of the ratio between the operating and the Debye temperatures. As a consequence in a cryogenic set-up like a dilution refrigerator this heat capacity could become so low that the increase of temperature due to the energy released by a particle in the absorber can be detected and measured by means of a suitable thermal sensor. The resolution of these detectors, even if still in their infancy, is already excellent. In X-ray spectroscopy made with bolometers of a milligram or less the FWHM resolution can be as low as 3 eV, more than an order of magnitude better than in any other detector. In the energy region of neutrinoless DBD the resolution with absorbers of masses up to a kg is comparable to or better than that of Ge diodes.

4˙3.2. Present results and future experiments.

Present results

The present results [11-14] on neutrinoless DBD are reported in table I with the corresponding limits on neutrino mass, where the large uncertainties on nuclear ma-

TABLE I. – *Present results on neutrinoless DBD and limits on neutrino mass (eV).*

Nucleus	T0n (y)	Sym	Rodin old	Civit	4	5	Rodin (New)
48Ca	>1.4x1022				22		
76Ge	>1.9x1025	.47	.55	.41	.97	.84	.36 ±.07
76Ge	>1.6x1025	.51	.60	.44	1.1	.91	.40±.08
76Ge	1.2x1025	.59	.69	.52	1.2	1.1	.46±.09
82Se	>2.1.x1023	2.2	2.9	1.8	2.8	3.7	1.9±2.0
100Mo	>5.8x1023	.97	2.7	1.1		11	1.1 ±.3
116Cd	>1.7x1023	2.4	3.5	1.4		2.7	2.3±.8
128Te	>7.7×1024	1.8	2.5		4.6	2.0	1.5±.6
130Te	>3x1024	.7	.85	.37	1.1	1.8	.16±.84
136Xe	>1.2x1024	2.9	2.0	.42	2.6		1.2±.5
150Nd	>1.2x1021	2.7	4.4		1.0	.84	2.3±1

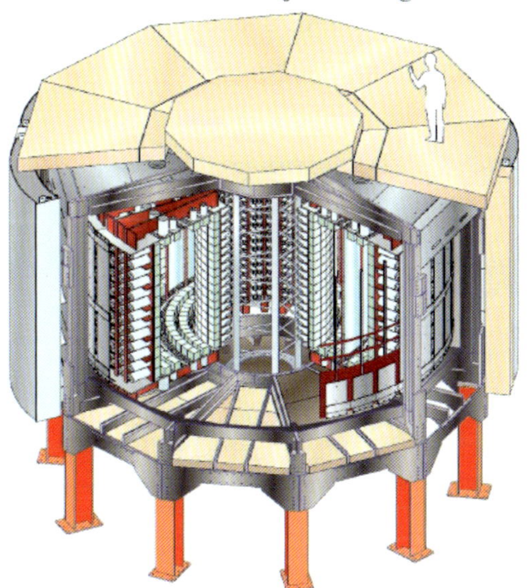

Fig. 12. – NEMO 3.

trix elements are taken into account. It can be seen that so far no experimental group has indicated the existence of neutrinoless DBD, with the exception of a subset of the Heidelberg-Moscow collaboration led by H. Klapdor-Kleingrothaus who claims the existence of this process in ^{76}Ge. This evidence is amply debated in the international arena.

NEMO 3 and CUORICINO

Two experiments are presently running with a sensitivity on neutrino mass comparable to the evidence reported by H. Klapdor-Kleingrothaus *et al.*: NEMO 3 and CUORICINO.

NEMO 3

It is a "source $\neq$ detector" experiment (fig. 12) presently running in a laboratory situated in the Frejus tunnel between France and Italy at a depth of $\sim$ 3800 meters of water equivalent (m.w.e). This experiment has yielded extremely good results on two-neutrino DBD of various nuclei. The limits on the neutrinoless channel of ^{100}Mo and ^{82}Se (table I) are already approaching the value of neutrino mass presented as evidence by Klapdor *et al.*

CUORICINO

It is at present the most sensitive running neutrinoless DBD experiment. It operates in the Laboratori Nazionali del Gran Sasso under a overburden of rock of $\sim$ 3500 m.w.e. (fig. 13). It consists in a column of 62 crystals of natural TeO$_2$ to search for neutrinoless

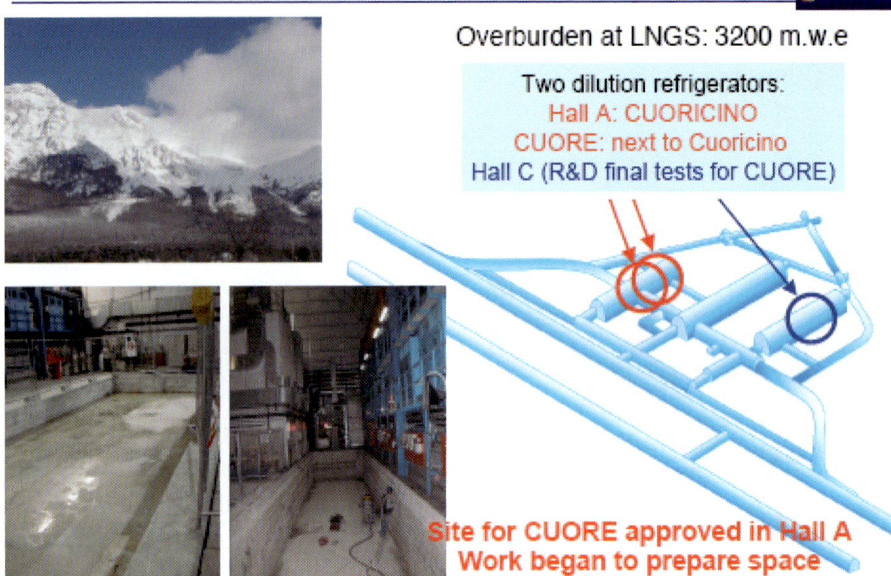

Fig. 13. – Location of CUORICINO and of R&D for CUORE in the Gran Sasso Laboratory.

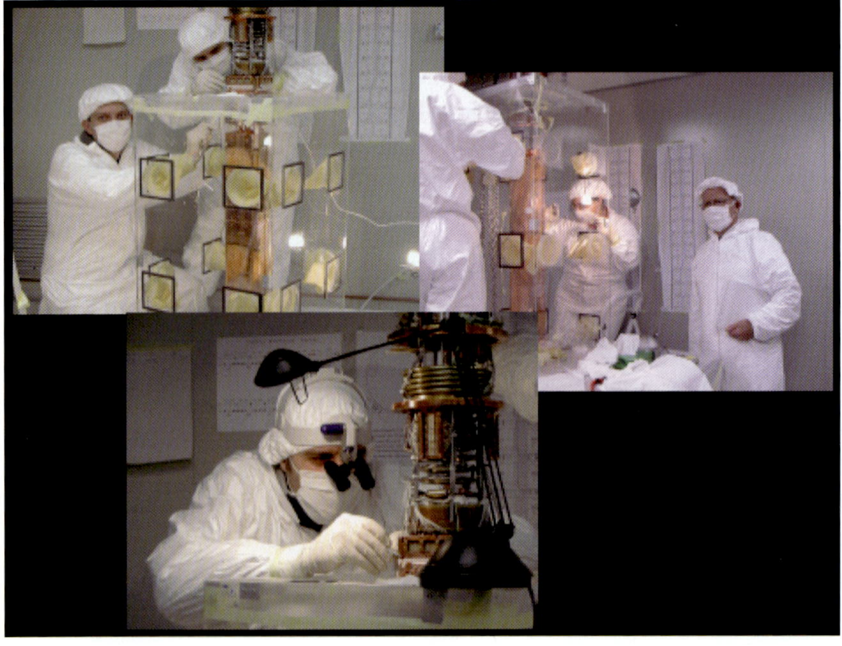

Fig. 14. – Mounting of CUORICINO.

TABLE II. – *Future experiments on Double Beta Decay.*

Name		%	$Q_{\beta\beta}$	% E	B c/y	T (year)	Tech	<m>
CUORE	^{130}Te	34	2533	90	3.5	1.8×10^{27}	Bolometric	9-57
GERDA	^{76}Ge	7.8	2039	90	3.85	2×10^{27}	Ionization	29-94
Majorana	^{76}Ge	7.8	2039	90	.6	4×10^{27}	Ionization	21-67
GENIUS	^{76}Ge	7.8	2039	90	.4	1×10^{28}	Ionization	13-42
Supernemo	^{82}Se	8.7	2995	90	1	$2 10^{26}$	Tracking	54-167
EXO	^{136}Xe	8.9	2476	65	.55	1.3×10^{28}	Tracking	12-31
Moon-3	^{100}Mo	9.6	3034	85	3.8	1.7×10^{27}	Tracking	13-48
DCBA-2	^{150}Nd	5.6	3367	80		1×10^{26}	Tracking	16-22
Candles	^{48}Ca	.19	4271	-	.35	3×10^{27}	Scintillation	29-54
CARVEL	^{48}Ca	.19	4271	-		3×10^{27}	Scintillation	50-94
GSO	^{160}Gd	22	1730	-	200	1×10^{26}	Scintillation	65-?
COBRA	^{115}Cd	7.5	2805				Ionization	
SNOLAB+	^{150}Nd	5.6	3367				Scintillation	

DBD of ^{130}Te. Its mass of 40.7 kg is more than an order of magnitude larger than in any other cryogenic set-up (fig. 14) No evidence is found for a peak in the region of neutrinoless DBD setting a 90% lower limit of 3×10^{24} years on the lifetime of neutrinoless DBD of ^{130}Te. The corresponding upper limit on $\langle m_\nu \rangle$ (0.16–0.84 eV) almost entirely covers the span of evidence coming from the claim of H. Klapdor-Kleingrothaus *et al.* (0.1–0.9 eV).

Future experiments

A list of proposed future experiments [11-14] is reported in table II with the adopted techniques and the expected background and sensitivity. Only one of them, CUORE, has been fully approved, while GERDA has been funded for its first preliminary version. These and a few others will be briefly described here.

GERDA and Majorana

Both these experiments (fig. 15) are based on the "classical" detection of neutrinoless DBD of ^{76}Ge in a "source = detector" approach with germanium diodes. They are logical continuations of the Heidelberg-Moscow and IGEX experiments, respectively. GERDA, already approved in its preliminary version, is going to be mounted in the Gran Sasso Underground Laboratory. An intense R&D activity is being carried out by the Majorana collaboration in view of the installation of this experiment. Its underground location has not yet been decided. The two experiments plan to join their forces in a future one ton experiment on ^{76}Ge.

MOON is based on the "source $\neq$ detector" approach to search for neutrinoless DBD of ^{100}Mo to be installed in the Oto underground laboratory in Japan. The set-up will be made (fig. 16) by thin sheets of enriched molybdenum interleaved with planes of scintil-

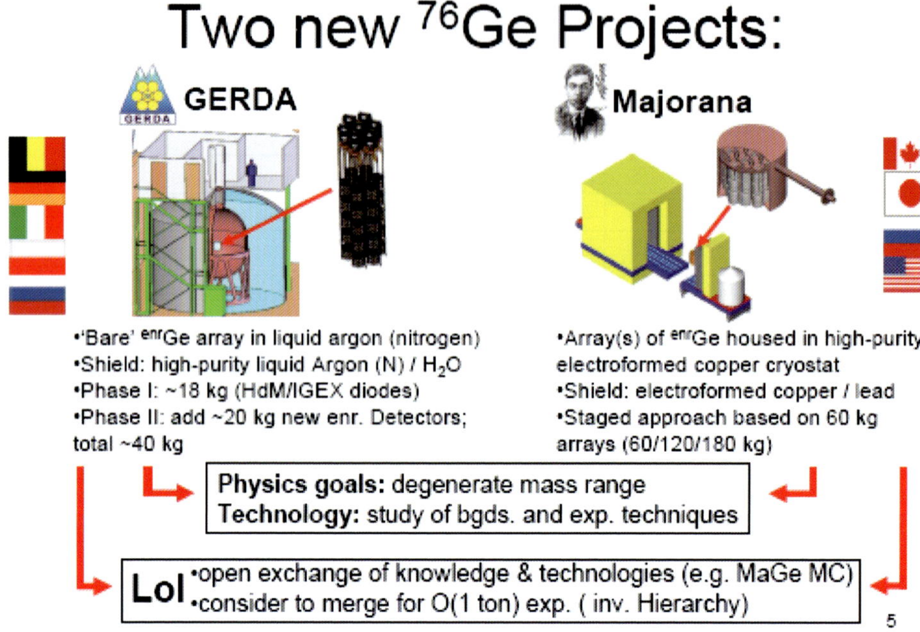

Fig. 15. – GERDA and Majorana.

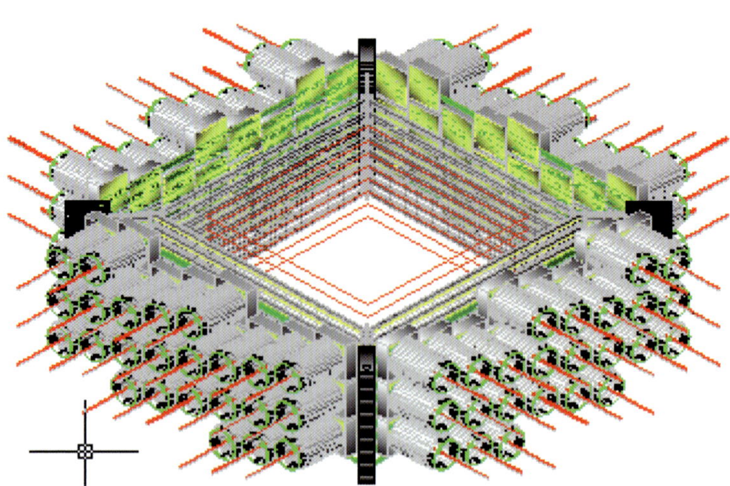

Fig. 16. – MOON I, a running prototype of MOON.

Fig. 17. – CUORE: an array of 988 crystals of TeO_2.

lating fibers. The experiment is also intended to detect the low-threshold interactions of solar neutrinos on ^{100}Mo leading to ^{100}Rb.

SUPERNEMO is also a "source $\neq$ detector" experiment mainly intended to search for neutrinoless DBD of ^{82}Se, to be installed in a not yet decided underground laboratory in Europe. The system is similar to the one adopted by NEMO 3, but with a considerably different geometry.

TABLE III. – *Possible thermal candidates for neutrinoless DBD.*

Compound	Isotopic abundance	Transition energy (keV)
^{48}CaF$_2$	.0187 %	4272
^{76}Ge	7.44 "	2038.7
^{100}MoPbO$_4$	9.63 "	3034
^{116}CdWO$_4$	7.49 "	2804
^{130}TeO$_2$	34 "	2528
^{150}NdF$_3$ ^{150}NdGaO$_3$	5.64 "	3368

XENON is an experiment to be carried out in Japan with a large mass of enriched Xenon based on scintillation to search for neutrinoless DBD of ^{136}Xe. Due to the large mass it will be also used to in a search for interactions of Dark Matter particles (WIMPS).

EXO is also intended to search for neutrinoless DBD of ^{136}Xe-^{136}Ba, but with a totally new approach: to search for DBD events by detecting with the help of LASER beams single Ba++ ions produced by the process. The option of liquid or gas Xenon and the underground location has not yet been decided, but a 100 kg litre liquid Xenon experiment without Ba tagging is going to operate soon in the WIPP underground laboratory in USA.

CUORE (for Cryogenic Underground Observatory of Rare Events) is the only second-generation experiment approved so far. It will consist in 988 crystals of natural TeO$_2$ arranged in 19 columns practically identical to the one of CUORICINO, with a total mass of about 750 kg (fig. 17). The experiment has already been approved by the Scientific Committee of the Gran Sasso Laboratory, by the Italian Institute of Nuclear Physics and by DOE and NSF. The basement for its installation has been prepared in Gran Sasso (fig. 13). As shown in table III ^{130}Te has been chosen for CUORE due to its high isotopic abundance, but the versatility of thermal detectors allows many other interesting, but expensive, double-beta active materials.

5. – Conclusions

Low-energy nuclear physics has contributed in a substantial way to the development of the study of weak interactions since the very beginning, and also after the so-called *particle astrophysics connection*. Since this year is the 50th anniversary of the discovery of parity non conservation in weak interactions, I have decided to start with this subject which still plays an important role in neutrino properties, which covers the rest of this lectures.

After 70 years the brilliant hypothesis of Ettore Majorana is still valid and is strongly supported by the discovery of neutrino oscillations which imply that the difference between the squared masses of two neutrinos of different flavours is finite. As a consequence at least one of the neutrinos has to be massive and the measurement of the neutrino mass becomes imperative. Double beta decay is at present the most powerful tool to obtain this result and also to clarify if the neutrino is a Majorana particle. The future second-generation experiments being designed, proposed and already in the case of CUORE under construction will allow in a few years to reach the sensitivity in the neutrino mass predicted by the results of oscillations in the inverse hierarchy scheme.

REFERENCES

[1] LEE T. D. and YANG C. N., *Phys. Rev.*, **104** (1956) 1.
[2] See, for instance, ADELBERGER E. G. and HAXTON W. C., *Annu. Rev. Nucl. Part. Sci.*, **35** (1985) 501.
[3] AJZENBERG-SELOVE F., *Nucl. Phys. A*, **490** (1989) 1 and references therein.
[4] GERICKE M. T. *et al.*, *Phys. Rev. C*, **74** (2006) 065503.

[5] AJZENBERG-SELOVE F., *Nucl. Phys. A*, **166** (1971) 1 and references therein.

[6] See, for instance, LINTZ M., GUÉNA J. and BOUCHIAT M.-A., *Eur. Phys. J. A*, **32** (2007) 525 and references therein.

[7] For a recent review including neutrino properties and recent results see: *Review of Particle Physics*, *J. Phys. G: Nucl. Part. Phys.*, **33** (2006) 1.

[8] FOGLI G. L., LISI E., MARRONE A. and PALAZZO A., *Prog. Part. Nucl. Phys. Phys.*, **57** (2006) 742 and references therein.

[9] GOEPPERT-MAYER M., *Phys. Rev.*, **48** (1935) 512.

[10] AALSETH C. *et al.*, arXiv:hep-ph/04123000.

[11] ZDESENKO Y., *Rev. Mod. Phys.*, **74** (2002) 663.

[12] AVIGNONE III F. T., KING III G. S. and ZDESENKO YU. G., *New J. Phys.*, **7** (2005).

[13] ELLIOTT S. R. and ENGEL J., *J. Phys. G: Nucl. Part. Phys.*, **30R** (2004) 183, hep-ph/0405078.

[14] AVIGNONE F. T., ELLIOTT S. R. and ENGEL J., *Double beta decay, Majorana neutrinos and neutrino mass*, submitted to *Rev. Mod. Phys.*

[15] FIORINI E. and NINIKOSKI T., *Nucl. Instrum. Methods*, **224** (1984) 83.

[16] TWERENBOLD D., *Rep. Prog. Phys.*, **59** (1996) 349.

[17] BOOTH N., CABRERA B. and FIORINI E., *Annu. Rev. Nucl. Part. Sci.*, **46** (1996) 471.

[18] ENNS C. (Editor), *Topics in Applied Physics*, Vol. **99** (Springer-Verlag, Germany) 2005, p. 453.

DOI 10.3254/978-1-58603-885-4-497

Experimental results on the GDR at finite temperature and in exotic nuclei

A. Bracco(*)

Dipartimento di Fisica dell'Università and INFN, Sezione di Milano
via Celoria 16, Milano, Italy

Summary. — Experiments on the gamma decay of the giant dipole resonance, addressing three different aspects of nuclear structure are here discussed. The three aspects are: i) the problem of the damping mechanisms at finite temperature, ii) the dipole radiation emitted by a dynamic dipole formed in the process leading to compound nuclei, and iii) the search of the pygmy strength in neutron-rich exotic nuclei. The experiments addressing the first two points were carried out at LNL using the beams of the Tandem-ALPI complex and the GARFIELD set-up including the HECTOR array. The compound nucleus mass region investigated was $A = 130$ at $T > 2\,\mathrm{MeV}$. The problem of searching the pygmy resonance was addressed with the Coulomb excitation technique at relativistic energies for the nucleus $^{68}\mathrm{Ni}$. The experiment was carried out at GSI using the RISING set-up. The experimental results in all cases are compared with the available predictions.

(*) E-mail: `angela.bracco@mi.infn.it`

1. – Introduction

The measurement of the gamma decay of the Giant Dipole Resonance (GDR) allows to investigate different aspects of nuclei at extreme conditions, going from the nuclear response at external excitations to the properties of reaction dynamics of heavy-ion collisions.

The gamma decay of the GDR in nuclei at finite temperature provides information on the damping mechanisms of this simple vibrational mode. In addition, it gives information on the nuclear shape at finite temperature and angular momentum due to the coupling of the dipole oscillation to the nuclear quadrupole deformation. The hot rotating nuclei are produced using fusion-evaporation reactions with heavy ions. The most recent results on this subject are presented and discussed in sect. **2**.

Another interesting aspect that can be investigated by measuring high-energy gamma in fusion-evaporation reactions with heavy ions is the formation of a dynamical electric dipole existing during the time in which the equilibration process takes place. The origin of the dynamic dipole is related to the fact that, in dissipative collisions, energy and angular momentum are quickly distributed among all single-particle degrees of freedom, while charge equilibration takes place on longer time scales. Consequently for charge asymmetric entrance channels one expects pre-equilibrium photon emission from the dipole oscillation in the isospin transfer dynamics at the time of CN formation. This problem is discussed in sect. **3**.

In the case of nuclei far from stability, the electric dipole ($E1$) response at energy around the particle separation energy is expected to have an enhanced strength as compared with the standard tail of the Lorentzian centered at $15\,\mathrm{MeV}$. There is presently interest in studying this strength around the particle binding energy not only to test mean-field theories but also because it affects reaction rates in astrophysical scenarios. The accumulation of $E1$ strength around the particle separation energy is commonly denoted as "pygmy" dipole resonance due to the smaller size of this strength in comparison with the "giant" dipole resonance dominating the $E1$ response and exhausting the Thomas-Reiche-Kuhn oscillator sum rule. The question of the evolution of the $E1$ strength in neutron nuclei far from stability is presently one of the most interesting topics since it provides information on the properties of neutron skins and consequently on the symmetry term relevant for the study of neutron stars. This topic is discussed in sect. **4** in connection with data concerning the nucleus ^{68}Ni.

2. – The width of the GDR at finite temperature

Extensive research has been made on the problem of the width of the Giant Dipole Resonance (GDR) at high temperature and angular momentum. However, the wealth of experimental data on this subject covers in most cases the interval of temperatures up to $\approx 2.5\,\mathrm{MeV}$. These data are based on the study of the γ-decay of hot rotating Compound Nuclei formed mainly using fusion-evaporation reactions. The measured systematics on the GDR width shows consistently a rapid increase of the GDR width and have been shown to provide an important testing ground for the theoretical models for damping of collective modes at finite temperature. In general [1-7] the change of the GDR width with angular momentum and temperature is found to be induced mainly by the large amplitude thermal fluctuations of the nuclear shape while quanta fluctuations do not seem to play a relevant role. In particular, the small amplitude quanta fluctuations, induced by nucleon-nucleon collisions could start affecting the GDR width at the highest temperature (3–4 MeV). It is clear that while at $T < 2$ MeV we have a rather good picture on the problem of the damping of the GDR, at higher temperatures the situation is more complex. In fact, the experimental study at $T > 2$ MeV could suffer from the problem of the incomplete thermalization of the nucleus, basic property that the nucleus must have when the dipole oscillation is built on it. This relevant question was raised in the works [8,9] concerning the study of the GDR width at finite temperature in Sn isotopes. In particular those works have shown that in the case of Sn at excitation energies $E^* > 150$ MeV the temperature of the γ-ray emitting systems had to be corrected with respect to that obtained using the kinematics of the complete fusion reaction because one had to take into account the pre-equilibrium emission. The measurement of Sn isotopes at temperature up to ≈ 2.5 MeV using the reaction ^{18}O + ^{100}Mo with $E_{\mathrm{beam}} = 122$–217 MeV was therefore particularly instructive. In fact, the analysis of the Light Charged Particle (LCP) spectra emitted in that reaction has shown that the pre-equilibrium contribution is sizable, corresponding, in the case of the highest bombarding energies, to a loss of excitation energy of approximately 20%. Based on that result corrections for the determination of the temperature of the emitting nucleus were applied also to previous data [8-14]. This reinterpretation of the existing data provided a picture of the temperature dependence of the GDR width no longer consistent with the previous one which indicated saturation effects in the damping mechanisms. Nevertheless, one weak point of the work on Sn is that only few data points were obtained in exclusive measurements and the correction of the excitation energy associated to the data point was measured only in the case of the reactions induced by Oxygen beams.

To obtain a better insight of the problem, further measurements were more recently made for another nucleus in the nearby mass region $A = 132$. For the chosen Ce isotopes a behavior similar to that of Sn is expected. The important feature of the measurements for Ce is that both the γ-decay, and also α and proton decay spectra, all in coincidence with the recoiling evaporation nuclei were obtained in the same experimental set-up [15-17].

The experiments concerning the GDR study in the hot Ce nuclei were performed at the INFN National Laboratory of Legnaro. The reaction, ^{64}Ni + ^{68}Zn at beam energy of 300, 400 and 500 MeV was used. The experimental set-up used in these experiments consists of the combination of 3 detector systems. The first is the GARFIELD array [18] for the measurement of charged particles, the second is the HECTOR set-up made of large volume BaF$_2$ detectors [19] and the third is a set of two Position Sensitive Parallel Plate Avalanche Counter telescopes (PSPPAC). A schematic drawing of the experimental set-up is shown in fig. 1.

The GARFIELD apparatus had a forward drift chamber, covering an angular range from $\theta = 29°$ to $\theta = 82°$ and 2π in ϕ. This drift chamber, filled with CF4 gas at low

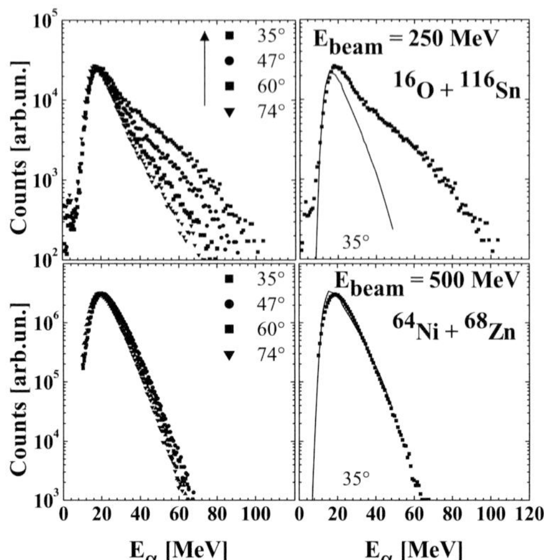

Fig. 1. – Left panels: measured α-spectra in the c.m. system at different angles. The lower panels show the data taken with the ^{64}Ni-beam ($E_{\mathrm{lab}} = 500\,\mathrm{MeV}$) and the upper panels show data with the ^{16}O-beam ($E_{\mathrm{lab}} = 250\,\mathrm{MeV}$). The right panels show the spectra specifically at $35°$ degrees. Both systems are leading to the same CN, namely ^{132}Ce having by kinematics and under the assumption of full thermalization the same excitation energy of 200 MeV. The continuous lines show the statistical model calculations for the decay of ^{132}Ce at $E^* = 200\,\mathrm{MeV}$.

pressure (50–70 mbar), is azimuthally divided into 24 sectors, and each sector consists of 8 E-ΔE telescopes, for a total of 192 telescopes. CsI(Tl) scintillation detectors, lodged in the same gas volume, were used to measure the residual energy E of the detected light ions. The calibration of the GARFIELD detectors was performed using elastically scattered ^{12}C and ^{16}O ions from a ^{181}Ta target at a number of bombarding energies in the interval from 6 to 20 MeV/u.

The recoiling heavy nuclei produced in the reactions were selected using two pairs of Position Sensitive Parallel Plate Avalanche Counters (PSPPACs). Each pair consists of two PSPPACs positioned the one behind the other and with a degrader (25 μm of Upilex) in between to stop the slow products. In this way the fusion-evaporation events were identify through the anti-coincidences between the two pairs of PSPPACs and through the Time-Of-Flight (TOF) measurement relative to the pulsed beam signal. The time resolution of the pulsed beam was better than 1 ns. The TOF measurement provided a precise discrimination between the fusion events and other reaction channels corresponding to reaction products with higher velocities.

HECTOR consists of 8 large ($14.5 \times 17\,\mathrm{cm}$) BaF$_2$ scintillators for the measurement of high energy γ-rays. In the present experiment the detectors were placed inside the large GARFIELD scattering chamber at backward angles between $125°$ and $160°$ degree at $\approx 30\,\mathrm{cm}$ from the target. At that distance from target it was possible to reject neutrons detected in BaF$_2$ by time-of-flight measurements. The BaF$_2$ detectors were calibrated using standard γ-ray sources and the 15.1 MeV γ-rays from the reaction $d(^{11}\text{B}, n\gamma)^{12}\text{C}$ at beam energy of 19.1 MeV.

The first important input for the statistical model calculations is the excitation energy of the compound nucleus. In fact, at the basis of the statistical model analysis of the high-energy gamma-ray spectra there is the assumption that the GDR vibration at finite temperature is built on a fully thermalised nucleus so that its gamma decay obeys statistical laws. For this reason it is important to examine the spectral shape of the alpha spectra measured with the same experimental conditions of the gamma-ray spectra. In the lower panels of fig. 1 a number of measured spectra of alpha-particles corresponding to different angles in the c.m. system are shown for the reaction $^{64}\text{Ni} + {}^{68}\text{Zn}$ at beam energy $E_{\text{lab}} = 500$ MeV. These spectra are compared with the corresponding spectra measured with the reaction $^{16}\text{O} + {}^{116}\text{Sn}$ at beam energy $E_{\text{lab}} = 250$ MeV (upper panels of fig. 1). Both systems are leading to the same CN, namely ^{132}Ce having by kinematics and under the assumption of full thermalization the same excitation energy of 200 MeV. However, only in the case of the reaction $^{64}\text{Ni} + {}^{68}\text{Zn}$ the alpha spectra are described by statistical model predictions, shown in fig. 1 with the continuous lines. In addition, the angular distribution of the emitted alpha-particles is consistent with statistical decay only for the $^{64}\text{Ni} + {}^{68}\text{Zn}$ reaction while that measured for the $^{16}\text{O} + {}^{116}\text{Sn}$ reaction, being strongly peaked at forward angles, has the typical behavior of the pre-equilibrium emission. The total energy loss in the excitation energy was found to be approximately 40 MeV [16, 17]. Therefore for the study of the GDR properties at finite temperature we have focused on the analysis of the high-energy gamma-rays from the $^{64}\text{Ni} + {}^{68}\text{Zn}$ reaction for which the alpha particle spectra show that the center-of-mass energy is entirely given to the equilibrated compound nucleus so that the giant dipole vibration can be assumed to be built on a fully thermalised system.

The high-energy gamma-ray spectrum is produced by the de-excitation of the compound nucleus and of all the other nuclei populated in its decay cascade [20, 21]. The GDR strength function was assumed to have the shape of a single Lorentzian function. The resonance width and centroid were treated as free parameters of the fit. In particular, the single Lorentzian strength function was centered at $E_{\text{GDR}} \approx 14$ MeV as in [22] and the GDR excitation was assumed to exhaust 100% of the energy-weighted sum rule (EWSR) strength. For the angular momentum of the compound nucleus we have used for all the three different excitation energies a triangular shape distribution with diffuseness corresponding to $2\,\hbar$. The maximum angular momentum of this distribution does not change with the bombarding energies in the case of the present experiment. In fact, the maximum angular momentum that the CN can sustain without fissioning has reached the largest possible value $L_{\text{max}} = 70\,\hbar$ already at bombarding energies lower than the present ones. For the level density description the Reisdorf formalism of Ignatyuk [23–25] was used with a value of the level density parameter a (MeV^{-1}) between $A/10$ and $A/9$ for $E^* < 100$ MeV. At higher excitation energies we used a level density parameter, as deduced from [26,27] which decreases linearly to $A/11$ up to $E^* < 170$ MeV and saturates down to $A/12.5$ for $E^* > 170$ MeV.

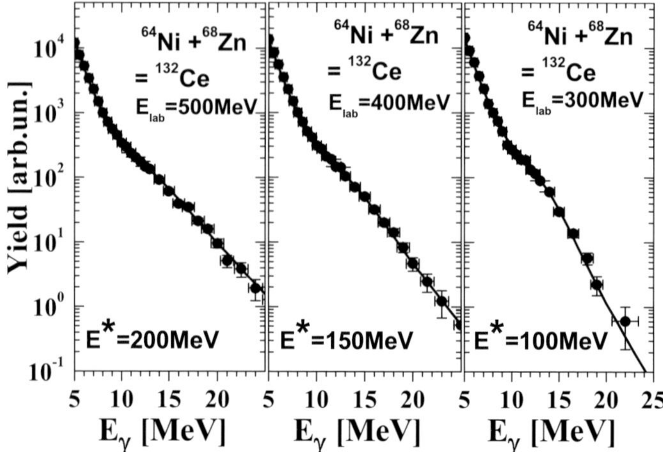

Fig. 2. – The measured (filled points) high-energy γ-ray spectra for ^{132}Ce at $E^* = 200$, 150 and 100 MeV. The corresponding predicted spectra were obtained with the statistical model and are shown with the full drawn lines. The calculations assume a fully thermalized compound nucleus and an average spin $\langle J \rangle$ of 45 $\hbar$. The resonance width and centroid were treated as free parameters of the fit.

The gamma-ray spectra calculated with the statistical model were folded with the response function of the BaF$_2$ array calculated using the GEANT [28] libraries and were then normalized at around 8 MeV. The width of the Lorentzian function was obtained from the best fit to the data using a χ^2 minimization procedure in the spectral region between 12–22 MeV. Because of the exponential nature of the γ-ray spectra, the χ^2 of this fit is dominated by the low-energy part and it is relatively insensitive to the high energy region. Consequently, the best-fitting GDR parameters were chosen to be those minimizing the χ^2 divided by the number of counts as, for example, in ref. [11]. The best-fitting statistical model calculations (full line) are shown in fig. 2 in comparison with the experimental results [15]. In order to display the measured and calculated spectra on a linear scale to emphasize the GDR region, the quantity $F(E_\gamma)Y_\gamma^{\exp}(E_\gamma)/Y_\gamma^{\text{cal}}(E_\gamma)$ was obtained and is displayed in fig. 3. In particular, $Y_\gamma^{\exp}(E_\gamma)$ is the experimental spectrum and $Y_\gamma^{\text{cal}}(E_\gamma)$ the best-fitting calculated spectrum, corresponding to the single Lorentzian function $F(E_\gamma)$. The best-fitting values deduced from the analysis of the GDR region correspond to a width $\Gamma_{\text{GDR}} = 8 \pm 1.5, 12.4 \pm 1.2$ and 14.1 ± 1.3 MeV at $E^* = 100, 150, 200$ MeV, respectively.

In order to compare the values of the GDR width extracted from the experiment with the theoretical predictions, one has to convert the excitation energy of the nucleus emitting the GDR γ-rays into temperature. In particular due to the nature or the measured spectrum one has to average the nuclear temperature over several decay steps [15]. To deduce the nuclear temperature T we have used the expression $T = 1/[\mathrm{d}(\ln(\rho))/\mathrm{d}E]$, as discussed in refs. [8, 29, 30], where ρ is the level density. The resulting values for the

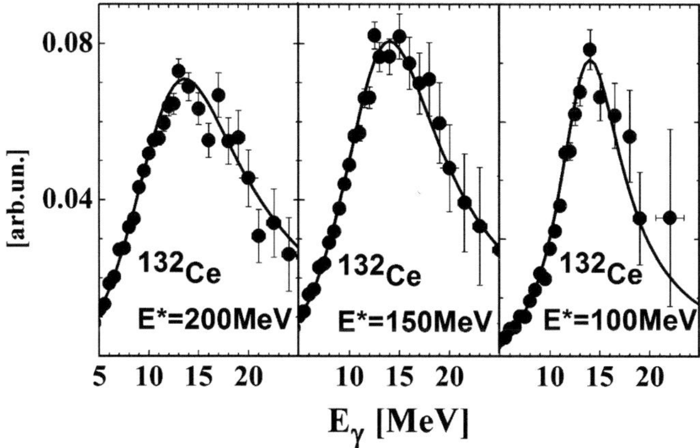

Fig. 3. – The quantity $F(E_\gamma)Y_\gamma^{\mathrm{exp}}(E_\gamma)/Y_\gamma^{\mathrm{cal}}(E_\gamma)$ is plotted for the measurement of ^{132}Ce at $E^* = 200, 150$ and $100\,\mathrm{MeV}$. $Y_\gamma^{\mathrm{exp}}(E_\gamma)$ is the experimental spectrum and $Y_\gamma^{\mathrm{cal}}(E_\gamma)$ the best fit calculated spectrum of fig. 2, corresponding to the single Lorentzian function $F(E_\gamma)$. The latter is shown in the figure with the continuum line.

present data are not substantially different from those calculated using the expression $T = [(E^* - E_{\mathrm{rot}} - E_{\mathrm{GDR}})/a]^{1/2}$, where E_{rot} is the rotational energy.

The measured GDR widths (extracted from the high-energy γ-ray spectra) consequently will not reflect the properties of the initial nucleus formed with the fusion reaction but rather an average over all the decay paths. Because the shape of the spectra emitted at different temperatures are rather different, it is important to make an appropriate average, which takes into account also the change in the spectral shape. This is equivalent to define an effective temperature corresponding to the average over a temperature region where there is a sensitivity in the fit with the statistical model of the experiment.

In order to compare the present results for ^{132}Ce with the existing ones for the Sn isotopes, we have determined also in this case the effective temperature with the same procedure.

The measured and predicted evolution of the GDR width for ^{132}Ce as a function of temperature is shown in top panel of fig. 4. The existing results for Sn are also presented (in the lower panel of fig. 4). The predictions for the Ce nucleus are for $\langle J \rangle = 45\,\hbar$, value corresponding to the reactions used for ^{132}Ce and for Sn are at $\langle J \rangle = 40\,\hbar$, value corresponding to the data on Sn. The measured widths of the GDR are plotted against the effective nuclear temperature of the compound system. In the case of the ^{132}Ce data the error bars in the width represents the statistical errors connected to the χ^2 minimization. The horizontal bar represents the average temperature range associated to 75% (lower value) and 25% (upper value) of the γ-yield. The neglected yield in the average represents the decay at the end of the CN cascade that is not sensitive to the GDR width because of its spectra shape. In the case of Sn the existing data are plotted at the values of the deduced effective temperatures and the error bars are those reported in [8].

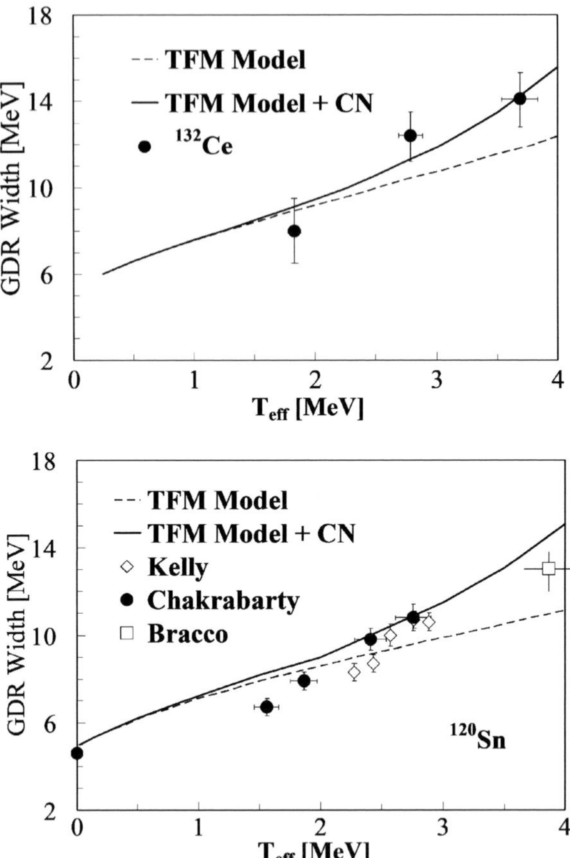

Fig. 4. – Comparison between the measured and predicted evolution of the GDR width as a function of temperature for the case of Ce (upper panel) and for Sn (lower panel). The predictions for the Ce nucleus are for $\langle J \rangle = 45\,\hbar$ and for Sn are at $\langle J \rangle = 40\,\hbar$. The measured widths of the GDR are plotted as a function of the effective nuclear temperature of the compound system. The thin continuous lines show the predictions of the thermal shape fluctuation model while the thick continuous lines include also the CN lifetime. For the Sn isotopes the zero temperature value is also shown.

The thin continuous lines show the predictions of the thermal shape fluctuation model while the thick continuous lines were obtained by adding to the thermal fluctuations also the CN lifetime. For the Sn isotopes data at temperatures larger than 2 MeV, the pre-equilibrium energy loss due to evaporation of light charged particles in the early stage of the fusion process was subtracted as reported in ref. [8]. No pre-equilibrium subtraction was made for the fusion reactions at lower excitation energies from the experiment of ref. [10]. For the Sn isotopes the zero temperature value is also shown in fig. 4.

The GDR strength function was calculated within the thermal shape fluctuation model by averaging the line shape corresponding to the different possible deformations. The averaging over the distribution of shapes is weighted with a Boltzmann factor $P(\beta, \gamma) \propto \exp[-F(\beta, \gamma)/T]$, where F is the free energy and T the nuclear temperature [2]. At each deformation point the width is $\Gamma(\beta, \gamma) = \Gamma_0(E_{\mathrm{GDR}}(\beta, \gamma)/E_0)^\delta$, where Γ_0 is the intrinsic width of the giant dipole resonance chosen equal to the zero temperature value, namely $4.5\,\mathrm{MeV}$. With this value of the intrinsic width one generally reproduce rather well the majority of the existing data at $T < 2.5\,\mathrm{MeV}$.

In both the Ce and Sn cases, the predictions account only partially for the GDR width increase measured at $T > 2.5$. Indeed by examining fig. 4 one can note that at $T > 2.5\,\mathrm{MeV}$ there is a discrepancy between the theory and experimental data. A possible explanation for this discrepancy could be related to the contribution of the lifetime of the compound nucleus which plays a role at these high temperatures. This effect was originally discussed in [31-33] and was here calculated for these cases. The thick continuous lines include the combined effect of thermal shape fluctuation plus the compound nucleus lifetime. The procedure adopted for summing the different contributions is that described in ref. [32], namely in the thermal fluctuation averaging at each point we use for the intrinsic width the expression $\Gamma_{\mathrm{GDR}} = \Gamma_0 + 2 * \Gamma_{CN_{\mathrm{evap}}}(T)$ where $\Gamma_{CN_{\mathrm{evap}}}(T)$ is the compound nucleus contribution at a given temperature [31]. Such calculations were made not only for Ce but also for Sn [36]. In this way it is possible to compare consistently the GDR width in these two cases.

A remarkable agreement between the experimental data and the predictions is found, for both, the $A = 130$ Ce and $A = 120$ Sn cases. From the present comparison one can also note that, as expected by the theory [2], for $T > 2\,\mathrm{MeV}$ there is no room for a significant increase of the intrinsic width Γ_0 with temperature [34], unless one unrealistically neglects the compound nucleus lifetime contribution to the total width.

The GDR width does not saturate at $T > 2.5\,\mathrm{MeV}$ but increases steadily with temperature at least up to $4\,\mathrm{MeV}$. Also the average nuclear deformation $\langle \beta \rangle$ [35] increases but much more steadily at $T > 2.5$, namely its increase is much less pronounced than that of the GDR width.

In summary, the GDR width in the two mass regions of Sn and Ce is found to increase consistently with temperature. Deformation effects and the sizable contribution of the intrinsic lifetime of the compound nucleus are the two combined mechanisms which explain the measured increase of the GDR width with temperature.

3. – Excitation of the dynamical dipole in heavy-ion fusion reactions

The gamma decay in heavy-ion reactions where the projectile and the target have different charge asymmetries is very interesting as it is affected by the value of the symmetry energy term of the equation of state at finite temperature. In fact, there is evidence that gamma-rays are emitted from a dynamical electric dipole which is formed in fusion reactions during the charge equilibration process. This emission is additional to that associated to the statistical decay from the thermally equilibrated compound nucleus. The origin of the dynamic dipole is related to the fact that, in dissipative collisions, energy and angular momentum are quickly distributed among all single-particle degrees of freedom,

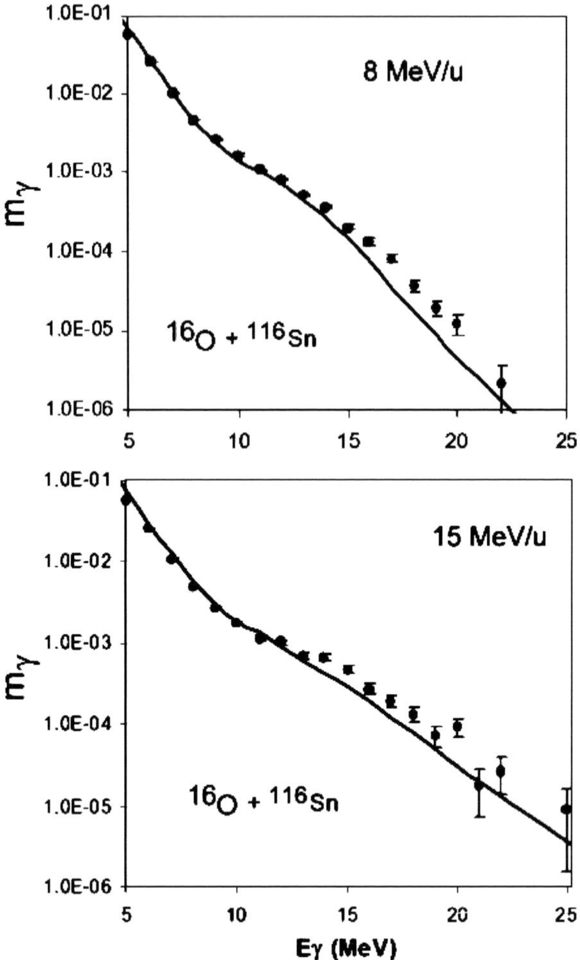

Fig. 5. – The energy spectra of gamma-rays measured with BaF_2 detectors at two different bombarding energies. The measured spectra are compared with statistical model calculations (shown with the continuous lines) using the parameters described in the text.

while charge equilibration takes place on larger time scales. Consequently for charge asymmetric entrance channels one expects pre-equilibrium photon emission from the dipole oscillation in the isospin transfer dynamics at the time of compound nucleus formation. A good way to test the models describing the pre-equilibrium emission of gamma rays as due to the dynamic dipole is to measure reactions characterized not only by a dipole moment in the entrance channel but also by the presence of sizable pre-equilibrium emission, the latter evidenced by charged particle emission. Here this problem is discussed in connection with the measurement of the reaction $^{16}O + ^{116}Sn$ (characterized by $(N/Z)_{proj} = 1$ and $(N/Z)_{targ} = 1.32$) at bombarding energies of 130 and 250 MeV.

The experiment was performed at LNL using beams from the Tandem-Alpi accelerator complex. As in the case of the experiments on Ce described in the previous section, the experimental set-up consisted of a combination of three detector systems: the first is the GARFIELD array [18] for the measurement of charged particles, the second is the HECTOR set-up made of large-volume BaF_2 detectors [19] for high-energy rays, and the third is a set of two Position Sensitive Parallel Plate Avalanche Counter telescopes (PSPPAC) to detect heavy ions. The measurement of the charged particles was important in this study to determine the presence of pre-equilibrium emission. In particular the ^{16}O-induced reaction has a sizable contribution from pre-equilibrium emission of alpha-particles [15]. A detailed study of the alpha and the proton spectra is presented in ref. [13].

The multiplicities of the high-energy gamma rays measured in coincidence with evaporation residues are shown in fig. 5. In the same figure the statistical model calculations made at the excitation energy $E^* = E_{CN} - E_{PE}$ (where E_{CN} is the CN excitation energy obtained from kinematics and E_{PE} is the energy removed by pre-equilibrium particle emission) are shown. We used as parameters of the GDR the values of E_{GDR} and Γ_{GDR} deduced from the measurement of $^{132}Ce^*$ made with the same experimental array, produced by the $^{64}Ni + ^{68}Zn$ reaction leading to a fully thermalized compound nucleus [15]. The response function of the detectors, calculated with the GEANT libraries was also folded to the calculations. One can note that, starting from a transition energy of 12 MeV, there is an excess yield as compared to the emission of the thermalized CN at excitation energy of 95 and 165 MeV.

The multiplicity of the measured excess yield, compared with the statistical model, is integrated from 8 to 20 MeV and shown in fig. 6. Both points correspond to the emission from a dynamical dipole with moment $D = 8.6\,\mathrm{fm}$ as obtained from the expression: $D = r_0(A_p^{1/3} + A_t^{1/3})/A * Z_p Z_t(N_t/Z_t - N_p/Z_p)$. The error bars of the data in fig. 6 include the statistical error and the uncertainty related to the energy lost in the pre-equilibrium emission. This uncertainty is mainly related to the emission of neutrons, which is not measured but deduced from systematics (see [17]). In the same figure predictions obtained within the BNV transport model are shown [37, 38]. It should be noted that, in spite of the rather good agreement between experiment and theory, the calculated spectral distribution is peaking 3–4 MeV lower than the measured one.

The present study of the high-energy gamma-rays produced in the asymmetric reaction $^{16}O + ^{116}Sn$ at two different bombarding energies has shown the existence of a pre-equilibrium gamma-ray emission due to a dynamic dipole moment, which is formed before full equilibration is reached. The intensity of the gamma-ray pre-equilibrium emission is rather well reproduced by a model based on the BNV model. If, on the one hand, the excitation mechanism seems to be understood, the spectral distribution of the gamma-rays is still debated: in fact, the average energy of the predicted spectra of the bremsstrahlung radiation is always 3–4 MeV lower than the measured one. Precise tests with stable beams are important before this problem can be attacked better in more asymmetric configurations as those that could be reached using the future generation of radioactive beams.

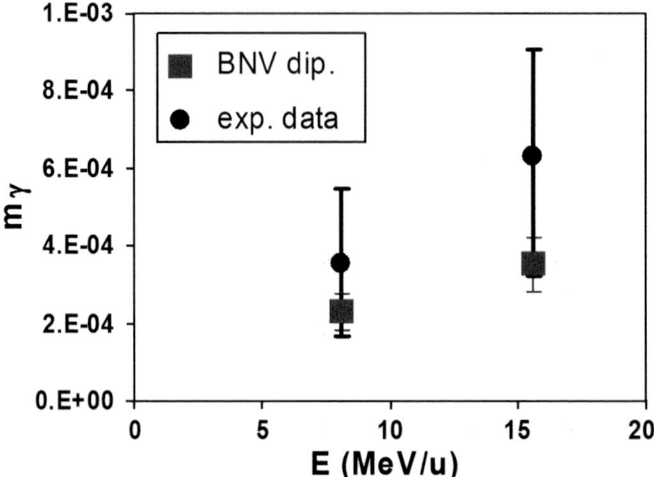

Fig. 6. – Multiplicity of the gamma-rays corresponding to pre-equilibrium emission. This represents the excess yield as compared to the statistical model. The error bars of the data include the statistical error and the uncertainty related to the energy involved in the pre-equilibrium emission. This uncertainty is mainly due to the emission of neutrons, that is deduced from models and not directly measured. The predictions based on calculations within the collisional model using the BNV approach are also shown.

4. – Search for the pygmy resonance in ^{68}Ni

In the case of nuclei far from stability, the electric dipole ($E1$) response at energy around the particle separation energy is expected to have an enhanced strength as compared with the standard tail of the Lorentzian centered at $\approx 15\,\mathrm{MeV}$. There is presently interest in studying this strength around the particle binding energy not only to test mean-field theories but also because it affects reaction rates in astrophysical scenarios. The accumulation of $E1$ strength around the particle separation energy is commonly denoted as "pygmy" dipole resonance due to the smaller size of this strength in comparison with the "giant" dipole resonance dominating the $E1$ response and exhausting the Thomas-Reiche-Kuhn oscillator sum rule. The question of the evolution of the $E1$ strength in neutron nuclei far from stability is presently one of the most interesting topics since it provides information on the properties of neutron skins and consequently on the symmetry term relevant for the study of neutron stars. So far data, obtained using neutron break-up reactions, are only available for neutron rich O and Sn isotopes [39, 40]. The unstable ^{68}Ni nucleus represents a good case to search for pygmy structures being this nucleus located in the middle of the long isotopic Ni chain having at the extremes the doubly magic ^{48}Ni and ^{78}Ni, while being experimentally accessible with the present radioactive beam facilities.

The search for the pygmy dipole resonance was performed by measuring the gamma decay following Coulomb excitation of a ^{68}Ni beam at $600\,\mathrm{MeV}/u$. At this energy the

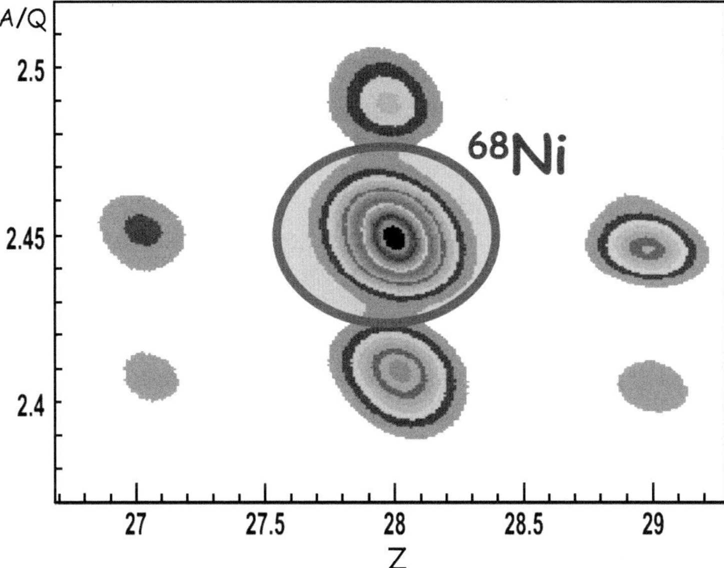

Fig. 7. – The figure shows the charge (horizontal axis) and the A/Q values (vertical axis) of the ions produced by fragmentation of ^{86}Kr at 900 MeV/u and transported through the Fragment Separator (FRS) to finally imping on the secondary Au target around which the reaction products are measured.

dipole excitation dominates over other excitation modes [41]. This is the first experiment to measure the dipole response through its gamma decay at such high energies. The radioactive ^{68}Ni beam was produced by fragmentation of the primary ^{86}Kr beam delivered, at 900 MeV/u, by the SIS synchrotron at GSI, with an intensity of 1010 particles per spill. The spill was approximately 6 seconds long with a period of 10 seconds. The primary beam was fragmented on a ^{9}Be target with a thickness of 4 g/cm^2. The ^{68}Ni ions were selected using the Fragment Separator (FRS) which was set in order to select and transport a cocktail beam of which ^{68}Ni nuclei accounted approximately for 30%, as shown in fig. 7. The transported ions were made to react on a secondary Au target. The particle identification after the target was performed by the CATE calorimeter [42, 43] which, in the present experiment, consisted of nine position sensitive Si detectors coupled to four 6-cm thick CsI scintillators, arranged to equally share the intensity of the incident beam. A total of approximately $3 \cdot 10^7$ ^{68}Ni events were collected. The gamma-ray emission from Coulomb-excited ^{68}Ni was measured using the RISING array [43] located at the final focus of the FRS spectrometer. The gamma detection array consisted of 15 HPGe cluster detectors of the Euroball array [44], each formed by 7 germanium crystals, located in rings of 16°, 33° and 36°, of 7 HPGe segmented clusters from the Miniball array [45] (at 90°), and of 8 BaF$_2$ detectors from the HECTOR array [19, 46] at 88° and at 142°. Only events with gamma multiplicity equal to one in which the ions were stopped in the

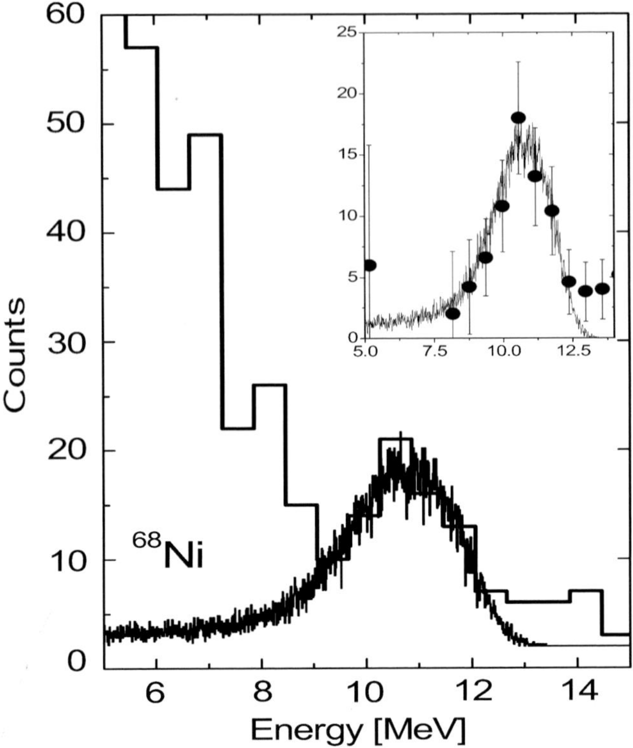

Fig. 8. – Energy spectrum of gamma-rays measured with the BaF_2 detectors at $90°$, after Doppler correction. The line is a simultation of the detector response calculated with GEANT. The inset shows the peak after background subtraction.

CATE calorimeter were further analysed. The peak in the energy spectrum measured with CATE corresponding to ^{68}Ni has a mass resolution of approximately 1%, which is sufficient to discriminate between different masses.

In order to disentangle the contribution of the pygmy dipole resonance a series of conditions were applied to the gamma-ray energy spectrum. The main constraint is the detection of ^{68}Ni, before and after the interaction on the Au target. A condition on the time spectrum measured with the different types of detectors was also requested. The gamma-ray energy spectrum was constructed after applying a Doppler correction on an event-by-event base. The trajectory of the incident secondary beam produced by fragmentation are reconstructed by tracing the particles using the beam tracking detectors of the FRS system. A peak structure centered between 10–11 MeV is visible in the gamma-ray energy spectrum measured by BaF_2 detectors, as shown in fig. 8. The width of this peak is much larger than the intrinsic energy resolution of the detector. A simulation was made for the detector response for gamma-rays of 10.5 MeV using the GEANT libraries. We included in the simulation the experimental conditions which contribute to the broadening of lines corresponding to a given gamma-ray transition. The experimental broadening is dominated by the opening of the detectors and by the straggling in the target, the first giving rise to a distribution of energies due to the variation of incident angles and the second a variation of the velocity and hence of the v/c. The result of the GEANT simulation is shown in fig. 8 with a line superimposed to the measured peak. In the inset of the figure the peak structure is shown after subtracting a linear background. A peak structure at the same energy is observed also in the gamma-ray energy spectrum measured with the HPGe detectors of RISING at forward angles. It should be noted that model predictions for the E1 strength for this nucleus also show an enhanced strength in this energy region. This is true for both predictions obtained within the RPA approach [47] and the ones using the relativistic mean-field theory [48].

5. – Conclusion

In this contribution I have presented results of experiments on the gamma decay of the giant dipole resonance addressing three different problems: i) the width at finite temperature, ii) the dynamic dipole formation in fusion reactions, and iii) the search of the pygmy resonance in exotic nuclei.

The problem of the GDR width was discussed for the temperature interval between 2 and $4\,\mathrm{MeV}$ in connection with the Ce nuclei (with $A \approx 130$). In fact, in this temperature region the data are scarce and there is the interesting question on wether or not the width saturates. The analysis of the new data shows that the GDR width does not saturate at $T > 2.5\,\mathrm{MeV}$ but increases steadily with temperature at least up to $4\,\mathrm{MeV}$. This behaviour is consistent with the one found for the Sn isotopes. However, in the case of the Ce data the thermalization was well determined experimentally by measuring simultaneously in the same experiment light charged particle spectra and γ-rays. Deformation effects and the sizable contribution of the intrinsic lifetime of the compound nucleus are the two combined mechanisms which explain the measured increase of the GDR width with temperature. More exclusive studies such as that of Ce should be made also in other mass regions including more exotic ones, or in other rotational frequency regimes to further test nuclear structure under extreme temperature condition and to learn more about nuclear deformation.

The study of the high-energy gamma-rays emitted in the asymmetric reaction $^{16}\mathrm{O} + {}^{116}\mathrm{Sn}$ at two different bombarding energies has shown the existence of a pre-equilibrium gamma-ray emission due to a dynamic dipole moment, which is formed before full equilibration is reached. The intensity of the gamma-ray pre-equilibrium emission is rather well reproduced by a model based on the BNV model. If, on the one hand, the excitation mechanism seems to be understood, the spectral distribution of the gamma-rays is still debated: in fact, the average energy of the predicted spectra of the bremsstrahlung radiation is always ≈ 3–$4\,\mathrm{MeV}$ lower than the measured one. Precise tests with stable beams are important before this problem can be attacked better in more asymmetric configurations as those that could be reached using the future generation of radioactive beams.

The search of the pygmy resonance with the gamma-ray emission was made for the exotic nucleus ^{68}Ni using the Coulomb excitation technique. The measured gamma-ray spectrum shows a peak centered around 10–11 MeV, corresponding to an enhanced strength as compared to that of the tail of the standard GDR lorentzian function. Such excess is also predicted by calculations of two different types, one based on the RPA approach and the other obtained using relativistic mean field theories. This first evidence of the pygmy resonance in a measurement using the virtual photon scattering technique opens possibilities for more systematics studies in the future.

$$* \ * \ *$$

The authors acknowledge all the collaborators of the RISING experiment (see the author list of ref. [46]) and of the GARFIELD experiment (see the author list of ref. [15]) whose work has been essential for the success of the experiments. We thank the LNS-Catania theoretical group and M. DI TORO for providing the BNV-Code and fruitful discussions.

REFERENCES

[1] KUSNEZOV D. and ORMAND W. E., *Phys. Rev. Lett.*, **90** (2003) 042501 and references therein.

[2] BORTIGNON P. F., BRACCO A. and BROGLIA R. A., *Giant Resonances: Nuclear Structure at Finite Temperature* (Harwood Academic Publishers, Amsterdam) 1998.

[3] GALLARDO M. *et al.*, *Nucl. Phys. A*, **443** (1985) 415.

[4] CAMERA F. *et al.*, *Nucl Phys. A*, **572** (1994) 401.

[5] MATIUZZI M. *et al.*, *Phys. Lett. B*, **364** (1995) 13.

[6] DONATI P. *et al.*, *Phys. Lett. B*, **383** (1996) 15.

[7] LE FAOU J. H. *et al.*, *Phys. Rev. Lett.*, **72** (1994) 3321.

[8] KELLY M. P. *et al.*, *Phys. Rev. Lett.*, **82** (1999) 3404.

[9] KELLY M. P. *et al.*, *Phys. Rev. C*, **56** (1997) 3201.

[10] CHAKRABARTY D. R. *et al.*, *Phys. Rev. C*, **36** (1987) 1886.

[11] BRACCO A. *et al.*, *Phys. Rev. Lett.*, **62** (1989) 2080.

[12] VOIJTECH R. J. *et al.*, *Phys. Rev. C*, **40** (1989) R2441.

[13] ENDERS G. *et al.*, *Phys. Rev. Lett.*, **69** (1992) 249.

[14] HOFFMAN H. J. *et al.*, *Nucl. Phys. A*, **571** (1994) 301.

[15] WIELAND O. *et al.*, *Phys. Rev. Lett.*, **97** (2006) 012501.

[16] KRAVCHUK V. L. *et al.*, *IWM2005, SIF Conf. Proc.*, Vol. **91** (Società Italiana di Fisica, Bologna) 2006, p. 33.

[17] BARLINI S. *et al.*, submitted to *Phys. Rev. C* (2007).

[18] GRAMEGNA F. *et al.*, *Nucl. Instrum. Methods A*, **389** (1997) 474.

[19] KMIECIK M. *et al.*, *Phys. Rev. C*, **70** (2004) 064317 references therein.

[20] PULHOFER F., *Nucl. Phys. A*, **280** (1977) 267.

[21] DIOSZEGI I., *Phys. Rev. C*, **64** (2001) 019801 and references therein.

[22] PIERROUTSAKOU D. *et al.*, *Phys. Rev. C*, **71** (2005) 054605.

[23] REISDORF W., *Z. Phys. A*, **300** (1981) 227.

[24] IGNATYUK A. V. *et al.*, *Sov. J. Phys.*, **21** (1975) 255.

[25] DIOSZEGI I. *et al.*, *Phys. Rev. C*, **63** (2001) 047601.

[26] De J. N. *et al.*, *Phys Rev. C.*, **57** (1998) 1398 and references therein.

[27] Gonin M. *et al.*, *Nucl. Phys. A*, **495** (1989) 139c.

[28] Brun R. *et al.*, CERN Report No. CERN-DD/EE/84-1.

[29] Kmiecik M. *et al.*, *Nucl. Phys. A*, **674** (2000) 29.

[30] Thoenessen M., Riken Review no. 23 (July 1999).

[31] Ormand W. E. *et al.*, *Nucl. Phys. A*, **614** (1997) 217.

[32] Chomaz Ph., *Phys. Lett. B*, **347** (1995) 1.

[33] Chomaz Ph., *Nucl. Phys. A*, **569** (1994) 569.

[34] Bracco A. *et al.*, *Phys Rev. Lett.*, **74** (1995) 3748.

[35] Matiuzzi M. *et al.*, *Nucl. Phys. A*, **612** (1997) 262.

[36] Bracco A. *et al.*, *Mod. Phys. Lett. A*, **22** (2007) 2479.

[37] Baran V. *et al.*, *Nucl. Phys. A*, **679** (373) 2001.

[38] Baran V. *et al.*, *Phys. Rev. Lett.*, **87** (2001) 182501.

[39] Leistenschneider A. *et al.*, *Phys. Rev. Lett.*, **86** (2001) 5442; Tryggestad E. *et al.*, *Phys. Lett. B*, **541** (2002) 32.

[40] Adrich P. *et al.*, *Phys. Rev. Lett.*, **95** (2005) 132501.

[41] Glashmacher T., *Annu. Rev. Nucl. Part. Sci.*, **48** (1998) 1.

[42] Lozeva R. *et al.*, *Nucl. Instrum. Methods A*, **562** (2006) 298.

[43] Wollersheim H. J. *et al.*, *Nucl. Instrum. Methods A*, **537** (2005) 637.

[44] de Angelis G., Bracco A. and Curien D., *Eur. News*, **34** (2003) 181; Simpson J., *Z. Phys. A*, **358** (1997) 139.

[45] Reiter P. *et al.*, *Nucl. Phys. A*, **701** (2002) 209c.

[46] Bracco A. *et al.*, *Acta Phys. Pol. B*, **38** (2007) 1229.

[47] Sarchi D. *et al.*, *Phys. Lett. B*, **601** (2004) 27; Colò G., private communications.

[48] Vretner D. *et al.*, *Nucl. Phys. A*, **692** (2001) 496; Liang J. *et al.*, *Phys. Rev. C*, **75** (2007) 054320.

DOI 10.3254/978-1-58603-885-4-515

Microscopic study of multiphonon excitations in nuclei

N. Lo Iudice, F. Andreozzi and A. Porrino

Dipartimento di Scienze Fisiche, Università di Napoli Federico II and INFN
Monte S. Angelo, via Cintia, I-80126 Napoli, Italy

F. Knapp and J. Kvasil

Institute of Particle and Nuclear Physics, Charles University
V Holešovičkách 2, CZ-18000 Praha 8, Czech Republic

1. – Introduction

Mean-field approaches, especially Random-Phase Approximation (RPA), provide an elegant, physically transparent, microscopic formalism for investigating collective excitations in nuclei [1, 2]. RPA, for instance, not only describes the global properties of giant resonances but explains in part also their fine structure by accounting for the decay of the collective mode to single-particle levels (Landau damping) [3]. RPA, however, being basically a harmonic approximation, cannot account for anharmonic features such as the collisional damping, responsible for the so-called spreading width, and is unable to describe multiphonon spectra whose evidence is growing rapidly.

At low energy, indeed, recent fluorescence scattering experiments have detected low-lying double-quadrupole, double-octupole and mixed quadrupole-octupole multiplets in nearly spherical heavy nuclei [4]. Other experiments, combining different γ-ray spectroscopy techniques, have discovered a class of multiphonon quadrupole excitations, interpreted within the Interacting Boson Model (IBM) as proton-neutron (F-spin) mixed symmetry states [5-7]. The same experiments have even detected triple-quadrupole multiplets [6, 7]. In deformed nuclei, a recent γ cascade experiment [8] could disentangle the $M1$ from the $E\lambda$ de-excitations providing evidence of a scissors mode [9, 10] built on excited states. A series of (p, t) transfer reaction experiments have populated an impressive number of 0^+ states up to $3\,\mathrm{MeV}$ [11-14]. At high energy, several experiments, exploiting different kinds of reactions, have established the existence of a double giant dipole resonance [15, 16].

The study of all these spectra and their inherent anharmonicities calls for theoretical approaches which go beyond mean-field theories. The ones already available are of phenomenological nature, like the IBM [17], or are natural microscopic extensions of the RPA, like the nuclear field theory [18, 19] or the Quasiparticle-Phonon Model (QPM) [20].

The IBM is an elegant and versatile algebraic approach which describes the low-lying properties of nuclei in terms of s $(L = 0)$ and d $(L = 2)$ bosons. It has been adopted with success for systematics of the low-energy multiphonon spectra throughout the whole periodic table [17]. Its s and d bosons have a simple underlying microscopic structure. They can be considered as highly correlated pairs of alike nucleons coupled to $L = 0$ and $L = 2$, respectively. This establishes a link with the boson expansion techniques developed long ago [21-23]. The IBM, indeed, can be considered a phenomenological realization of a fermion-boson mapping [24] according to the Marumori prescriptions [22].

The fermion-boson mapping idea inspires the mentioned microscopic approaches, namely the nuclear field theory [18, 19] and the QPM [20]. The first has been especially suitable for characterizing the anharmonicities of the vibrational spectra and the spreading widths of the giant resonances, while the QPM has been extensively adopted for the study of the fine structure of multiphonon excitations [25-27].

Much less exploited are other microscopic multiphonon approaches based on the iterative solution of equations of motion [28, 29].

Moving along the lines of these latter approaches, we have developed a new equation-of-motion method [30] which generates iteratively a basis of multiphonon states, built out of phonons constructed in the Tamm-Dancoff approximation (TDA) [1]. The basis so obtained is highly correlated and makes the task of diagonalizing the nuclear Hamiltonian much easier.

Here, we first outline the mean-field theories pointing out their harmonic character. We, then, give a very brief account of the QPM analyzing its virtues and limitations.

We finally illustrate our equation of motion phonon method (EMPM) and show how it can be implemented exactly in the specific case of $^{16}\mathrm{O}$, a nucleus of highly complex shell structure, which represents a severe test for any microscopic approach.

2. – Collective modes in Tamm-Dancoff and random phase approximations

In macroscopic models one first singles out a collective coordinate α_λ, of multipolarity λ, describing for instance a proton-neutron relative displacement $(\lambda = 1)$ or a quadrupole $(\lambda = 2)$ or octupole $(\lambda = 3)$ shape vibration. One then defines a conjugate momentum π_λ, and writes a harmonic oscillator (HO) Hamiltonian in terms of these coordinates. This Hamiltonian is more conveniently expressed in terms of creation $(O_\lambda^\dagger)$ and annihilation

(O_λ) boson operators. In such a form, it is immediate to derive the eigenvalue equation

$$(1) \qquad [H, O_\lambda^\dagger] = \hbar\omega_\lambda O_\lambda^\dagger.$$

The HO vacuum $|0\rangle$ defines the nuclear ground state. The collective mode is described by a single state, the first excited HO state $|\lambda\rangle = O_\lambda^\dagger|0\rangle$, of energy ω_λ, which collects the whole strength of the collective coordinate α_λ.

This purely harmonic model can only account qualitatively for the gross features of nuclear collective excitations. Indeed, the collective HO Hamiltonian is not immediately related to the nuclear Hamiltonian H, which includes many two-body pieces of complex structure. Nonetheless, we will see that collective and microscopic approaches are intimately correlated.

Let us write the nuclear Hamiltonian H in second quantization

$$(2) \qquad H = \sum_i \epsilon_i a_i^\dagger a_i + \frac{1}{4} \sum_{ijkl} V_{ijkl}\, a_i^\dagger a_j^\dagger a_l a_k \,,$$

where ϵ_i are the single-particle energies, V_{ijkl} the antisymmetrized matrix elements of the nucleon-nucleon interaction, $a_i^\dagger$ (a_i) the creation (annihilation) particle operators with respect to the nucleon vacuum.

We now assume that the single-particle energies have been obtained self-consistently in Hartree-Fock (HF) approximation and define the HF ground state $|\rangle$ as the particle-hole (ph) vacuum

$$(3) \qquad \begin{aligned} a_p^\dagger|\rangle &= |p\rangle, & a_p|\rangle &= 0, & \epsilon_p &> \epsilon_F, \\ a_{\bar h}|\rangle &= |h^{-1}\rangle, & a_h^\dagger|\rangle &= 0, & \epsilon_h &< \epsilon_F, \end{aligned}$$

where ϵ_F is the Fermi energy determined by the last HF filled orbit, $|p\rangle$ and $|h^{-1}\rangle$ are single-particle and single-hole states, and $\bar h$ denotes time reversal.

The most straightforward microscopic mechanism for constructing collective states is provided by the TDA. This consists just in solving the nuclear eigenvalue problem

$$(4) \qquad H|\lambda\rangle = E_\lambda|\lambda\rangle$$

in a restricted space spanned by ph states. One, thus, obtains the eigenvalue equations

$$(5) \qquad \sum_{p'h'} A(ph;p'h')\, X^\lambda(p'h') = (E_\lambda - E_0)\, X^\lambda(ph),$$

where the matrix A is given by

$$(6) \qquad A(ph;p'h') = (\epsilon_p - \epsilon_h)\delta_{pp'}\delta_{hh'} + V_{ph'h'p}\,.$$

The eigenstates have the simple form

$$(7) \qquad |\lambda\rangle = \sum_{ph} X_{ph}(\lambda) a_p^\dagger a_h |\rangle.$$

If we remain within the ph space and adopt a HF basis, we can write the eigenvalue eq. (4) in the HO form

$$(8) \qquad [H, O_\lambda^\dagger]|\rangle = \omega_\lambda O_\lambda^\dagger |\rangle,$$

where

$$(9) \qquad O_\lambda^\dagger = \sum_{ph} X_{ph}(\lambda) a_p^\dagger a_h$$

and $\omega_\lambda = E_\lambda - E_{\mathrm{HF}}$ is the excitation energy with respect to the HF unperturbed ground state.

The HO-like eigenvalue equation can be generalized by replacing the HF vacuum $|\rangle$ with the lowest eigenstate of the full Hamiltonian H, namely the nuclear ground state $|0\rangle$ [1,2]. In fact, one can always define operators O_λ and $O_\lambda^\dagger$ satisfying the conditions

$$(10) \qquad O_\lambda|0\rangle = 0, \qquad |\lambda\rangle = O_\lambda^\dagger |0\rangle$$

for any eigenstate $|\lambda\rangle$ of H. Under the above constraints, the operators O_λ and $O_\lambda^\dagger$ satisfy the following HO-like equations

$$(11) \qquad \left[H, O_\lambda^\dagger\right]|0\rangle = \omega_\lambda O_\lambda^\dagger|0\rangle = (E_\lambda - E_0)O_\lambda^\dagger|0\rangle.$$

RPA consists in solving the above eigenvalue equations in a restricted space spanned by ph states, assuming that $|0\rangle$ is the true, highly correlated, ground state. Under this assumption, the boson-like operators take the form

$$(12) \qquad O_\lambda^\dagger = \sum_{ph} \left(Y_{ph}^\lambda a_p^\dagger a_h - Z_{ph}^\lambda a_h^\dagger a_p\right).$$

The explicit eigenvalue equations are obtained by expanding the commutator in eq. (11) and linearizing the resulting expression. It is necessary, at this stage, to make the so-called Quasi-Boson-Approximation (QBA). This is the basic approximation of RPA and consists in using the Hartree-Fock ph vacuum $|\rangle$, instead of the correlated one $|0\rangle$, to actually compute the quantities

$$(13) \qquad \langle 0|a_p^\dagger a_{p'}|0\rangle \sim \langle|a_p^\dagger a_{p'}|\rangle = 0,$$

$$(14) \qquad \langle 0|a_h^\dagger a_{h'}|0\rangle \sim \langle|a_h^\dagger a_{h'}|\rangle = \delta_{hh'}.$$

Under this approximation, one obtains

$$(15) \qquad \begin{vmatrix} A & B \\ B^* & A^* \end{vmatrix} \begin{vmatrix} Y_\lambda \\ Z_\lambda \end{vmatrix} = \hbar\omega_\lambda \begin{vmatrix} Y_\lambda \\ -Z_\lambda \end{vmatrix},$$

where A is nothing but the TDA matrix (6) and B takes into account the correlations of the ground state and is given by

$$(16) \qquad B(ph; p'h') = V_{pp'hh'}.$$

The RPA states $|\lambda\rangle = O_\lambda^\dagger |0\rangle$ are orthonormalized according to

$$(17) \qquad \langle\lambda|\lambda'\rangle = \langle 0|O_\lambda O_{\lambda'}^\dagger|0\rangle = \langle 0|[O_\lambda, O_{\lambda'}^\dagger]|0\rangle \cong \langle|[O_\lambda, O_{\lambda'}^\dagger]|\rangle = \delta_{\lambda\lambda'},$$

where, once again, use of the QBA has been made. More explicitly, the orthonormalization condition yields

$$(18) \qquad \sum_{ph}(Y_{ph}^{\lambda*}Y_{ph}^{\lambda'} - Z_{ph}^{\lambda*}Z_{ph}^{\lambda'}) = \delta_{\lambda\lambda'}.$$

The transition amplitudes for the generic one-body operator W are given by

$$(19) \qquad \langle\lambda|W|0\rangle = \langle 0|[O_\lambda, W]|0\rangle \cong \langle|[O_\lambda, W]|\rangle = \sum_{ph}(Y_{ph}^{\lambda*}W_{ph} + Z_{ph}^{\lambda*}W_{hp}),$$

where the QBA has been made in order to obtain the final expression.

For open shell nuclei, one has to move from a particle to a quasi-particle formalism by means of the canonical Bogolyubov transformation

$$(20) \qquad \begin{aligned} \alpha_k^\dagger &= u_k a_k^\dagger - v_k a_{\bar{k}}, \\ \alpha_k &= u_k a_k - v_k a_{\bar{k}}^\dagger. \end{aligned}$$

The HF vacuum $|\rangle$ is replaced by a new one defined by the BCS state $|\tilde{0}\rangle$.

The quasi-particle RPA (QRPA) states have the form

$$(21) \qquad |\lambda\rangle = O_\lambda^\dagger |0\rangle,$$

where the phonon operator $O_\lambda^\dagger$ is given by

$$(22) \qquad O_\lambda^\dagger = \sum_{kl}\{Y_{kl}^\lambda \alpha_k^\dagger \alpha_l^\dagger - Z_{kl}^\lambda \alpha_l \alpha_k\}.$$

The QBA consists now in replacing in actual calculations the exact correlated ground state with the BCS vacuum $|\tilde{0}\rangle$. This approximation is also used to derive the QRPA transition amplitudes

$$(23) \qquad \langle \lambda | W | 0 \rangle = \langle 0 | [O_\lambda, W] | 0 \rangle \simeq \langle \tilde{0} | [O_\lambda, W] | \tilde{0} \rangle$$
$$= \sum_{\alpha > \beta} (Y_{\alpha\beta}^{\lambda^*} W_{\alpha\beta} + Z_{\alpha\beta}^{\lambda^*} W_{\beta\alpha})(u_\alpha v_\beta + \tau v_\alpha u_\beta),$$

where $\tau = +1$ or -1 according that the operator W is even or odd under time-reversal.

Clearly, the RPA equations have a much richer structure than the purely harmonic one (1). Indeed, they yield not one but many eigenstates. Only one state, however, is collective and gets a large fraction of the strength. The remaining strength is distributed among the other states (Landau damping). These have basically a single-particle or quasi-particle nature and, therefore, are non-collective.

It is to be pointed out, once more, that RPA, though more general than TDA, is not exact but relies on the QBA.

While accounting for the Landau damping, neither TDA or RPA can describe the fragmentation induced by the coupling of the mode with complex configurations or the anharmonicities associated to the multiphonon excitations. The study of these phenomena requires approaches which go beyond mean-field theories.

3. – A well-established multiphonon approach: The quasi-particle–phonon model

In the QPM [20], one adopts a Hamiltonian of the following structure:

$$(24) \qquad H = H_0 + V_{pp} + V_{ph}.$$

H_0 is the unperturbed one-body Hamiltonian composed of a kinetic term plus a Woods-Saxon potential, V_{pp} and V_{ph} are the two-body potentials of general separable form, acting in the particle-particle and particle-hole channels, respectively. V_{pp} consists of a monopole plus a sum of λ multipole proton-proton and neutron-neutron pairing potentials. V_{ph} is composed of a sum of proton-proton, neutron-neutron and proton-neutron separable potentials of different multipolarity λ including, among others, quadrupole-quadrupole and octupole-octupole pieces.

The QPM procedure goes through several steps. One first transforms the particle $a_q^\dagger$ ($a_{\bar{q}}$) into quasi-particle $\alpha_q^\dagger$ ($\alpha_{\bar{q}}$) operators by making use of the Bogolyubov canonical transformation (20).

In the second step, one solves the RPA equations (15), which, in the case of separable forces, become simple dispersion equations. One, thus, obtains the RPA energy spectrum and the QRPA phonon operators (22). Few of them result to be a coherent linear combination of many quasi-particle pairs. Most of the RPA states, instead, are pure two–quasi-particle configurations. Only a small subset of these non-collective phonons, together with the collective ones, is included in the actual calculations.

In the third step, one writes the Hamiltonian (24) in terms of quasi-particle and RPA phonon operators. One obtains a Hamiltonian of the quasi-particle–phonon form

$$(25) \qquad H_{QPM} = \sum_\lambda \omega_\lambda O_\lambda^\dagger O_\lambda + H_{vq},$$

where the last piece is a quasi-particle–phonon coupling term which mixes states with different numbers of phonons.

In the fourth step, one brings the quasi-particle–phonon Hamiltonian to diagonal form. This is done by using the variational principle with a trial wave function

$$(26) \qquad \Psi_\nu = \sum_i R_i(\nu) O_i^\dagger \Psi_0 + \sum_{i_1 i_2} P_{i_1 i_2}(\nu) \left[O_{i_1}^\dagger \otimes O_{i_2}^\dagger \right] \Psi_0,$$

where Ψ_0 represents the phonon vacuum state and ν labels the specific excited state. Sometimes, the QPM wave function includes up to three phonons.

The method has been extensively and successfully adopted to describe both low- and high-energy multiphonon excitations in spherical as well as deformed nuclei. It has greatly contributed to clarify the microscopic structure of the mixed-symmetry states [25] and of the double giant dipole resonance [26]. The method has also been crucial in the understanding of the nature of the 0^+ states observed in large abundance in deformed nuclei [27].

Though very successful, the QPM has some limitations. It is suitable only for a Hamiltonian of separable form and is based on the QBA. One consequence of such an approximation is that the correlations in the ground state are accounted for only virtually.

Very recently, we have developed a new multiphonon approach which is free of the QBA and accounts explicitly for the correlations in the ground state. The method is in principle exact and completely equivalent to shell model.

4. – A new multiphonon approach: An equation-of-motion phonon method

The goal of the EMPM is to construct a basis composed of multiphonon states $\mid n, \beta \rangle$, built out of TDA phonons, to be used for diagonalizing the nuclear Hamiltonian in a space which is the direct sum of subspaces with $n = 0, 1, 2, \ldots, N$ phonons. The most obvious basis would be

$$(27) \qquad \mid n, \beta \rangle = \mid \nu_1, \nu_2, \ldots, \nu_n \rangle = O_{\nu_1}^\dagger O_{\nu_2}^\dagger \ldots O_{\nu_n}^\dagger \mid \rangle,$$

where $O_{\nu_i}^\dagger$ is a TDA phonon given by eq. (9). Such a basis, however, is of no practical use. Constructing and diagonalizing the Hamiltonian using these states would be simply prohibitive. We have therefore to find ways of circumventing a brute force approach.

The way out is suggested by the structure of the TDA state itself. Let us suppose in fact that the $(n-1)$-phonon states $|n-1,\alpha\rangle$, of energy $E_\alpha^{(n-1)}$, are known. We can then adopt for the n-phonon subspace a basis composed of the states

$$(28) \qquad b_{ph}^\dagger |n-1,\alpha\rangle = a_p^\dagger a_h |n-1,\alpha\rangle,$$

having put $b_{ph}^\dagger = a_p^\dagger a_h$. For $n=1$, we have simply $b_{ph}^\dagger |n-1=0\rangle = b_{ph}^\dagger |\rangle = |ph^{(-1)}\rangle$, namely the ph basis states used for obtaining the TDA eigenstates $|n=1,\beta\rangle$.

$4^\cdot 1$. *Equations of motion.* – Guided by the $n=1$ TDA case, we require that, also for $n>1$, the basis (28) brings the Hamiltonian H to diagonal form within the n-phonon subspace obtaining the states $|n,\beta\rangle$ of energy $E_\beta^{(n)}$. Having in mind this goal, and exploiting the fact that the ph operators $a_p^\dagger a_h$ couple only subspaces which differ at most by one TDA phonon, we derive the following equations of motion:

$$(29) \qquad \langle n;\beta| \, [H, b_{ph}^\dagger] \, |n-1;\alpha\rangle = (E_\beta^{(n)} - E_\alpha^{(n-1)}) \langle n;\beta|b_{ph}^\dagger|n-1;\alpha\rangle.$$

We then write the Hamiltonian in second quantized form and expand the commutator $[H, a_p^\dagger a_h]$ on the left-hand side of the equation. After a linearization procedure, we obtain for the n-phonon subspace the eigenvalue equation

$$(30) \qquad \sum_{\gamma p'h'} A_{\alpha\gamma}^{(n)}(ph;p'h') \, X_{\gamma\beta}^{(n)}(p'h') = E_\beta^{(n)} \, X_{\alpha\beta}^{(n)}(ph),$$

where

$$(31) \qquad X_{\alpha\beta}^{(n)}(ph) \equiv \langle n;\beta|b_{ph}^\dagger|n-1;\alpha\rangle$$

are the vector amplitudes and the matrix $A^{(n)}$ is given by

$$(32) \qquad A_{\alpha\gamma}^{(n)}(ph;p'h') = \delta_{hh'}\delta_{pp'}\delta_{\alpha\gamma}^{(n-1)} \left[(\epsilon_p - \epsilon_h) + E_\alpha^{(n-1)} \right]$$
$$+ \sum_{h_1} V_{p'h_1 h'p}\, \rho_{\alpha\gamma}^{(n-1)}(h_1 h) - \sum_{p_1} V_{p'hh'p_1}\, \rho_{\alpha\gamma}^{(n-1)}(pp_1)$$
$$+ \frac{1}{2}\delta_{hh'} \sum_{p_1 p_2} V_{p'p_1 pp_2}\rho_{\alpha\gamma}^{(n-1)}(p_1 p_2) - \frac{1}{2}\delta_{pp'} \sum_{h_1 h_2} V_{hh_1 h'h_2}\big\{\rho_{\alpha\gamma}^{(n-1)}(h_1 h_2)\big\}.$$

The quantity

$$(33) \qquad \rho_{\alpha\gamma}^{(n)}(kl) = \langle n;\gamma|a_k^\dagger a_l|n;\alpha\rangle$$

is the density matrix and plays a crucial role. It is, in fact, seen to weight the particle-hole, particle-particle and hole-hole interaction.

The simple structure of the equations is to be stressed. It is also to be pointed out that the above matrix $A^{(n)}$ contains only density matrices defined within the $(n-1)$-phonon subspace. For them, as we shall see, recursive relations hold which allow an iterative solution of eq. (30). We can indeed start from the ph vacuum $\mid n = 0\rangle = \mid\rangle$ and solve the equations step by step up to a given number of phonons. In the first step, $n = 1$, since the $(n-1)$-phonon subspace contains only the ph vacuum $\mid\rangle$, the density matrices appearing in eq. (32) assume the values

$$\rho^{(0)}(pp') = \langle\mid a_p^\dagger a_{p'} \mid\rangle = 0, \qquad \rho^{(0)}(hh') = \langle\mid a_h^\dagger a_{h'} \mid\rangle = \delta_{hh'},$$

so that the matrix $A^{(n=1)}$ becomes just the TDA Hamiltonian matrix (6). Thus, our method represents the most natural extension of the TDA to multiphonon spaces.

Before implementing the iterative process, however, we must solve the redundancy problem we would run into, if we just pretended to solve the eigenvalue equations in the form given by eq. (30)

$4^{\cdot}2.$ *Overcompleteness of the basis and removal of the redundancy.* – The basis states we have adopted are of the factorized form $b_{ph}^\dagger\mid n - 1; \alpha\rangle = (a_p^\dagger a_h)\mid n - 1; \alpha\rangle$, where the ph operator $a_p^\dagger a_h$ acts on each phonon state $\mid n - 1; \alpha\rangle$ as a whole and not on its Fermion constituents. It is therefore impossible to enforce the Pauli principle between the ph and the fermion constituents of the phonon piece $\mid n - 1; \alpha\rangle$. In the absence of such a constraint, the number of ph states is determined independently of the phonon components $\mid n - 1; \alpha\rangle$ and vice versa.

It follows that the $b_{ph}^\dagger\mid n - 1; \alpha\rangle$ form an overcomplete set of linearly dependent states, yielding a redundant number of eigensolutions with spurious admixtures in each eigenvector.

In order to eliminate this redundancy problem and obtain the correct number of eigenstates, we have to formulate an eigenvalue problem of general type. Let us denote by N_r the total number of states $b_{ph}^\dagger\mid n - 1; \alpha\rangle$. It is always possible to expand the state to be determined $\mid n; \beta\rangle$ in terms of these redundant N_r states

$$(34) \qquad \mid n; \beta\rangle = \sum_{\alpha\,ph} C_{\alpha\beta}^{(n)}(ph)\, b_{ph}^\dagger \mid n - 1; \alpha\rangle.$$

Using this formula in eq. (31) we get

$$(35) \qquad\qquad X = \mathcal{D}C,$$

where $\mathcal{D}$ is the overlap or metric matrix defined by

$$(36) \qquad \mathcal{D}_{\alpha\beta}^{(n)}(ph; p'h') = \langle n - 1; \beta \mid b_{p'h'} b_{ph}^\dagger \mid n - 1; \alpha\rangle.$$

Upon insertion of the relation (35), eq. (30) is transformed into an eigenvalue equation of general form

$$(37) \qquad A\mathcal{D}C = E\mathcal{D}C.$$

This eigenvalue equation, however, is ill defined. The matrix $\mathcal{D}$ is singular with respect to inversion. Its determinant is necessarily vanishing, since the vectors $b_{ph}^{\dagger} \, | \, n-1; \alpha \rangle$ are not linearly independent.

The traditional prescriptions adopted to overcome this problem are based on the straightforward diagonalization of $\mathcal{D}$. This yields a number of vanishing eigenvalues equal to the number of redundant states. Thus, one can form a basis of linear independent states out of the eigenstates of $\mathcal{D}$ with non-vanishing eigenvalues [31].

We have avoided the direct diagonalization of $\mathcal{D}$, which is time consuming, by adopting an alternative method based on the Choleski decomposition [30]. This provides a fast and efficient recipe for extracting a basis of linear independent states $b_{ph}^{\dagger} \, | n - 1; \alpha \rangle$ spanning the physical subspace of the correct dimensions $N_n < N_r$. One, then, uses this basis to extract a $N_n \times N_n$ minor $\bar{\mathcal{D}}$ from the overlap matrix $\mathcal{D}$. This matrix $\bar{\mathcal{D}}$ has a non vanishing determinant and, therefore, can be inverted. By left multiplication in the N_n-dimensional subspace we get from eq. (37)

$$(38) \qquad \left[\bar{\mathcal{D}}^{-1}(A\mathcal{D})\right] C = HC = EC.$$

Once the matrix multiplication in the first member has been performed, the solution of the equation determines only the coefficients $C_{\alpha\beta}^{(n)}(ph)$ of the N_n-dimensional physical subspace. The remaining redundant $N_r - N_n$ coefficients are undetermined and, therefore, can be safely put equal to zero. The eigenvalue problem is thereby formulated correctly. To obtain the solutions we need, first, to compute the matrices A and $\mathcal{D}$ and extract the minor $\bar{\mathcal{D}}$.

This brings us to the second major difficulty of our problem. In general, the calculation of the metric matrix $\mathcal{D}$ is a highly non-trivial task. Elaborated diagrammatic techniques and complex iterative procedures have been envisaged to this purpose [28, 29, 32, 33].

Our EMPM makes such a calculation almost trivial. It provides, in fact, the simple formulas

$$(39) \qquad \mathcal{D}_{\alpha\beta}^{(n)}(ph; p'h') = \sum_{\gamma} \left[\delta_{pp'}\delta_{\gamma\beta} - \rho_{\gamma\beta}^{(n-1)}(pp')\right] \rho_{\alpha\gamma}^{(n-1)}(hh'),$$

where the matrix densities are given by the recursive relations

$$(40) \qquad \rho_{\alpha\beta}^{(n)}(p_1 p_2) = \sum_{ph\gamma\delta} C_{\alpha\gamma}^{(n)}(ph) \, X_{\delta\beta}^{(n)}(p_1 h) \left[\delta_{pp_2}\delta_{\gamma\delta} - \rho_{\gamma\delta}^{(n-1)}(pp_2)\right],$$

$$(41) \qquad \rho_{\alpha\beta}^{(n)}(h_1 h_2) = \sum_{ph\gamma\delta} C_{\alpha\gamma}^{(n)}(ph) \left[\delta_{hh_1}\delta_{\gamma\delta} X_{\gamma\beta}^{(n)}(ph_2) - X_{\delta\beta}^{(n)}(ph) \, \rho_{\gamma\delta}^{(n-1)}(h_1 h_2)\right].$$

The above formulas make also the calculation of the matrix $A^{(n)}$ straightforward.

We have thus accomplished all necessary steps for solving the generalized eigenvalue problem (37). These can be summarized as follows:

1) Assuming that the eigensolutions in the $(n-1)$-phonon subspace are known, construct the matrix $A^{(n)}$ through eq. (32).

2) Compute the metric matrix $\mathcal{D}$.

3) Use the Choleski decomposition method to extract a set of N_n linearly independent states and construct the $N_n \times N_n$ non-singular matrix $\bar{\mathcal{D}}$.

4) Solve the generalized eigenvalue equation in the N_n-dimensional physical subspace to obtain the correct number of eigenvalues and the expansion coefficients $C_{\alpha\beta}^{(n)}(ph)$ of the corresponding eigenvectors.

5) Compute within the n-phonon subspace the amplitudes $X_{\alpha\beta}^{(n)}(ph)$ using eqs. (35) and the density matrix $\rho_{\alpha\beta}^{(n)}(kl)$ using the recursive relations (40) and (41).

6) $X_{\alpha\beta}^{(n)}(ph)$ and $\rho_{\alpha\beta}^{(n)}(kl)$ are the new entries for the equations of motion in the $(n+1)$-phonon subspace.

The iterative process is clearly outlined. To implement it, we have just to start with the lowest trivial 0-phonon subspace, the ph vacuum, and, then, solve the eigenvalue problem step by step within each n-phonon subspace following the prescriptions 1 to 6.

4'3. *Solution of the full eigenvalue problem.* – Such a procedure yields a multiphonon basis which reduces the Hamiltonian to diagonal blocks, mutually coupled by off-diagonal terms

$$(42) \qquad H = \sum_{n\alpha} E_\alpha^{(n)} \mid n,\alpha \rangle\langle n,\alpha \mid + \sum_{n\alpha n'\beta} \mid n',\beta \rangle\langle n',\beta \mid H \mid n,\alpha \rangle\langle n,\alpha \mid,$$

where $n' = n\pm 1, n\pm 2$, since only subspaces differing from each others by two phonons, at most, are coupled. The off-diagonal pieces are given by recursive formulas and, therefore, easily computed.

The Hamiltonian matrix is therefore sparse, being composed of diagonal blocks of large dimensions and of off-diagonal pieces coupling only diagonal blocks differing by one and two phonons (fig. 1). Its sparsity property makes the diagonalization of H quite straightforward even in a large space.

The eigenvalues and eigenvectors are obtained with no approximations. They are given by

$$(43) \qquad \mid \Psi_\nu \rangle = \sum_{n,\alpha} C_\alpha^{(\nu)} \mid n,\alpha \rangle.$$

Solving iteratively eq. (34) for each state $\mid n,\alpha \rangle$, we get

$$(44) \qquad \mid \Psi_\nu \rangle = \sum_{n=0,N} \sum_{\nu_1..\nu_n} C_{\nu_1,\nu_2,.....,\nu_n}^{(\nu)} \mid n;\nu_1,\nu_2,.....\nu_n \rangle,$$

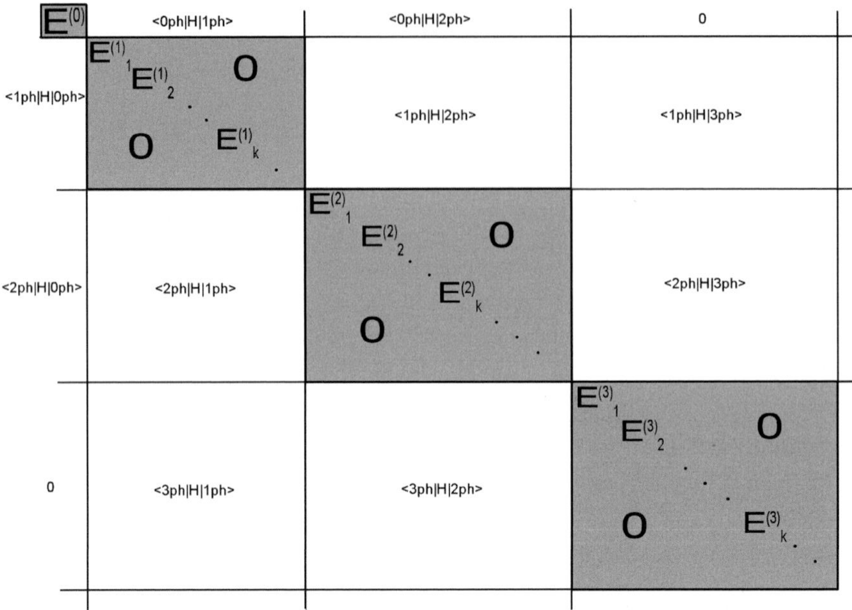

Fig. 1. – Block structure of the Hamiltonian matrix in the multiphonon representation.

where ν_i labels TDA states. The coefficients $C^{(\nu)}_{\nu_1,\nu_2,....,\nu_n}$ are very involved sums of products of expansion coefficients. This proves that the EMPM states $| \Psi_\nu \rangle$, including the ground state, are highly correlated.

Using the wave functions (43), we can compute the transition amplitudes

$$(45) \qquad \mathcal{M}(if) = \langle \Psi_{\nu_f} | \mathcal{M} | \Psi_{\nu_i} \rangle$$

of the one-body operator

$$(46) \qquad \mathcal{M} = \sum_{kl} \mathcal{M}_{kl} a_k^\dagger a_l,$$

obtaining

$$(47) \qquad \mathcal{M}(if) = \langle \Psi_{\nu_f} | \mathcal{M} | \Psi_{\nu_i} \rangle = \sum_{kln_in_f\alpha_i\beta_f} \mathcal{M}_{kl} C^{(\nu_i)}_{n_i\alpha_i} C^{*(\nu_f)}_{n_f\beta_f} \rho^{(n_in_f)}_{\alpha_i\beta_f}(kl).$$

These amplitudes involve the density matrix, which can be easily computed. Their composite structure reflects the fact that all correlations are accounted for explicitly in the EMPM wave functions and not virtually as in RPA.

5. – A numerical implementation of the method

In order to show that and how the EMPM can be implemented exactly, we have used ^{16}O as our test ground. The low-lying excitations of this nucleus are known to have a highly complex ph structure since the pioneering work of Brown and Green [34] and, even nowadays, represent a benchmark for all microscopic calculations.

Indeed, the low-energy positive parity spectrum of this nucleus was studied in a shell model calculation which included up to $4p - 4h$ and $4\hbar\omega$ configurations [35]. The same spectrum was studied very recently within a no-core shell model and an algebraic symplectic shell model [36]. In both approaches, the model space was enlarged so as to include all configurations up to $6\hbar\omega$.

We have included all ph configurations up to $3\hbar\omega$, which limit our phonon space up to $n = 3$. Such a space is considerably smaller than the one adopted in shell model. This is sufficient for our purpose, which is to show that our method can be implemented exactly. Moreover, our approach generates at once the whole spectrum of positive and negative parity states. This allows the study of low-lying as well as high-energy spectroscopic properties and, in particular, the fine structure of the giant resonances. To this purpose, we used a modified harmonic-oscillator one-body Nilsson Hamiltonian [37] plus a bare G-matrix deduced from the Bonn-A potential [38].

We have adopted the method of Palumbo [39] to separate the intrinsic from the center-of-mass motion. This method was applied to standard shell model by Glockner and Lawson [40] and, since then, widely adopted in nuclear-structure studies. It consists of adding a HO Hamiltonian in the center-of-mass coordinates multiplied by a coupling constant. If all configurations up to $3\hbar\omega$ are included, as in our case, each eigenfunction of the full Hamiltonian gets factorized into a intrinsic and center-of-mass components. For a large enough coupling constant, the center-of-mass excited states are pushed high up in energy, leaving at the low physical energies only the intrinsic states, namely the eigenfunctions with the center of mass in the ground state. Thus, a complete separation of the center of mass from the intrinsic motion is achieved.

$5{}^{\cdot}1.$ *Energy levels and transition probabilities.* – Being the space confined to $3\hbar\omega$, the ground state contains correlations up to two phonons only. These account for about 20% of the state, while the remaining 80% pertains to the ph vacuum. These values are reasonably close to the estimates of the no-core symplectic shell model. This yielded about 60% for the $0p - 0h$, 20% for $2p - 2h$ and 20% for the other more complex configurations [36], excluded from our restricted space.

As shown in fig. 2, the effect of the multiphonon configurations on the negative parity spectrum (right panel) is very important. The coupling with the two phonons pushes down the one-phonon levels. Since, however, the energy shifts on the excited levels are much smaller than the one induced on the ground state, the gap between excited and ground-state levels increases with respect to the one-phonon case. The agreement with experiments improves drastically as we include the three-phonon configurations.

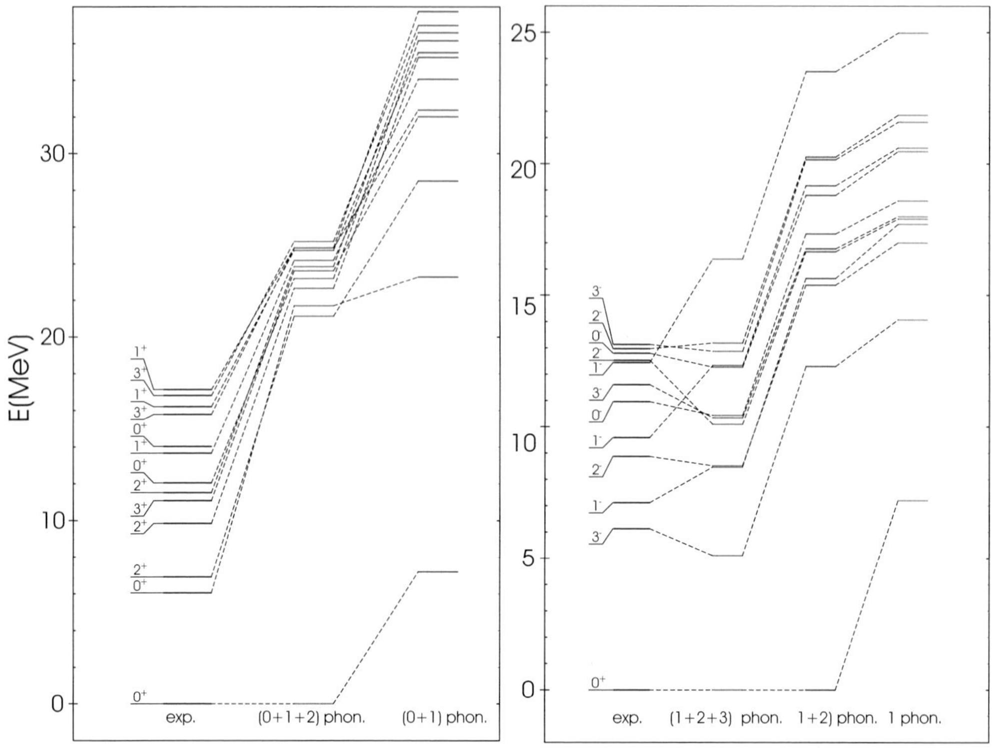

Fig. 2. – Low-lying positive (left panel) and negative (right panel) parity spectrum in ^{16}O. The energy of the one-phonon ground state, namely the ph vacuum, is relative to the one of the correlated 0^+_{gs}, assumed coincident with the experimental ground-state level.

These are quite effective in bringing all the states down within the energy range of the corresponding experimental levels [41-43].

Because of the coupling with the two-phonon subspace, the positive parity excited levels get compressed (left panel of fig. 2). They, however, remain too high in energy with respect to the ground state to reach a reasonable agreement with the experimental spectrum. This was largely expected since most of the low-lying positive parity levels are known to be dominated by $4p - 4h$ configurations at least. These states are outside our space confined to configurations up to $3\hbar\omega$. This upper energy limit excludes also the presence of 3-phonon configurations since, in the positive parity case, these states have at least an energy of $4\hbar\omega$.

Let us now investigate the anharmonicities induced by the multiphonon configurations on some selected giant resonances. We have studied isoscalar and isovector dipole and quadrupole transitions. To this purpose, we have computed the strength function [44]

$$(48) \qquad \mathcal{S}(\lambda, \omega) = \sum_\nu B_\nu(\lambda)\, \delta(\omega - \omega_\nu) \approx \sum_\nu B_\nu(\lambda)\, \rho_\Delta(\omega - \omega_\nu),$$

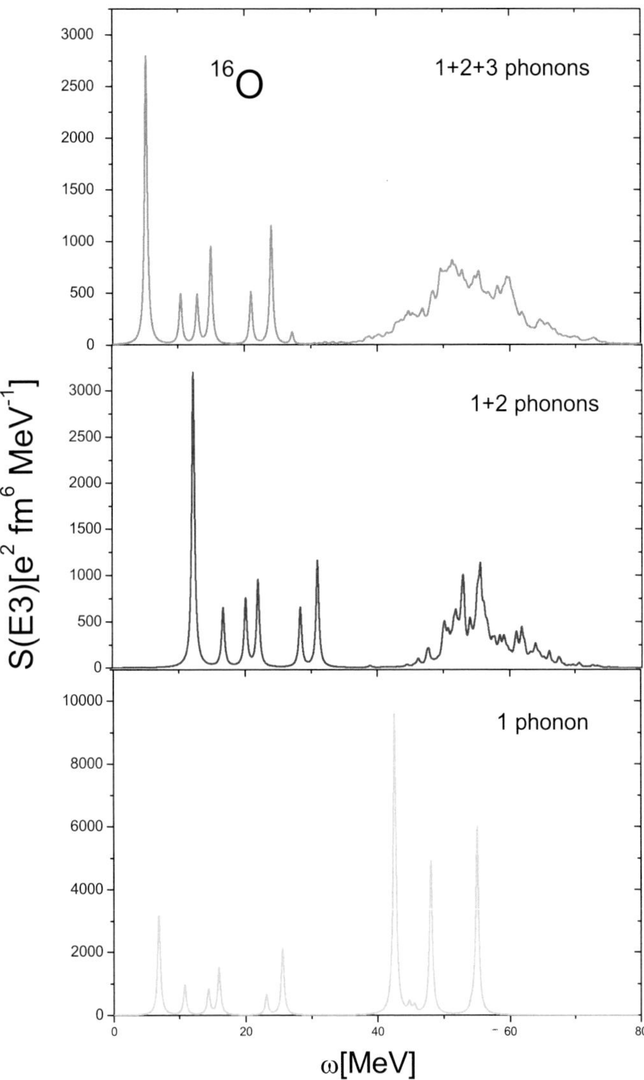

Fig. 3. – $E3$ strength distributions in ^{16}O.

where ω is the energy variable, ω_ν the energy of the transition of multipolarity λ from the ground to the ν_{th} excited state $\Psi_\lambda^{(\nu)}$ of spin $J = \lambda$, given by eq. (43), and

$$(49) \qquad \rho_\Delta(\omega - \omega_\nu) = \frac{\Delta}{2\pi} \frac{1}{(\omega - \omega_\nu)^2 + (\Delta/2)^2}$$

is a Lorentzian of width Δ, which replaces the δ function as a weight of the reduced

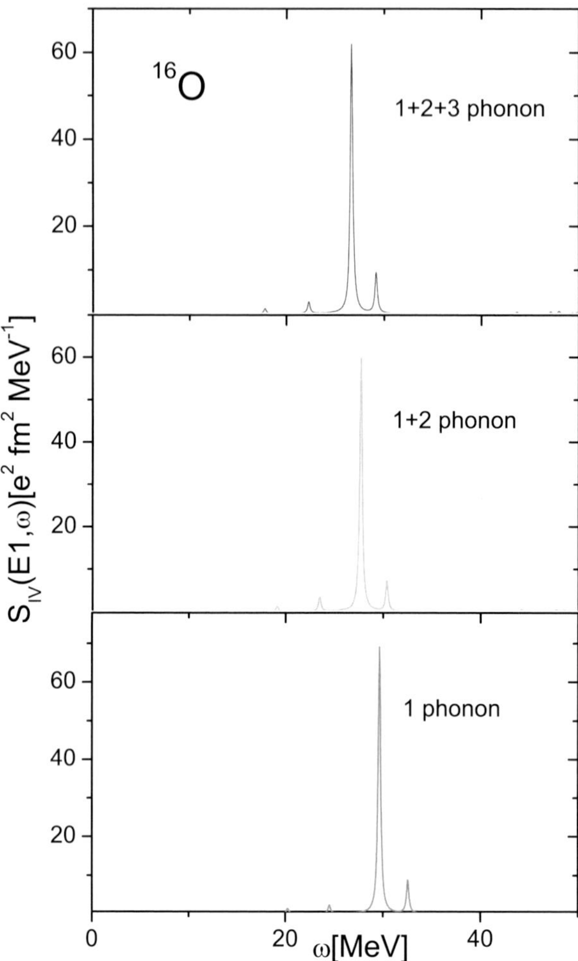

Fig. 4. – Isovector $E1$ strength distributions in ^{16}O.

transition probability

$$(50) \qquad B^{(\nu)}(\lambda) = \sum_{\mu} \left| \langle \Psi_{\lambda\mu}^{(\nu)} | \mathcal{M}(\lambda\mu) | 0 \rangle \right|^2.$$

The standard $E\lambda$ operators promoting the electric transitions are

$$(51) \qquad \mathcal{M}(E\lambda\mu) = \frac{e}{2} \sum_{i=1}^{A} (1 - \tau_3^i) r_i^\lambda Y_{\lambda\mu}(\hat{r}_i),$$

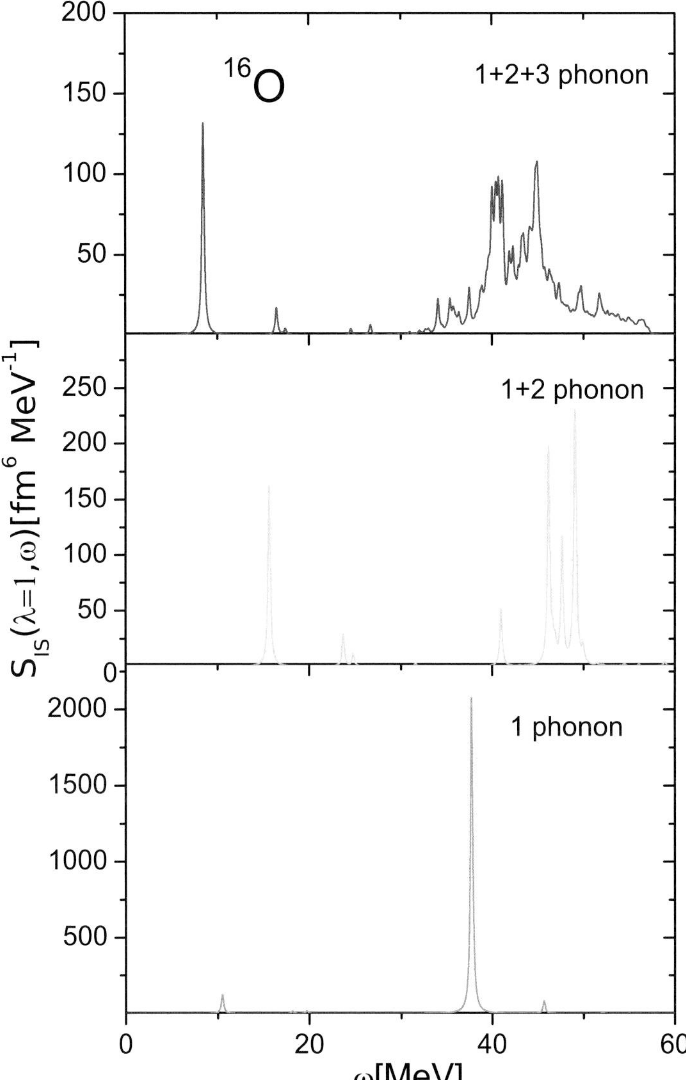

Fig. 5. – Isoscalar $E1$ strength distributions in ^{16}O.

where $\tau_3 = 1$ for neutrons and $\tau_3 = -1$ for protons.

On the other hand, the $\tau = 0$ component of the $\lambda = 1$ operator (51) is just proportional to the nuclear center-of-mass coordinate and, therefore, excites the center-of-mass spurious mode. Our calculation yields exactly vanishing strengths for all transitions promoted by this operator proving that the spurious center-of-mass mode has been completely removed.

An isoscalar dipole mode exists, nonetheless. It is known as squeezed dipole mode, or giant dipole isoscalar resonance, and is excited by the operator

$$(52) \qquad \mathcal{M}_{\mathrm{IS}}(\lambda = 1\mu) = \sum_{i=1}^{A} r_i^3 Y_{1\mu}(\hat{r}_i).$$

One should notice the absence of corrective terms generally included in order to eliminate the spurious contribution due to the center-of-mass excitation. Such a term is not necessary in our approach which guarantees a complete separation of the center of mass from the intrinsic motion.

Figure 3 shows the distribution of the $E3$ strengths over a very large energy interval. The two-phonon configurations have a damping and spreading effect and push the spectrum up in energy. The strength is further redistributed and shifted downward by the three-phonon configurations.

It is interesting to investigate the phonon structure of some of these 3^- states. While the low-lying ones are mainly one-phonon states, those contributing to the main peaks at high energy are dominated by three-phonon components.

The isovector $E1$ response $S_{\mathrm{IV}}(E1, \omega)$ is only slightly affected by the multiphonon configurations (fig. 4), an indication that the isovector giant dipole resonance mode is basically harmonic.

More dramatic is the effect of the multiphonon excitations on the isoscalar $E1$ response. The strength gets spread over a much larger energy range as we include these more complex configurations (fig. 5). Such a spreading was expected. Indeed, the isoscalar giant dipole resonance is due to ph excitations of $3\hbar\omega$. This is also the energy of many $2p - 2h$ as well as $3p - 3h$ configurations which are therefore to be included in a consistent description of the mode.

6. – Conclusions

The proposed EMPM yields eigenvalue equations of simple structure, valid for a Hamiltonian of general form. Recursive formulas hold for all quantities. It is, thus, possible to solve the equations of motion iteratively to generate a multiphonon basis. The matrix representing the nuclear Hamiltonian in such a basis is sparse. It can, therefore, be brought easily to diagonal form. The method is exact not only in principle but also in its actual numerical implementation.

We have, indeed, applied the method to ^{16}O without making any approximation in the calculation. Such an exact treatment has forced us to confine ourselves to a space which included up to three phonons and $3\hbar\omega$. This space is sufficient for our illustrative purposes, but too restricted to describe exhaustively and faithfully all spectroscopic properties of a complex nucleus like ^{16}O.

On the other hand, the highly correlated nature of the phonon states suggests that we do not need to include all of them in actual calculations but the *relevant ones*. It is, in fact, conceivable that most TDA phonons are non-collective and unnecessary. Thus, a drastic truncation of the space should be feasible with little detriment to the accuracy of the solutions.

In this perspective, we are developing a new formulation which makes the truncation procedure efficient and, at the same time, accurate. The new formulation gives us unambiguous criteria for selecting the relevant TDA phonons and discard the others, thereby reducing drastically the number of TDA phonons coming into play. Such a severe cut should make possible the access to phonon subspaces with a quite large number N of phonons and, thus, render the approach competitive with shell model even in describing the low-energy spectroscopic properties.

The method, however, goes beyond shell model. In fact, it allows to include in the TDA phonons ph configurations of arbitrarily high energy not at reach of shell model calculations. Moreover, as we have seen in our illustrative example, it enables us to study the fine structure of giant resonances at low and high energies as well.

Several generalizations are still possible. The method can be reformulated so as to include RPA phonons. This extension, however, might be unnecessary since the method, already in its present TDA formulation, yields an explicitly correlated ground state.

A formulation of the method in terms of quasi-particle rather than particle-hole states is also straightforward and especially useful. It allows, indeed, to study anharmonicities and multiphonon excitations in open shell nuclei not easily accessible to shell model methods.

$$* \ * \ *$$

Work supported in part by the Italian Ministero della Istruzione Università e Ricerca (MIUR), the Czech grant agency under the contract No. 202/06/0363 and the Czech Ministry of Education (contract No. VZ MSM 0021620859).

REFERENCES

[1] ROWE D. J., *Nuclear Collective Motion* (Methuen London) 1970.
[2] RING P. and SCHUCK P., *The Nuclear Many/Body Problem* (Springer-Verlag, New York) 1980.
[3] BORTIGNON P. F., BRACCO A. and BROGLIA R. A., *Giant Resonances* (Harwood Academic Publishers, Amsterdam) 1998.
[4] KNEISSL M., PITZ H. H. and ZILGES A., *Prog. Part. Nucl. Phys.*, **37** (1996) 439.
[5] PIETRALLA N. *et al.*, *Phys. Rev. Lett.*, **83** (1999) 1303.
[6] FRANSEN C. *et al.*, *Phys. Rev. C*, **71** (2005) 054304.
[7] KNEISSL M., PIETRALLA N. and ZILGES A., *J. Phys. G: Nucl. Part. Phys.*, **32** (2006) R217.
[8] KRTICKA M. *et al.*, *Phys. Rev. Lett.*, **92** (2004) 172501.
[9] LO IUDICE N. and PALUMBO F., *Phys. Rev. Lett.*, **41** (1978) 1532.
[10] BOHLE D. *et al.*, *Phys. Lett. B*, **137** (1984) 27.
[11] LESHER S. R. *et al.*, *Phys. Rev. C*, **66** (2002) 051305(R).
[12] WIRTH H.-F. *et al.*, *Phys. Rev. C*, **69** (2004) 044310.
[13] BUCURESCU D. *et al.*, *Phys. Rev. C*, **73** (2006) 064309.
[14] MEYER D. A. *et al.*, *Phys. Rev. C*, **74** (2006) 044309.

[15] Frascaria N., *Nucl. Phys. A*, **482** (1988) 245c.

[16] Auman T., Bortignon P. F. and Hemling H., *Annu. Rev. Nucl. Part. Sci.*, **48** (1998) 351.

[17] For a review see Arima A. and Iachello F., *Adv. Nucl. Phys.*, **13** (1984) 139.

[18] Bohr A. and Mottelson B. R., *Nuclear Structure*, Vol. **II** (Benjamin, New York) 1975 and references therein.

[19] Bortignon P. F., Broglia R. A., Bes D. R. and Liotta R., *Phys. Rep.*, **30** (1977) 305 and references therein.

[20] Soloviev V. G., *Theory of Atomic Nuclei: Quasiparticles and Phonons* (Institute of Physics, Bristol) 1992.

[21] Belyaev S. T. and Zelevinsky V. G., *Nucl. Phys.*, **39** (1962) 582.

[22] Marumori T., Yamamura M. and Tokunaga A., *Prog. Theor. Phys.*, **31** (1964) 1009.

[23] For a review see Klein A. and Marshalek E. R., *Rev. Mod. Phys.*, **63** (1991) 375.

[24] Otsuka T., Arima A. and Iachello F., *Nucl. Phys. A*, **309** (1978) 1.

[25] Lo Iudice N. and Stoyanov Ch., *Phys. Rev. C*, **62** (2000) 047302; **65** (2002) 064304.

[26] Ponomarev V. Yu., Bortignon P. F., Broglia R. A. and Voronov V. V., *Phys. Rev. Lett.*, **85** (2000) 1400.

[27] Lo Iudice N., Sushkov A. V. and Shirikova N. Yu., *Phys. Rev. C*, **70** (2004) 064316; **72** (2005) 034303.

[28] Pomar C., Blomqvist J., Liotta R. J. and Insolia A., *Nucl. Phys. A*, **515** (1990) 381.

[29] Grinberg M., Piepenbring R., Protasov K. V. and Silvestre-Brac B., *Nucl. Phys. A*, **597** (1996) 355.

[30] Andreozzi F., Lo Iudice N., Porrino A., Knapp F. and Kvasil J., *Phys. Rev. C*, **75** (2007) 044312.

[31] Rowe D. J., *J. Math. Phys.*, **10** (1969) 1774.

[32] Liotta R. J. and Pomar C., *Nucl. Phys. A*, **382** (1982) 1.

[33] Protasov K. V., Silvestre-Brac B., Piepenbring R. and Grinberg M., *Phys. Rev. C*, **53** (1996) 1646.

[34] Brown G. E. and Green A. M., *Nucl. Phys.*, **75** (1966) 401.

[35] Haxton W. C. and Johnson C. J., *Phys. Rev. Lett.*, **65** (1990) 1325.

[36] Dytrych T., Sviratcheva K. D., Bahri C., Draayer J. P. and Vary J. P., *Phys. Rev. Lett.*, **98** (2007) 162503.

[37] Ring P. and Schuck P., *The Nuclear Many-Body Problem* (Springer-Verlag, New York) 1980.

[38] Machleidt R., *Adv. Nucl. Phys.*, **19** (1989) 189.

[39] Palumbo F., *Nuc. Phys.*, **99** (1967) 100.

[40] Glockner D. H. and Lawson R. D., *Phys. Lett. B*, **53** (1974) 313.

[41] Tilley D. R., Weller H. R. and Cheves C. M., *Nucl. Phys. A*, **564** (1993) 1.

[42] Berman D. L. and Filtz S. C., *Rev. Mod. Phys.*, **47** (1975) 713.

[43] Lui Y.-W., Clark H. L. and Youngblood D. H., *Phys. Rev. C*, **64** (2001) 064308.

[44] See for instance Kvasil J., Lo Iudice N., Nesterenko V. O. and Kopal M., *Phys. Rev. C*, **58** (1998) 209.

DOI 10.3254/978-1-58603-885-4-535

$E0$ decay of the first excited 0^+ state in ^{156}Dy

G. Lo Bianco

Dipartimento di Fisica, Università di Camerino - Via Madonna delle Carceri
I-62032 Camerino, Italy
INFN, Sezione di Perugia - Via Pascoli, I-06123 Perugia, Italy

Summary. — Following an introduction on the general properties of $E0$ transitions and their significance in nuclear structure, an experiment is described in which the branching ratio between the $E0$ and $E2$ decays of the 0_2^+ state in ^{156}Dy was measured.

1. – $E0$ transitions in nuclei

Let us consider the electromagnetic field in vacuum. If we introduce electromagnetic potentials

$$\vec{A}(\vec{r},t), \qquad \varphi(\vec{r},t)$$

and impose the Coulomb gauge

$$\nabla \cdot \vec{A} = 0, \qquad \varphi = 0,$$

the electric and magnetic fields are given by

$$\vec{E} = -\frac{\partial \vec{A}}{\partial t}, \qquad \vec{B} = \nabla \times \vec{A}.$$

In free space in the case of a monochromatic field

$$\vec{A}(\vec{r}, t) = \vec{A}_0(\vec{r})e^{i\omega t}$$

the space-dependent factor is a solution of the homogeneous Helmholtz equation

$$\nabla^2 \vec{A}_0 + \left(\frac{\omega}{c}\right)^2 \vec{A}_0 = 0.$$

It can then be shown (see, *e.g.*, ref. [1]) that the vector potential can be expressed by the following multipole expansion:

$$\vec{A}_0(\vec{r}) = \sum_{jm} \sum_{l=j-1}^{j+1} j_l(kr)\vec{X}_{jlm}(\vartheta, \varphi),$$

where $j_l(kr)$ is a spherical Bessel function and the vector spherical harmonix is defined by

$$\vec{X}_{jlm} = \sum_{m_l \mu}(lm_l1\mu \mid jm)Y_{lm_l}(\vartheta, \varphi)\vec{e}_\mu$$

in terms of the ordinary spherical harmonix functions and of the unit vectors of the spherical basis.

The terms of the multipole expansion with $j = l$,

$$\vec{A}_{jm}{}^M = j_j(kr)\vec{X}_{jjm}(\vartheta, \varphi)$$

have parity $(-1)^l$ and are called magnetic multipoles; the terms with $j = l \pm 1$ have the opposite parity and can be regrouped to form the electric (transverse) and the longitudinal linear combinations.

In free space the longitudinal term is eliminated by the condition

$$\nabla \cdot \vec{A} = 0.$$

In the presence of charges or currents the vector potential must satisfy the inhomogeneous Helmholtz equation. Solutions may again be expressed in terms of the multipole expansion and the longitudinal terms may still be neglected far away from the sources (in the wave zone) but must be considered in the near zone, where they are, in particular, responsible for the interaction between the electrons and the nucleus that gives rise to internal conversion.

Because of the transverse nature of the electromagnetic field in the wave zone (or, in other terms, because the photon has only two polarization states) electric and magnetic multipoles with $j = 0$ do not exist and monopole ($E0$) electromagnetic transitions are only possible via emission of conversion electrons.

Monopole transitions take place between states having the same spin I and the same parity. If I is not zero, $E0$ transitions compete with $E2$ transitions, but for $I = 0$ they are the only possibility.

Church and Weneser [2] showed that $E0$ transition probabilities can be factorized into electronic and nuclear terms:

$$W = \Omega \rho^2 \,,$$
$$\rho = \sum_p \int \psi_f^* \left[\left(\frac{r_p}{R}\right)^2 + \ldots \right] \psi_i \, d\tau,$$

where the sum is over the protons, R is the nuclear radius and additional terms inside the brackets are usually neglected.

A theoretical estimate of the transition probability was obtained by Rasmussen [3] in the framework of the collective geometrical model. In the case of vibrations in a deformed nucleus the reduced transition probability is given by

$$\rho^2 = \frac{9}{8\pi^2} Z^2 R^2 \beta^2 \hbar (BC)^{-1/2} \,,$$

where β is the deformation parameter and B and C are the parameters of the vibrational Hamiltonian.

It is convenient to introduce a dimensionless ratio of the reduced $E0$ and $E2$ transition probabilities which is independent of B, C, R and Z:

$$X(E0/E2; 0_2^+ \to 0_1^+) = \frac{\rho^2(E0; 0_2^+ \to 0_1^+) e^2 R^4}{B(E2; 0_2^+ \to 2_1^+)} \,,$$

where R is the nuclear radius; in the model of Rasmussen the ratio is $4\beta^2$.

Empirically $\rho^2(E0; 0_2^+ \to 0_1^+)$ is large in the transitional region. This fact was explained by strong mixing of states with different deformation [4] or (within the interacting boson model) with different number of d bosons [5]. An extended discussion on $E0$ transitions and their interpretation was published by Woods [6].

The $X(5)$ model of Iachello [7] provides an analytical description of the phase transition between spherical and deformed nuclei; the position of the 0_2^+ is particularly significant.

Can this model predict the value of $X(E0/E2 : 0_2^+ \to 0_1^+)$?

In order to address that question we decided to measure the $X(E0/E2; 0_2^+ \to 0_1^+)$ ratio in ^{156}Dy, which is a candidate $X(5)$ nucleus [8], even if the variation of the $B(E2)$ values with spin is probably closer to the rotational-model prediction. A partial level scheme of ^{156}Dy is shown in fig. 1.

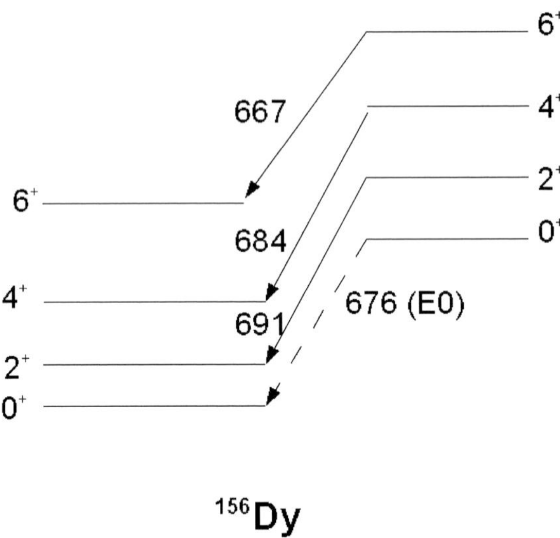

Fig. 1. – Partial level scheme of ^{156}Dy.

2. – Experimental procedures

The experiment was performed at the Laboratori Nazionali del Sud of INFN (Catania) in March 2007. Levels in ^{156}Dy were populated by the ε-decay chain

$$^{156}\text{Er } (19.6\,\text{m}) \rightarrow {}^{156}\text{Ho } (56\,\text{m}) \rightarrow {}^{156}\text{Dy}.$$

^{156}Er was produced by the ^{148}Sm(^{12}C, 4n) reaction at 73 MeV. The beam was switched on and off with a period of 1 h and the main data was collected during the beam-off periods. Measurements were also performed with the beam on for calibration purposes.

γ-rays were measured by a coaxial high-purity germanium detector and electrons by a "mini-orange" spectrometer. Such instrument (fig. 2) consists of a set of permanent magnets and of a lithium-drifted silicon detector cooled at liquid-nitrogen temperature. A central absorber stops low-energy electrons and greatly reduces the X-ray flux reaching the detector.

The energy resolution of the mini-orange spectrometer is entirely due to the properties of the silicon detector while the magnetic field produced by the permanent magnets has the only purpose of transporting the electrons to the detector. The efficiency by which electrons are collected on the silicon detector is, however, strongly energy dependent. The position and the width of the transmission window depend on the type and number of the magnets and on the distance between the magnets and the source and between the magnets and the silicon detector.

The efficiency of the mini-orange spectrometer as a function of the electron energy

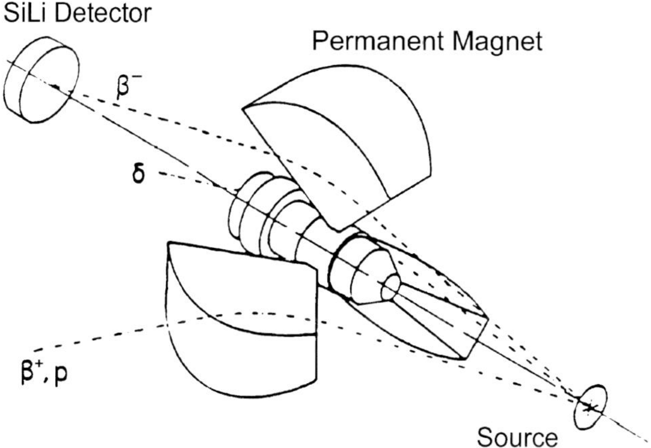

Fig. 2. – Schematic view of the mini-orange spectrometer.

can be evaluated from the measured areas of the electron peaks corresponding to lines whose relative γ-ray intensities are known. Indicating by ε_e the efficiency of the electron spectrometer, by A_e the area of the electron peak, by I_γ the γ-ray intensity, by α the conversion coefficient and by B the electron binding energy, we have for two lines of energy $E1$ and $E2$:

$$\frac{\varepsilon_e(E_1 - B)}{\varepsilon_e(E_2 - B)} = \frac{A_e(E_1 - B)}{A_e(E_2 - B)} \times \frac{\alpha(E_2)I_\gamma(E_2)}{\alpha(E_1)I_\gamma(E_1)} .$$

If the γ-ray relative intensities are not known they can be derived from the areas of the peaks in the γ spectrum by measuring the relative efficiency of the γ-ray detector using a standard source.

3. – Results

Figure 3 shows the relevant part of the summed γ-ray and conversion-electron spectra collected during all the beam-off periods.

The arrow in lower panel shows the energy of the expected $E0$ peak; the corresponding arrow in the upper panel shows the position where a peak should appear in the γ-ray spectrum if the supposed $E0$ transition were instead a K line of higher multipolarity in one of the dysprosium isotopes. After a careful inspection of the γ-ray spectrum we could also exclude a significant contribution from K lines of any identified contaminant or from L lines.

The X ratio defined by Rasmussen can be obtained from the branching ratio

$$q^2 = \frac{W_{e,K}(E0, 0_2^+ \to 0_1^+)}{W_{e,K}(E2, 0_2^+ \to 2_1^+)}$$

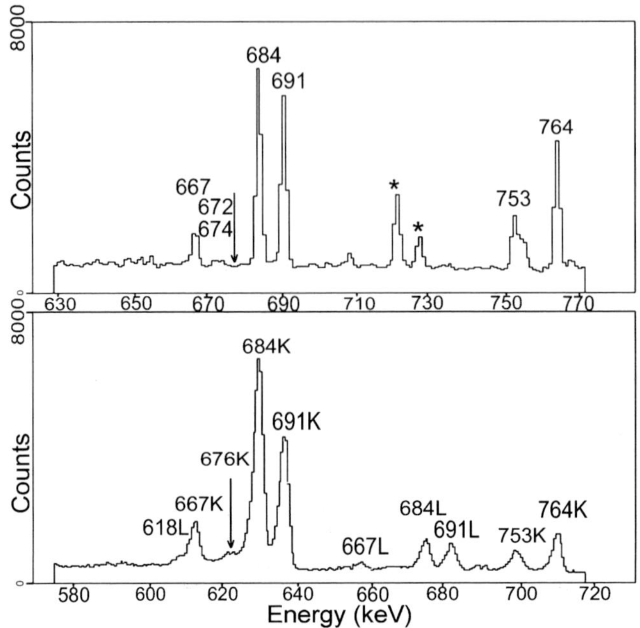

Fig. 3. – Partial off-beam γ-ray (upper panel) and conversion-electron spectra (a star indicates contaminant lines that do not originate from the irradiated target).

using the formula [9]

$$X(E0/E2; 0_2^+ \to 0_1^+) = 2.54 \times 10^9 A^{4/3} E_\gamma^5 q^2 \alpha_K(E2)/\Omega_K \,.$$

The quantity

$$q^2 \alpha_K(E2) = \frac{W_{e,K}(E0; 0_2^+ \to 0_1^+)}{W_\gamma(E2; 0_2^+ \to 2_1^+)}$$

can be obtained from our data if we choose as a reference a line whose γ-ray intensity and conversion coefficient are known. Indicating by $A_{e,K}$ the area of the K-electron peak and by $\varepsilon_{e,K}$ the efficiency for K electrons, we have

$$W_{e,K}(E0; 0_2^+ \to 0_1^+) = C \frac{A_{e,K}(E0)}{\varepsilon_{e,K}(E0)}$$

and

$$W_{e,K}(R) \equiv \alpha_K(R) W_\gamma(R) = C \frac{A_{e,K}(R)}{\varepsilon_{e,K}(R)} \,,$$

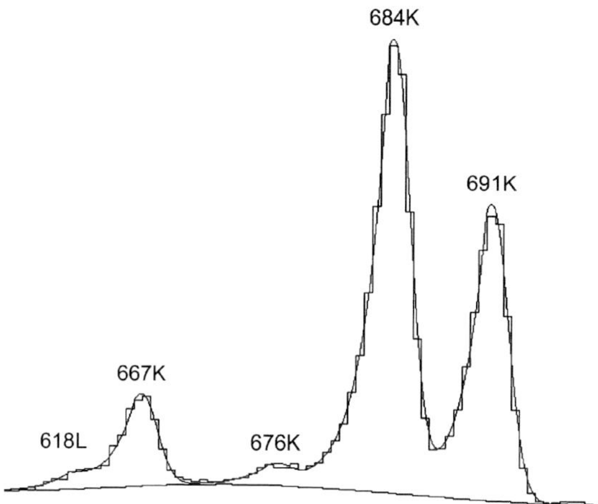

Fig. 4. – Fit of a part of the conversion-electron spectrum.

where R indicates the reference line and C is a constant. We have therefore

$$q^2 \alpha_K(E2) = \frac{A_{e,K}(E0)}{A_{e,K}(R)} \times \frac{\varepsilon_{e,K}(R)}{\varepsilon_{e,K}(E0)} \times \alpha_K(R) \times \frac{W_\gamma(R)}{W_\gamma(E2; 0_2^+ \to 2_1^+)} .$$

A convenient reference is the K-electron line of the 684 KeV $(4_2^+ \to 4_1^+)$ transition. The ratio of γ-ray transition probabilities can be obtained from the intensities measured by Caprio *et al.* [10], who populated the levels of ^{156}Dy in exactly the same way as we did.

The relative efficiency of the electron spectrometer at the energies of the reference line and of the $E0$ transition can be obtained from an internal calibration using the formula that was given in sect. **2**. Convenient calibration points are obtained from the K-electron lines of the 667 keV $(6_2^+ \to 6_1^+)$ and 691 keV $(2_2^+ \to 2_1^+)$ transitions.

Conversion coefficients of transitions in ^{156}Gd were measured by Gromov *et al.* [11] following ε decay and by de Boer *et al.* [12] following the (p, 4n) and (α, 4n) reactions.

Figure 4 shows a fit of the relevant portion of the experimental electron spectrum where peaks were assumed to have a shape described by the combination of a Gaussian function and an exponential low-energy tail. In drawing the assumed background a contamination from the K line of the 674 keV transition in ^{157}Ho was taken into account.

From the fit we obtained the areas of the four peaks that were used in the analysis, that correspond to the K-electron lines of the 667, 676 ($E0$), 684 and 691 keV transitions.

The relative efficiency of the mini-orange spectrometer for the 667 and 691 K lines was computed from the measured areas, the intensities of ref. [10] and the conversion coefficients of ref. [11]. The efficiency for the $E0$ and the 684 K transitions was obtained assuming a linear dependence.

The quantities that fully describe the four transitions for the purpose of our analysis are collected in table I.

Table I. – *Properties of the transitions that are important in the present analysis.*

Transition	$J_i^\pi \to J_f^\pi$	$I_\gamma^{(a)}$	$\alpha_K^{(b)}$	$A_{e,K}$	$\varepsilon_{e,K}$
667 K	$6_2^+ \to 6_1^+$	1.92 (10)	0.058 (3)	6872 (344)	1.056 (119)$(^+)$
$E0$	$0_2^+ \to 0_1^+$			1346 (175)	1.035 (74)
684 K	$4_2^+ \to 4_1^+$	13.3 (9)	0.043 (3)	32886 (822)	1.016 (35)
691 K	$2_2^+ \to 2_1^+$	10.4 (5)	0.036 (2)	21889 (657)	$1(^{++})$

$(^a)$ Normalized to $I_\gamma(2_1^+ \to 0_1^+) = 100$, from ref. [10].

$(^b)$ From ref. [11] (assuming 6% error), $(^+)$ calibration point, $(^{++})$ calibration reference point.

Using for the intensity of $0_2^+ \to 2_1^+$ transition the value of ref. [10],

$$I_\gamma(538\,\text{keV}) = 0.86\ (12),$$

we obtain

$$q^2 \alpha_K(E2) = 0.027\ (6)$$

and, using the theoretical conversion coefficient $\alpha_K(E2; 538\,\text{keV}) = 0.01019\ (14)$, we have

$$q^2 = 2.62\ (60).$$

Finally, we computed the electronic factor using the method of Kantele [13] obtaining $\Omega_K = 4.05 \times 10^{10}\,\text{s}^{-1}$, that inserted in the formula for the X ratio gives

$$X(E0/E2; 0_2^+ \to 0_1^+) = 0.063\ (15).$$

4. – Conclusion

The $X(E0/E2; 0_2^+ \to 0_1^+)$ ratio in a single nucleus cannot at the present time be compared to the result of $X(5)$ calculations, because in that model the $E0$ and the $E2$ operators are both proportional to a constant and the relationship between the two constants is not known. While it can be hoped that further theoretical work will solve the problem, a meaningful test of the $X(5)$ predictions can be obtained by measuring the $X(E0/E2; 0_2^+ \to 0_1^+)$ value in a series of nuclei, so that the ratio of the two constants can be fixed to reproduce one of the experimental values.

$$* \quad * \quad *$$

The experiment was performed in collaboration with L. A. Atanasova (Sofia), D. L. Balabanski (Camerino and Sofia), N. Blasi (Milano), K. Gladnishki (Camerino and Sofia), S. Nardelli (Camerino) and A. Saltarelli (Camerino).

REFERENCES

[1] EISENBERG J. M. and GREINER W., *Excitation Mechanisms of the Nucleus* (North Holland) 1970.
[2] CHURCH E. L. and WENESER J., *Phys. Rev.*, **103** (1956) 1035.
[3] RASMUSSEN J. O., *Nucl. Phys.*, **19** (1960) 85.
[4] HEYDE K. and MEYER R. A., *Phys. Rev. C*, **37** (1988) 2170.
[5] VON BRENTANO P. *et al.*, *Phys. Rev. Lett.*, **93** (2004) 152502.
[6] WOOD J. L. *et al.*, *Nucl. Phys. A*, **651** (1999) 323.
[7] IACHELLO F., *Phys. Rev. Lett.*, **87** (2001) 052502.
[8] DEWALD A. *et al.*, *J. Phys. G*, **31** (2005) S1427.
[9] KIBEDI T. *et al.*, *Nucl. Phys. A*, **567** (1994) 183.
[10] CAPRIO M. A. *et al.*, *Phys. Rev. C*, **66** (2002) 054310.
[11] GROMOV K. Y. *et al.*, *Acta Phys. Pol. B*, **7** (1976) 507.
[12] DE BOER F. W. N. *et al.*, *Nucl. Phys. A*, **290** (1977) 173.
[13] KANTELE J., *Nucl. Instrum. Methods A*, **271** (1988) 625.

DOI 10.3254/978-1-58603-885-4-545

Proton-neutron interactions, collectivity and DFT calculations

R. B. CAKIRLI

Department of Physics, University of Istanbul - Istanbul, Turkey

A. W. Wright Nuclear Structure Laboratory, Yale University - New Haven, CT, 06520, USA

Summary. — The structure of nuclei can often be understood in terms of interactions of valence particles. Changes in shell structure are dominated by the proton-neutron interaction. The experimental interaction between the last two protons and last two neutrons, δV_{pn}, is discussed, and some realistic calculations are shown. It is also stressed that collectivity depends on the p-n interaction strengths.

1. – Introduction

A key issue in the study of atomic nuclei is to understand how their structure varies with N and Z. A useful approach is to study the interactions of the valence particles (protons and/or neutrons). Of course, this is not sufficient; however, it gives a good starting point to obtain basic ideas about structure and structural evolution.

If a nucleus has no valence particles, it has a spherical shape. With increasing number of valence particles (both protons and neutrons), the shape of the nucleus changes from spherical to deformed. This can be seen from empirical $E(2_1^+)$ values. They are very low ($< 100\,\mathrm{keV}$) for heavy deformed nuclei and high ($> 800\,\mathrm{keV}$) for spherical nuclei. Figure 1 shows this. Tin has 50 protons which is a magic number: Yet, $E(2_1^+)$ does not

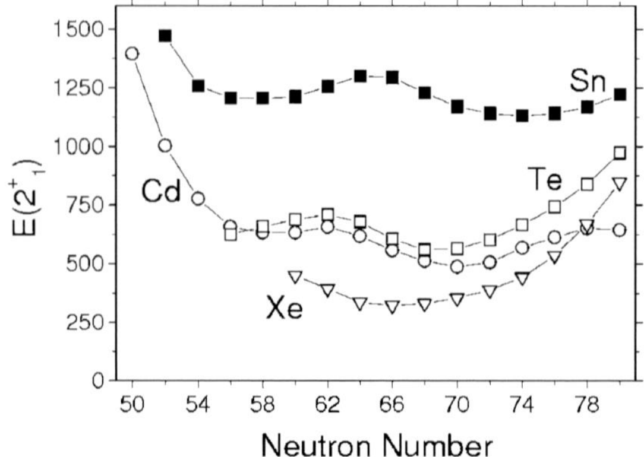

Fig. 1. – First $E(2^+)$ level energies against neutron number for $_{48}$Cd, $_{50}$Sn, $_{52}$Te and $_{54}$Xe.

change much although it has valence neutrons. When we add two protons, that is $_{52}$Te, $E(2^+_1)$ suddenly decreases for the same neutron numbers. It is similar for $_{54}$Xe and $_{48}$Cd as $_{52}$Te. This shows that $E(2^+_1)$ is a good observable to obtain an idea of the shape of a nucleus. Another is $R_{4/2} = E(4^+_1)/E(2^+_1)$. If a nucleus is near doubly magic, $R_{4/2}$ ratio is less than 2. If it is around 2, the nucleus is spherical and exhibits collective vibrations, and it is deformed if it is 3.33.

Changes in the shape of the nucleus are related to the proton-neutron (p-n) interaction. For example, for the $A = 150$ region, $R_{4/2}$ values are given in fig. 2. As clearly seen, there are two different trends. To make these two different trends clear, two different symbols are used in fig. 2. Open squares show $N \geq 90$ nuclei and the stars have $N \leq 88$.

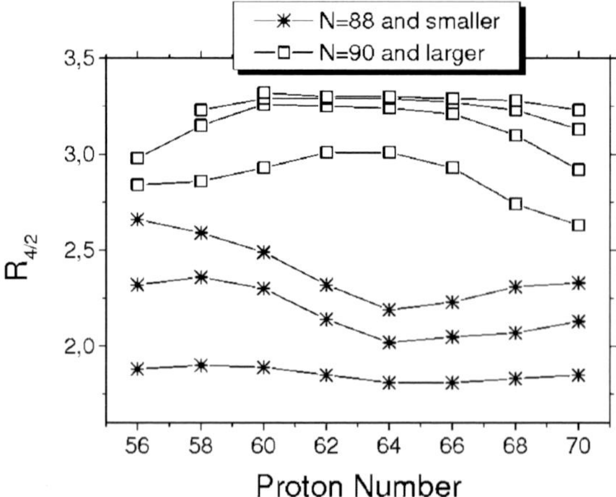

Fig. 2. – $R_{4/2}$ ratios against proton number for $N = 84$–96.

Not only are the different patterns obvious but also the $R_{4/2}$ ratios show a dip at $Z = 64$ for $N \leq 88$ although they have a maximum for $N \geq 90$ nuclei. This is evidence that $Z = 64$ is a magic number for $N \leq 88$ nuclei in which the $R_{4/2}$ ratio is a minimum, and $Z = 64$ is mid-shell when $R_{4/2}$ is a maximum, that is, for $N \geq 90$ nuclei. We directly see that changing shell structure is related to the p-n interaction. The interaction also plays a key role in the onset of collectivity and deformed shapes of nuclei. Therefore, it is useful to study p-n interactions. We can understand the magic numbers, single-particle energies, development of configuration mixing, phase/shape transitions from a study of p-n interactions [1-3].

Since nucleonic interactions are important, we would like to have some empirical measures of them. This is possible using binding energies because they reflect the interactions. In 1989 J. Y. Zhang *et al.* [4], described *double differences of binding energies* which give the interaction between the last 2 protons and last 2 neutrons, δV_{pn}. We can isolate specific nucleonic interactions with this method. δV_{pn} is given by

$$(1) \quad \delta V_{\mathrm{pn}}(Z, N) = \frac{1}{4} \left[\{ B(Z, N) - B(Z, N - 2) \} - \{ B(Z - 2, N) - B(Z - 2, N - 2) \} \right].$$

This is an average interaction between Z^{th} and $(Z - 1)^{\mathrm{th}}$ with N^{th} and $(N - 1)^{\mathrm{th}}$ nucleons.

2. – Methods and results

We obtained empirical δV_{pn} values for nuclei in the major shell region $Z = 50$–82, $N = 82$–126 using experimental masses [5] and eq. (1). The results are shown in fig. 3. In this figure, the δV_{pn} values are coded with colors. Red-reddish colors are higher δV_{pn} values than blue-bluish colors. We see that the δV_{pn} values are higher along the diagonal line (from Te at $N = 84$ to Pb at $N = 126$) and lower further away from the diagonal line. Note that there are no data for the lower right quadrant. In addition, there are some other unknown δV_{pn} values elsewhere.

3. – Interpretation

It is easy to understand these results from simple shell model arguments. δV_{pn} can be thought of in terms of the orbits of the last 2 proton and last 2 neutron orbits [6]. For heavy mass nuclei, these shells start with high j (angular momentum), low n (principal quantum number) and end with low j, high n. Large p-n interactions occur if the overlap between proton and neutron wave functions is large, that is, if the protons and neutrons have similar j and n quantum numbers. If protons and neutrons do not fill similar orbits, the overlap will be small; therefore, we expect low p-n interactions. From this, we expect the largest p-n interactions for similar proton and neutron orbits, that is, for similar fractional filling for given shells. For example, for the $Z = 50$–82, $N = 82$–126 shells, we expect the largest p-n interactions along the diagonal in fig. 3 (left).

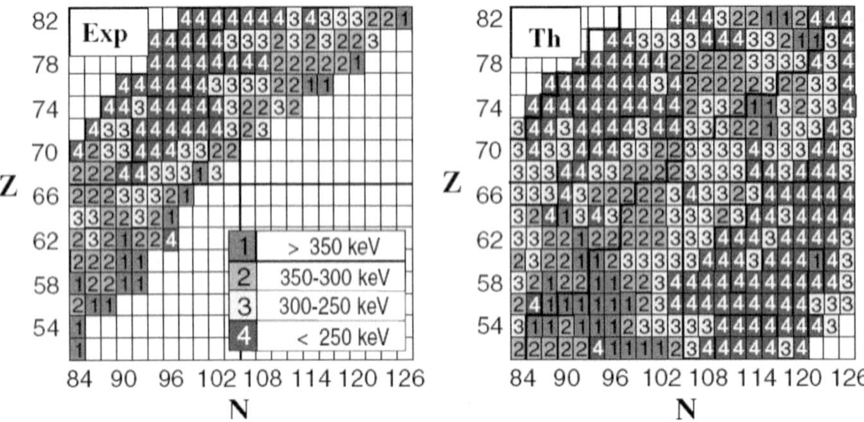

Fig. 3. – Empirical (left) and theoretical (right) δV_{pn} values for the $Z = 50$–82, $N = 82$–126 shells. δV_{pn} values were extracted using the 2003 mass evaluation [5] (based on ref. [7]).

In order to attempt to understand these interactions microscopically and to test advanced approaches to realistic calculations of nuclear structure and binding, M. Stoitsov, W. Nazarewicz and W. Satula carried out extensive calculations using Density Functional Theory (DFT) for over 1000 nuclei. The results for δV_{pn} for the $Z = 50$–82, $N = 82$–126 shells are shown in fig. 3 (right) [7]. As seen, calculations work very well compared to the empirical δV_{pn} values. Along the diagonal line, we see more red-reddish colors and blue-bluish colors are dominant further away from the line. The calculations are also shown for the lower right quadrant although there is not even one empirical δV_{pn} value. This

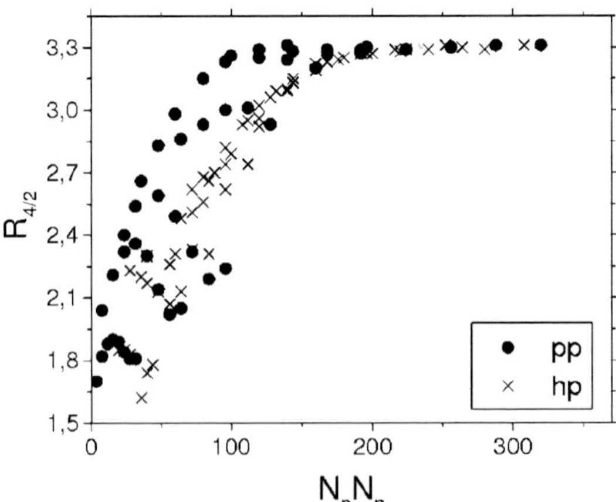

Fig. 4. – Known $R_{4/2}$ values for $Z = 50$–82, $N = 82$–104 nuclei against $N_{\mathrm{p}}N_{\mathrm{n}}$ values (based on ref. [8]).

is a useful guide for future mass measurements. For example, to obtain δV_{pn} of ^{210}Pb, we need to know ^{210}Pb, ^{208}Pb, ^{208}Hg and ^{206}Hg binding energies. In the literature we only lack the ^{208}Hg experimental binding energy. This measurement has been planned at CERN-ISOLDE-ISOLTRAP.

There is another interesting point in fig. 3 (right): the dominant blue colors in the lower right quadrant. This result is expected. Similarly as explained above, in this quadrant, protons will be at the beginning of the shell (high j, low n), and neutrons will be at the end of the shell (low j, high n). Thus, there should be small overlap in proton and neutron wave functions so we expect small δV_{pn} values.

We can learn one more important thing from fig. 3. It is seen that the lower left and upper right quadrants have higher δV_{pn} values than upper left and lower right. Thus, the lower left (particle-particle or pp) quadrant has stronger p-n interactions than the upper left (hole-particle or hp) quadrant. This is directly related with the growth of collectivity. Figure 4 shows $R_{4/2}$ ratios for pp (solid circle) and hp (cross) regions against $N_{\mathrm{p}}N_{\mathrm{n}}$. As seen, the pp region nuclei show a faster growth of collectivity than the hp region nuclei. This is the first direct correlation of observed growth rates of collectivity with empirical p-n interaction strengths [8].

$$* \ * \ *$$

I would like thank R. F. Casten for his limitless help. I also would like to thank W. Nazarewicz and M. Stoitsov so much for providing us useful calculations. I also owe thanks to E. A. McCutchan for her ideas. Otherwise, ref. [8] would not exist. Work supported by the US DOE Grant No. DE-F602-91-ER-40609.

REFERENCES

[1] Van Isacker P., Warner D. D. and Brenner D. S., *Phys. Rev. Lett.*, **74** (1995) 4607.
[2] Heyde K., *Phys. Lett. B*, **155** (1985) 303.
[3] Federman P. and Pittel S., *Phys. Lett. B*, **69** (1977) 385; **77** (1978) 29.
[4] Zhang J.-Y. *et al.*, in *Proceedings of the International Conference on Contemporary Topics in Nuclear Structure Physics Cocoyoc, Book of Abstracts,* Unpublished, Mexico (1988) p. 109.
[5] Audi G., Wapstra A. H. and Thibault C., *Nucl. Phys. A*, **729** (2003) 337.
[6] Cakirli R. B., Brenner D. S., Casten R. F. and Millman E. A., *Phys. Rev. Lett,* **94** (2005) 092501.
[7] Stoitsov M., Cakirli R. B., Casten R. F., Nazarewicz W. and Satula W., *Phys. Rev. Lett.*, **98** (2007) 132502.
[8] Cakirli R. B. and Casten R. F., *Phys. Rev. Lett.*, **96** (2006) 132501.

DOI 10.3254/978-1-58603-885-4-551

Detection of fast neutrons and digital pulse-shape discrimination between neutrons and γ-rays

P.-A. SÖDERSTRÖM

Department of Nuclear and Particle Physics, Uppsala University - 751 21 Uppsala, Sweden

Summary. — The basic principles of detection of fast neutrons with liquid-scintillator detectors are reviewed, together with a real example in the form of the neutron wall array. Two of the challenges in neutron detection, discrimination of neutrons and γ-rays and identification of cross talk between detectors due to neutron scattering, are briefly discussed, as well as possible solutions to these problems. The possibilities of using digital techniques for pulse-shape discrimination are examined. Results from a digital and analog versions of the zero crossover algorithm are presented. The digital pulse-shape discrimination is shown to give, at least, as good results as the corresponding analogue version.

1. – Introduction

In experimental studies of the structure of exotic nuclei far from the line of β stability it is important to accurately and efficiently be able to identify the nuclei produced in the reactions. One of the most common types of reactions used is the heavy-ion fusion-evaporation reaction, in which a number of neutrons, protons and/or α-particles are evaporated, and exotic nuclei are produced with very small cross-sections. The protons and α-particles are for example detected by highly efficient silicon arrays. For studies of proton-rich nuclei, the clean detection of the number of emitted neutrons in each

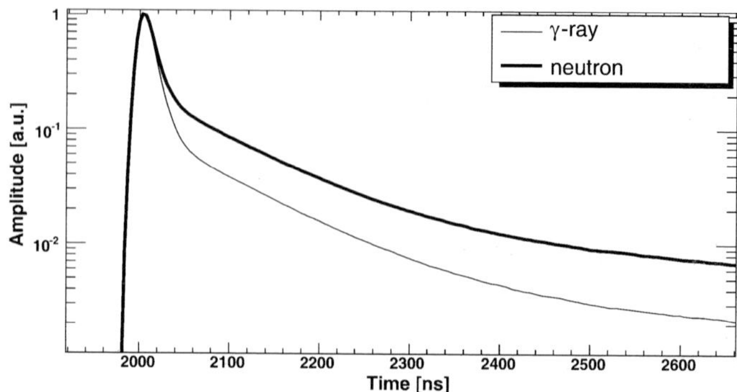

Fig. 1. – Idealized pulse shapes from a BC501 liquid scintillator from a γ-ray and a neutron interaction.

reaction, is of utmost importance. Due to the uncharged nature of the neutrons, they are however much more difficult to detect with high efficiency than the light charged particles. The future neutron detection arrays will operate at radioactive ion-beam facilities, like SPIRAL2 [1] and FAIR [2]. The demand on these arrays will be much larger due to the intense γ-ray background originating from the radioactive beam itself.

2. – Neutron detection

The most common fast neutron([1]) detector for in in-beam γ-ray spectroscopy experiments is the organic, liquid-scintillator detector. Detection of fast neutrons in an organic scintillator is an indirect process, in which the neutrons deposit energy mainly by elastic scattering with the protons in the liquid. The energy of the recoil protons can have values from zero up to the incoming neutron energy, depending on the scattering angle. The recoil proton excites the organic molecules into either singlet, S_i, or triplet states, T_i, of order i. When the molecule de-excites and cascades down the states it will emit photons, and the decay $S_1 \rightarrow S_0$ will be detected as scintillation light in the photomultiplier tube, as a fast component of the pulse. If two molecules in a triplet state collide, they can change their configuration to one S_0 and one S_1 state. The decay of the S_1 state contributes with a slow component to the pulse [3, 4]. The amount of light produced in this process is proportional to the ionization density of the interacting particle. Hence a proton (neutron) interaction will give a larger contribution of this slow component compared to an electron (γ-ray). See fig. 1.

An existing and so far very successful, neutron detector array for use in nuclear structure experiments is the Neutron Wall [5]. It consists of 15 hexagonal detectors and one pentagonal, which are assembled into a closely packed array covering about 1π of the solid angle, and mounted in the forward hemisphere of the set-up. The 16 detectors are in

---

([1]) In this paper, fast neutrons in the energy range 0.2 to 10 MeV are considered.

turn divided into 50 segments and filled with a liquid scintillator of the type BC501A [6]. The total volume of the liquid in the neutron wall is 150 litre. The neutron wall was originally designed for experiments together with EUROBALL [7], but since 2005 it has been located at GANIL where it has been used together with EXOGAM [8]. The total neutron efficiency of the neutron wall is about 25% in symmetrical fusion-evaporation reactions. The Pulse-Shape Discrimination (PSD) electronics used are NIM units of the type NDE202 [9] based on the zero cross-over discrimination method.

3. – Cross talk

Since the recoil proton only gets a fraction of the full energy of the neutron, the neutron will in many cases have substantial kinetic energy left after the interaction, but a velocity in some random direction. Because of this there exists a probability that the neutron will scatter into a neighboring detector and interact. This can cause quite severe errors in the counting of the number of neutrons emitted in the reaction. Several methods attempting to correct for this exist [10, 11]. For the neutron wall a method based on the time-of-flight difference between the different segments has shown to give good results [11]. The idea of this rejection technique is that the time difference between the signals of two separate neutrons, on average, is shorter than the time it takes for a scattered neutron to travel between two detectors.

4. – Discrimination between γ-rays and neutrons

Another important property of a neutron detector is the ability to discriminate between detection of γ-rays and neutrons. It has also been shown that even a small amount of γ-rays misinterpreted as neutrons dramatically reduce the quality of the cross-talk rejection [11]. The differences in pulse shape make it possible to distinguish between neutrons and γ-rays that hit the detector, and several methods to do this are available. Two well-known techniques are the zero crossover method where the signal is shaped into a bipolar pulse from which the zero crossing is extracted. This shaping can for example be done via a RC-CR integrator and differentiator network [12], or by using double delay lines [13]. Another technique is the charge comparison method, where the charge in the fast component is integrated and compared to the charge in the slow component [14]. Both of these methods are well studied and have been compared using analogue electronics [15].

5. – Digital pulse-shape discrimination

When developing the next generation of neutron detector arrays, it will be desirable to use digital instead of analogue electronics. Some of the advantages of using digital electronics is the possibility to store the data and do very complex data treatment off-line, as well as doing advanced digital PSD on-line. For example, instead of using the regular charge comparison method one can enhance the discrimination properties by using a

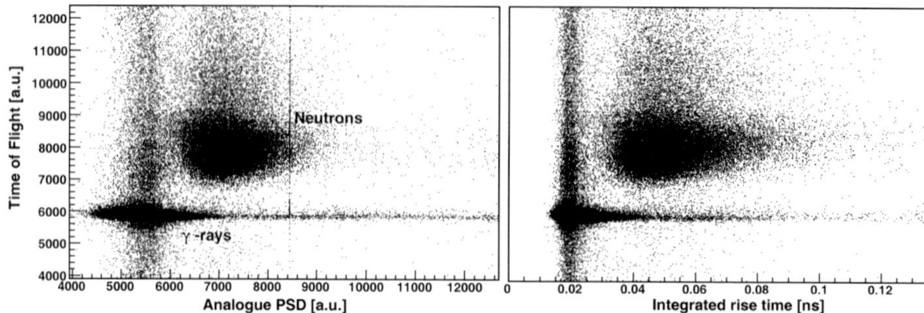

Fig. 2. – Neutron and γ-ray separation using analogue (left panel) and digital (right panel) methods in the energy range of 90 to 680 keV for γ-rays and 150 to 1130 keV for recoil protons. The vertical and horizontal γ-ray distributions are random coincidences pile-up events, respectively.

suitable weight [16] for the integral of the slow component, instead of the very limited modifications to the integration that is available in the analogue case.

The digital PSD method has been tested using an experimental set-up consisting of a ^{252}Cf source, a neutron detector similar to the neutron wall detectors and a sampling ADC system running at 300 MHz and 14 bits. As an example, the implementation of a digitized version of the zero crossover method, the integrated rise-time, is shown in fig. 2 together with the analogue separation. In this method of separation, instead of taking the detour with pulse shaping, the information was directly obtained from integrating the pulse and using the time difference between the 10% and 72% pulse height, which gives information equivalent to the zero crossover method [17]. The analogue separation is obtained by one of the neutron wall PSD units [9]. By using well-known analog algorithms the results can easily be compared with their analogue counterparts, and thus be used for benchmarking of the effectiveness of the digital electronics, especially with respect to the bit resolution and sampling frequencies of the ADC, a work that is currently in progress [18]. As seen in fig. 2, it is possible to get, at least, as good separation with this set-up as when using analogue electronics.

6. – Conclusion and outlook

A short introduction to the principles of neutron detection by organic liquid scintillation counting has been presented, as well as an example of an existing neutron detection array that has been extensively used. Two of the great challenges in neutron detection, cross talk and neutron-γ discrimination, were also briefly described.

It has been shown that one can get, at least, as good a separation of neutrons and γ-rays using digital and analogue versions of the same type of algorithms. By taking full advantage of the potential of digital PSD it might be possible to get a much better discrimination. For example, nice separation has recently been obtained with a folding approach [19] and using a curve fitting method [20]. Other possible improvements for the future neutron detector arrays might be the development of new detector materials, like solid plastic scintillators with PSD properties, or deuterated scintillators with better energy response than standard liquid scintillators.

$$* \; * \; *$$

Thanks to my supervisor J. NYBERG for all the help, J. LJUNGVALL for the TNT2 library, and to R. WOLTERS for the preparation of the set-up. I would also like to thank the organizers for a very rewarding summer school. And of course my fellow students for a great time.

REFERENCES

[1] GALES S., this volume, p. 79.
[2] AUMANN T., this volume, p. 57.
[3] KNOLL G., *Radiation Detection and Measurements*, 3rd edition (Wiley) 1999.
[4] BIRKS J. B., *The Theory and Practice of Scintillation Counting* (Pergamon) 1964.
[5] SKEPPSTEDT Ö. *et al.*, *Nucl. Instrum. Methods A*, **421** (1999) 531. http://nsg.tsl.uu.se/nwall/
[6] Saint-Gobain Crystals, France, BC501A data sheet.
[7] SIMPSON J., *Z. Phys. A*, **358** (1997) 139.
[8] AZAIEZ F., *Nucl. Phys. A*, **654** (1999) 1003c. http://www.ganil.fr/exogam/
[9] The NDE202 was built by WOLSKI D. and MOSZYŃSKI M. *et al.* at The Andrzej Soltan Institute for Nuclear Studies, Swierk, Poland.
[10] CEDERKÄLL J. *et al.*, *Nucl. Instrum. Methods A* , **385** (1997) 166.
[11] LJUNGVALL J., PALACZ M. and NYBERG J., *Nucl. Instrum. Methods A*, **528** (2004) 741.
[12] ROUSH M. L., WILSON M. A and HORNYAK W. F., *Nucl. Instrum. Methods*, **31** (1964) 112.
[13] ALEXANDER T. K. and GOULDING F. S., *Nucl. Instrum. Methods*, **13** (1961) 244.
[14] BROOKS F. D., *Nucl. Instrum. Methods*, **4** (1959) 151.
[15] WOLSKI D. *et al.*, *Nucl. Instrum. Methods A*, **360** (1995) 584.
[16] GATTI E. and DE MARTINI F., in *Nuclear Electronics, Proceedings of International Conference at Belgrade*, Vol. II (IAEA, Vienna) 1962, pp. 265-276.
[17] RANUCCI G., *Nucl. Instrum. Methods A*, **354** (1995) 389.
[18] SÖDERSTRÖM P.-A., NYBERG J. and WOLTERS R., to be published in *Nucl. Instrum. Methods A*.
[19] KORNILOV N. V. *et al.*, *Nucl. Instrum. Methods A*, **497** (2003) 467.
[20] MARRONE S. *et al.*, *Nucl. Instrum. Methods A*, **490** (2002) 299.

International School of Physics "Enrico Fermi"

Villa Monastero, Varenna

Course CLXIX

17–27 July 2007

"Nuclear Structure far from Stability: New Physics and New Technology"

Directors

Aldo COVELLO
Dipartimento di Scienze Fisiche
Università di Napoli "Federico II"
Complesso Universitario Monte S. Angelo
Via Cintia
80126 Napoli
Italy
Tel.: ++39-081-676263
Fax: ++39-081-676263
covello@na.infn.it

Francesco IACHELLO
Center for Theoretical Physics
Sloane Laboratory
217 Prospect Street
Yale University
New Haven, CT 06520-8120
USA
Tel.: +1-203-4326944
Fax: ++1-203-4325741
francesco.iachello@yale.edu

Renato Angelo RICCI
INFN, Laboratori Nazionali di Legnaro
Via Romea 4
35020 Legnaro (PD)
Italy
Tel.: ++39-049-8068311
Fax: ++39-049-641925
Renato.A.Ricci@lnl.infn.it

Scientific Secretary

Giuseppe MAINO
ENEA C.R. "E. Clementel"
Via don Fiammelli 2
40128 Bologna
Italy
Tel.: ++39-051-6098206
Fax: ++39-051-6098352
giuseppe.maino@bologna.enea.it

Lecturers

Baha BALANTEKIN
Department of Physics
University of Wisconsin
1150 University Avenue
Madison, WI 53706
USA
Tel.: ++1-608-2637931
Fax: ++1-608-2630800
baha@physics.wisc.edu

Richard CASTEN
Wright Nuclear Structure Laboratory
Yale University
PO Box 208124
New Haven, CT 06520-8124
USA
Tel.: ++1-203-4326174
Fax: ++1-203-4323522
rick@riviera.physics.yale.edu

Hans FELDMEIER
GSI
Planckstrasse 1
64291 Darmstadt
Germany
Tel.: ++49-6159-712744
Fax: ++49-6159-712990
h.feldmeier@gsi.de

Karlkeinz LANGANKE
GSI
Planckstrasse 1
64291 Darmstadt
Germany
Tel.: ++49-6159-712747
Fax: ++49-6159-712990
k.langanke@gsi.de

Petr NAVRATIL
LLNL
PAT Directorate L-414
PO Box 808
Livermore, CA 94551
USA
Tel.: ++1-925-4243725
Fax: ++1-925-4225940
navratil1@llnl.gov

Takaharu OTSUKA
Department of Physics and Center
for Nuclear Study
University of Tokyo
7-3-1 Hongo, Bunkyo-ku
Tokyo 113-0033
Japan
Tel.: ++81-3-8122111 ext 4134
Fax: ++81-3-56849642
otsuka@phys.s.u-tokyo.ac.jp

Steven PIEPER
Physics Division
Argonne National Laboratory
Building 203
Argonne, IL 60439
USA
Tel.: ++1-630-2524232
Fax: ++1-630-2526008
spieper@anl.gov

Pieter VAN ISACKER
GANIL
BP 55027
F-14076 Caen Cedex 5
France
Tel.: ++33-231-454565
Fax: ++33-231-454421
isacker@ganil.fr

Seminar Speakers

Thomas AUMANN
GSI
Planckstrasse 1
64291 Darmstadt
Germany
Tel.: ++49-6159-711408
Fax: ++49-6159-712809
T.Aumann@gsi.de

Ettore FIORINI
Dipartimento di Fisica
Università di Milano Bicocca
Edificio U2
Piazza della Scienza 3
20126 Milano
Italy
Tel.: ++39-02-64482424
Fax: ++39-02-64482585
ettore.fiorini@mib.infn.it

Sidney GALÉS
GANIL
BP 55027
F-14076 Caen Cedex 5
France
Tel.: ++33-231-454591
Fax: ++33-231-454665
gales@ganil.fr

Silvia LENZI
Dipartimento di Fisica
Università di Padova
Via Marzolo 8
35131 Padova
Italy
Tel.: ++39-049-8277180
Fax: ++39-049-8762641
silvia.lenzi@pd.infn.it

Nicola LO IUDICE
Dipartimento di Scienze Fisiche
Università di Napoli "Federico II"
Complesso Universitario Monte S. Angelo
Via Cintia
80126 Napoli
Italy
Tel.: ++39-081- 676157
Fax: ++39-081-676346
loiudice@na.infn.it

Tohru MOTOBAYASHI
Heavy Ion Nuclear Physics Laboratory
RIKEN
2-1 Wako
Saitama 351-0198
Japan
Tel.: ++81-4-84624449
Fax: ++81-4-84624464
motobaya@riken.jp

Karsten RIISAGER
CERN
CH-1211 Geneva 23
Switzerland
Tel.: ++41-22-7675825
Fax: ++41-22-7678990
karsten.riisager@cern.ch

Aurora TUMINO
INFN
Laboratori Nazionali del Sud
Via Santa Sofia 62
95123 Catania
Italy
Tel.: ++39-095-542385
Fax: ++39-095-542252
tumino@lns.infn.it

Jan VAAGEN
Department of Physics and Technology
University of Bergen
Allegaten 55
5007 Bergen
Norway
Tel.: ++47 -55-582724
Fax: ++47-55-589440
jan.vaagen@ift.uib.no

Students

Lucrezia AUDITORE
INFN, Gruppo Collegato di Messina
Dipartimento di Fisica
Università di Messina
Contrada Papardo, Salita Sperone 31
98166 Messina
Italy
Tel.: ++39-090-393654 int.24
Fax: ++39-090-395004
lauditore@unime.it

R. Burcu CAKIRLI
Department of Physics
University of Istanbul
Vezneciler
Istanbul
Turkey
Tel.: ++90-21-24555700
Fax: ++90-21-2519
burcu@wnsl.physics.yale.edu

Anna CORSI
Dipartimento di Fisica
Università di Milano
Via Celoria 16
20133 Milano
Italy
Tel.: ++39-348-0524526
Fax: ++39-02-50317295
anna.corsi@email.it

Antonio DI NITTO
Dipartimento di Scienze Fisiche
Università di Napoli "Federico II"
Via Cinthia
80126 Napoli
Italy
Tel.: ++39-081-676236
Fax: ++39-081-676904
dinitto@na.infn.it

Francesca GIACOPPO
INFN
Laboratori Nazioni del Sud
Via S. Sofia 62
95123 Catania
Italy
Tel.: ++39-095-542390
giacoppo@ct.infn.it
f_giacoppo@libero.it

Philip M. GORE
Department of Physics
Vanderbilt University
629 Country Club Lane
Nashville, TN 37205
USA
Tel.: ++1-615-3566470
p.m.gore.phys@earthlink.net

Andrea GOTTARDO
INFN
Laboratori Nazioanli di Legnaro
Viale dell'Università 2
35020 Legnaro (PD)
Italy
Tel.: ++39-0444-585312
andrea.gottardo@lnl.infn.it

Gabriela ILIE
Department of Nuclear Physics
University of Koln
Zulpicherstrasse 77
50937 Koln
Germany
Tel.: ++49-0221-4702796
Fax: ++49-0221-4705168
ilie@ikp.uni-koeln.de

Elizabeth F. JONES
Department of Physics
Vanderbilt University
629 Country Club Lane
Nashville, TN 37205
USA
Tel.: ++1-615-3568848
e.f.jones.phys@earthlink.net

Michal MACEK
Department of Particle
and Nuclear Physics
University of Charles in Prague
v Holesovickach 2
18000 Prague
Czech Republic
Tel.: ++420-22191-2483
Fax: ++420-22191-2434
macek@ipnp.troja.mff.cuni.cz

Marina MANGANARO
Dipartimento di Fisica
Università di Messina
Contrada Papardo, Salita Sperone 31
98166 Messina
Italy
Tel.: ++39-090-6765024
Fax: ++39-090-395004
manganarom@unime.it

Marco MAZZOCCO
Dipartimento di Fisica
Università di Padova
Via Marzolo, 8
35131 Padova
Italy
Tel.: ++39-049-8277181
Fax: ++39-049-8277181
marco.mazzocco@pd.infn.it

Daniele MONTANARI
Dipartimento di Fisica
Università degli Studi di Milano
Via Celoria, 16
20133 Milano
Italy
Tel.: ++39-349-8426899
Fax: ++39-02-50317295
daniele.montanari@mi.infn.it

Francesca Manuela MORTELLITI
Dipartimento di Fisica
Università di Messina
Contrada Papardo, Salita Sperone 31
98166 Messina
Italy
Tel.: ++39-090-6765024
Fax: ++39-090-395004
fmortelliti@alice.it

Sara NARDELLI
Dipartimento di Fisica
Università di Camerino
Via Madonna delle Carceri 9
62032 Camerino (MC)
Italy
Tel.: ++39-347-9449064
Fax: ++39-0737-402853
sara.nardelli@unicam.it

Giovanni RICCIARDETTO
INFN
Sezione di Catania
Viale A. Doria 6
95125 Catania
Italy
Tel.: ++39-349-1458149
giovanni.ricciardetto@ct.infn.it

Lorenza ROVERSI
ENEA C.R. "E. Clementel"
Via don Fiammelli 2
40129 Bologna
Italy
Tel.: ++39-051-6098206
Fax: ++39-051-6098359
lorenza.roversi@bologna.enea.it

Domenico SANTONOCITO
INFN
Laboratori Nazionali del Sud
Via S. Sofia 62
95123 Catania
Italy
Tel.: ++39-095-542353
santonocito@lns.infn.it

Maria Letizia SERGI
INFN
Laboratori Nazionali del Sud
Via S. Sofia 62
95123 Catania
Italy
Tel.: ++39-095-542397
Fax: ++39-095-542335
sergi@lns.infn.it

Par-Anders SODERSTROM
Nuclear and Particle Physics
University of Uppsala
P.O.Box 535
75121 Uppsala
Sweden
Tel.: ++46-18-4717617
Par-Anders.Soderstrom@tsl.uu.se

Pavel STRANSKY
Department of Particle
and Nuclear Physics
University of Charles in Prague
v Holesovickach 2
18000 Prague 8
Czech Republic
Fax: ++420-22191-2434
stransky@ipnp.troja.mff.cuni.cz

Observers

Thomas MATERNA
Institut Laue-Langevin
University of Grenoble
6 rue Jules Horowitz
BP 156
38042 Grenoble
France
Tel.: ++33-476-207511
Fax: ++33-476-207777
materna@ill.fr

Luigi VANNUCCI
INFN
Laboratori Nazionali di Legnaro
Viale dell'Università 2
35020 Legnaro (PD)
Italy
Tel.: ++39-049-8068633
Fax: ++39-049-641925
vannucci@lnl.infn.it

PROCEEDINGS OF THE INTERNATIONAL SCHOOL
OF PHYSICS "ENRICO FERMI"

---

[1]This course belongs to the NATO ASI Series C, Vol. 460 (Kluwer Academic Publishers).